T0134976

Handbook of Big Geospatial Data

Martin Werner • Yao-Yi Chiang
Editors

Handbook of Big Geospatial Data

Editors
Martin Werner
Professorship Big Geospatial Data Management
Technical University of Munich (TUM)
Munich, Germany

Yao-Yi Chiang
Spatial Sciences Institute
University of Southern California
Los Angeles, CA, USA

ISBN 978-3-030-55464-4 ISBN 978-3-030-55462-0 (eBook)
https://doi.org/10.1007/978-3-030-55462-0

This Springer imprint is published by the registered company Springer Nature Switzerland AG
The registered company address is: Gewerbestrasse 11, 6330 Cham, Switzerland

Preface

Spatial data form the backbone of many science and engineering investigations and applications. For example, spatial data are fundamental to construction and engineering, social sciences and life sciences, art and visualization, and many more fields. In spatial data, the geometry component captures the concept of shape, presence, movement, and similarity about things in space. In this book, we focus on a particular type of spatial data: data that are suitable to understand various aspects of the physical surroundings, either in a local environment or for the entire Earth. Such data are known as geospatial data, as they are geometric data linked to things in the physical world. Specifically, we focus on big geospatial data, a subdomain within spatial data, which considers the computing technologies for handling large datasets of geometry representing things on Earth (hence "geo").

Due to the advent of mobile computing and GPS (global positioning system) technologies, the location of mobile devices, including smartphones, cars, and assets such as ships, containers, and more, can be determined without a high cost. The implication of these mobile location data is potent. Still, it requires careful considerations and technologies to understand and exploit the positive potential and mitigate the risks of the negative potential of such data. On the one hand, such information can be used to optimize logistics, increase the efficiency of transportation systems in cities, and enhance personalized experiences for individual users. On the other hand, ubiquitous location information can produce undesired applications, including deanonymizing individuals, tracking user behavior unwillingly, and micro-targeting commercials with fake news, for example, micro-targeting fake news that trick people into spending money they would not have spent or (much worse) deceive people to vote for certain politicians and political parties.

In addition to mobile location data, public and private satellites and in situ measurements are continuously generating geospatial databases of unprecedented sizes and accuracy. High-quality satellite missions with global coverage at limited resolutions, like Landsat and Copernicus, provide us a global view of our planet with pretty short update times (a few days) between individual acquisitions. High-resolution satellites can provide centimeter-level geospatial observations with typically either cheaper sensors building some constellation or less frequent visits of the same place on Earth. Airborne machines (plane, helicopter, drone) and street-level measurement campaigns (streetview, urban mapping) provide targeted, highly precise data describing selected aspects of the physical and natural environment. Typically, the amount of collected/sensed data is so large, and we do not yet understand the impact of the existence of these collections at all. For example, what is the best use case for such data? Can we solve some of our most pressing challenges, like feeding the world, fighting climate change, and fighting poverty, using increased precision and detailed views of the Earth? Or are we going to experience malicious use of these data first, such as predictive policing affected by biases present in such observations limiting civil rights?

Yet another domain generating huge amounts of big geospatial data that have provenly affected our democratic elections at least once (though there is a debate about how much actually) is the domain of social media. Social media datasets with location information represent the connection of where people are (accessible from their smartphones) with what they want to communicate (content) and to whom (their social network). This domain generates impressive amounts of data and enables positive and negative use cases.

The previous paragraphs have shown that our societies are currently facing a significant challenge: we are collecting huge amounts of location data and do not yet understand the impact. We hope that the positive use of these data will outweigh the risks, but we cannot be sure. Therefore, research on the technologies for making use of spatial data is crucial, especially in the context of big data where the sheer amount of data allows for advanced processing, including machine learning and data mining. However, spatial data are highly heterogeneous, and, therefore, many interdisciplinary islands do exist, each of which is having their ideals, languages, problems, and cultures. The various fundamental theories of spatial computing come from both computer science and spatial science, which presents a challenge for working with spatial data as there are not many truly interdisciplinary collections of knowledge that can serve as a starting point. Very often, these fundamental theories can be hidden from the applications, since they are not at all easy to access or understand for the users who approach spatial computing from an applied setting. Also, there are many domains in which spatial data are generated and processed for specific applications. This leads to the fact that significant developments in spatial computing are being published in separate domains, not directly accessible or transferrable to other domains. As a result, the same techniques often are developed multiple times.

In the context of the ACM SIGSPATIAL GIS conference in 2017, we concluded that a collection of a diverse set of articles from the various subfields of our community with strong ties to spatial computing would be beneficial for many research communities that use spatial data. This collection will provide a forum in which our research topics can be discussed and made accessible across boundaries. In this setting, we have asked our colleagues to provide book chapters in which various topics of spatial computing are explained in a didactic setting while still allowing for presenting new research results. The selected chapters are organized into several sections to provide an overview of big geospatial data. Each chapter represents a unique contribution from expert researchers within their own traditionally isolated disciplines and provides a medium for them to write from their view of the topic. These disciplines include mathematics, computational geometry, machine learning, statistics, remote sensing, smart cities, indoor navigation, politics, big data, and social media analytics, to name a few. We hope that the book will further facilitate interdisciplinary dialogue so that all disciplines working with big geospatial data can understand each other a bit better.

In summary, this book is intended to be an initial contribution to shed light on how spatial data could improve our lives and, at the same time, how they could harm our societies. The book consists of contributions of more than 70 authors writing a total of more than 600 pages aiming at helping to increase the interdisciplinary understanding of big geospatial data for future collaboration and evolution of the field.

Overview of the Book

We organized the book into five topic areas aiming to differentiate the nature of the chapters and the mindset of the authors. The first Part, *Spatial Computing Systems and Applications,* introduces the topic of big geospatial data by presenting example systems and domains from the wide area of big geospatial data. This includes processing systems like IBM PAIRS and GeoSpark, as well as domain-dependent applications like Indoor Mapping or Disaster Response using satellite data. The second Part, *Trajectories, Event, and Movement Data*, collects various chapters on this interesting and challenging type of spatiotemporal data. This includes chapters on two lately introduced extensions of geospatial trajectories with semantics and, more generally, toward multi-attribute trajectories and application examples from urban road map extraction. It concludes with a hands-on chapter on the explorative analysis of big trajectory datasets. The third Part, *Statistics, Uncertainty, and Data Quality*, focuses on issues arising from uncertainty, methods to combat uncertainty from a database research perspective, and an introduction to statistical models for spatial datasets. The fourth Part, *Information Retrieval from Multimedia Spatial Datasets*, provides a broad range of chapters related to information retrieval, either from text and social media, bridging the domains of big geospatial data and classical information retrieval, or from spatial datasets, such as historic map archives. The fifth Part, *Governance, Infrastructures, and Society*, collects chapters on more societal and systemic questions related to big geospatial data. Two chapters explore the state of the art of the governmental big geospatial data infrastructures in Europe (INSPIRE) and the United States (the National Spatial Data Infrastructure). They are augmented with a discussion of how decision-makers can be integrated into spatial data warehouse architectures and how semantic graphs can be used to deal with spatial evolution, for example, changing polygonal subdivision of urban regions due to growth.

In summary, this book provides a comprehensive view of the field of big geospatial data. In addition to covering a wide range of topics, three clusters of chapters stand out: multimedia information retrieval, trajectories, and moving data, as well as societal and organizational aspects of infrastructures. This is in line with the current research trends and needs.

For the next generation of big geospatial data systems, we need to overcome the traditional approach of collecting data for local analysis. Instead, we need to envision strong infrastructures for bringing the analysis code directly to the data saving energy and cost. However, this contains challenges related to reproducibility in science and to the business models of both spatial data companies and national agencies that have been used to earn money on request and to provide "dumb" data instead of services. In addition, urbanization and mobility are tightly interlinked societal challenges: how can we better understand and optimize our urban spaces, and how can we sustainably fulfill the mobility demand of people and goods? Trajectory computing is a very hard subfield of spatial computing that is essential to answering these questions in a data-driven way.

At this point, we want to thank all authors of papers and all colleagues who helped with this book project in one way or another. We hope that the book will be useful to grow our community of spatial computing. Now, enjoy the book!

Munich, Germany Martin Werner

Los Angeles, USA Yao-Yi Chiang

Contents

Part I Spatial Computing Systems and Applications

1 IBM PAIRS: Scalable Big Geospatial-Temporal Data and Analytics As-a-Service 3
Siyuan Lu and Hendrik F. Hamann

2 Big Geospatial Data Processing Made Easy: A Working Guide to GeoSpark 35
Jia Yu and Mohamed Sarwat

3 Indoor 3D: Overview on Scanning and Reconstruction Methods 55
Ville V. Lehtola, Shayan Nikoohemat, and Andreas Nüchter

4 Big Earth Observation Data Processing for Disaster Damage Mapping 99
Bruno Adriano, Naoto Yokoya, Junshi Xia, and Gerald Baier

5 Spatial Data Reduction Through Element-of-Interest (EOI) Extraction 119
Samantha T. Arundel and E. Lynn Usery

6 Semantic Graphs to Reflect the Evolution of Geographic Divisions 135
C. Bernard, C. Plumejeaud-Perreau, M. Villanova-Oliver, J. Gensel, and H. Dao

Part II Trajectories, Event and Movement Data

7 Big Spatial Flow Data Analytics 163
Ran Tao

8 Semantic Trajectories Data Models 185
Maria Luisa Damiani

9 Multi-attribute Trajectory Data Management 199
Jianqiu Xu

10 Mining Colocation from Big Geo-Spatial Event Data on GPU 241
Arpan Man Sainju and Zhe Jiang

11 Automatic Urban Road Network Extraction From Massive GPS Trajectories of Taxis 261
Song Gao, Mingxiao Li, Jinmeng Rao, Gengchen Mai, Timothy Prestby, Joseph Marks, and Yingjie Hu

12 Exploratory Analysis of Massive Movement Data 285
Anita Graser, Melitta Dragaschnig, and Hannes Koller

Part III Statistics, Uncertainty and Data Quality

13 Spatio-Temporal Data Quality: Experience from Provision of DOT Traveler Information 323
Douglas Galarus, Ian Turnbull, Sean Campbell, Jeremiah Pearce, and Leann Koon

14 Uncertain Spatial Data Management: An Overview 355
Andreas Züfle

15 Spatial Statistics, or How to Extract Knowledge from Data 399
Anna Antoniuk, Miryam S. Merk, and Philipp Otto

Part IV Information Retrieval from Multimedia Spatial Datasets

16 A Survey of Textual Data & Geospatial Technology 429
Jochen L. Leidner

17 Harnessing Heterogeneous Big Geospatial Data 459
Bo Yan, Gengchen Mai, Yingjie Hu, and Krzysztof Janowicz

18 Big Historical Geodata for Urban and Environmental Research 475
Hendrik Herold

19 Harvesting Big Geospatial Data from Natural Language Texts 487
Yingjie Hu and Benjamin Adams

20 Automating Information Extraction from Large Historical Topographic Map Archives: New Opportunities and Challenges 509
Johannes H. Uhl and Weiwei Duan

Part V Governance, Infrastructures and Society

21 The Integration of Decision Maker's Requirements to Develop a Spatial Data Warehouse 525
Sana Ezzedine, Sami Yassine Turki, and Sami Faiz

22 Smart Cities 563
Mayank Kejriwal

23 The 4th Paradigm in Multiscale Data Representation: Modernizing the National Geospatial Data Infrastructure 589
Barbara P. Buttenfield, Lawrence V. Stanislawski, Barry J. Kronenfeld, and Ethan Shavers

24 INSPIRE: The Entry Point to Europe's Big Geospatial Data Infrastructure 619
Marco Minghini, Vlado Cetl, Alexander Kotsev, Robert Tomas, and Michael Lutz

Part I
Spatial Computing Systems and Applications

Chapter 1
IBM PAIRS: Scalable Big Geospatial-Temporal Data and Analytics As-a-Service

Siyuan Lu and Hendrik F. Hamann

1.1 Introduction

Traditional geographic information systems connect data with geolocations (e.g., weather, maps etc.). Since their first computerized instantiations in 1960 (Clarke 1986) such systems have been widely used to process and analyze (mostly) static, geo-coded "vector" data (=points, lines and polygons). Traditional GIS is a central technology to geospatial analytics which is a fast growing market projected to reach 96 billion USD by 2025 at a 12.9% CAGR (Compound Annual Growth Rate) (https://www.marketsandmarkets.com/PressReleases/geospatial-analytics.asp). However, GIS is at an inflection point for mainly two reasons: On the technology side, the backends of traditional GIS are hitting serious scalability limits as a result of the emergence of "mega" big data in the form of geo-coded imagery (e.g., from drones and satellites) (Tang and Shao 2015; Bouwmeester and Guo 2010), time-series IoT (Internet of Things) (Gubbi et al. 2013; Weber 2016), LiDAR (Light Detection and Ranging) (Dubayah and Drake 2000) or RaDAR (Radio Detection and Ranging) data. By way of example, the European Space Agency (ESA) produces more than 10 TeraBytes of satellite data in a single day (Petiteville n.d.). Ten TeraBytes cannot by handled by most GIS backends. The growth of GIS data generation is expected to be exponential, considering the emergence of new platforms for data collections such as drones (Dubayah and Drake 2000) or nanosatellites (Bouwmeester and Guo 2010) or new sensor types such as hyperspectral LiDAR (Hakala et al. 2012).

On the application side, GIS users are now looking more and more to take full advantage of these ever-growing, ubiquitous new data sources leveraging the

S. Lu (✉) · H. F. Hamann
IBM T. J. Watson Research Center, Yorktown Heights, NY, USA
e-mail: lus@us.ibm.com

M. Werner, Y.-Y. Chiang (eds.), *Handbook of Big Geospatial Data*,
https://doi.org/10.1007/978-3-030-55462-0_1

Dimensionality / Data type	Spatial	Temporal	Spatial and temporal
Vector	Traditional GIS		
Raster			
Vector and Raster			Future GIS

Small data
Big data
Small data
Big data

Fig. 1.1 Transformation of traditional GIS

latest advances of machine-learning and artificial intelligence with the goal to operationalize GIS use cases (Chen et al. 2011; Yuan 2009). Examples of such "geospatial-temporal" use cases are plentiful and cut across different industries ranging from the energy and utility industry (when and where to trim vegetation to avoid costly outages), agribusiness (when and where to buy or sell agricultural commodities), insurance (when and where are the highest risk assets) to governments (when and where to optimally respond to a natural disaster).

Figure 1.1 illustrates the transformation of traditional GIS from a static, mostly vector-based, planning tool to an operational, real-time technology, which can process all kinds of different data at scale. An example for this is the application of smart meter monitoring (advanced metering infrastructure = AMI) in the context of renewable energy management (Resch et al. 2014).

The technical challenges for the transformation as depicted in Fig. 1.1 are at least twofold. On a more practical level, scalable integration of data from different data sources is still a major bottleneck, where often more than 90% of all effort is spent on data pre-processing, curation and integration. Most use cases require a combination of different data whether this is raster (or imagery), vector (points, lines, and polygons), or time-series information. It is well known that such data can be highly complex, with hundreds of different formats, resolutions, projections and reference systems.

On a more fundamental level though, while such multi-modal data integration can be very difficult, it is arguably much more challenging to do this at scale. Many of the emerging geospatial-temporal data sets, which users seek exploiting are simply too big to be moved or downloaded in time to be useful for an operational application. By way of example, the daily 10 TeraBytes from the European Space Agency (at 100 MB/s – read speed of a hard disk drive) takes more than a day to "move" from a storage device to the memory of a processor for subsequent computation.

The facts that (i) many of new emerging geospatial-temporal data sets (LiDAR, RaDAR, imagery, time-series) are too big to be moved and (ii) most use cases require the integration of multiple data sets, leads to the notion of data gravity. Data gravity means that big data tends to attract more data – in the same way a bigger mass attracts a smaller mass – and with that, big data attracts more compute and applications. Most traditional GIS "move" data to the application or analytics and thus they are inherently limited in terms of how much can be processed and exploited. To be more specific, the database backend of GIS must become much more powerful to cope with these challenges, where in the future, analytics and data must be directly collocated.

The solution to these data gravity challenges involves many technologies. First, given the size of the data and the fact that many users require the same big data sets (such as weather) for their different applications, a shared, often cloud-based system becomes more economical, which can be used remotely as a service. Other key technologies may include HDFS (Hadoop Distributed File System) (Lam 2010; Zikopoulos and Eaton 2011), which allows a scalable distributed storage layer exemplified by key-value stores such as HBase (Dimiduk et al. 2013; Harter et al. 2014a) to be combined with a highly parallel processing layer using frameworks such as MapReduce (Dean and Ghemawat 2008; Ekanayake et al. 2008). This in turn enables processing of very large data sets by pushing analytics tasks "into" queries and thus avoiding data movement.

By way of comparison, GIS systems even today rely often on relational database systems such as Postgres, mostly for vector data and/or file-based storage for raster data. It is well known that relational databases have difficulties in scaling to data sizes beyond a few tens of TeraBytes. The use of file-based storage comes with other major drawbacks. Often users need to assemble different images thereby dealing with different timestamps, resolutions, map projections etc. Even in simple cases where a user wants to extract a time series from multiple satellite observations for the same location, one would have to download and open often thousands of files to extract the right information. Ironically, the inability to perform analytical tasks within the data and without downloading often leads to more data, where data providers compute ahead of time more derivatives of the raw data (such as vegetation index from hyperspectral satellite data).

To address the aforementioned challenges recently the IBM PAIRS Geoscope (Physical Analytics Integrated Data Repository & Services) was introduced (Klein et al. 2015; Lu et al. 2016), which unlike most systems does not use relational database systems or file-based storage (object or cold store). PAIRS is based on a distributed, highly parallel, key-value big data system with a big, ready-made catalog of carefully indexed, diverse, and continually updated geospatial-temporal information (of both spatial and/or temporal vector and imagery data) in the cloud, enabling scalable access to complex queries and machine learning-based analytics and AI to run without the need for downloading data.

PAIRS provides several benefits to the users. Firstly, PAIRS *allows access to PetaBytes of geospatial-temporal data sets at low cost*. That is because many users require the same data sets (e.g., weather, satellite etc.) and analytics capabilities and thus the shared services of PAIRS are much more efficient and cost effective. Second and as we will discuss in more detail below, *PAIRS drastically accelerates the analytics by reducing the time to insights* when retrieving and analyzing geospatial-temporal information – whether PAIRS (i) just provides AI-ready curated data, or (ii) returns results from search and analytics queries involving multiple data sets (by filtering, aggregating, applying mathematical functions etc.) or (iii) provides platform services for custom analytics without downloading the data or (iv) enables clients to integrate their own data thereby allowing them to exploit, analyze or monetize their data along with the PetaBytes of already curated data. Finally, and thirdly, due to the technology's unique scalability, *PAIRS enables users – often for the first time – to scale and operationalize their respective geospatial-temporal use cases*.

PAIRS is not the only technology for geospatial information which leverages a combination of key-value store with a distributed parallel big data system to overcome scalability limits. GeoMesa and GeoWave are two exciting and innovative, open-source research projects using a similar idea (Hughes et al. 2015; Whitby et al. 2017). By way of comparison, GeoMesa and GeoWave designs are primarily centered on vector data, while PAIRS complements this capability by focusing on raster data. In addition, PAIRS aims to provide end-to-end functionality from data curation to customizable "in data" analytics which a user can directly use without performing deployment or configuration optimization. Such "as-a-service" nature of PAIRS is reflected in its architecture and user experience as discussed next.

1.2 PAIRS Architecture Overview

Key components of the PAIRS architecture are shown in Fig. 1.2. In overview, PAIRS has four main components: (i) an *ingestion/data curation engine*, (ii) a massive *distributed compute and data store based on HDFS/HBase*, (iii) an *analytics and data platform*, which enables users via (iv) an *interface* to interact with the system.

(i) The *ingestion/data curation engine* includes data cleaning, filtering, re-projecting, resampling. It is a highly tuned C++ program compatible with a large variety of geospatial-temporal data in over 200 formats built on top of GDAL/OGR (Warmerdam 2008) and PDAL (Contributors 2018). During the ingestion process, all imagery data is remapped onto a set of nested resolution layers and to a common projection and reference system. Details were described previously (Klein et al. 2015; Lu et al. 2016). The data curation engine can process with today's infrastructure (over 100 servers, ~4000 cores, ~30 TB of RAM, and ~500GB/sec total network switching bandwidth) more than 50 TeraBytes per day. Routinely, PAIRS curates more than 15 TeraBytes per day and has subscribed to many agencies such as NASA, ECMWF, ESA, NOAA etc. for continuous, near real-time data ingestion.

All data uploaded are indexed to a *massive distributed compute and data store based on HDFS/HBase*. In this key-value store, every raster data layer is modeled as (x, y, t, θ – value), i.e. value as a function of x (longitude), y (latitude), t (time),

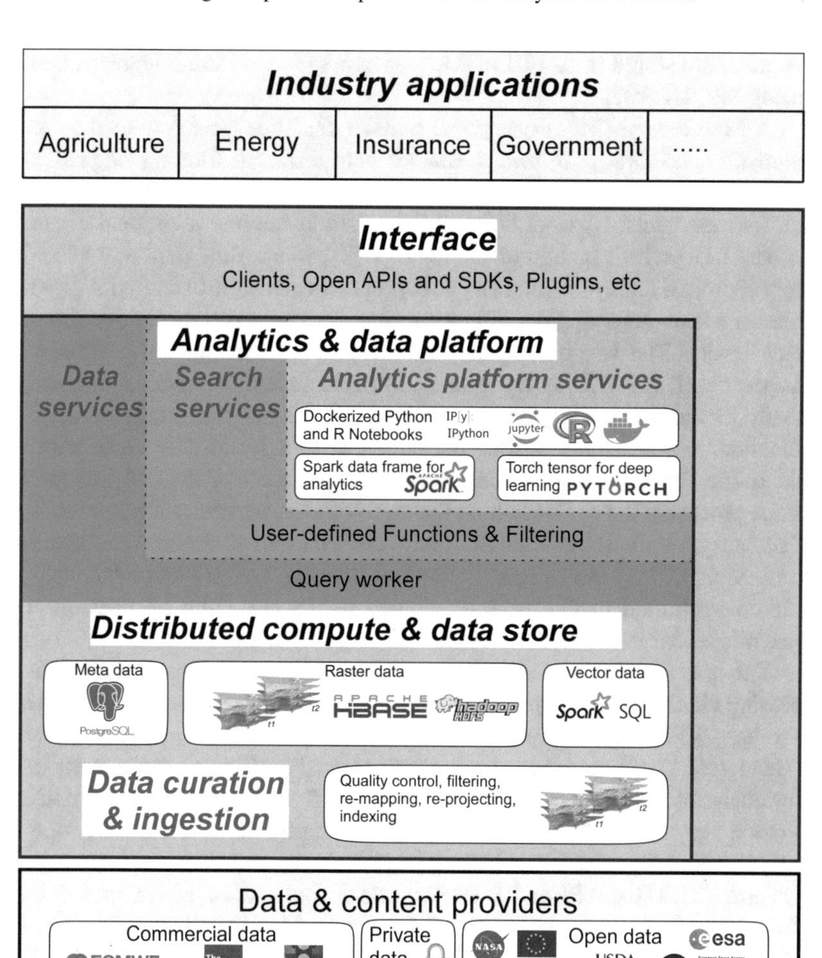

Fig. 1.2 Overview of PAIRS architecture

and additional dimensions θ. Here θ represents any additional dimensions, other than x, y, t, which are required to uniquely define the value. For example, additional dimensions may include vertical elevation (for 3D atmospheric data) or forecast lead time, Δt (between the issue of a weather forecast and the actual forecasted time) or a sensor ID. A distinguishing feature of PAIRS is that all data are ready for use without a data staging or preparation step. Unlike many other technologies, PAIRS uses object and cold store *only* for archiving data which have already been ingested into the key-value store. All PAIRS data are organized in layers, where each layer is linked in space and time. Layers can be access controlled (visualization only, read, write, admin) according to the privileges of user groups. In addition to PetaBytes of raster data stored, vector data (discrete points, polygons, typically much smaller in volume) are stored in PAIRS in Postgres or a key-value store which can be queried using SPARK SQL.

A MapReduce (M/R) query and built-in analytics engine is at the core of the *analytics and data platform*. It enables data retrieval, filtering, logical joins and complex math of a layer or between different layers. A basic PAIRS query is based on four elements: (i) what (specifies the data layers and additional dimensions if needed), (ii) where (geographical region), (iii) when (time period, maybe different for different data layers), and (iv) post-processing or built-in analytics (aggregation, mathematical computation, filtering). The query syntax is unified for different data layers. The key differences with respect to conventional platforms are the following: (i) A PAIRS query returns physical and logically organized data which is ready for analytics. This contrasts with a conventional platform's "search for data" function, which simply returns the reference to a set of files containing relevant data. The PAIRS data store design ensures that most of the data required by the same processor are co-located on the same cluster, which minimizes the burden of data reorganization. (ii) A PAIRS query takes care of commonly encountered post-processing tasks, such as aggregation and filtering, which often effectively reduces the data returned to the users (compared to the raw data) by over one order of magnitude. More detailed examples will be given below.

The query results are available as files for download, for visualization or processing via OGC web map service (WMS) (Consortium n.d.-a) and web processing service (WPS) (Consortium n.d.-b) which are served from a set of geo-servers (Henderson 2014), or as Pandas data frames or SPARK data frames ready to be used for analytics without downloading. For this, a Docker encapsulated ("dockerized") Python Jupyter Notebook with access to the data frame can be readily "spun" up.

All interactions with PAIRS are available to end users via an *interface* of an open RESTful API and a PAIRS client application, which includes a query GUI (graphical user interface) and Python Jupyter Notebooks. Two screenshots of the PAIRS GUI are shown in Fig. 1.3. A freemium version of this PAIRS client is available at this reference (https://ibmpairs.mybluemix.net). Further updates of the PAIRS Client will be made including user-enabled uploading and 3D visualization. An initial version of the PAIRS API is available at this reference (https://pairs.res.ibm.com/tutorial/). For the convenience of Python users, an open sourced PAIRS SDK wrapping API functionalities is available at (https://github.com/ibm/ibmpairs) or from pip or conda Python package management system.

Multiple PAIRS deployment models including SaaS, on premise, or hybrid can co-exist to accommodate clients' focus, e.g. business user, data distributor, application developer. In addition, PAIRS supports multiple data protection schemes to accommodate full data and/or analytics privacy protection, residency requirements, and flexible selective data sharing.

While we have described PAIRS from an architectural point of view it is equally important to understand PAIRS from a users' perspective. From a users' perspective, PAIRS can be a (i) *data curation service*, where users upload their geospatial-temporal data into PAIRS. The benefit of that is that the user's data becomes

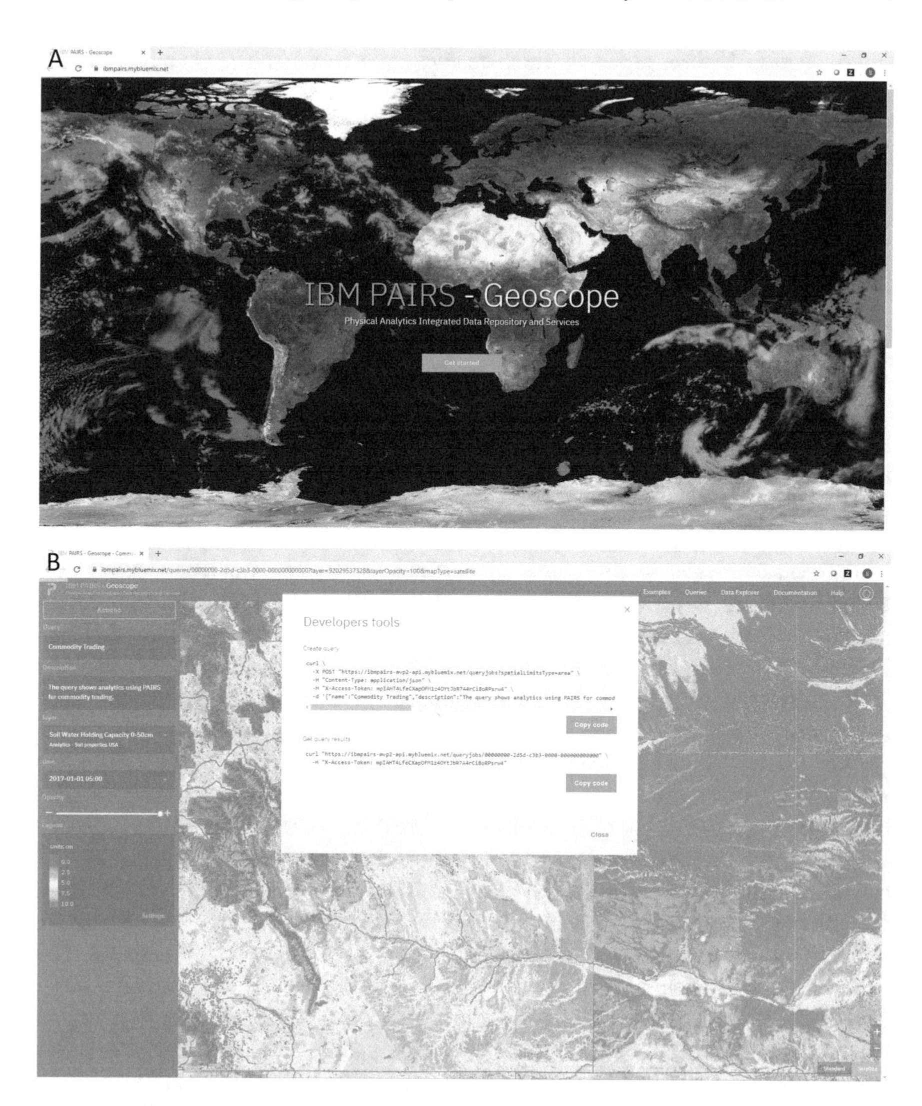

Fig. 1.3 (a) PAIRS Client landing page and (b) screenshot of the PAIRS query UI

query-able with all the already curated PAIRS data. Many users exploit PAIRS as a smart (ii) *data service*, where for example, time-series information from satellite data is being requested at multiple points. On the next level PAIRS enables a (iii) *search or discovery service* for geospatial-temporal data. More specifically, users can query PAIRS to identify or find all locations in a certain geographic area, which meet a couple of requirements. For example, show me all areas in the United States, where the population is larger than 1000 people per square mile and the temperature will be below freezing in the next 5 days (for heating energy consumption estimation). Finally, a user can fully leverage the different

(iv) *analytics platform services*, which enables users to customize analytics without downloading the information first. The anticipation of such PAIRS usage patterns from a *users' perspective* dictates the design of PAIRS key-value store which we detail next.

1.3 Key-Value Store Design and Performance

At the core of PAIRS is its data store based on HBase on top of Hadoop (Dimiduk et al. 2013; Harter et al. 2014a). For brevity, in the following discussion we focus on the implementation of the raster data store. Interested readers may refer to GeoMesa (Hughes et al. 2015) and GeoWave (Whitby et al. 2017) for the implementation of big vector data store. In HBase all data abstractly can be thought of as being stored as multi-level ordered key-value pairs on a distributed system, which extends over many data nodes (region servers) controlled by a master. Each region server hosts several consecutive key-value pair sections. Such sections are referred to as regions. Using MapReduce (M/R) and SPARK respectively, queries and analytical tasks are executed, which may access multiple regions on different region servers of HBase in parallel, thereby providing excellent scalable performance (Dean and Ghemawat 2008; Ekanayake et al. 2008). Unlike relational databases, carefully tuned key-value stores are scalable to hundreds of PetaBytes (DeCandia et al. 2007).

While details vary, fundamental to the implementation based on key-value store is how to effectively translate or index multi-dimensional geospatial-temporal data (at least 3 dimensions x, y, t) to a one-dimensional key so that optimal and balanced performance of writing/reading is achieved for different types of raster data. The salient design decisions of the PAIRS key-value store are summarized in Table 1.1 and actual implementation is provided in Table 1.2. The design decisions are motivated by the anticipated read/write patterns of the geospatial-temporal data encoded in the key-value store, and importantly, how to efficiently handle raster data at both of the two extreme ends of the spectrum. As depicted in Fig. 1.4, on one end, there are cases with data of very high spatial resolution but only a few timestamps, such as the high-resolution aerial imagery from the US Department of Agriculture (USDA) National Agriculture Imagery Program (NAIP) dataset. The spatial resolution of this dataset is 0.5–1 m but data is only available every other year. On the other end, there is data of lower spatial resolution with many timestamps. For example, weather forecasts often come with hourly resolution but are of few kilometers' spatial resolution.

Moreover, merely reading the key-value store to retrieve data is not enough. As discussed earlier, the retrieved data must be organized logically and physically in a way that readily enables downstream data analytics (i.e. analytics that proceeds without major reshuffling of the data, which can be a bottleneck in a public cloud environment).

In PAIRS, each dataset (such as a satellite imagery product) is represented as an HBase table (HTable), which conceptually is a hierarchical, three-level key-value

Table 1.1 Summary of PAIRS design decisions

Design decision	Rationale
Use a fixed coordinate reference system and a fixed set of nested grids.	To enable analytics-ready data at the cost of reprojection errors.
Employ a multi-aspect row key to encode spatial and temporal information.	To enable efficient parallelized data processing for large queries.
Supercells (32 × 32 pixels) as the value of the key-value store.	To reduce the storage overhead of the keys and the computional overhead of the key-value operations.
Temporal hash in the HBase row keys.	To mitigate the resorting of the HBase key-value stores during ingestion of data layers of high temporal frequency.
Spatial hash in the HBase row keys.	To mitigate HBase region server "hot-spotting" when Ingesting data layers of high spatial resolution.
Construct a set of coarse-grained overview layers for each data layers.	To enable (1) effective retrieval of available timestamps at given locations, (2) rapid preview of the results of large queries, and (3) acceleration of queries which require the filtering of data layer values.

Table 1.2 Design of the PAIRS key-value store

	Key			Value
Row key [128 bits]	Column		Version timestamp	2^N × 2^N pixel super cell
	Column family	Column qualifier		
[4 bits reserved] + [4 bits temporal hash] + [4 bits spatial hash] + [52 bits spatial] + [16 bits reserved] + [48 bits temporal key]	Data layer	Additional dimensions	Not used	Typical 32 × 32 pixels

store – the three keys being row key, column family and column qualifier. The key-value store is ordered by the three keys with row keys being at the top of the hierarchy. PAIRS employs a key design as shown in Table 1.2. The highest level HBase row key encodes space and time (i.e. x, y, t). The second level column family encodes a data layer (such as a band of a satellite imagery product). The third level column qualifier encodes any additional dimensions. For example, atmospheric data usually comes in at different altitude. The altitude information is stored as one of the additional dimensions. Weather forecasts may also involve a forecast lead time (e.g. forecast is for 1 day or 10 days ahead), the forecast lead time can be stored as another additional dimension. In the following, we note a few salient features of the raster key-value store design.

To encode the location information of incoming data by spatial keys, a predetermined map projection and spatial resolution are necessary. PAIRS is designed primarily as a cloud hosted data and analytics service for industrial applications. We also anticipate many of its users may come from non-geospatial background. Thus,

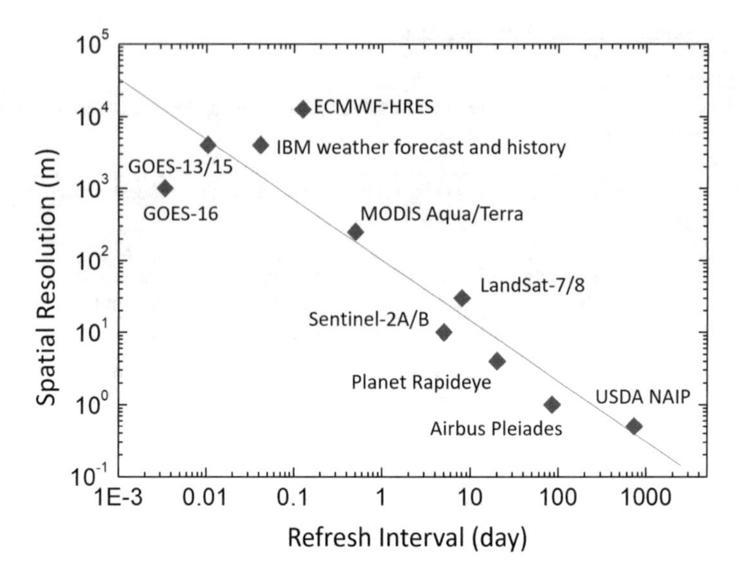

Fig. 1.4 Spatial and temporal resolutions of different raster data sets

PAIRS adopted the WGS 84 projection in favor of its simplicity. The inefficiency and non-convergence problem of WGS 84 near the poles are typically of less importance for most conceivable industrial use cases but this can be addressed by alternative projections. A fixed and nested resolution hierarchy (spatial resolutions) as shown in Table 1.3 is adopted. The grid size reduces by a factor of 2 when resolution increases one level. During ingestion, all data is re-sampled to the next higher resolution level, e.g., data from a satellite with 1.0 m resolution is re-sampled to level 26 (0.89 m at equator). While such implementation increases the data volume, it has the advantage that all information is linked and thus PAIRS can provide very fast "contextual" information (e.g. from different satellites with different resolutions) compared to other systems. Queries including multiple layers of geospatial-temporal data – e.g. "show me all areas in the Middle East where the accumulated precipitation in the last week was lower than 0.2 mm and the population density is larger than 1,000 people/km^2" – are orders of magnitude faster because no re-sampling of the data is required at query time.

Temporal and spatial hash in row keys Naively, one might use a key with location and time information and then a z-order to ensure that the data from the same location is stored close on the same part of the disk. However, this creates "hotspots" when reading/writing always hits the same server (Vohra 2016). To overcome this, PAIRS introduces a special spatial and temporal hash in the beginning of the key to achieve an improved and balanced performance for data layers at the two extreme ends of Fig. 1.4.

First consider a data layer with low spatial resolution but high temporal refresh rate such as the GOES-16 satellite data (5 minutes refreshing interval). Because the

Table 1.3 Shows the PAIRS grid size. The different rows show the grid size in latitude/longitude degree ($\Delta\theta$,$\Delta\varphi$)

Levels	$\Delta\theta$, $\Delta\phi$ [degree]	Δy Δx ($\theta = 0°$)	Δx ($\theta = 40°$)
31	2.5e-7	2.78 cm	2.09 cm
30	5e-7	5.56 cm	4.19 cm
29	1e-6	11.1 cm	8.37 cm
...	...	...	...
25	1.6e-5	1.78 m	1.34 m
...	...	...	...
20	5.12e-4	57.0 m	42.9 m
...	...	...	...
10	0.524288	58.3 km	43.9 km

main temporal key comes after the spatial key, writing data with a new timestamp to a spatial location means inserting (in contrast to appending) new rows into the HTable. Doing such insertion frequently is computationally quite expensive because within the HTable one must re-sort and re-compact to keep the key-store ordered (Harter et al. 2014b). To mitigate this problem, we have introduced a four-bit temporal hash. We note that GeoMesa employs a very similar temporal hash. This hash ensures that only a small part of the HTable must be re-sorted and re-compacted as new timestamps get inserted. Tests have shown that this temporal hash improves the data ingestion/curation process by more than 10x.

A different problem is encountered when ingesting and querying a data layer of high spatial resolution but very low temporal refreshing such as the NAIP data (2 years refresh rate). In such case, because the tailing temporal key has only a few different values, when querying a polygon area or ingesting a new image tile, one will be effectively reading/ writing a set of continuous keys of HBase which usually are hosted on the same region server. This causes the aforementioned and well-known issue of "region hot-spotting" (Vohra 2016). To overcome this difficulty, a 4 bits spatial hash is introduced after the temporal hash. This hash ensures that reading or writing of a large spatial area is parallelized on multiple regions to avoid "hot-spotting".

Supercell as values Moreover, PAIRS uses supercells, which are arrays of 32 × 32 pixels, as the value of the key-value store. In this way the storage taken by the key (16 bytes) becomes negligible compared to the value (4 KiloBytes for pixels of 4-byte float type). Reading/writing each key-value pair then processes 1024 pixels at once, significantly enhancing performance. Our benchmarking showed over 50× improvement compared to one pixel per key-value pair.

PAIRS aims to achieve high performance for both "big" queries of raster data for a "large" area and "small" queries of point locations. Indeed, the profiling of PAIRS queries (Fig. 1.5) indicates a large fraction of queries are the "small" ones for point location, typically 1–100 KiloBytes in size. We found empirically that supercells made of 32 × 32 pixels is a good size as it enables balanced performance. With a 32 × 32 pixel supercell, the time required to seek a key is already much less than the time to reading/writing each value (involves 1024 pixels, 4 KiloBytes for a pixel of 4

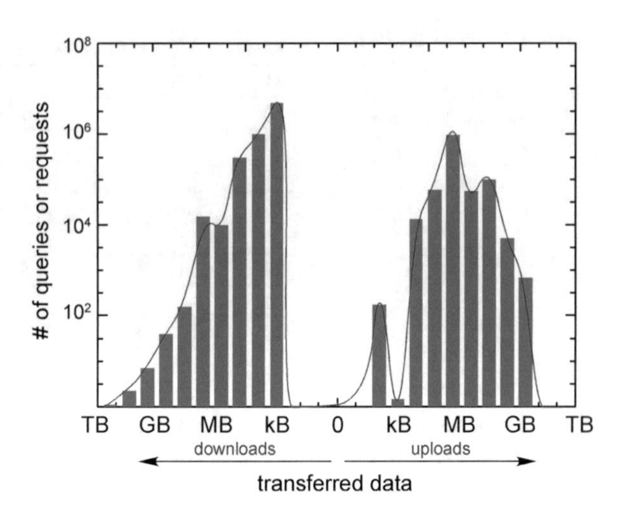

Fig. 1.5 PAIRS user behavior sampled from more than 7M requests between 01/01 and 12/31 2017

Byte float type). Thus for "big" spatial queries, the effect of further reducing the time to seek keys by increasing the supercell size is only marginal. On the other hand, for queries of point locations, even though we need to retrieve a supercell (1024 pixels) for a single pixel value, the performance is not significantly degraded either. This is because even with a 32×32 supercell, the time for data retrieval is still about 100 microsecond or less, insignificant compared to the overhead (establishing https connection, logging the query etc.).

The overview layer key-value stores The discussed key design favors data retrieval from point spatial locations for a period, which matches the preferred user behavior (see Fig. 1.5). In this case, the starting and ending row keys for the point locations can be readily determined. One may simply retrieve all the row keys in between (retrieving all the key-values between a starting and an ending key is called a "scan" operation). In contrast, to retrieve data for an area for a period (or a set of time periods) is more problematic. The reason is that for any given spatial key, one does not necessarily know what temporal keys exist. For example, in satellite imagery, different parts of the earth are imaged at different times. Without prior timestamp information, one will have to either read out all the timestamps possible or first scan each spatial location to know what keys are available. In key-value store operations, such a "scan" operation often takes on more than 1 millisecond, which is much too slow compared to merely retrieving the value for a known key (called a "get" operation, typically on the order of 1 microsecond).

To overcome such difficulty, the PAIRS innovation includes the introduction of multi-level overview layers as illustrated in Fig. 1.6. An overview layer uses a similar key-value store structure (Table 1.2) as the main layer, except that its spatial resolution is coarse-grained by a factor of two per level-up. At selected overview levels, statistics of the supercells are stored. For example, each pixel of the fifth level overview layer stores the mean, min, and max values of the corresponding

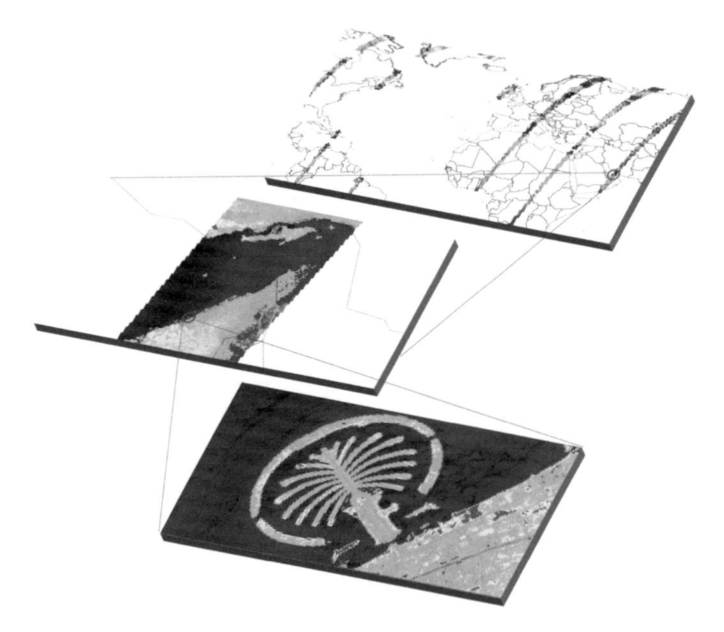

Fig. 1.6 Landsat-8 (NIR band) illustrating the relation of main layer (bottom), fifth level overview layer (middle) and tenth level overview layer (top)

32×32 pixel supercell in the main layer ($1024\times$ reduction in the number of pixels). Similarly, the tenth level overview layer stores the mean, min, max values of the corresponding 32×32 supercell in the fifth level overview layer.

To retrieve data for an area for a time period (or a set of time periods), one first scans the overview layer to obtain the timestamps available for the area. This enables one to pre-calculate all the row keys needed to retrieve in the main layer, leading to much faster data retrieval using the "get" operation (instead of the "scan" operation).

Moreover, the overview layer also brings the added benefit that it enables rapid overview visualization of the data layers and accelerated data filtering (e.g. "get data where temperature is below freezing"), because if the filtering condition can be ascertained by the overview layer, the retrieval of unnecessary data layers can be completely skipped.

PAIRS queries of point locations are usually served in real-time (with hundreds of milliseconds response times). In contrast, for queries of areas, Fig. 1.7 is a useful way to characterize their performance. In Fig. 1.7 the time for retrieving data is plotted as a function of data size. More complex queries would apparently change the curve. Since writing output to disk is the most expensive part of a query, query time can reduce substantially if a query involves data reduction (see examples in the next section).

From Fig. 1.7 one observes that if little data is retrieved, the performance is limited by *latency*, which is determined by the overhead of logging, authorization of

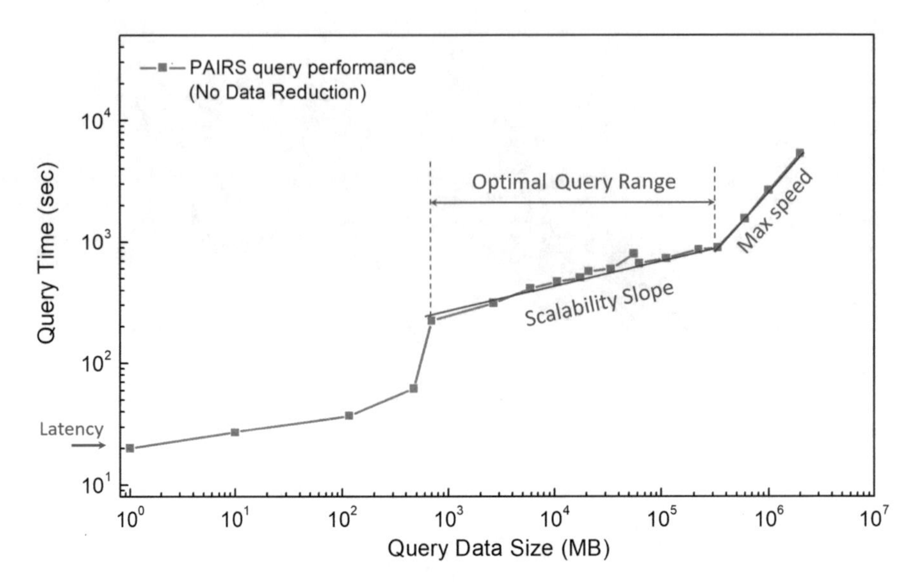

Fig. 1.7 PAIRS query times as a function of query data size

a query, and queueing for the availability of resources before starting a MapReduce job. As the processing time increases with query size, latency becomes negligible. Within an *optimal query size range*, the time is only weakly dependent on the size because the number of parallel mappers (in MapReduce) processing the query increases with its size. The slope in Fig. 1.7 in this regime characterizes the *scalability*. For a typical user, the PAIRS system has an optimal query size range from 0.5 to 500 GB. The scalability slope is ~0.2 because the number of mappers scales sub-linearly with the query size for a typical user. Beyond the optimal query size, the number of mappers reaches an upper bound, and the query time scales about linearly with size. The slope of the linear relation defines the max query *speed*, which is currently ~400 MB/s for a typical user. The maximum *throughput* is the largest possible query size limited by the memory available to hold the result for in-memory analytics, which is ~2 TB for a typical user.

1.4 PAIRS User Experience

The implantation and performance of the PAIRS key-value store discussed above enables a paradigm-shifting user experience for many geospatial-temporal use cases when compared to a conventional system. Indeed, the conventional usage of geospatial-temporal data at scale is convoluted. For example, a user intends to run analytics on satellite imagery using a conventional system such as the US Geological Survey (USGS) EarthExplorer (https://earthexplorer.usgs.gov/). One first selects an area-of-interest (AOI) and a time range to obtain a list of image files of the relevant satellite tiles. Then these image files are downloaded and processed for the user's task at hand. Suppose one is to obtain a time series of near-infrared surface reflectance from ESA's Sentinel-2 satellite for a particular region over many years. In such a scenario, often hundreds of tiles will need to be downloaded and opened/sought to extract the pixel(s) corresponding to the region of interest. Such a task gets increasingly complicated as more data sources become involved – e.g. we need temperature in addition to surface reflectance and/or we want to use data from other satellites such as LandSat or MODIS.

The PAIRS design emphasizes that the platform relieves the user from performing such data reorganization and provides a simple and unified experience regardless of the details of the original data. The PAIRS API and GUI capabilities are detailed in its documentation (https://pairs.res.ibm.com/tutorial/). In this section we illustrate how the PAIRS provides a new user experience. For generality, we discuss the PAIRS query examples using the native PAIRS REST API. Often the open sourced PAIRS Python SDK (short PAW = PAIRS Geoscope RESTful API Wrapper) (https://github.com/ibm/ibmpairs) which wraps the native PAIRS API offers more convenient interaction with PAIRS.

1.4.1 Data Service

The simplest service which PAIRS offers is the *data service*. For example, to obtain the time series of surface reflectance as well as temperature, one simply sends a POST request to PAIRS with a JSON payload (query_json). The sample Python code snippet is:

```
import pandas as pd, requests
pairs_auth = ('<username>', '<password>') # your username and password here.
query_json = {
    "layers" : [
        {
            "id" : "49361" # near IR surface reflectance, Sentinel-2 L2 band 8
        },
        {
            "id" : "49257" # 2m temperature, TWC gCOD hourly weather
        }
    ],
    "spatial" : {"type" : "point",  "coordinates" : ["41.213", "-73.798"]},
    "temporal" : {"intervals" : [
        {"start" : "2017-01-01T00:00:00Z", "end" : "2019-10-31T00:00:00Z"}
    ]}
}

# send a POST request containing query_json to PAIRS API endpoint.
api_response = requests.post(
    'https://pairs.res.ibm.com/v2/query', auth = pairs_auth, json = query_json
)
# convert the response json into a pandas dataframe
pairs_data = pd.DataFrame(api_response.json()['data'])
```

The query above requests about 3 years of near-infrared surface reflectance (Sentinel-2 band 8, PAIRS data layer id = 49361) and 2 m temperature (global

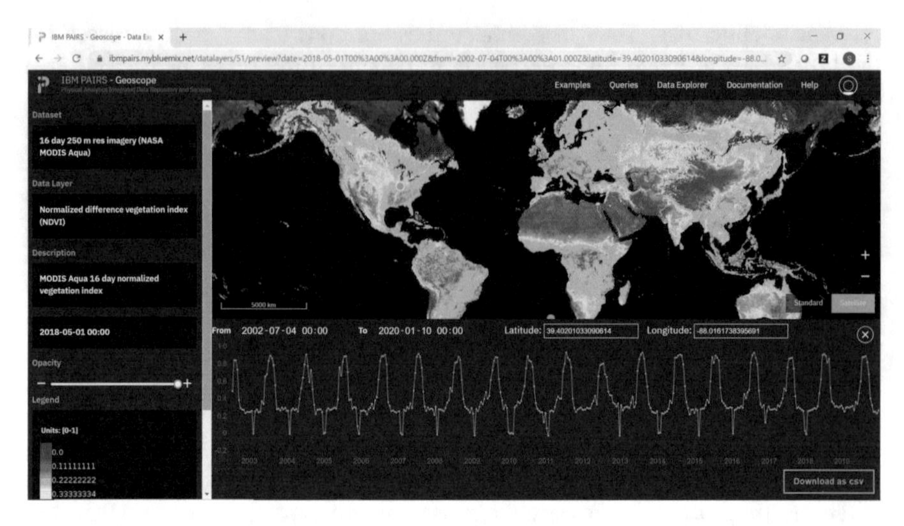

Fig. 1.8 PAIRS GUI (Geoscope) enables the real-time visualization of the content of data layers in PAIRS. This example shows global NDVI from MODIS Aqua on 2015-11-25 and a 20-year time series at latitude 37.3°/longitude −89.1°

weather history hourly temperature from TWC (The Weather Company, an IBM business), PAIRS data layer id = 49257) for a location somewhere in New York with the coordinates 41.213/−73.798 degree (latitude/longitude). PAIRS responds with a JSON (api_response) with about 100 records of surface reflectance and about 24,000 records of hourly temperature typically within a few hundreds of milliseconds. The last line of the code snippet above converts the JSON into a Pandas data frame for downstream analytics. Beyond point location query, a user may specify a query JSON with the spatial part representing a bounding box or a multi-polygon. In such case PAIRS returns either a set of geotiffs or CSVs (latitude, longitude, timestamp, value) for the queried area.

As the GUI counterpart to such query capability, a user can in real-time visualize the content of a data layer in PAIRS GUI as shown in Fig. 1.8. This includes (1) picking a timestamp to visualize a data layer as a color map and (2) picking a location and time range to visualize a time-series.

The performance of the query is the benefit of the key-value store design of PAIRS. Take the surface reflectance and temperature timeseries query above for example, given the key design shown in Table 1.2, the location and time range specified in the query can be directly translated to a set of starting and ending keys for scanning the HBase to retrieve data. In contrast, in a conventional file-based system, one would be forced to open and seek tens of thousands of files (each for different timestamps) to retrieve data, which limits performance. It may also be obvious to the readers that the visualization of the data layer content on the GUI relies on the construction of the overview layers (Fig. 1.6). Thus, at a given zoom level, the PAIRS overview layers can supply data to the GUI at the appropriately coarse-grained resolution.

1.4.2 Search or Discovery Service

More sophisticated than the data service, PAIRS enables a user to push spatial and/or temporal aggregation, filtering, math computation, and basic geospatial transformation into a query, which we refer to as the *search or discovery service*. Let us use the following query as an example. The example computes the summer 2018 max temperature for all the corn fields averaged for all states in the contiguous US (CONUS).

```
query_json = {
    "layers" : [
        {
            "id" : "92", # PRISM daily maximum temperature
            "temporal" : {"intervals" : [
                {"start" : "2018-06-01T00:00:00Z", "end" : "2018-08-31T00:00:00Z"}
            ]},
            "aggregation" : "Max"
        },
        {
             "id" : "111", # USDA cropscape
            "temporal" : {"intervals" : [{
                "start" : "2018-01-01T00:00:00Z", "end" : "2018-12-31T00:00:00Z"}
            ]},
            "aggregation" : "Mean", # collapse crop type in the temporal range into a single value
            # filter out spatial area for which crop type equals 1 (corn per USDA designation)
            "filter": {"expression" : "EQ 1"}, "output" : "false"
        }
    ],
     "spatial" : {
         "type" : "poly",
         "aoi" : "24", # polygon of Contiguous US
         # list of polygon id for 48 states
         "aggregation" : {"aoi": [121, 123, 124, … , 130, 131, 133, 134, … , 171]}
     },
     # "temporal" below is irrelevant as it is overridden by "temporal" within the data layers above
    "temporal" : {"intervals" : [
        {"start" : "2018-06-01T00:00:00Z", "end" : "2018-09-30T00:00:00Z "}
    ]}
}
```

There is quite a lot going on in the example. To begin we are requesting data for spatial area CONUS (“aoi” : “24”) and for two data layers: daily maximum temperature from PRISM dataset (“id” : “92”), and crop type from USDA cropscape dataset (“id” : “111”). The spatial resolution of PRISM is ~4 km (PAIRS level 14), while crop type is ~30 m (PAIRS level 21). PAIRS automatically samples the temperature data to match the crop type which is the highest resolution data layer of this query. Moreover, for each of the two data layers we use a different temporal range. We requested temporal range 06/01/2018 to 09/30/2018 for temperature and requested PAIRS to apply max aggregation for the time period (“aggregation” : “Max”) to obtain the highest temperature during the time period for each pixel. Separately for crop type, the temporal is 01/01/2018 to 12/31/2018. Temporal aggregation (“aggregation” : “Mean”) is applied to collapse crop type within the temporal range into a single value. A filter (“filter”: {“expression” : “EQ 1”}) is applied, which request PAIRS to retrieve data for only spatial areas with crop type =1 (i.e. corn per US department of agriculture convention). Finally, a spatial aggregation is specified with a list of polygon id’s for the 48 contiguous US states (“spatial”: {“type”: “poly”, “aoi”: “24”, “aggregation”: {“aoi”: [121, 123, 124, 125,

… … 171]}}).PAIRS thus spatially aggregates the corn field temperature by states and provides an output file in the default CSV format.

A salient character of such a query is that all the computations are performed in parallel in the PAIRS cluster. While there is a large amount of data being processed, a user merely retrieves a CSV with 48 rows in which each row contains the 2018 spatially averaged summer max temperature of corn fields for one state.

We can now take such query capability to answer some less trivial questions. Say we are interested in the impact of global warming on agriculture, thus would like to know which part of the croplands in the northern hemisphere have seen a substantial summer daily maximum temperature (Tmax) rise of over 1.5 °C in the last 40 years. The sample query is below.

```
query_json = {
    "layers" : [
        {
            # data layer id 49188 is daily maximum temperature (Tmax) at 2 m above surface
            # virtual layer "Y2018" is the mean summer Tmax of Jun to Aug 2018. Same below.
            "alias" : "Y2018","temporal" : {"intervals" : [{"start" : "2018-06-01",\
            "end" : "2018-08-31"}]}, "id" : 49188, "aggregation" : "Mean",
            "output" : "false" # PAIRS does not write output for this layer
        },
        {
            # virtual layer "Y2017" is the mean summer Tmax of 2017.
            "alias" : "Y2017","temporal" : {"intervals" : [{"start" : "2017-06-01",\
            "end" : "2017-08-31"}]}, "id" : 49188, "aggregation" : "Mean", "output" : "false"
        },

        … … # not showing "Y1980" to "Y2016" due to space limitation

        {
            # virtual layer "Y1979" is the mean summer Tmax of 1979.
            "alias" : "Y1979","temporal" : {"intervals" : [{"start" : "1979-06-01",\
            "end" : "1979-08-31"}]}, "id" : 49188, "aggregation" : "Mean", "output" : "false"
        },
        {
            # the mean summer Tmax difference between 2009 to 2018 and 1979 to 1998
            "alias" : "TempDiff",
            "expression" : " ($Y2018 + $Y2017 + $Y2016 + $Y2015 + $Y2014 \
                           + $Y2013 + $Y2012 + $Y2011 + $Y2010 + $Y2009)/10 \
                           - ($Y1988 + $Y1987 + $Y1986 + $Y1985 + $Y1984 \
                           + $Y1983 + $Y1982 + $Y1981 + $Y1980 + $Y1979)/10",
            "filter" : {"expression" : "GT 1.5"} # filter out pixels of value greater than 1.5
        },
        {
            "alias" : "crop_fraction",
            "temporal" : {"intervals": [{"snapshot" : "2017-01-01"}]},
            "id" : 49307, # crop fraction at 250 m resolution, survey of timestamp 2017-01-01
            "aggregation" : "Mean",
            "filter" : {"expression" : "GT 0.5"} #  filter where crop fraction > 50%
        }
    ],
     "spatial" : {
        # bbox north hemisphere, latitude 0 to 80 deg north and longitude -179.9 to 179.9 deg east
         "type" : "square", 'coordinates': [0,-179.9, 80, 179.9]
     },
     # "temporal" below is overridden by "temporal" within the data layers above
    "temporal" : {"intervals" : [
        {"start" : "1976-01-01", "end" : "2018-12-31"}
    ]}
}
```

In this example, we are requesting data for the northern hemisphere between latitude −0 to 80 degree north and longitude −179.9 to 179.9 degree east (defined by ‘coordinates’: [0, −179.9, 80, 179.9]). A number of user-defined “intermediate” layers are created by "mean" aggregation of Tmax (PAIRS data layer id = 49188). For example layer "Y2018"

```
{
    # data layer id 49188 is daily maximum temperature (Tmax) at 2 m above surface
    # virtual layer "Y2018" is the mean summer Tmax of Jun to Aug 2018. Same below.
    "alias" : "Y2018","temporal" : {"intervals" : [{"start" : "2018-06-01",\
    "end" : "2018-08-31"}]}, "id" : 49188, "aggregation" : "Mean",
    "output" : "false" # PAIRS does not write output for this layer
},
```

represents the mean Tmax in summer (June to August) 2018. Note that "output" : "false" instructs PAIRS to not write output, thus the intermediate layer stays only in memory.

Based on the intermediate layers, a user defined function (UDF)

```
{
    # the mean summer Tmax difference between 2009 to 2018 and 1979 to 1998
    "alias" : "TempDiff",
    "expression" : " ($Y2018 + $Y2017 + $Y2016 + $Y2015 + $Y2014 \
                   + $Y2013 + $Y2012 + $Y2011 + $Y2010 + $Y2009)/10 \
                   - ($Y1988 + $Y1987 + $Y1986 + $Y1985 + $Y1984 \
                   + $Y1983 + $Y1982 + $Y1981 + $Y1980 + $Y1979)/10",
    "filter" : {"expression" : "GT 1.5"} # filter out pixels of value greater than 1.5
},
```

computes the mean summer daily maximum temperature difference between 2009 to 2018 and 1979 to 1988. The temperature difference is subsequently filtered by "filter": {"expression" : "GT 1.5"}, i.e. selecting the pixels of temperature rise over 1.5 °C. Moreover, a filter using crop fraction (data layer id 49307) selects the pixels in which the crop fraction percentage is over 50%.

The result of the query is shown in Fig. 1.9 below, which concludes that Europe croplands had the most notable summer daily maximum temperature rise in the last 40 years.

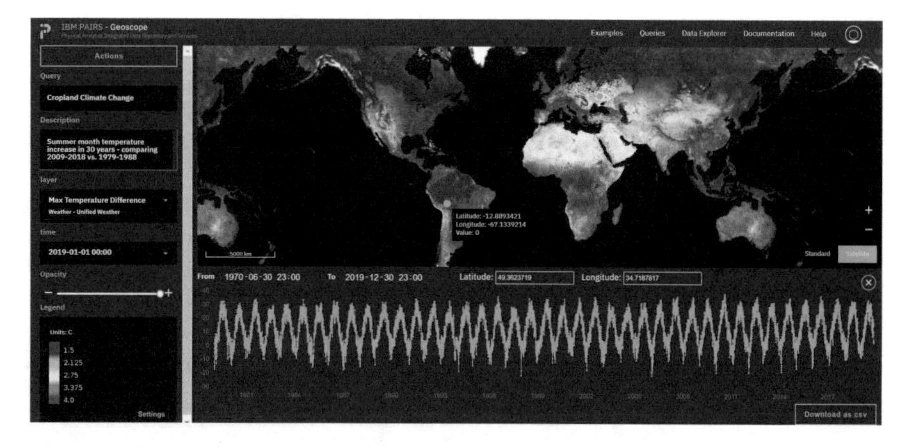

Fig. 1.9 A PAIRS query showing northern hemisphere croplands in which mean summer daily maximum temperature has risen over 1.5 °C comparing 1979 to 1988 and 2008 to 2018. Note that Europe stands out as the most affected region

The query above involves over 1800 timestamps of temperature and 6e9 spatial grid points (northern hemisphere at around 250 m meter resolution, PAIRS level 18 for crop fraction). This combination represents ~1e13 spatial-temporal grid points. Such a task, starting from raw data gathering may conventionally take a data scientist from days to weeks to complete. In contrast, it took a single query and around 60 seconds to execute on PAIRS, showcasing the processing power and user experience on the PAIRS platform. While the UDF employed in the query is simplistic involving only arithmetic operations, PAIRS UDF supports common mathematical and logical operators and functions. More sophisticated analytics which can be pushed into UDFs includes regression models and decision trees. Some examples can be found in this reference (https://github.com/IBM/ibmpairs/tree/master/examples).

In addition to the functionalities discussed above, a PAIRS query may also include common geospatial processing including coarse graining, contouring, as well as customized functions such as delineating trees from satellite images (see section below on vegetation management) etc. The list of built-in functionalities is continually evolving. For the latest refer to documentation (https://pairs.res.ibm.com/tutorial/).

1.4.3 Analytics Platform Service

Finally, it is anticipated that a query (*discovery service*) by itself may not be able to perform all the analytics a user may want to. In such cases, it is expected that a PAIRS query would have substantially reduced the amount of data via aggregation, filtering etc. as discussed earlier. The last mile of customized analytics beyond query capability is handled by *analytics platform service*. A user may request a Python Jupyter Notebook on the PAIRS cluster or an IBM Watson Studio Notebook which contains the query result as data frame(s) to be launched. In API mode, a user makes the request using the id of a completed query and gets in the response a unique URL for the notebook. In the GUI, a user clicks on the "generate Jupyter Notebook" button in the "Actions" menu as shown in the screenshot (Fig. 1.10). For resource management and access control purposes the notebook is dockerized (Merkel 2014). In the Jupyter Notebook, the user can take advantage of all the latest modules of Python including PyTorch for deep learning and SPARK for scalable processing. For privileged users, a big query result will be returned as a SPARK data frame instead of a usual Pandas data frame. A SPARK data frame is distributed throughout the memory of the PAIRS cluster when possible. Using PySpark, the data frame may be accessed from the Jupyter Notebook through a set of RESTful APIs orchestrated by an Apache Livy server.

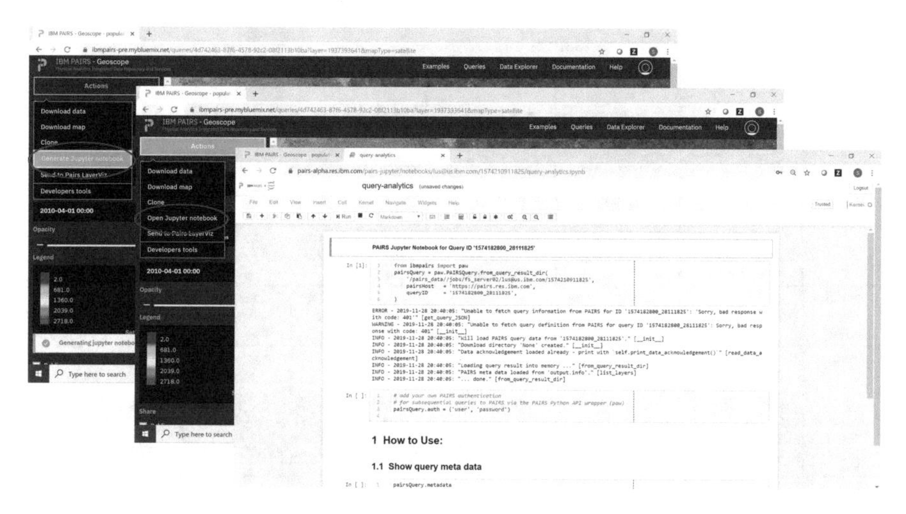

Fig. 1.10 GUI interface by which a user may spin up a Jupyter Notebook on the PAIRS cluster from a query result. The Jupyter Notebook is pre-loaded with the query result as dataframe(s)

1.5 Selected Industry Applications

Following the discussion of PAIRS architecture and user experience above, we present next a couple of selected geospatial-temporal use cases whose solutions were developed using PAIRS in the past few years. Geospatial-temporal use cases are, generally speaking "What-When-and-Where" type of applications and are naturally plentiful cutting across multiple industries and sectors (Table 1.4), such as government (how, when and where to respond to a disaster such as a hurricane?), retail (what, when and where to promote a product?), finance (when and where to buy and sell what commodities), agriculture (when and where to apply the right amount of fertigation), or energy (when, where and how much renewable energy will be generated?). While sometimes such PAIRS enabled applications are described as being "on top" of PAIRS, we note that this notion is misleading. It is better to refer to such applications as ones "within" PAIRS as they exploit the in-data computation capabilities as discussed in the previous sections.

1.5.1 PAIRS Enabled Improvements in Weather Forecasting

One of the useful applications of PAIRS is weather forecasting, which is an old field of science (i.e., meteorology) but still very actively researched. For one, weather impacts literally every aspect of our lives and the economy, and for another, weather is highly complex with many aspects of the underlying physical phenomena not quite understood. It is thus well known that not every weather model provides

Table 1.4 Exemplary PAIRS industrial use cases

Industry	Finance	Utility	Agriculture	Insurance	Retail	Public
Example	Commodity trading	Vegetation management	Decision support for agriculture	Develop risk models (Flood, Fire)	Supply change	Disaster response
Example Queries / Questions	How much crop is planted? How much corn will be produced in Argentina?	Where do trees infringe on utility assets? When to schedule tree trimming?	What is the best crop to plant? When to apply fertilizers, pesticides?	Where are my assets at risk? At what time of the year is the risk the highest ?	Where, when and what to ship? Where should I promote a product?	Who is being impacted by a hurricane? What is the best emergency response?
PAIRS Data Layers	Weather, land class, soil, satellite	Weather, satellite, power line data	Weather, land class, soil, satellites	Climate, vegetation, traffic, census	Weather, socio-economic data, store locations	Weather, climate, satellites, map, socio-economic data
Example	Early Crop Recognition	Tree Identification	Precision Agriculture	Flooding Risk	Optimal store locations	Emergency Response

the same forecasts and accuracy. By way of example, this becomes very evident during extreme weather events such as hurricanes, where multiple weather models can project very different pathways (Brennan and Majumdar 2011). The difference in the forecasts is naturally most pronounced for longer term forecasts (going beyond 10 days) and thus we focus in the following on such long-term forecasts. Specifically, we discuss briefly how PAIRS can be used to improve the accuracy of such forecasts using its scalable big data processing capabilities by leveraging state-of-the-art machine-learning techniques. While the discussion will be focused on weather, it should be noted that the general framework, presented below, is applicable to "consolidate" between different forecasts or prediction modalities, which is a common challenge. For example, the presented framework could be used to consolidate the information received from different IoT sensor systems, which measure similar or related but not agreeing parameters.

Table 1.5 shows a selected list of data sets available in PAIRS, which are relevant for improving the accuracy of long-term forecasting. This includes weather station data from RAWS (=Remote Automatic Weather Stations) (Horel and Dong 2010; https://raws.nifc.gov/), the ISD (=Integrated Surface Database) from NOAA (National Oceanic and Atmospheric Administration) (Smith et al. 2011; https://www.ncdc.noaa.gov/isd), and NOAA's Surface Radiation network (SurfRad) (Augustine et al. 2008; https://www.esrl.noaa.gov/gmd/grad/surfrad/index.html). PAIRS also includes data from GPS-RO (=Radio Occultation) (Kuo et al. 2004; https://www.cosmic.ucar.edu/), which is a technique for measuring atmospheric parameters from space. There are also outputs from several extended or long range weather forecast models available, including NOAA's CFS v2 (=Climate Forecasting System) (Saha et al. 2006), as well as extended range forecasts from ECMWF (=European Centre for Medium-Range Weather Forecasts) (https://www.ecmwf.int/en/forecasts/datasets/set-vi) and from JMA (=Japanese Meteorology Agency) (Japan Meteorological Agency 2013). Note that Table 1.5 is only a rough estimate about the total amount of data content one may be able

Table 1.5 Selected data sets in PAIRS including long-term weather models and weather station data

	Weather station			Satellite	Model		
Source	RAWS	NOAA ISD	NOAA SurfRad	GPS-RO (Cosmic I)	NOAA CFS v2	ECMWF Extended	JMA Extended
Type	Vector	Vector	Vector	Vector	Raster	Raster	Raster
Coverage	US	Global	US	Global	Global	Global	Practically Global
Spatial resolution	Point ~2200 sites	Point ~30,000 sites	Point 7 sites	Point ~1000 ROs per day	0.5 deg	0.4 deg	1 deg
Temporal resolution	<1 h	1 h	1 min	~1 min per ROs	6 h	6 h	24 h
Forecasting horizon	NA	NA	NA	NA	0–6 months	0–46 days	0–30 days
Forecasting issuance	NA	NA	NA	NA	4 times per day	Twice per week	Once per week
Ensemble forecast	NA	NA	NA	NA	4 members	51 members	50 members
Estimated date size	~4 MB/day	~30 MB/day	~2 MB/day	~9 MB/day	~1 TB/day	~400 GB/day	~1 GB/day

to retrieve from the data sources from a user's perspective. It is not about the internal complexity and data processing necessary to produce the user accessible outputs. For example, the ECMWF extended range forecasts are for 46 days ahead at 6 hourly resolution (185 timestamps) and for a 0.4 degree global grid (globally ~4e5 grid points). On the order of 100 parameters and/or pressure levels and 51 ensemble members are available from the forecasts. The forecasts are issued twice a week. Assuming parameter values are stored as four bytes floating-point numbers, we estimate that, phenomenologically, the daily data content is around $185 \times 4e5 \times 100 \times 51 \times 2 / 7 \times 4 \sim 400$ GB/day.

Table 1.5 highlights the complexity of the data integration. For example, the forecast models not only differ in the underlying physics and assumptions which are used to generate them but also provide data at different spatial and temporal resolutions and cover different forecasting horizons. By virtue of PAIRS, the output from all these different models are "automatically" harmonized, integrated and spatially linked. Note that although many relevant datasets, such as ECMWF weather reanalysis etc. are omitted in Table 1.5, the amount of data listed already amounts to over 500 TeraBytes annually. Clearly to exploit all that data within a reasonable processing time, a scalable big data platform such as PAIRS is required.

As for any machine-learning task, the analytics includes two steps (training and deployment). In the training, first, an error analysis is performed to identify the most important features, where historical forecasts are compared with actual measurements from high quality weather station. Because PAIRS allows quick and scalable access to data, this can be followed by a very granular functional analysis of variance (*FANOVA*) (Hooker 2007), which identifies zeroth, first and second order errors of the forecasted parameters (such as temperature, precipitation rate etc.). By way of example, the zeroth order of a temperature forecast is just a bias, while the first order error depends on one feature and the second order error on two features and so on. Such features can include other forecasted parameters. An example of a second order error of a 30-day temperature forecast from NOAA CFSv2 member 1 is shown in Fig. 1.11, where it is compared to a class I weather station from the SurfRad network (here for Bondville, IL, 40.05192°N, 88.37309°W). In this example, the second order error is a function of wind speed and solar irradiance. As shown in Fig. 1.11, FANOVA reveals that for this specific location the NOAA CFSv2 model on average *overpredicts* the temperature if the forecasted wind speed and solar irradiance are high. However, the same model *underpredicts* the temperature if the wind speed is low and solar irradiance is high. Clearly and as shown in Fig. 1.11, different regimes of weather categories can be identified. We note that the second order error does not only depend on the forecast location, but it is also a function of forecast horizon (e.g., 30 days ahead vs 60 days ahead), and forecast parameter (e.g., temperature, precipitation). While we only show in Fig. 1.11 four weather categories, many others are identified using FANOVA for different features, forecast horizons, locations etc.

Next we train an individual machine learning (ML) for each weather situation, forecast horizon, location and forecast parameter. Best results have been achieved

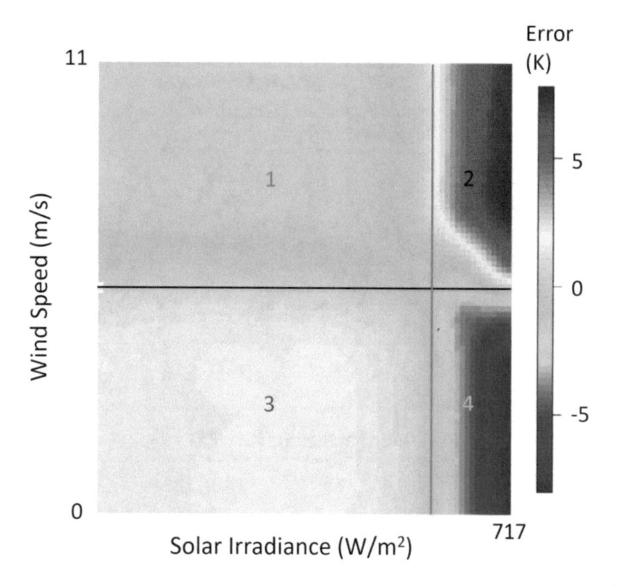

Fig. 1.11 Second order error of a 30-day ahead temperature NOAA forecast for Bondville, IL, USA (here as a function of wind speed and solar irradiance)

using ensemble learning methods for regression (random forest) although other ML methods show good results as well. One may even run multiple ML models in parallel and adopt a multi-expert learning system which dynamically picks the most accurate ML method based on recent performance. The key, however, is to have on-demand access to data frames of training label (ground truth) and a large number of features so that the important parameters for different weather situations and different ML methods can be selected.

A typical training involves querying over 3 years of historical training data from around 100,000 point locations globally, which means around 1 TeraBytes have to be processed for each forecast variable – a nontrivial task. For this purpose, a specific data assembly module is used. The module uses an XML file as the template to construct complex PAIRS queries, manages those queries, and reorganizes the query results into training or forecasting data frames. As noted in Table 1.5, NOAA CFS forecasts are issued four times per day, while ECMWF and JMA extended range forecasts are issued twice and once a week, respectively. Thus, one of the roles of the data assembly module is to pick different lead times (Δt between the issue of a forecast and the actual forecasted time) for the different forecasting models in the PAIRS query so that the latest forecasts of the models are selected.

In the deployment step, after data assembly and classifying the respective weather categorization, we apply the specific trained machine-learning model for this case, for each forecast parameter, location, and forecast horizon if applicable. Figure 1.12 shows an example of such a forecast, which nicely illustrates the power of the approach and how a scalable platform such as PAIRS can help to develop such fine-grained ML models. We show in Fig. 1.12 the best long-term forecast from the four members of NOAA's CFSv2 model (in red) for a location with a high quality

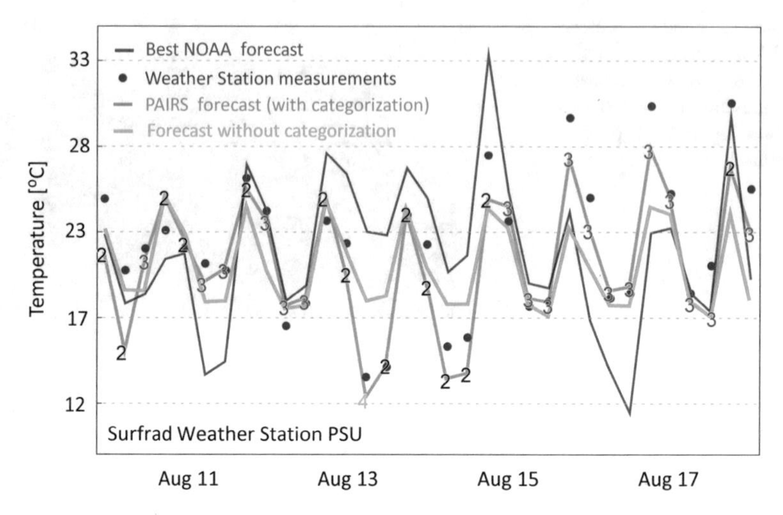

Fig. 1.12 Comparison of a 30 day ahead temperature forecasts from the "best" NOAA forecasts and a PAIRS forecast (with and without categorization) with measurements from a SurfRad weather station at Pennsylvania State University (year 2015)

weather station at Pennsylvania State University (40.72012°N, 77.93085°W). The temperature measurements are shown in blue. Because forecasts were available every 6 hours the comparison between forecast and measurement is performed for the same time interval. As becomes evident the NOAA CFS forecast provides moderate accuracy. For comparison we have plotted forecasts, where we used machine-learning without categorization (green) and with categorization (blue). While the machine-learnt forecasts without categorization show improvements over NOAA CFS forecast it tends to be "biased towards the mean", which is a common pitfall of certain machine-learning approaches. Clearly, as shown in Fig. 1.12, PAIRS big data capabilities, which enable specific machine-learning for each weather category, this problem can be mitigated and overall the best accuracy can be achieved. For reference in this plot we show four corresponding weather categories (labeled from 1 to 4).

While the data in Fig. 1.12 shows just a snapshot for a single location and forecast parameter (i.e., temperature) we show in Fig. 1.13 the mean absolute error (MAE) results for 7 locations in the US with class 1 weather stations (Bondville, IL (40.05192°N, 88.37309°W), Boulder, CO (40.12498°N, 105.23680°W), Desert Rock, NV (36.62373°N, 116.01947°W), Fort Peck, MT (48.30783°N, 105.10170°W), Goodwin Creek, MS (34.2547°N, 89.8729°W), Penn State, PA (40.72012°N, 77.93085°W), Sioux Fall, SD (43.73403°N, 96.62328°W)) for wind speed and temperature, respectively in comparison with the four NOAA CFS member models (for the duration from 06/20/15 – 09/20/15). Figure 1.13 shows improvements in MAE of 30% over the best NOAA CFS ensemble model member.

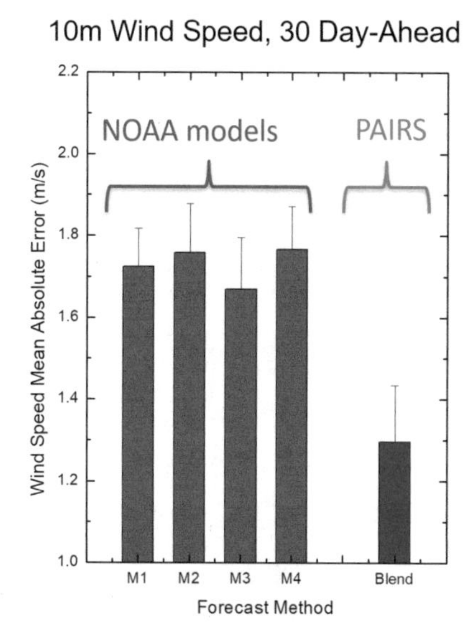

Fig. 1.13 Mean absolute error (MAE) of the PAIRS model for wind speed and temperature for seven locations in the US with class 1 weather stations (Surfrad) in comparison with the 4 NOAA CFS ensemble model members (for the duration from 06/20/15 – 09/20/15)

1.5.2 Vegetation Management

Besides weather forecasting, PAIRS recently attracted many applications in the electric utility industry. In many parts of the world electrical powerlines are above ground. For example, in the US alone, the electrical grid includes 200,000 miles of transmission and 5.5 million miles of distribution lines and almost all of it is above ground, where it can come in contact with vegetation. However, this is a safety hazard and can cause major outages or even spark wildfires. Consequently, many electric utilities have complex vegetation management programs, which include regular pruning, brush removal, herbicide applications, etc., to prevent the vegetation from interfering with the utility assets and overgrowing of the conductors. Naturally such line clearance programs are not only difficult and complex but most importantly very costly with vegetation management being the largest preventive maintenance expense for many utilities (Hollenbaugh and Champagne 2006). By way of example, it has been reported that San Diego Gas & Electric must trim more than 450,000 trees regularly (Rodgers 2014).

It is well known that tree growth can vary tremendously by species and other environmental conditions such as weather, soil etc. (McPherson and Peper 2012). However, the lack of monitoring capabilities, which could provide actionable insights in where and when vegetation management is required, leaves many utilities

Table 1.6 Different data sets in PAIRS relevant for vegetation management

	Other	Satellite		Weather		Soil	
Source	LiDAR	NAIP	Sentinel II	NAM	GFS	SURGO	Soilgrid
Type	Vector	Raster	Raster	Raster	Raster	Vector	Raster
Coverage	Local	US	Global	US	Global	US	Global
Spatial resolution	<0.1 m	0.5 m	10 m	5 km	0.25 deg	Point	250 m
Temporal resolution	NA	2 y	Weekly	1 h	3 h	NA	NA
Forecasting horizon	NA	NA	NA	0–60 h	6–192 h	NA	NA
Estimated data size	NA	~80 TB/year	~12 TB/day	~0.6 TB/day	~1 TB/day	~1 TB	~2 TB

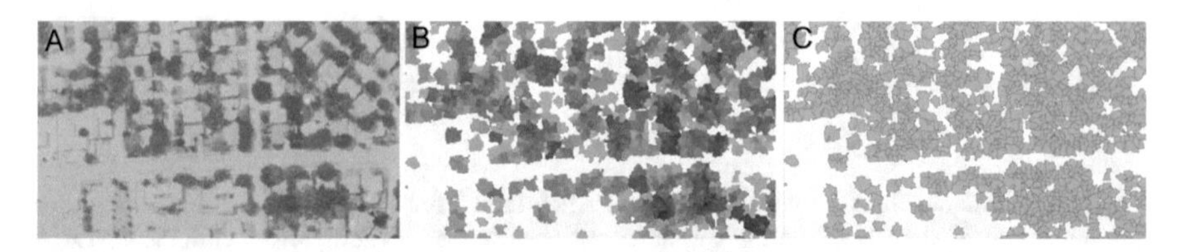

Fig. 1.14 Shows the basic process of tree delineation using the normalized difference vegetation index (NDVI): (**a**) raw NDVI, (**b**) result of smart thresholding, (**c**) vectorization to obtain the outline of tree canopies

with no other choice than regular maintenance schedules for their programs, which is naturally non-optimal and adds to the already very large cost.

In the following we discuss briefly how a vegetation management solution has been developed using PAIRS big data processing capabilities. Table 1.6 shows selective data sets which are available in PAIRS and are relevant to understand and monitor the progression of vegetation. As in the previous example, multiple very large data sources with different data types, resolutions etc. must be integrated.

Key to the vegetation management solution in PAIRS is the combination of high and low-resolution (spatial) hyperspectral aerial/satellite imagery. The high-resolution imagery, for example, the NAIP dataset in US at 0.5–1 m spatial resolution (USDA NAIP n.d.), enables the computation of a vegetation base layer. In some cases, the base layer can also be derived from LiDAR (light detection and ranging) data sets, which can be easily processed in PAIRS (Klein 2019). Both LiDAR and/or NAIP data sets can be used to estimate and delineate the canopy size as illustrated in Fig. 1.14. The computation of the vegetation from NAIP is based on the normalized difference vegetation index (NDVI), which is the normalized difference between the red and near infrared band of the imagery. In some cases, some additional data layers (i.e., land use, OpenStreet map etc.) can be leveraged to improve the tree identification process.

The difference between different types of vegetation can be inferred from a time series of remote observations of the vegetation index, which is often not available

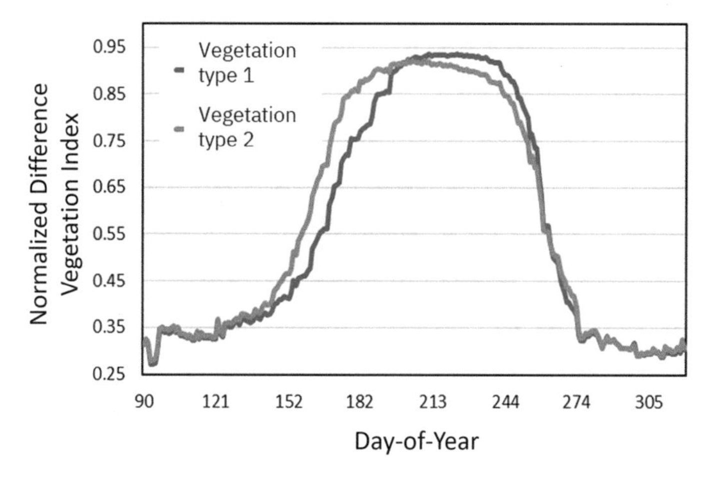

Fig. 1.15 Time-series of normalized difference vegetation index (NDVI) for two different vegetation types

at such high spatial resolution. However, the Sentinel-2 satellite from the European space agency (ESA) provides NDVI data at 10 m spatial resolution every 5 days if clouds are not interfering (ESA Sentinel-2 n.d.). An alternative data source, which is not listed in Table 1.6, but available in PAIRS is the LandSat-8 dataset from the USGS (USGS LandSat n.d.). Figure 1.15 shows a time series of NDVI data for two different vegetation types. In combination with ground truth data, which can also be obtained from LiDAR scans, such information can then be used to identify the vegetation type and tree species as applicable.

Combining tree type identification with consecutive high-resolution imagery or LiDAR scans, one can further estimate the tree growth using the canopy size. The basic relationship between tree canopy size and tree growth is shown in Fig. 1.16 (McPherson and Peper 2012) for different weather conditions.

Finally, we show in Fig. 1.17 the results from such an analysis with the delineated vegetation and tree height in the vicinity of a power line. While the model is simplistic, initial validations have shown that this model can provide between 80–90% accuracy (Klein 2019).

1.6 Conclusion and PAIRS Resources

To conclude, IBM Physical Analytics Integrated Data Repository and Services (PAIRS) is a big data and analytics service platform designed to support complex industrial applications which require analytics of a wide range of geospatial-temporal data. PAIRS features highly scalable (PetaBytes scale) storage of curated data and processing close to the data for advanced analytics and offers a unified user interface and user experience independent of the source of such data. It substantially

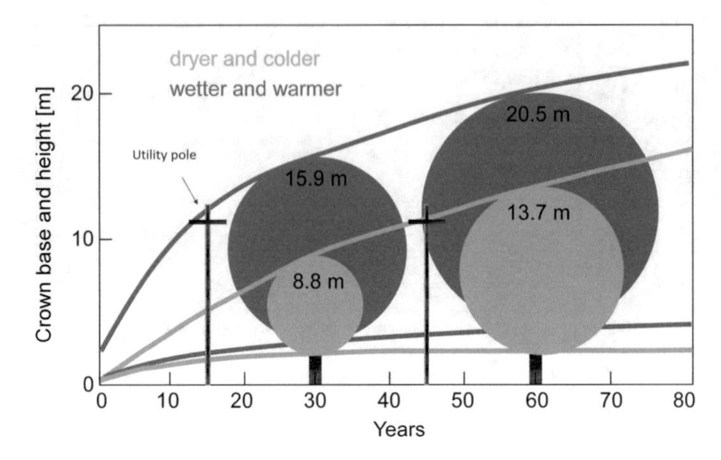

Fig. 1.16 Shows the relationship between tree height and canopy size for different environmental conditions for a green ash tree. (Rendered from McPherson and Peper 2012)

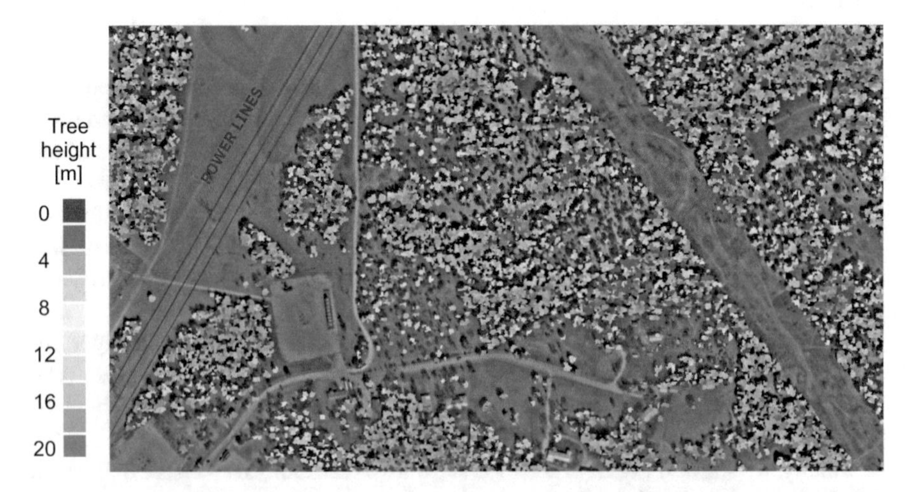

Fig. 1.17 Tree distribution around power lines after processing high resolution aerial and LiDAR data

reduces users' data management burden and, in many use cases, enables the users to drastically accelerate "data-to-sights" through its ability to compare, combine, filter, sort and display multiple large data sets simultaneously for correlation discovery and change detection.

Some useful PAIRS resources are provided below.

- Freemium GUI: https://ibmpairs.mybluemix.net
- API tutorial and reference: https://pairs.res.ibm.com/tutorial/
- Data documentation: https://ibmpairs.mybluemix.net/data-explorer
- Python SDK and sample Jupyter Notebooks (open source): https://github.com/ibm/ibmpairs

Acknowledgement The authors thank Bruce Elmegreen, Fernando Marianno, Ildar Khabibrakhmanov, Michael Schappert, Conrad Albrecht, Marcus Freitag, Johannes Schmude, Levente Klein, Wang Zhou, Carlo Siebenschuh, Xiaoyan Shao, Rui Zhang, Norman Bobroff, Patrick Dantressangle, Simon Laws, Steffan Taylor, Robert Goodman, Ivan Milman, Kelvin Shank, Ismael Faro, Axel Hernandez Ferrera, Daniel Wolfson and many other colleagues for their work on PAIRS and the useful discussions.

References

Augustine JA et al (2008) An aerosol optical depth climatology for NOAA's national surface radiation budget network (SURFRAD). J Geophys Res Atmos 113(D11)

Bouwmeester J, Guo J (2010) Survey of worldwide pico-and nanosatellite missions, distributions and subsystem technology. Acta Astronaut 67(7–8):854–862

Brennan MJ, Majumdar SJ (2011) An examination of model track forecast errors for hurricane Ike (2008) in the Gulf of Mexico. Weather Forecast 26(6):848–867

Chen J et al (2011) Exploratory data analysis of activity diary data: a space–time GIS approach. J Transp Geogr 19(3):394–404

Clarke KC (1986) Advances in geographic information systems. Comput Environ Urban Syst 10(3–4):175–184

Consortium OG.. Web map service

Consortium OG.. Web processing service

Contributors P, PDAL Point Data Abstraction Library 2018

Dean J, Ghemawat S (2008) MapReduce: simplified data processing on large clusters. Commun ACM 51(1):107–113

DeCandia G et al (2007) Dynamo: Amazon's highly available key-value store. In: ACM SIGOPS operating systems review. ACM

Dimiduk N et al (2013) HBase in action. Manning Shelter Island

Dubayah RO, Drake JB (2000) Lidar remote sensing for forestry. J For 98(6):44–46

ECMWF ENS extended forecast. https://www.ecmwf.int/en/forecasts/datasets/set-vi

Ekanayake J, Pallickara S, Fox G (2008) Mapreduce for data intensive scientific analyses. In: 2008 IEEE fourth international conference on eScience. IEEE

ESA Sentinel-2. https://sentinel.esa.int/web/sentinel/missions/sentinel-2

Geospatial analytics market worth 86.32 billion USD by 2023. https://www.marketsandmarkets.com/PressReleases/geospatial-analytics.asp

Gubbi J et al (2013) Internet of Things (IoT): a vision, architectural elements, and future directions. Futur Gener Comput Syst 29(7):1645–1660

Hakala T et al (2012) Full waveform hyperspectral LiDAR for terrestrial laser scanning. Opt Express 20(7):7119–7127

Harter T, et al. Analysis of {HDFS} Under HBase: a Facebook messages case study. In: Proceedings of the 12th {USENIX} conference on file and storage technologies ({FAST} 14). 2014a

Harter T et al (2014b) Analysis of HDFS under HBase: a facebook messages case study. In: FAST

Henderson C (2014) Mastering GeoServer. Packt Publishing Ltd

Hollenbaugh R, Champagne B. Utility vegetation management: the key driver of system reliability. 2006

Hooker G (2007) Generalized functional anova diagnostics for high-dimensional functions of dependent variables. J Comput Graph Stat 16(3):709–732

Horel JD, Dong X (2010) An evaluation of the distribution of remote automated weather stations (RAWS). J Appl Meteorol Climatol 49(7):1563–1578

Hughes JN et al (2015) Geomesa: a distributed architecture for spatio-temporal fusion. In: Geospatial informatics, fusion, and motion video analytics V. International Society for Optics and Photonics

IBM PAIRS application examples. https://github.com/IBM/ibmpairs/tree/master/examples
IBM PAIRS freemium. https://ibmpairs.mybluemix.net
IBM PAIRS tutorial and reference manual. https://pairs.res.ibm.com/tutorial/
IBM PAIRS SDK. https://github.com/ibm/ibmpairs
Japan Meteorological Agency, 2013: Outline of the Operational Numerical Weather Prediction at the Japan Meteorological Agency. Appendix to WMO Tech. Progress Rep. on the Global Data-Processing and Forecasting System and Numerical Weather Prediction, 188 pp. Available online at http://www.jma.go.jp/jma/jma-eng/jma-center/nwp/outline2013-nwp/index.htm
Klein LJ (2019) N-dimensional geospatial data and analytics for critical infrastructure risk assessment. IEEE Big Data conference, in press
Klein LJ et al (2015) PAIRS: a scalable geo-spatial data analytics platform. In: 2015 IEEE international conference on Big Data (Big Data). IEEE
Kuo YH et al (2004) Inversion and error estimation of GPS radio occultation data. J Meteorol Soc Jpn 82(1B):507–531
Lam C (2010) Hadoop in action. Manning Publications Co
Lu S et al (2016) IBM PAIRS curated big data service for accelerated geospatial data analytics and discovery. In: 2016 IEEE international conference on Big Data (Big Data). IEEE
McPherson EG, Peper PJ (2012) Urban tree growth modeling. J Arboricult Urban For 38(5):175–183
Merkel D (2014) Docker: lightweight linux containers for consistent development and deployment. Linux J 2014(239):2
National Interagency Fire Center (NIFC) Remote Automated Weather Stations (RAWS). https://raws.nifc.gov/
NOAA ISD. https://www.ncdc.noaa.gov/isd
NOAA Surfrad. https://www.esrl.noaa.gov/gmd/grad/surfrad/index.html
Petiteville I,. personal communication
Resch B et al (2014) GIS-based planning and modeling for renewable energy: challenges and future research avenues. ISPRS Int J Geo Inf 3(2):662–692
Rodgers D (2014) Trimming the cost of tree trimming. T&D World Magazine
Saha S et al (2006) The NCEP climate forecast system. J Clim 19(15):3483–3517
Smith A, Lott N, Vose R (2011) The integrated surface database: recent developments and partnerships. Bull Am Meteorol Soc 92(6):704–708
Tang L, Shao G (2015) Drone remote sensing for forestry research and practices. J For Res 26(4):791–797
UCAR COSMIC. https://www.cosmic.ucar.edu/
USDA NAIP (National Agriculture Imagery Program). https://www.fsa.usda.gov/programs-and-services/aerial-photography/imagery-programs/naip-imagery/
USGS EarthExplorer. https://earthexplorer.usgs.gov/
USGS LandSat. https://www.usgs.gov/land-resources/nli/landsat
Vohra D (2016) Defining the row keys. In: Apache HBase Primer. Springer, pp 117–119
Warmerdam F (2008) The geospatial data abstraction library. In: Open source approaches in spatial data handling. Springer, pp 87–104
Weber RM (2016) Internet of Things becomes next big thing. J Financ Serv Prof 70(6)
Whitby MA, Fecher R, Bennight C (2017) Geowave: utilizing distributed key-value stores for multidimensional data. In: International symposium on spatial and temporal databases. Springer
Yuan M (2009) Challenges and critical issues for temporal GIS research and technologies. In: Handbook of research on geoinformatics. IGI Global, pp 144–153
Zikopoulos P, Eaton C (2011) Understanding big data: analytics for enterprise class hadoop and streaming data. McGraw-Hill Osborne Media

Chapter 2
Big Geospatial Data Processing Made Easy: A Working Guide to GeoSpark

Jia Yu and Mohamed Sarwat

2.1 Introduction

In the past decade, the volume of available geospatial data increased tremendously. Such data includes but not limited to: weather maps, socio-economic data, and geo-tagged social media. Moreover, the unprecedented popularity of GPS-equipped mobile devices and Internet of Things (IoT) sensors has led to continuously generating large-scale location information combined with the status of surrounding environments. For example, several cities have started installing sensors across the road intersections to monitor the environment, traffic and air quality. Making sense of the rich geospatial properties hidden in the data may greatly transform our society. This includes many subjects undergoing intense study: (1) Climate analysis: that includes climate change analysis (N. R. C. Committee on the Science of Climate Change 2001), study of deforestation (Zeng et al. 1996), population migration (Chen et al. 1999), and variation in sea levels (Woodworth et al. 2011), (2) Urban planning: assisting government in city/regional planning, road network design, and transportation/traffic engineering, (3) Commerce and advertisement (Dhar and Varshney 2011): e.g., point-of-interest (POI) recommendation services. These data-intensive spatial analytics applications highly rely on the underlying database management systems (DBMSs) to efficiently manipulate, retrieve and manage data.

There has been a flurry of spatial database systems (Spatial DBMSs) that provide spatial data types, query operators and index structures to handle spatial data based on the Open Geospatial Consortium standards (http://www.opengeospatial.org/). Some of these systems opt to extend existing relation databases (PostGIS,

J. Yu (✉) · M. Sarwat
Arizona State University, Tempe, AZ, USA
e-mail: jiayu2@asu.edu; msarwat@asu.edu

M. Werner, Y.-Y. Chiang (eds.), *Handbook of Big Geospatial Data*,
https://doi.org/10.1007/978-3-030-55462-0_2

http://postgis.net/), key – value stores (RocksDB, Rocksdb – spatial indexing. https://rocksdb.org/) or document-based databases (MongoDB, Mongodb – geospatial. https://www.mongodb.com/). Others are dedicated to standalone data systems (ArcGIS, https://www.arcgis.com/index.html; QGIS, https://qgis.org/) which have more granular control over the internal data organization. Even though the aforementioned systems offer full support for spatial data access, they suffer from the scalability issue when handling large-scale spatial data. That happens because the massive scale of available spatial data hinders making sense of it when using traditional spatial query processing techniques.

On the other hand, researchers and practitioners have been using cluster computing systems such as Hadoop MapReduce (Apache. Hadoop, http://hadoop.apache.org/) and Apache Spark (https://spark.apache.org) to process data at scale. However, the existing systems do not provide native support for spatial data. Users have to write tedious code or even inefficient algorithms to conduct spatial analysis applications. Recent works Eldawy and Mokbel (2015), Xie et al. (2016), and Yu et al. (2019) attempt to extend these systems to bolster spatial indices, spatial query operators and so on.

In this chapter, we mainly focus on GeoSpark, a cluster computing system that extends the core engine of Apache Spark and Spark SQL to process geospatial data at scale. This chapter is organized as follows: it first gives a brief introduction of the existing cluster computing systems and explains basic concepts that will be used in the other sections. It then explains different components of GeoSpark system including Spatial RDDs, and spatial queries. Section 2.6 explains how to write spatial data analytics applications in GeoSpark.

2.2 Background

2.2.1 Cluster Computing Systems

Over the years, researchers and practitioners have designed several different cluster computing systems to address the scalability issue on big data. These systems do not come with the functionalities of spatial data processing by default. However, learning their internals is very beneficial for the readers to understand how people extend the core models to support geospatial data. This section briefly summarizes the key concepts proposed in the state-of-art cluster computing systems.

Big Table and its derivatives Google Big Table (Chang et al. 2008) is a distributed key-value store which builds on top of Google File System (Ghemawat et al. 2003). Each record in Big Table contains a composite key and a string value. The composite key comprises three parts: row key, column key and timestamp. Among them, the row key and column key work in conjunction to identify the physical location of the record (the location on a particular data partition). The timestamp describes when the record is received and it also allows for versioning and garbage collection. In

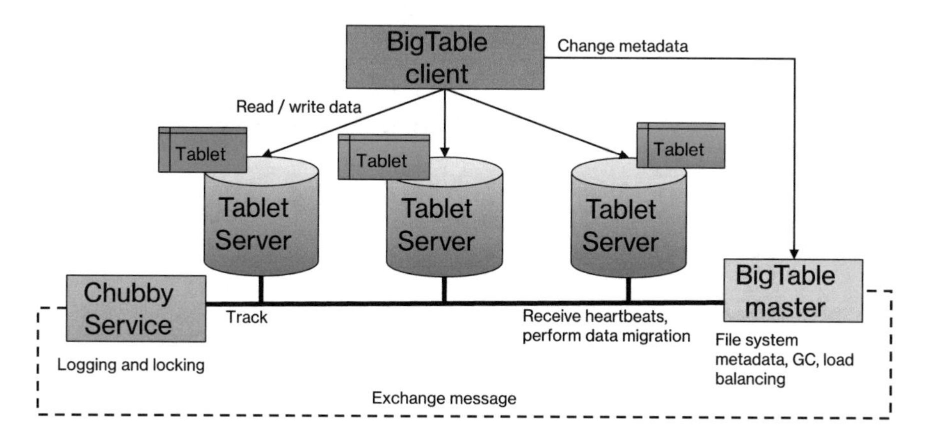

Fig. 2.1 Google Big Table

Big Table, the records at the same location may have several versions based on timestamps. A table in Big Table is split into many small tablets across the cluster (see Fig. 2.1), each of which contains a subset of data according to the row keys. The system tries to utilize the data locality such that any key search will only touch a few tablets. Although Google never open-sources Big Table, the emergence of this system has motivated a couple of open-source alternatives such as Apache Hbase (http://hbase.apache.org/) and Apache Accumulo (https://accumulo.apache.org/). These derivatives follow the same design philosophy of Big Table but use Hadoop Distributed File System (HDFS) (Shvachko et al. 2010) as the underlying file system because Google File System is for in-house use only.

Hadoop MapReduce The Hadoop MapReduce system (Apache. Hadoop, http://hadoop.apache.org/) is an open-source implementation of Google MapReduce model (Dean and Ghemawat 2008). The MapReduce model is inspired by the decades-old Map and Fold operations in functional programming languages such as S (Becker 1984) and R (Ihaka and Gentleman 1996) but extends the idea to the cluster computing environment by incorporating fault tolerance, task scheduling and so on. It is designed to process large datasets with a distributed algorithm on a cluster. This system gives a simple abstraction of complex distributed programs and hides the details of parallelization, fault-tolerance, data distribution and load balancing. A MapReduce program usually consists of three phases (see Fig. 2.2): Map, Shuffle, and Reduce. Among them, the Map and Reduce phases can run user-defined functions. The map function takes an input value key/value pair and generates a set of intermediate key/value pairs. The reduce function will merge the values of that key to a smaller set of values, based on the logic written by the user. A complex program may need to repeat the three phases, Map, Shuffle, and Reduce, many times and the intermediate data between two phases are persisted on local disk. The MapReduce system will execute the algorithm by scheduling tasks to

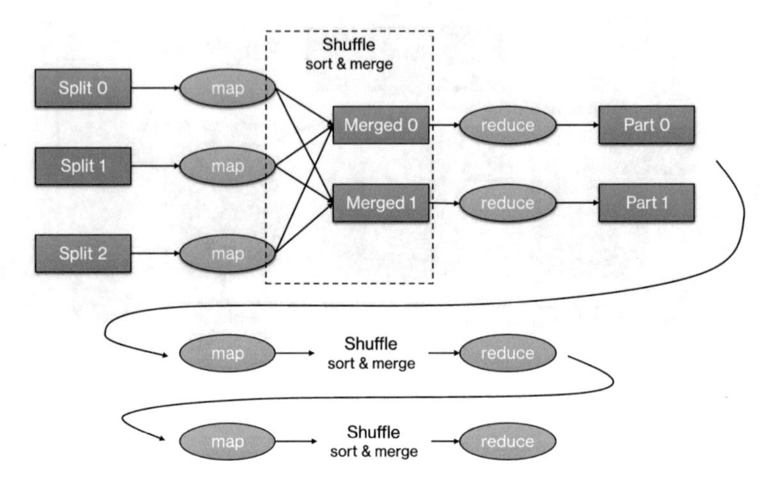

Fig. 2.2 A MapReduce model

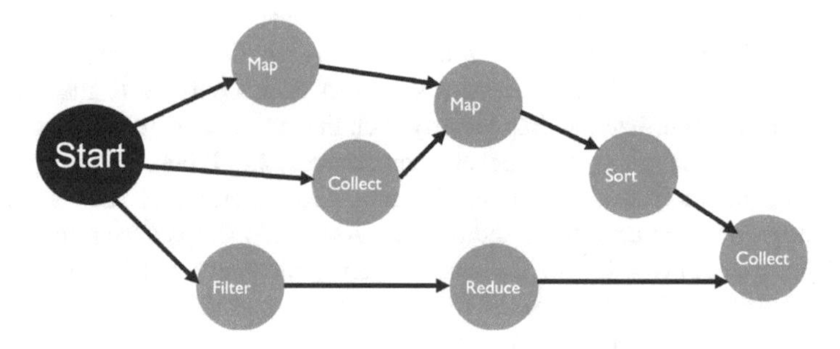

Fig. 2.3 Spark Directed Acyclic Graph

distributed machines, running different tasks in parallel, managing data shuffle and providing redundancy and fault tolerance.

Apache Spark The Spark system is a distributed general-purpose cluster computing framework which allows users to easily write distributed programs without being involved in the details of parallelism. It also can tolerate faults and scale out to many commodity machines. It is an implementation of Resilient Distributed Datasets (RDD) (Zaharia et al. 2012). RDD is an immutable distributed collection of in-memory objects. Each RDD is built using parallelized transformations (filter, join or groupBy). For fault tolerance, Spark rebuilds lost data on failure using lineage: each RDD remembers how it was built from other datasets (through transformations) to recover itself. The lineage among RDDs is represented as a Directed Acyclic Graph which consists of a set of RDDs (points) and directed Transformations (edges) (Fig. 2.3).

There are two transformations can be applied to RDDs, (1) narrow transformation: Each partition of the parent RDD is used by at most one partition of the child RDD. A Narrow transformation does not incur any data shuffle. Examples are map and filter. (2) Wide transformation: Each partition of the parent RDD is used by multiple child partitions. An example is join. A wide transformation will introduce data shuffle. The dependency between the parent and child is called wide dependency.

Directed Acyclic Graph (DAG) scheduler is deemed one of the most important components of Apache Spark that conducts stage-oriented scheduling. After a complex job is submitted to Spark, the scheduler computes a Directed Acyclic Graph for this job and divides the job into a set of stages based on this graph. This way, Spark can maximize the utilization of in-memory intermediate data and avoid unnecessary data shuffle (aka data transfer among nodes in the cluster).

2.2.2 Spatial Queries

Spatial queries are the basic building blocks of spatial analytics applications. With neat and concise APIs, users will be able to assemble complex applications which fit in their scenarios.

Spatial range query A spatial range query (Pagel et al. 1993) returns all spatial objects that lie within a geographical region. For example, a range query may find all parks in the Phoenix metropolitan area or return all restaurants within one mile of the user's current location. In terms of the format, a spatial range query takes a set of spatial objects and a polygonal query window as input and returns all the spatial objects which lie in the query area.

Spatial join Spatial join queries (Patel and DeWitt 1996) are queries that combine two datasets or more with a spatial predicate, such as distance and containment relations. There are also some real scenarios in life: tell me all the parks which have lakes and tell me all of the gas stations which have grocery stores within 500 feet. Spatial join query needs two sets of spatial objects as inputs. It finds a subset from the cross product of these two datasets such that every record satisfies the given spatial predicate.

Spatial K Nearest Neighbors (KNN) query Spatial KNN query takes a query center point, a spatial object set as inputs and finds the K nearest neighbors around the center points. For instance, a KNN query finds the 10 nearest restaurants around the user.

Spatial query language The state-of-art distributed geospatial data systems usually provide neat APIs and hide the underlying system details from the users. This transparency is actually an important design principle because most of the geospatial data scientists are not experts in distributed systems. The existing systems can be put into two categories:

- **High-level declarative language.** Some systems allow users to write their applications using declarative languages which is more flexible and user-friendly. GeoSpark (Yu et al. 2019) and ESRI Tools (ESRI, Esri tools for hadoop, https://esri.github.io/gis-tools-for-hadoop/) implement Spatial SQL API in Apache Spark and Hadoop MapReduce, respectively. Both of them follow the SQL-MM3 standard (Ashworth 2016) which is widely used in spatial databases such as PostGIS (http://postgis.net/). Spatial SQL has a syntax similar to the regular SQL but equips many spatial data types (e.g., point, polygon, ...) and functions (e.g., ST_Contains, ST_Within, ...). Apache Pig develops its own high-level language called Pig Latin, on top of Hadoop MapReduce, which has some key properties: ease of programming, optimization opportunities, and extensibility. Pigeon (Eldawy and Mokbel 2014) is a Spatial MapReduce language which extends Pig to support spatial data processing. It implements spatial functionalities via User-Defined Functions such that existing query operators such as Filter, GroupBy and Join can directly work with the operators. It implements the Open Geospatial Consortium (OGC) standard (http://www.opengeospatial.org/) (e.g., Cross, Overlap). SpatialHadoop (Eldawy and Mokbel 2015) ships with Pigeon by default.
- **Operational language.** Almost all of the existing systems provide APIs in different operational languages such as Java, Python, and Scala. Writing spatial analysis applications in such languages has a higher learning curve but these APIs are helpful for smart application-dependent optimizations.

2.3 Overview

GeoSpark[1] is a cluster computing framework that can process geospatial data at scale. It extends the core engine of Apache Spark (https://spark.apache.org) and SparkSQL to support spatial data types, distributed spatial indices, distributed spatial data partitioning and distributed spatial queries. As depicted in Fig. 2.4, GeoSpark allows users to issue queries using the out-of-box Spatial SQL API and RDD API. The RDD API provides a set of interfaces written in operational programming languages including Scala, Java, Python and R. The Spatial SQL interfaces offers a declarative language interface to the users so they can enjoy more flexibility when creating their own applications. These SQL API implements the SQL/MM Part 3 (Ashworth 2016) standard which is widely used in many existing spatial databases such as PostGIS (http://postgis.net/) (on top of PostgreSQL). In particular, GeoSpark put the available Spatial SQL functions into three categories: (1) Constructors: create a geometry type column (2) Predicates: evaluate whether a spatial condition is true or false. Predicates are usually used in WHERE clauses, HAVING clauses and so on (3) Geometrical functions: perform a specific geo-

[1]GeoSpark website: https://datasystemslab.github.io/GeoSpark/

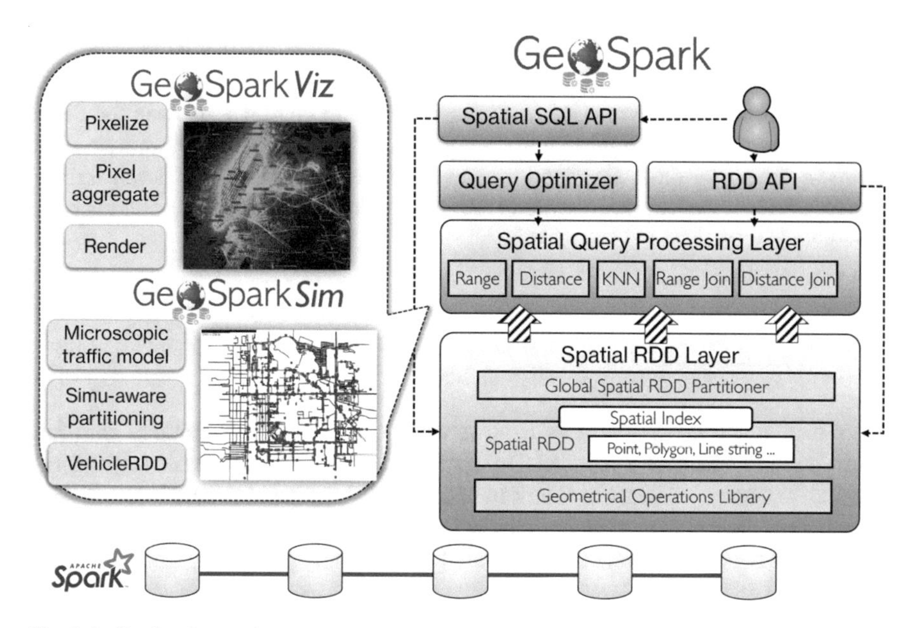

Fig. 2.4 GeoSpark overview

metrical operation on the given inputs. These functions can produce geometries or numerical values such as area or perimeter.

Spatial Resilient Distributed Dataset layer extends RDD, the core data structure in Apache Spark, to accommodate big geospatial data in a cluster. It consists of data partitions that are distributed across the Spark cluster. A Spatial RDD can be created by RDD transformation or be loaded from a file that is stored on permanent storage. This layer provides a number of APIs which allow users to read heterogeneous spatial object from various data formats. Moreover, Spatial RDDs equip distributed spatial indices and distributed spatial partitioning to speed up spatial queries.

Spatial query processing layer supports a number of spatial queries which harness distributed spatial indices and distributed spatial partitioning in Spatial RDDs and parallelize the workload using efficient distributed query algorithms. This layer provides standards APIs for the most-widely used spatial query operators, such as spatial range query, spatial join query, spatial KNN query.

GeoSparkViz is an extension of GeoSpark which provides native support of general cartographic design. It encapsulates the main steps of the geospatial map visualization to a set of massively parallized RDD transformation in Apache Spark. The visualization operators in GeoSparkViz directly take Spatial RDDs as input and generate high-resolution maps (i.e., scatter plot and heat map) using the Spark cluster. GeoSparkViz, together with GeoSpark, offers users a holistic system to data management and visualization on spatial data in the same cluster and avoid the unnecessary overhead of transferring intermediate data between two isolated systems.

GeoSparkSim is another GeoSpark extension which upholds large-scale road network traffic simulation. It allows users to generate trajectories data for numerous vehicles over any arbitrary road network. The system employs microscopic traffic models, such as traffic lights, light changing and car following, to produce more realistic trajectory data. GeoSparkSim works in conjunction with GeoSpark and GeoSparkViz to deliver an end-to-end data system to simulate, analyze and visualize urban traffic data.

2.4 Spatial RDD Layer

GeoSpark loads geospatial data from a variety of data sources, chops data into partitions and assemble Spatial RDDs. Spatial RDD in GeoSpark intuitively extends the RDD structure in Apache Spark to accommodate geospatial data in the cluster. The system equips several techniques to improve the geospatial data storage in the cluster and hence accelerate the distributed spatial query processing in the cluster.

2.4.1 Supported Spatial Data Sources

In the past, researchers and practitioners have developed a number of geospatial data formats for different purposes. However, the heterogeneous sources make it extremely difficult to integrate geospatial data together. For example, WKT format is a widely used spatial data format that stores data in a human readable tab-separated-value file. Shapefile is a spatial database file which includes several sub-files such as index file, and non-spatial attribute file. In addition, geospatial data usually possesses different shapes such as points, polygons and trajectories.

Supported data formats Currently, GeoSpark can read WKT, WKB, GeoJSON, Shapefile, and NetCDF/HDF format data from different external storage systems such as local disk, Amazon S3 and Hadoop Distributed File System (HDFS) to Spatial RDDs.

Supported geometry types Spatial RDDs now can accommodate seven types of spatial data including Point, Multi-Point, Polygon, Multi-Polygon, Line String, Multi-Line String, GeometryCollection, and Circle. Moreover, spatial objects that have different shapes can co-exist in the same Spatial RDD because GeoSpark adopts a flexible design which generalizes the geometrical computation interfaces of different spatial objects.

2.4.2 *Spatial RDD Built-In Geometrical Library*

It is quite common that spatial data scientists need to exploit some geometrical attributes of spatial objects in GeoSpark, such as perimeter, area and intersection. Spatial RDD equips a built-in geometrical library to perform geometrical operations at scale so the users will not be involved into sophisticated computational geometry problems. Currently, GeoSpark provides over 20 different functions in this library and put them in two separate categories

Regular geometry functions are applied to every single spatial object in a Spatial RDD. For every object, it generates a corresponding result such as perimeter or area. The output must be either a regular RDD or Spatial RDD.

Geometry aggregation functions are applied to a Spatial RDD for producing an aggregate value. It only generates a single value or spatial object for the entire Spatial RDD. For example, GeoSpark can compute the bounding box or polygonal union of the entire Spatial RDD.

2.4.3 *Spatial RDD Partitioning*

Data partitioning is an important concept in Spark due to its significant impact on the performance. By default, Spark tries to read data into RDD from the nodes that are close to it. If the input data is already partitioned like the files in Hadoop Distributed File System (HDFS), Spark will use the same RDD partitions to hold the HDFS partitions and these RDD partitions are put at the same place where HDFS partitions stay. This way, Spark can create RDD without introducing data shuffle. If the input data is not partitioned such as the data file on local disk or from Amazon S3, Spark will use the default Hash partitioner to partition the data automatically and distribute the partitions across the cluster. However, these partitioning methods do not take into account the spatial proximity which has a dramatic impact on the query speed.

GeoSpark equips a data partitioning method which is tailored to spatial data processing in a cluster. Data in Spatial RDDs are partitioned according to the spatial data distribution and nearby spatial objects are very likely to be put into the same partition. The effect of spatial partitioning is two-fold: (1) when running spatial queries that target at particular spatial regions, GeoSpark can speed up queries by avoiding the unnecessary computation on partitions that are not spatially close. Figure 2.5 shows the spatial partitioning grids currently supported by GeoSpark. (2) it can chop a Spatial RDD to a number of data partitions which have similar number of records per partition. This way, GeoSpark can ensure the load balance and avoid stragglers when performing computation in the cluster.

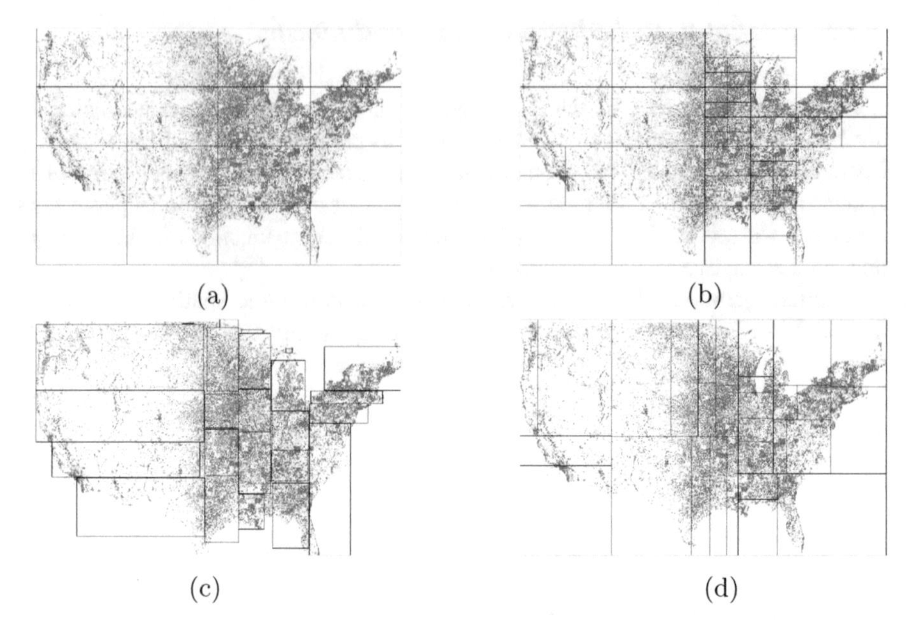

Fig. 2.5 Grids generated by SRDD spatial partitioning techniques. (**a**) SRDD partitioned by uniform grids. (**b**) SRDD partitioned by Quad-Tree. (**c**) SRDD partitioned by R-Tree. (**d**) SRDD partitioned by KDB-Tree

2.4.4 Spatial RDD Index

Spatial indices such as R-Tree and Quad-Tree are widely used in spatial data systems to accelerate the query processing. This is because spatial indices internally leverage a hierarchical structure to cluster spatial objects according to the spatial proximity. Given a query window, the indices quickly navigate the predicate to corresponding tree nodes in a top-down fashion. However, a traditional index on the entire Spatial RDD can cost an additional 15% storage overhead (Yu and Sarwat 2016, 2017). No single machine can afford such storage overhead when the data scale becomes large. Moreover, a single-machine spatial index cannot leverage the cluster to parallelize the workload.

GeoSpark proposes a distributed spatial index to index Spatial RDDs in the cluster (see Fig. 2.6). This distributed index consists of two parts (1) global index: is stored on the master machine and generated during the spatial partitioning phase. It indexes the bounding box of partitions in Spatial RDDs. The purpose of having such a global index is to prune partitions that are guaranteed to have no qualified spatial objects. (2) local index: is built on each partition of a Spatial RDD. Since each local index only works on the data in its own partition, it can have a small index size. Given a spatial query, the local indices in the Spatial RDD can speed up queries in parallel.

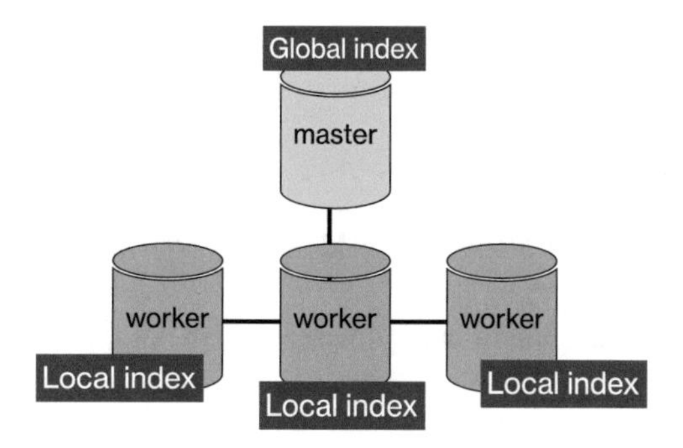

Fig. 2.6 Distributed spatial index in GeoSpark

2.4.5 Spatial RDD Customized Serializer

When distributed databases transfer objects across machines (e.g., data shuffle), all objects have to be first serialized in byte arrays. The receiver machines will put the received data chunk in memory and then de-serialize the data. Many distributed spatial databases use generic serializers such as Kryo which can provide a compact representation of simple objects (e.g., integers). However, for spatial objects that possess very complex geometrical shapes, the generic serializer may lead to large-scale data shuffled across the network and tremendous memory overhead across the cluster.

GeoSpark provides a customized serializer for spatial objects and spatial indexes. The proposed serializer can serialize spatial objects and indices into compressed byte arrays. This serializer is faster than the widely used kryo serializer and has a smaller memory footprint when running complex spatial operations, e.g., spatial join query. When converting spatial objects to a byte array, the serializer follows the encoding and decoding specification of Shapefile (ESRI 1998). The detailed specification is given below:

- **Byte 1** specifies the type of the spatial object. Each supported spatial object type has a unique ID in GeoSpark.
- **Byte 2** specifies the number of sub-objects in this spatial object.
- **Byte 3** specifies the type of the first sub-object (only needed for GeometryCollection, other types don't need this byte).
- **Byte 4** specifies the number of coordinates (n) of the first sub-object. Each coordinate is represented by two double type (8 bytes * 2) data X and Y.
- **Byte 5 – Byte 4+16*n** stores the coordinate information.
- **Byte 16*n+1** specifies the number of coordinates (n) of the second sub-object...
- **Until the end** Here all sub-objects have been serialized.

The serializer can also serialize and deserialize local spatial indices, such as Quad-Tree and R-Tree. For serialization, it uses the Depth-First Search (DFS) to traverse each tree node following the pre-order strategy (first write current node information then write its children nodes). For de-serialization, it will follow the same strategy used in the serialization phase. The de-serialization is also a recursive procedure. When serialize or de-serialize every tree node, the index serializer will call the spatial object serializer to deal with individual spatial objects.

2.5 Spatial Query Processing Layer

The spatial query processing layer in GeoSpark equips a set of distributed spatial query algorithms to uphold various spatial query operators. Currently, range query, distance query, K Nearest Neighbors (KNN) query, range join query, and distance join query are supported. The query processing layer intuitively leverages the spatial indices and data partitioning in the Spatial RDD layer to speed up the query execution.

2.5.1 Spatial Range Query

A spatial range query takes as input a query window which can be polygon or rectangle, and a Spatial RDD. It usually runs fast and does not incur a data shuffle. GeoSpark completes this query using a parallelized filter transformation in Apache Spark. The steps are: (1) Broadcast the query window to each machine in the cluster and create a spatial index on each Spatial RDD partition if necessary. (2) For each Spatial partition, if a spatial index is created, use the query window to query the spatial index. Otherwise, check the spatial predicate between the query window and each spatial object in the Spatial RDD partition. If the spatial predicate holds true, the algorithm adds the spatial object to the result set.

2.5.2 Spatial K Nearest Neighbors (KNN) Query

A spatial KNN query takes as input a query point, a Spatial RDD and a number. The straightforward way to perform this query in a Spark cluster is to compute the distance among the query point and all other spatial objects, conduct a global sorting, and take the top K spatial objects that have the shortest distances to the query point. However, this approaches requires a global sorting which is slow and incurs a huge data shuffle. GeoSpark adopts a distributed top-k algorithm which runs in two phases (1) selection phase: for each Spatial RDD partition, GeoSpark calculates the distances from the given query point to each spatial object, then maintains a local priority queue on the distances. Such a queue contains the nearest K objects around the given query point and becomes a partition of the new intermediate Spatial RDD. For the indexed Spatial RDDs, GeoSpark can query the local indexes in partitions to accelerate the distance calculation. (2) sorting phase: each partition in the RDD produced in the previous step contains k spatial objects which are the candidates of the global top k nearest neighbors. In the sorting phase, GeoSpark sorts spatial objects in this RDD and takes global top k nearest neighbors. This sorting only works on a few spatial objects that are candidates for the global K nearest neighbors such that the sorting overhead is reduced.

2.5.3 Spatial Join Query

A spatial join query in GeoSpark takes as input two Spatial RDDs, A and B. Both of them should be partitioned by the same spatial partitioning grid file. One of the Spatial RDD can have local indices. GeoSpark develops a join algorithm called GSjoin (Yu et al. 2019) which leverages the spatial data partitioning and indices built in Spatial RDDs. The algorithm zips two Spatial RDDs partition by partition, according to spatial proximity. In every partition, for spatial objects from Spatial RDD A and B, the algorithm computes their spatial relations. If the spatial object from A overlaps the other one from B, the algorithm keeps them in the final results. Finally, the algorithm removes the duplicated points and returns the result to other operations in the Spark.

2.6 Perform Spatial Data Analytics in GeoSpark

This section details the usage of GeoSpark RDD and SQL APIs and showcases the system. In particular, it illustrates several GeoSpark queries and gives a complete example about how to interact with Zeppelin. The example code in this section is written in Scala but also works for Java.[2]

2.6.1 Run Queries Using RDD APIs

The section outlines the steps to create Spatial RDDs and run spatial queries using GeoSpark RDD APIs. The example code is written in Scala but also works for Java.

Set up dependencies Before starting to use GeoSpark, users must add the corresponding package to their projects as a dependency. For the ease of managing dependencies, the binary packages of GeoSpark are hosted on the Maven Central Repository which includes all JVM based packages from the entire world. As long as the projects are managed by popular project management tools such as Apache

[2]Example code: https://github.com/jiayuasu/GeoSparkTemplateProject

Maven and sbt, users can easily add GeoSpark by adding the artifact id in the project specification file such as POM.xml and build.sbt.

Initiate SparkContext Any RDD in Spark or GeoSpark must be created by SparkContext. Therefore, the first task in a GeoSpark application is to initiate a SparkContext. The code snippet below gives an example. In order to use GeoSpark custom spatial object and index serializer, users must enable them in the SparkContext.

```
1 val conf = new SparkConf()
2 conf.setAppName("GeoSparkExample")
3 // Enable GeoSpark custom Kryo serializer
4 conf.set("spark.serializer", classOf[KryoSerializer].getName)
5 conf.set("spark.kryo.registrator", classOf[
     GeoSparkKryoRegistrator].getName)
6 val sc = new SparkContext(conf)
```

Create a Spatial RDD Spatial objects in a SpatialRDD is not typed to a certain geometry type and open to more scenarios. It allows an input data file which contains mixed types of geometries. For instance, a WKT file might include three types of spatial objects, such as LineString, Polygon and MultiPolygon. Currently, GeoSpark can load data in many different data formats. This is done by a set of file readers such as WktReader and GeoJsonReader. For example, users can call ShapefileReader to read ESRI Shapefiles (ESRI 1998).

```
1 val spatialRDD = ShapefileReader.readToGeometryRDD(sc, filePath)
```

Transform the coordinate reference system GeoSpark doesn't control the coordinate unit (i.e., degree-based or meter-based) of objects in a Spatial RDD. When calculating the distance between two coordinates, GeoSpark simply computes the Euclidean distance. In practice, if users want to obtain the accurate geospatial distance, they need to transform coordinates from the degree-based coordinate reference system (CRS), i.e., WGS84, to a planar coordinate reference system (i.e., EPSG: 3857). GeoSpark provides this function to the users such that they can perform this transformation to every object in a Spatial RDD and scale out the workload using a cluster.

```
1 // epsg:4326: is WGS84, the most common degree-based CRS
2 val sourceCrsCode = "epsg:4326"
3  // epsg:3857: The most common meter-based CRS
4 val targetCrsCode = "epsg:3857"
5 objectRDD.CRSTransform(sourceCrsCode, targetCrsCode)
```

Build the distributed spatial index Users can call APIs to build a distributed spatial index on the Spatial RDD. Currently, GeoSpark provides two types of spatial indexes, Quad-Tree and R-Tree, as the local index on each partition. The code of this step is as follows:

```
spatialRDD.buildIndex(IndexType.QUADTREE, false) // Set to true
    only if the index will be used join query
```

Write a spatial range query A spatial range query takes as input a range query window and a Spatial RDD and returns all geometries that intersect/are fully covered by the query window. Assume the user has a Spatial RDD. He or she can use the following code to issue a spatial range query on this Spatial RDD. The output format of the spatial range query is another Spatial RDD.

```
val rangeQueryWindow = new Envelope(-90.01, -80.01, 30.01, 40.01)
 /*If true, return gemeotries intersect or are fully covered by
    the window; If false, only return the latter. */
val considerIntersect = false
// If true, it will leverage the distributed spatial index to
    speed up the query execution
val usingIndex = false
var queryResult = RangeQuery.SpatialRangeQuery(spatialRDD,
    rangeQueryWindow, considerIntersect, usingIndex)
```

Write a spatial KNN query A spatial K Nearest Neighbor query takes as input a K, a query point and a Spatial RDD and finds the K geometries in the RDD which are the closest to the query point. If the user has a Spatial RDD, he or she then can perform the query as follows. The output format of the spatial KNN query is a list which contains K spatial objects.

```
val geometryFactory = new GeometryFactory()
val pointObject = geometryFactory.createPoint(new Coordinate
    (-84.01, 34.01)) // query point
val K = 1000 // K Nearest Neighbors
val usingIndex = false
val result = KNNQuery.SpatialKnnQuery(objectRDD, pointObject, K,
    usingIndex)
```

Write a spatial join query A spatial join query takes as input two Spatial RDDs A and B. For each object in A, finds the objects (from B) covered/intersected by it. A and B can be any geometry type and are not necessary to have the same geometry type. Spatial RDD spatial partitioning can significantly speed up the join query. Three spatial partitioning methods are available: KDB-Tree, Quad-Tree and R-Tree. Two Spatial RDDs must be partitioned by the same spatial partitioning grid file. In other words, If the user first partitions Spatial RDD A, then he or she must use the data partitioner of A to partition B. The example code is as follows:

```
// Perform the spatial partitioning
objectRDD.spatialPartitioning(joinQueryPartitioningType)
queryWindowRDD.spatialPartitioning(objectRDD.getPartitioner)
// Build the spatial index
val usingIndex = true
queryWindowRDD.buildIndex(IndexType.QUADTREE, true) // Set to
    true only if the index will be used join query
val result = JoinQuery.SpatialJoinQueryFlat(objectRDD,
    queryWindowRDD, usingIndex, considerBoundaryIntersection)
```

2.6.2 Run Queries Using SQL APIs

The page outlines the steps to manage spatial data using the Spatial SQL interface of GeoSpark. The SQL interface follows SQL/MM Part3 Spatial SQL Standard (Ashworth 2016). In order to use the system, users need to add GeoSpark as the dependency of their projects, as mentioned in the previous section.

Initiate SparkSession Any SQL query in Spark or GeoSpark must be issued by SparkSession, which is the central scheduler of a cluster. To initiate a SparkSession, the user should use the code as follows:

```
1 var sparkSession = SparkSession.builder()
2 .appName("GeoSparkExample")
3 // Enable GeoSpark custom Kryo serializer
4 .config("spark.serializer", classOf[KryoSerializer].getName)
5 .config("spark.kryo.registrator", classOf[GeoSparkKryoRegistrator
    ].getName)
6 .getOrCreate()
```

Register SQL functions GeoSpark adds new SQL API functions and optimization strategies to the catalyst optimizer of Spark. In order to enable these functionalities, the users need to explicitly register GeoSpark to the Spark Session using the code as follows.

```
1 GeoSparkSQLRegistrator.registerAll(sparkSession)
```

Create a geometry type column Apache Spark offers a couple of format parsers to load data from disk to a Spark DataFrame (a structured RDD). After obtaining a DataFrame, users who want to run Spatial SQL queries will have to first create a geometry type column on this DataFrame because every attribute must have a type in a relational data system. This can be done via some constructors functions such as ST_GeomFromWKT. After this step, the users will obtain a Spatial DataFrame. The following example shows the usage of this function.

```
1 SELECT ST_GeomFromWKT(wkt_text) AS geom_col, name, address
2 FROM input
```

Transform the coordinate reference system Similar to the RDD APIs, the Spatial SQL APIs also provide a function, namely ST_Transform, to transform the coordinate reference system of spatial objects. It works as follows:

```
1 SELECT ST_Transform(geom_col, "epsg:4326", "epsg:3857") AS
    geom_col
2 FROM spatial_data_frame
```

Write a spatial range query GeoSpark Spatial SQL APIs have a set of predicates which evaluate whether a spatial condition is true or false. ST_Contains is a classical function that takes as input two objects A and returns true if A contains B. In a given SQL query, if A is a single spatial object and B is a column, this becomes a spatial range query in GeoSpark (see the code below).

```
1 SELECT *
2 FROM spatial_data_frame
3 WHERE ST_Contains (ST_Envelope(1.0,10.0,100.0,110.0), geom_col)
```

Write a spatial KNN query To perform a spatial KNN query using the SQL APIs, the user needs to first compute the distance between the query point and other spatial objects, rank the distances in an ascending order and take the top K objects. The following code finds the 5 nearest neighbors of Point(1, 1).

```
1 SELECT name, ST_Distance(ST_Point(1.0, 1.0), geom_col) AS
    distance
2 FROM spatial_data_frame
3 ORDER BY distance ASC
4 LIMIT 5
```

Write a spatial join query A spatial join query in Spatial SQL also uses the aforementioned spatial predicates which evaluate spatial conditions. However, to trigger a join query, the inputs of a spatial predicate must involve at least two geometry type columns which can be from two different DataFrames or the same DataFrame. The following query involves two Spatial DataFrames, one polygon column and one point column. It finds every possible pair of <polygon, point> such that the polygon contains the point.

```
1 SELECT *
2 FROM spatial_data_frame1 df1, spatial_data_frame2 df2
3 WHERE ST_Contains(df1.polygon_col, df2.point_col)
```

Perform geometrical operations GeoSpark provides over 15 SQL functions for geometrical computation. Users can easily call these functions in their Spatial SQL query and GeoSpark will run the query in parallel. For instance, a very simple query to get the area of every spatial object is as follows:

```
1 SELECT ST_Area(geom_col)
2 FROM spatial_data_frame
```

Aggregate functions for spatial objects are also available in GeoSpark. They usually take as input all spatial objects in the DataFrame and yield a single value. For example, the code below computes the union of all polygons in the Data Frame.

```
1 SELECT ST_Union_Aggr(geom_col)
2 FROM spatial_data_frame
```

2.6.3 Interact with GeoSpark via Zeppelin Notebook

Although Spark bundles interactive Scala and SQL shells in every release, these shells are not user-friendly and not possible to do complex analysis and charts. Data scientists tend to run programs and draw charts interactively using a graphic

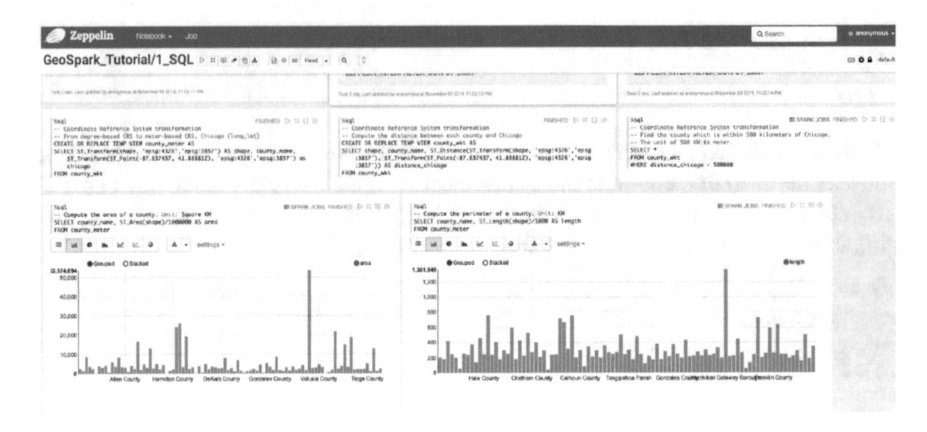

Fig. 2.7 Run GeoSpark Spatial SQL in Apache Zeppelin web-based notebook

interface. Starting from 1.2.0, GeoSpark provides a Helium plugin tailored for Apache Zeppelin (Foundation AS, Apache zeppelin. https://zeppelin.apache.org/) web-based notebook. Users can perform spatial analytics on Zeppelin web notebook (see Fig. 2.7) and Zeppelin will send the tasks to the underlying Spark cluster.

Users can create a new paragraph on a Zeppelin notebook and write code in Scala, Python or SQL to interact with GeoSpark. Moreover, users can click different options available on the interface and ask GeoSpark to render different charts such as bar, line and pie over the query results. For example, Zeppelin can visualize the result of the following query as a bar chart and show that the number of landmarks in every US county.

```
1 %sql
2 SELECT C.name, count(*)
3 FROM US_county C, US_landmark L
4 WHERE ST_Contains(C.geom_col, L.geom_col)
5 GROUPBY C.name
```

Another example is to find the area of each US county and visualize it on a bar chart (see Fig. 2.7). The corresponding query is as follows. This actually leverages the geometrical functions offered in GeoSpark.

```
1 %sql
2 SELECT C.name, ST_Area(C.geom_col) AS area
3 FROM US_county C
```

References

Ashworth M (2016) Information technology – database languages – SQL multimedia and application packages – part 3: Spatial, standard, International Organization for Standardization, Geneva

Becker RA, Chambers, JM (1984) S: an interactive environment for data analysis and graphics. CRC Press

Chang F, Dean J, Ghemawat S, Hsieh WC, Wallach DA, Burrows M, Chandra T, Fikes A, Gruber RE (2008) Bigtable: a distributed storage system for structured data ACM Trans Comput Syst 26(2):4:1–4:26

Chen C, Burton M, Greenberger E, Dmitrieva J (1999) Population migration and the variation of dopamine D4 receptor (DRD4) allele frequencies around the globe. Evol Hum Behav 20(5):309–324

Dean J, Ghemawat S (2008) MapReduce: simplified data processing on large clusters. Commun ACM 51:107–113

Dhar S, Varshney U (2011) Challenges and business models for mobile location-based services and advertising. Commun ACM 54(5):121–128

Eldawy A, Mokbel MF (2014) Pigeon: a spatial mapreduce language. In: IEEE 30th international conference on data engineering, ICDE 2014, Chicago, 31 Mar–4 Apr 2014, pp 1242–1245

Eldawy A, Mokbel MF (2015) Spatialhadoop: a mapreduce framework for spatial data. In: Proceedings of the IEEE international conference on data engineering, ICDE, pp 1352–1363

ESRI E (1998) Shapefile technical description. An ESRI White Paper

Ghemawat S, Gobioff H, Leung S (2003) The google file system. In: Proceedings of the ACM symposium on operating systems principles, SOSP, pp 29–43

Ihaka R, Gentleman R (1996) R: a language for data analysis and graphics. J Comput Graph Stat 5(3):299–314

N. R. C. Committee on the Science of Climate Change (2001) Climate change science: an analysis of some key questions. The National Academies Press, Washington, DC

Pagel B-U, Six H-W, Toben H, Widmayer P (1993) Towards an analysis of range query performance in spatial data structures. In: Proceedings of the twelfth ACM SIGACT-SIGMOD-SIGART symposium on principles of database systems, PODS'93

Patel JM, DeWitt DJ (1996) Partition based spatial-merge join. In: Proceedings of the ACM international conference on management of data, SIGMOD, pp 259–270

Shvachko K, Kuang H, Radia S, Chansler R (2010) The hadoop distributed file system. In: 2010 IEEE 26th symposium on mass storage systems and technologies (MSST). IEEE, pp 1–10

Woodworth PL, Menéndez M, Gehrels WR (2011) Evidence for century-timescale acceleration in mean sea levels and for recent changes in extreme sea levels. Surv Geophys 32(4–5):603–618

Xie D, Li F, Yao B, Li G, Zhou L, Guo M (2016) Simba: efficient in-memory spatial analytics. In: Proceedings of the ACM international conference on management of data, SIGMOD

Yu J, Sarwat M (2016) Two birds, one stone: a fast, yet lightweight, indexing scheme for modern database systems. Proc Int Conf Very Large Data Bases, VLDB 10(4):385–396

Yu J, Sarwat M (2017) Indexing the pickup and drop-off locations of NYC taxi trips in postgresql – lessons from the road. In: Proceedings of the international symposium on advances in spatial and temporal databases, SSTD, pp 145–162

Yu J, Zhang Z, Sarwat M (2019) Spatial data management in Apache Spark: the GeoSpark perspective and beyond. GeoInformatica 23(1):37–78

Zaharia M, Chowdhury M, Das T, Dave A, Ma J, McCauly M, Franklin MJ, Shenker S, Stoica I (2012) Resilient distributed datasets: a fault-tolerant abstraction for in-memory cluster computing. In: Proceedings of the USENIX symposium on networked systems design and implementation, NSDI, pp 15–28

Zeng N, Dickinson RE, Zeng X (1996) Climatic impact of Amazon deforestation? A mechanistic model study. J Climate 9:859–883

Chapter 3
Indoor 3D: Overview on Scanning and Reconstruction Methods

Ville V. Lehtola, Shayan Nikoohemat, and Andreas Nüchter

3.1 Introduction

Accurate three-dimensional (3D) data is called for creating accurate reconstructions of indoor spaces, i.e., application-suitable digital twins of these spaces. For instance, the global indoor 3D laser scanner market accounted for 3.79 billion in 2017 (Businesswire 2019). The purpose of reconstruction is roughly dividable into two types: schematic models for engineering purposes or visual models that are intended for a broader audience than just engineers. On the one hand, schematic applications include performing change detection between building information models (BIM) and as-built data, and the related planning and monitoring of construction processes and building conditions. On the other hand, visually-appealing virtual models are useful for facility management, supporting high-level decision making, real-estate brokering and marketing, displaying cultural and historical heritage, and other applications. The schematic and virtual properties of digital 3D models can also be combined. Indoor models of public buildings, e.g., airports and shopping malls, can be used to assist indoor navigation and location-based services. Concerning the public sector, construction permit processes may be sped up by applying automated model checkers into these digital models – before and after the construction. Furthermore, decision making on city planning can be facilitated when plans are made for indoor or underground public places. Such places are common, for example, near metro stations and in northern countries where winters are cold.

V. V. Lehtola (✉) · S. Nikoohemat
ITC Faculty, University of Twente, Enschede, The Netherlands
e-mail: ville.lehtola@iki.fi

A. Nüchter
Robotics and Telematics, University of Würzburg, Würzburg, Germany
e-mail: andreas@nuechti.de

M. Werner, Y.-Y. Chiang (eds.), *Handbook of Big Geospatial Data*,
https://doi.org/10.1007/978-3-030-55462-0_3

Fig. 3.1 Point cloud of Startup Sauna entrance at Aalto University. (Reproduced from Lehtola et al. 2017)

Perhaps surprising to a common man, the scanning and modeling of building interiors and exteriors are two different things. The reader may reflect on this while they proceed. The activities for reconstructing building exteriors were already well-known when the interest towards the indoor spaces was taking its first steps (Musialski et al. 2013).

The creation of indoor 3D models from scanned data was mainly a curiosity before 2010s, and was done without modern mobile mapping methods. The reconstruction of indoor models relied on 3D point clouds obtained from terrestrial laser scanning (TLS), see Fig. 3.1, or on classical photogrammetry, specifically, bundle adjustment. On one hand, TLS scanning required professional level high-cost equipment and post-processing software for the lidar data, which made the process impossible to automate. On the other hand, digital RGB images taken with calibrated cameras were employed to find similar features from images and then triangulate the 3D geometry from these images. After sparse matching, dense matching and reconstruction techniques were employed, e.g. those based on voxelization (Furukawa et al. 2009). Considering automated processing, this bundle adjustment-based technique required professional knowledge of camera calibration from the user and had problems with lighting, textures, and the complex geometry of the indoor environments (Lehtola et al. 2014). These initial techniques however brought an initial sense of success and with the development in miniaturization of sensors (lidars, MEMS INS), they sparked a boom of interest in indoor scanning and reconstruction.

The indoor scanning problem has been a hot topic throughout the 2010s, seeing many different scanning systems being designed (Lehtola et al. 2017). The problem itself was approached from several directions. First, the positioning procedure of the mobile mapping system that traditionally relied on global navigation satellite systems (GNSS) was 're-designed' to operate in interior spaces. This meant disabling the GNSS receiver and using only the inertial sensor to navigate. This so-called (pedestrian) dead reckoning[1] technique however results into a rapidly increasing uncertainty about the position of the sensor system, because the inertial sensor drift rate is an unknown function with respect to time. In other words, even if the inertial sensor drift is calibrated at an instant of time, that drift changes at a certain rate. Now, this change rate could also be measured and calibrated away, but because the change rate is unknown, the calibration does not last. With a navigation grade inertial measurement unit (IMU), i.e. equipment with a very small drift change rate, Trimble was able to design a pushcart system in 2012. This unit however remains to be a test system due to the high costs of such an IMU (>20,000 euros). Hence, the key in indoor scanning is the robustness of the positioning method. The position of the scanning system, when developed as a function of time, becomes the traversed path of motion, i.e. the trajectory.

After the bundle adjustment, the TLS, and the dead reckoning methods led into shortcomings, the research focus was intensified in mobile systems and trajectory-based methods. There, the basic idea is to track the position and heading (i.e. pose) of the sensor system as a function of time in 3D relative coordinates. The pose updates are done using the overlaps in optical data, that is for example keeping record on déjà-vu's, or technically, features that have been seen before. In robotics, this is known as the simultaneous localization and mapping (SLAM).

This book chapter is written as follows. We shall begin by considering the properties of indoor environments and what problems they pose for scanning and reconstruction (Sect. 3.2). Then we discuss how can these spaces can be understood by computers, i.e., map representations (Sect. 3.3). The development of the indoor scanning techniques are reflected on the introduced problems and we list some prominent mobile mapping methods (Sect. 3.4). Based on these and given an application, the reader should to be able to identify the scanning challenges related to that application and then be able to select a suitable indoor mobile mapping system for that application. Furthermore, in order to give the reader a basic understanding in how the indoor mapping systems perform simultaneous localization and mapping (SLAM), we describe the algorithm based on iterative closest points (ICP) (Sect. 3.5). This description (along with the cited works) allows – in principle – for the reader to construct their own indoor mobile scanning system.[2] We expect, however, that most readers do not construct their own systems but are instead interested in the functionality of the existing systems and their development in the creation of point clouds. The reconstruction of indoor spaces (Sect. 3.6) covers the necessary step of semantically segmenting the created point cloud and the following step on turning this labeled point cloud into a meshed model. Hence, by indoor 3D reconstruction we are referring to the process of generating a meshed model which is exportable to one of the standard formats such as IFC (industry

[1]The terms dead reckoning and pedestrian dead reckoning are used in the field of positioning and navigation.

[2]Note that ready open source SLAM codes are also available, e.g. https://github.com/googlecartographer/ (Hess et al. 2016).

foundation classes) or IndoorGML (Chen and Clarke 2017). In other words, the point cloud that is obtained from scanning is replaced by a mesh that consists of continuous geometrical shapes such as planes. In Sect. 3.7, we review applications. The book chapter ends with a discussion on future trends (Sect. 3.8) and a list of exercises for students (Sect. 3.9).

A common thread of this book chapter, as the reader will discover, is that the scanning trajectory is of critical importance in each of the steps towards the final 3D model, i.e. an application-suitable digital twin of the indoor space. The concept of trajectory, i.e. the path that the scanning system has traveled, is at the very core of mobile mapping and we highlight that it is important to understand what it stands for as it is exploited not only in scanning but also in reconstruction steps of the indoor spaces. To this end, we need to review some terminology.

3.1.1 Terminology

The development in indoor mobile mapping has heritage in multiple fields of science. Hence, there are several words that bear a similar or identical meaning. A systematic review of the scanning terminology is listed in Table 3.1. Additionally, there are some apparent ambiguities that need to be clarified. It is important to differentiate between *relative positioning*, where a map with an internal coordinate system is created,[3] and *absolute positioning*, where a map with geographic coordinates is created (typically using a GNSS receiver). Here, the term *map* follows from robotics (definition in Sect. 3.3) and does not refer to a cartographic map. When relative positioning such as SLAM techniques are used to create a map, this map can be *geo-referenced*. Geo-referencing is a surveying term that means that the internal coordinate system of a map or image is transformed into a geographic coordinate system, typically WGS82 (World Geodetic System). In other words, after an indoor space is mapped, the obtained map may be connected onto an outdoor map to form a seamless indoor-outdoor transition in the map. Finally, the range *precision* of

Table 3.1 Typical terminology related to (indoor) point cloud registration comes from different disciplines of science (term data registration is sometimes also used). The symbols $(xyz\theta\phi\kappa)$ correspond to 3 Cartesian coordinates and 3 Euler angles

Discipline	Equipment	Term	Mathematical equivalent
Laser scanning	lidar	(uses terms below)	
Photogrammetry	Digital camera	Orientation	$(xyz\theta\phi\kappa)$
Computer vision	Digital camera	External calibration	$(xyz\theta\phi\kappa)$
Robotics	Robot (or sensor)	Pose (or posture)	$(xyz\theta\phi\kappa)$
Navigation	GNSS receiver	Position, heading	(xyz), $(\theta\phi\kappa)$

[3]Typically, the origin is chosen to be at the start point of scanning, i.e. $(x, y, z) = (0, 0, 0)$.

lidars is around some millimeters, so positioning from dense scans using SLAM almost never results in issues with precision. Instead, there are problems related to erroneous scan registrations which are discussed in Sect. 3.5.5 and which we refer to with the word accuracy or accurate data.

3.2 Properties of Indoor Environments and Identification of Scanning and Reconstruction Problems

Indoor spaces may be dark or over-illuminated. They can be colorful or lacking texture. These properties of indoor spaces have an immediate impact on the functionality of sensors. Sensor capabilities with respect to different conditions in indoor environments are detailed in Table 3.2. Note that only the most commonly used sensors are included. The range of a sensor is important when large indoor facilities are scanned. In addition to sensors, there are further things to consider.

Every object and feature within an interior space has a specific purpose, as they have been designed by humans. These objects and features come in different sizes, see Fig. 3.2. There are, in fact, a lot of objects in which people do not normally pay attention, and some of these may have geometrically complex shapes. Some are small, such is the width of an electric wire, and some are big, such is a room. The magnitude of sizes varies from the order of one centimeter to dozens or even hundreds of meters. In other words, the characteristic length scale of interior spaces spans four orders of magnitude. We call this as a *multi-scale problem* (Lehtola et al. 2017).

The multi-scale problem sets apparently conflicting criteria to the design of the indoor mapping system. On one hand, the sampling resolution should be large to be able to account for the smallest details, but on the other it should be sparse to make covering large spaces computationally tractable. However, the fast accumulation of data from large resolution may be dealt with sophisticated data distillation techniques. Hence, an ideal system designed for three-dimensional (3D) indoor reconstruction has a sampling rate that can account for the smallest details, but is able to do efficient data distillation so that even the largest interior spaces may

Table 3.2 Optical sensor capabilities with respect to different conditions in indoor environments

Conditions	RGB (camera)	RGB-D (range camera)	lidar	RGB and lidar
Nominal	Y	Y	Y	Y
Weak textures	N	Y	Y	Y
Dark	N	N	Y	Y
Direct light or sunlight	N	N	Y	Y
Advantage	Textures (and geometry)	Textures and geometry	Geometry	Textures and geometry
Range	Unlimited	6–10 m	30–100 m	30–100+ m

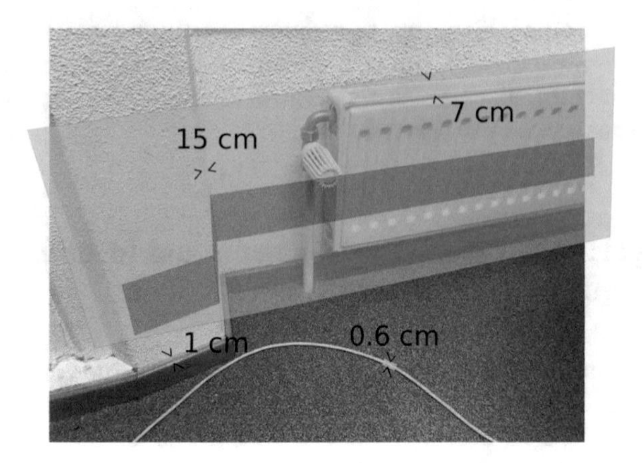

Fig. 3.2 Multi-scale problem. Each object and feature in indoor environments serves a specific purpose for which it has been put there. These span a multitude of length scales, for example, a network cable has 0.6 cm thickness, while the thickness of the radiator is one order of magnitude larger, i.e. 7 cm. The building itself can span a distance of hundreds of meters. Success in separating the objects of these different scales depends on the precision of the data and the models used. A coarse assumption of a rectangular room leads to the elimination of these features, depicted with a cyan plane. Instead, using a piece-wise planar model shown with red planes allows for the recovery of the different objects

be covered. Usually for applications, it is important that the level of detail stays the same regardless of the size of the building. Hence, in practice, the application determines the properties that the measurement system should fulfill.

The measuring geometry of indoor data is very different from traditional remote sensing, where the Earth is viewed from above, and from 3D scanning of single objects, since of two restrictions. First, scanning techniques must account for not being able to see the surrounding indoor space in one snapshot, as sensors typically have a field of view that does not cover 360 degree rotation around two directions. Second, the sensor trajectory is more restricted and difficult measuring geometries that may lead to registration problems are encountered for example in narrow doorways. In contrast, an air-borne scanning system can be freely flown above the Earth or a studio-system freely moved around the single object that is 3D scanned.

Indoor environments are highly convoluted spaces. In topological sense, they can be thought to resemble Swiss cheeses, i.e. bulks with multiple carved holes. Such a bulk can be discretized with an occupancy grid for optical 'mining' (see Sect. 3.3). One typical problem that is encountered is the difficulty in distinguishing the points captured from the two different sides of a thin wall. Another is distinguishing between an opening caused by missing data and an opening caused by an existing window.

Occlusions are abundant indoors, since often an object or one part of the area to be scanned blocks another part of the area to be scanned, see Fig. 3.3. Outdoors, when large platforms may be used, the problem can be alleviated by fusing sensor data from different platform locations (Schneider et al. 2010). Indoors, the sensor

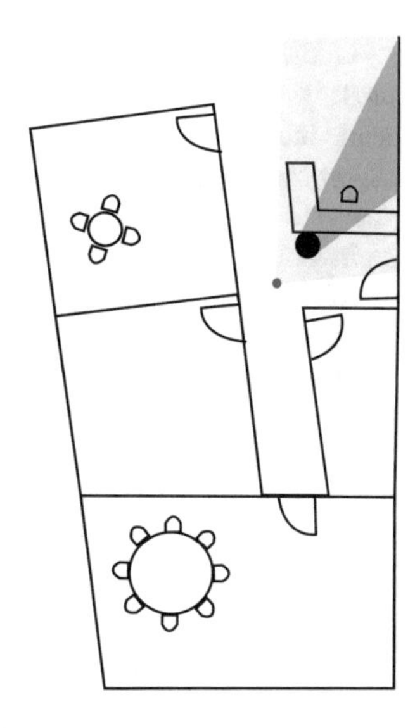

Fig. 3.3 Occlusion is a common problem indoors. Some parts of the scene occlude some other parts of the scene. Obvious examples of occluding objects include static constructs such as pillars and corners, but there are also dynamic objects such as furniture and doors. Here, a pillar is occluding a part of the view of the scanning system located at the red spot (visible area shown in blue, occluded area in grey)

systems are purposefully smaller and this approach is less feasible. These occlusions can be then overcome with footwork, i.e. a thorough scan of the indoor space, which is likely to require an online interface from which the operator can see what part of the areas need further scanning. This need for an interface has partly lead the development and design of commercial indoor scanning products.

Dynamic occlusions and static occlusions are two different things. If a scan is planned, it needs to be taken into account that indoor spaces are often full of people and objects. In technical sense, people that move around may be referred to as dynamic occlusions while objects that do not move present static occlusions. Dynamic occlusions can be detected by performing scans of the same environment at different instances of time and then comparing the obtained point clouds. If only one instance of time is used, then dynamic occlusions may not be easily detected. Static occlusions need to be treated in the reconstruction phase, for example, oftentimes a priori knowledge of the environment is used to fill in the gaps (Sect. 3.6). Different measures to detect changes in indoor MLS point clouds are discussed in Lehtola et al. (2017).

Reflection is the 'evil twin' of occlusion. Reflections of optical rays may occur from transparent surfaces, e.g. glass, or shiny surfaces, e.g. metal. With digital images, light sources are probably the most common cause for reflections from surfaces. As another simple example, the first return of a laser beam is backscattered from a window and the second one follows from the beam hitting something solid beyond that window. These are a typical cause of outliers in the indoor 3D data. One straightforward way of eliminating these is to use a threshold value to omit returns that have a low intensity value. However, this is not always feasible for automated methods, as the threshold value depends on multiple factors and hence may appear arbitrary.

Outliers in indoor 3D data may be considerably harder to eliminate than the ones present in 3D object data because objects have a simple (convex hull) topology while indoor spaces usually do not. In other words, while outlier points inside a 3D object are harmless, they are a problem inside a room. Also, airborne scanned laser data that forms a surface with height differences usually is easier to de-noise than an indoor 3D point cloud that contains empty spaces inside.

In indoor 3D scanning, all surfaces are explicit surfaces in contrast to object 3D scanning. When scanning separate objects, e.g. by moving a camera around them, it is typically assumed that the surface of that object does not contain any holes, i.e. that the surface is implicit. This assumption greatly facilitates the reconstruction, because then a coherent surface without holes is always recovered. However, this assumption must be relaxed for indoor spaces, because for example windows (or arbitrary decorations) form holes on the walls (or other surfaces). This, that all indoor surfaces are explicit, makes the reconstruction process significantly harder than what it is for single objects. Data that is missing due to scanning occlusions or due to incomplete scan coverage must then be identified, and dealt with. The identification of this missing data is plausible with e.g. machine learning techniques that can benefit from the consistence of the existing data to create an estimate for occluded shapes and textures (Sect. 3.6).

List of problems or challenges identified in indoor 3D scanning and reconstruction is then as follows

- Optical sensor challenges as in Table 3.2 (S)
- Multi-scale problem: objects of different size (S,R)
- Occlusions from the measurement geometry in a highly convoluted space (S)
- Dynamic occlusions (S) and static occlusions (R)
- Reflections and outliers (R)
- Convoluted space with explicit surfaces (S,R)

Note that some of these problems are typically solved in either the scanning (S) or the reconstruction (R) phase. This depends on the problem characteristics. The reader should keep these in mind when reading the following sections.

3.3 Map Representations

Computers (or robots) understand the indoor spaces differently than humans. In their memory, they form a map. How the map looks like is explained in a while. First, consider the following procedure where a mobile mapping system gathers optical data of the environment, while being propagated forward by a human operator:

Initially, the map does not exist. It is generated from optical observations. In this process, we can see that there are two important concepts, positioning and map expansion. These are intertwined. The captured optical data can be transformed to expand a map, if and only if the platform movement is known. That is, if the platform can be accurately positioned with respect to time. In outdoor

Result: Map (and other scanned data)
Initialize new map from the first scan;
while *Scanning* **do**
  Observe new data;
  if *Match between new data and the stored map is found* **then**
    Update the position of the system on the map;
    Expand the map with the new data (e.g. Figure 1.4) ;
  end
end

environments, global navigation satellite system (GNSS) receivers are typically employed to provide absolute positions for a mobile mapping system in a so-called PVT format (position, velocity, precise time).[4] However, as the GNSS signals are not available indoors, relative positioning methods must be employed. This means that the generated map is employed to localize the system on it. Therefore, this is called simultaneous localization and mapping (SLAM, Sect. 3.5). In other words, estimating the trajectory of the platform and estimating the map is the very same problem, that is, the problem has a dualistic nature.

The map itself may have a variety of forms, see Table 3.3. A map can be 2D or 3D. However, the localization method used to construct it may be limited to 2D even if it outputs a 3D map. A good example of such system would be a multi-sensor system that utilizes a 2D lidar to perform the localization but has a 3D lidar or digital cameras to capture data. In some instances, these systems or methods are referred to output 2.5D maps. The 2.5D stands for two and half dimensions meaning that the outputted point clouds are 3D but there are some limitations in the method, e.g. restricting the mapping of two stories on top of each other into the same map.

Point-based maps are point clouds, such as Fig. 3.1, which are extended as the scanning continues. The benefits of these maps are that the point density can be allowed to vary from dense to sparse. The point density is stored and may be utilized later to evaluate the uncertainties in the scan result. Furthermore, point-based maps lack the discretization error that is present when the space is discretized into voxels or when planes are used to represent the space. Their limitation, however, is that they are not infinitely dense.

The voxel maps (or occupancy grids), such as Fig. 3.4, consist of cells of a given size, e.g. $5\,\text{cm}^3$, that are labeled as either occupied or unoccupied (or unexplored). For example, the map updating could go as follows: if during a scan a point is observed and that point resides in a voxel, then that voxel is marked as occupied. Also, the voxels residing along the line of sight to that point are also marked as unoccupied. A voxel map does not, however, have to be binary. It can also be probabilistic, for example see OctoMap (Hornung et al. 2013). Then the voxel cells are do not have binary states of being either occupied or unoccupied, but

[4]Navipedia of European Space Agency: https://gssc.esa.int/navipedia/

Table 3.3 Map representations used when scanning indoor environments

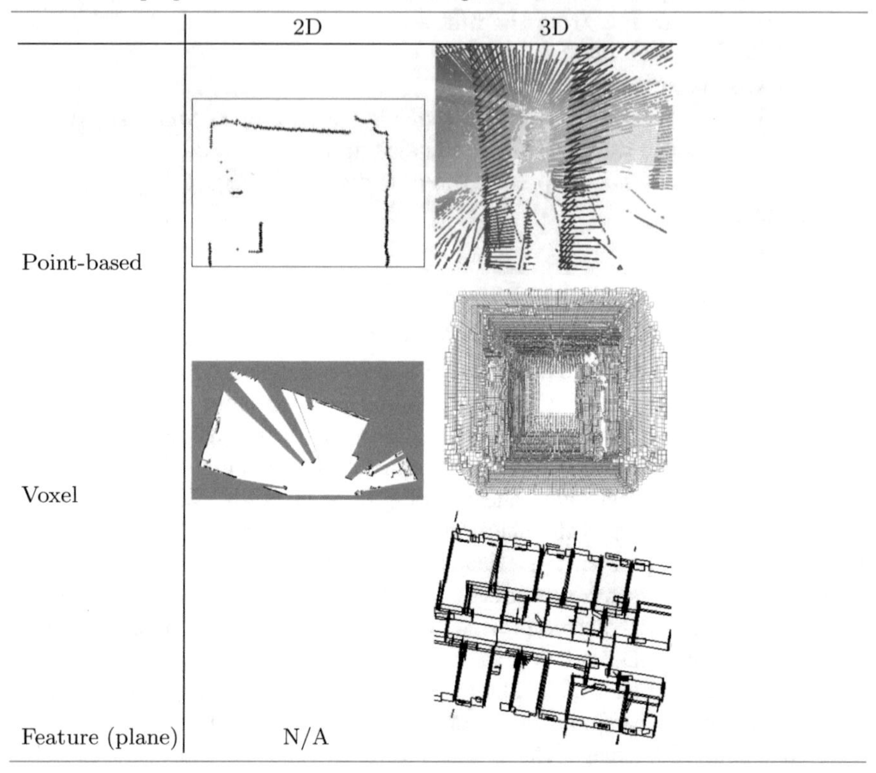

	2D	3D
Point-based		
Voxel		
Feature (plane)	N/A	

have a probability of being full (or empty). Note that in the reconstruction phase[5] (Sect. 3.6), the voxels are converted into a binary (occupied or empty) format, while here they may have unexplored or probabilistic states. Occupancy grids offer a straightforward way to represent the scanned space and are powerful in 2D, where they offer a computationally light way for keeping track of dense scans. In 3D, however, this beneficial property is severely countered by the rapidly increasing amount of (empty) voxels. The usefulness of a 3D voxel map hence easily suffers from the amount of memory required to span a large volume, because this demand increases as $O(N^3)$.

Planes are commonly used to represent floors, walls, and ceilings in indoor reconstruction (Sect. 3.6). Hence, planar features are beneficial in that they may allow for the SLAM algorithm to output models that are close to ones obtained from reconstruction (Grant et al. 2019; Karam et al. 2019). Further discussion related to SLAM algorithms is in Sect. 3.5. Other geometrical features may also be used instead or in conjunction with planar features.

[5]In reconstruction literature, voxel maps are also referred to as Manhattan world approximation.

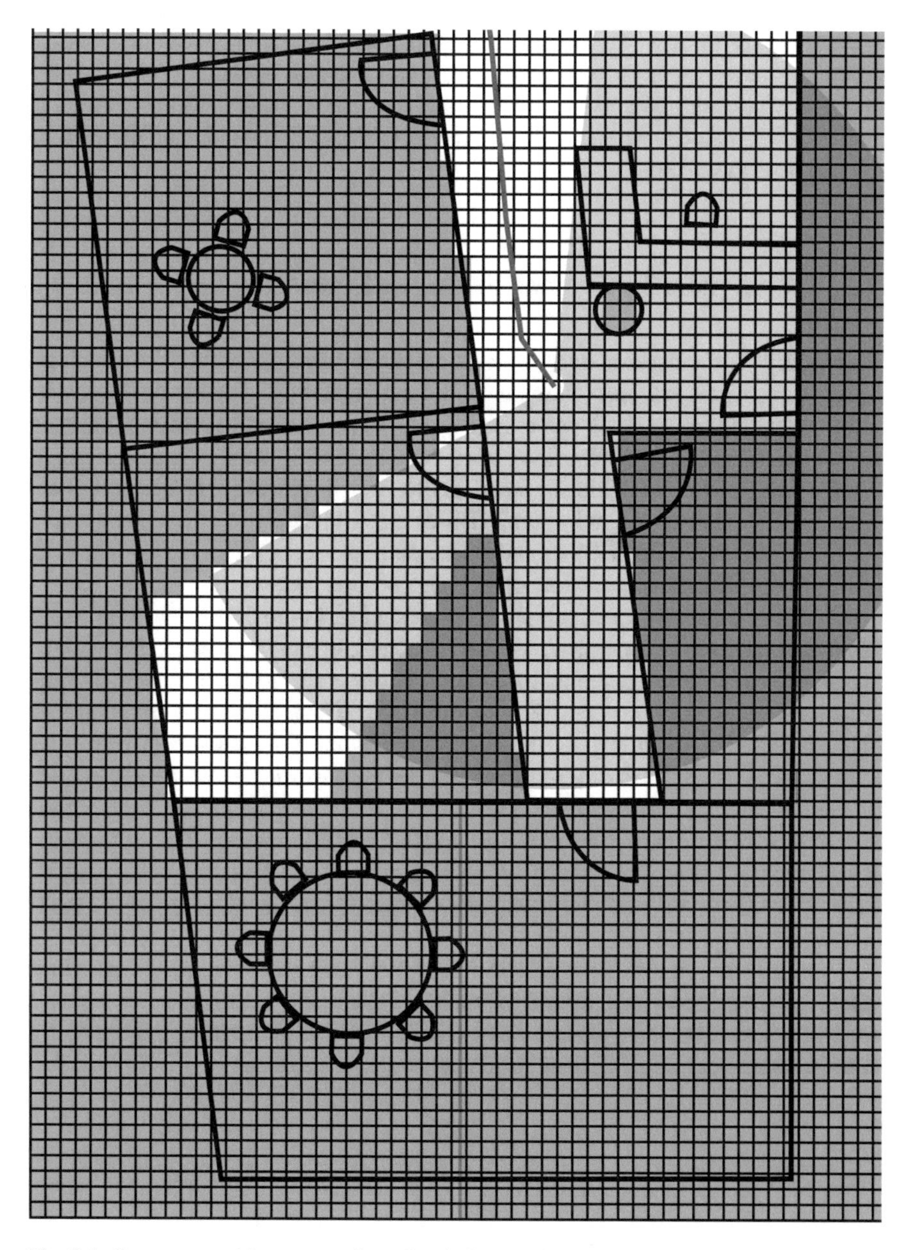

Fig. 3.4 Occupancy grid representation of an indoor environment: grey areas are not explored, white areas are empty, except those with lines. Red path is the past trajectory of the mobile mapping system. Red arc displays the field of view of the scanner. One room door is open and the scanner can partly see inside that room. Other doors are closed

Multiple maps may be created and benefited from. If the scanning system provides a map that is updated online while the operator is walking, it shows the operator which areas are yet unexplored. This is helpful, so that all the rooms and corners of the indoor environment get covered in the final detailed map that will be the output of the scanning.

3.4 Development of Indoor Scanning Systems

The history of the development of indoor scanning systems is briefly visualized in Fig. 3.5. First conscious attempts to capture whole 3D indoor environments were concluded using RGB cameras (Fig. 3.5c) and structure from motion techniques (Furukawa et al. 2009). Soon after, consumer-level depth cameras (RGB-D, Fig. 3.5a were found suitable for some limited mapping tasks but their weakness remains to be a very limited range[6] (Du et al. 2011). Almost simultaneously, a backpack platform with cameras and laser scanners was put together by Liu et al. (2010) to enable the capture of accurate geometry and textures.

The following wave of development consisted of improving the way of usage of one scanline (or 2D) lidars. A single 2D lidar was used in conjunction with an inertial sensor in a system called Zebedee (Bosse et al. 2012). Another 2D lidar system used a rotation encoder (Zhang and Singh 2014). Ultimately, a mobile 2D lidar was used without any other sensors to capture 3D indoor data (Fig. 3.5e, Lehtola et al. 2015, 2016). In other words, the a-priori model for the ego-motion required to start registering the data was successfully relaxed at a later stage of the SLAM processing. Lauterbach et al. (2015) (Fig. 3.5f) combined a 2D laser scanner in conjunction with a 3D scanner on a backpack system. Here, data from the 3D scanner could augment the 2D trajectory from an initial localization from the 2D scanner into full 3D.

The third wave of development consists of multi-line scanners (Fig. 3.5g). These relatively inexpensive but potent scanners enabled the emergence of multiple commercial systems. The systems based on these multi-line scanners include for example hand-held (Fig. 3.5h) and backpack (Fig. 3.5i) systems.

3.4.1 Single Sensor Methods and Multi-sensor Systems

Single sensor methods are important in understanding the possibilities and limitations of the sensors. Their study helps designing multi-sensor systems with optimal combinations of sensors that complement each other. In the following, we list selected state-of-the-art systems and methods.

[6]Today the commercial RGB-D cameras have a range of only up to 10 m.

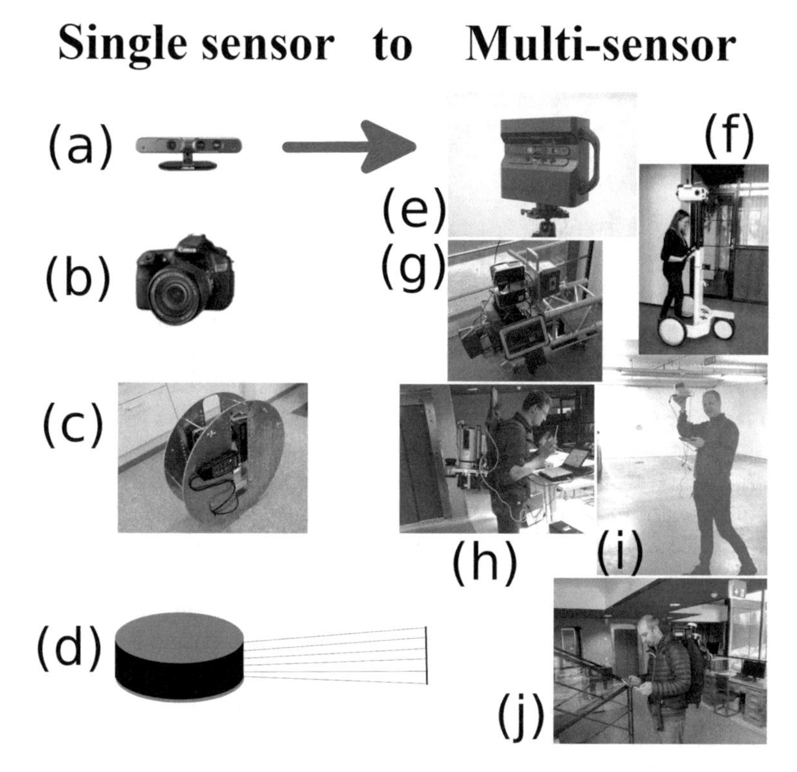

Fig. 3.5 Development of indoor 3D scanning platforms, from single sensor systems (**a**, **b**, **c**, **d**) to multi-sensor systems (**e**, **f**, **g**, **h**, **i**, **j**). These selected example systems are (**a**) RGB-D camera (Kinect), (**b**) RGB camera, (**c**) rolling 2D scanner (Aalto VILMA), (**d**) a multi-line laser scanner, (**e**) 3x RGB-D cameras (Matterport), (**f**) a multi-sensor trolley (NavVis), (**g**) a multi-sensor pushcart (FGI Slammer), (**h**) backpack with a 2D scanner and a 3D scanner (From University of Würzburg), (**i**) hand-held system (Kaarta stencil), and (**j**) multi-sensor backpack (Leica Pegasus). Many more systems exist. (**e**, **f**, **g**, **h**, **i**, **j**) are reproduced from Lehtola et al. (2017) (CC)

Single sensor methods operate on the data from

- Digital RGB and RGB-D (depth) cameras: visual SLAM, see e.g. Mur-Artal and Tardós (2017) and review by Taketomi et al. (2017). Individual frames from a video feed are matched so that the movement of the camera can be estimated.
- 2D lidar (Lehtola et al. 2016). Note that it is very challenging to reconstruct a 3D model out of 2D lidar data, and therefore these lidars are usually employed in a multi-sensor system.
- Multi-line lidar (Moosmann and Stiller 2011; Grant et al. 2019). The multi-line lidar outputs a scan that already has some 3D geometry, and the overlap from different scans can be employed in 3D scan registration for SLAM (see Sect. 3.5).

Multi-sensor systems are common in mobile mapping of indoor spaces. Our careful estimate is that there are dozens of different multi-sensor systems. Those who are interested in quantitative measures should refer to Lehtola et al. (2017), where the performance of eight different systems has been compared. Three of these are further analyzed in Tucci et al. (2018). Here, we divide multi-sensor systems into human-carriable systems and mobile platforms.

3.4.1.1 Carriable Systems

The simplest multi-sensor systems consist of a 2D lidar and an inertial sensor, e.g. Zebedee scanner (Bosse et al. 2012), or a 2D lidar and an angular decoder, e.g. LOAM (Zhang and Singh 2014). These additional sensors provide digital a-priori knowledge about the motion of the scanner that can be utilized in a prediction algorithm to make registration of the data feasible. The prediction step is especially important in providing information about the extra dimension, when a 3D reconstruction is attempted using the data from a 2D scanner. It is also worthwhile in keeping track of fast rotations when using a 3D scanner that has a low frequency (of some 10–20 Hz) in capturing scan lines, see e.g. Velas et al. (2018).

RGB cameras are also used in minimalist systems. Using an inertial measurement unit (IMU) in conjunction with a RGB camera allows for solving the absolute scale of the camera network (Nützi et al. 2011) and it provides robustness against sudden rotations where the camera system would otherwise lose track (Concha et al. 2016).

Lidar backpack systems can mount several sensors and have a combination of lidars, cameras, and IMUs. One of the first backpack systems had 3 cameras, 3 Hokyuo lidars, and one IMU (Liu et al. 2010). Since then, backpacks are seeing more 3D lidars such as Riegl VZ-400 (Lauterbach et al. 2015) and multi-line Velodyne scanners (Blaser et al. 2018).

3.4.1.2 Mobile Platforms

Mobile platforms roll on wheels, having space to mount multiple sensors to ensure a full capture of the environment. Also, they offers some advantages related to the predictability of the platform movement. For example, the localization may be conducted in 2D. Numerous experimental pushcart or trolley platforms have been assembled. For example, Radler is an instrumented surveyor's wheel that uses low-cost sensors, a 2D laser scanner and an IMU, to create 3D point clouds (Borrmann et al. 2018), while the FGI scanner (Kaijaluoto et al. 2015) (Fig. 3.5g) is more cumbersome to move, but consists of solely state-of-the-art high-end sensors. One of the known commercial platforms is NavVis (2016), see Fig. 3.5f.

3.4.1.3 Micro Aerial Vehicles

Micro Aerial Vehicles (MAVs) offer maneuverability and flexibility in mapping indoor spaces. They have, for example, been used in inventing warehouses (Eudes et al. 2018) and mapping caves (Kaul et al. 2016). One captivating scene of autonomous MAV mapping can be found in the movie Prometheus. In buildings, however, the use of MAVs is limited by that they cannot open doors and therefore are restricted by closed doors.

3.5 Iterative Closest Point SLAM

The positioning of indoor mobile mapping systems is performed using simultaneous localization and mapping (SLAM) techniques, since satellite-based positioning is unavailable indoors. We offer an example on SLAM techniques in the form of ICP-based SLAM, but the reader should be aware that there are other SLAM techniques as well. The ICP algorithm is used for matching new observations against the stored map, after which the map expansion can be done and the system position may be updated.

3.5.1 The ICP Algorithm

The ICP algorithm is the de-facto baseline for all other algorithms. The complete algorithm was invented at the same time in 1991 by Besl and McKay, by Chen and Medioni and by Zhang. The method is called the *Iterative Closest Points (ICP) algorithm.*

Given two independently acquired sets of 3D points, $\hat{M}$ (model set) and $\hat{D}$ (data set) which correspond to a single shape, we want to find the transformation $(\mathbf{R}, \mathbf{t})$ consisting of a rotation matrix $\mathbf{R}$ and a translation vector $\mathbf{t}$ which minimizes the following cost function:

$$E(\mathbf{R}, \mathbf{t}) = \frac{1}{N} \sum_{i=1}^{N} ||\mathbf{m}_i - (\mathbf{R}\mathbf{d}_i + \mathbf{t})||^2 , \tag{3.1}$$

All corresponding points can be represented in a tuple $(\mathbf{m}_i, \mathbf{d}_i)$ where $\mathbf{m}_i \in M \subset \hat{M}$ and $\mathbf{d}_i \in D \subset \hat{D}$. Two things have to be calculated: First, the corresponding points, and second, the transformation $(\mathbf{R}, \mathbf{t})$ that minimizes $E(\mathbf{R}, \mathbf{t})$ on the basis of the corresponding points. The ICP algorithm uses closest points as corresponding points. A sufficiently good starting guess, i.e. that the matched point sets are quite similarly oriented already, enables the ICP algorithm to converge to the correct minimum, see Fig. 3.6.

Current research in the context of ICP algorithms mainly focuses on fast variants of ICP algorithms (Rusinkiewicz and Levoy 2001). If the input are 3D meshes then a point-to-plane metric can be used instead of Eq. (3.1). Minimizing using a

Fig. 3.6 Registration of 3D scans. The scanned scene shows the Domshof in Bremen. Left: 3D point cloud, Right: Bird eye's view. Top: Initial registration based on rough estimates. Middle: Result after 5 iterations of ICP. Below: Final registration after ICP has terminated

point-to-plane metric outperforms the standard point-to-point one, but requires the computation of normals and meshes in a pre-processing step.

The computation of closest points is the most expensive step in the ICP algorithm. Using the optimized k-d trees the cost for finding the closest point to a given query point is at average in the order of $O(\log N)$ (Friedman et al. 1977), thus the overall cost is $O(N \log N)$ (expected time). Note: N can be very large (Elseberg et al. 2013). Improvements to k-d tree search have been presented by Elseberg et al. (2012). They include approximate k-d tree search (Greenspan and Yurick 2003), registration using $d2$-trees (Mitra et al. 2004) and cached k-d tree search (Nüchter et al. 2007).

3.5.2 Computing Optimal Poses

Four algorithms are currently known that solve the error function (3.1) in closed form (Lorusso et al. 1995). The difficulty of this minimization is to enforce the orthonormality constraint for the rotation matrix $\mathbf{R}$. Three of these algorithms separate the computation of the rotation $\mathbf{R}$ from the computation of the translation $\mathbf{t}$. These algorithms compute the rotation first and afterward the translation is derived using the rotation. For this separation, two point sets M' and D' have to be computed, by subtracting the mean of the points that are used in the matching:

$$\mathbf{c}_m = \frac{1}{N}\sum_{i=1}^{N} \mathbf{m}_i, \qquad \mathbf{c}_d = \frac{1}{N}\sum_{i=1}^{N} \mathbf{d}_i \tag{3.2}$$

and

$$M' = \{\mathbf{m}'_i = \mathbf{m}_i - \mathbf{c}_m\}_{1,\ldots,N}, \qquad D' = \{\mathbf{d}'_i = \mathbf{d}_i - \mathbf{c}_d\}_{1,\ldots,N}. \tag{3.3}$$

After replacing Eqs. (3.2) and (3.3) in the error function, $E(\mathbf{R}, \mathbf{t})$ Eq. (3.1) becomes:

$$\begin{aligned} E(\mathbf{R}, \mathbf{t}) &= \frac{1}{N}\sum_{i=1}^{N} ||\mathbf{m}'_i - \mathbf{R}\mathbf{d}'_i - \underbrace{(\mathbf{t} - \mathbf{c}_m + \mathbf{R}\mathbf{c}_d)}_{=\tilde{\mathbf{t}}}||^2 \qquad (3.4) \\ &= \frac{1}{N}\sum_{i=1}^{N} ||\mathbf{m}'_i - \mathbf{R}\mathbf{d}'_i||^2 - \frac{2}{N}\tilde{\mathbf{t}} \cdot \sum_{i=1}^{N} (\mathbf{m}'_i - \mathbf{R}\mathbf{d}'_i) + \frac{1}{N}\sum_{i=1}^{N} ||\tilde{\mathbf{t}}||^2. \end{aligned}$$

In order to minimize the sum above, all terms have to be minimized. The second sum is zero, since all values refer to centroid. The third part has its minimum for $\tilde{\mathbf{t}} = \mathbf{0}$ or

$$\mathbf{t} = \mathbf{c}_m - \mathbf{R}\mathbf{c}_d.$$

Therefore the algorithm has to minimize only the first term, and the error function is expressed in terms of the rotation only:

$$E(\mathbf{R}, \mathbf{t}) \propto \sum_{i=1}^{N} \left|\left|\mathbf{m}'_i - \mathbf{R}\mathbf{d}'_i\right|\right|^2. \tag{3.5}$$

1. The first method was developed in 1987 by Arun, Huang, and Blostein. The rotation $\mathbf{R}$ is represented as an orthonormal 3×3 matrix. The optimal rotation is calculated by $\mathbf{R} = \mathbf{V}\mathbf{U}^T$. Here the matrices $\mathbf{V}$ and $\mathbf{U}$ are derived by the singular value decomposition $\mathbf{H} = \mathbf{U}\mathbf{\Lambda}\mathbf{V}^T$ of a cross correlation matrix $\mathbf{H}$. This 3×3 matrix $\mathbf{H}$ is given by

$$\mathbf{H} = \sum_{i=1}^{N} \mathbf{m}_i'^T \mathbf{d}'_i = \begin{pmatrix} S_{xx} & S_{xy} & S_{xz} \\ S_{yx} & S_{yy} & S_{yz} \\ S_{zx} & S_{zy} & S_{zz} \end{pmatrix}, \tag{3.6}$$

where $S_{xx} = \sum_{i=1}^{N} m'_{x,i} d'_{x,i}$, $S_{xy} = \sum_{i=1}^{N} m'_{x,i} d'_{y,i}$, ...
2. The second method is similar to the previous method and was independently developed in 1988 by Horn, Hilden and Negahdaripour. Again, a correlation Matrix $\mathbf{H}$ according to Eq. (3.6) is calculated. Afterwards a so-called polar decomposition is computed, i.e., $\mathbf{H} = \mathbf{P}\mathbf{S}$, where $\mathbf{S} = (\mathbf{H}^T\mathbf{H})^{1/2}$. For this polar decomposition, Horn et al. (1988) define a square root of a matrix. Let $\mathbf{H}$, $\mathbf{S}$ and $\mathbf{P}$ the matrices as described above. Then the optimal rotation is given by

$$\mathbf{R} = \mathbf{P} = \mathbf{H}\left(\frac{1}{\sqrt{\lambda_1}}\mathbf{u}_1\mathbf{u}_1^T + \frac{1}{\sqrt{\lambda_2}}\mathbf{u}_2\mathbf{u}_2^T + \frac{1}{\sqrt{\lambda_3}}\mathbf{u}_3\mathbf{u}_3^T\right),$$

where $\{\lambda_i\}$ are the eigenvalues and $\{\mathbf{u}_i\}$ the corresponding eigenvectors of the matrix $\mathbf{H}^T\mathbf{H}$ (Horn et al. 1988).
3. The third method finds the transformation for the ICP algorithm by using unit quaternions. This method was invented in 1987 by Horn. The rotation represented as unit quaternion $\dot{\mathbf{q}}$, that minimizes Eq. (3.1), corresponds to the largest eigenvalue of the cross covariance matrix $\mathbf{N} =$

$$\begin{pmatrix} (S_{xx} + S_{yy} + S_{zz}) & (S_{yz} + S_{zy}) & (S_{zx} + S_{xz}) & (S_{xy} + S_{yx}) \\ (S_{yz} + S_{zy}) & (S_{xx} - S_{yy} - S_{zz}) & (S_{xy} + S_{yx}) & (S_{zx} + S_{xz}) \\ (S_{zx} + S_{xz}) & (S_{xy} + S_{yx}) & (-S_{xx} + S_{yy} - S_{zz}) & (S_{yz} + S_{zy}) \\ (S_{xy} + S_{yx}) & (S_{yz} + S_{zy}) & (S_{zx} + S_{xz}) & (-S_{xx} - S_{yy} + S_{zz}) \end{pmatrix}.$$

4. The fourth solution method for minimizing Eq. (3.1) uses so-called dual quaternions. This method was developed by Walker et al. in 1991. Unlike the first three methods covered so far the transformation is found in a single step. There is

no need to apply the trick with centroids to compute the rotation in a separate fashion. Here, the optimal transformation consisting of a rotation and translation is again a solution of the eigenvalue problem of a 4×4 matrix function that is built from corresponding point pairs.

The closed-form solutions discussed so far are all non-linear, since they need an eigenvector/eigenvalue solver, e.g., in case of using the third method, a quartic equation must be solved (Horn 1987).

For SLAM applications it is necessary to have a notion of the uncertainty of the poses calculated by the registration algorithm. The following is the extension of the probabilistic approach first proposed by Lu and Milios (1997) to 6 DoF. This extension is not straightforward, since the matrix decomposition, i.e., Eq. (3.8) cannot be derived from first principles. For a more detailed description of the extension refer to Borrmann et al. (2008a,b). In addition to the pose $\mathbf{X}$, the pose estimate $\bar{\mathbf{X}}$ and the pose error $\Delta\mathbf{X}$ are required.

The positional error of a scan at its pose $\mathbf{X}$ is described by:

$$E = \sum_{i=1}^{m} \|\mathbf{X} \oplus \mathbf{d}_i - \mathbf{m}_i\|^2 = \sum_{i=1}^{m} \|\mathbf{Z}_i(\mathbf{X})\|^2$$

Here, $\oplus$ is the compounding operation that transforms a point $\mathbf{d}_i$ into the global coordinate system. For small pose errors $\Delta\mathbf{X}$, E can be linearized by use of a Taylor expansion:

$$\begin{aligned}\mathbf{Z}_i(\mathbf{X}) &\approx \bar{\mathbf{X}} \oplus \mathbf{d}_i - \mathbf{m}_i - \nabla\mathbf{Z}_i(\bar{\mathbf{X}})\Delta\mathbf{X} \\ &= \mathbf{Z}_i(\bar{\mathbf{X}}) - \nabla\mathbf{Z}_i(\bar{\mathbf{X}})\Delta\mathbf{X}\end{aligned}$$

Utilizing the matrix decomposition $\mathbf{M}_i\mathbf{H}$ of $\nabla\mathbf{Z}_i(\bar{\mathbf{X}})$ that separates the pose $\mathbf{X}$, which is contained in $\mathbf{H}$ from the points $\mathbf{m}_i$ and $\mathbf{d}_i$, which are contained in $\mathbf{M}_i$:

$$\mathbf{Z}_i(\mathbf{X}) \approx \mathbf{Z}_i(\bar{\mathbf{X}}) - \mathbf{M}_i\mathbf{H}\Delta\mathbf{X}$$

Appropriate decompositions are given for the Euler angles, quaternion representation and the Helix transform in the following paragraphs. Because $\mathbf{M}_i$ is independent of the pose, the positional error E is approximated as:

$$E \approx (\mathbf{Z} - \mathbf{M}\mathbf{H}\Delta\mathbf{X})^T(\mathbf{Z} - \mathbf{M}\mathbf{H}\Delta\mathbf{X}),$$

where $\mathbf{Z}$ is the concatenation of all $\mathbf{Z}_i(\bar{\mathbf{X}})$ and $\mathbf{M}$ the concatenation of all $\mathbf{M}_i$'s.

E is minimized by the ideal pose:

$$\bar{\mathbf{E}} = (\mathbf{M}^T\mathbf{M})^{-1}\mathbf{M}^T\mathbf{Z}$$

and its covariance is given by

$$\mathbf{C} = s^2(\mathbf{M}^T\mathbf{M}),$$

where s^2 is the unbiased estimate of the covariance of the identically, independently distributed errors of $\mathbf{Z}_i$:

$$s^2 = (\mathbf{Z} - \mathbf{M}\bar{\mathbf{E}})^T(\mathbf{Z} - \mathbf{M}\bar{\mathbf{E}})/(2m - 3). \tag{3.7}$$

Note that $\bar{\mathbf{E}}$ is the minimum for the linearized pose $\mathbf{H}\Delta\mathbf{X}$. To obtain the optimal $\mathbf{X}$ the following transformation is performed:

$$\mathbf{X} = \bar{\mathbf{X}} - \mathbf{H}^{-1}\bar{\mathbf{E}},$$
$$\mathbf{C} = (\mathbf{H}^{-1})\mathbf{C}(\mathbf{H}^{-1})^T.$$

The representation of pose $\mathbf{X}$ in Euler angles, as well as its estimate and error is as follows:

$$\mathbf{X} = \begin{pmatrix} t_x \\ t_y \\ t_z \\ \theta_x \\ \theta_y \\ \theta_z \end{pmatrix}, \bar{\mathbf{X}} = \begin{pmatrix} \bar{t}_x \\ \bar{t}_y \\ \bar{t}_z \\ \bar{\theta}_x \\ \bar{\theta}_y \\ \bar{\theta}_z \end{pmatrix}, \Delta\mathbf{X} = \begin{pmatrix} \Delta t_x \\ \Delta t_y \\ \Delta t_z \\ \Delta\theta_x \\ \Delta\theta_y \\ \Delta\theta_z \end{pmatrix}.$$

The matrix decomposition $\mathbf{M}_i\mathbf{H} = \nabla\mathbf{Z}_i(\bar{\mathbf{X}})$, i.e., the Jacobian, is given by:

$$\mathbf{H} = \begin{pmatrix} 1 & 0 & 0 & 0 & \bar{t}_z\cos(\bar{\theta}_x) + \bar{t}_y\sin(\bar{\theta}_x) & \bar{t}_y\cos(\bar{\theta}_x)\cos(\bar{\theta}_y) - \bar{t}_z\cos(\bar{\theta}_y)\sin(\bar{\theta}_x) \\ 0 & 1 & 0 & -\bar{t}_z & -\bar{t}_x\sin(\bar{\theta}_x) & -\bar{t}_x\cos(\bar{\theta}_x)\cos(\bar{\theta}_y) - \bar{t}_z\sin(\bar{\theta}_y) \\ 0 & 0 & 1 & \bar{t}_y & -\bar{t}_x\cos(\bar{\theta}_x) & \bar{t}_x\cos(\bar{\theta}_y)\sin(\bar{\theta}_x) + \bar{t}_y\sin(\bar{\theta}_y) \\ 0 & 0 & 0 & 1 & 0 & \sin(\bar{\theta}_y) \\ 0 & 0 & 0 & 0 & \sin(\bar{\theta}_x) & \cos(\bar{\theta}_x)\cos(\bar{\theta}_y) \\ 0 & 0 & 0 & 0 & \cos(\bar{\theta}_x) & -\cos(\bar{\theta}_y)\sin(\bar{\theta}_x) \end{pmatrix}. \tag{3.8}$$

and

$$\mathbf{M}_i = \begin{pmatrix} 1 & 0 & 0 & 0 & -d_{y,i} & -d_{z,i} \\ 0 & 1 & 0 & d_{z,i} & d_{x,i} & 0 \\ 0 & 0 & 1 & -d_{y,i} & 0 & d_{x,i} \end{pmatrix}.$$

As required, $\mathbf{M}_i$ contains all point information while $\mathbf{H}$ expresses the pose information. Thus, this matrix decomposition constitutes a pose linearization similar to those proposed in the preceding sections. Note that, while the matrix decomposition is arbitrary with respect to the column and row ordering of **H**, this particular description was chosen due to its similarity to the 3D pose solution given by Lu and Milios (1997).

3.5.3 Marker and Feature-Based Registration

Sometimes the ICP algorithm does not properly converge from the starting guess and is attracted into a local minimum. To avoid these issues with starting guess in the ICP framework, marker based registration uses defined artificial or natural landmarks as corresponding points. This manual data association ensures that by minimizing Eq. (3.1) the scans are registered at the correct location. Iterations are no longer required. Feature based algorithms, like using SIFT features, automatically extract the 3D position of natural features and do not need any iterations nor manual interference for registration (Böhm and Becker 2007).

While registering several 3D data sets using the ICP algorithm or marker and feature-based registration techniques, errors sum up. These errors are due to imprecise measurements and small registration errors. Therefore, globally consistent scan matching algorithm aim at reducing these errors.

3.5.4 ICP-Based SLAM

Chen and Medioni (1992) aimed at globally consistent range image alignment when introducing an incremental matching method, i.e., all new scans are registered against the so-called metascan, which is the union of the previously acquired and registered scans. This method does not spread out the error and is order-dependent.

Bergevin et al. (1996), Stoddart and Hilton (1996), Benjemaa and Schmitt (1997); Benjemaas and Schmitt (1998), and Pulli (1999) present iterative approaches. Based on networks representing overlapping parts of images, they use the ICP algorithm for computing transformations that are applied after all correspondences between all views have been found. However, the focus of research is mainly 3D modeling of small objects using a stationary 3D scanner and a turn table; therefore, the used networks consist mainly of one loop (Pulli 1999), where the loop closing has to be smoothed. These solutions are locally consistent algorithms that stick to the mentioned analogy of the spring system (Cunnington and Stoddart 1999), whereas true globally consistent algorithms minimize the error function in one step. A probabilistic approach was proposed by Williams et al. (1999), where each scan point is assigned a Gaussian distribution in order to model the statistical errors made by laser scanners. This causes high computation time due to the large amount of data in practice. Krishnan et al. (2000) presented a global registration algorithm that minimizes the global error function by optimization on the manifold of 3D rotation matrices.

The n-scan registration using linearization allows us to compute global optimal poses in one step given point correspondences between adjacent scans. These scans are given by a graph, where each link, $j \rightarrow k$ denotes a set of point pairs, i.e., closest points. Following the notation of ICP, scan j serves as the model set, while scan k serves as data set. Next we present four novel linear methods for the parameterization of the rotation.

For an uncertainty-based global point cloud registration method or SLAM method, the 2-scan case, discussed above is extended. Under the assumption that two poses $\mathbf{X}'_j$ and $\mathbf{X}'_k$ are related by the linear error metric $\mathbf{E}'_{j,k}$ we wish to minimize the Mahalanobis distance that describes the global error of all the poses:

$$\begin{aligned} W &= \sum_{j \rightarrow k} (\bar{\mathbf{E}}_{j,k} - \mathbf{E}'_{j,k})^T \mathbf{C}_{j,k}^{-1} (\bar{\mathbf{E}}'_{j,k} - \mathbf{E}'_{j,k}) \\ &= \sum_{j \rightarrow k} (\bar{\mathbf{E}}_{j,k} - (\mathbf{X}'_j - \mathbf{X}'_k)) \mathbf{C}_{j,k}^{-1} (\bar{\mathbf{E}}'_{j,k} - (\mathbf{X}'_j - \mathbf{X}'_k)). \end{aligned} \tag{3.9}$$

The error between two poses is modeled by the Gaussian distribution $(\bar{\mathbf{E}}_{j,k}, \mathbf{C}_{j,k})$. In matrix notation, W becomes:

$$W = (\bar{\mathbf{E}} - \mathbf{H}\mathbf{X})^T \mathbf{C}^{-1} (\bar{\mathbf{E}} - \mathbf{H}\mathbf{X}).$$

Here $\mathbf{H}$ is the signed incidence matrix of the pose graph, $\bar{\mathbf{E}}$ is the concatenated vector consisting of all $\bar{\mathbf{E}}'_{j,k}$ and $\mathbf{C}$ is a block-diagonal matrix comprised of $\mathbf{C}_{j,k}^{-1}$ as submatrices. Minimizing this function yields new optimal pose estimates. The minimization of $\mathbf{W}$ is accomplished via the following linear equation system:

$$\begin{aligned} (\mathbf{H}^T \mathbf{C}^{-1} \mathbf{H}) \mathbf{X} &= \mathbf{H}^T \mathbf{C}^{-1} \bar{\mathbf{E}} \\ \mathbf{B}\mathbf{X} &= \mathbf{A}. \end{aligned}$$

The matrix $\mathbf{B}$ consists of the submatrices

$$\mathbf{B}_{j,k} = \begin{cases} \sum_{k=0}^{n} \mathbf{C}_{j,k}^{-1} & (j = k) \\ \mathbf{C}_{j,k}^{-1} & (j \neq k). \end{cases}$$

The entries of $\mathbf{A}$ are given by:

$$A_j = \sum_{\substack{k=0 \\ k \neq j}}^{n} \mathbf{C}_{j,k}^{-1} \bar{\mathbf{E}}_{j,k}.$$

In addition to $\mathbf{X}$, the associated covariance of $\mathbf{C}_X$ is computed as follows:

$$\mathbf{C}_X = \mathbf{B}^{-1}$$

The actual positional error of two poses $\mathbf{X}_j$ and $\mathbf{X}_k$ is not linear:

$$E_{j,k} = \sum_{i=1}^{m} \left\| \mathbf{X}_j \oplus \mathbf{d}_i - \mathbf{X}_k \oplus \mathbf{m}_i \right\|^2 = \sum_{i=1}^{m} \left\| \mathbf{Z}_i(\mathbf{X}_j, \mathbf{X}_k) \right\|^2 .$$

Analogous to the simple 2-scan case the linearized pose difference $\mathbf{E}'_{j,k}$ is obtained by use of a Taylor expansion of $\mathbf{Z}_i(\mathbf{X}_j, \mathbf{X}_k)$:

$$\mathbf{Z}_i(\mathbf{X}_j, \mathbf{X}_k) \approx \mathbf{Z}_i(\bar{\mathbf{X}}_j, \bar{\mathbf{X}}_k) - \left(\nabla_{\mathbf{X}_j} \mathbf{Z}_i(\bar{\mathbf{X}}_j, \bar{\mathbf{X}}_k) \Delta \mathbf{X}_j - \nabla_{\mathbf{X}_k} \mathbf{Z}_i(\bar{\mathbf{X}}_j, \bar{\mathbf{X}}_k) \Delta \mathbf{X}_k \right).$$

Here, $\nabla_{\mathbf{X}_j}$ refers to the derivative with respect to $\mathbf{X}_j$. Utilizing the same matrix decomposition $\mathbf{M}_i\mathbf{H}$ of $\nabla \mathbf{Z}_i(\bar{\mathbf{X}})$ as in the 2-scan case $\mathbf{Z}_i(\mathbf{X}_j, \mathbf{X}_k)$ is approximated as:

$$\mathbf{Z}_i(\mathbf{X}_j, \mathbf{X}_k) \approx \mathbf{Z}_i(\bar{\mathbf{X}}_j, \bar{\mathbf{X}}_k) - \mathbf{M}_i \mathbf{E}'_{j,k},$$

where $\mathbf{E}'_{j,k}$ is the linear error metric given by:

$$\begin{aligned} \mathbf{E}'_{j,k} &= (\mathbf{H}_j \Delta \mathbf{X}_j - \mathbf{H}_k \Delta \mathbf{X}_k) \\ &= (\mathbf{X}'_j - \mathbf{X}'_k). \end{aligned}$$

$\mathbf{E}'_{j,k}$ is linear in the quantities $\mathbf{X}'_j$ that will be estimated by the algorithm. Again, the minimum of $\mathbf{E}'_{j,k}$ and the corresponding covariance are given by

$$\begin{aligned} \bar{\mathbf{E}}_{j,k} &= (\mathbf{M}^T\mathbf{M})^{-1}\mathbf{M}^T\mathbf{Z} \\ \mathbf{C}_{j,k} &= s^2(\mathbf{M}^T\mathbf{M}). \end{aligned}$$

Here $\mathbf{Z}$ is the concatenated vector consisting of all $\mathbf{Z}_i = \bar{\mathbf{X}}_j \oplus \mathbf{d}_i - \bar{\mathbf{X}}_k \oplus \mathbf{m}_i$.

Note that the results have to be transformed in order to obtain the optimal pose estimates, just like in the 2-scan case.

$$\begin{aligned} \mathbf{X}_j &= \bar{\mathbf{X}}_j - \mathbf{H}_j^{-1}\mathbf{X}'_j, \\ \mathbf{C}_j &= (\mathbf{H}_j^{-1})\mathbf{C}_j^X(\mathbf{H}_j^{-1})^T. \end{aligned}$$

Note that SLAM techniques have not been developed for indoor applications only, as applications for SLAM exist also in undersea, space, underground, and forest environments. We recommend interested readers to get acquainted with probabilistic techniques such as Kalman and particle filtering in SLAM (see e.g. Thrun et al. 2005).

3.5.5 *Assessing the SLAM Errors*

The Mahalanobis distance of Eq. (3.9) offers a formulation for a computational error that can be minimized to match all poses. However, the minimized residual of this error can not be straightforwardly interpreted to assess the quality of the final point cloud. For example, it does not take a stance on whether the ICP algorithm has been attracted into local minima leading to failed pose matching, which in turn may lead into serious distortions in the final point cloud. In other words, the residual from Eq. (3.9) can be very small even if the final point cloud is nonsense. This is because of the dualistic nature of the SLAM problem: the pose errors are transformed onto errors in the observed 3D shape of the environment. The errors in the final point cloud are thus a function of all data (as these were used to estimate the poses), and include sensor errors, system calibration errors, and quality and extent of observation overlap. For the case of any SLAM (also ICP), the errors can therefore be

- quantitatively assessed only if reference data is available, from the trajectory or from the point cloud
- qualitatively assessed with the human eye, which is commonly used in 3D visualization, from the point cloud
- assessed against a-priori knowledge, e.g. geometric rules that require all walls to be planar or such that require all corners to be straight.

Point cloud to point cloud comparisons can be conducted by using a measure for control points, point subsets, or whole point clouds (Lehtola et al. 2017). Different measures are summarized in Table 3.4. The choice depends on the properties of the scanned object, i.e. whether it can change shape, and whether the point cloud has already been smoothed, e.g. filtered for outliers. Shape change is a property often related to human 3D body scanning but a room with swinging doors could also be considered with these measures. Smoothing is usually an integrated part of commercial products, meaning that if the data is captured with such a product, the output is a smoothed point cloud.

Once a model is reconstructed (see Sect. 3.6) from the scanned point cloud, the assessment becomes more straightforward. It is then a model to model comparison. The straightforwardness follows from that the measures such as completeness, that the model covers the reference, and correctness, that the model does not contain anything extra with respect to the reference, can be defined (Tran et al. 2019). However, the cave-at here is that then the total errors are a function of not only the SLAM process but also the reconstruction process. This may make the assessment appear oblique if the cause of errors is of interest. In industry, the reconstructed

Table 3.4 Metrics for point cloud to point cloud comparison. DEM stands for digital elevation models. (Adapted from Lehtola et al. 2017)

	Smoothed point cloud	Non-smoothed point cloud
Rigid object	L_p norms Hausdorff measure Examples: Scanned 3D objects, DEM	L_1 and L_2 norms with cutoff radius Examples: Raw point clouds
Non-rigid object	Gromov-Hausdorff Gromov-Wasserstein Examples: Shape changing objects	N/A

models are in standard formats (see Sect. 3.6) and are validated with commercial model checkers.

In mobile mapping, a real time map from an online SLAM is sometimes used to direct the operator when data is gathered. However, pose errors are typically larger for online SLAM than offline SLAM, since less data is used for overlap computation in the online versions to keep the computational load tractable. In turn, the offline, or post-processing, SLAM algorithms can optimize over all data. Such is the minimization of the Mahalanobis distance of Eq. (3.9) and such are also the so-called graphSLAM techniques that are based on graphs representing all observations (Grisetti et al. 2010).

3.6 Indoor 3D Reconstruction

By indoor 3D reconstruction we are referring to the process of generating a mesh model which is exportable to one of the standard formats such as IFC (industry foundation classes) or IndoorGML (Chen and Clarke 2017). In other words, the point clouds are replaced by a mesh that consists of continuous geometrical shapes such as planes and boundary representations (B-Rep). During a reconstruction process, a successful composition of walls is the most important factor because it defines the main layout of the interiors. However, some approaches are contented by providing a volumetric model of the interiors without an explicit representation of walls. The reconstruction process here includes the data segmentation step, where the point cloud is divided into rooms and subspaces. Note that some of the room segmentation methods explained next directly result in a final mesh model (e.g. cell decomposition), while some just assign labels to points (e.g. mathematical morphology) and require another method for the creation of the mesh.

3.6.1 Space Subdivision and Room Segmentation

Space subdivision is referred to the problem of dividing the space into semantic subspaces. Another term used in the literature for space subdivision is room segmentation. However, there are slight differences between the concepts of a room and a subspace. A room is separated from other rooms by permanent structures such as walls, floors and ceilings and there should be an opening (e.g. door) to connect two rooms. A subspace can represent a room or part of a room, for example a meeting area which is separated from the rest of that room by temporary partitions. When spaces are physically separated by permanent structures, space subdivision is equivalent to room segmentation. Several important remarks need to be considered when dealing with space subdivision:

1. The space subdivision can be done in 2D (Bormann et al. 2016), in 2.5D (Ikehata et al. 2015) and in full 3D (Mura et al. 2016).
2. The space subdivision can be done with Manhattan-World assumptions (Khoshelham and Díaz-Vilariño 2014) or without it (Ochmann et al. 2019). In Manhattan-World assumption, walls are assumed to be perfectly vertical and perpendicular to each other.
3. The trajectory of the acquisition device, in case of mobile laser scanners, can be a valuable data source for the space subdivision (Elseicy et al. 2018; Nikoohemat et al. 2018).

In the following, the most common space subdivision methods in the literature are presented along with their limitations.

Mathematical morphology The input data is converted into a 2D grid, which is essentially an image (with pixels), and or into a 3D voxel grid. The pixels (or voxels) are labeled as occupied (not accessible) or empty (accessible). A morphological erosion is applied on empty pixels which causes the occupied pixels (e.g. walls) to grow and the empty pixels to either vanish or get separated, if they had a weak connection. Then a connected component analysis is run, which identifies all connected segments of empty space. Each empty segment represents a room candidate. Finally, a morphological dilation is applied on the generated room segments to grow until the border of the room meets the occupied space (Bormann et al. 2016; Nikoohemat et al. 2018). See Fig. 3.7. The limitation of the morphology approach is that one has to make a selection for the pixel size to provide a good trade-off between computational cost and accuracy of the room topology. Obviously, smaller pixel size represents a better accuracy of the room topology but becomes computationally expensive and needs more iteration to converge to the correct number of rooms.

Delaunay Triangulation/Voronoi Diagram (Bormann et al. 2016; Turner et al. 2014): For this method, the input is either a set of points representing wall samples or a 2D grid of occupied and empty pixels (similar to the morphology case). Delaunay triangulation is run on the input producing a set of triangles that connect

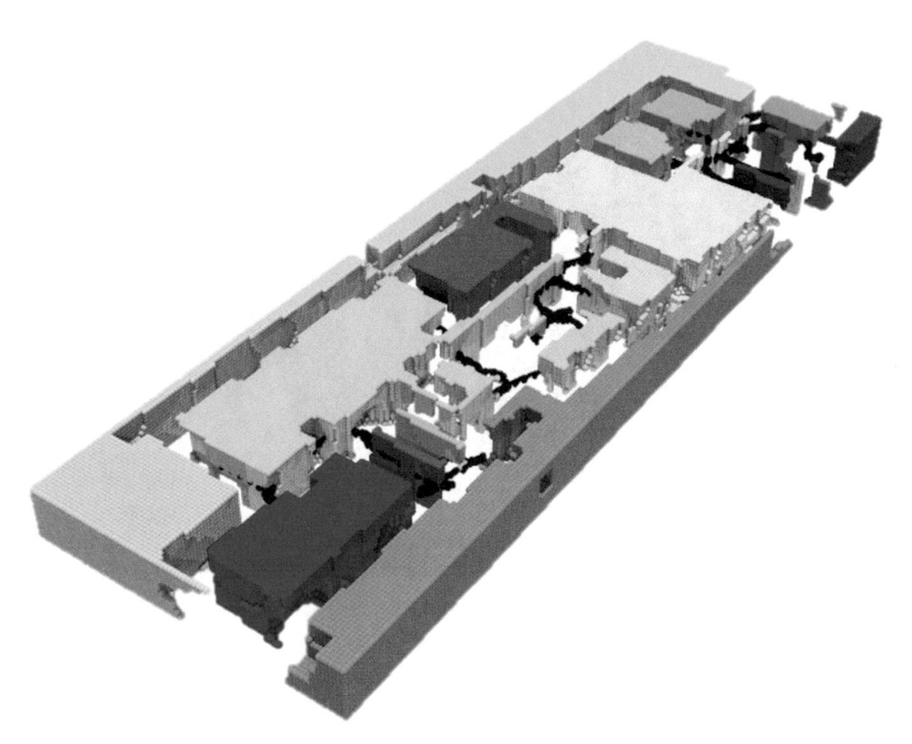

Fig. 3.7 Room segmentation by mathematical morphology. Each color is one segment. (Reproduced from Nikoohemat et al. 2018)

the wall sample points of the input data, see Fig. 3.8. Then the triangles are labeled as inside or outside using the line of sight established from the scanner trajectory. If an intersection between the line of sight and the triangle is encountered, then the triangle is labeled as inside. Inside triangles are used as room seeds. For each triangle, a circumcircle is generated (i.e. the unique circle that passes through each corner point of that triangle). It is assumed that the circumcircles with the highest overlap belong to the same room, and only one of these is stored. The initial set of room seeds then equals to the largest remaining circumcircles. The result is a rough location of each room and an initial number of rooms. Finally, candidate rooms are merged under two conditions: (i) if they share a large perimeter with another room and (ii) if they share a border which is too large to be a door. This, however, results in over-segmentation in the long corridors. Delaunay triangulation is mainly implemented in 2D and then extended to 2.5D. It is not an ideal method for true 3D modeling.

Cell decomposition is perhaps the most used approach in the literature. It consists of three steps, see Fig. 3.9. (i) The input data is converted into a set of lines (2D) or planes (3D). (ii) Lines or planes are elongated so that they intersect with the bounding box limits of the modeled space. This process generates a 2D or 3D cell complex, where each cell is represented by a piece-wise planar polyhedron

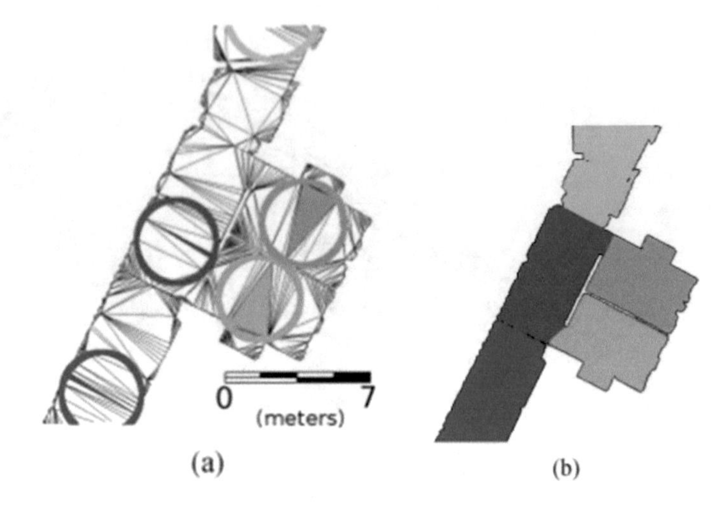

Fig. 3.8 (**a**) Delauney triangulation. The shown circles with the longest circumcircles form seeds for rooms. (**b**) Obtained room segmentation. (Reproduced from Turner et al. 2014 with permission)

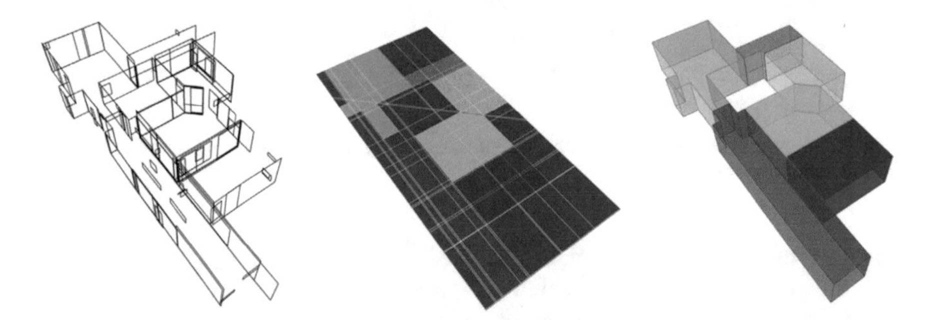

Fig. 3.9 Room segmentation by cell decomposition. Steps in (i) left, (ii) middle, and (iii) right image explained in text are visualized here. (Reproduced from Mura et al. 2014)

or a convex cell. Additionally, each cell is labeled as an inside cell or an outside cell. Typically, this is done using line-of-sight ray tracing techniques. (iii) Cells with the inside label are clustered and merged to form individual rooms. This is the step where methods show most differences, especially in the way room seeds are clustered, for example, with Markov clustering (Mura et al. 2016), Integer Linear Programming (Ochmann et al. 2019), or graph cut optimization (Oesau et al. 2014). This that there has been multiple different approaches to the optimization of the cell clustering step tells that it is not a straightforward problem, which may be considered as a limitation.

MLS trajectory-based method Mobile laser scanner data is registered into a 3D point cloud using SLAM techniques. From that process, for each point in the point cloud, we can obtain the point on the trajectory from which that particular

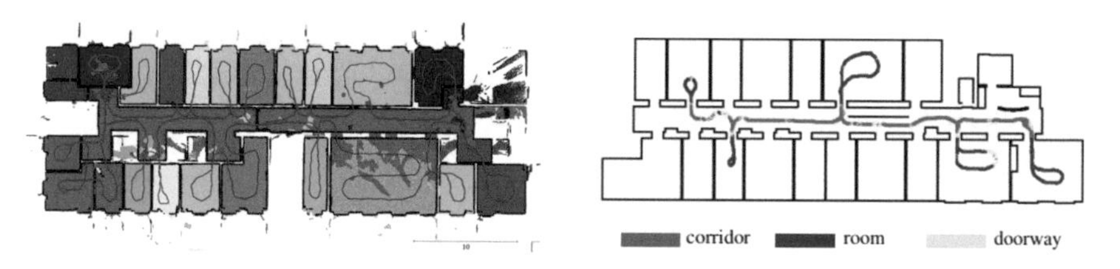

Fig. 3.10 Room segmentation by using the trajectory of a mobile laser scanning system. This method relies on doorway (opening) detection. Left: Elseicy et al. (2018), right: Mozos (2010, p.31)

measurement was done. In order to do room segmentation, first the trajectory is divided to segments using the locations, where a door is detected from the data, as division points. The location of the doors can be extracted by intersecting the trajectory with wall candidates, which are lines in 2D and planes in 3D. Obviously, only doors which were entered during the scanning can be identified and used to segment the trajectory. In addition, one can consider that each loop in the trajectory is a possible room candidate, see Fig. 3.10. By applying both criteria (door location and loops) the trajectory is segmented more robustly in that it takes into account also the cases where the entryway is larger or otherwise different than a standard door. Then the subset of the point cloud captured from that trajectory segment can be collected (using e.g. the time attribute in both point clouds and the trajectory). One limitation of this approach is the existence of openings, meaning that from the scanner positions residing inside a room parts of other spaces can also be scanned if there are openings to those spaces. This causes the room topology to become inaccurate near doors and windows. Another limitation concerns the correct detection of loops, for example if the scanning operator enters one room and exits from another door a loop is not formed. Similarly, if the operator makes an unnecessary loop in a big hall or a corridor it results in an over-segmentation of the space (Elseicy et al. 2018; Mozos 2010).

Machine learning methods bring in the possibility of naming the rooms by their functions. This so-called semantic labeling of rooms goes one step further than plain room segmentation. Machine learning methods such as random forest, adaBoost and conditional random field are used to cluster the detected planes or super segments into rooms with a function (e.g. corridor, kitchen, bedroom, etc.) (Bassier and Vergauwen 2019; Bormann et al. 2016; Mozos 2010). For this kind of methods, however, training data needs to be created and that training data should represent different types of rooms. Another novel approach is using deep learning (Convolutional Neural Network), which needs an even larger training dataset but in return is able to produce, for example, floor plans from a large set of RGB-D samples (Liu et al. 2018). Hence, the effort of creating a method is partly replaced by an effort to create a set of labeled data to be used for training. On the other hand, if the method can take sensor data as input, in a so-called end-to-end network fashion, this can be seen advantageous.

Other approaches for room segmentation such as graph-based methods, enclosure of the spaces, and shape grammar are referred to in the bibliography (Murali et al. 2017; Nikoohemat et al. 2019; Tran et al. 2018).

3.6.2 Reconstruction of Walls

Walls can be reconstructed as infinitely-thin planar planes, see Fig. 3.9, or as volumetric objects that have a non-zero thickness. Oftentimes the choice depends on the application, for example, in BIM models walls are volumetric objects. Given a point cloud, there are several approaches to separate walls form the furniture and to create the correct wall arrangement. Some of these approaches were explained in Sect. 3.6.1 on room segmentation, because these two problems are related. Room segmentation determines the room layout and consequently the wall arrangement. Furthermore, there is another approach that is based on constructive solid geometry (CSG) (Xiao and Furukawa 2014). It begins by slicing the point clouds to horizontal layers and looking for primitives such as rectangles in 2D or cuboids in 3D. Rectangle primitives are the most common shapes in the architecture. By exploiting a so-called free-space constraint and optimizing an objective function, the best arrangement of the primitives is selected, which means that some of the rectangles are merged into one another. Finally, the 2D CSG models are elongated in the third dimension to make them volumetric, and the objects stacked on top of each other form the sought-after 3D model with the correct wall arrangement. This approach generates volumetric walls but it does not create a topology for rooms. That is, the geometrical objects are created in their respective positions but the relations between these objects, e.g. how they form rooms, need to be established with additional means.

3.6.3 Grammar Approach

Grammar represents the rules of how things should be expressed. Ill-defined grammar or bad grammar results in ambiguities and misunderstandings. A natural language, English for example, would not function without grammar, because it would not make sense. Grammar is also widely applied in programming languages (see e.g. Chomsky 1959). Another types of grammar, ones intended for 3D models, can also be defined. For example, shape grammars have been used as modeling techniques in architecture, computer graphic and engineering (Stiny and Gips 1971). These were decades before the introduction of 3D scanning data. Later, grammar approaches that exploit the regularity and repetitive pattern of architectures were employed to generate models that are not based on real data, for urban designers or for computer games (Wonka et al. 2003).

Considering scanned 3D data, data-based reconstruction activities started from building façade reconstruction for city models (e.g. Müller et al. 2006; Musialski et al. 2013). Soon after, grammar was harnessed to benefit indoor 3D reconstruction (Becker et al. 2013; Boulch et al. 2013) with the founding idea that data-based indoor 3D reconstruction would yield BIM models (see Sect. 3.7). Simultaneously, automatically generated 2D floor plans could also be extracted (Ikehata et al. 2015). Floor plan generation is important, for example, in Americas as there most homes lack floor plans. Latest studies present BIM-aimed approaches where grammar rules are used iteratively to fit the best representations (parametric models with semantics) into the data (Becker et al. 2015; Tran et al. 2018). Point clouds have been established as the main input data format for grammar based indoor reconstruction.

The reconstruction of walls, or segmentation of rooms, is the goal also when we are using grammar. A grammar is a set of components such as terminals, non-terminals, rules, axioms and attributes. Rules define how the non-terminals (e.g. shape) should be transformed to terminals during the model creation. Rules generally are defined by an expert and this is the challenging part of the grammar. Some of the common rules in shape grammar are merge, split and a set of transformations. Attributes can be color, texture, material and labels. Attributes are not always part of the grammar unless we are using an attribute grammar.

One basic idea to apply grammar in indoor reconstruction is to define a primitive shape such as a cuboid as an axiom and fit the cuboid to the data (e.g. a point cloud). For this purpose a parametric shape can be defined and placed aligned with the axis in the data. If the data is axis-aligned then a Manhattan World can be assumed and fitting the cuboid to the data becomes easy. By placing, scaling, and transforming more cuboids the model can be reconstructed. Then, based on the fact that whether there are enough points to support the faces of each cuboid as valid walls, the cuboids can be merged or split to form rooms. Useful information may additionally be obtained from the locations of doorways which can be used to justify whether some cuboid faces should be added to split the cuboids (e.g. one cuboid per room). Similar approach is used by Ikehata et al. (2015) and Khoshelham and Díaz-Vilariño (2014). One limitation of using grammar lies in assuming Manhattan-World structures, as the rooms are considered to be in a grid. Another limitation is the need of an expert who defines the rules. However, as a line of research, it is possible to attempt to learn the rules form a large training data.

3.6.4 *Detection and Reconstruction of Openings*

Detection of openings (e.g. doors and windows) from the scanned data is important so that these openings can be successfully reconstructed. This adds an important level of detail to the model. It turns out that in most of indoor environments the structural surfaces are actually occluded (see Sect. 3.2), and as a consequence there are holes, for example, on the walls, which necessarily do not represent openings. The challenge then is to discriminate between these holes caused by occlusion and

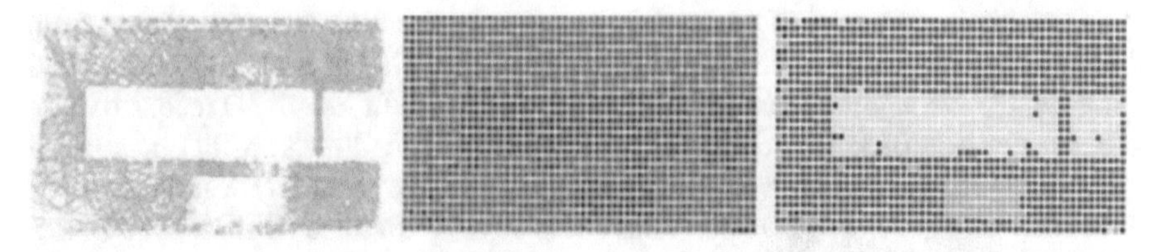

Fig. 3.11 Left: A window in a point cloud. Middle: Ray casting divides the voxels into occupied (red) and unoccupied (blue) classes. Right: Results after Eq. (3.10) include also occluded (green) and opening (yellow) classes. (Reproduced from Nikoohemat et al. 2017)

the holes which represent windows or doorway openings, for instance. We discuss two types of opening detection methods. First, ray casting can be used generally for opening detection. Second, trajectory-based methods are used for door detection.
classes

1. Ray casting is illustrated in Fig. 3.11. Initially, an image is generated for each wall surface, defined by a surface plane. The optical axis of the image is taken parallel to the normal vector of the surface plane. The image is bounded by a bounding box encompassing the wall surface. All the pixels are labeled by an initial label L_0. The objective then is to label the pixels on the surface to opening, occluded or occupied. A ray is cast from each scanner position S_i to each point P_i in the point cloud and the ray intersects with a surface at the intersection point I_j. Then two distances can be calculated from these three points: $D_{S_i-P_i}$, which is the distance between the scanner position and the measured point. $D_{S_i-I_j}$, which is the distance between scanner position and the intersection point. By comparing these two distances, we have

$$\text{Label} = \begin{cases} \text{Occlusion, if} D_{Si-Pi} < D_{S_i-I_j} \\ \text{Occupied, if} D_{Si-Pi} = D_{S_i-I_j} \\ \text{Opening, if} D_{Si-Pi} > D_{S_i-I_j} \end{cases} \tag{3.10}$$

Note that some of the pixels remain with the label L_0 and in the image on the right they are shown in blue. After labeling each surface, the openings can be distinguished from occlusions and the borders of the openings can be extracted. Adan and Huber (2011) use a support vector machine (SVM) method to further reconstruct the border of the opening in each surface. Additionally, note that ray casting can also be done in 3D by using a set of voxels, which enables the treatment of more complex geometries for openings. Finally, a common drawback of any occlusion reasoning method is that when an opening is partially occluded it is hard to reconstruct the border between the opening and the occlusion.

2. Using the Trajectory: When indoor data is collected with a mobile laser scanning system, the trajectory of the laser scanner intersects doorways, see Fig. 3.12. Nikoohemat et al. (2017) exploits this fact to detect the doorways. Given a point cloud, the 3D space is voxelized and the voxels are labeled as empty and occupied (depending if there are points in the voxel). For doorway detection, the

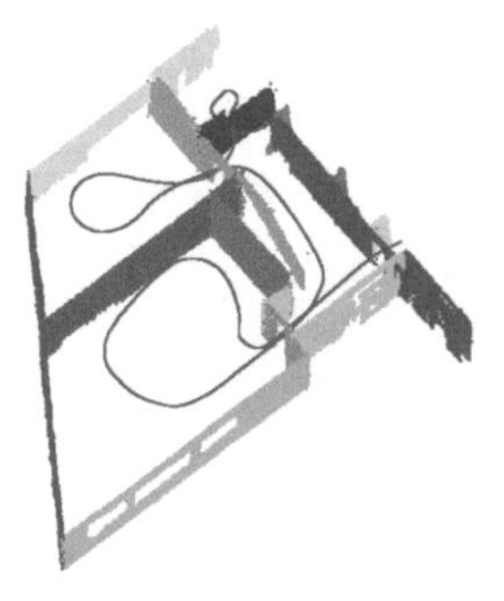

Fig. 3.12 (**a**) Snapshot from inside a point cloud with the trajectory from a mobile mapping system shown with a line. (**b**) Reconstruction of openings with the help of a known trajectory. ((**a**) is reproduced from Nikoohemat et al. 2018)

goal is to find the voxels at the center and on the top of the door frame, which gives us an approximate location of the doorway and its orientation. Note that this approach does not require any knowledge about the walls. Three criteria are checked for each voxel. It represents the center of a doorway if (i) above it, there are several occupied voxels, (ii) it resides close to the trajectory, e.g. distance is within 15 cm, and (iii) it is surrounded by empty voxels when considering a short radius (of e.g. 30 cm). The first criterion scouts for the top of the door frame. The second one is self-explanatory. The third criterion implies that the door center is in the middle of an opening (i.e. not a door that was closed). When identifying the voxel candidates for the door center, the voxels on top of the door center (considering a standard door height) are similarly identified as top of the door. For identifying closed doors, the empty voxels in the third criterion are replaced by occupied voxels, because we expect the center of a closed door is occupied by points. One of the benefits of this approach is that the doorways can be identified regardless of whether the doors are being opened or closed during the measurement. This approach has later been extended by others (Elseicy et al. 2018; Staats et al. 2019).

This method can be simplified if the wall surfaces are known. For example, each place where the trajectory intersects a wall is a probable doorway candidate. Also, the orientation of the door can be derived from the wall normal vector. Obviously, the limitation in using the trajectory knowledge is that there is no guarantee that any untraversed doorways would be detected (e.g. with closed doors).

3.6.5 Reconstructing Occluded Data by Machine Learning

When an object is between the measurement instrument and another object, the first object occludes the other object (see Sect. 3.2). In a point cloud, this is seen as missing data (or holes) that are shaped as a shadow of the first object. For 2D images, inpainting is a well-known technique to restore missing pixels (so-

Fig. 3.13 Left: Point cloud with occlusion. Right: Disocclusion by inpainting. (Reproduced from Xiong et al. 2013)

Fig. 3.14 Inpainting a 3D point cloud separately for geometry and colors and then combining the results. (Reproduced from Väänänen and Lehtola 2019)

called disocclusion in computer vision). For 3D data, a common approach is to take snapshots from the point cloud and then apply some known 2D techniques to fill the missing data. Xiong et al. (2013) use a 3D Markov Random Field inpainting algorithm for disocclusion, shown in Fig. 3.13. Essentially, the algorithm uses planar patches that are labeled as wall, ceiling, floor, or clutter, and the characteristics of opening shape and location for each patch class are learned using machine learning techniques. This learning allows for enough prior knowledge in order to distinguish between occlusion and openings, so that occlusions are filled but openings are left as they are.

Arbitrary geometries and textures can also be patched up, see Fig. 3.14. Väänänen and Lehtola (2019) train the patches separately for geometry and colors with a generative adversarial network (GAN) for each pre-defined class in the point cloud. Their method is noteworthy in that there is no need for external data, i.e. each point cloud can be inpainted per se, and the inpainting is independent of both the occlusion shape and cause. There are limitations as well. The patch size is essentially the size of the band-aid that is laid on the occluded holes, and covering large areas well is difficult.

3.7 Applications

As we have seen, indoor 3D data is important as reconstruction material and that reconstruction aims for industry standardized formats (see Sect. 3.6) in building information modeling (BIM). Figure 3.15 displays applications in facility management, asset management, and construction. Other applications for BIM include real estate brokering and various planning activities. The planning, building, and operating actions can be thought to form a so-called BIM cycle. The BIM-supported decisions relate either to the construction (or renovation) phase, during which the indoor spaces are physically modified, or to the operational phase, during which the indoor spaces are being used. The cycle is closed when the planning is revisited with the scanned as-built data and other data gathered while operating functions inside the building.

In the construction phase, for example, making spaces that are energy efficient brings savings in the fixed costs. This can be achieved through the integration of thermal data (Lagüela et al. 2013) onto 3D models which allows for insulation planning and energy conservation. Also, daylight simulations on the other hand allow for optimization of the electrical illumination (Díaz-Vilariño et al. 2014).

On the other hand, designing the spaces so that they facilitate the activities increases the efficiency of operations, which brings savings in the operational costs. Importantly, 3D models allow for detailed modeling of future operations, which in turn enables better advance planning before construction. As a specific example, hospitals are interested in the smooth flow of people and equipment and in tracking the room occupancy data. Indoor routing for pedestrians can be planned with the help of navigation graphs (Flikweert et al. 2019). Also, BIM helps with inventory data such as equipment serial numbers and make and model data so that it can be connected to room data and accessed when needed. See Fig. 3.15.

Automated model checkers are by themselves a commercial application.[7] Model checkers bring economic savings in altering the ways that problems are handled in construction. One classic example is the overlap of pipelines, meaning that a pipeline is planned to run in a space occupied already for another purpose such as another pipeline. Conventionally, such overlaps were detected in the field and solved by ad hoc methods in the order of occurrence. With 3D planning and model checking, such problems are detected already in the planning phase, and therefore better solutions are plausible. Concerning the public sector, these model checkers can be programmed to test and validate whether e.g. private 3D building plans comply with state regulations.

[7]For example, https://www.solibri.com/how-it-works

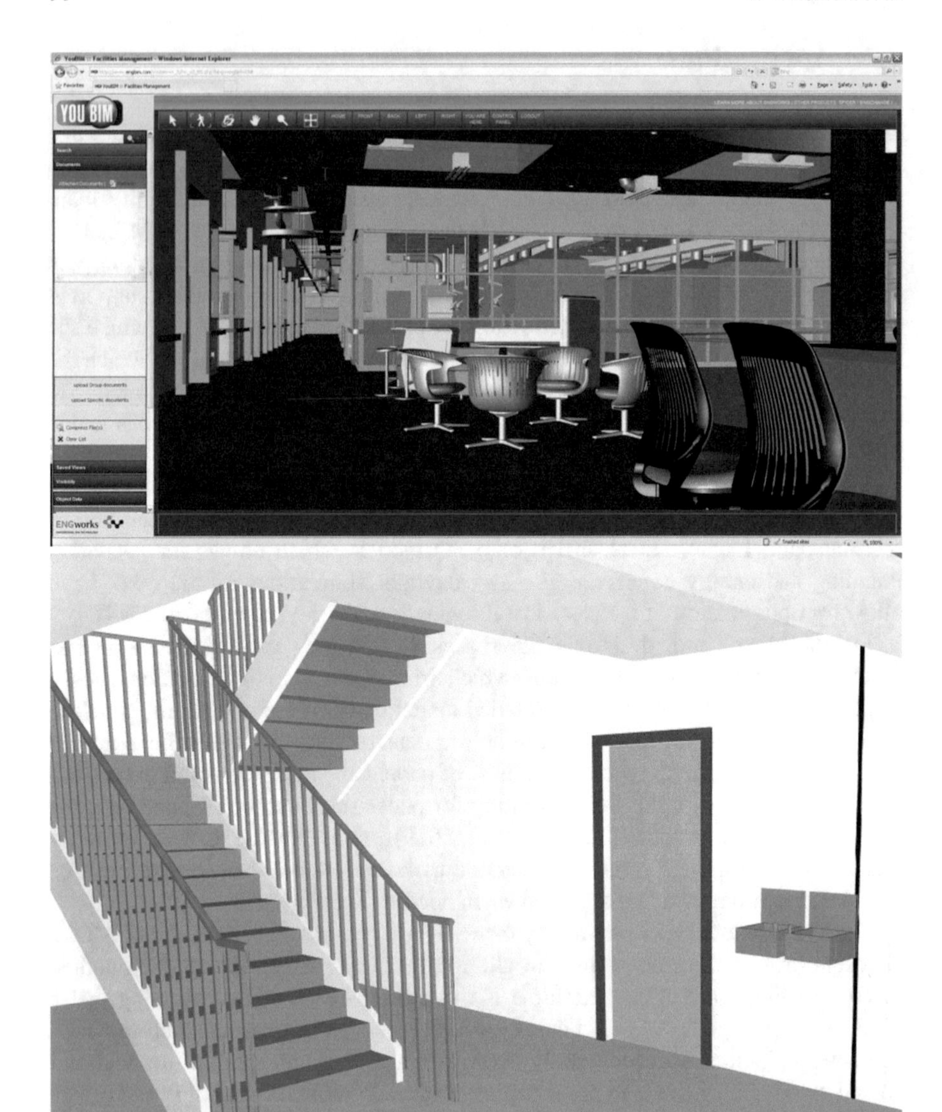

Fig. 3.15 3D BIM models for industry applications, e.g. facility and asset management (top, courtesy of Engworks) and construction (bottom, courtesy of Youbim, youbim.com). Building information modeling (BIM) is important also for various other planning applications

Imagine that your company would possess a large real-estate portfolio. To make most profit out of it, your company would benefit from having a detailed understanding of this ownership. Once the company has the understanding, it can be matched to meet the needs of clients. Further questions may then emerge: are the indoor spaces adjustable for different functions? What are the cost estimates for the adjustments? What functions would provide the most income? Is the portfolio missing something? Asset management is also important. Example assets are lights, fire extinguishers, tables, chairs, etc. Your company is asking: are all the assets that are supposed to be there in place? Are there some synergies inside the portfolio related to the procurement of new assets? Can maintenance be optimized?
Concerning the public sector, consider the vast amount of indoor spaces owned by state ministries, health services, schools, public transportation systems, and even universities. Same questions apply.

3.8 Future Trends

At the time of writing, we see the future trends in scanning, in reconstruction, and in society, as follows.

In scanning, several indoor mobile mapping systems (e.g. commercial systems Kaarta Stencil, Paracosm PX-80, Leica Pegasus: backpack) are relying on lidars that capture multiple scan lines simultaneously. This is because these multiline scanners offer robustness in the SLAM registration process. The authors expect that these multiline scanners will be upgraded into solid-state lidars which have no micro-mechanical moving parts and therefore offer more robustness in mobile use and further miniaturization possibilities for these platforms. New applications are likely to emerge. Single photon techniques are also interesting, since they require less energy to operate than traditional pulsed beams. Their cave-at is background illumination, mainly from sunlight, which may not manifest as overly restrictive in indoor environments (Lehtola et al. 2019).

In reconstruction, one of the main problems encountered is that the trouble in the capture of point clouds often leads into imperfections of the scanned surfaces. Some of these imperfections follow from the scanning geometry and from visual occlusions and manifest themselves as holes and missing observations. The authors expect that machine learning based methods that learn from the intact surfaces to tailor covering patches for the holes and missing observations shall become even more popular in the near future. Also, note that the BIM models in Fig. 3.15 are reconstructed from mobile mapping data and abide industry standards. The difference between these standardized BIM models and the reconstructed models shown in Sect. 3.6 is that there is an extra step in between. The reconstruction into a standardized BIM format typically requires manual effort as it is very hard to automate reliably, but this is about to change.

In society, the shift from 2D to 3D will eventually be completed. At the time of writing, 3D cadaster registration has been started, for example in Northern Europe, to allow for non-surface (underground) properties. In planning, private businesses are leading the way. Building information modeling (BIM) is widely used for construction planning by large construction companies, especially in Northern Europe. One major hindrance still standing in front of these models are the official processes of the state and the municipalities, such as construction permit admission processes. Oftentimes these processes cannot be set in motion with the 3D models, even if they abide an industrial standard, but the official process requires that floorplans need to be extracted into a conservative 2D form and stored in portable document format (PDF). The work on standards is therefore important (e.g. Zlatanova et al. 2016).

Finally, the authors hope that more open data sets would become available for indoor method testing in addition to the ISPRS dataset (Khoshelham et al. 2017).

3.9 Exercises for Students

The following exercises require the reading of this book chapter, some of the cited work, and – some thinking.

- How does the scanning process differ for building interiors and exteriors?
- How does the reconstruction process differ for building interiors and exteriors?
- Write a definition for the trajectory of an indoor mobile mapping system, using mathematical expressions when necessary. Let t denote time.
- An indoor space has been scanned. Give an example on how the available trajectory can be utilized to benefit the reconstruction of an indoor model.
- What are occlusions and how can they be avoided and/or dealt with to obtain watertight models?
- Give two examples on how machine learning can be benefited from in indoor scanning and/or reconstruction.
- Which representation contains more information about the indoor environment: a scanned point cloud or a reconstructed meshed model? Why?

References

Adan A, Huber D (2011) 3D reconstruction of interior wall surfaces under occlusion and clutter. In: 2011 international conference on 3D imaging, modeling, processing, visualization and transmission. IEEE, pp 275–281

Arun KS, Huang TS, Blostein SD (1987) Least square fitting of two 3-d point sets. IEEE Trans Pattern Anal Mach Intell 9(5):698–700

Bassier M, Vergauwen M (2019) Clustering of wall geometry from unstructured point clouds using conditional random fields. Remote Sens 11(13):1586
Becker S, Peter M, Fritsch D, Philipp D, Baier P, Dibak C (2013) Combined grammar for the modeling of building interiors. ISPRS Ann Photogramm Remote Sens Spat Inf Sci 4:1–6
Becker S, Peter M, Fritsch D (2015) Grammar-supported 3D indoor reconstruction from point clouds for "as-built" bim. ISPRS Ann Photogramm Remote Sens Spat Inf Sci 2:17–24
Benjemaa R, Schmitt F (1997) Fast global registration of 3D sampled surfaces using a multi-Z-buffer technique. In: Proceedings IEEE international conference on recent advances in 3D digital imaging and modeling (3DIM'97), Ottawa
Benjemaas R, Schmitt F (1998) A solution for the registration of multiple 3D point sets using unit quaternions. In: Computer vision – ECCV'98, vol 2, pp 34–50
Bergevin R, Soucy M, Gagnon H, Laurendeau D (1996) Towards a general multi-view registration technique. IEEE Trans Pattern Anal Mach Intell (PAMI) 18(5):540–547
Besl P, McKay N (1992) A method for registration of 3-D shapes. IEEE Trans Pattern Anal Mach Intell (PAMI) 14(2):239–256
Blaser S, Cavegn S, Nebiker S (2018) Development of a portable high performance mobile mapping system using the robot operating system. ISPRS Ann Photogramm Remote Sens Spat Inf Sci 4(1):13–20
Bormann R, Jordan F, Li W, Hampp J, Hägele M (2016) Room segmentation: survey, implementation, and analysis. In: 2016 IEEE international conference on robotics and automation (ICRA). IEEE, pp 1019–1026
Borrmann D, Jörissen S, Nüchter A (2018) Radler-a radial laser scanning device
Borrmann D, Elseberg J, Lingemann K, Nüchter A, Hertzberg J (2008a) Globally consistent 3D mapping with scan matching. J Robot Auton Syst (JRAS) 56(2):130–142. https://robotik.informatik.uni-wuerzburg.de/telematics/download/ras2007.pdf
Borrmann D, Elseberg J, Lingemann K, Nüchter A, Hertzberg J (2008b) The efficient extension of globally consistent scan matching to 6 DoF. In: Proceedings of the 4th international symposium on 3D data processing, visualization and transmission (3DPVT'08), Atlanta, pp 29–36. https://robotik.informatik.uni-wuerzburg.de/telematics/download/3dpvt2008.pdf
Bosse M, Zlot R, Flick P (2012) Zebedee: design of a spring-mounted 3-D range sensor with application to mobile mapping. IEEE Trans Robot 28(5):1104–1119
Boulch A, Houllier S, Marlet R, Tournaire O (2013) Semantizing complex 3D scenes using constrained attribute grammars. In: Proceedings of the eleventh eurographics/ACMSIGGRAPH symposium on geometry processing. Eurographics Association, pp 33–42
Böhm J, Becker S (2007) Automatic marker-free registration of terrestrial laser scans using reflectance features. In: Proceedings of 8th conference on optical 3D measurment techniques, Zurich, pp 338–344
Businesswire (2019) Global indoor 3D laser scanner market outlook, 2017–2026. https://www.businesswire.com/news/home/20191009005303/en/Global-Indoor-3D-Laser-Scanner-Market-Outlook
Chen J, Clarke KC (2017) Modeling standards and file formats for indoor mapping. In: GISTAM, pp 268–275
Chen Y, Medioni G (1991) Object modelling by registration of multiple range images. In: Proceedings of the IEEE conference on robotics and automation (ICRA'91), Sacramento, pp 2724–2729
Chen Y, Medioni G (1992) Object modelling by registration of multiple range images. Image Vis Comput 10(3):145–155
Chomsky N (1959) On certain formal properties of grammars. Inf Control 2(2):137–167
Concha A, Loianno G, Kumar V, Civera J (2016) Visual-inertial direct slam. In: 2016 IEEE international conference on robotics and automation (ICRA). IEEE, pp 1331–1338
Cunnington S, Stoddart A (1999) N-view point set registration: a comparison. In: Proceedings of the 10th British machine vision conference (BMVC'99), Nottingham. citeseer.nj.nec.com/319525.html

Díaz-Vilariño L, Lagüela S, Armesto J, Arias P (2014) Indoor daylight simulation performed on automatically generated as-built 3D models. Energy Build 68:54–62

Du H, Henry P, Ren X, Cheng M, Goldman DB, Seitz SM, Fox D (2011) Interactive 3D modeling of indoor environments with a consumer depth camera. In: Proceedings of the 13th international conference on ubiquitous computing. ACM, pp 75–84

Elseberg J, Magnenat S, Siegwart R, Nüchter A (2012) Comparison on nearest-neigbour-search strategies and implementations for efficient shape registration. J Softw Eng Robot (JOSER) 3(1):2–12. https://robotik.informatik.uni-wuerzburg.de/telematics/download/joser2012.pdf

Elseberg J, Borrmann D, Nüchter A (2013) One billion points in the cloud – an octree for efficient processing of 3D laser scans. ISPRS J Photogramm Remote Sens (JPRS) Special Issue Terr 3D Model 76:76–88. https://robotik.informatik.uni-wuerzburg.de/telematics/download/isprs2012.pdf

Elseicy A, Nikoohemat S, Peter M, Elberink S (2018) Space subdivision of indoor mobile laser scanning data based on the scanner trajectory. Remote Sens 10(11):1815

Eudes A, Marzat J, Sanfourche M, Moras J, Bertrand S (2018) Autonomous and safe inspection of an industrial warehouse by a multi-rotor mav. In: Field and service robotics. Springer, Cham, pp 221–235

Flikweert P, Peters R, Díaz-Vilarino L, Voûte R, Staats B (2019) Automatic extraction of a navigation graph intended for indoorgml from an indoor point cloud. ISPRS Ann Photogramm Remote Sens Spat Inf Sci 4(2/W5):271–278

Friedman JH, Bentley JL, Finkel RA (1977) An algorithm for finding best matches in logarithmic expected time. ACM Trans Math Softw 3(3):209–226

Furukawa Y, Curless B, Seitz SM, Szeliski R (2009) Reconstructing building interiors from images. In: 2009 IEEE 12th international conference on computer vision, pp 80–87, iD: 1

Grant WS, Voorhies RC, Itti L (2019) Efficient velodyne slam with point and plane features. Auton Robots 43(5):1207–1224

Greenspan M, Yurick M (2003) Approximate K-D tree search for efficient ICP. In: Proceedings of the 4th IEEE international conference on recent advances in 3D digital imaging and modeling (3DIM'03), Banff, pp 442–448

Grisetti G, Kummerle R, Stachniss C, Burgard W (2010) A tutorial on graph-based slam. IEEE Intell Transp Syst Mag 2(4):31–43

Hess W, Kohler D, Rapp H, Andor D (2016) Real-time loop closure in 2D lidar slam. In: 2016 IEEE international conference on robotics and automation (ICRA). IEEE, pp 1271–1278

Horn BKP (1987) Closed–form solution of absolute orientation using unit quaternions. J Opt Soc Am A 4(4):629–642

Horn BKP, Hilden HM, Negahdaripour S (1988) Closed–form solution of absolute orientation using orthonormal matrices. J Opt Soc Am A 5(7):1127–1135

Hornung A, Wurm KM, Bennewitz M, Stachniss C, Burgard W (2013) Octomap: an efficient probabilistic 3D mapping framework based on octrees. Auton Robots 34(3):189–206

Ikehata S, Yang H, Furukawa Y (2015) Structured indoor modeling. In: Proceedings of the IEEE international conference on computer vision, pp 1323–1331

Kaijaluoto R, Kukko A, Hyyppä J (2015) Precise indoor localization for mobile laser scanner. ISPRS Int Arch Photogramm Remote Sens Spat Inf Sci XL-4/W5:1–6. https://www.int-arch-photogramm-remote-sens-spatial-inf-sci.net/XL-4-W5/1/2015/

Karam S, Vosselman G, Peter M, Hosseinyalamdary S, Lehtola V (2019) Design, calibration, and evaluation of a backpack indoor mobile mapping system. Remote Sens 11(8):905

Kaul L, Zlot R, Bosse M (2016) Continuous-time three-dimensional mapping for micro aerial vehicles with a passively actuated rotating laser scanner. J Field Robot 33(1):103–132

Khoshelham K, Díaz-Vilariño L (2014) 3D modelling of interior spaces: learning the language of indoor architecture. Int Arch Photogramm Remote Sens Spat Inf Sci 40(5):321

Khoshelham K, Vilariño LD, Peter M, Kang Z, Acharya D (2017) The ISPRS benchmark on indoor modelling. Int Arch Photogramm Remote Sens Spat Inf Sci 42:367–372

Krishnan S, Lee PY, Moore JB, Venkatasubramanian S (2000) Global registration of multiple 3D point sets via optimization on a manifold. In: Eurographics symposium on geometry processing

Lagüela S, Díaz-Vilariño L, Martínez J, Armesto J (2013) Automatic thermographic and rgb texture of as-built bim for energy rehabilitation purposes. Autom Constr 31:230–240

Lauterbach H, Borrmann D, Heß R, Eck D, Schilling K, Nüchter A (2015) Evaluation of a backpack-mounted 3D mobile scanning system. Remote Sens 7(10):13753–13781

Lehtola VV, Kurkela M, Hyyppä H (2014) Automated image-based reconstruction of building interiors–a case study. Photogramm J Finl 24(1):1–13

Lehtola VV, Virtanen J-P, Kukko A, Kaartinen H, Hyyppa H (2015) Localization of mobile laser scanner using classical mechanics. ISPRS J Photogramm Remote Sens 99(0):25–29. http://www.sciencedirect.com/science/article/pii/S0924271614002585

Lehtola VV, Virtanen J-P, Vaaja MT, Hyyppä H, Nüchter A (2016) Localization of a mobile laser scanner via dimensional reduction. ISPRS J Photogramm Remote Sens 121:48–59

Lehtola V, Kaartinen H, Nüchter A, Kaijaluoto R, Kukko A, Litkey P, Honkavaara E, Rosnell T, Vaaja M, Virtanen J-P et al (2017) Comparison of the selected state-of-the-art 3D indoor scanning and point cloud generation methods. Remote Sens 9(8):796

Lehtola V, Hyyti H, Keränen P, Kostamovaara J (2019) Single photon lidar in mobile laser scanning: the sampling rate problem and initial solutions via spatial correlations. Int Arch Photogramm Remote Sens Spat Inf Sci 42:91–97

Liu T, Carlberg M, Chen G, Chen J, Kua J, Zakhor A (2010) Indoor localization and visualization using a human-operated backpack system. In: 2010 international conference on indoor positioning and indoor navigation. IEEE, pp 1–10

Liu C, Wu J, Furukawa Y (2018) Floornet: a unified framework for floorplan reconstruction from 3D scans. In: Proceedings of the European conference on computer vision (ECCV), pp 201–217

Lorusso A, Eggert D, Fisher R (1995) A comparison of four algorithms for estimating 3-D rigid transformations. In: Proceedings of the 4th British machine vision conference (BMVC'95), Birmingham, pp 237–246. citeseer.nj.nec.com/lorusso95comparison.html

Lu F, Milios E (1997) Globally consistent range scan alignment for environment mapping. Auton Robots 4:333–349

Mitra NJ, Gelfand N, Pottmann H, Guibas L (2004) Registration of point cloud data from a geometric optimization perspective. In: Scopigno R, Zorin D (eds) Eurographics symposium on geometry processing, pp 23–32

Moosmann F, Stiller C (2011) Velodyne slam. In: 2011 IEEE intelligent vehicles symposium (IV). IEEE, pp 393–398

Mozos ÓM (2010) Semantic labeling of places with mobile robots, vol 61. Springer, Berlin/Heidelberg

Müller P, Wonka P, Haegler S, Ulmer A, Van Gool L (2006) Procedural modeling of buildings. In: ACM transactions on graphics (Tog), vol 25. ACM, pp 614–623

Mur-Artal R, Tardós JD (2017) Orb-slam2: an open-source slam system for monocular, stereo, and RGB-D cameras. IEEE Trans Robot 33(5):1255–1262

Mura C, Mattausch O, Villanueva AJ, Gobbetti E, Pajarola R (2014) Automatic room detection and reconstruction in cluttered indoor environments with complex room layouts. Comput Graph 44:20–32

Mura C, Mattausch O, Pajarola R (2016) Piecewise-planar reconstruction of multi-room interiors with arbitrary wall arrangements. In: Computer graphics forum, vol 35. Wiley Online Library, pp 179–188

Murali S, Speciale P, Oswald MR, Pollefeys M (2017) Indoor scan2bim: building information models of house interiors. In: 2017 IEEE/RSJ international conference on intelligent robots and systems (IROS). IEEE, pp 6126–6133

Musialski P, Wonka P, Aliaga DG, Wimmer M, van Gool L, Purgathofer W (2013) A survey of urban reconstruction. Comput Graph Forum 32(6):146–177

NavVis (2016) Digitizing indoors – NavVis. http://www.navvis.com. Accessed: 20 Oct 2016

Nikoohemat S, Peter M, Elberink SO, Vosselman G (2017) Exploiting indoor mobile laser scanner trajectories for semantic interpretation of point clouds. ISPRS Ann Photogramm Remote Sens Spat Inf Sci 4:355–362

Nikoohemat S, Peter M, Oude Elberink S, Vosselman G (2018) Semantic interpretation of mobile laser scanner point clouds in indoor scenes using trajectories. Remote Sens 10(11):1754

Nikoohemat S, Diakité A, Zlatanova S, Vosselman G (2019) Indoor 3D modeling and flexible space subdivision from point clouds. ISPRS Ann Photogramm Remote Sens Spat Inf Sci 4:285–292

Nüchter A, Lingemann K, Hertzberg J (2007) Cached *k*-D tree search for ICP algorithms. In: Proceedings of the 6th IEEE international conference on recent advances in 3D digital imaging and modeling (3DIM'07), Montreal, pp 419–426. https://robotik.informatik.uni-wuerzburg.de/telematics/download/3dim2007.pdf

Nützi G, Weiss S, Scaramuzza D, Siegwart R (2011) Fusion of IMU and vision for absolute scale estimation in monocular slam. J Intell Robot Syst 61(1–4):287–299

Ochmann S, Vock R, Klein R (2019) Automatic reconstruction of fully volumetric 3D building models from oriented point clouds. ISPRS J Photogramm Remote Sens 151:251–262

Oesau S, Lafarge F, Alliez P (2014) Indoor scene reconstruction using feature sensitive primitive extraction and graph-cut. ISPRS J Photogramm Remote Sens 90:68–82

Pulli K (1999) Multiview registration for large data sets. In: Proceedings of the 2nd international conference on 3D digital imaging and modeling (3DIM'99), Ottawa, pp 160–168

Rusinkiewicz S, Levoy M (2001) Efficient variants of the ICP algorithm. In: Proceedings of the third international conference on 3D digital imaging and modellling (3DIM'01), Quebec City, pp 145–152

Schneider S, Himmelsbach M, Luettel T, Wuensche H-J (2010) Fusing vision and lidar-synchronization, correction and occlusion reasoning. In: 2010 IEEE intelligent vehicles symposium. IEEE, pp 388–393

Staats B, Diakité A, Voûte R, Zlatanova S (2019) Detection of doors in a voxel model, derived from a point cloud and its scanner trajectory, to improve the segmentation of the walkable space. Int J Urban Sci 23(3):369–390

Stiny G, Gips J (1971) Shape grammars and the generative specification of painting and sculpture. In: IFIP congress (2), vol 2

Stoddart A, Hilton A (1996) Registration of multiple point sets. In: Proceedings of the 13th IAPR international conference on pattern recognition, Vienna, pp 40–44

Taketomi T, Uchiyama H, Ikeda S (2017) Visual slam algorithms: a survey from 2010 to 2016. IPSJ Trans Comput Vis Appl 9(1):16

Thrun S, Burgard W, Fox D (2005) Probabilistic robotics. The MIT Press, Cambridge, MA

Tran H, Khoshelham K, Kealy A, Díaz-Vilariño L (2018) Shape grammar approach to 3D modeling of indoor environments using point clouds. J Comput Civil Eng 33(1):04018055

Tran H, Khoshelham K, Kealy A (2019) Geometric comparison and quality evaluation of 3D models of indoor environments. ISPRS J Photogramm Remote Sens 149:29–39

Tucci G, Visintini D, Bonora V, Parisi E (2018) Examination of indoor mobile mapping systems in a diversified internal/external test field. Appl Sci 8(3):401

Turner E, Cheng P, Zakhor A (2014) Fast, automated, scalable generation of textured 3D models of indoor environments. IEEE J Sel Top Signal Process 9(3):409–421

Väänänen P, Lehtola V (2019) Inpainting occlusion holes in 3D built environment point clouds. Int Arch Photogramm Remote Sens Spat Inf Sci 42:393–398

Velas M, Spanel M, Hradis M, Herout A (2018) Cnn for IMU assisted odometry estimation using velodyne lidar. In: 2018 IEEE international conference on autonomous robot systems and competitions (ICARSC). IEEE, pp 71–77

Walker MW, Shao L, Volz RA (1991) Estimating 3-D location parameters using dual number quaternions. CVGIP Image Underst 54:358–367

Williams J, Bennamoun M (1999) Multiple view 3D registration using statistical error models. In: Vision modeling and visualization

Wonka P, Wimmer M, Sillion F, Ribarsky W (2003) Instant architecture, SIGGRAPH '03: ACM SIGGRAPH 2003 Papers July 2003 pp 669–677 vol 22. ACM Headquarters. https://doi.org/10.1145/1201775.882324

Xiao J, Furukawa Y (2014) Reconstructing the worlds museums. Int J Comput Vis 110(3):243–258

Xiong X, Adan A, Akinci B, Huber D (2013) Automatic creation of semantically rich 3D building models from laser scanner data. Autom Constr 31:325–337

Zhang Z (1992) Iterative point matching for registration of free–form curves. Technical Report RR-1658, INRIA–Sophia Antipolis, Valbonne Cedex. citeseer.nj.nec.com/zhang92iterative.html

Zhang J, Singh S (2014) Loam: lidar odometry and mapping in real-time. In: Robotics: science and systems conference (RSS), vol 2, p 9

Zlatanova S, Van Oosterom P, Lee J, Li KJ, Lemmen C (2016) Ladm and indoorgml for support of indoor space identification. ISPRS Ann Photogramm Remote Sens Spat Inf Sci 4:257–263

Chapter 4
Big Earth Observation Data Processing for Disaster Damage Mapping

Bruno Adriano, Naoto Yokoya, Junshi Xia, and Gerald Baier

4.1 Monitoring Disasters from Space

Earth observation has been receiving considerable attention in disaster management in recent years. As such, the imaging capability of national or international earth observation missions has been improving steadily. Also, driven by technology innovation in New Space, the number of satellites has been increasing dramatically. Satellite constellations enable high-frequency data acquisition, which is often required in disaster monitoring and rapid response.

In the last two decades, enormous efforts have been made in international cooperative projects and services for sharing and analyzing satellite imagery in emergency response. Some representative ones are listed below.

- International Charter 'Space and Major Disasters':[1] The International Charter 'Space and Major Disasters' is an international collaboration among space agencies and companies (e.g., Maxar and Planet Labs) to support disaster response activities by providing information and products derived from satellite data. The charter was initiated by the European Space Agency (ESA) and the French space agency (CNES), came into operation in 2000, and activated 601 times for 125 countries supported by 17 charter members with 34 satellites as of April 1, 2019.

[1] https://disasterscharter.org

B. Adriano · N. Yokoya (✉) · J. Xia · G. Baier
RIKEN Center for Advanced Intelligence Project, Tokyo, Japan
e-mail: bruno.adriano@riken.jp; naoto.yokoya@riken.jp; junshi.xia@riken.jp; gerald.baier@riken.jp

M. Werner, Y.-Y. Chiang (eds.), *Handbook of Big Geospatial Data*,
https://doi.org/10.1007/978-3-030-55462-0_4

- UNOSAT:[2] UNOSAT is a technology-intensive programme of the United Nations Institute for Training and Research (UNITAR) to provide satellite imagery analysis and solutions to the UN system and its partners for decision making in critical areas, including humanitarian response to natural disasters. UNOSAT was established in 2001 and the Humanitarian Rapid Mapping service of UNOSAT was launched in 2003 and contributed to 28 humanitarian response to natural disasters in 22 countries in 2018.
- Sentinel Asia:[3] The Sentinel Asia initiative is a voluntary basis international collaboration among space agencies, disaster management agencies, and international agencies to support disaster management activities in the Asia-Pacific region by applying remote sensing and Web-GIS technologies. Sentinel Asia was initiated by the Asia-Pacific Regional Space Agency Forum (APRSAF) in 2005 and its members consist of 93 organizations from 28 countries/regions and 16 international organizations. In 2018, there were 25 emergency observation requests and disaster response activities are supported by 8 data provider nodes and 48 data analysis nodes.
- Copernicus Emergency Management Service (Copernicus EMS):[4] Copernicus EMS provides geospatial information for emergency response to disasters as well as prevention, preparedness, and recovery activities by analyzing satellite imagery. Copernicus EMS is coordinated by the European Commission as one of the key services of the European Union's Earth Observation programme Copernicus. The two Mapping services of Copernicus EMS (i.e., Rapid Mapping, Risk and Recovery Mapping) started operation since April 2012 and 349 mapping activations have been conducted as of April 3, 2019.

Owing to the development of hardware, big earth observation data is now available from various types of satellites and imaging sensors. Large volume and a wide variety of earth observation data promote new applications but also raise challenges in understanding satellite imagery for disaster response. In this book chapter, we summarize recent advances and challenges in the processing of big earth observation data for disaster management.

4.2 Earth Observation Satellites

Over the last decades, the number of earth observation satellites has steadily increased, providing an unprecedented amount of available data. This includes optial (multi- and hyperspectral) images (e.g., Fig. 4.1b) and also synthetic aperture radar (SAR) images (e.g., Fig. 4.1e, f). Regarding disaster response, the sheer number

[2]https://unitar.org/unosat/

[3]https://sentinel.tksc.jaxa.jp

[4]https://emergency.copernicus.eu/mapping/

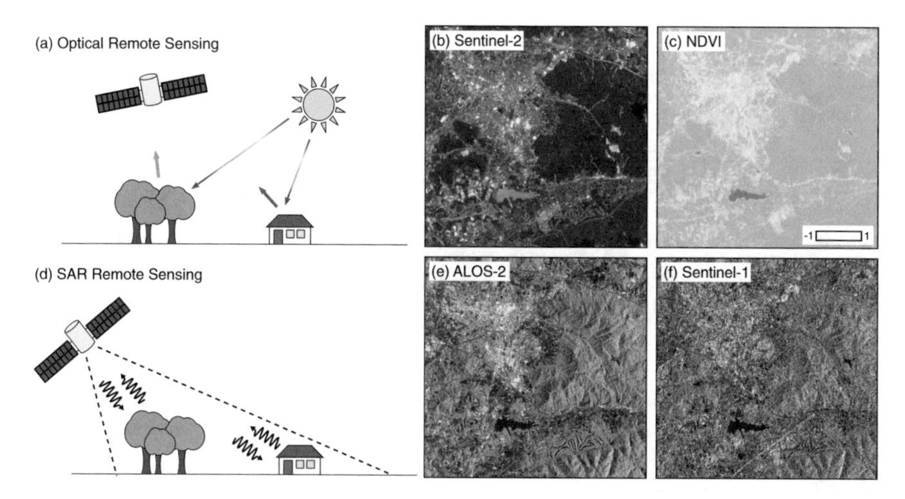

Fig. 4.1 (**a**) Illustration of optical remote sensing. (**b**) Sentinel-2 imagery. (**c**) NDVI derived from Sentinel-2 data. (**d**) Illustration of SAR remote sensing. (**e**) ALOS-2 (L-band) imagery. (**f**) Sentinel-1 (C-band) imagery

of satellites ensures quick post-event acquisitions and often, due to the regular acquisition patterns of many satellite missions, the availability of a recent pre-event image. In the following paragraphs, we provide a summary of current and future earth observation satellite missions and how they benefit mapping damages and the extent of disasters.

4.2.1 *Optical Satellite Missions*

Table 4.1 shows the list of optical satellite missions. An explosive amount of data has become available in the last decade. For instance, only Sentinel-2 satellites acquire over one petabyte per year. Data policies are different depending on resolution: datasets from moderate-resolution satellites (e.g., Landsat-8 and Sentinel-2) are freely available and those from very high-resolution satellites (e.g., Pleiades and WorldView-3) are commercial. For emergency responses, even some commercial satellite images are openly distributed through special data programs (e.g., Open Data Program[5] for WorldView images and Disaster Data Program[6] for PlanetScope).

Optical remote sensing records the solar radiation reflected from the surface in visible, near-infrared, and short-wave infrared ranges as illustrated in Fig. 4.1a.

[5]https://www.digitalglobe.com/ecosystem/open-data

[6]https://www.planet.com/disasterdata/

Table 4.1 Current optical satellite missions

Satellite mission	Best GSD	Swath width	# of Bands	Revisit cycle	Launch date
Landsat 8	15 m	185 km	11	16 days	2013
Sentinel-2 A/B	10 m	290 km	13	5 days	2015
SPOT 7	1.5 m	60 km	5	26 days	2014
WorldView-3	0.31 m	13.1 km	29	daily	2014
Pleiades	0.5 m	20 km	5	daily	2011
PlanetScope	3 m	16.4 km	4	daily	2016
Gaofen-2	0.8 m	45 km	5	5 days	2014

Reflected spectral signatures allow us to discriminate different types of land covers. Owing to its similar characteristics to human vision, optical imagery is straightforward to analyze for damage recognition. A pair of pre- and post-disaster optical images are commonly used to detect pixel-wise or object-wise changes and identify damage levels of affected areas. In particular, if there is any clear change in the normalized difference vegetation index (NDVI) (e.g., Fig. 4.1c) or the normalized difference water index (NDWI) due to landslides or floods, affected areas can be detected easily and accurately.

Optical satellite imaging systems have evolved in terms of spatial, temporal, and spectral resolutions. Spatial and temporal resolutions are critical for disaster damage mapping. Improvement of temporal resolution has been achieved by forming satellite constellations. For instance, the revisit cycle of Sentinel-2 is five days and it was accomplished by a constellation of twin satellites (i.e., Sentinel-2 A and B). An extreme example is PlanetScope: the daily acquisition is possible for the entire globe with a constellation of 135+ small satellites (i.e., Droves). The evolution in temporal resolution allows disaster damage mapping within a day under good weather conditions.

Spatial resolution is another key factor to ensure accuracy of disaster damage mapping. Medium-resolution satellites such as Landsat-8 and Sentinel-2 are sufficient for mapping large-scale changes of the surface due to floods, landslides, wildfires, and volcanos. High-resolution satellites data are necessary particularly when analyzing damages in urban areas. Visual interpretation in emergency response relies on sub-meter satellite imagery such as Pleiades and WorldView to identify building damages.

The major drawback of optical satellites is that they cannot acquire images when affected areas are covered by clouds. Because of this limitation, in many real cases, datasets from different sensors are only available before and after disasters in a few days after disasters. Integration and fusion of multisensor data sources are crucial to deliver map products of disaster damages.

Table 4.2 Current and future SAR missions. Resolutions and swath widths depend on the acquisition mode. The table lists the maximum resolution and the corresponding swath width

Satellite mission	Max. resolution	Swath width	Band	Launch date
TerraSAR-X/TanDEM-X	1×0.25 m	5 km	X	2007/2010
COSMO-SkyMed	1×1 m	10 km	X	2007/2008/2010
RADARSAT-2	3×1 m	18 km	C	2007
KOMPSAT-5	1 m	5 km	X	2013
ALOS-2	3×1 m	25 km	L	2014
Sentinel-1 A/B	5×5 m	80 km	C	2014/2016
Gaofen-3	1×1 m	10 km	C	2016
NovaSAR-S	6 m	20 km	S	2018
PAZ	1×0.25 m	5 km	X	2018
SAOCOM 1A	10 m	40 km	L	2018
ICEYE X2	1×1 m	10 km	X	2018
COSMO-SkyMed 2nd Gen.	0.5×0.5 m	7 km	X	2019
RADARSAT Constellation	3×1 m	20 km	C	2019
ALOS-4	3×1 m	35 km	L	2020
Capella	0.5×0.5 m	10 km	X	2020
Synspective	1×1 m	10 km	X	2020
NISAR	6×8 m	240 km	L	2021

4.2.2 SAR Satellite Missions

Unlike optical imagery, SAR sensors have the advantage that they are undisturbed by clouds, making them invaluable for responding to disasters due to their reliable image acquisition schedule. Table 4.2 lists current and future SAR missions, together with their highest resolution modes, the corresponding swath widths, their frequency bands and launch dates. All of these satellites also have lower resolution acquisition modes with increased spatial coverage. As can be seen from Table 4.2, even moderately large areas can easily result in multiple GB of data if several sensors are used and acquisitions before and after an event are collected.

As a quite recent development, several startup companies (ICEYE, Capella and Synspective) announced plans to create constellations of dozens of comparatively small and cheap satellites, that enable frequent and short notice acquisitions. Such constellations would produce a wealth of data, compounding the need, both for big data systems and algorithms.

The following publications provide more verbose information for the respective SAR satellites and list additional references. Morena et al. (2004) for RADARSAT-2, Lee (2010) for KOMPSAT-5 and Werninghaus and Buckreuss (2010) for TerraSAR-X, TanDEM-X and the essentially identical PAZ satellite (Suri et al. 2015). Torres et al. (2012) describes ESA's Sentinel-1 satellites, Bird et al. (2013) NovaSAR-S, Caltagirone et al. (2014) COSMO-SkyMed, Rosenqvist et al. (2014)

SAOCOM, and Sun et al. (2017) Gaofen-3. Future SAR missions are covered in Rosen et al. (2017) for NISAR, Motohka et al. (2017) for ALOS-4, and finally De Lisle et al. (2018) introduces the RADARSAT constellation mission. Technical details and developments regarding small SAR satellite constellations are given in Farquharson et al. (2018) and Obata et al. (2019).

Many satellites have acquisition modes where the resolution suffices to detect changes and damages for individual buildings. In any case, large scale destructions, caused by earthquakes (Karimzadeh et al. 2018), wildfires (Tanase et al. 2010; Verhegghen et al. 2016) landslides or flooding (Martinis et al. 2018) can be observed by all sensors. We cover these in greater detail in Sects. 4.4.1, 4.4.2 and 4.4.3. Here we introduce the reader to SAR image formation and how these characteristics are applicable for disaster damage mapping. For a more thorough introduction we advice the interested reader to consult (Moreira et al. 2013).

SAR sensors emit electromagnetic waves and measure the reflected energy (see Fig. 4.1d), called backscatter, which depends on the geometric and geophysical properties of the target. This renders the SAR sensors sensitive to different kinds of land cover but also physical parameters, such as soil moisture. In addition, depending on the SAR's operating frequency, parts of the electromagnetic wave also penetrate the surface and image layers below the uppermost land cover.

Just like visible light, microwaves are polarized, and the polarimetric composition of reflected waves depends on the imaged targets' geometric and physical properties. These polarimetric signatures permit further analysis and classification of the imaged area.

Inside one SAR resolution cell, i.e. pixel, numerous elemental scatterers reflected the impinging electromagnetic wave. The superposition of all these reflections make up the received signal at the SAR sensors. Between two SAR acquisitions changes of the elemental scatterers can be estimated, providing a direct measure of differences, the so-called coherence.

All of these properties: backscatter, polarimetric composition, and coherence are useful when analysing disaster-struck areas.

Some newer SAR satellite systems, namely PAZ, NovaSAR-S and the RADARSAT constellation, are additionally equipped with automatic identification system (AIS) receivers, enabling them to track shipping traffic. In most countries AIS transceivers are mandatory for vessels above a certain size. AIS is an additional data source that could be exploited for responding to disasters affecting ships.

4.3 Land Cover Mapping

Map information is necessary in all phases of disaster management. Mapping of buildings and roads is essential for rescue, relief, and recovery activities. The map information is generally well maintained in the developed countries; however, it is not the case for developing countries, particularly where uncontrolled urbanization is happening, and thus there is high demand for the automatic update of map information from satellite imagery at a large (e.g., country) scale.

Mapping of buildings, roads, and land cover types is one of the key applications using satellite imagery. Global land cover maps at a high resolution have been derived from satellite data in the last decade. Global Urban Footprint (GUF) was created with a ground sampling distance of 12 m by the German Aerospace Center by processing 180,000 TerraSAR-X and TanDEM-X scenes (Esch et al. 2013). The GUF data was released in 2012, freely available at a full resolution for any scientific use and also open to any nonprofit applications at a degraded resolution of 84 m. GlobeLand30 is the first open-access and high-resolution land cover map comprising 10 land cover classes for the years from 2000 to 2010 by analyzing more than 20,000 Landsat and Chinese HJ-1 satellite images (Jun Chen et al. 2015). In 2014, China donated the GlobeLand30 data to the United Nations to contribute to global sustainable development and climate change mitigation.

Recently, building and road mapping technologies that apply machine and deep learning to high-resolution satellite imagery have been dramatically improved. For instance, Ecopia U.S. Building Footprints powered by DigitalGlobe (currently a part of Maxar) has been released in 2018 as the first precise, GIS-ready building footprints dataset covering the entire United States produced by semi-automated processing based on machine learning. The 2D vector polygon dataset will be updated every six months using latest DigitalGlobe big satellite image data to ensure up-to-date building footprint information. Going beyond 2D is the next standard in the field of urban mapping. 3D reconstruction and 3D semantic reconstruction using large-scale satellite imagery have been receiving particular attention in recent years.

Benchmark datasets and data science competitions have been playing key roles in advancing 2D/3D mapping technologies. Representative benchmark datasets are listed below.

- **SpaceNet**:[7] SpaceNet is a repository of freely available high-resolution satellite imagery and labeled training data for computer vision and machine learning research. SpaceNet was initiated by CosmiQ Works, DigitalGlobe, and NVIDIA in 2016. SpaceNet building and road extraction competitions were organized with over 685,000 building footprints and 8000 km of roads from large cities in the world (i.e., Rio de Janeiro, Las Vegas, Paris, Shanghai, Khartoum).
- **DeepGlobe**:[8] DeepGlobe is a challenge-based workshop initiated by Facebook and DigitalGlobe as conjunction with CVPR 2018 to promote research on machine learning and computer vision techniques applied to satellite imagery and bridge people from the respective fields with different perspectives. DeepGlobe was composed of three challenges: road extraction, building detection, and land cover classification. The building detection challenge used the SpaceNet data; the road extraction and land cover classification challenges used images sampled

[7] https://spacenetchallenge.github.io/

[8] http://deepglobe.org/

from the DigitalGlobe Basemap +Vivid dataset. The road extraction challenge dataset comprises images of rural and urban areas in Thailand, Indonesia, and India, whereas the land cover classification challenge focuses on rural areas (Demir et al. 2018).

- **BigEarthNet**:[9] The BigEarthNet archive was constructed by the Technical University of Berlin and released in 2019. The archive is a large scale dataset composed of 590,326 Sentinel-2 image patches with land cover labels. BigEarthNet was created from 125 Sentinel-2 tiles covering 10 countries of Europe and the corresponding labels were provided from CORINE Land Cover database. BigEarthNet advances research for the analysis of big earth observation data archives.
- **2019 IEEE GRSS Data Fusion Contest**:[10] 2019 IEEE GRSS Data Fusion Contest, organized by the Image Analysis and Data Fusion Technical Committee (IADF TC) of the IEEE Geoscience and Remote Sensing Society (GRSS) and the Johns Hopkins University (JHU), promoted research in semantic 3D reconstruction and stereo using machine learning and satellite images. The contest was composed of four challenges: three of them are simultaneous estimation of land cover semantics and height information from single-view, pairwise, and multi-view satellite images, respectively; the last one is 3D point cloud classification. The contest used high-resolution satellite imagery and airborne LiDAR data over Jacksonville and Omaha, US (Le Saux et al. 2019).

One major challenge in land cover mapping is the generalization ability. Most of training data was prepared for a limited number of countries and cities. Trained models for such data do not always work globally due to different characteristics of structures. The technical focus has been on how to ensure the generalization ability between different cities (Yokoya et al. 2018). To exploit the capability of machine learning and maximize the mapping accuracy, the simplest approach is to increase training data. Many mapping projects have been progressing in developing countries through annotation efforts by local people (e.g., Open Cities Africa[11]). Collaborative mapping based on crowdsourced data represented by OpenStreetMap plays a major role in creating training data. The synergy of openly available big earth observation data, crowdsourcing-based annotations, and machine learning technologies will accelerate the land cover mapping capability for the entire globe.

[9] http://bigearth.net/

[10] http://www.grss-ieee.org/community/technical-committees/data-fusion/2019-ieee-grss-data-fusion-contest/

[11] https://opencitiesproject.org/

4.4 Disaster Mapping

4.4.1 Flood Mapping

Besides the above-mentioned international cooperative projects and services for the disaster response in the introduction part, flood mapping systems are also availability.

- Global flood detection system.[12] The objective of this system is to detect and map major river floods using daily passive microwave remote sensing sensors (AMSR2 and GPM).
- NASA Global flood detection system.[13] This system adapts real-time TRMM Multi-satellite Precipitation Analysis (TMPA) and Global Precipitation Measurement (GPM) Integrated Multi-Satellite Retrievals.
- Tiger-Net.[14] ESA supports the African with earth observation for monitoring water resource (including flood mapping) through the satellites of ESA.
- Dartmouth flood observatory.[15] It was founded in 1993 at Dartmouth College, Hanover, NH USA and moved to the University of Colorado, INSTAAR in 2010. They have used all the available satellite datasets (optical and SAR) to estimate the flood inundation map using change detection methods.
- DLR flood service.[16] Sentinel-1 and TerraSAR-X SAR datasets are used to extract the flooding maps using a fully automatic chains (i.e., pre-processing, auxiliary datasets collection, initialized classification and post-processing) via a web-client.

For flood mapping, SAR images are the better choice compared to the optical and UAV images, as clouds are penetrated by electromagnetic waves and do not corrupt the resulting image. Usually, due to the lower reflectance in optical and lower backscattering in SAR datasets, water bodies are easily detected. Two traditional but efficient methods are usually utilized (seen in Fig. 4.2). The first one is to apply the change detection methods between pre- and post-flood images and then use the filters (e.g., morphological closing and opening) to remove the noise. This kind of techniques is suitable to detect the flood area using single source datasets, such as Landsat series (Chignell et al. 2015), ENVISAT ASAR (Schlaffer et al. 2015), and Sentinel SAR (Li et al. 2018).

The second one is to extract the water bodies using classification methods (water and non-water areas) and indexes (listed in Table 4.3) from pre- and post-flood images. Then, the flood area is produced by analyze the changes between the water

[12]http://www.gdacs.org/flooddetection/

[13]https://disasters.nasa.gov/datasets/global-flood-monitoring-system

[14]http://www.tiger-net.org/

[15]https://floodobservatory.colorado.edu/index.html

[16]https://www.dlr.de/eoc/en/desktopdefault.aspx/tabid-12939/22596_read-51634/

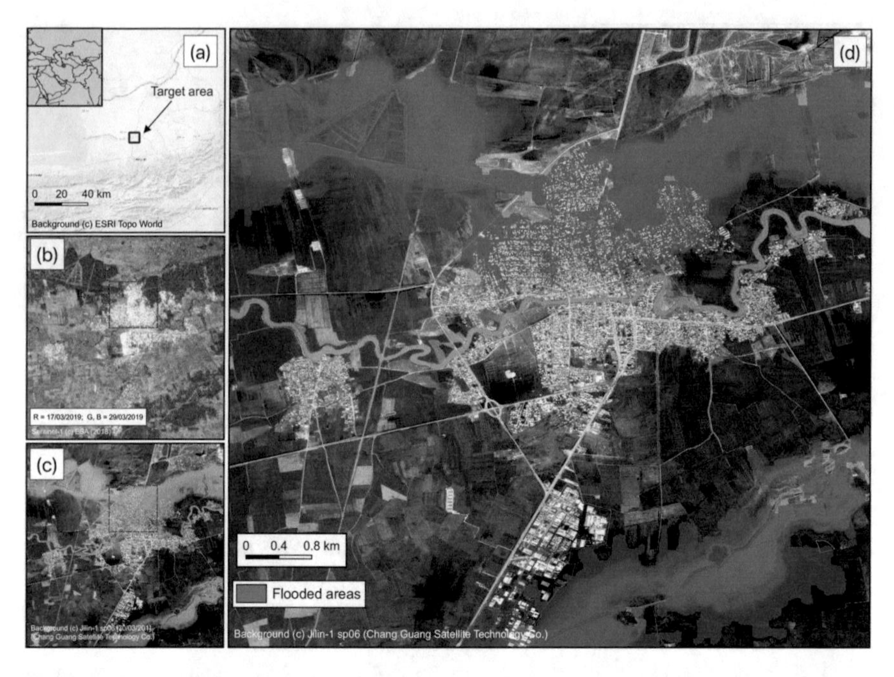

Fig. 4.2 Flood detection methods. (**a**) Location map of the target area. (**b**) False composite color image of Sentinel-1 SAR (R: pre-event, G, B: post-event). (**c**) Post-event high-resolution optical image (Jilin-1 sp06). (**d**) Mapping of flooded areas

Table 4.3 Water indices with their equations and sources for optical datasets

Indices	Equation	Source
NDWI	NDWI = (Green-NIR)/(Green+NIR)	Mcfeeters (1996)
MNDWI	MNDWI = (Green-SWIR)/(Green+SWIR)	Xu (2006)
AWEI	AWEI_{NSH} = 4(Green-SWIR1)-(0.25NIR+2.75*SWIR2)	Feyisa et al. (2014)
	AWEI_{SH} = Blue+2.5Green-1.5(NIR+SWIR1)-0.25SWIR2	

NDWI normalized difference water index, *MNDWI* modification of normalised difference water index, *AWEI* automated water extraction index, $AWEI_{NSH}$ AWEI in non-shadow area, $AWEI_{SH}$ AWEI in shadow area

bodies of two periods. Tong et al. (2018) have applied the support vector machine and the active contour without edges model for extracting water from Landsat 8 and COMSO-SkyMed and then mapped the flood using image difference method.

Technical challenges and future directions are list as follows:

1. Mapping flood in small specific area. Very high resolution remote sensing provide an opportunity to monitoring the flood in a small scale (e.g., downtown area). However, water is always mixed by the shadow areas. To separate the shadow from water body will improve the performance of flood monitoring.
2. Developing more computationally efficient and robust method without considering spatial resolution, spectral signature, or viewing angle. Normalized

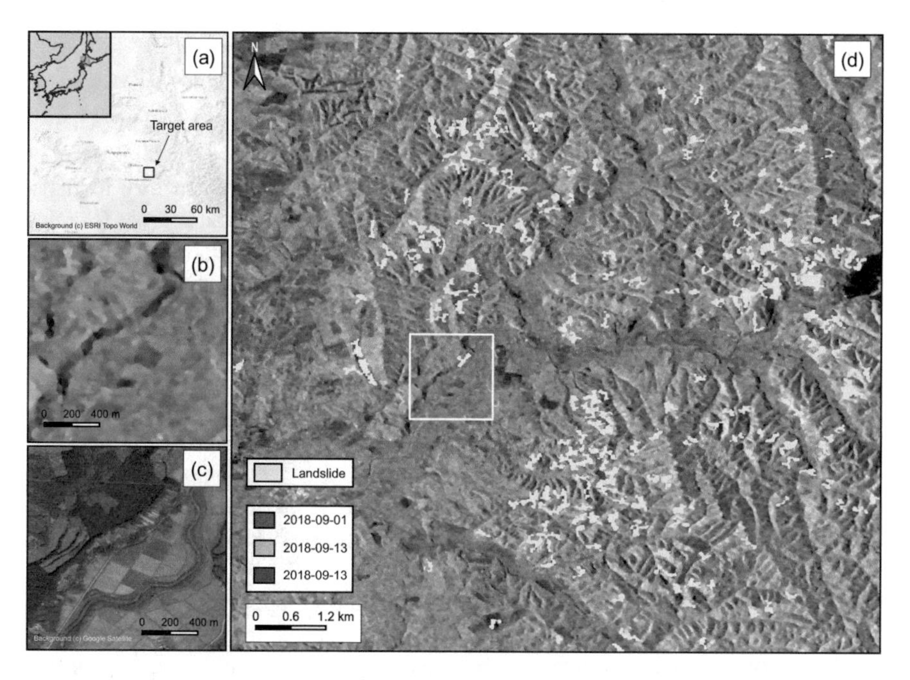

Fig. 4.3 (**a**) Location map of the target area. (**b**) False composite color image of Sentinel-1 SAR (R: pre-event, G, B: post-event). (**c**) Post-event high-resolution optical image (Jilin-1 sp06). (**d**) Mapping of flooded areas

Difference Flood Index (NDFI) (Cian et al. 2018), which is computed using multi-temporal statistics of SAR images, will give us the inspirations.

3. Flood detection via satellite and social media by deep learning. Satellite images can provide large scale flooding information, however, we should wait for the datasets. Social media can provide real-time information. A proper way should be found to integrate the information derivied from satellite images and social multimedia. Interested reader can read more details in http://www.multimediaeval.org/mediaeval2018/.

Here, a typical example of combining medium-resolution SAR (i.e., Sentinel-1) and high-resolution optical (i.e., Jilin-1 sp06) datasets to detect the flood areas in Iran is shown in Fig. 4.3. Due to the coarse resolution of Sentinel-1, the small flood areas in the city center (red rectangle areas in Fig. 4.3b) could not be detected by using only Sentinel-1 images. However, it can be identified by the high-resolution optical images. Thus, the final flooded mapping is the combination of the city flooded areas extracted by high-resolution and the non-city flood areas generated by Sentinel-1.

4.4.2 Landslide Mapping

Landslide disasters are frequently triggered by heavy rains and earthquakes (Martelloni et al. 2012; Tanyaş et al. 2019). These deadly events can cause a large number of fatalities (Intrieri et al. 2019). As a result, there have been several efforts to map a global subjectively of occurrence using big earth observation data sources (Stanley and Kirschbaum 2017). These activities take advantage of the relationship between landslides and four main variables such as topography slope computed from global topography models (SRTM, ASTER GDEM), land cover, rainfall data, and seismic activity (NASA Goddard Space Flight Center 2007; Muthu and Petrou 2007; Kirschbaum et al. 2010, 2015; Kirschbaum and Stanley 2018). These techniques are mainly based on models that integrate all variables using heuristic functions to evaluate the possibility of landslide occurrence. These models can map the landslide susceptibility on a continental scale (approximately 1 km^2), regional, and local scale with a resolution of few hundred meters. These studies provide an overview of the landslide hazard and can be used for mitigation and preparation activities before these disasters occurred.

Differently, earth observation data is also applied for mapping landslide damages in smaller scales focusing on particular events. Visual interpretation methods employ very-high-resolution optical imagery acquired from either space- or airbone platforms. Although these approaches provide high-reliability on the damage assessment, their applicability is often restricted by the availability of suitable images such as cloud-free and good-illumination conditions. It is also important to notice that these techniques require huge human efforts for damage interpretation, specially in case of rapid disaster response.

Change detection models, on the other hand, use a set of images acquired before and after the disaster to evaluate the damages. The land cover changes estimated from multi-temporal optical imagery is used for delineating the extent of landslides. Furthermore, spectral indexes (e.g. normalized vegetation and soil index) are also employed for landslide mapping (Rau et al. 2014; Lv et al. 2018; Yang et al. 2013; Zhuo et al. 2019; Ramos-Bernal et al. 2018). Integration of high-resolution digital terrain models allows estimation of landslide-induced damages such as debris and land scars distribution in the affected area (Dou et al. 2019; Bunn et al. 2019). Similarly to visual interpretation approach, the availability of suitable multi-temporal image datasets firmly bound the deployment of these techniques.

In the case of SAR data that has almost all-weather acquisition conditions, mapping techniques take advantage of the side-looking nature of these sensors. The two properties of SAR data, intensity, and phase information of the backscattered signal are exploited for detecting landslide damages. The later is widely applied for monitoring and mapping seismic-induced landslides (Cascini et al. 2009; Kalia 2018). Interferometric SAR (InSAR) analysis using detail DEM data provide the spatial distribution and displacement fields of the ground movement (Riedel and Walther 2008; Rabus and Pichierri 2018; Amitrano et al. 2019). Furthermore, time-series InSAR models allow landslide monitoring of slow-movement landslides (Kang et al. 2017). On the other hand, change detection techniques, using

SAR intensity images, are also powerful means to estimate the spatial distribution of landslide damages (Shi et al. 2015). For instance, texture features computed from multi-temporal datasets shows good correlation with the areas affected by landslides (Darvishi et al. 2018; Mondini et al. 2019). Furthermore, in case of disaster response where rapid geolocations of affected areas are crucial for rescue efforts, change detection based on intensity information has great applicability because of low computation time and direct manipulation of geocoded images. For instance, on September 6, last year, the 2018 Hokkaido Eastern Iburi Earthquake caused several landslides distributed in an extensive area (Yamagishi and Yamazaki 2018). Figure 4.4 shows a repid landslide mapping (yellow segments) using a combination of pixel- and object-based change detection analysis, proposed by Adriano et al. (2020), of a pre- and post-event Sentinel-1 intensity images acquired on September 1 and 13, 2018, respectively.

Recently, machine learning algorithms together with earth observation data are applied to detect landslide areas. Application of well establish classifiers such as support vector machine and ensemble learning models are used to identify landslide areas from optical, SAR intensity, and SAR coherence images (Bui et al. 2018; Park et al. 2018; Burrows et al. 2019). Furthermore, deep neural networks are

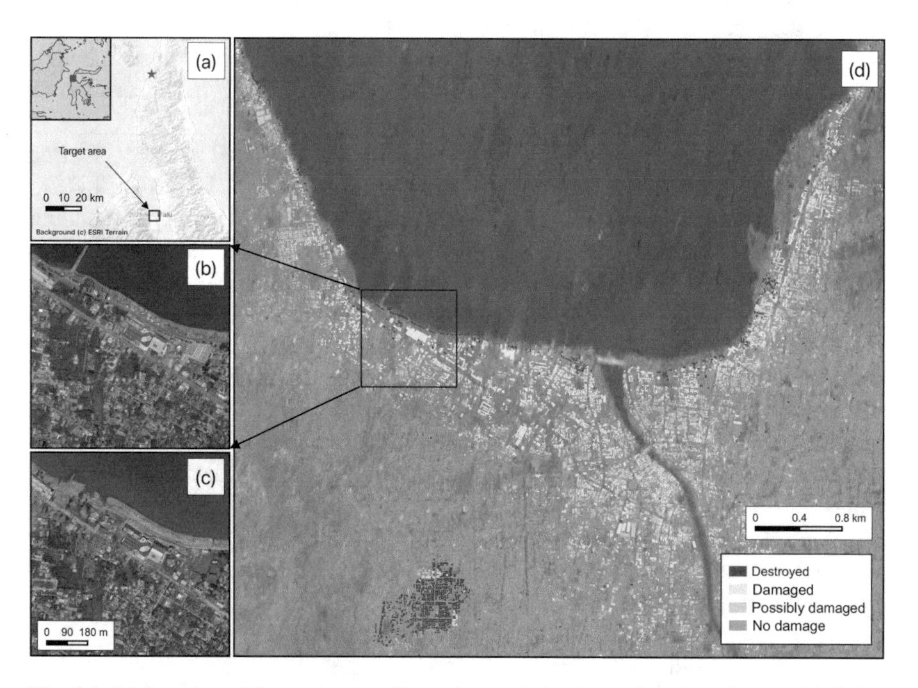

Fig. 4.4 (**a**) Location of the target area. The red start shows the earthquake epicenter (**b**) Color-composed image from pre- and post-event Sentinel-1 intensity images (R: pre-event, G, B: post-event). (**c**) Google Satellite imagery corresponding to the same area shown in *b*. (**d**) Landslide mapping results using multi-temporal Sentinel-1 imagery. Background image corresponds to the color-composed RGB image

also employed to map the landslide detection (Ghorbanzadeh et al. 2019; Wang et al. 2019). These approaches focused on high-resolution remote sensing imagery and landslide influencing features such as DEM data, land cover, and rainfall information.

4.4.3 Building Damage Mapping

Assessing the building damage in the aftermath of major disasters, such as earthquakes, tsunamis, and typhoons, are crucial for post-disaster rapid and efficient relief activities. In this context, earth observation data is a good alternative for damage mapping because satellite imagery can observe large scenes from remote or inaccessible affected areas (Matsuoka and Yamazaki 2004). Based on the evolution of sensor platforms and their spatial resolution, damage mapping can be divided into two parts. Initial applications for building damage recognition were based on change detection analysis of moderate-resolution, mainly using sensor launched in the late 90's such as the Landsat-7 Satellite and the European Remote Sensing (ERS-1) SAR satellite, optical and SAR imagery. These applications relied on the interpretation of texture and linear correlation features computed from pre- and post-event datasets. Besides, due to their relative low spatial resolution (about $30\,m^2$), these methods were efficiently applied for building damage mapping in a block-scale (Yusuf et al. 2001; Matsuoka and Yamazaki 2005; Kohiyama and Yamazaki 2005).

The following generation of high-resolution optical and SAR imagery, starting in early 2000s such as the QuickBird, GeoEye-1, TerraSAR-X, COSMO-SkyMed satellites, contribute to developing frameworks for detail mapping of building damage. These methods, besides of change detection techniques, implemented sophisticated pixel- and object-based image processing algorithms for damage recognition (Miura et al. 2016; Tong et al. 2012; Brett and Guida 2013; Gokon et al. 2015; Ranjbar et al. 2018). Moreover, taking advantage of very-high-resolution datasets, sophisticated frameworks were implemented to extract building damage using only post-event images (Gong et al. 2016). Most of these methodologies rely on specific features of SAR data. For instance, some studies analyzed the polarimetric characteristics of radar backscattering that are correlated with building damage patterns observed in SAR images (Yamaguchi 2012; Chen and Sato 2013). Furthermore, SAR platforms such as the Sentinel-1 and ALOS-2 repeatedly acquired images constructing large time-series datasets. Phase coherence computed from multi-temporal SAR acquisitions can provide important characteristics of the degree of changes in urban areas in the case of earthquake-induced damage (Yun et al. 2015; Olen and Bookhagen 2018; Karimzadeh et al. 2018).

Recently, advanced machine learning algorithms are implemented using multi-temporal and multi-source remote sensing data for mapping building damage. These methodologies learn from limited but properly labeled samples of damaged buildings to assign a level on the whole affected area (Endo et al. 2018). A recent example, Adriano et al. (2019) used an ensemble learning classifier on

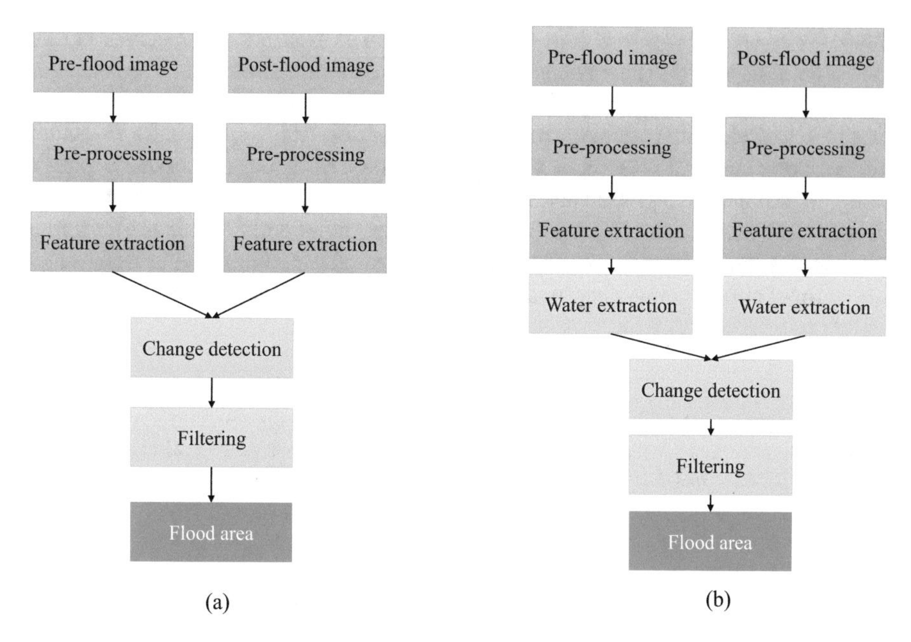

Fig. 4.5 (**a**) Location of the target area. The red start shows the earthquake epicenter (**b**) Pre-event WorldView-3 image. (**c**) Post-event WorldView-3 images. (**d**) Damage mapping results using multi-sensor and multi-temporal remote sensing data. Background image corresponds to the pre-event Sentinel-1 SAR image capture on May 26, 2018

SAR and optical datasets to map the building damage following the 2018 Sulawesi Earthquake-Tsunami in Palu, Indonesia. Their methodology successfully classified three levels of building damage with an overall accuracy greater than 90% (Fig. 4.5). Furthermore, their implemented framework provided a reliable thematic map after only after three hours of acquired all raw remote sensing datasets.

4.5 Conclusion and Future Lines

Open data policy in earth observation and international cooperation in emergency responses have expanded practical use of image and signal processing techniques for rapid disaster damage mapping. In this chapter, we have reviewed earth observation systems available for disaster management and showcased recent advances in land cover mapping, flood mapping, landslide mapping, and building damage mapping.

Although human visual interpretation is still required to determine the level of detailed building damages, it takes a long time to acquire high-resolution images and conduct visual interpretation. One possible future direction is to construct training data on past disasters via human visual interpretation and develop machine learning models that can respond quickly to unknown disasters. Another challenge is that there are many cases where data can not be obtained from the same sensor before and after a disaster (He and Yokoya 2018). How to extract disaster-induced changes from multisensor and possibly heterogeneous data sources before and after disasters is a practical problem in damage mapping. Furthermore, it is important for the entire disaster management process to verify the accuracy of damage assessment results using in-situ data. Integration and fusion of earth observation data with ground-shot images and text information available online (e.g., news and SNS) is also a future subject. On the basis of the remote sensing image and signal processing technology and human expert knowledge, machine learning technologies have the potential to accelerate the accuracy and speed of damage mapping from big earth observation data.

Acknowledgments If you want to include acknowledgments of assistance and the like at the end of an individual chapter please use the `acknowledgement` environment – it will automatically be rendered in line with the preferred layout.

References

Adriano B, Xia J et al (2019) Multi-source data fusion based on ensemble learning for rapid building damage mapping during the 2018 Sulawesi earthquake and tsunami in Palu, Indonesia. Remote Sens 11(7). ISSN: 2072–4292. https://doi.org/10.3390/rs11070886. https://www.mdpi.com/2072-4292/11/7/886

Adriano B, Yokoya N et al (2020) A semiautomatic pixel-object method for detecting landslides using multitemporal ALOS-2 intensity images. Remote Sens 12(3). issn: 2072-4292. https://doi.org/10.3390/rs12030561. https://www.mdpi.com/2072-4292/12/3/561

Amitrano D et al (2019) Long-term satellite monitoring of the Slumgul- lion landslide using space-borne synthetic aperture radar sub-pixel offset tracking. Remote Sens 11(3):369. https://doi.org/10.3390/rs11030369

Bird R et al (2013) NovaSAR-S: a low cost approach to SAR applications. In: Conference proceedings of 2013 Asia-Pacific conference on synthetic aperture radar (APSAR), pp 84–87

Brett PTB, Guida R (2013) Earthquake damage detection in urban areas using curvilinear features. IEEE Trans Geosci Remote Sens 51(9):4877–4884. ISSN: 0196-2892. https://doi.org/10.1109/TGRS.2013.2271564. http://ieeexplore.ieee.org/lpdocs/epic03/wrapper.htm?arnumber=6565347

Bui DT et al (2018) Landslide detection and susceptibility mapping by AIRSAR data using support vector machine and index of entropy models in Cameron Highlands, Malaysia. Remote Sens 10(10). ISSN: 20724292. https://doi.org/10.3390/rs10101527

Bunn MD et al (2019) A simplified, object-based framework for efficient landslide inventorying using LIDAR digital elevation model derivatives. Remote Sens 11(3). ISSN: 20724292. https://doi.org/10.3390/rs11030303

Burrows K et al (2019) A new method for large-scale landslide classification from satellite radar. Remote Sens 11(3):237. https://doi.org/10.3390/rs11030237

Caltagirone F et al (2014) The COSMO-SkyMed dual use earth observation program: development, qualification, and results of the commissioning of the overall constellation. IEEE J Sel Top Appl Earth Obs Remote Sens 7(7):2754–2762. ISSN: 1939–1404. https://doi.org/10.1109/JSTARS.2014.2317287

Cascini L, Fornaro G, Peduto D (2009) Analysis at medium scale of low-resolution DInSAR data in slow-moving landslide-affected areas. ISPRS J Photogramm Remote Sens 64(6):598–611. ISSN: 09242716. https://doi.org/10.1016/j.isprsjprs.2009.05.003

Chen J et al (2015) Global land cover mapping at 30 m resolution: a POK-based operational approach. ISPRS J Photogramm Remote Sens 103:7–27

Chen S-W, Sato M (2013) Tsunami damage investigation of built-up areas using multitemporal spaceborne full polarimetric SAR images. IEEE Trans Geosci Remote Sens 51(4):1997. ISSN: 0196-2892. https://doi.org/10.1109/TGRS.2012.2210050. http://ieeexplore.ieee.org/lpdocs/epic03/wrapper.htm?arnumber=6353568

Chignell SM et al (2015) Multi-temporal independent component analysis and landsat 8 for delineating maximum extent of the 2013 Colorado front range flood. Remote Sens 7(8):9822–9843. ISSN: 2072-4292. https://doi.org/10.3390/rs70809822. http://www.mdpi.com/2072-4292/7/8/9822

Cian F, Marconcini M, Ceccato P (2018) Normalized difference flood index for rapid flood mapping: taking advantage of EO big data. Remote Sens Environ 209:712–730. ISSN: 0034-4257. https://doi.org/10.1016/j.rse.2018.03.006. http://www.sciencedirect.com/science/article/pii/S0034425718300993

Darvishi M et al (2018) Sentinel-1 and ground-based sensors for continuous monitoring of the corvara landslide (South Tyrol, Italy). Remote Sens 10(11):1781. https://doi.org/10.3390/rs10111781

De Lisle D et al (2018) RADARSAT constellation mission status update. In: EUSAR 2018; 12th European conference on synthetic aperture radar, pp 1–5

Demir I et al (2018) DeepGlobe 2018: a challenge to parse the Earth through satellite images. In: The IEEE conference on computer vision and pattern recognition (CVPR) workshops

Dou J et al (2019) Evaluating GIS-based multiple statistical models and data mining for earthquake and rainfall-induced landslide susceptibility using the LiDAR DEM. Remote Sens 11(6):638. ISSN: 2072-4292. https://doi.org/10.3390/rs11060638. https://www.mdpi.com/2072-4292/11/6/638

Endo Y et al (2018) New insights into multiclass damage classification of tsunami-induced building damage from SAR images. Remote Sens 10(12):2059. ISSN: 2072-4292. https://doi.org/10.3390/rs10122059. http://www.mdpi.com/2072-4292/10/12/2059

Esch T et al (2013) Urban footprint processor – fully automated processing chain generating settlement masks from global data of the TanDEM-X mission. IEEE Geosci Remote Sens Lett 10(6):1617–1621

Farquharson G et al (2018) The capella synthetic aperture radar constellation. In: IGARSS 2018–2018 IEEE international geoscience and remote sensing symposium, pp 1873–1876. https://doi.org/10.1109/IGARSS.2018.8518683

Feyisa GL et al (2014) Automated water extraction index: a new technique for surface water mapping using Landsat imagery. Remote Sens Environ 140:23–35. ISSN: 0034-4257. https://doi.org/10.1016/j.rse.2013.08.029. http://www.sciencedirect.com/science/article/pii/S0034425713002873

Ghorbanzadeh O et al (2019) Evaluation of different machine learning methods and deep-learning convolutional neural networks for landslide detection. Remote Sens 11(2). ISSN: 20724292. https://doi.org/10.3390/rs11020196

Gokon H et al (2015) A method for detecting buildings destroyed by the 2011 Tohoku earthquake and tsunami using multitemporal TerraSAR-X data. IEEE Geosci Remote Sens Lett 12(6):1277–1281. ISSN: 1545-598X. https://doi.org/10.1109/LGRS.2015.2392792. http://ieeexplore.ieee.org/lpdocs/epic03/wrapper.htm?arnumber=7042770

Gong L et al (2016) Earthquake-induced building damage detection with post-event sub-meter VHR terrasar-X staring spotlight imagery. Remote Sens 8(11):1–21. ISSN: 20724292. https://doi.org/10.3390/rs8110887

He W, Yokoya N (2018) Multi-temporal sentinel-1 and -2 data fusion for optical image simulation. ISPRS Int J GeoInf 7(10). ISSN: 2220-9964. https://doi.org/10.3390/ijgi7100389. https://www.mdpi.com/2220-9964/7/10/389

Intrieri E, Carlà T, Gigli G (2019) Forecasting the time of failure of landslides at slope-scale: a literature review. Earth-Sci Rev. ISSN: 00128252. https://doi.org/10.1016/j.earscirev.2019.03.019. https://linkinghub.elsevier.com/retrieve/pii/S001282521830518X

Kalia AC (2018) Classification of landslide activity on a regional scale using persistent scatterer interferometry at the Moselle Valley (Germany). Remote Sens 10(12). ISSN: 20724292. https://doi.org/10.3390/rs10121880

Kang Y et al (2017) Application of InSAR techniques to an analysis of the Guanling landslide. Remote Sens 9(10):1–17. ISSN: 20724292. https://doi.org/10.3390/rs9101046

Karimzadeh S et al (2018) Sequential SAR coherence method for the monitoring of buildings in Sarpole-Zahab, Iran. Remote Sens 10(8):1255. ISSN: 2072-4292. https://doi.org/10.3390/rs10081255. http://www.mdpi.com/2072-4292/10/8/1255

Kirschbaum D, Stanley T (2018) Satellite-based assessment of rainfall-triggered landslide hazard for situational awareness. Earth's Future 6(3):505–523. ISSN: 23284277. https://doi.org/10.1002/2017EF000715

Kirschbaum DB et al (2010) A global landslide catalog for hazard applications: method, results, and limitations. Nat Hazards 52(3):561–575. ISSN: 0921030X. https://doi.org/10.1007/s11069-009-9401-4

Kirschbaum D, Stanley T, Zhou Y (2015) Spatial and temporal analysis of a global landslide catalog. Geomorphology 249:4–15. ISSN: 0169555X. https://doi.org/10.1016/j.geomorph.2015.03.016

Kohiyama M, Yamazaki F (2005) Damage detection for 2003 Bam, Iran, earthquake using Terra-ASTER satellite imagery. Earthquake Spectra 21(S1):267–274. https://doi.org/10.1193/1.2098947

Le Saux B et al (2019) 2019 data fusion contest [technical committees]. IEEE Geosci Remote Sens Mag 7(1):103–105. ISSN: 2168–6831. https://doi.org/10.1109/MGRS.2019.2893783

Lee S (2010) Overview of KOMPSAT-5 program, mission, and system. In: 2010 IEEE international geoscience and remote sensing symposium, pp 797–800. https://doi.org/10.1109/IGARSS.2010.5652759

Li Y et al (2018) An automatic change detection approach for rapid flood mapping in Sentinel-1 SAR data. Int J Appl Earth Obs Geoinf 73:123–135. ISSN: 0303-2434. https://doi.org/10.1016/j.jag.2018.05.023. http://www.sciencedirect.com/science/article/pii/S0303243418302782

Lv ZY et al (2018) Landslide inventory mapping from bitemporal high-resolution remote sensing images using change detection and multiscale segmentation. IEEE J Sel Top Appl Earth Obs Remote Sens 11(5):1520–1532. ISSN: 21511535. https://doi.org/10.1109/JSTARS.2018.2803784

Martelloni G et al (2012) Rainfall thresholds for the forecasting of landslide occurrence at regional scale. Landslides 9(4):485–495. ISSN: 1612–5118. https://doi.org/10.1007/s10346-011-0308-2

Martinis S, Plank S, Ćwik K (2018) The use of Sentinel-1 time-series data to improve flood monitoring in arid areas. Remote Sens 10(4). ISSN: 2072-4292. https://doi.org/10.3390/rs10040583. https://www.mdpi.com/2072-4292/10/4/583

Matsuoka M, Yamazaki F (2004) Use of satellite SAR intensity imagery for detecting building areas damaged due to earthquakes. Earthquake Spectra 20(3):975. ISSN: 87552930. https://doi.org/10.1193/1.1774182. http://link.aip.org/link/EASPEF/v20/i3/p975/s1%7B%5C&%7DAgg=doi

Matsuoka M, Yamazaki F (2005) Building damage mapping of the 2003 Bam, Iran, earthquake using Envisat/ASAR intensity imagery. Earthquake Spectra 21(S1):S285. ISSN: 87552930. https://doi.org/10.1193/1.2101027. http://link.aip.org/link/EASPEF/v21/iS1/pS285/s1%7B%5C&%7DAgg=doi

Mcfeeters SK (1996) The use of the Normalized Difference Water Index (NDWI) in the delineation of open water features. Int J Remote Sens 17(7):1425–1432

Miura H, Midorikawa S, Matsuoka M (2016) Building damage assessment using high-resolution satellite SAR images of the 2010 Haiti earthquake. Earthquake Spectra 32(1):591–610. ISSN: 8755-2930. https://doi.org/10.1193/033014EQS042M

Mondini A et al (2019) Sentinel-1 SAR amplitude imagery for rapid landslide detection. Remote Sens 11(7):760. ISSN: 2072-4292. https://doi.org/10.3390/rs11070760. https://www.mdpi.com/2072-4292/11/7/760

Moreira A et al (2013) A tutorial on synthetic aperture radar. IEEE Geosci Remote Sens Mag 1(1):6–43. ISSN: 2168-6831. https://doi.org/10.1109/MGRS.2013.2248301
Morena LC, James KV, Beck J (2004) An introduction to the RADARSAT-2 mission. Can J Remote Sens 30(3):221–234. https://doi.org/10.5589/m04-004
Motohka T et al (2017) Status of the advanced land observing satellite-2 (ALOS-2) and its follow-on L-band SAR mission. In: 2017 IEEE international geo-science and remote sensing symposium (IGARSS), pp 2427–2429. https://doi.org/10.1109/IGARSS.2017.8127482
Muthu K, Petrou M (2007) Landslide-Hazard mapping using an expert system and a GIS. IEEE Trans Geosci Remote Sens 45(2):522–531. ISSN: 0196-2892
NASA Goddard Space Flight Center (2007). The Global Landslide Catalog. http://web.archive.org/web/20080207010024/. http://www.808multimedia.com/winnt/kernel.htm (visited on 30 Sept 2010)
Obata T et al (2019) The development status of the first demonstration satellite of our commercial small synthetic aperture radar satellite constellation. In: AIAA/USU conference on small satellites, pp 1–4
Olen S, Bookhagen B (2018) Mapping damage-affected areas after natural hazard events using Sentinel-1 coherence time series. Remote Sens 10(8):1272. ISSN: 2072-4292. https://doi.org/10.3390/rs10081272. http://www.mdpi.com/2072-4292/10/8/1272
Park S-J et al (2018) Landslide susceptibility mapping and comparison using decision tree models: a case study of Jumunjin Area, Korea. Remote Sens 10(10):1545. https://doi.org/10.3390/rs10101545
Rabus B, Pichierri M (2018) A new InSAR phase demodulation technique developed for a typical example of a complex, multi-lobed landslide displacement field, Fels Glacier Slide, Alaska. Remote Sens 10(7). ISSN: 20724292. https://doi.org/10.3390/rs10070995
Ramos-Bernal RN et al (2018) Evaluation of unsupervised change detection methods applied to landslide inventory mapping using ASTER imagery. Remote Sens 10(12). ISSN: 20724292. https://doi.org/10.3390/rs10121987
Ranjbar HR et al (2018) Using high-resolution satellite imagery to provide a relief priority map after earthquake. Nat Hazards 90(3):1087–1113. ISSN: 0921-030X. https://doi.org/10.1007/s11069-017-3085-y
Rau J, Jhan J, Rau R (2014) Semiautomatic object-oriented landslide recognition scheme from multisensor optical imagery and DEM. IEEE Trans Geosci Remote Sens 52(2):1336–1349. ISSN: 0196-2892
Riedel B, Walther A (2008) InSAR processing for the recognition of landslides. Adv Geosci 14:189–194. ISSN: 1680-7359. https://doi.org/10.5194/adgeo-14-189-2008. https://www.adv-geosci.net/14/189/2008/
Rosen P et al (2017) The NASA-ISRO SAR (NISAR) mission dual-band radar instrument preliminary design. In: 2017 IEEE international geoscience and remote sensing symposium (IGARSS), pp 3832–3835. https://doi.org/10.1109/IGARSS.2017.8127836
Rosenqvist A et al (2014) A brief overview of the SAOCOM Integrated Mission Acquisition Strategy (IMAS). In: 1st ESA SAOCOM companion satellite workshop, ESA ESTEC
Schlaffer S et al (2015) Flood detection from multi-temporal SAR data using harmonic analysis and change detection. Int J Appl Earth Obs Geoinf 38:15–24. ISSN: 0303-2434. https://doi.org/10.1016/j.jag.2014.12.001. http://www.sciencedirect.com/science/article/pii/S0303243414002645
Shi X et al (2015) Landslide deformation monitoring using point-like target offset tracking with multi-mode high-resolution TerraSAR-X data. ISPRS J Photogramm Remote Sens 105:128–140. ISSN: 09242716. https://doi.org/10.1016/j.isprsjprs.2015.03.017
Stanley T, Kirschbaum DB (2017) A heuristic approach to global landslide susceptibility mapping. Nat Hazards 87(1):145–164. ISSN: 15730840. https://doi.org/10.1007/s11069-017-2757-y
Sun J, Yu W, Deng Y (2017) The SAR payload design and performance for the GF-3 mission. Sensors 17(10). ISSN: 1424–8220. https://doi.org/10.3390/s17102419. http://www.mdpi.com/1424-8220/17/10/2419

Suri S et al (2015) TerraSAR-X/PAZ constellation: CONOPS, highlights and access solution. In: 2015 IEEE 5th Asia-Pacific conference on synthetic aperture radar (APSAR), pp 178–183. https://doi.org/10.1109/APSAR.2015.7306183

Tanase MA et al (2010) TerraSAR-X data for burn severity evaluation in mediterranean forests on sloped terrain. IEEE Trans Geosci Remote Sens 48(2):917–929. ISSN: 0196-2892. https://doi.org/10.1109/TGRS.2009.2025943

Tanyaş H et al (2019) Rapid prediction of the magnitude scale of landslide events triggered by an earthquake. Landslides 16(4):661–676. ISSN: 1612-5118. https://doi.org/10.1007/s10346-019-01136-4

Tong X, Hong Z et al (2012) Building-damage detection using pre- and post-seismic high-resolution satellite stereo imagery: a case study of the May 2008 Wenchuan earthquake. ISPRS J Photogramm Remote Sens 68:13–27. ISSN: 09242716. https://doi.org/10.1016/j.isprsjprs.2011.12.004. http://linkinghub.elsevier.com/retrieve/pii/S0924271611001584

Tong X, Luo X et al (2018) An approach for flood monitoring by the combined use of Landsat 8 optical imagery and COSMO-SkyMed radar imagery. ISPRS J Photogramm Remote Sens 136:144–153. ISSN: 0924-2716

Torres R et al (2012) GMES Sentinel-1 mission. In: Remote sensing of environment 120. The Sentinel missions – new opportunities for science, pp 9–24. ISSN: 0034-4257. https://doi.org/10.1016/j.rse.2011.05.028. http://www.sciencedirect.com/science/article/pii/S0034425712000600

Verhegghen A et al (2016) The potential of Sentinel satellites for burnt area mapping and monitoring in the Congo Basin Forests. Remote Sens 8(12). ISSN: 2072-4292. https://doi.org/10.3390/rs8120986. https://www.mdpi.com/2072-4292/8/12/986

Wang Y, Fang Z, Hong H (2019) Comparison of convolutional neural networks for landslide susceptibility mapping in Yanshan County, China. Sci Total Environ 666:975–993. ISSN: 18791026. https://doi.org/10.1016/j.scitotenv.2019.02.263

Werninghaus R, Buckreuss S (2010) The TerraSAR-X mission and system design. IEEE Trans Geosci Remote Sens 48(2):606–614. ISSN: 0196-2892. https://doi.org/10.1109/TGRS.2009.2031062

Xu H (2006) Modification of normalised difference water index (NDWI) to enhance open water features in remotely sensed imagery. Int J Remote Sens 27(14):3025–3033

Yamagishi H, Yamazaki F (2018) Landslides by the 2018 Hokkaido Iburi-Tobu earthquake on September 6. Landslides 15(12):2521–2524. ISSN: 1612-5118. https://doi.org/10.1007/s10346-018-1092-z

Yamaguchi Y (2012) Disaster monitoring by fully polarimetric SAR data acquired with ALOS-PALSAR. Proc IEEE 100(10):2851–2860. ISSN: 0018-9219. https://doi.org/10.1109/JPROC.2012.2195469. http://ieeexplore.ieee.org/document/6205771/

Yang W, Wang M, Shi P (2013) Using MODIS NDVI time series to identify geographic patterns of landslides in vegetated regions. IEEE Geosci Remote Sens Lett 10(4):707–710. ISSN:1545-598X. https://doi.org/10.1109/LGRS.2012.2219576

Yokoya N et al (2018) Open data for global multimodal land use classification: outcome of the 2017 IEEE GRSS Data Fusion Contest. IEEE J Sel Top Appl Earth Obs Remote Sens 11(5):1363–1377

Yun S-H et al (2015) Rapid damage mapping for the 2015 M w 7.8 Gorkha earthquake using synthetic aperture radar data from COSMO-SkyMed and ALOS-2 satellites. Seismological Res Lett 86(6):1549–1556. ISSN: 0895-0695. https://doi.org/10.1785/0220150152. http://srl.geoscienceworld.org/lookup/doi/10.1785/0220150152%20. https://pubs.geoscienceworld.org/srl/article/86/6/1549-1556/315478

Yusuf Y, Matsuoka M, Yamazaki F (2001) Damage assessment after 2001 Gujarat earthquake using Landsat-7 satellite images. J Indian Soc Remote Sens 29(1):17–22. ISSN: 0974-3006. https://doi.org/10.1007/BF02989909

Zhuo L et al (2019) Evaluation of remotely sensed soil moisture for land-slide hazard assessment. IEEE J Sel Top Appl Earth Obs Remote Sens 12(1):162–173. ISSN: 1939-1404

Chapter 5
Spatial Data Reduction Through Element-of-Interest (EOI) Extraction

Samantha T. Arundel and E. Lynn Usery

5.1 Introduction

Big Data are characterized as any vast and complex data collection that is difficult to manage with traditional practices. Big data containing geo-referenced attributes, which are viewed as big geospatial data, are transforming the geospatial industries into data-driven disciplines (Graham and Shelton 2013; Kitchin 2013; Miller and Goodchild 2015). Most geospatial big data currently derive from remotely sensed imagery via satellite, airborne and ground vehicles, or geo-enabled social media platforms like Twitter and Facebook. Big geospatial data suffer from issues related to the five 'Big Vs' – variety, volume, velocity, veracity and value, which respectively describe the great diversity of data types, the tremendous volume of data, the speed at which new data are produced, the reliability of the data, and the degree to which the data may actually be utilized (Emani et al. 2015; Lee and Kang 2015).

Whereas the benefits, including analytics, visualization, and knowledge discovery, of big data are reported elsewhere (Lee and Kang 2015; Li et al. 2013), complications associated with big geospatial data analysis have recently attracted considerable attention. The belief that more data results in higher accuracy has led to the collection of vast amounts and varieties of data, despite the cost of collection and storage (Chen and Zhang 2014; Goodchild 2013). Although big spatial data are currently available to the public via data portals and cyberinfrastructures, little use is made of them because they often fail to meet specific study requirements in their raw form. In fact, data collection continues to be a significant element in the design of geographical research. Instead, it may be wise to emphasize better exploitation

S. T. Arundel (✉) · E. L. Usery
Center of Excellence for Geospatial Information Science, U.S. Geological Survey, Rolla, MO, USA
e-mail: sarundel@usgs.gov

M. Werner, Y.-Y. Chiang (eds.), *Handbook of Big Geospatial Data*,
https://doi.org/10.1007/978-3-030-55462-0_5

of existing datasets. Utilizing big data in their raw formats results in formidable challenges to visualization, data mining, data query and analysis. To meet these challenges, raw data must be reduced to features, or elements of interest (EOI) must be extracted from them, without losing essential characteristics. In general, however, such features are scarce, and their quality is unsatisfactory in big data analytics (Lecun et al. 2015; Lekamalage et al. 2013; Najafabadi et al. 2015). This "big but valueless" predicament has become a significant obstruction to the beneficial use of big geospatial data.

EOI hail from two general data sources: (1) the mapping and remote sensing fields, which include surveying; and (2) social media and location-based data sources, which include the Internet of Things (IoT). We refer to the first type as active spatial data sources, in that data are actively collected for spatial applications, whereas the second type is considered passive, where data are 'collected' for other reasons but happen to be spatially enabled.

Remotely sensed data (RSD) in the spatial domain, and information derived from them, are an exceptional example of geospatial Big Data. Recently, the volume of data available to remote sensing consumers has seen such a rapid increase that a shift in the traditional way they are handled is needed. Big Data are often unstructured or collected for different purposes, whereas RSD are designed to be highly structured to mine significant information from the planetary surface and its atmosphere about an object, value, state or condition. In the case where imagery is collected to understand objects and their spatial patterns, efficient and timely extraction of those objects from the base imagery is essential for data reduction, which in turn lessens the infrastructure required to support the data.

Social media and location-based data, including IoT, form a large class of Big Data with location as a secondary attribute, often derived from the data content or context. Data sources and applications of location-based data include the IoT, fleet management, volunteered geographic information (VGI), governmental activities such as INSPIRE in the European Union, and cadastral information. These data are continually being updated, sent in real-time or video streams, some with GPS locations, but most with their place of origin derived from the static location of the sensors or the message content. Identification of EOI in these data rely heavily on data mining and knowledge discovery of similarities in content to interpret environmental or social patterns. The breadth of applications for these Big Geospatial Data spans urban monitoring, human movement, natural and human-created event processing, predictive maintenance for fleet management (Killeen et al. 2019) and many others.

5.2 Methods to Obtain EOI from Georeferenced Big Data

This section presents current state-of-the-art methods to create EOI from some types of georeferenced big data.

5.2.1 *Methods Commonly Used in the Remote Sensing and Mapping Fields*

5.2.1.1 Pixel-Based Methods

Satellite and other remotely-sensed images are viewed as a collection of samples each of which is an individual pixel created by optical to electrical conversion to generate a brightness value in a number of different bands of the electromagnetic spectrum. The analysis, definition, and classification of EOI rely on individual pixel values and the methods to extract EOI are called pixel-based methods. Each pixel location has a unique value for each band of wavelengths, such as, blue, green, red, and infrared, that are acquired. The pixel values are assigned based on the brightness and spectral reflectance. A collection of these values for a specific geographic feature forms a "spectral signature" and spatial location for the feature. These collections can be created by the user in a supervised classification mode, in which the user identifies groups of pixels that spectrally and spatially identify a geographic feature of known type. A series of these collections of pixels form a set of sample signatures that can be used by an automatic classifier, such as nearest neighbor, minimum distance, or maximum likelihood, to group unidentified pixels into the defined classes. An alternative is an unsupervised approach in which the user allows the computer algorithm to determine clusters of pixels with common spectral and spatial signatures. Although these clusters may represent geographic features, it is likely the user must aggregate clusters and define the features of interest. A similar classification from these clusters can be obtained and then interpreted to represent geographic features such as land use classification. These pixel-based methods have been extensively researched and are described in detail in general reference textbooks on remote sensing and image classification (Jensen 2015; Lillesand et al. 2015).

Pros and Cons of Pixel-Based Methods in Big Data

Pixel-based methods have a long history in analysis of remotely sensed big data beginning with satellite images with coarse spatial resolutions of 79x57 meters (m) from the original Landsat Multispectral data, 30x30 m from Landsat Thematic Mapper data, and 10 m from SPOT data. The accuracy of the methods depends on the spatial resolution and classification system used. For example, land cover, often the target product of the analysis, usually is classed in multilevel classification systems, such as the USGS Anderson land cover classes, and could usually attain an 85% accuracy for most images. These methods were optimized on the basis of aggregation of Earth surface materials inside the pixel ground area. Thus, in a scene in which vegetation and concrete occurred together, they could not be separated and were classed together. Sub-pixel classifiers were developed to handle fractionated land covers within individual pixels. These mixed pixel classifiers use techniques that estimate the abundance of different materials within each pixel and although they can successfully produce subpixel classifications, they are less well developed and understood than pixel-based methods. High resolution remotely sensed images, those with 1 m or higher spatial resolution, appeared in the 1990s and led to failure of many of the pixel-based methods. The diversity of Earth reflectance in high resolution images mitigated the pixel-based approach which depends on aggregation of Earth materials in a single pixel. The need for an alternative to analyze high spatial resolution images, including newer sensors such as lidar, led to segmentation of the images into objects with common reflectance characteristics and object-based image classification.

5.2.1.2 Object-Based Methods

Object-based image classification with a geographical element (GEOBIA) studies spatial elements and their relationships by defining and analyzing image objects – collections of adjoining pixels that share a common property (Castilla and Hay 2008; Blaschke 2010). These image objects become the rudimentary entity of analysis rather than an individual pixel. GEOBIA can generally be simplified into two processes: segmentation and classification. Segmentation is a process that groups similar pixels into objects, based on chosen components like spectral properties, numerical values, size, shape, texture, context and geometry (Blaschke 2010). Many segmentation algorithms exist, but in the natural sciences, multiresolution segmentation is used regularly due to its ability to simultaneously produce objects of various sizes (scales) in a single image (Baatz and Schäpe 2000). This capability performs well in spatial applications because similar features may vary in size, as do geologic faults or hydrological water bodies. Once fundamental objects are created, GEOBIA classification uses selected properties to group them into categories.

Multiple image bands, images and other spatial datasets, including vector data, can be fused in GEOBIA to create and classify objects. An example is the use of elevation, aspect, slope and proximity to streams to find specific vegetation types (Kim et al. 2010). Each layer not only provides its own context, but the spatial relationships within and between them can also be evaluated. Objects hold proximity and distance relationships between neighbors. As another example, objects with high negative curvature adjacent to ridges can be classified as glacial cirques (Arundel 2016). Another important advantage of GEOBIA over traditional pixel-based approaches is GEOBIA's ability to assimilate semantics into the translation of objects into everyday features, relying on the insight of the user (Blaschke and Strobl 2001). Thus, GEOBIA may avail many improvements over pixel-based methods (Blaschke et al. 2014; Chen et al. 2018).

Big data challenges for GEOBIA are typically related to imagery of high resolution covering large areas, reaching limits of RAM, CPU, disk space and I/O speed. As in pixel-wise approaches, one way to improve the processing of these images is by tiling them into smaller chunks, processing each chunk separately and then merging the results. Whereas pixel- or neighborhood-wise procedures result in identical output using this process, most GEOBIA segmentation routines must consider relationships between pixels within a much broader spatial context (Michel et al. 2015). To address this issue, automated methods are being explored to extract smaller regions of interest (ROI) in a meaningful way to support spatial image segmentation (Gonçalves et al. 2019). Through appropriate sampling design, ROIs should be representative of the desired processes, and thus should allow independent processing.

Research into improving big data analysis has also recently centered on enhancing the automation of GEOBIA procedures. One topic of focus has been developing techniques to assist the selection of the scale at which neighboring pixels are evaluated, known as the scale parameter (Cánovas-García and Alonso-Sarría 2015; Drăguţ et al. 2010; Ming et al. 2015), as well as other input parameters (Gonçalves et al. 2019; Kim et al. 2008; Liu et al. 2012). Tools that consider the research question to automatically determine the appropriate resolution and select data based on it offer possible future solutions to the big data problem (Chen et al. 2018). Parallel processing, recently introduced to GEOBIA algorithms, also promises better handling of big spatial data (Scrucca 2013).

5.2.1.3 Machine Learning

Machine learning (ML) is a principal subset of artificial intelligence focused on algorithms that permit computers to develop responses based on empirical data, as compared to imposing reactions up front. It is an essential counterpart to conventional techniques like geostatistics. The most unmistakable property of machine learning is its ability to uncover information and automatically choose behaviors based on that knowledge through the use of nonlinear, adaptive, robust and universal tools for pattern extraction and data modeling (Chen and Zhang 2014). Although in theory they should be capable of modeling any process, learning machines must be structured appropriately and their parameters (called hyperparameters) tuned properly. Careful selection and testing of these two elements give rise to the majority of machine learning research today (Kanevski et al. 2008).

Two important categories of machine learning algorithms are Statistical Learning Theory such as kernel-based methods like Support Vector Machines and Support Vector Regression, and Artificial Neural Networks of various construction (Li et al. 2016). Being data driven (black/grey boxes), such techniques rely substantially on the quality and quantity of data. Variograms are often used to regulate the quality of data analysis and modeling, and tune hyperparameters (Kanevski and Maignan 2004).

As datasets have increased in resolution, variety, temporality and size, they have revealed weaknesses in conventional approaches to handling them (Li et al. 2016; Shekhar et al. 2012). In many cases, geospatial questions are answered with multivariate, highly dimensional and scale-dependent data for which prediction becomes quite demanding (Leuenberger and Kanevski 2015). One advantage to machine-learning algorithms is their nonparametric capabilities – i.e. lack of dependence on specific data distributions, their flexibility in the testing data they can accept as input, and their general aptitude for modeling complex data types with multi-dimensional feature space. These data types are modeled with higher accuracy using leaning machines than when addressed with traditional parametric techniques (Maxwell et al. 2018).

5.2.2 Methods to Analyze Social Media and Location-Based Data

These big geospatial data sources include advanced sensors and devices, such as mobile (smart) phones, health monitors, air quality sensors, sound sensors, and many others connected to the Internet. Except for smart phones, many of these devices are collectively referred to as the Internet of Things. These sensors measure time, motion, temperature, noise, movement of humans and machines, vibration, temperature, humidity and other chemical and physical changes in the environment. Many sensors are permanently or semi-permanently placed geographically and form sensor networks that broadcast to the Internet. The term social sensing was applied by Liu et al. (2015) to big social data types including taxi trajectories, mobile phone records, social media and social network data, smart card records in public transportation systems, and others. Yao and Li (2018) define a class of data called big spatial vector data that includes many of these social media and location-based data types. These data are commonly represented as points, lines, and polygons (Shekhar et al. 2011). The processing and extraction of EOI from these big geospatial data sources use algorithms for hotspot and anomaly detection, often implemented in parallel and cloud computing environments (Yang et al. 2017). The primary methods for EOI extraction include data mining and knowledge discovery methods, data analytics, especially visual analytics, and machine learning.

5.2.2.1 Data Mining

Data mining is identifying novel, interesting and actionable patterns in data. Knowledge discovery is sometimes equated with data mining. Data mining techniques can be used with social media data to determine community and group detection, influence propagation, diffusion of information topic detection and monitoring, behavior analysis and market research (Barbier 2011). Data mining methods have been used with social big data to analyze social interactions, health records, phone logs, and digital traces (Boyd and Crawford 2012). The methods use clustering of similar data responses to identify hotspots and anomalies. These then become the EOI for these data, identifying conversation topics, intensity of discussion, and ultimately evaluation of subject matter opinions expressed and social and cultural results.

Classification is a common technique for social data analysis, and similar to land remote sensing data analysis, techniques for mining social data include supervised and unsupervised methods. Supervised approached depend on a-priori class labels whereas unsupervised approaches characterize data based on internal data patterns. In both cases a set of training data is prepared and then an algorithm to classify the full dataset is implemented. Clustering is a common unsupervised approach used in data mining of social data. Clustering techniques do not depend on labels in the data, but rather converge patterns of similarity. Additional methods for mining for social data include association rules, Bayesian classification (Cui et al. 2019), rule-based classifiers, support vector machines, text mining, link analysis, and multi-relational data mining (Barbier 2011).

5.2.2.2 Data Analytics

Data analytics for big geospatial data include statistical analysis methods such as simulation, classification, and common visual representations for both analysis and results. Strategies to achieve this integration include abstraction in which data are defined with a representation similar in form to the semantics (meaning), while hiding details. Aggregation is used to summarize information about a certain bounded region. Visual analytics are analytical reasoning methods with interactive visual interfaces that rely on the tight integration of visualization and analytics of big data. This visualization works especially well for geospatial data because these data a include representation of space (and time) along with the phenomena. Visual analytics seeks to determine explicit and latent relationships in data and interpret how these relationships inform analytic tasks. Goals in visual analytics are to synthesize information, detect the expected and discover the unexpected and unknown, provide assessments, and communicate the assessments effectively. Key steps in the process include data transformation, visual mapping/layout, model-based analysis and user interactions (Keim et al. 2008).

For big social and location-based data, goals of analytics are often determination of public opinion, perception, sentiment, and market analysis. The methods involve location tracking and transit analysis, as with taxi trip records, polling of public sentiment, as with political commentary using Twitter feeds or intensity and needs during hazard events and disasters. Concept-level analysis from social text uses keyword spotting, lexical affinity, and statistical methods (Cambria et al. 2013). The tracking and opinion usually come from the public at large and methods of aggregation and statistical analysis are used to refine the data into clusters of similar phenomena that can be interpreted.

5.2.2.3 Machine Learning

As with remote sensing data, high resolution social media and location–based data are being processed and analyzed using machine learning and artificial intelligence methods. These methods use a series of algorithms that enable computers to identify patterns in data and classify them in clusters. The algorithms detect objects by using high volumes of labeled training data in a learning mode. Semantic signatures are sometimes used with machine learning and geographic artificial intelligence techniques to analyze the digital trace of human activity (Janowicz et al. 2019). Digital traces are produced and collected through mobile devices, sensor fields, and IoT devices and thus are passive data. Data collected and processed may include natural language posts and text, human movement tracers, video, sound, and sensor feeds of environmental data. The volume and heterogeneity of the data make them particularly suited to analysis in geographic artificial intelligence. Specific machine learning methods for social media analytics, domain adaptation, sentiment analysis, and link prediction are provided by Hayat et al. (2019). Martin et al. (2018) provide an example of convolutional neural networks to analyze human mobility and behavior. All these methods require extensive, high-quality well-labeled training data for successful implementation.

5.3 Use Cases in the Active and Passive Big Data Spatial Realms

5.3.1 Active Use Cases

Components of the geospatial terrain obtained from digital elevation models present an important use case of EOI extraction (Fig. 5.1). This research has focused on the basic terrain elements *peaks, pits* and *passes* of surfaces (Ehsani and Quiel 2008; Takahashi 2006); slope, aspect and curvature (Csillik et al. 2015; Eisank and Drăguț 2010), physiographic units (Drăguţ and Eisank 2012; Gerçek et al. 2011), general landform segmentation and classification (Camiz and Poscolieri 2015; Graff and Usery 1993), and specific landform types such as mountains (Miliaresis and Argialas 2002; Sinha and Mark 2010) and cirques (Anders et al. 2009; Arundel 2016).

Drăguţ and Eisank (2012) used object-based image analysis to classify ~90 m resolution Shuttle Radar Topography Mission (SRTM) data, resampled to ~1 km, into topographic or terrain classes for the globe. The original 90 m dataset, assuming the globe is approximately 510 million km^2, is covered by 57 billion pixels, requiring about a half terabyte (TB) in 8-byte storage. The resulting objects represent the terrain classes along with additional descriptors at three scale levels – small, medium and large. These 8-byte vector files are 94.6 megabytes (MB), 216 mg, and 395 MB, respectively (L. Drăguț, personal communication, November 26, 2019). The source 90 m SRTM dataset is 1250 times larger than even the highest resolution object dataset, and the resampled 1 km SRTM (~4 gigabytes (GB)) is ten times larger. Thus, data reduction using EOI extraction from digital elevation models can be quite remarkable.

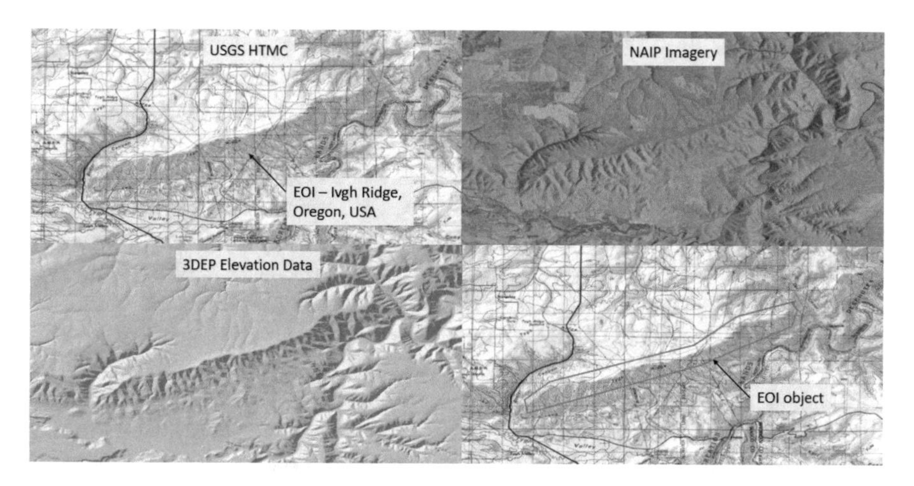

Fig. 5.1 An example of data reduction through element-of-interest extraction. Three raster datasets (USGS historical topographic map, NAIP imager and 3DEP elevation data) are reduced to one vector object

Community acceptance of the dataset was high, with 88% of experts rating the dataset as moderately to very useful. Suggested dataset applications fall mainly within geomorphology, landscape ecology, geology, soil science, ecology, hydrology and agriculture. For example, Robinne et al. (2018) estimated global wildfire-water risks based on the classification. The methodological procedures have been employed and expanded in the same fields (e.g. Manfré et al. 2015).

Vegetation mapping as a use case is a broad field that benefits greatly from EOI extraction from various datasets. Over the decades, various indices, like the Normalised Difference Vegetation Index (NDVI) and the Simple Ratio NIR/RED have been developed and deployed to reduce remote sensing imagery and other data sources to species or community level vegetation maps (Xie et al. 2008). This work includes more complex elements of interest like vegetation height, biomass, soil moisture, leaf area index and chlorophyll density/concentration (Huesca et al. 2019; Khadim et al. 2019; Koyama et al. 2019; Lang et al. 2019). Large remote sensing datasets, as well as those encompassing other physical elements like temperature, precipitation and soil moisture, can be terribly unwieldy in size, especially when analyzed together across large spatial realms.

Huang et al. (2017) implemented a procedure to extend vegetation heights from ICESat/GLAS (70 m^2 resolution) using MODIS' 250 m^2 resolution Vegetation Continuous Fields (VCF) product, the Nadir BRDF-Adjusted (Bidirectional reflectance distribution function) reflectance product (500 m^2) and Leaf-area Index (LAI) product (1 km^2); WORLDCLIM near-surface air temperature, precipitation, and seasonal fluctuation data (1 km^2), and SRTM elevations, slopes, aspects and the Compound Topographic Index (CTI) data (90 m^2) to create complete coverage of China's vegetation heights. These datasets covering China's area of ~10 million km^2 (approximately 20 billion pixels) add to over 5 TB, assuming an 8-byte storage system. The resulting 500 m^2 resolution vegetation height product is about 160 gb, which is less than 1/30th the size of the input datasets.

This ensuing product can in turn be used directly in land cover assessment (Gong et al. 2013), understanding carbon sequestration (Smith and Reid 2013), agriculture (Stanton et al. 2017), biomass estimations (Qi et al. 2019), and biodiversity studies (Adhikari et al. 2020). Future applications of data reduction through vegetation EOI extraction will include developing new strategies to integrate country and continent-wide datasets like the China vegetation heights into other large-area datasets to improve land cover classifications, global change time-series, and global and regional climate model predictions.

As lidar coverage becomes more complete at higher resolutions, and additional sensors on more platforms are launched for Earth observations, the need for methods to reduce and integrate datasets will continue to increase. Machine learning techniques will likely continue to expand their role in synthesizing these data into functional products.

5.3.2 *Passive Use Cases*

Human mobility – Social data provide a basis for human mobility research and tracking of human interactions. These data support measures of human conditions, for example a happiness index (Mitchell et al. 2013) and demographics (Li et al. 2013). Other uses include sentiment processing and representation of place semantics (Liu et al. 2015).

Building identification – With VGI from Open Street Map and natural color images from Worldview II, Yuan et al. (2018) used convolutional neural networks to identify and classify building footprints. Hecht et al. (2015) also used neural networks to identify building footprints from several big geospatial data sources including topographic databases, raster maps and digital landscape models. Jiang et al. (2015) used point-of-interest data and data mining methods to identify and classify land use at high resolution from social networks and VGI.

Event Processing – The analysis of big geospatial data for events uses data from social network and social media platforms such as Twitter, Facebook, Flicker, FourSquare, and many others. The analytical results of processing the data streams allow interpretation of social and political sentiment, food and restaurant choice and selection, and environmental awareness, among others. These data also allow prediction of event activity, success, and failure. Traffic trajectories from taxi and other public carrier records can also be used for event processing. Complex event processing (CEP), which handles many different data streams, has become a major component of the use of big social data and specific platforms for CEP such as Oracle CEP and Esper (Lee and Kang 2015).

Smart Cities – The availability of big data from the IoT has allowed the development and enhancement of the smart cities concept. Use cases for big data in smart cities include transportation with traffic monitoring and individual mobility identifying areas of congestion with alternative solutions such as rerouting; energy consumption and use in which these big data analytics provide the ability to manage individual lights or entire power grids; public health identifying areas for spread of disease, using preventive measures, and smart medical record keeping for diagnosis and patient care; governance with open citizen participation, environment, and security with advanced sensing and spotting of anomalies in the big data streams of pollution, crime and other activities. These smart city approaches are aimed at a better quality of life for urban citizens resulting from more efficient transportation, better planning of living and work spaces, and better decision support for many urban problems (Al Nuaimi et al. 2015).

5.4 Conclusion

Big spatial data reduction through element-of-interest extraction applies an established field of research to challenging new horizons. Both active and passive data collection domains supply input to EOI methods, which are processed through extremely varied means like object-based image analysis, data mining and machine learning. The applications to which this research applies are just as diverse, including vegetation mapping, terrain feature extraction, building identification, understanding human mobility, predicting events and developing Smart Cities.

Future research can be motivated by asking the question: "what spatial queries remain that need automated methods to answer them?" The answer to this question is many and varied. The greatest challenge is recreating the human reasoning mechanism that can easily answer a question like "which streams drain the North Rim of the Grand Canyon to the Colorado River?" given a topographic map. To answer this query requires the extraction of many EOIs, including the Grand Canyon, the North Rim and the Colorado River, its tributaries within the Grand Canyon, and their relations. Research to return appropriate responses to these queries should include exploration of best methods to handle extremely big data, particularly those using parallel and distributed processing and high-performance computing, and understanding of the feasibility of cloud, quantum and biological computing in EOI extraction. The development of additional and refined methods to better extract spatial EOI from big data should include all procedural tracks. Great progress can be gained by adapting existing advanced EOI techniques in fields like medicine and image recognition to the geospatial realm. There is no doubt that the discipline is perfectly posed to make rapid advances in the near future.

References

Adhikari H, Valbuena R, Pellikka PKE, Heiskanen J (2020) Mapping forest structural heterogeneity of tropical montane forest remnants from airborne laser scanning and Landsat time series. Ecol Indic 108:105739. https://doi.org/10.1016/j.ecolind.2019.105739

Al Nuaimi E, Al Neyadi H, Mohamed N, Al-Jaroodi J (2015) Applications of big data to smart cities. J Internet Serv Appl 6:1–15. https://doi.org/10.1186/s13174-015-0041-5

Anders NS, Seijmonsbergen AC, Bouten W (2009) Multi-scale and object-oriented image analysis of high-res LiDAR data for geomorphological mapping in Alpine mountains. Geomophometry 2009:61–65

Arundel ST (2016) Pairing semantics and object-based image analysis for national terrain mapping – a first-case scenario of cirques. In: Kerle N, Gerke M, Lefevre S (eds) GEOBIA 2016: solutions and synergies. University of Twente Faculty of Geo-Information and Earth Observation (ITC), Enscede

Baatz M, Schäpe A (2000) Multiresolution Segmentation – an Optimization Approach for High Quality Multi-Scale Image Segmentation. In: Strobl J, Blaschke T, Griesebner, G (eds) Proceedings of Angewandte Geographische Informationsverarbeitung XII. Wichmann, Heidlelberg, pp. 12–23

Barbier G (2011) Social network data analytics. Soc Netw Data Anal. https://doi.org/10.1007/978-1-4419-8462-3

Blaschke T (2010) Object based image analysis for remote sensing. ISPRS J Photogramm Remote Sens 65:2–16. https://doi.org/10.1016/j.isprsjprs.2009.06.004

Blaschke T, Strobl J (2001) What's wrong with pixels? Some recent developments interfacing remote sensing and GIS. Geo-Inf-Syst 14:12–17

Blaschke T, Hay GJ, Kelly M, Lang S, Hofmann P, Addink E, Queiroz Feitosa R, van der Meer F, van der Werff H, van Coillie F, Tiede D (2014) Geographic object-based image analysis – towards a new paradigm. ISPRS J Photogramm Remote Sens 87:180–191. https://doi.org/10.1016/j.isprsjprs.2013.09.014

Boyd D, Crawford K (2012) Critical questions for big data: provocations for a cultural, technological, and scholarly phenomenon. Inf Commun Soc 15:662–679. https://doi.org/10.1080/1369118X.2012.678878

Cambria E, Rajagopal D, Olsher D, Das D (2013) Big social data analysis. Big Data Comput:401–414. https://doi.org/10.1201/b16014-19

Camiz S, Poscolieri M (2015) Geomorpho: a methodology for the classification of terrain units. In: Jasiewicz J, Zwoliński Z, Mitasova H, Hengl T (eds) Geomorphometry for geosciences. Adam Mickiewicz University, Poznań, pp 149–152

Cánovas-García F, Alonso-Sarría F (2015) A local approach to optimize the scale parameter in multiresolution segmentation for multispectral imagery. Geocarto Int 30. https://doi.org/10.1080/10106049.2015.1004131

Castilla G, Hay GJ (2008) Image objects and geographic objects. In: Object-based image analysis. Springer, Berlin/Heidelberg, pp 91–110

Chen CLP, Zhang C (2014) Data-intensive applications, challenges , techniques and technologies: a survey on Big Data. Inf Sci 275:314–347

Chen G, Weng Q, Hay GJ, He Y (2018) Geographic object-based image analysis (GEOBIA): emerging trends and future opportunities. GISci Remote Sens 55:159–182. https://doi.org/10.1080/15481603.2018.1426092

Csillik O, Evans IS, Drăguţ L (2015) Transformation (normalization) of slope gradient and surface curvatures, automated for statistical analyses from DEMs. Geomorphology 232:65–77. https://doi.org/10.1016/j.geomorph.2014.12.038

Cui K, Jiang Y, Li Y, Pfoser D (2019) A vocabulary recommendation method for spatiotemporal data discovery based on Bayesian network and ontologies. Big Earth Data 3:220–231. https://doi.org/10.1080/20964471.2019.1652431

Drăguţ L, Eisank C (2012) Automated object-based classification of topography from SRTM data. Geomorphology 141–142:21–33. https://doi.org/10.1016/j.geomorph.2011.12.001

Drăguţ L, Tiede D, Levick SR (2010) ESP: a tool to estimate scale parameter for multiresolution image segmentation of remotely sensed data. Int J Geogr Inf Sci 24:859–871. https://doi.org/10.1080/13658810903174803

Ehsani AH, Quiel F (2008) Geomorphometric feature analysis using morphometric parameterization and artificial neural networks. Geomorphology 99:1–12. https://doi.org/10.1016/j.geomorph.2007.10.002

Eisank C, Drăguț LD (2010) Detecting characteristic scales of slope gradient. In: Car A, Griesebner G, Strobl J (eds) Geospatial Crossroads@GI_Forum'10. Wichman, Berlin, pp 48–57

Emani CK, Cullot N, Nicolle C (2015) ScienceDirect understandable big data: a survey. Comput Sci Rev 17:70–81. https://doi.org/10.1016/j.cosrev.2015.05.002

Gerçek D, Toprak V, Strobl J (2011) Object-based classification of landforms based on their local geometry and geomorphometric context. Int J Geogr Inf Sci 25:1011–1023. https://doi.org/10.1080/13658816.2011.558845

Gonçalves J, Pôças I, Marcos B, Mücher CA, Honrado JP (2019) SegOptim—a new R package for optimizing object-based image analyses of high-spatial resolution remotely-sensed data. Int J Appl Earth Obs Geoinf 76:218–230. https://doi.org/10.1016/j.jag.2018.11.011

Gong P, Wang J, Yu L, Zhao Y, Zhao Y, Liang L, Niu Z, Huang X, Fu H, Liu S, Li C, Li X, Fu W, Liu C, Xu Y, Wang X, Cheng Q, Hu L, Yao W, Zhang H, Zhu P, Zhao Z, Zhang H, Zheng Y, Ji L, Zhang Y, Chen H, Yan A, Guo J, Yu L, Wang L, Liu X, Shi T, Zhu M, Chen Y, Yang G, Tang P, Xu B, Giri C, Clinton N, Zhu Z, Chen J, Chen J (2013) Finer resolution observation and monitoring of global land cover: first mapping results with Landsat TM and ETM+ data. Int J Remote Sens 34:2607–2654. https://doi.org/10.1080/01431161.2012.748992

Goodchild MF (2013) The quality of big (geo) data. Dialogues Hum Geogr 3:280–284. https://doi.org/10.1177/2043820613513392

Graff LH, Usery EL (1993) Automated classification of generic terrain features in digital elevation models. Photogramm Eng Remote Sensing 59:1409–1417

Graham M, Shelton T (2013) Geography and the future of big data, big data and the future of geography. Dialogues Hum Geogr 3:255–261

Hayat MK, Daud A, Alshdadi AA, Banjar A, Abbasi RA, Bao Y, Dawood H (2019) Towards deep learning prospects: insights for social media analytics. IEEE Access 7:36958–36979. https://doi.org/10.1109/ACCESS.2019.2905101

Hecht R, Meinel G, Buchroithner M (2015) Automatic identification of building types based on topographic databases – a comparison of different data sources. Int J Cartogr 1:18–31. https://doi.org/10.1080/23729333.2015.1055644

Huang H, Liu C, Wang X, Biging GS, Chen Y, Yang J, Gong P (2017) Mapping vegetation heights in China using slope correction ICESat data, SRTM, MODIS-derived and climate data. ISPRS J Photogramm Remote Sens 129:189–199. https://doi.org/10.1016/j.isprsjprs.2017.04.020

Huesca M, Riaño D, Ustin SL (2019) Spectral mapping methods applied to LiDAR data: application to fuel type mapping. Int J Appl Earth Obs Geoinf 74:159–168. https://doi.org/10.1016/j.jag.2018.08.020

Janowicz K, Gao S, Mckenzie G, Hu Y (2019) GeoAI: spatially explicit artificial intelligence techniques for geographic knowledge discovery and beyond. Int J Geogr Inf Sci 00:1–12. https://doi.org/10.1080/13658816.2019.1684500

Jensen JR (2015) Introductory digital image processing: a remote sensing perspective, 4th edn. Prentice Hall Press, Upper Saddle River

Jiang S, Alves A, Rodrigues F, Ferreira J, Pereira FC (2015) Mining point-of-interest data from social networks for urban land use classification and disaggregation. Comput Environ Urban Syst 53:36–46. https://doi.org/10.1016/j.compenvurbsys.2014.12.001

Kanevski M, Maignan M (2004) Analysis and modelling of spatial environmental data. EPFL Press, Lausanne

Kanevski M, Pozdnukhov A, Timonin V (2008) Machine learning algorithms for geospatial data. Appl Soft Tools. Iemss.org 320–327

Keim D, Andrienko G, Fekete JD, Görg C, Kohlhammer J, Melançon G (2008) Visual analytics: definition, process, and challenges. LectNotes ComputSci(including Subser Lect Notes Artif Intell Lect Notes Bioinformatics) 4950 LNCS, 154–175. https://doi.org/10.1007/978-3-540-70956-5_7

Khadim FK, Su H, Xu L, Tian J (2019) Soil salinity mapping in Everglades National Park using remote sensing techniques and vegetation salt tolerance. Phys Chem Earth 110:31–50. https://doi.org/10.1016/j.pce.2019.01.004

Killeen P, Ding B, Kiringa I, Yeap T (2019) IoT-based predictive maintenance for fleet management. 2nd international conference on emerging data and industry 4.0 (EDI40) Leuven, Belgium

Kim M, Madden M, Warner T (2008) Estimation of optimal image object size for the segmentation of forest stands with multispectral IKONOS imagery. Object-Based Image Anal. 291–307. https://doi.org/10.1007/978-3-540-77058-9_16

Kim M, Madden M, Xu B (2010) GEOBIA vegetation mapping in Great Smoky Mountains National Park with spectral and non-spectral ancillary information. Photogramm Eng Remote Sensing 76:137–149. https://doi.org/10.14358/PERS.76.2.137

Kitchin R (2013) Big data and human geography: opportunities, challenges and risks. Dialogues Hum Geogr 3:262–267. https://doi.org/10.1177/2043820613513388

Koyama CN, Watanabe M, Hayashi M, Ogawa T, Shimada M (2019) Mapping the spatial-temporal variability of tropical forests by ALOS-2 L-band SAR big data analysis. Remote Sens Environ 233:111372. https://doi.org/10.1016/j.rse.2019.111372

Lang N, Schindler K, Wegner JD (2019) Country-wide high-resolution vegetation height mapping with Sentinel-2. Remote Sens Environ 233:111347. https://doi.org/10.1016/j.rse.2019.111347

Lecun Y, Bengio Y, Hinton G (2015) Deep learning. Nature 521:436–444. https://doi.org/10.1038/nature14539

Lee J, Kang M (2015) Geospatial Big Data: challenges and opportunities ⋆. Big Data Res 2:74–81. https://doi.org/10.1016/j.bdr.2015.01.003

Lekamalage L, Kasun C, Zhou H, Huang G, Vong CM (2013) Representational learning with extreme learning machine for big data. IEEE Intell Syst 28:31–34

Leuenberger M, Kanevski M (2015) Extreme learning machines for spatial environmental data. Comput Geosci 85:64–73. https://doi.org/10.1016/j.cageo.2015.06.020

Li W, Goodchild MF, Church R (2013) An efficient measure of compactness for two-dimensional shapes and its application in regionalization problems. Int J Geogr Inf Sci 27:1227–1250. https://doi.org/10.1080/13658816.2012.752093

Li S, Dragicevic S, Castro FA, Sester M, Winter S, Coltekin A, Pettit C, Jiang B, Haworth J, Stein A, Cheng T (2016) Geospatial big data handling theory and methods: a review and research challenges. ISPRS J Photogramm Remote Sens 115:119–133. https://doi.org/10.1016/j.isprsjprs.2015.10.012

Lillesand T, Kiefer RW, Chipman J (2015) Remote Sensing and Image Interpretation, Seventh Edition. John Wiley & Sons, Inc, Hoboken pp 736

Liu Y, Bian L, Meng Y, Wang H, Zhang S, Yang Y, Shao X, Wang B (2012) Discrepancy measures for selecting optimal combination of parameter values in object-based image analysis. ISPRS J Photogramm Remote Sens 68:144–156. https://doi.org/10.1016/j.isprsjprs.2012.01.007

Liu Y, Liu X, Gao S, Gong L, Kang C, Zhi Y, Chi G, Shi L (2015) Social sensing: a new approach to understanding our socioeconomic environments. Ann Assoc Am Geogr 105:512–530. https://doi.org/10.1080/00045608.2015.1018773

Manfré LA, de Albuquerque Nóbrega RA, Quintanilha JA (2015) Regional and local topography subdivision and landform mapping using SRTM-derived data: a case study in southeastern Brazil. Environ Earth Sci 73:6457–6475. https://doi.org/10.1007/s12665-014-3869-2

Martin H, Bucher D, Suel E, Zhao, Perez-Cruz F, Raubal M, Martin H, Bucher D, Suel E, Zhao P, Perez-Cruz F (2018) Graph Convolutional Neural Networks for Human Activity Purpose Imputation from GPS-based Trajectory Data 1–6. https://doi.org/10.3929/ethz-b-000310251

Maxwell AE, Warner TA, Fang F (2018) Implementation of machine-learning classification in remote sensing: an applied review. Int J Remote Sens. https://doi.org/10.1080/01431161.2018.1433343

Michel J, Youssefi D, Grizonnet M (2015) Stable mean-shift algorithm and its application to the segmentation of arbitrarily large remote sensing images. IEEE Trans Geosci Remote Sens 53:952–964. https://doi.org/10.1109/TGRS.2014.2330857

Miliaresis GC, Argialas DP (2002) Quantitative representation of mountain objects extracted from the global digital elevation model (GTOPO30). Int J Remote Sens 23:949–964. https://doi.org/10.1080/01431160110070690

Miller HJ, Goodchild MF (2015) Data-driven geography. GeoJournal 80:449–461. https://doi.org/10.1007/s10708-014-9602-6

Ming D, Li J, Wang J, Zhang M (2015) Scale parameter selection by spatial statistics for GeOBIA: using mean-shift based multi-scale segmentation as an example. ISPRS J Photogramm Remote Sens 106. https://doi.org/10.1016/j.isprsjprs.2015.04.010

Mitchell L, Frank MR, Harris KD, Dodds PS, Danforth CM (2013) The geography of happiness: connecting Twitter sentiment and expression, demographics, and objective characteristics of place. PLoS One 8. https://doi.org/10.1371/journal.pone.0064417

Najafabadi MM, Villanustre F, Khoshgoftaar TM, Seliya N, Wald R, Muharemagic E (2015) Deep learning applications and challenges in big data analytics. J Big Data 2:1. https://doi.org/10.1186/s40537-014-0007-7

Qi W, Saarela S, Armston J, Ståhl G, Dubayah R (2019) Forest biomass estimation over three distinct forest types using TanDEM-X InSAR data and simulated GEDI lidar data. Remote Sens Environ 232:111283. https://doi.org/10.1016/j.rse.2019.111283

Robinne FN, Bladon KD, Miller C, Parisien MA, Mathieu J, Flannigan MD (2018) A spatial evaluation of global wildfire-water risks to human and natural systems. Sci Total Environ 610–611:1193–1206. https://doi.org/10.1016/j.scitotenv.2017.08.112

Scrucca L (2013) GA: a package for genetic algorithms in R. J Stat Softw 53:1–37. https://doi.org/10.18637/jss.v053.i04

Shekhar S, Evans MR, Kang JM, Mohan P (2011) Identifying patterns in spatial information: a survey of methods. Wiley Interdiscip Rev Data Min Knowl Discov 1:193–214. https://doi.org/10.1002/widm.25

Shekhar S, Gunturi V, Evans MR, Yang K (2012) Spatial Big-data challenges intersecting mobility and cloud computing. In: Proceedings of the eleventh ACM international workshop on data engineering for wireless and mobile access, MobiDE'12. ACM, New York, pp 1–6. https://doi.org/10.1145/2258056.2258058

Sinha G, Mark DM (2010) Cognition-based extraction and modelling of topographic eminences. Cartogr Int J Geogr Inf Geovisualization 45:105–112. https://doi.org/10.3138/carto.45.2.105

Smith R, Reid N (2013) Carbon storage value of native vegetation on a subhumid-semi-arid floodplain. Crop Pasture Sci 64:1209–1216. https://doi.org/10.1071/CP13075

Stanton C, Starek MJ, Elliott N, Brewer M, Maeda MM, Chu T (2017) Unmanned aircraft system-derived crop height and normalized difference vegetation index metrics for sorghum yield and aphid stress assessment. J Appl Remote Sens 11:026035. https://doi.org/10.1117/1.jrs.11.026035

Takahashi S (2006) Algorithms for extracting surface topology from digital elevation models. Topol Data Struct Surfaces Introd Geogr Inf Sci 31–51. https://doi.org/10.1002/0470020288.ch3

Xie Y, Sha Z, Yu M (2008) Remote sensing imagery in vegetation mapping: a review. J Plant Ecol 1:9–23. https://doi.org/10.1093/jpe/rtm005

Yang C, Huang Q, Li Z, Liu K, Hu F (2017) Big Data and cloud computing: innovation opportunities and challenges. Int J Digit Earth 10:13–53. https://doi.org/10.1080/17538947.2016.1239771

Yao X, Li G (2018) Big spatial vector data management: a review. Big Earth Data 2:108–129. https://doi.org/10.1080/20964471.2018.1432115

Yuan J, Roy Chowdhury PK, McKee J, Yang HL, Weaver J, Bhaduri B (2018) Exploiting deep learning and volunteered geographic information for mapping buildings in Kano, Nigeria. Sci Data 5:1–8. https://doi.org/10.1038/sdata.2018.217

Chapter 6
Semantic Graphs to Reflect the Evolution of Geographic Divisions

C. Bernard, C. Plumejeaud-Perreau, M. Villanova-Oliver, J. Gensel, and H. Dao

6.1 Introduction

All around the world, directives or laws are enacted to open up data to citizens. For instance, the *Open Data Directive*[1] in Europe sets up a legal framework to make public sector information widely accessible and reusable. Indeed, according to the European Commission, allowing public sector data to be re-used should foster the participation of citizens in political and social life, and increase the transparency of public policies. Publishing data on the Web is one way to achieve this open data movement. Thus, public institutions in the world are facing the challenge of publishing data on the Open Data Web, on behalf of governments or other political organizations. As a consequence, the volume of data coming from the public sector is growing rapidly and the Web has been subject to a series of transformations over the last years, moving from the *Documents Web* (as Web pages interlinked by hyperlinks) to the *Data Web* also called the *Linked (Open) Data (LD or LOD) Web*. Today, almost every political authority (State, region, etc.) in the world has an

[1]https://ec.europa.eu/digital-single-market/en/open-data

C. Bernard (✉) · M. Villanova-Oliver · J. Gensel
Univ. Grenoble Alpes, CNRS, Grenoble INP, LIG, Grenoble, France
e-mail: camille.bernard@univ-grenoble-alpes.fr;
marlene.villanova-oliver@univ-grenoble-alpes.fr; Jerome.gensel@univ-grenoble-alpes.fr

C. Plumejeaud-Perreau
LIttoral ENvironnement et Sociétés (LIENSs) - UMR, La Rochelle, France
e-mail: christine.plumejeaud-perreau@univ-lr.fr

H. Dao
Department of Geography and Environment Geneva, University of Geneva, Geneva, Switzerland
e-mail: hy.dao@unige.ch

M. Werner, Y.-Y. Chiang (eds.), *Handbook of Big Geospatial Data*,
https://doi.org/10.1007/978-3-030-55462-0_6

Open Data Portal in order to centralize data and make it accessible to citizens (e.g., the opendata.swiss portal,[2] the U.S. Government's open data portal,[3] the European Union Data Portal,[4] etc.).

Major actors in this process are (National) Statistical Agencies ((N)SAs) and Mapping Agencies which create and disseminate official statistics and geographic information (such as the administrative or electoral boundaries) in order to monitor their jurisdiction and sub-jurisdictions. Official statistics measure diverse socio-economic or natural phenomena that occur and evolve on these jurisdictions (e.g., demography). The expression *geo-coded statistics* or *territorial statistics* is used to designate such statistics, meaning that data refer to a territorial reference system, using alphanumerical identifiers of *geographic areas* (also called geographic units or statistical areas) (Eurostat 2001). All these geographic areas are organized into what is called a *Territorial Statistical Nomenclature* (TSN), built by (N)SAs in order to observe a territory at several nested geographic division levels (e.g., regions, districts, sub-districts levels). These artifact TSNs are built for statistical purpose, although they usually correspond to an electoral or administrative structure. Numerous TSNs exist throughout the world. For instance, the one from Eurostat,[5] called the *Nomenclature of Territorial Units for Statistics* (NUTS),[6] provides four nested subdivisions levels of the European Union (EU) territory, for the collection of EU regional statistics (State members, major Regions, basic Regions, and small Regions levels).

Territorial statistics are of utmost importance for policy-makers to conduct various analyses upon their jurisdiction, through time and space. For instance, using data available at two or more periods in time, they can observe the evolution of the unemployment rate in a given administrative region. These observations and the analysis of these observations are essential prerequisites for political decision-making (spatial planning, public health policies and preventive measures, etc.). Thus, there is a strong demand from governments, organizations and researchers regarding time-series of these territorial statistical data. One unexpected underlying problem is that past territorial data cannot be compared to more recent data if the geographic areas observed have changed in the meantime i.e., data collected in different versions of a TSN are not directly comparable because the observed geographic areas are potentially not the same areas anymore. Territorial changes are very frequent in Europe (for instance in France, in 2016, administrative regions have been merged into greater regions) or in the U.S.A. through a well-known process called *gerrymandering* or, more broadly *redistricting*. Territorial changes lead to broken time-series and are source of both misinterpretations, and statistical

---

[2]https://opendata.swiss/

[3]https://www.data.gov/

[4]https://data.europa.eu/euodp/en/home?

[5]The European Statistical Office that provides official statistics to the NSAs of the European Union member states.

[6]http://ec.europa.eu/eurostat/web/nuts/overview

biases when not properly documented. For instance, the identifiers of the geographic areas in the nomenclature do not necessarily change after a territorial redistribution, even if the areas have changed their name or merged with a neighbor. Then, the main drawback of using such identifiers for areas lies in their lack of consistency through time and space, as they may designate a region that has changed over time in its boundaries or/and name. To address this problem, statistical services often transfer former statistical data into the latest version of the TSN. Hence, statistical data sets do not contain traces of territorial changes. However, this non-evolving view hampers a good understanding of the territory life itself. Indeed, changes of the areas are not meaningless because they are decided or/and voted by an authority pursuing some objective. Thereby, solutions for representing different versions of the geographic divisions, and their evolution are to be proposed in order to enhance the understanding of territorial dynamics, providing statisticians, researchers, citizens with descriptions to comprehend the motivations and the impact of changes on territorial data (on electoral results for instance). In fact, providing an explicit representation of territorial changes through times is a prerequisite to a reliable analysis of time-series of statistical data. Therefore, it is crucial to keep and enrich such information about territorial changes with metadata and other resources available on the Web that may contribute to explain the changes (e.g., societal reasons, historical events). The fundamental question we address in this chapter is *how to provide statisticians, researchers or citizens with semantic descriptions of changes and lifelines of evolving geographic areas in order to comprehend the evolution of the territories over time?*

Our solution is to adopt Semantic Web technologies and Linked Open Data (LOD) (in order to comply with Open Data directives among others) representation for the description of the geographic divisions, and of their evolution over time. This guaranties, among others, the syntactic and semantic interoperability between systems exchanging statistical and geometric datasets about geographic divisions. We propose a solution to handle the whole TSN data life cycle on the LOD Web: from the modeling of geographic areas and of their changes, to the automatic detection of the changes then exploitation of these descriptions on the LOD Web, using SPARQL requests. Our system embeds two ontologies, TSN Ontology and TSN-Change Ontology, we have designed for an unambiguous description, in time and space, of the geographic structures and of their changes over time. The knowledge graphs we generate improve the understanding of territorial dynamics, providing policy-makers, technicians, researchers and general public with fine-grained semantic descriptions of territorial changes to conduct various accurate and traceable analyses.

In this chapter, we first introduce the current state of territorial statistics and geographic divisions on the Open Data Web, and issues behind approaches that erase traces of territorial changes over time. In Sect. 6.3, we investigate solutions for changes representation using the Semantic Web technologies. In Sect. 6.4, we present spatiotemporal solutions for the description of evolution of geographic areas over time. We focus on solutions offered in the Semantic Web. In Sect. 6.5, we present our solution that provides researchers, citizens with Linked Open geospatial representation of the areas evolution over time. In Sect. 6.6, we conclude by considering other data whose evolution could be describe in order to further comprehend territorial dynamics over time.

6.2 Context

In this section, we describe in more detail the current states of territorial statistical information on the Web. We focus on three main issues regarding this information composed of both statistics and geographic data: (1) statistical and geographic data are not fully interconnected on the Web, (2) due to change in the territorial units over time, time-series of socio-economic data are often broken, (3) then statistical agencies use to estimate data in the latest partition versions and then, this approach erase traces of territorial changes.

6.2.1 Not Fully Interconnected Data

Although socio-economic or environmental, etc. statistics (produced by NSAs) and geographic information (produced by National Mapping Agencies) have a strong connection because statistics measure some observed phenomenon that occurs on a territory, most of the time, on the Open Data Web they are available in separated data sets. They are generally available in different formats, using different Web Services (SDMX REST Web services for the statistics,[7] and OGC Web services (WFS,[8] WMS,[9] CSW[10]...) for the geographic information). And, even when they are available in the same format (CSV or XML for instance), most of the time the two types of information are described using different data models that are not interconnected. The only link that is made between these siloed data sets relies on the identifier (a numeric or alphanumeric identifier) of the geographic area that the statistical observation refers to. And, as explained in Sect. 6.1, the main drawback of using such identifiers for areas lies in their lack of accuracy in time and space, as they may designate a region that has changed over time in its boundaries or/and name. Even if sometimes the units are recodified after a major change, the fact remains that in both cases (recoding or not), the statistical series referring to the identifiers are broken, as explained in Sect. 6.2.2.

[7]For instance, the REST SDMX API of Eurostat gives access to the Eurostat data (https://ec.europa.eu/eurostat/web/sdmx-web-services/rest-sdmx-2.1

[8]https://www.opengeospatial.org/standards/wfs

[9]https://www.opengeospatial.org/standards/wms

[10]https://www.opengeospatial.org/standards/cat

6.2.2 Broken Time-Series

The boundaries of geographic areas defined by humans evolved over time, because of reforms, electoral concerns, alliances or conflicts between human groups. This leads to broken statistical series since past data about a given statistical indicator (e.g., unemployed number, income distribution, life expectancy) can no more be compared to more recent data if the geographic areas observed have changed. This problem, well known as the *Change of Support Problem* (COSP) (Openshaw and Taylor 1979; Gotway Crawford and Young 2005; Howenstine 1993), describes the phenomenon where data collected in different TSNs or versions of a TSN are not comparable due to potential differences between the geographic areas (for instance, after the split or merge of two regions) used as supports for the collected data. Consequently, territorial changes are a clear obstacle to the comparability of socio-economic data over time, as this is only possible if data are estimated in the same geographical divisions. Changes of geographic areas are, most of the time, more complex than a split or merge of areas, and very numerous the more we go down the territorial hierarchy (changes in the municipalities for instance).

6.2.3 Removal of Changes

As seen above, territorial statistics are often simply not comparable over time due to potential differences between the geographic areas at two periods of time. To cope with this problem, methods such as aggregation, disaggregation (e.g., Simple Areal Weighted), and areal interpolation (e.g., Inverse Distance Weighting, Kriging method) can be used to transfer data into another TSN or version of a TSN or to abolish the boundaries between areas by using spatial smoothing (Flowerdew 1991; Wang 2014; Anderson et al. 2012; Plumejeaud et al. 2010). However, these methods mitigate traces of changes, and thus do not help to understand the various governance and planning choices behind these territorial changes (Plumejeaud et al. 2011). It is crucial to keep and enrich such information about territorial changes for several reasons: (1) it avoids errors in the analysis of statistics when the areas keep their identifier while their boundaries have changed; (2) the nature of the change the areas undergone (e.g., split, merge, redistribution of the areas) helps in estimating data in a new geographic division. Then, it helps in constructing long time-series and analyzing the evolution of a territory, using the latest boundaries; (3) conversely estimating data, such as electoral results, in former electoral areas helps in analyzing, for instance, electoral votes as if there was no redistricting and then observe the influence of the new areas on results; and finally, (4) knowing which area has changed, because of some event or law helps to understand which and why changes occurred. For instance, the fusion of the French regions in 2016 was acted by a law that aims at reducing the number of administrative levels for cost saving purposes.

In the next section, we investigate solutions offered by the Open Data Web to the three issues identified (not fully interconnected data, broken time-series, removal of changes). In particular, we investigate solution relying on Semantic Web technologies, semantic data being a particular form of Open Data.

6.3 Towards a Change in Representation with the Semantic Web

6.3.1 Open Data

Considering the Open Data challenge the (N)SAs have to meet, it should be first noted that there are different degrees of data openness, depending on the data format chosen by institutions. This data format determines what can be made with data available on the Web and how they can be linked with other resources on the Web.

Tim Berners-Lee, the inventor of the Web and Linked Data initiator, suggested a deployment scheme for Open Data (Fig. 6.1) that starts from, at least, the publication of data as PDF data on the Web, then as structured data (e.g., within an Excel file instead of an image scan of a table, so that a computer program may extract the value of each cell). The next step is to make data available in a non-proprietary format (e.g., comma-separated values instead of Excel). Data reach 4 stars when each of them (e.g., each cell of a table) may be identified uniquely using a *Unique Resource Identifier* (URI), so that people can point at data using their URI. The Open data process ends when linking each URI with each other i.e., each cell data being identified uniquely by a URI is connected to other data on the Web using a *link* (also identified by a URI). Indeed, the more data are linked, the more users can discover new facts, by going from one node in the distributed Web graph database to another. Such a link between two resources on the Web is called a *triple* (subject-predicate-object) (Fig. 6.2). Most of the time, the *Resource Description Framework* (RDF) standard syntax is used to write these triples, called RDF triples. For instance, from the data resource "Helsinki", identified by the URI http://dbpedia.org/resource/

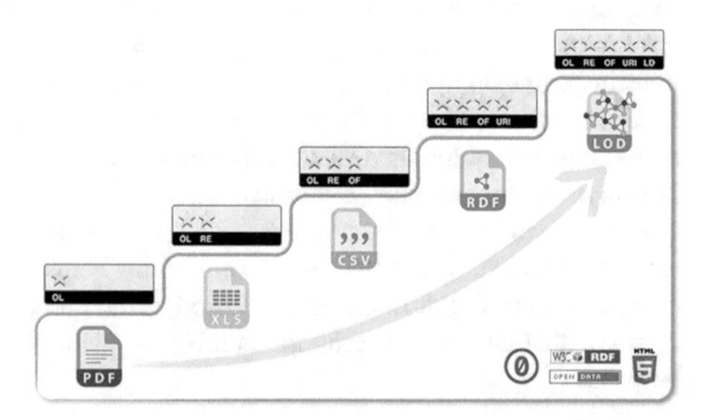

Fig. 6.1 Tim Berners-Lee, the inventor of the Web and Linked Data initiator, suggested a 5-star deployment scheme for Open Data (Hausenblas 2012)

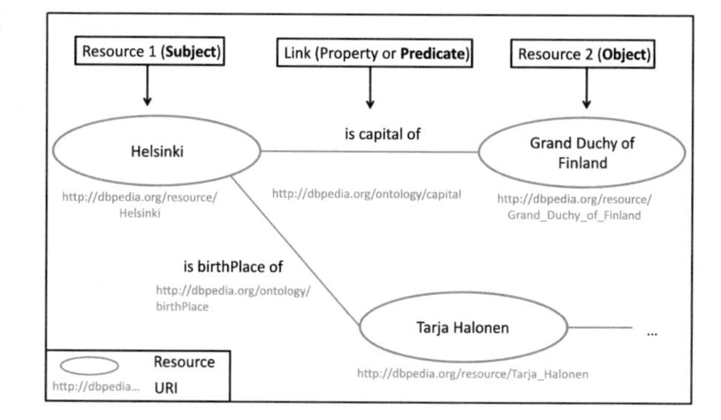

Fig. 6.2 Example of Linked Resources on the Web using URI and the RDF syntax for their identification and representation

Helsinki on the Linked Open Data (LOD) Web, users discover famous people born in Helsinki by visiting linked nodes.

Hence, nowadays for the institutions the challenge is to publish data on the LOD Web, which implies a significant change from traditional data infrastructures to approaches based on new Web standard formats and tools dedicated to the sharing of Linked Data (Tandy et al. 2017).

6.3.2 Linked Data

For the SAs, the stakes behind adopting the LOD approach and technologies are crucial. Indeed, statistical indicators linked to historical, environmental, etc. information or linked with each other (in time and space) allow multi-criteria analyzes of the territories. As a matter of fact, analysts may explore correlations, causalities and understand the territory under analysis. So far, SAs share their statistics at data set level, meaning that users have to download the whole data set (most of the time in a tabular file that list all the values for each geographic areas of the TSN), even if they are interested in only one indicator or one value of the file to perform their analysis. Similarly, to address a specific problem, analysts have to download multiple isolated data sets (containing one or several indicator(s)), each of them resolving only a part of their problem. SAs are thus far from disseminating atomic data that can be automatically processed, reused and from which new facts can be inferred.

Using the LOD paradigm, each statistical data set available on the Web is identified by a URI, each of them representing a resource, and each indicator, each data and metadata composing the data set is identified by a URI as well: everything being a resource node in the distributed Web graph database. Thus, data sets as LOD are no longer isolated and monolithic instances, they are immersed into the Web as graph(s). They are all interconnected at a finer granularity, and so the indicators and the statistical values can be as well. Each statistical value is a resource on the Web, linked to the geographic area where the observation was made. Since all values and areas are addressable and queryable, users can search for data concerning a particular region, and cross data from multiple providers.

6.3.3 Semantic Data

LOD technologies foster syntactic interoperability, and, most of all, semantic interoperability between systems. Indeed, for each data being published on the LOD Web, it is necessary to define to which set of things – real-world objects, events, situations or abstract notion – it belongs, by linking it to a concept defining this set. As part of LOD, *ontologies* are documents that formally define these concepts and their relations (Berners-Lee et al. 2001). They help both people and machines to communicate, supporting the share of semantics and not only of syntax (Alevxander and Staab 2001). Thus, on the LOD Web, data but also data semantics is explicitly defined and shared through *Ontologies* or *Vocabularies*. Hence, using RDF triples, the syntax of data is homogenized, data can be transferred from one system to another (as explained in Sect. 6.3.1), and because of the explicit semantics, systems "understand" data they receive and can determine the appropriate process to be applied to data (e.g., data tagged as geospatial data may be shown on a map). Thus, the term Semantic Web is also used to denote on LOD data sets, ontologies and technologies related to them.

More recently the term *knowledge graph* emerges to designate LOD graph (containing both data sets and ontologies) because data in the graph bear formal semantics, which can be used to interpret data and infer new facts (Ontotext 2018). A knowledge graph can be seen as a network of several data sets and ontologies which are relevant to a specific domain, and upon which a reasoner can operate in order to infer new knowledge (Ehrlinger and Wöß 2016). Even if the term is borrowed from a commercial data graph, the *Google Knowledge Graph*, it is now used to denote also on openly available graphs, such as DBpedia, YAGO, and Freebase (Paulheim 2017).

Thus, there are many benefits for (N)SAs in using Semantic Web technologies for their statistics or geospatial data:

- users may navigate from one data set to another or even from one finer piece of information to another;
- data and metadata are interlinked;
- systems "understand" data they receive and can determine the appropriate process to apply to data;
- data are put in context, as they are linked to other resources on the LOD Web;
- each statistical indicator and value become addressable;
- using a network of ontologies and data sets, a knowledge graph representing the territorial dynamics over time may be constructed, combining several statistical data sets from various disciplines (e.g., environment, socio-economy, ethology, transports), at different instants in time, observed in different geographic divi-

sions and versions of these divisions. Also, historical events or other contextual information could be mobilized in order to explain the changes that occur over time, such as the changes in boundaries.

From this knowledge graph, one can build intelligent tools for the restitution of these very disparate data which taken as a whole may help in understanding the complexity of dynamic phenomena. These tools might also be able to infer new data (such as the estimation of unemployment values after a redistricting), using information on the nature of the indicator and on the nature of the territorial change for accurate estimation.

6.3.4 Linked Open Geospatial Data

Spatial data on the Web are impacted by the transformation of the Web (from the Web of documents to the Web of data) led by projects such as *LinkedGeoData* (Auer et al. 2009) and groups, such as *Spatial Data on the Web Working Group*[11] (a group that gathers members from the World Wide Web Consortium (W3C) and the Open Geospatial Consortium (OGC)). Software tools, such as the *Geotriples Software*[12] emerge. Using this software, one can transform geospatial files (here *shapefile*) into RDF triples. However, Semantic Web technologies have not yet being adopted by every one because they imply, as noticed by the *Spatial Data on the Web Working Group*,[13] "*a significant change of emphasis from traditional Spatial Data Infrastructures (SDI)*" (Tandy et al. 2017).

Regarding geographic divisions, even if some agencies published their areas as LOD (i.e., using the RDF syntax), in most cases these geospatial data are only available online as ESRI ®shapefiles i.e., a format for geospatial data (the specification of the format are open). If we refer to the Tim Berner-Lee deployment scheme for Open Data (see Fig. 6.1), geographic divisions deployment, being available as geospatial vector files, are at level 3. Obviously, they are not yet available as Linked (Open) Data.

In the next section, we investigate the existing solutions through the Semantic Web to describe the evolution of resources over time, including geospatial ones.

[11]https://www.w3.org/2015/spatial/wiki/Main_Page

[12]http://geotriples.di.uoa.gr/

[13]https://www.w3.org/2015/spatial/wiki/Main_Page

6.4 Modeling Geospatial Changes in the Semantic Web

6.4.1 Identity and Changes

As explained in the previous sections, the descriptions of territorial changes over time are critical for the interpretation of statistical data produced over several versions of a geographic area. The description of the world dynamics should help in understanding the events behind changes and the motivations for partitioning a territory. In Beller (1991) and Claramunt and Thériault (1995), it is stated that reproducing the dynamics of spatiotemporal processes consists in describing changes and relationships between entities, considering the events behind changes and the facts which enable the observation of these changes. On these fundamentals, authors in Del Mondo et al. (2013) notice that modeling these dynamics of the world requires modeling the entities themselves but must of all, the *spatial*, *spatiotemporal* and *filiation relations* between them (e.g., topological relation, topological relation considered in time, ancestor/descendant relation). Authors in Del Mondo et al. (2010) define *Spatiotemporal entities* as abstractions of the real world that have a fixed identity and a type (for instance counties, cities, lakes are types of entities). Spatiotemporal entities can also have time-dependent properties, such as their geometry that can vary in different time instants. The expressions *fixed identity* and *a property that can vary* bring to attention the fact that even if an object change, its identity may endure over time.

The term *identity* requires further definition, since in the context of evolving object, one may wonder, as noticed in Harbelot et al. (2013), *"How far can an entity vary before losing its identity?"*. In philosophy, this issue is often illustrated by *The Ship of Theseus* Greek legend: the Ship of Theseus was rebuilt entirely, over centuries, as every plank was broken one by one. Is the ship still the Ship of Theseus after all planks being replaced? In Fearon (1999), a philosophical sense of identity is provided: "*the identity of a thing (not just a person) consists of those properties or qualities in virtue of which it is that thing. That is, if you changed these properties or qualities, it would cease to be that thing and be something different.*"

This issue of entities that change and remain or not the same entities is a problem that the database community faced many years ago. However, ontology development is a more decentralized process than database schema development since a change in one ontology may affect the other ontologies that use it. In Noy and Klein (2004), it is stated further that in the context of ontologies, one has to distinguish between "changes in the domain or changes in the real world" and "changes in conceptualization". Attention should be paid to the way the LOD technologies deal with changes over time. The RDF Specification (Cyganiak et al. 2014) states that the RDF data model is atemporal since RDF graphs are static snapshots of information. However, RDF graphs can express information about events and temporal aspects, given appropriate vocabulary terms (i.e., ontological concepts). Whenever possible, it is recommended to use existing vocabularies (such as OWL-Time[14]) to achieve semantic interoperability. Another concern addressed by the W3C (Cyganiak et al. 2014) is the possibility to express different states of a resource over time through different RDF graphs enclosed in one resource. However, authors in Stefanidis et al. (2014) warn about possible duplication of data from one version to another. Then, this approach can rapidly increase the space memory requirements. An alternative solution is to store only one version (e.g., the first one) and the differences between two consecutive versions (called *deltas* in software engineering).

Another field to explore is precisely the field of software engineering and the task called *Software Configuration Management (SCM)* for managing the evolution of large software systems (Tichy 1988). Within this field, a *version* is defined as a state of an evolving item and different types of version are identified (Conradi and Westfechtel 1998) such as: *revisions* that are versions intended to supersede their predecessor; *variants* that are versions intended to coexist. Another core concept of SCM is the *version model*. A *version model* defines the items to be versioned and the *delta*, which is the difference between two consecutive versions (Conradi and Westfechtel 1998).

To summarize, a model that reflects territorial dynamics should address a set of prerequisites: (1) the modeling of spatiotemporal entities, (2) the representation of their relations over time (2.1) and space (2.2) and in filiation (2.3), (3) a definition of what makes the identity of these entities over time, (4) the description of their differences (the *delta*) and changes over time, and (5) the consideration of events behind changes.

6.4.2 Modeling Changes

In this subsection, we focus on Semantic Web models and standards dedicated to the representation of territorial dynamics.

6.4.2.1 Standard Space and Time Ontologies

Nowadays, any spatiotemporal modeling of objects on the LOD Web necessarily implies two standards and fundamentals ontologies that are the *GeoSPARQL* and *OWL-Time* ontologies.

GeoSPARQL is an OGC standard that supports two main actions: representing and querying geospatial data on the Semantic Web. It defines an ontology for representing spatial data in RDF, and an extension to the SPARQL query language for semantic data, for querying spatial data.[15] The GeoSPARQL namespace is

[14]https://www.w3.org/TR/owl-time/

[15]http://www.opengeospatial.org/standards/geosparql

available from the URI http://www.opengis.net/ont/geosparql# (the nickname of this URI is the prefix `geo`). It provides access to the GeoSPARQL ontological concepts. The GeoSPARQL ontology is made of three main classes (Perry and Herring 2012): `geo:SpatialObject` defined as the super class of every feature or geometry that can have a spatial representation; `geo:Feature` defined as the super class of every feature; `geo:Geometry` defined as the super class of every geometry. GeoSPARQL defines topological relations between `geo:SpatialObject`. For instance, the relation `geo:sfTouches` "exists if the subject SpatialObject spatially touches the object SpatialObject".[16] Therefore, using the OGC GeoSPARQL concepts one may describe the topological relations between entities, fulfilling the prerequisite 2.2 regarding the modeling of the territorial dynamics.

The *Time Ontology in OWL* (also called *OWL-Time*) is a W3C Recommendation (Cox and Little 2017) defined as an ontology that "*provides a vocabulary for expressing facts about topological (ordering) relations among instants and intervals, together with information about durations, and about temporal position including date-time information.*". It supports the set of interval relations defined by Allen (1983). Therefore, using the W3C OWL-Time ontology one may represent the temporal relation between entities, another prerequisite 2.1 to the modeling of the territorial dynamics i.e., being able to temporally characterize geographic subdivisions, both in an absolute (through dates) and a relative (using order relation) ways.

Using both the OWL-Time and GeoSPARQL Ontologies, one may describe the boundaries of areas at a specific time interval (prerequisite 1) and the topological relations as well as the topological relation considered in time between spatial objects (prerequisites 2.1 and 2.2). We will see in the next sub-section how to model the filiation relations between entities (e.g., ancestor/descendant relation) (prerequisite 2.3), which sometimes raises the issues of the identity of entities (prerequisite 3) that change over time (prerequisite 4).

6.4.2.2 Fundamentals for the Modeling of Evolving Geospatial Entities

In Khoshafian and Copeland (1986), the authors examine the evolution of objects in the context of database systems. They identify several possible states of an object that evolves over time, over periods of existence and non-existence that are: (a) non-existence without history, (b) create, (c) recall, (d) destroy, (e) continue existence, (f) eliminate, (g) forget, (h) reincarnate, and (i) non-existence with history.

In Renolen (1997), objects are described as a series of consecutive *static states* and transitions (i.e., *changing states*). *Events* are defined as transitions with zero duration (i.e., when objects change suddenly), whereas other objects may change continuously. The author presents a notation called the *history graph* to represent changes of an entity over time. It consists in creating a series of consecutive versions

[16]http://www.opengis.net/ont/geosparql#sfTouches

(i.e., static states), and transitions (i.e. changing states) of an entity that represent its history over time.

The graph approach is of particular interest in this context of representing entities (i.e., vertices), and all their relations (i.e., edges) with other entities (descendants or neighbors). The *graph model for spatiotemporal evolution* of Del Mondo et al. (2013) argues also in this direction. The authors define *Spatiotemporal graphs* as entities (i.e., vertices) related by spatial and filiation binary relations (i.e., edges). In Del Mondo et al. (2010), two filiation relation cases are defined (prerequisite 2.3): a filiation of type *Continuation* is when *the first entity is the same as the second entity (e.g., one person at two times)*, i.e., entities maintained their identity; a filiation of type *Derivation* is when *the first entity creates (possibly with others) the second entity (e.g., a parent of a child)*.

Considering once again the example of the ship of Theseus: does the ship of Theseus at time 0 has a relation of type *Continuation* with the ship of Theseus at time 1? What if we consider the relation between the ship at time 0 and time 100, all the planks having being replaced? In other word, it still the same ship or a different one? In the literature, there is no unique answer to this question, the response varies from one chosen definition of what makes the identity of the ship to another, and more broadly it depends on modeling choices. In the Sect. 6.4.4, we present different models and ontologies that address this issue of change to the identity of entities (prerequisites 3 and 4).

6.4.3 Ontological Approaches for the Modeling of Evolving Entities

6.4.3.1 Versioning Approach

The terms *Versioning* and *Version Control* are most of the time used in the context of computer programming and, more broadly, for the management of modifications to text document. As previously explained, in the field of *Software configuration management (SCM)* (Tichy 1988) a *version* is defined as a state of an evolving item.

Version-Difference Spatiotemporal model

Adopting the approach of *Software configuration management (SCM)*, the authors in Huibing et al. (2005) introduce the *Version-Difference Spatiotemporal model* developed with the requirement of using historical information in order to analyze changes of spatial objects over time. Information systems based on this model store "the current state of an object (called the *default version*) and all its historic changes (called *version differences*)." Then, using a reconstruction operator and the change descriptions, all the states of an entity over time can be obtained.

PAV and DC-terms Ontologies

The *Provenance, Authoring and Versioning Ontology* (PAV) (URI: http://purl.org/pav/), and the Dublin Core Metadata Initiative Terms (DC-terms) (URI: http://purl.

org/dc/terms/) also adopt the term *Version*. They focus on entities that are published on the Web: DC-terms provides terms to describe resources on the Web with metadata, whereas the PAV ontology is more devoted to the agents contributing in Web resources: contributors, authors, curators and digital artifact creators. Also, PAV provides terms for tracking provenance of digital entities that are published on the Web (Ciccarese et al. 2013). Both PAV and DC-terms use the predicate `hasVersion` to point to a resource that is a version, edition, or adaptation of the described resource. The PAV ontology provides the following definition of the property *hasVersion*: "*This property is intended for relating a non-versioned or abstract resource to several versioned resources, e.g. snapshots.* The term *Snapshot* is found in several models (Renolen 1996; Grenon and Smith 2004),

6.4.3.2 SNAP and SPAN Approach

The *Basic Formal Ontology* (BFO) is an upper ontology based on the approach of Grenon and Smith (2004) that accounts for both the static SNAP and the dynamic SPAN entities, while most of the existing models account only for one type at a time.

Ontologies for *continuants* are called *SNAP*. The terms 'continuant' or 'endurant' are used interchangeably for those *entities that have continuous existence and a capacity to endure (persist self-identically) through time even while undergoing different sorts of changes (e.g., a person, the planet Earth)* (Grenon and Smith 2004).

Ontologies for *occurrents* are called *SPAN*. Two different terms are used to refer to occurrent entities (e.g., a smile, the passage of a rainstorm over a forest) (Grenon and Smith 2004): (1) *processes* that are occurrent entities which persist (perdure) in time i.e., they are not instantaneous. The term *perdurant* is also used for these occurrents; and (2) *events* that are occurrent entities which exhaust themselves in single instants of time.

The BFO framework addresses both the continuant and occurrent entities: it is a SNAP-SPAN framework. Whereas these two alternative views have traditionally been considered as incompatible, the authors argue that, as reality is essentially dynamic, an ontology must be capable of accounting for spatial reality both synchronically (entities that exist at a time) and diachronically (how the things unfold through time). They introduce the notion of Trans-Ontological relation. A trans-Ontological relation is a "*relation between entities that are constituents of distinct ontologies.*" The SNAP-SNAP Trans-Ontologies are ontologies that depict the world over time as a succession of temporally separated snapshots. Changes from one SNAP to another are described and may belong to one of the three main types that are:

- qualitative change: for instance, the color of a table becomes tarnished over time, yet there is still something which remains the same;

- locational change: for instance, "in one SNAP ontology the cup is on the table, in a later SNAP ontology it is on the floor. The cup underwent location change";
- substantial change: for instance, it is when a substance is divided up so as to produce a plurality of substances.

The SNAP-SPAN Trans-Ontologies are ontologies that depict the *life* or *history* of entities over time. The lifeline of an entity is a SPAN object, and the entity itself is a SNAP object. The two approach SNAP-SNAP and SNAP-SPAN may also be combined in order to depict both the changes of a SNAP entity between two periods of time and its evolution process over time (i.e., its life). Then, the authors in Grenon and Smith (2004) address in particular the geographic objects, and dissociate between:

- *SNAP Geographical Object Ontology* from which the subcategory "*Boundaries and Geographical Regions*" is one of the 5 major subcategories to geographical SNAP entities. The authors explain that administrative boundaries are SNAP *fiat* objects, that is to say, they are constructed object that exist according to an administrative, social or political convention (in opposition to *bona fide* objects that correspond to "natural" objects, such as natural boundaries like mountains) (Smith and Varzi 2000; Mathian and Sanders 2014).
- *SPAN Geographical Process Ontology* that may be of two types depending on the processes described: physical processes or social processes. For instance, change of the administrative boundaries of regions is a social process.

To conclude, as remarked by Felix (2011) "*Since most geographical ontologies still take the view of a static world, Grenon and Smith developed the Basic Formal Ontology (BFO), an upper ontology that accounts for both the static SNAP and the dynamic SPAN entities.*" In the following sub-section, we present another perdurantist approach that focuses on the representation of a relation (between entities) that changes over time, as most of the time *relationships are diachronic, i.e. they vary with time*. This approach is called *Ontology for fluents*.

6.4.3.3 Ontologies for Fluents Approach

Based on the *perdurantist* approach (from D.K. Lewis philosopher), *Ontologies for fluents* propose a way to represent in OWL relationships between entities that change with time (Welty et al. 2006). In the *fluents* approach, entities are four dimensional, the 4th dimension being time (Batsakis et al. 2017). In the 4D view, all entities are perdurant (i.e., an individual has a succession of distinct temporal parts throughout its existence). Thus, all entities have temporal parts and can be thought of intuitively as four dimensional *spacetime worms* whose temporal parts are slices of the *worm* (Sider 2001; Welty et al. 2006). In Welty et al. (2006), the authors present an example that explains their 4D fluents approach: in this approach, statements such as "*Joe walked into the room*" (i.e., the relationship between Joe and the room at one time), are represented as "*a temporal part of Joe walked into a temporal part*

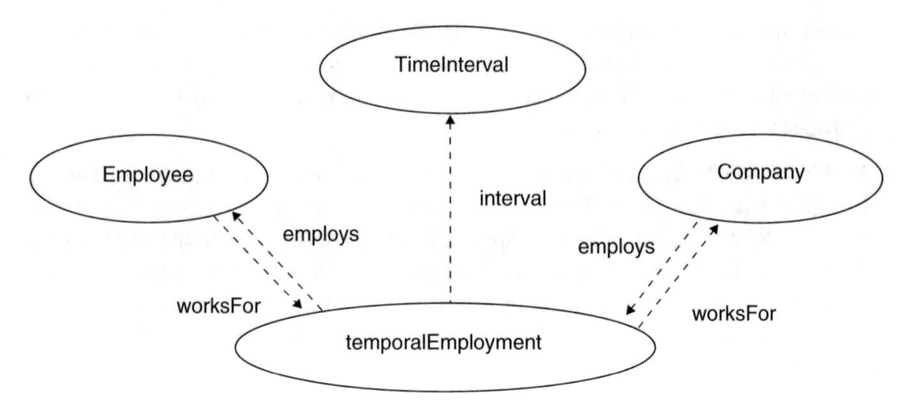

Fig. 6.3 Example of N-ary Relations from Batsakis et al. (2017)

of the room". The main issue they address is the representation of relations between entities that change over time. They call *Fluents* relations that hold within a certain time interval and not in others. The example of the relation between an employee and a company is a common example in the literature to explain relationships that change over time: the relationship is true for a defined period of time only, since the person has not always been an employee of this company. There are different ways to address this problem in ontologies. We present two of them here:

- the *reification* solution that reify the relationships into an object, as shown in Fig. 6.3. "*Reification is a general purpose technique for representing n-ary relations using a language such as OWL that permits only binary relations.*"(Batsakis and Petrakis 2011). Although this approach requires only one additional object for every temporal relation, it suffers from redundancy of the properties (e.g., *employs, worksFor*) that participate in the reification of the relationships.
- the 4D-fluents (perdurantist) approach where objects in time are represented by *TimeSlices*, and their temporary relationships are described between these timeslices. The main advantage of this approach is the possibility to describe changes of the entities on the timeslices sub-objects. However, this approach suffers from proliferation of objects, as remarked by Batsakis et al. (2017) since it introduces two additional objects (e.g., *EmployeeTimeSlice* and *CompanyTimeSlice*) for each temporal relation (instead of one in the case of N-ary relations (e.g., *TemporalEmployment*)). The 4D fluents (perdurantist) approach offers a new way to answer the *identity* question over time, exposed previously through the Ship of Theseus story. In this approach, the Ship of Theseus life is represented through one *space-time worm* ship which has temporal parts (timeslices): a part at year zero, a part at year 1, ..., a part at year 1000 (Sider 2001) (Fig. 6.4).

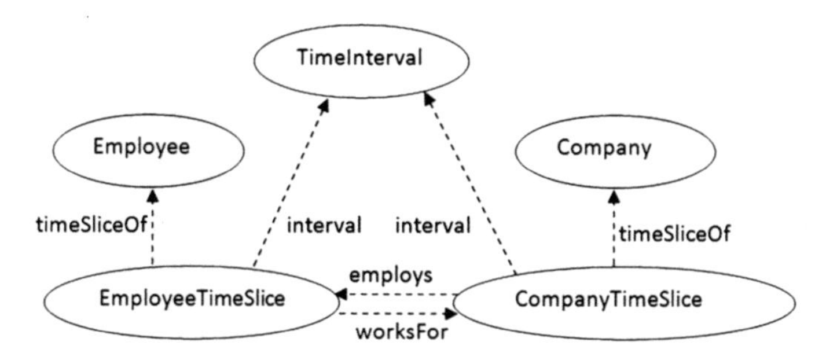

Fig. 6.4 Example of 4D fluents from Batsakis et al. (2017)

We have presented several ontologies that model the evolution of entities over time. In the next sub-section, we investigate the ontologies that model in particular the evolution of geospatial entities.

6.4.4 *Ontological Approaches for the Modeling of Evolving Geospatial Entities*

The 4D fluents approach, combined to the *OWL-Time* ontology (in order to assign time points or interval to a *TimeSlice*), is often used in the LOD Web to represent the evolution of resources (Batsakis and Petrakis 2011), including geospatial ones (Harbelot et al. 2015; Tran et al. 2015). For instance, the *Continuum Spatiotemporal Model* represents the evolution of parcels in time, adopting the 4D fluents approach of Welty et al. (2006). In Harbelot et al. (2015), parcels are described as perdurant entities that have two types of attributes: identity attributes and other attributes, valid for a period of time (semantic, spatial, temporal components). If the value assigned to attributes that hold the identity change, then a new parcel is created. Whereas, if the other attributes change, a new sub-object, i.e. *TimeSlice*, linked to the main one is created: it holds all the attributes of the parcel that may change over time (Welty et al. 2006). Using the *Continuum Spatiotemporal Model*, the authors construct lineage of land-parcels through time (Harbelot et al. 2015). Two attributes of the land-parcels are taken into account to determine whether the filiation link throughout the versions of the parcels of the CORINE COoRdinate INformation on the Environment Land Cover (CLC) data set (from the European Environmental Agency[17]) are of type *continuation* or *derivation*: the polygon

[17]Available online from the Copernicus Web site https://land.copernicus.eu/pan-european/corine-land-cover

geometry of the parcels and the nature of the land cover (e.g., agricultural area, forest area, construction site, etc.) defined in the *Corine Land Cover Nomenclature*.[18]

The main drawback of this approach is that the semantics of lineage and changes are both hold by only one predicate that links two parcels or time-slices of a parcel over time (for instance, the predicate "deforestation" describes a filiation relation of type derivation between two parcels and it describes also a change of the nature of the land cover of the parcel). If one wants to extend this approach in order to describe some other changes such as an identifier change, a name change, an inhabitant number change, etc., then, whenever an attribute changes, a link has to be drawn between the two parcels or time-slices of the parcel. Also, the direct links created between two time-slices in case of *Derivation* situations are questionable. Indeed, in many cases, direct links might not be relevant in the context of statistics. For instance, let us consider *Redistribution* events, where the identifiers and the boundaries of several areas are modified simultaneously in a way that it is difficult to define the nature of the change (e.g., in terms of *merge* or *split*). A solution here could be to determine the set of impacted areas and the set of created ones, then to link them all through a node that allows to switch from one version to another.

In these cases, the *Change Bridge* approach of Kauppinen and Hyvönen (2007) has to be considered. In Kauppinen et al. (2008), the authors introduce the notion of *Change Bridge* to chain former territories (e.g., *East Germany* and *West Germany*) to the following ones (e.g., *Germany*), using the *Change Vocabulary* for the characterization of changes. It consists in indirectly linking *input* and *output* elements that are involved in a change event, through a *Change node* that describes the nature of the territorial change. The *Change Vocabulary*[19] of Kauppinen et al. (2008) and Kauppinen and Hyvönen (2007) is a lightweight spatiotemporal vocabulary made of two properties (*before*, *after*) and five classes for the description of changes (*Establishment*, *Merge*, *Split*, *Namechange*, *Changepartof*). The *Change Bridge* approach has been implemented in order to create the *Finnish Spatiotemporal Ontology*, an *ontology time series* of the Finnish municipalities over the time interval 1865–2007 (Kauppinen et al. 2008). The modeling approach presented in Kauppinen and Hyvönen (2007) and Kauppinen et al. (2008) is criticized by Lacasta et al. (2014) as it may lead to a proliferation of instances. Actually, the complete status of the area has to be described before and after the change event.

The approach from Lacasta et al. (2014) consists in describing changes in political subdivisions, called Jurisdictional Domains, at the property level (each property can have its own time span). The authors propose to minimize the description and to avoid duplication of data. This prevents from the creation of new instances after each change. They present in López-Pellicer et al. (2008, 2012) an ontology schema that combines, in a single model, the political structure, the spatial components, and the temporal evolution of areas. However, this approach requires

[18]Please consult the nomenclature of CLC classes from the Copernicus Project Web site at https://land.copernicus.eu/Corinelandcoverclasses.eps.75dpi.png/

[19]http://linkedearth.org/change/ns/

the use of a written dictionary of changes as input, listing every individual change. This requirement may be seen as a drawback for current statistical information systems because the listing by hand of toponym and geospatial changes takes times. Automating the change detection and annotation allows for processing any geographic subdivisions in the world (using the same semantics for the description of changes), whether or not there exists a catalog of changes.

In the next section, we present our 4D perdurantist approach that offers a model for a fine description of changes that occurred between a series of versions of geographic divisions and automates their detection and description. The originality of this approach lies in the fact that it automatically describes territorial changes at each level of the hierarchy (e.g., States, districts, sub-districts, municipalities subdivisions). An algorithm has been designed for this and implemented in a framework. It also seeks for changes to be linked when they impact nested areas. The framework automatically creates catalogs of changes that are published on the LOD Web. These catalogs allow double readings of changes: a *horizontal reading* of the RDF graph provides a view on the lineage of each area over time; while a *vertical reading* of the graph gives to see the propagation of a change event through the divisions levels.

6.5 Contributions

We adopt a descriptive approach for the territorial dynamics, strongly defended within the community of geoinformation science (Claramunt and Thériault 1995; Wachowicz 2003; Del Mondo et al. 2010; Harbelot et al. 2015). As far as we know, this is the first time that this approach is used in the context of statistics in order to describe the evolution processes of geographic areas, and of the geographic division level these areas belong to. The geographic hierarchies we process are standard nomenclatures, called *Territorial Statistical Nomenclatures* (TSNs) (as explained previously in Sect. 6.1), built by (N)SAs to observe a territory at several nested levels (e.g., regions, districts, sub-districts geographic subdivisions). TSNs change over time to reflect on political, administrative or population changes. Statisticians create a “snapshot” of the TSN and amend the snapshot at the end of a specified period of stability (e.g., three years at least for the Eurostat European TSN, 10 years for the U.S. *Census Tract*) to reflect changes in the regional breakdown of a territory.

Our approach is multiscalar and allows to zoom out to visualize in a global way the main changes of the TSN geographic hierarchy, version to version. Also, it allows to zoom in to visualize all the sub-changes, until the name change of the smallest geographic area of the structure. We focus on links between areas (hierarchical links, spatiotemporal links, filiation links) and automatic description of these links on the LOD Web.

Most of the time, in statistics, all traces of change and evolution of the areas are erased. While, in our approach, according to Grasland and Madelin (2006), we consider the *Change of Support Problem* no longer as a “Problem” but as a

"Potential" that provides information about data and territories. Rather than erasing each previous version of a TSN, we preserve it together with the new one. Our challenge is to automatically identify, describe and contextualize every change event that has occurred between two consecutive TSN versions. Our objective is to find: *What* (the name of a district, its boundaries ...), *Where* and *When* changes occurred on the studied area, and above all, *How* areas have changed (i.e., the nature of the change(s) e.g., a fusion of areas). It also aims at discovering *Who* is responsible for the change in boundaries, *Why* (because of a reform?) the change occurs.

To this end, we adopt the Semantic Web technologies for the description of both the areas and their changes. Thus, we make sure that the semantics of changes are understood in the same way, by humans and software agents, that share the same concepts and definitions. Another benefit is the LOD Web, which allows us to connect and contextualize the change events by connecting to historical or law data sets on the LOD Web. Figure 6.5 below presents, in a simplified way, the kind of RDF graph we obtain automatically. Here, the chosen example takes place in France where the administrative regions have been merged to create great regions. These changes have been officially adopted in January 2015.[20] The illustration presents a trivial example of fusion of two regions (Auvergne and Rhône-Alpes) which leads to a new bigger one called Auvergne-Rhône-Alpes.

We have designed and implemented the Theseus Framework in order to supervise the automatic creation and publication of such RDF graphs on the LOD Web. It is called *Theseus* by reference to the identity philosophical issue raised by the *Ship of Theseus* that changes over time. This framework has been designed for the Statistical Agencies, statisticians, or researchers who wish to publish on the LOD Web successive versions of their geographic areas, as well as change and similarity descriptions between the versions i.e., filiation links between the features throughout the versions. The framework encapsulates two ontologies, called TSN-Ontology and TSN-Change Ontology (available from the URIs http://purl.org/net/tsn# and http://purl.org/net/tsnchange#) (Bernard et al. 2018b). They are designed for the unambiguous identification of the geographic areas in time and space, and for the description of their filiation links (comprising similarity and change descriptions) over time on the LOD Web. Our proposal is built on an original combination of approaches we have introduced in Sect. 6.4.2. We adopt the *perdurantist* approach of ontologies for *fluents*, for the description of the geographic areas that vary in time. Also, we adopt the *Change Bridges* approach for managing the union of successive *time-slices* of the areas. Theseus also embeds our TSN Semantic Matching Algorithm (Bernard et al. 2018a). It is designed to automate both the detection and the description on the LOD Web of territorial similarities and changes among various TSNs.

By exploiting the distributed LOD Web, our descriptions of geographic areas and of their changes can be enriched by looking for other resources in order to understand more deeply the evolution of the territory. Different data sets exist

[20]Law No 2015-29 of January 16th, 2015 https://www.legifrance.gouv.fr/eli/loi/2015/1/16/INTX1412841L/jo/texte

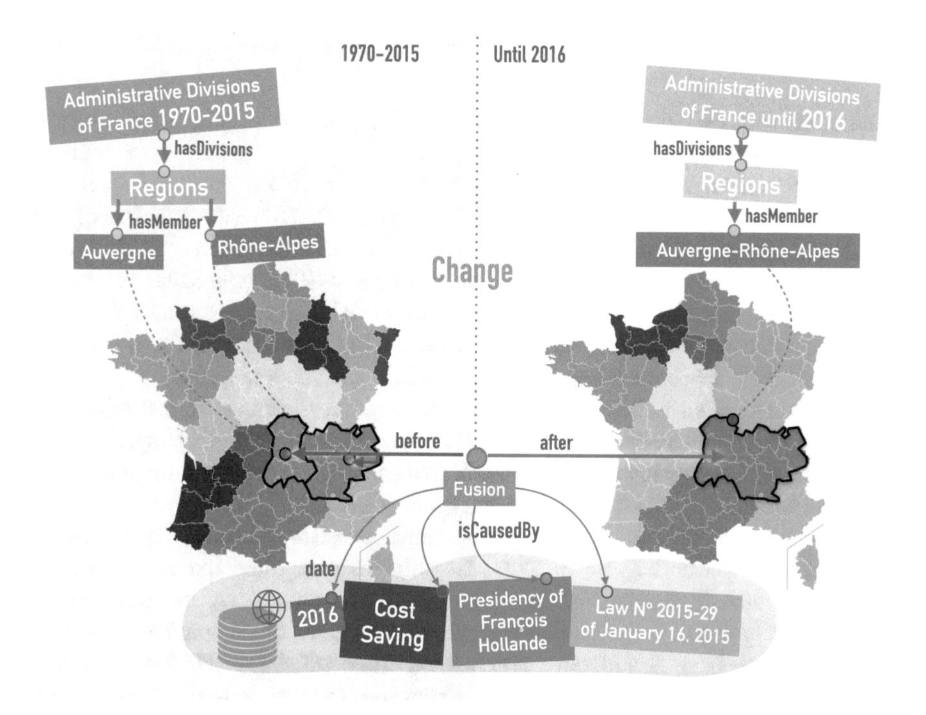

Fig. 6.5 Simplified illustration of the RDF Graph representing territorial changes generated with the Theseus Framework

on the LOD cloud, but the ones we found the most relevant in our context are those from DBPedia[21] and WikiData.[22] Indeed, they are generalist and provide encyclopedia-style information, such as historical data. Then, it may be possible to find the historical cause of a territorial changes in these data sets. All together the software modules contribute to the publication on the LOD Web of TSNs semantic history graphs. Those latter constitute catalogs of evolving geographic areas that enhance the understanding of territorial dynamics, providing statisticians, researchers, citizens with descriptions to comprehend the motivations and the impact of changes. Our framework is then a step towards knowledge graph of evolving territorial statistics that link several ontologies and data sets: RDF Data Cube data sets, geospatial TSN data sets, (historical) event data sets, law data sets (for instance, in Europe, the European Legislation Identifier (ELI)[23] provides Web identifiers (URIs) for legal information)...from which one could build intelligent tools for the analysis of the territorial dynamics over time. These tools are capable of inferring new data (such as estimating statistical values in a new TSN version).

[21]https://wiki.dbpedia.org/

[22]https://www.wikidata.org/wiki/Wikidata:Main_Page

[23]https://eur-lex.europa.eu/eli-register/about.html

6.6 Conclusion and Perspectives

In this chapter, we have presented issues behind changes of the geographic divisions over time, specially when the geographic areas serve as a reference for counting statistical values. We have highlighted the importance of keeping track of former geographical divisions and furthermore, the importance of describing their evolution and their changes over time to comprehend the territorial dynamics.

Then, we have presented a state of the art of the existing approaches to describe the evolution of geographic areas on the Semantic Web. Since the geographic areas and their updates are numerous and frequent (in order to reflect the real evolution of the world), it becomes very important to adopt a versioning, or a perdurantist approach dedicated to the description of the processes leading the evolution. The spatiotemporal and semantic approaches we have presented in this chapter may be useful in many domains (socio-economic, archaeological, cultural, linguistic, ...) in order to correctly use data in time and space.

We have also presented our approach which consists in describing the lifeline and the changes undergone in TSNs in order to strengthen the consistency of associated statistical data. We have proposed two ontologies, an algorithm to automatically detect and describe lineages and changes of geographic areas over time, and we have tested our approach on three official TSN: The European *Nomenclature of Territorial Units for Statistics* (NUTS) (versions 1999, 2003, 2006, and 2010) from the *Eurostat Statistical Institute*; The *Switzerland Administrative Units* (SAU), from The *Swiss Federal Statistical Office*, that describes the cantons, districts and municipalities of Switzerland in 2017 and 2018; The *Australian Statistical Geography Standard* (ASGS), built by the *Australian Bureau of Statistics*, composed of seven nested subdivisions of the Australian territory, in versions 2011 and 2016. The created RDF graphs are available at http://purl.org/steamer/nuts, http://purl.org/steamer/sau, and http://purl.org/steamer/asgs.

Generalizing this approach to historical map data on the Web may improve search engines for geographic data, offering the possibility to search by obsolete names for instance. Also, applying our approach to other kind of geographical divisions, such as non nested divisions (urban (metropolitan) areas), or other geospatial objects (network) and giving to see their changes on a map will considerably help policy makers in the analyses of their jurisdictions evolution over time.

References

Allen JF (1983) Maintaining knowledge about temporal intervals. Commun ACM 26(11):832–843

Anderson CW et al (2012) Quantitative methods for current environmental issues. Springer Science & Business Media, London

Auer S, Lehmann J, Hellmann S (2009) LinkedGeoData: adding a spatial dimension to the web of data. In: Bernstein A et al (ed) The semantic web ISWC 2009: 8th international semantic web conference, ISWC 2009, Chantilly, 25–29 Oct 2009. Proceedings. Springer, Berlin/Heidelberg, pp 731–746. ISBN: 978-3-642-04930-9. https://doi.org/10.1007/978-3-642-04930-9_46

Batsakis S, Petrakis EGM (2011) SOWL: a framework for handling spatio-temporal information in OWL 2.0. In: International workshop on rules and rule markup languages for the semantic web. Springer, pp 242–249
Batsakis S et al (2017) Temporal representation and reasoning in OWL 2. Semant Web 8(6):981–1000. https://doi.org/10.3233/SW-160248. Gangemi A (ed)
Beller A (1991) Spatial/temporal events in a GIS. Proc GIS/LIS 91(57):4
Bernard C et al (2018a) An ontology-based algorithm for managing the evolution of multi-level territorial partitions. In: Proceedings of the 26th ACM SIGSPATIAL international conference on advances in geographic information systems, SIGSPATIAL'18, Seattle. ACM, Washington, pp 456–459. ISBN: 978-1-4503-5889-7. https://doi.org/10.1145/3274895.3274944
Bernard C et al (2018b) Modeling changes in territorial partitions over time: ontologies TSN and TSN-change. In: Proceedings of the 33rd annual ACM symposium on applied computing, SAC'18. ACM, Pau, pp 866–875. ISBN: 978-1-4503-5191-1. https://doi.org/10.1145/3167132.3167227
Berners-Lee T et al (2001) The semantic web. Sci Am 284(5):28–37
Ciccarese P et al (2013) PAV ontology: provenance, authoring and versioning. J Biomed Semant 4(1):37
Claramunt C, Thériault M (1995) Managing time in GIS an event-oriented approach. In: van Rijsbergen CJ, Clifford J, Tuzhilin A (eds) Recent advances in temporal databases. Springer, London, pp 23–42. https://doi.org/10.1007/978-1-4471-3033-8_2. ISBN: 978-3-540-19945-8 978-1-4471-3033-8. (Visited on 01 Feb 2017)
Conradi R, Westfechtel B (1998) Version models for software configuration management. ACM Comput Surv (CSUR) 30(2):232–282. (Visited on 30 May 2016)
Cox S, Little C (2017) Time ontology in OWL – W3C recommendation 19 Oct 2017. https://www.w3.org/TR/owl-time/. (Visited on 08 Jan 2019)
Cyganiak R, Wood D, Lanthaler M (2014) RDF 1.1 concepts and abstract syntax. https://www.w3.org/TR/rdf11-concepts/. (Visited on 01 Mar 2017)
Del Mondo G et al (2010) A graph model for spatio-temporal evolution. J UCS 16(11):1452–1477
Del Mondo G et al (2013) Modeling consistency of spatio-temporal graphs. Data Knowl Eng 84:59–80. ISSN: 0169023X. https://doi.org/10.1016/j.datak.2012.12.007. http://linkinghub.elsevier.com/retrieve/pii/S0169023X12001188
Ehrlinger L, Wöß W (2016) Towards a definition of knowledge graphs. In: SEMANTiCS (Posters, Demos, SuCCESS), vol 48
Eurostat (2001) Manual of concepts on land cover and land use information systems. OCLC: 464871883. Office for Official Publications of the European Communities, Luxembourg. ISBN: 978-92-894-0432-7
Fearon JD (1999) What is identity (as we now use the word)? Unpublished manuscript, Stanford University, Stanford
Flowerdew R (1991) Data integration: statistical methods for transferring data between zonal systems. In: In handling geographical information: methodology and potential applications, pp 39–54
Gantner F (2011) A spatiotemporal ontology for the administrative units of Switzerland. Theses, University of Zurich – Switzerland
Gotway Crawford CA and LJ Young (2005) Change of support: an inter-disciplinary challenge. In: Geostatistics for environmental applications. Springer, Berlin/Heidelberg, pp 1–13
Grasland C, Madelin M (2006) Modifiable area unit problem. ESPON. https://halshs.archives-ouvertes.fr/halshs-00174241
Grenon P, Smith B (2004) SNAP and SPAN: towards dynamic spatial ontology. Spat Cogn Comput 4(1):69–104. ISSN: 1387-5868, 1542–7633. https://doi.org/10.1207/s15427633scc0401_5. (Visited on 14 Feb 2019)
Harbelot B, Arenas H, Cruz C (2013) Continuum: a spatiotemporal data model to represent and qualify filiation relationships. In: Proceedings of the 4th ACM SIGSPATIAL international workshop on geoStreaming. ACM, pp 76–85. (Visited on 15 Dec 2015)

Harbelot B, Arenas H, Cruz C (2015) LC3: a spatio-temporal and semantic model for knowledge discovery from geospatial datasets. Web Semant Sci Serv Agents World Wide Web 35:3–24. http://www.sciencedirect.com/science/article/pii/S1570826815000840. (Visited on 19 Apr 2017)

Hausenblas M (2012) 5-star Open Data. (Visited on 08 Feb 2019)

Howenstine E (1993) Measuring demographic change: the split tract problem. Prof Geogr 45(4):425. ISSN: 00330124

Huibing W et al (2005) Modeling spatial-temporal data in version-difference model, p 4

Kauppinen T, Hyvönen E (2007) Modeling and reasoning about changes in ontology time series. In: Ontologies. Springer, Boston, pp 319–338. (Visited on 03 Feb 2016)

Kauppinen T, Väätäinen J, Hyvönen E (2008) Creating and using geospatial ontology time series in a semantic cultural heritage portal. Springer, Berlin/New York. (Visited on 11 Apr 2016)

Khoshafian SN, Copeland GP (1986) Object identity. Association for computing machinery. New York, NY, USA, vol 21(11). ISSN: 0362–1340. https://doi.org/10.1145/960112.28739

Lacasta J et al (2014) Population of a spatio-temporal knowledge base for jurisdictional domains. Int J Geogr Inf Sci 28(9):1964–1987. ISSN: 1365-8816, 1362-3087. (Visited on 18 Jan 2017)

López-Pellicer FJ et al (2008) Administrative units, an ontological perspective. In: Advances in conceptual modeling–challenges and opportunities. Springer, pp 354–363. https://doi.org/10.1007/978-3-540-87991-6_42 (Visited on 07 Apr 2016)

López-Pellicer FJ et al (2012) An ontology for the representation of spatiotemporal jurisdictional domains in information retrieval systems. Int J Geogr Inf Sci 26(4):579–597. ISSN: 1365-8816, 1362-3087. (Visited on 18 Jan 2017)

Maedche A, Staab S (2001) Learning ontologies for the semantic web. In: Proceedings of the second international conference on semantic web-volume 40. CEUR-WS.org, pp 51–60. (Visited on 01 July 2016)

Mathian H, Sanders L (2014) Spatio-temporal approaches: geographic objects and change process. Wiley, Hoboken

Noy NF, Klein M (2004) Ontology evolution: not the same as schema evolution. Knowl Inf Syst 6(4):428–440. ISSN: 0219-1377, 0219-3116. https://doi.org/10.1007/s10115-003-0137-2. (Visited on 08 Feb 2016)

Ontotext (2018) What is a knowledge graph?| Onotext en-US. https://www.ontotext.com/knowledgehub/fundamentals/what-is-a-knowledge-graph/. (Visited on 19 Feb 2019)

Openshaw S, Taylor PJ (1979) A million or so correlation coefficients: three experiments on the modifiable areal unit problem. Stat Appl Spat Sci 21:127–144

Paulheim H (2017) Knowledge graph refinement: a survey of approaches and evaluation methods. Semant Web 8(3):489–508

Perry M, Herring J (2012) OGC GeoSPARQL – a geographic query language for RDF data, p 75

Plumejeaud C et al (2010) Transferring indicators into different partitions of geographic space. In: Taniar D et al (eds) Computational science and its applications ICCSA 2010. Lecture notes in computer science, vol 6016. Springer, Heidelberg, pp 445–460. ISBN: 978-3-642-12155-5

Plumejeaud C et al (2011) Spatio-temporal analysis of territorial changes from a multi-scale perspective. Int J Geogr Inf Sci 25(10):1597–1612. (Visited on 07 Apr 2016)

Renolen A (1996) History graphs: conceptual modeling of spatio-temporal data. Gis Front Bus Sci 2:46

Renolen A (1997) Conceptual modelling and spatiotemporal information systems: how to model the real world. In: ScanGIS, vol 97, pp 1–22

Sider T (2001) Four-dimensionalism: an ontology of persistence and time. Oxford University Press on Demand, Oxford

Smith B, Varzi AC (2000) Fiat and bona fide boundaries. Philos Phenomenol Res 60(2):401–420

Stefanidis K, Chrysakis I, Flouris G (2014) On designing archiving policies for evolving RDF datasets on the web. In: Yu E et al (eds) Conceptual modeling: 33rd international conference, ER 2014, Atlanta, 27–29 Oct 2014. Proceedings. Springer International Publishing, Cham, pp 43–56. https://doi.org/10.1007/978-3-319-12206-9_4. ISBN: 978-3-319-12206-9

Tandy J, Van Den Brink L, Barnaghi P (2017) Spatial data on the web best practices – OGC and W3C recommendation. https://www.w3.org/TR/sdw-bp. (Visited on 28 Nov 2017)
Tichy W (1988) Software configuration management overview. Citeseer. (Visited on 21 June 2016)
Tran B-H et al (2015) A semantic mediator for handling heterogeneity of spatio-temporal environment data. In: Research conference on metadata and semantics research. Springer, pp 381–392. (Visited on 02 May 2017)
Wachowicz M (2003) Object-oriented design for temporal GIS. Research monographs in GIS. Taylor & Francis. ISBN: 9780203212394
Wang F (2014) Quantitative methods and socio-economic applications in GIS. CRC Press. (Visited on 29/ June 2017)
Welty C, Fikes R, Makarios S (2006) A reusable ontology for fluents in OWL. In: FOIS, vol 150, pp 226–236. (Visited on 05 May 2017)

Part II
Trajectories, Event and Movement Data

Chapter 7
Big Spatial Flow Data Analytics

Ran Tao

7.1 Introduction

Spatial flows, also referred to as origin-destination flows, represent meaningful interaction activities between regions, organizations, and individuals. Flow activities widely exist in our daily life, for example, migration flows, commuting flows, trade flows, and information flows. For the past few decades, studying various spatial patterns and the decision processes behind flow phenomena have been an interesting topic in geography, regional science, animal ecology, environmental science, physics, and urban planning (Farmer and Oshan 2017).

Spatial flow data requires specially-designed analytical methods given its dyadic nature and unique characteristics. A flow event consists of a pair of corresponding points. Besides the locational information of the two endpoints, direction, distance or flow length, flow type, and flow volume often bear the essential information and deserve special treatment during the analysis. Unlike trajectory data, flow data is more abstract as it does not account for the actual path between endpoints (Tao et al. 2017). This makes flow more comparable to the edges in a network, where topological relationship is the key to gaining insight of the data. Moreover, discrete flows of fine spatial resolution can be easily converted to aggregated flows by grouping the ones originated from or destinated to the same upper-level spatial unit, which calls for different types of analysis and introduces issues such as the modifiable areal unit problem (MAUP).

The recent data evolution and the widespread adoption of location-aware technologies such as the GPS-enabled smartphones amass flow data at individual level, along with much finer spatiotemporal granularity and abundant semantic

R. Tao (✉)
School of Geosciences, University of South Florida, Tampa, FL, USA
e-mail: rtao@usf.edu

M. Werner, Y.-Y. Chiang (eds.), *Handbook of Big Geospatial Data*,
https://doi.org/10.1007/978-3-030-55462-0_7

information. The increasing availability of big spatial flow has brought us with unprecedented opportunities to study all kinds of spatial interaction phenomena from new perspectives. For example, to optimize transportation planning by identifying spatial territories of multi-type taxi flows (Tao and Thill 2019b); to uncover inter-urban movement patterns from individual trip flows embedded in online travel blogs (Gao et al. 2019) and social media check-in data (Liu et al. 2014); to improve our understanding of the communication space by studying information flows spreading among cellphone users (Gao et al. 2013). In the meanwhile, great intellectual challenges need to be overcome in order to extract meaningful information from the big flow data and visualize it in a readily comprehensive manner. Therefore, developing data-driven methods specific for big spatial flow data has become pressing for understanding the dynamics of spatial interaction phenomena across scientific disciplines (Yan and Thill 2009).

This chapter introduces a collection of the latest analytical methods and techniques tailored for big spatial flow data. Important literatures are reviewed from several perspectives: flow mapping and geovisualization, spatial data mining, spatial statistics and spatial regression. One representative method of each category is selected for elaboration. Available toolkits are introduced. An overview of the current works and an outlook for the future directions are summarized in the end.

7.2 Flow Mapping & Geovisualization

Flow mapping can serve as the first step of exploring a flow dataset, as well as the last step of presenting the analytical results. However, visualizing a large flow dataset can be challenging, and a higher degree of abstraction is needed (Andrienko et al. 2008). In a flow map, flows are commonly represented by a number of straight or curved lines connecting origin and destination locations (Zhu and Guo 2014). Accompanied with well-designed color schemes, labels, or symbols, it can be used as a visual analytic method to represent the dynamics of movement between two pair-wise interacting geographical regions (Cao et al. 2015).

The first known map of spatial flows can be traced back to 1837 in which Lt. Harness depicted bidirectional traffic flows between major Irish urban centers (Marble et al. 1997). While the first experiment of flow mapping (migration flows) with the assistance of a computer was done by Tobler (1987). Since then, considerable efforts have been made to design new layout of flow maps, to increase the manageable data size, to enhance the drawing speed, to emphasize important thematic information, and to integrate user-friendly features such as interactive selection and brushing.

Notwithstanding, scholars have soon found that flow mapping is much more complex than a pure cartographic technique. Problems emerge even when mapping a relatively small dataset from today's perspective (Marble et al. 1997). Severe visual cluttering caused by massive intersections and overlapping of flows easily turns the map unreadable. The reason is that unlike mapping point or polygon data which are discrete spatial objects, mapping flows is to visually represent the dynamic processes or relationships between two sets of geographical locations, which can easily reach a massive size. Given that an increasing amount of flow data are collected at the individual level, flow mapping becomes even a more challenging task. A number of geovisualization approaches have been developed in recent years to visualize big flow data.

7.2.1 *Flow Aggregation*

As a pioneer in flow mapping, Tobler (1987) suggested that information aggregation and removal is an important part of identifying patterns through visualization. Tobler (1987) observed that 75% of migration flow connections on the small side contain less than 25% of the flow volume. Therefore, filtering out the low-value flows while preserving the high-value ones is a plausible solution. However, the choice of which flows to keep or to remove is arbitrary and can result in loss of key information. An alternative way is to aggregate flows that share common origin or destination, for example aggregating county level flows to the state level. Tobler (2004) generated a series of flow maps of migration originated from California between 1995 and 2000 (e.g. Fig. 7.1a), in which straight lines with arrows represent migration flows with the line width and arrow correspond to flow magnitude and flow direction, respectively. Aggregation can dramatically improve the clarity of a flow map, but it still comes short to show many flows that intersect with each other.

When accurate coordinates are available, endpoint aggregation can be processed with other techniques. Andrienko and Andrienko (2011) utilized a point clustering method to group flow origins and destinations before drawing the flow maps. It works for visualizing individual flows as it takes advantage of the high spatial resolution. Guo (2009) proposed another approach to aggregate locations into regions based on the flow topology with a graph partitioning technique. This approach also manages to discover the natural regions from massive individual

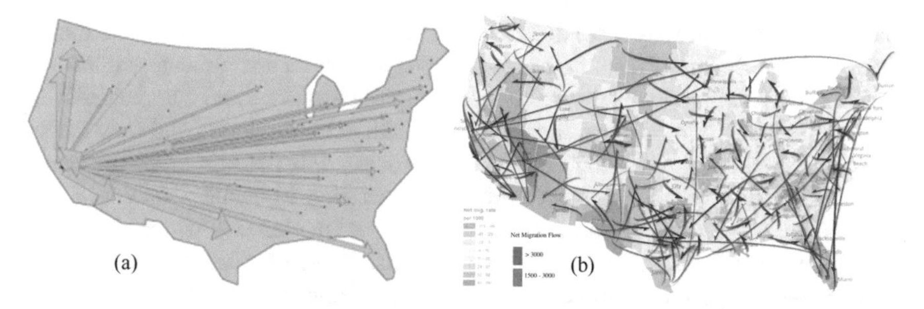

Fig. 7.1 (**a**) From CA migration from Tobler (2004); (**b**) Top 200 migration flows from Guo and Zhu (2014)

flows instead of using pre-defined political boundaries. While endpoint location aggregation is effective at reducing visual clutter, it bears the modifiable areal unit problem (MAUP) as selecting a "perfect" geographic scale or region to aggregate the endpoints is impossible. Aggregating to big regions would result in the loss of short-distance flows whose origin and destination locate within the same region, while aggregating to small regions may not remove clutter effectively. Even when there is an appropriate scale for aggregation, applying the same scale everywhere may smooth out much of the interesting local spatial structure of the spatially heterogeneous flow data.

A similar but different strategy is to aggregate flows themselves instead of the endpoints. This type of strategy first performs a flow grouping approach, for example, hierarchical flow clustering method (Zhu and Guo 2014), hot flow detection method (Tao and Thill 2016a), hierarchical and density-based flow clustering method (Tao et al. 2017), and flow-based density-estimation technique (Guo and Zhu 2014; Zhu et al. 2019). The grouped flows are then visualized as the final map. Figure 7.1b shows such a map of the selected top 200 migration flows in the U.S., after aggregated by a flow-based density estimation method (Guo and Zhu 2014). In contrast with Fig. 7.1a, the aggregated flows in Fig 7.1b are not restricted to the predefined spatial scale such as state. Moreover, its curved flow symbols with gradually varied width can visualize a large number of flows with different origins or destinations at the same time. The choropleth map in the background helps convey additional information such as the net migration rate. Notwithstanding, the emphasis of this line of works is usually on the flow grouping approach rather than the cartographic design. In other words, flow maps in this case serve as the presentation of the final analytical results.

7.2.2 Edge Bundling

Another type of flow mapping method is called edge bundling, i.e. to bundle nearby flows together in order to minimize edge crossing and to improve visual clarity. Compared with previous methods, edge bundling methods make use of geometric characteristics of flows (edges) directly. The result flow maps are usually no longer straight line-based graphs but in the road-map-style. While it significantly improves the overall visual clarity, it inevitably compromises accuracy and completeness of the original data. For example, the endpoint locational information and flow length information can be lost during the bundling process. Phan et al. (2005) presented an edge bundling method using hierarchical clustering to create a flow tree that connects a source (the root) to a set of destinations (the leaves), while preserving branching substructure across flow maps with different roots that share a common set of nodes. Later on, improvements of the edge-bundling framework have been done by grouping links based on their intersections in the Delaunay triangulation of the endpoints (Qu et al. 2006), through a control mesh generation method that can better capture the underlying graph patterns (Cui et al. 2008), and through a

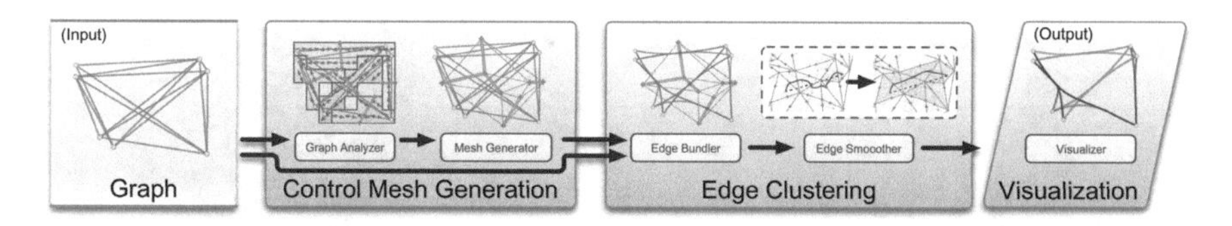

Fig. 7.2 Framework of the edge-bundling approach from Cui et al. (2008)

self-organizing approach modeling edges as flexible springs that can attract each other (Holten and Van Wijk 2009). Most recently, Yang et al. (2019) extended the application scenarios of edge bundling to three-dimensional flow maps.

Here I mainly introduce the edge-bundling approach by Cui et al. (2008), which is one of the most highly cited works and considered as an important foundation for later related studies. Figure 7.2 shows the three major steps: control mesh generation, edge cluster, and the final visualization. Explanations of each step are provided.

A preparation step is needed to convert the input flow dataset into a topological graph (e.g. Fig. 7.3a), so that each flow is treated as an edge in the following algorithm. This conversion from flow to graph is obvious for aggregated flows such as city-to-city travel flows, but it takes an extra prior step for discreate flows such as taxi pick-up and drop-off flows. This prior step is to build the network topology by grouping the endpoints based on Thiessen polygon or predefined administrative boundaries. Once the preparation is done, the algorithm automatically generates a control mesh to cage the entire graph. First, regular grid cells are created based on the bounding box for the input flow data. Kernel Density Estimator is then applied to decide the primary direction of each cell. Adjacent cells with similar primary directions are merged as a larger region, such as the polygons with red outlines in Fig. 7.3b. The primary direction after merging is the weighted average of the cells. Next, some mesh edges are placed perpendicular to each region's primary direction, e.g. the green edges in Fig. 7.3c. Based on these edges and their vertices, the final meshes are constructed by using Constrained Delaunay triangulation (Fig. 7.3d). With the mesh generated, the next step is to compute control points, based on which the flows are clustered. There are many intersection points (red dots shown in Fig. 7.3e) between the flows and the mesh edges created in the previous step. Applying the K-means clustering, the centers of a bunch of nearby intersection points are located as the control points. Forcing the flows to pass through these control points, the edges are bundled together (Fig. 7.3f).

The last step is to add cartographic designs to the bundled edges for a visually pleasing map. Adding opacity is a great way to show the density of the clustered flows, while adding colors can symbolize an additional flow attribute of user's choice. For instance, Fig. 7.4a shows the original map of the 9,798 migration flows between 1,790 cities in the contiguous United States. The map is too cluttered to present any useful information except for the city locations. Figure 7.4b is the result of this method. The flows sharing similar migration route, i.e. adjacent origins and

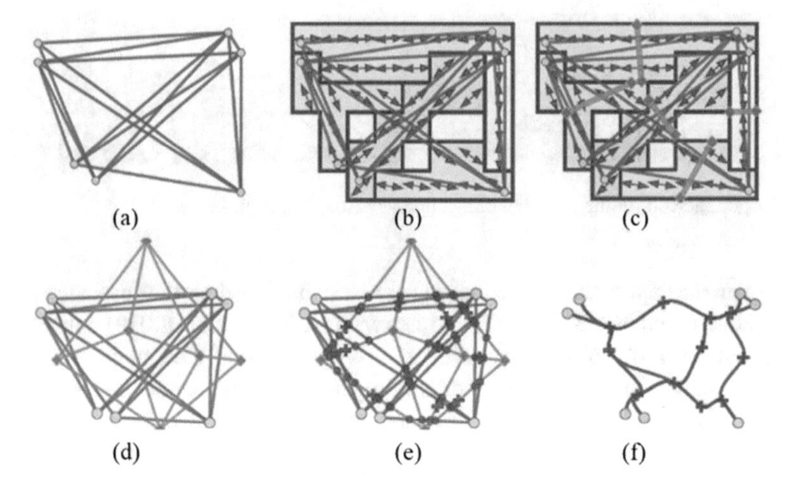

Fig. 7.3 Automatic mesh generation and edge clustering: (**a**) a graph; (**b**) grid the graph and merge the grids based on their calculated primary direction; (**c**) set some mesh edges perpendicular to the blocks' primary direction; (**d**) graph with a control mesh; (**e**) the intersections and the control points; (**f**) the merged graph from Cui et al. (2008)

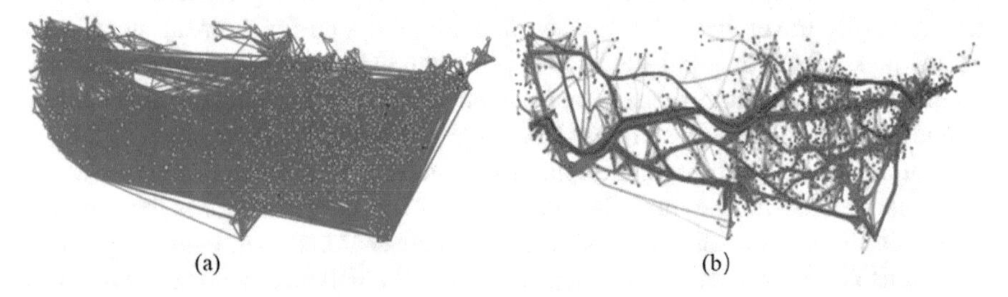

Fig. 7.4 (**a**) The original map and (**b**) the result map of U.S. migration flows from Cui et al. (2008)

destinations or similar migration direction along the route, are bundled together in a much cleaner and informative flow map. The color scheme from red to blue corresponds to the gross migration value from the highest to the lowest. And the edge width corresponds to the varied volume of migration flows at each segment of the edge.

7.2.3 Visual Analytics and Tools

The purpose of visualization is not only to depict the geographic entities, but also as a powerful exploratory tool to discover underlying stories behind the data. Therefore, visual analytic methods are frequently applied in studying big flow data. Yan and Thill (2009) use self-organizing maps (SOM) to uncover the structure of air transport system from air traffic flows. Thill (2011) illustrates the relative space of migration with cartograms. In those distorted maps, destination states are allocated further or closer to the origins, in coordination with different amounts of migrants.

Developing handy tools for visualization and visual analytics of flow data is an enduring important task. Beginning with Tobler's Flow Mapper (Tobler 1987, 2004, various software) applications have been developed for this purpose. Well-known toolkits to visualize flows in an interactive fashion include Glennon's Flow Data Model Tools in a series of ArcGIS 9 Visual Basic for Applications (VBA) macros (Glennon 2005), Flow Mapping with Graph Participating and Regionalization (Guo 2009), and jFlowMap (Boyandin et al. 2010). Commercial software such as Gephi, VisIt, and Mapbox also have flow mapping capability. Nowadays it is the trend that more applications are web-based because it has advantages such as light, portable, and highly-interactive. Some popular open-source flow mapping tools or libraries include: d3.ForceBundle (https://github.com/upphiminn/d3.ForceBundle); the interactive maps of arc and line data on https://kepler.gl/; flowmap.blue (https://flowmap.blue/) by lya Boyandin; and the FlowMapper plugin of QGIS 2 by Cem Gulluoglu (https://plugins.qgis.org/plugins/FlowMapper/).

7.3 Spatial Data Mining Methods

Spatial data mining (SDM) is the application of data mining techniques to spatial data, in order to discover previously unknown, but interesting and potentially useful patterns from high volume and heterogenous spatial datasets. Due to the nature of the geographic space and the data complexity, SDM of flow data holds uniqueness from several aspects. First, flow data is born with internal spatial dependence. While spatial dependence is a type of spatial properties commonly exist between nearby spatial objects, flow data hold an additional internal spatial dependence between each corresponding origin and destination. Therefore, treating the endpoints separately or condensing a flow object as a point object will ruin this basic property and end up with erroneous results. Second, flow data have unique spatial and geometric properties that increase the difficulty of SDM. Often viewed as a vector line that connects two endpoints, flow data need special treatment for even basic data analysis such as finding spatial neighbors (Tao and Thill 2019a) and calculating distance (Tao and Thill 2016a). The path between the two endpoints is usually abstracted as a straight or curved line since the actual route is unknown, unlike trajectory data. However, it is risky to simply ignore it during SDM, because it may embed important information such as potential geographic barriers and political boundaries that can hinder the flow process. Third, flows are spatially continuous. The flows of people, shipments, or information do not stop at the destination. Instead, it is highly likely that a flow object is just one segment of a complex and dynamic spatial process. For example, a proportion of travelers from city A to city B will continue traveling to a third city or return to city A. Therefore, it is important to have a comprehensive and systematic view of the entire flow process rather than taking individual flow objects as irrelevant events.

7.3.1 Spatial Outlier Detection

Spatial outlier detection is defined as the technique to extract a spatially referenced object whose spatial or non-spatial attributes appear to be inconsistent with other objects within its spatial neighborhood (Shekhar et al. 2003). It has been proved useful to pick out outlier flows based on OD locations, flow length, start and end time, and flow volume. Liu et al. (2010) classified taxi drivers by their income, and relate this to their driving habits mined from 48 million taxi pick-up and drop-off flows: such as operation time, average length of single trip, activity space coverage, capability of avoiding congestion, etc. Interestingly they found the secrets of top-earning (outliers) taxi drivers: long operation time; good sense of business (short time intervals between trips); knowledge to avoid congestion; and preference on the fastest path rather than shortest or longest ones. Fang et al. (2012) identified critical transportation links in Wuhan, China such as major road bridges, also from the taxi flows standing out from a large volume. Three exploratory analysis functions were developed to examine and visualize flows in an integrated spatial and temporal environment, and alternative travel paths for those bridges are identified.

7.3.2 Flow Clustering

Another SDM technique that has been commonly used is cluster analysis. Treating flows as vector lines, scholars have migrated various point-based clustering techniques to the flow context. The classical K-means algorithms have been proved very effective with respect to multi-location spatial data (Ossama et al. 2011; Genolini and Falissard 2010). Density-based clustering methods such as DBSCAN (Ester et al. 1996), OPTICS (Ankerst et al. 1999), and their variants have also been adjusted to flow data (Nanni and Pedreschi 2006; Lee et al. 2007; Zhu and Guo 2014; Tao and Thill 2016b) as density-based methods are the most suitable for discovering clusters of arbitrary shapes and filtering out noise. The key of such methods is to define a set of distance functions tailored for line-segment that can measure both positional and directional differences. Hierarchical clustering can also be used for flows. For instance, Zhu and Guo (2014) developed an approach that can generalize flows to different hierarchical levels and has the potential to support multi-resolution flow mapping. In general, spatial flow methods are designed to group observations into "clusters" based on their similarity. Unlike directly aggregating flows to predefined regions, cluster analysis methods are able to identify groups of similar flows that are not limited within predefined boundaries. Therefore, the impact of uneven density levels or ad hoc zoning definition of flow endpoints can be well handled. As discussed earlier, cluster analysis methods are frequently combined with visualization techniques when analyzing spatial flows, as the extracted flow clusters are essential information that deserves visual emphasis on the map. It is worth mentioning that there is another family of flow clustering method that root in

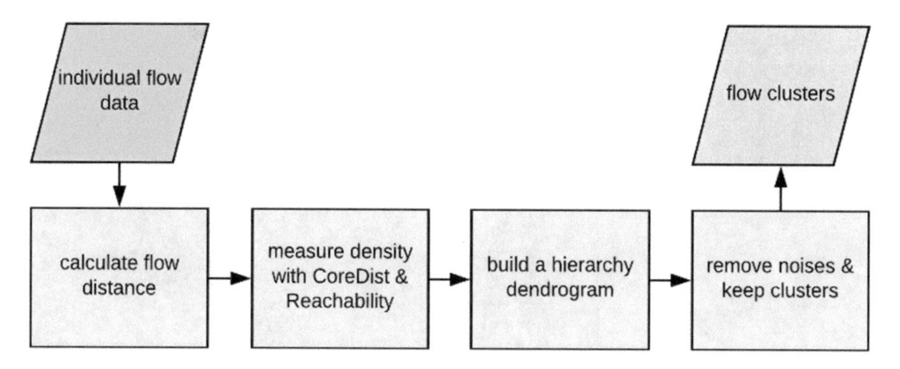

Fig. 7.5 The flowchart of FlowHBDSCAN

spatial statistics (to be introduced in section 4) rather than SDM. The difference is that the statistics-based methods focus on detecting the spatial clustering effects, i.e. concentration of flow objects or flow value in a certain region, while the SDM-based methods aim at grouping the similar or nearby flow objects together.

I select FlowHDBSCAN to give a detailed introduction here because it integrates the advantages of both hierarchical clustering method and density-based clustering method. It is effective to cluster individual flows with fine spatial resolution while solving common problems like MAUP, loss of spatial information, and uneven distribution or hoc zoning definition of flow endpoints. Moreover, it can reveal the internal hierarchical data structure of the extracted flow clusters. Last but not least, its sole-parameter design improves its ease to use by avoiding arbitrary parameterization. Figure 7.5 illustrates the general steps of this method.

The input data is in the form of vector lines ideally with accurate endpoint locations, e.g. Fig. 7.6. The first step is to build an N by N flow distance matrix with the distance measure as Equation (7.1), which integrates all the spatial elements of a flow including endpoint locations, length, and direction (implicitly).

$$FDist_{ij} = \sqrt{\frac{\alpha\left[\left(x_i - x_j\right)^2 + \left(y_i - y_j\right)^2\right] + \beta\left[\left(u_i - u_j\right)^2 + \left(v_i - v_j\right)^2\right]}{\left(L_i L_j\right)^{\gamma}}}.$$

$$or\ simplify\ as : FDist_{ij} = \sqrt{\left(\alpha {d_O}^2 + \beta {d_D}^2\right) / \left(L_i L_j\right)^{\gamma}}. \tag{7.1}$$

FDist$_{ij}$ is the distance between flow F_i (from O_i (x_i, y_i) to D_i (u_i, v_i)) and flow F_j (from O_j (x_j, y_j) to D_j (u_j, v_j)); d_O and d_D are the Euclidean distances between the two origins and two destinations, respectively; L_i and L_j are flow lengths; α and β are designed to control the relative importance of origins and destinations ($\alpha > 0$; β

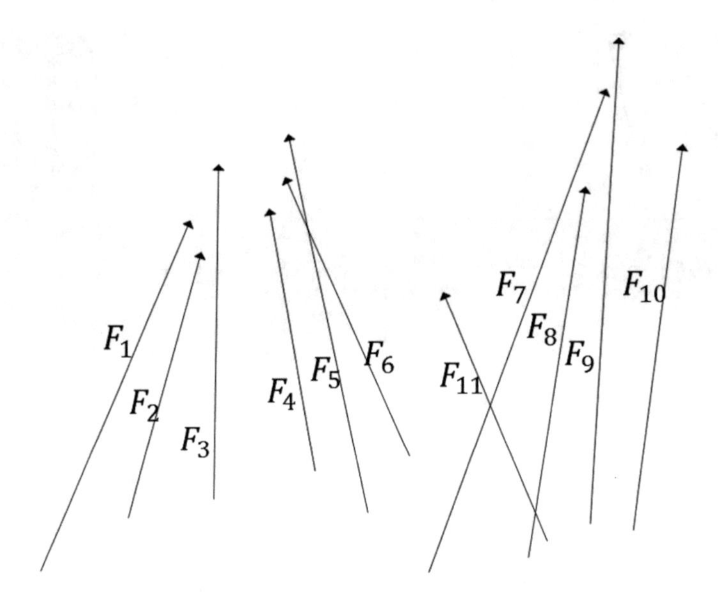

Fig. 7.6 Sample set of flows from Tao et al. (2017)

> *0*; $\alpha + \beta = 2$; by default $\alpha = \beta = 1$); γ controls the impact of flow length (by default $\gamma = 1$).

The next step is to calculate two density measures, namely CoreD with Equation (7.2) and MReachD with Equation (7.3). Both measures are borrowed from classical density-based clustering methods such as DBSCAN (Ester et al. 1996) and OPTICS (Ankerst et al. 1999). CoreD, or core distance, is calculated for each flow based on flow distances and one parameter, namely the minimum cluster size MinFlows. The smaller CoreD is, the more likely the flow belongs to a cluster. In Fig. 7.6, $CoreD_1$ is smaller than $CoreD_{11}$ when MinFlows = 3, which suggests F_1 has a higher likelihood than F_{11} to be a cluster member. MReachD, or mutual reachability distance, is calculated for each pairs of flows to measure the relative separation from one flow to another. In Fig. 7.6, $MReachD_{1,\,2}$ is smaller than $MReachD_{1,\,4}$, which means F_1 has a higher likelihood to be in the same cluster with F_2 rather than F_4.

$$CoreD_i = FDist_{i,(MinFlows-1)th\ nearest\ neighbor\ of\ i} \tag{7.2}$$

$$MReachD_{ij} = \max\left(CoreD_i, CoreD_j, FDist_{ij}\right) \tag{7.3}$$

Starting from the next step, the algorithm transits from a density-based clustering approach to a hierarchical-based one. A minimum spanning tree (MST) is built by connecting each flow with its minimal MReachD to another flow. Figure 7.7a illustrates the MST with respect to the sample data in Fig. 7.6. Some unitless numbers (1–4, and) are used here to reflect the magnitudes of MReachD. Flows

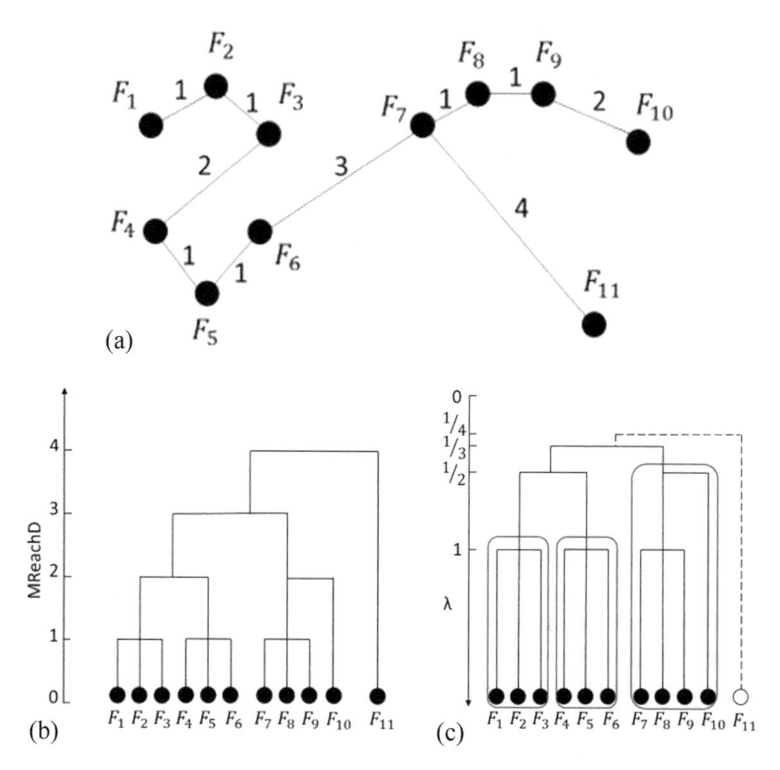

Fig. 7.7 (**a**) Minimum spanning tree (MST); (**b**) dendrogram; (**c**) pruned dendrogram with identified clusters and noises from Tao et al. (2017)

that are connected by short edges, such as F_1, F_2 and F_3, are more likely to form a cluster. Sort the MST by an increasing order of MReachD value, a dendrogram is obtained (Fig. 7.7b). While it illustrates the hierarchical structure of the entire flow dataset, a further step is needed to distinguish vertices belonging to a cluster from noises.

Traverse the dendrogram from the top. At the split of every MReachD level: drop the descendant set that has fewer members than MinFlows as noises, then keep traversing the rest. Setting MinFlows = 3 in the example of Fig. 7.7b, flow F_{11} is dropped at the first split of MReachD = 4. At the next split of MReachD = 3, both descendant sets are preserved as they have more than three flows. Keeping traversing the dendrogram all the way down to the bottom, the flows dropped at a high MReachD level such as F_{11} are likely to be noises, while the ones preserved till the end such as F_1, F_2, F_3, F_4, F_5, F_6, F_7, F_8, and F_9 are likely to be cluster members.

The last step is to calculate the cluster stability index (Equation 7.4) to extract the final clusters. Taking F_{10} in Fig. 7.7b as an example, it is not an obvious choice to remove it as a noise or to join a four-member cluster with F_7, F_8, and F_9. Unclear situation also applies to (F_1, F_2, F_3, F_4, F_5, and F_6), as it is debatable to split them into small clusters or keep them as a whole. Cluster stability is introduced for comparing and deciding between such ambiguous situations.

$$S(C_i) = \sum_{F_j \in C_i} \lambda_{stay}(F_j) = \sum_{F_j \in C_i} \left(\lambda_{end}(F_j) - \lambda_{begin}(F_j)\right) \tag{7.4}$$

$S(C_i)$ is the cluster stability of cluster C_i; $\lambda = 1/MReachD$, $\lambda_{begin}(F_j)$ and $\lambda_{end}(F_j)$ correspond to the smallest and largest λ value that flow F_j belongs to cluster C_i, respectively. And $\lambda_{stay}(F_j)$ means the range of λ value in between. According to the cluster stability index, F_1, F_2, F_3, F_4, F_5, and F_6 separate as two three-member clusters, while F_{10} sticks with F_7, F_8, and F_9 as a four-member cluster, as denoted by the red bounding boxes in Fig. 7.7c.

Figure 7.8a shows the result map of flowHDBSCAN using a real-world eBay online trade dataset that contains 8,607 flows connecting each seller and buyer (Tao et al. 2017). In total 39 clusters are extracted between popular location pairs between eBay buyers and sellers, while the rest of the flows (in grey color) are discriminated as noises. Most of the clusters are between big cities. The pattern shows that physical distance is not an impedance in this online trade example, as there are some coast-to-coast clusters. The hierarchical structure reveals that inside some of the big clusters, there exist lower-level smaller clusters. For instance, the cluster from the Greater New York area to the San Francisco Bay area can be broken down into two smaller clusters, based on their different destination locations in the northern part and southern part of the Bay area (Fig. 7.8b and c).

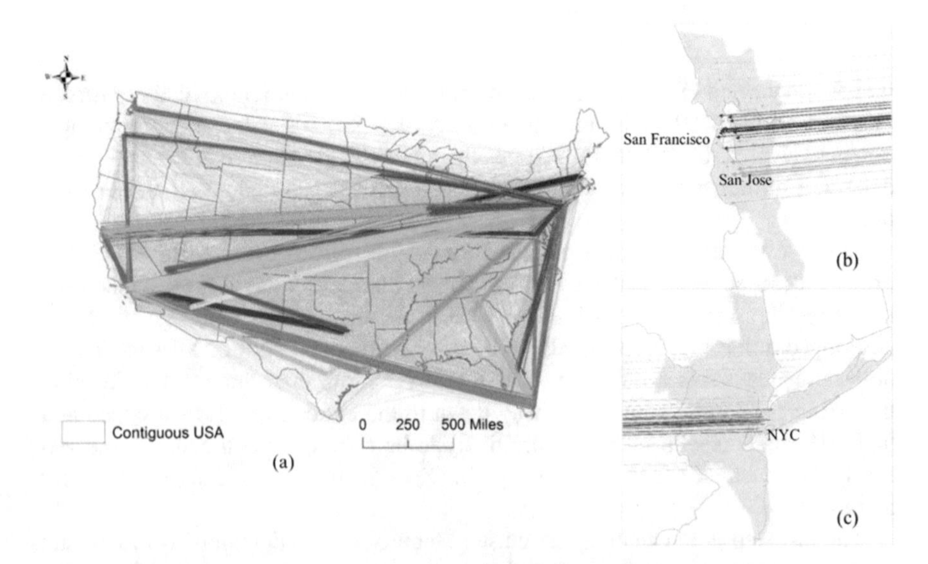

Fig. 7.8 (**a**) Map of eBay trade flow clusters; (**b**) flow clusters end in San Francisco Bay area; (**c**) flow clusters originated from the Greater New York area from Tao et al. (2017)

7.4 Spatial Statistical Methods

While SDM techniques can discover knowledge from large databases, an important question is whether it is possible to derive some understanding, explore relationships and develop hypotheses associated with observed movements and spatial interactions (Murray et al. 2011). In order to further examine these questions in a confirmatory way, spatial statistics are favored owing to their ability to establish inferential properties.

7.4.1 Spatial Patterns Detection

The preponderance of the literature on spatial point pattern analysis treats each point as an event independent from all the others. Spatial flow data, however, encompass at least two points (polygons), one corresponding to the origin location (region) or start of the flow and one for the destination location (region) of the flow. Flow data, therefore, differ fundamentally from point data or polygon data and methods designed to handle the points and polygons cannot be directly applied to flow data. Some endeavors have been undertaken in previous research to fill this gap. Berglund and Karlström (1999) applied the G_i statistics (Getis and Ord 1992; Ord and Getis 1995) to identify local spatial association in flow data. Although several different spatial weight matrices were proposed in this paper to address spatial non-stationarity, only the simplest binary spatial weight matrix based on identical origins or destinations was implemented, which certainly limits its usage. Lu and Thill (2003) proposed an ad hoc and partially qualitative approach in which they apply point cluster detection methods to analyze origin and destination points separately and combine the two sets of results via a relationship table to conclude on the patterns exhibited by the flows. Related issues such as sensitivity to scale and neighborhood definition were discussed in their later work (Lu and Thill 2008).

While decomposing one-dimensional flows into zero-dimensional points can considerably simplify the problem, this approach would inevitably overlook the simultaneity of some critical information, such as flow direction and flow length. Murray et al. (2011) departed from this approach by combining exploratory spatial data analysis and confirmatory circular statistics to analyze the similarities of flow direction and length. However, they sacrifice the actual locational information in the process so that little knowledge on spatial relationships between movements can be extracted. More recently, Liu et al. (2015) extended both global and local Moran's I statistics to a flow context, considering movement distances and directions at once. Nonetheless, their approach is still based on the spatial proximity relationship of either set of end points rather than entire vectors. Therefore, it remains within the scope of measuring spatial autocorrelation of vectors/flows in parts rather than as a whole. Tao and Thill (2016a) came up with Flow K-function, which upgrades classic "hot spot" detection to the level of "hot flow" detection, so that both global and local patterns of flow's spatiotemporal distribution can be measured. The method not only considers flow characteristics, i.e. end points, length, and direction, but also builds on proper measurement of spatial proximity relationship between entire flows. Tao and Thill (2019b) further developed the Flow Cross K-function, which is a bivariate flow analytical method that detects spatial dependency between two types of flows.

Here I mainly introduce the flowAMOEBA (Tao and Thill 2019a), which is a data-driven and bottom-up spatial statistical method for identifying anomalous high- or low-value flow values, for instance, very large volume of travelers between two regions. As a spatial statistical method, flowAMOEBA is mainly an ESDA approach that concentrates on analyzing both the spatial distribution but also the attribute distribution of flow data. The results of flowAMOEBA can be used for further confirmatory analysis, e.g. spatial interaction modeling, which will be introduced in Section 4.2.

Technically, flowAMOEBA is an extension to the method called AMOEBA (Aldstadt and Getis 2006), which identifies clusters of high- or low-value from areal data. flowAMOEBA targets aggregated flows as oppose to individual flows. Therefore, in the preparation stage, flows that sharing identical origin and destination need to be aggregated as one flow by summing up of their numeric values. While the OD matrix of a flow dataset can easily reach an enormous size, it is indeed very common to observe large proportion of the matrix elements as null-value OD pairs. Taking the county-to-county migration flow in the Contiguous U.S. as an example, from year 2010 to 2014 there are less than five percent of the total 9,656,556 OD pairs have at least one migrant. Therefore, only the non-zero flows are taken into the algorithm. Another common characteristic of flow dataset is the heavy-tailed distribution of flow values, in other words, the majority of the non-zero flows have low value. Using a head-tail break to select the flows with relatively large absolute value into the seed pool can significantly reduce the computing time without compromising the result quality.

The first step is to identify the neighbors of every flow based on flow neighborhood definition. A flow's neighbors can be identified based on its endpoints' contiguity. Taking flow *a* in Fig. 7.9 as an example, flow *b* shares the same destination zone and their origin zones are contiguous. low *b'* is in a similar standing with the same origin as flow *a* and an adjacent destination. Both flow *b* and flow *b'* are considered as neighbor flows to flow *a*. Flow *c* represents another situation where both its origin and destination are contiguous to flow *a*. Flow *c* is also seen as a neighbor to flow *a*, but with a longer distance compared with flow *b* and flow *b'*. In a less strict definition, it is possible to extend neighborhood of flow *a* to include second-order neighbors such as flow *d*. But here only flow *b*, flow *b'*, flow *c*, and flows in their equivalent situations are considered as the neighbors of flow *a*, all other flows such as flow *d* and flow *e* are not.

Next, select an arbitrary flow from the preselected pool of seeds. The seed flow is the first member of a flow ecotope, which is the group of the seed flows and some of its neighbors that can expand in space to potentially form a flow cluster. Calculate the G_i^* statistic (Getis and Ord 1992; Ord and Getis 1995) of the seed-only flow ecotope with Equation (7.5).

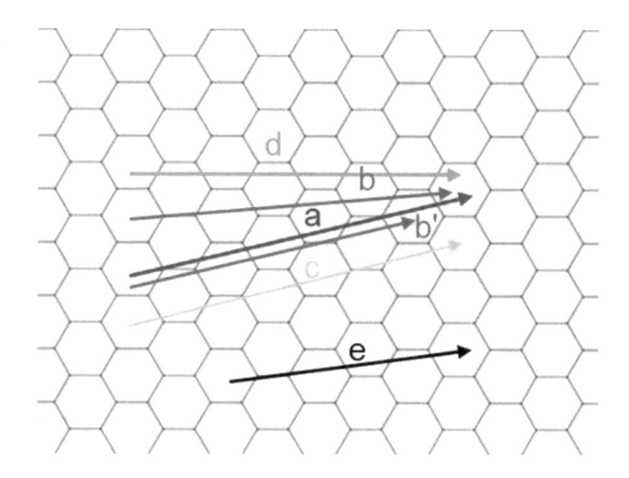

Fig. 7.9 Flow neighborhood definition

$$G_i^* = \left(\sum_{j=1}^{N} w_{ij}x_j - \overline{x}\sum_{j1}^{N} w_{ij}\right) / S\sqrt{\frac{N\sum_{j=1}^{N} {w_{ij}}^2 - \sum_{j1}^{N} {w_{ij}}^2}{N-1}}, \quad \text{where}$$
$$S = \sqrt{\frac{\sum_{j=1}^{N} {x_j}^2}{N} - \overline{x}^2}. \quad (7.5)$$

The classical G_i^*statistic is used to measure the concentration of high or low values at a given location i. The spatial weight w_{ij} is set as 1 if flow j neighbors flow i, otherwise 0. N is the total number of flows. x_j is the value of flow j. $\overline{x}$ is the mean value of all flows.

A search and expand process starts from the seed flow towards its neighbors. Traverse the neighbors one at a time. if the G_i^* statistic of the flow ecotope increases after including a neighbor, keep it. Move on to the next till all neighbors of the seed flow are traversed. In Fig. 7.10, flow i is selected as the seed flow, and its G_i^* statistic is calculated. The grid cells filled with red stripe lines represent the origins and destinations of flow i's neighbors.

Keep expanding the flow ecotope by traversing the neighbors of the newly joined members in the previous step. Again, include only those can increase the G_i^* statistic of the flow ecotope. Do not stop the expansion until the G_i^* statistic reaches the maximum absolute value. The expansion from each seed will result in a stable flow ecotope, which is a flow cluster of high absolute values if it passes statistical significant test later on. Repeat the above steps until a flow cluster is obtained by growing from every seed flow in the preselected seed pool. Figure 7.11 illustrates all identified potential flow clusters originated from different seeds.

The final step is statistical significance test. Conduct a 1,000-time Monte-Carlo simulation by randomly permutating the flow values. Keep the potential flow clusters that pass the statistical significance test as the final outcome. An application of flowAMOEBA with real dataset can be found in Gong et al. (2018). Figure 7.12

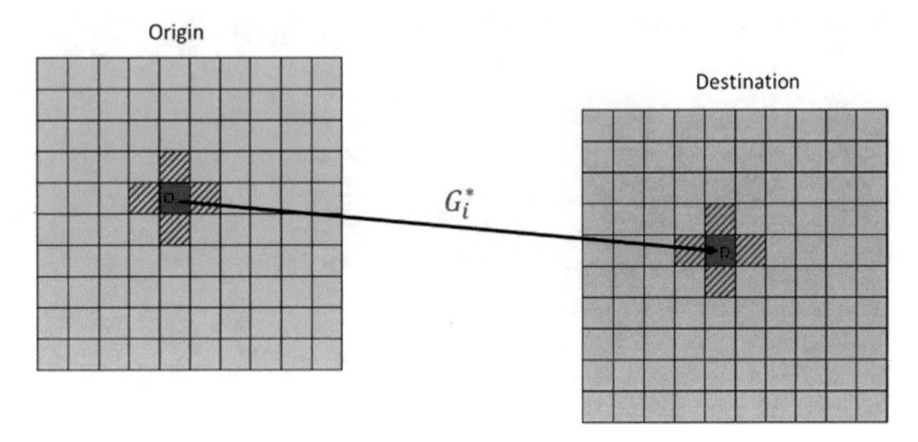

Fig. 7.10 The initial stage of flow ecotope expansion

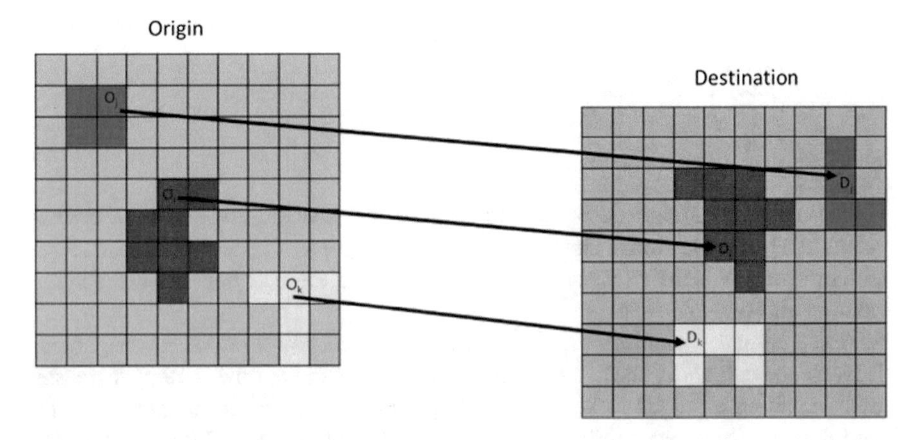

Fig. 7.11 The final stage of flow ecotope expansion

shows one of the results of flowAMOEBA. It detects high volume of daily trips between some cities in Northeast China and Sanya, a mid-size tourist city in the south. The result reflects the popularity of Sanya as a destination for people from Northeast China. Given the raw flow data is about trips in April, the story behind can be more interesting than a seasonal "winter birds" migration for vocational purpose. Socioeconomic and political factors are probably playing a role here, considering the ongoing shrinking-city phenomena in Northeast China.

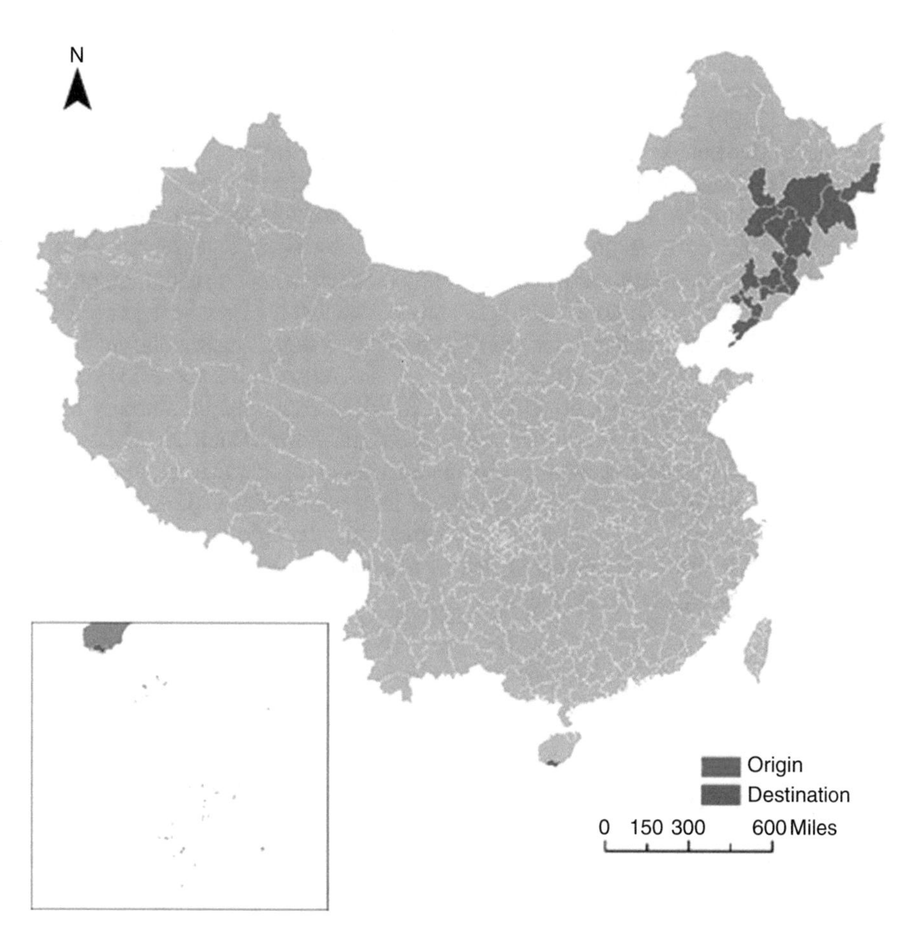

Fig. 7.12 High volume of daily trips between Northeast China cities and Sanya from Gong et al. (2018)

7.4.2 From Patterns to Spatial Interaction Models

The spatial statistical approaches listed above still belong to the broad ESDA family as they aim at exploring the data and detect patterns such as spatial clustering, spatial autocorrelation, and spatial heterogeneity. However, it is still one step short from taking these findings to confirmatory studies in order to reach solid explanatory conclusions. Taking spatial dependence as an example, many researchers have found that this type of spatial effects exists among OD flow data and specifically refer to it as network autocorrelation (Black 1992; Griffith 2007; LeSage and Pace 2008; Chun 2008). Implementing classical spatial interaction models, for example, the gravity model (Zipf 1946); the competing destinations model (Fotheringham 1983; Haynes and Fotheringham 1984); the random utility model (Block and Marschak 1959); and the radiation model (Simini et al. 2012), without taking account of this effect tends to result in incorrect parameter estimation and unsound conclusions as it violates one of the key assumptions of those models, i.e. independence of observations. LeSage and Pace (2008) overcame this problem by proposing spatial weight structures that model dependence among OD flows that are consistent with standard spatial autoregressive models. The spatial weight structures consist of three spatial connectivity matrices capturing origin, destination, and origin-to-destination dependences. Griffith and Chun came up with another solution by using eigenfunction-based filters for accommodating spatial autocorrelation effects within a spatial interaction model (Griffith 2007; Chun 2008; Chun and Griffith 2011). They have proved the effectiveness of using eigenvector filtering to improve spatial interaction modeling in various application cases. Except for common migration flows, scenarios such as journey-to-work commuting flows (Griffith 2009), interregional commodity flows (Chun et al. 2012), and space-time crime incidents flows (Chun 2014) have all been successfully tested. Accounting for spatial autocorrelation in traditional spatial interaction modeling provides an example of incorporation of exploratory spatial statistical considerations into confirmatory hypothesis-testing research. More efforts should be made along this research line to discover the usefulness of interesting findings of emerging ESDA methods.

7.5 Conclusion

The increasing availability of big spatial flow has brought both challenges and opportunities to study spatial interaction phenomena. One of the keys is to develop data-driven methods tailored to big spatial flow data. First, flow mapping and geovisualization are critical to discover and present essential information from the usually cluttered flow maps. Representative techniques include edge bundling (Cui et al. 2008), which bundle the nearby flows on the map together based on their spatial and geometric characteristics, is elaborated in this chapter. Second, various data mining techniques are designed to discover interesting spatiotemporal flow patterns. Among these techniques, clustering analysis is particularly useful to quickly gain insights from massive individual flows. A recent method called flowHBSCAN is introduced step by step. It groups similar flows in space by combining the advantages of hierarchical-based clustering method and density-based clustering method. Third, spatial statistics have been updated to extract patterns of big flow data as well as to explain flow activities and the relevant factors with confirmatory regression models. A clear advantage of statistical methods is that they thoroughly consider flow type and flow values, in addition to the spatial distribution of flow objects. A bottom-up strategy called flowAMOBEA (Tao and Thill 2019a) is selected to introduce in detail. It is designed for identifying and delineating corresponding OD regions, between which there exist an anomalous amount of spatial interaction activities.

Besides the above-mentioned aspects, cyberinfrastructure and high-performance computing (HPC) are commonly incorporated with existing methods to handle the origin-destination matrix of flow data as well as to boost the computing efficiency. However, the way to apply the powerful computing infrastructure and techniques to flow data analysis does not clearly distinguish from other geocomputing applications, therefore, they are not discussed in particular. Last but not least, because spatial flows can be converted to topological graphs, or network data, it is common and sometimes beneficial to apply network analysis to study flow data. For example, Gao et al. (2013) used the Louvain algorithm (Blondel et al. 2008) to extract communities from cellphone call flows. Chin and Wen (2015) applied PageRank (Brin and Page 1998) to rank the relative importance of locations in a spatial flow network. To sum, while we keep developing new flow-exclusive analytical methods to better suit its unique characteristics, we should be aware of its similarities to other types of geospatial data and look out for applicable methods from a broader range. After all, the best method is the one that can help fulfill the research goal most effectively, whether it is flow-specific or not.

References

Aldstadt J, Getis A (2006) Using AMOEBA to create a spatial weights matrix and identify spatial clusters. Geographical Analysis 38(4):327–343

Andrienko N, Andrienko G (2011) Spatial generalisation and aggregation of massive movement data. IEEE Trans Vis Comput Graph 17:205–219

Andrienko G, Andrienko N, Dykes J, Fabrikant SI, Wachowicz M (2008) Geovisualization of dynamics, movement and change: key issues and developing approaches in visualization research. SAGE Publications Sage UK, London

Ankerst M, Breunig MM, Kriegel HP, Sander J (1999) OPTICS: ordering points to identify the clustering structure. ACM Sigmod Rec 28(2):49–60

Berglund S, Karlström A (1999) Identifying local spatial association in flow data. J Geogr Syst 1(3):219–236

Black WR (1992) Network autocorrelation in transport network and flow systems. Geographical Analysis 24(3):207–222

Block HD, Marschak J (1959) Random orderings and stochastic theories of response (No. 66). Cowles Foundation for Research in Economics, Yale University.

Blondel VD et al (2008) Fast unfolding of communities in large networks. J Stat Mech Theory Exp 10:1–12

Boyandin I, Bertini E, Lalanne D (2010) Using flow maps to explore migrations over time. In: Geospatial visual analytics workshop in conjunction with the 13th AGILE international conference on geographic information science 2(3).

Brin S, Page L (1998) The anatomy of a large-scale hypertextual web search engine. Comput Netw ISDN Syst 30(1–7):107–117

Cao G, Wang S, Hwang M, Padmanabhan A, Zhang Z, Soltani K (2015) A scalable framework for spatiotemporal analysis of location-based social media data. Computers, Environment and Urban Systems 51:70–82

Chin WCB, Wen TH (2015) Geographically modified PageRank algorithms: identifying the spatial concentration of human movement in a geospatial network. PLoS One 10(10):e0139509

Chun Y (2008) Modeling network autocorrelation within migration flows by eigenvector spatial filtering. Journal of Geographical Systems 10(4):317–344

Chun Y (2014) Analyzing space-time crime incidents using eigenvector spatial filtering: an application to vehicle burglary. Geographical Analysis 46(2):165–184

Chun Y, Griffith D (2011) Modeling network autocorrelation in space–time migration flow data: an eigenvector spatial filtering approach. Ann Assoc Am Geogr 101(3):523–536
Chun Y, Kim H, Kim C (2012) Modeling interregional commodity flows with incorporating network autocorrelation in spatial interaction models: an application of the US interstate commodity flows. Computers, Environment and Urban Systems 36(6):583–591
Cui W, Zhou H, Qu H, Wong PC, Li X (2008) Geometry-based edge clustering for graph visualization. IEEE Trans Vis Comput Graph 14(6):1277–1284
Ester M, Kriegel HP, Sander J, Xu X (1996) A density-based algorithm for discovering clusters in large spatial databases with noise. Kdd 96(34):226–231
Fang Z, Shaw S, Tu W, Li Q, Li Y (2012) Spatiotemporal analysis of critical transportation links based on time geographic concepts: a case study of critical bridges in Wuhan, China. J Transp Geogr 23:44–59
Farmer C, Oshan T (2017) Spatial interaction. The geographic information science & technology body of knowledge (4th quarter 2017 edition), John P. Wilson (ed.). https://www.ucgis.org/gis-t-body-of-knowledge
Fotheringham AS (1983) A new set of spatial-interaction models: the theory of competing destinations. Environ Plan A Econ Space 15(1):15–36
Gao S, Liu Y, Wang Y, Ma X (2013) Discovering spatial interaction communities from mobile phone data. Transactions in GIS 17(3):463–481
Gao Y, Ye C, Zhong X, Wu L, Liu Y (2019) Extracting spatial patterns of intercity tourist movements from online travel blogs. Sustainability 11(13):3526
Genolini C, Falissard B (2010) KmL: K-means for longitudinal data. Comput Stat 25(2):317–328
Getis A, Ord J (1992) The analysis of spatial association by use of distance statistics. Geographical Analysis 24(3):189–206
Glennon JA (2005) Flow data model tool for ArcGIS 9.0. Department of Geography, University of California, Santa Barbara. Flow tool available at http://www.alanglennon.com/flowtools/
Gong Z, Tao R, Ma Q, Kan C (2018) Exploratory analysis of inter-city human activity interactions for shrinking cities in Northeast China with flowAMOEBA. Proceedings of the 26th GIScience research UK conference, Leicester, UK, 17–20 Apr 2018
Griffith D (2007) Spatial structure and spatial interaction: 25 years later. Rev Reg Stud 37(1):28–38
Griffith D (2009) Modeling spatial autocorrelation in spatial interaction data: empirical evidence from 2002 germany journey-to-work flows. Journal of Geographical Systems 11(2):117–140
Guo D (2009) Flow mapping and multivariate visualization of large spatial interaction data. IEEE Trans Vis Comput Graph 15(6):1041–1048
Guo D, Zhu X (2014) Origin-destination flow data smoothing and mapping. IEEE Trans Vis Comput Graph 20(12):2043–2052
Haynes KE, Fotheringham AS (1984) Gravity and spatial interaction models, vol 2. Sage, Beverly Hills
Holten D, Van Wijk JJ (2009) Force-directed edge bundling for graph visualization. Comput Graph Forum 28:983–990
Lee JG, Han J, Whang KY (2007) Trajectory clustering: a partition-and-group framework. In Proceedings of the 2007 ACM SIGMOD international conference on management of data. 593–604.
LeSage JP, Pace RK (2008) Spatial econometric modeling of origin-destination flows. J Reg Sci 48(5):941–967
Liu L, Andris C, Ratti C (2010) Uncovering cabdrivers' behavior patterns from their digital traces. Computers, Environment and Urban Systems 34(6):541–548
Liu Y, Sui Z, Kang C, Gao Y (2014) Uncovering patterns of inter-urban trip and spatial interaction from social media check-in data. PLoS One 9(1):e86026
Liu Y, Tong D, Liu X (2015) Measuring spatial autocorrelation of vectors. Geographical Analysis 47(3):300–319
Lu Y, Thill J-C (2003) Assessing the cluster correspondence between paired point locations. Geographical Analysis 35(4):290–309

Lu Y, Thill J-C (2008) Cross-scale analysis of cluster correspondence using different operational neighborhoods. Journal of Geographical Systems 10(3):241–261
Marble DF, Gou Z, Liu L, Saunders J (1997) Recent advances in the exploratory analysis of interregional flows in space and time. In: Kemp Z (ed) Innovations in GIS 4. Taylor & Francis, London, pp 75–88
Murray A, Liu Y, Rey SJ, Anselin L (2011) Exploring movement object patterns. Ann Reg Sci 49(2):471–484
Nanni M, Pedreschi D (2006) Time-focused clustering of trajectories of moving objects. J Intell Inf Syst 27(3):267–289
Ord J, Getis A (1995) Local spatial autocorrelation statistics: distributional issues and an application. Geographical Analysis 27(4):286–306
Ossama O, Mokhtar H, El-Sharkawi M (2011) Clustering moving objects using segments slopes. Int J Database Manag Syst 3(1):35–48
Phan D, Xiao L, Yeh R, Hanrahan P (2005) Flow map layout. IEEE Symp Inf Vis:219–224
Qu H, Zhou H, Wu Y (2006) Controllable and progressive edge clustering for large networks. Proc Symp Graph Draw:399–404. Springer, Berlin, Heidelberg
Shekhar S, Lu CT, Zhang P (2003) A unified approach to detecting spatial outliers. GeoInformatica 7(2):139–166
Simini F, González MC, Maritan A, Barabási AL (2012) A universal model for mobility and migration patterns. Nature 484(7392):96
Tao R, Thill JC (2016a) Spatial cluster detection in spatial flow data. Geogr Anal 48(4):355–372
Tao R, Thill JC (2016b) A density-based spatial flow cluster detection method. In: The ninth international conference on geographic information science (GIScience2016) short paper proceedings. Montreal, Canada. pp 288–291
Tao R, Thill JC (2019a) flowAMOEBA: identifying regions of anomalous spatial interactions. Geographical Analysis 51(1):111–130
Tao R, Thill JC (2019b) Flow cross K-function: a bivariate flow analytical method. Int J Geogr Inf Sci 33(10):2055–2071
Tao R, Thill JC, Depken C, Kashiha M (2017) Flow HDBSCAN: a hierarchical and density-based spatial flow cluster analysis method. In: Proceedings of UrbanGIS'17:3rd ACM SIGSPATIAL workshop on smart cities and urban analytics. Redondo Beach
Thill JC (2011) Is spatial really that special? A tale of spaces. In: Information fusion and geographic information systems. Springer, Berlin, Heidelberg, pp 3–12
Tobler WR (1987) Experiments in migration mapping by computer. Am Cartogr 14:155–163
Tobler WR (2004) Movement mapping. http://csiss.ncgia.ucsb.edu/clearinghouse/FlowMapper
Yan J, Thill J-C (2009) Visual data mining in spatial interaction analysis with self-organizing maps. Environ Plan B Plan Design 36(3):466–486
Yang G, Ma K, Yuan X, Li J, Lu Q (2019) Expectation-based 3D edge bundling. Multimed Tools Appl 78(24):35099–35118
Zhu X, Guo D (2014) Mapping large spatial flow data with hierarchical clustering. Trans GIS 18(3):421–435
Zhu X, Guo D, Koylu C, Chen C (2019) Density-based multi-scale flow mapping and generalization. Comput Environ Urban Syst 77(2):101359
Zipf GK (1946) The P 1 P 2/D hypothesis: on the intercity movement of persons. Am Sociol Rev 11(6):677–686

Chapter 8
Semantic Trajectories Data Models

Maria Luisa Damiani

8.1 Introduction

Semantic trajectories is a major paradigm for representation of the individual movement. Different from spatial trajectories, which are primarily intended to represent the continuous movement of an entity in a coordinated space, e.g. the path of a vehicle, semantic trajectories are grounded on the idea of movement as *context evolution*, where the context can be any set of features characterizing the situation of the entity in time (Dey 2001).

Semantic trajectories can represent many different kinds of movement data, such as, series of geo-tagged tweets posted by an individual, series of activities performed in a time frame, paths of vehicles augmented with sensor data on e.g., weather conditions. Abstractly, a semantic trajectory can be seen as a finite sequence of states, where a state is a snapshot of the context in which the movement takes place, typically specifying the individual location together with non-spatial attributes whose values changes in time. The transitions from one state to the next describe the object movement. Actually, the notion of state is instrumental in abstracting a common ground from the variety of existing semantic trajectories representations. A state can be seen as consisting of two components, a time interval I, and a description A of the context during I. With little abuse of terminology, we refer to A as *annotation*. Structurally, a semantic trajectory, is a series of states or *units*, i.e. $(I_1, A_1) \ldots (I_n, A_n)$.

The paradigm of semantic trajectories encompasses different models, which differ for level of abstraction and purpose. This article attempts to provide a few insights into major research themes and controversial aspects, without any ambition

M. L. Damiani (✉)
University of Milan, Milan, Italy
e-mail: maria.damiani@unimi.it

M. Werner, Y.-Y. Chiang (eds.), *Handbook of Big Geospatial Data*,
https://doi.org/10.1007/978-3-030-55462-0_8

for completeness. The remainder of the article is structured as follows. In order to put the discussion into a proper perspective, Sect. 8.2 overviews the origins and evolution of the paradigm, reporting as well a few considerations on the relationship between spatial and semantic trajectories. Section 8.3 elaborates on a simple meta-model of semantic trajectory, to illustrate core concepts. Section 8.4 presents a possible classification of semantic trajectories models based on their purpose. The article ends up with a discussion on research directions.

8.2 Preliminaries

8.2.1 Historical Perspective

For a better understanding of the current research context, we start from some historical notes. Spatial trajectories is the first prominent representation of movement data. Rooted in GPS technology, this representation is at the heart of the Moving Object database model developed in early 2000 (Güting et al. 2000).

In the second half of the first decade, novel types of sequential data, other than GPS trajectories, became available, including telecommunication data (e.g., Call Detail Record – CDR data), and series of activities, possibly resulting from a knowledge discovery process. Following this trend, the GeoPKDD project (Giannotti and Pedreschi 2008), a research project funded by the European Union and carried out by a number of research groups spread across Europe, coined the term *semantic trajectory*, first appeared in 2007 (Alvares et al. 2007). The core idea was to enrich spatial trajectories with application-dependent information, possibly extracted from raw mobility data and supplementary data sources through the use of knowledge discovery methods. In the same period, similar concerns were raised in other projects carried out in Asia and US (Zheng et al. 2008; Liu et al. 2006). Overall, the research in this period is mostly focused on analytical methods and methodologies related to the conceptual modeling and construction of semantic trajectories, e.g. Yan et al. (2013). For a survey of early research, we refer the reader to Parent et al. (2013).

A more recent stream of research focuses on the data engineering aspects, in particular the management of enriched trajectories through a database system. This line of research is motivated by the growing availability of big enriched trajectory data, heterogeneous and voluminous, calling for efficient processing methods. Especially the plethora of geo-social applications, developed in the last decade, has led to the collection of large amounts of data, e.g. check-in data, naturally organized as sequences of geo-referenced POIs, where a POI (Point of Interest) is a name, possibly accompanied by supplementary information, such as the facility type. Concerning data modeling, we witness the specification of rigorous trajectory models and the systematic development of operational solutions targeting the practical utilization of semantic trajectories. Two contrasting views, however, emerge, the one is application-driven and interprets semantic trajectories primarily as series of time-stamped POIs; the other is application-independent and targets development of generic trajectory models.

The view presented in this article is in line with the latter interpretation, that is a semantic trajectory is any representation of the movement as evolving context. In this sense, the term 'semantic trajectories' does not indicate a specific model, but rather a paradigm. In conclusion, semantic trajectories respond to the need of encompassing in a representation situational information made available by novel data acquisition technologies and applications.

8.2.2 Spatial vs. Semantic Trajectories

We elaborate a bit more on the difference between semantic and spatial trajectories:

Spatial trajectories Spatial trajectories are built on sequences of time-stamped points sampling the entity's movement in a geometric space $\mathcal{S}$, typically the Euclidean or the geographical space. Given a temporal domain $\mathcal{T}$, a spatial trajectory is the sequence: $T = (x_1, y_1, t_1) \ldots (x_n, y_n, t_n)$, with $(x_i, y_i) \in \mathcal{S}, t \in \mathcal{T}$. The underlying assumption is that the movement is continuous, namely the location changes smoothly in time, while the sampling rate is sufficiently high so as to ensure a good approximation of the actual movement. Spatial trajectories can be stored in a database, typically as values of suitable data types, and accessed through spatio-temporal query languages (Güting et al. 2000; Zheng and Zhou 2011). A standard model of spatial trajectories has been recently proposed by OGC (Open GeoSpatial Consortium) (OGC 2019).

Semantic trajectories From the data modeling perspective, we point out three major differences between spatial and semantic trajectories:

- Geometry vs. annotation. Spatial trajectories are grounded on the geometric representation of the object's location. In contrast, semantic trajectories rely on the notion of annotation, which encompasses multiple and diverse kinds of information, including spatial data as a special case. In this sense, the notion of semantic trajectory goes beyond the spatial context.
- Whole vs. part. Spatial trajectories describe the movement of an object as a whole. Therefore, attributes can be only attached to the entire trajectory and not to parts of it. For example, the attribute 'weather' can be only specified for the whole travel. By contrast, a semantic trajectory is an aggregation of smaller components, the units, while the relationship between the unit and the whole trajectory is the mereological relationship of parthood. Annotations can thus be attached to units, and that results in a finer-grained representation of the actual movement.
- Continuous vs. non-continuous. A spatial trajectory provides an approximate representation of a continuous movement taking place in a physical space. In con-

trast, semantic trajectories can describe a broader range of movements, including discrete and stepwise evolving movements. In this sense, semantic trajectories can provide a framework suitable for a broader spectrum of applications.

8.3 A Semantic Trajectory Meta-model

To exemplify the above concepts, we elaborate on the generic trajectory model sketched in the introductory section. The goal is to provide a simple meta-model sufficiently general to accommodate the diverse interpretations and facilitate the understanding of key concepts.

A simple meta-model A semantic trajectory T is a sequence of units, where the unit i-th specifies the temporal extent $I_i = [t_b^i, t_e^i]$ and the annotation A_i, i.e. $T = (I_1, A_1) \ldots (I_n, A_n)$. Trajectories have the following properties:

- A time interval can degenerate in a single instant, i.e. $t_b^i = t_e^i$. In that case the unit is an *event*. A simplified representation for the event is (t_i, A_i)
- Time periods $I_1, \ldots, I_n$ do not overlap, i.e. $\bigcap_i I_i = \emptyset$.
- There cannot exist two adjacent periods reporting the same annotation, i.e. every unit contains the maximal interval with identical annotation.
- There can exist temporal discontinuities (gaps), namely consecutive units are not necessarily adjacent in time.
- The annotations are application-dependent. Moreover, an annotation may include spatial data as a special case.

In the following, we show a few examples of annotations and how these can be expressed in terms of the generic model.

Annotations as simple values Annotations can indicate generic activities, such as tasks (e.g. working, shopping) and transportation means (e.g. train, taxi). Activities have normally a duration, moreover can be represented as attributes of simple type, e.g. string. The movement described by the trajectory has thus the meaning of stepwise evolution in a discrete space. A possible structure for the trajectory unit is: $u_i = (I_i, [l_1, \ldots, l_n])$ where I_i is the time interval and $[l_1, \ldots, l_n]$ the list of one or more values that, globally, form the annotation. A graphical example of trajectory annotated with strings, specifically reporting the transportation means used by an individual during a trip, is shown in Fig. 8.1.

Annotations as labeled points A labeled point is a named location. Semantic trajectories annotated with labeled points can describe, for example, the series of check-ins posted by the members of a geo-social network. In that case, the unit model is an *event* (in the sense given earlier), while the annotation can include the name of the POI, its category (and additional attributes), and the location coordinates. To generalize, a possible structure of the trajectory units is: $u_i = (t_i, [att_i^1, att_i^n, (x, y)_i])$ indicating: a time-stamp, a sequence of attributes, and a

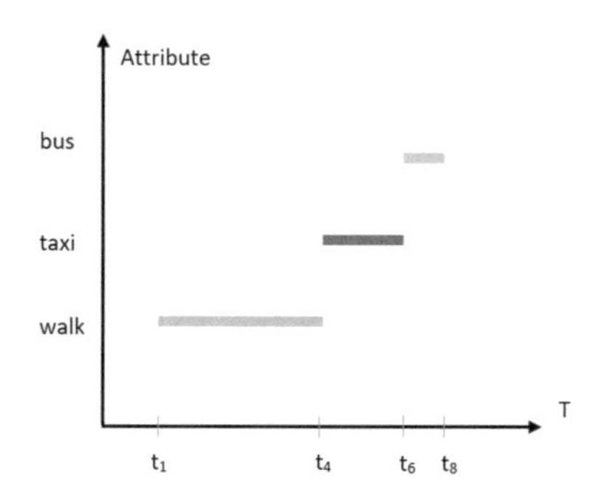

Fig. 8.1 Semantic trajectory as sequence of time-stamped strings

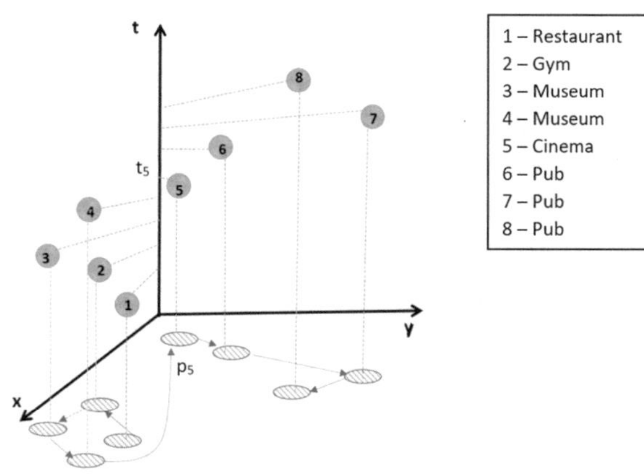

Fig. 8.2 A semantic trajectory as sequence of labeled points

coordinated point, respectively. The graphical representation of semantic trajectory as sequence of labeled points is shown in Fig. 8.2.

Annotations as spatial sub-trajectories with attributes In this case, spatial trajectories are split into parts, i.e. segments, based on some partitioning criteria, and each segment is assigned one or more attributes. Attributes can indicate for example the activity performed in the period and the weather conditions. A possible structure for the unit is: $u_i = (I_i, [l_i^1, \ldots, l_m^i], segm_i)$, where the pair $[l_i^1, \ldots, l_m^i], segm_i$ indicates the series of attributes, and the segment, respectively. A graphical example of trajectory is reported in Fig. 8.3.

8.4 Semantic Trajectory Data Models: A Purpose-Driven Taxonomy

The paradigm of semantic trajectories encloses a variety of trajectory models. In this section, we discuss a possible classification of the existing models that, either explicitly or implicitly, follow the semantic trajectory paradigm. Models are

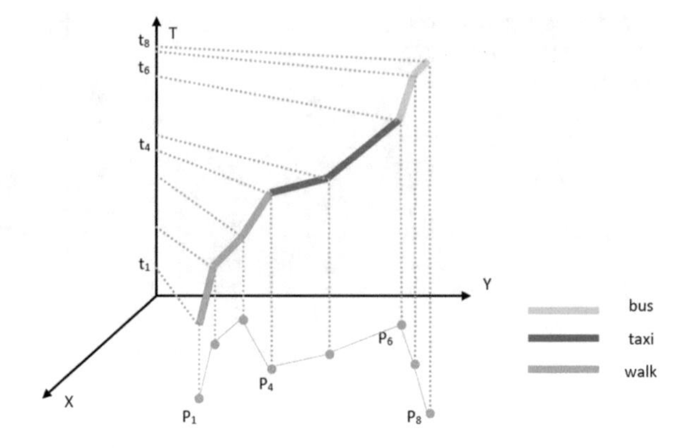

Fig. 8.3 A semantic trajectory as sequence of labeled spatial sub-trajectories

primarily characterized by the structural properties of trajectories. Such a structure is strongly related to the purpose of the representation. In particular, we identify four main purposes and thus classes, we refer to as:

- Conceptual representation
- Database logical model
- Query processing
- Data analytics

In the following, we describe those classes in more detail and show a few representative models.

8.4.1 *Conceptual Representation*

The semantic trajectory models of this class target generic and expressive representations and are exclusively defined at conceptual level. That is, models are built on user-oriented abstractions drawn from application practices and requirements, while the operational aspects, for example which operations can be performed over semantic trajectories, are not taken into account. This class comprises two main groups of models, those based on stop-and-move patterns and those based on episodes, respectively, described as follows.

Models based on stops-and-moves The seminal work in Spaccapietra et al. (2008) was inspired by the observation that moving individuals often alternate periods of relative stationarity (stops) with periods of mobility (move). Goal of the model was to capture that mobility behavior by annotating spatial trajectories with labels denoting stops and moves.

Stop-and-move is a pattern of practical relevance in a variety of phenomena. For example, in an urban setting, the stops can represent POIs, while the moves the

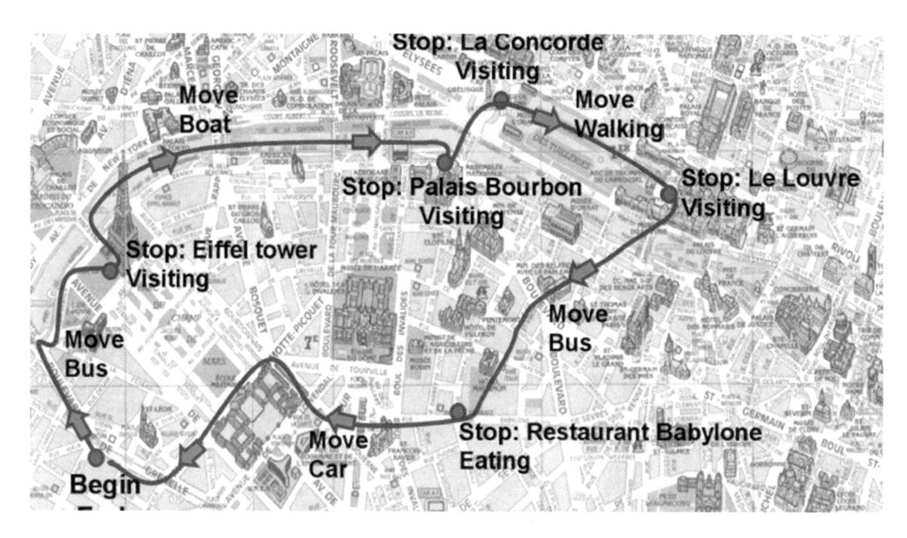

Fig. 8.4 A semantic trajectory consisting of stop and moves (Parent et al. 2013)

transportation means used by individuals for moving from one stop to another; in the field of animal ecology, the stops can indicate the animal home-ranges, while the moves the seasonal migrations. To convey the intuition, an example of trajectory consisting of stops and moves is illustrated in Fig. 8.4. Following the traditional approaches to conceptual modeling, this model is expressed using a graphical notation. The substantial limitation of this model (and alike) is that it is built on a specific mobility pattern.

Models based on episodes The models of this group aim to overcome the above limitations, by introducing generic concepts not confined to any specific mobility pattern. The core idea is to enrich spatial trajectories with supplementary information, i.e. numerical and categorical attributes, which can regard either the single points of the trajectories, or the segments resulting from a prior operation of trajectory partitioning. In the latter case, the segments are called *episodes* and are associated with a time interval (Parent et al. 2013). In this view, stops and moves can be seen as specific types of episodes. The operation of segmentation turns the continuous movement, represented by the spatial trajectory, into a stepwise movement. A semantic trajectory can specify one or multiple segmentations. For example, the trajectory representing a travel can be partitioned based on the transportation means used by the individual and weather conditions. This generalization provides the basis for extended conceptual models, e.g. Bogorny et al. (2014) and Mello et al. (2019).

8.4.2 Database Logical Models

Another class of semantic trajectory models are those utilized in databases. In this case, the goal is not only to devise a rich representation, but also to provide appropriate techniques and languages for the manipulation and querying of trajectories through a database system. In general, the use of rich-content data models is paid in terms of complexity, usability, efficiency. Simply, the more expressive the representation, the more complex the operational system. Therefore, a key challenge is to find a trade-off between expressivity and cost of the solution. Another aspect of concern is how to embed the trajectory data model into an existing and extensible database system, so as to make the system usable in real applications. Two major models of this class are *symbolic trajectories*, and its evolution, *multi-attribute trajectories*, discussed next.

Symbolic trajectories Symbolic trajectories (Valdés et al. 2013; Damiani and Güting 2014; Güting et al. 2015) describe a movement that evolves step-wise, not necessarily in a physical space. In its basic form, a symbolic trajectory consists of a series of units: $(I_1, l_1), \ldots, (i_n, l_n)$ where I_j is a time interval and l_j a single label (i.e., a name) or, in alternative, a set of labels. Units can be also annotated with one or more places, where a place is a labeled point. An example of symbolic trajectory, reporting a series of activities (Damiani et al. 2015), is as follows:

```
([8:45 - 17:00] working)([17:00 - 18:30] shopping) (..)
```

Key feature of the model is the pattern-based query language, for the retrieval of trajectories based on pattern matching and rewriting. Matching is used to retrieve the trajectories satisfying a pattern, while rewriting is to extract or redefine parts of trajectories matching the given pattern. A pattern is sequential, namely it consists of a series of *simple patterns* that are to be matched in the given order. Moreover, patterns are defined using regular expressions, variables, and predicates. In particular, variables are used to verify predicates beyond the scope of a simple pattern (Damiani et al. 2015). For example a query containing a variable (X) and a condition on the variable (the condition following //) is the following:

```
Q: * (morning working) X(_ shopping)* //duration X.time> 2 * hour
```

In this example, the trajectories matching Q are those in which the working activity takes place in the morning and is followed by a shopping activity taking more than 2 hours. For the language syntax, we refer the reader to Güting et al. (2015).

Another feature is that the pattern language and the type system are integrated into an extensible moving object database (Güting et al. 2015). In particular, a symbolic trajectory is a value of type *mlabel* (or mlabels for sets of labels), or *mplace* (and mplaces), wherein the construct *m*type is a shortcut of *moving*(type), defining a mapping from the time domain to the domain of the given type. Hence, a value of type *mlabel* indicates a time-varying string. As an example, we can construct a relation describing the trips of individuals through an attribute of type *mlabel* specifying the road along which the user is traveling:

```
CREATE TABLE Trips (Id: int, Trip: mlabel)
```

This model suffers from a major limitation, that the annotation is exclusively textual.

Trajectories with multiple attributes This data model (Valdés and Güting 2019) responds to the limitations of symbolic trajectories, by extending such a model with annotations consisting of multiple attributes of arbitrary time-varying type. Time-varying types include *mpoint* (i.e. a time-varying point representing a spatial trajectory) and *mreal/mint* (i.e., a time varying numeric quantity). In essence, the idea is to annotate a time interval with a tuple of attribute values, where the attributes can be of different type, either spatial or not. As an example, the *Trips* table can have, as attributes, a spatial trajectory describing the actual path, the time-varying road name, and the time-varying speed limit along the road, as shown below:

```
CREATE TABLE Trips (Id: int, Path: mpoint, RoadName: mlabel,
                    SpeedLimit: mint)
```

In this example, the non-spatial attributes, i.e. road name and speed limit, are categorical attributes that evolve step-wise, inducing two different segmentations of the spatial trajectory. Sequential queries over multi-attribute trajectories are formulated through a slightly extended pattern-based language, in which variables are associated to tuple-based units.

8.4.3 *Query Processing*

The models in this class are instrumental to the efficient computation of selected types of queries over large datasets of trajectories. This class also includes intermediate representations, possible resulting from a pre-processing step. In this sense, these models are possibly defined ad-hoc. In the following, we outline two representative approaches dealing with different types of semantic trajectories and focusing on specific types of query.

Top-K spatial keyword queries over trajectories of labeled locations The overall goal is the efficient retrieval of semantic trajectories defined as *activity trajectories* (Zheng et al. 2016). An activity trajectory is a series of timestamped POIs, supplemented with textual keywords describing the activities undertaken at that place. Structurally, a trajectory consists of units representing events (in the sense of Sect. 8.3) annotated with a POI p and with a set of keywords $h_1, \ldots, h_m$, i.e., $u = (t, p, \{h_1, \ldots h_m\})$. Given a query q=$(x, y, tw, qw, a)$ specifying a location (x, y), a time interval tw, a set of keywords qw and a weight factor a used as preference, the query retrieves the k trajectories that are most similar to the query, based on a criteria of similarity encompassing spatial distance, keywords matching and places popularity. Example: the user located in p is willing to eat Japanese food, watch movie and go to a pub. The user does not know which places are more convenient to visit, thus search for the trajectories reporting similar travel experience by users moving nearby. Query processing is supported by an augmented R-tree index, called ITB-tree, supporting both spatial and keyword search.

Sequential range queries over labeled spatial sub-trajectories
In this case, the goal is to efficiently retrieve spatial trajectories augmented with textual annotations. Specifically, a semantic trajectory is a sequence of labeled segments, possibly resulting from a prior temporal overlay of a spatial trajectory with a symbolic trajectory (Güting et al. 2015). The units of the trajectory are triples (I_i, l_i, seg_i), where the pair l_i, seg_i consists of a spatial trajectory segment and a label, respectively (Issa and Damiani 2016).

A sequential range query q is a sequence of *simple* queries, $q = q_1, \ldots, q_n$, each specifying a range constraint on time, space and labels, which are to be solved in the given order. Example: find the users traveling by car or by bus and passing by region A in the morning and by region B in the evening. In this query, A/B define the spatial range, morning/evening the temporal range and car/bus the label range. The query semantics is defined as follows: a trajectory satisfies the query q if there exist n instants $t_1, \ldots, t_n$ with $t_i < t_{i+1}$ such that for every instant t_i, there exists a trajectory unit matching q_i (Issa and Damiani 2016). Importantly, the semantics, and thus the result of the query, does not depend on the distribution of the sample points in the spatial trajectory. Query processing is supported by an hybrid index, called IRWI (IR index with Trajectory Identifiers), an augmented 3D R-Tree integrating space, time and label search. Moreover the input query is processed by evaluating concurrently every simple query of the sequence during the traversal of the IRWI tree.

8.4.4 Data Analytics

The last class of trajectory models are those employed for data analytics purposes. These models are application dependent, namely the structural properties of the model do not result from an abstraction effort, but rather are drawn from the application problem. These trajectories are, therefore, grounded on real application needs, and, for that reason, interesting for the discussion. Since this is a broad area, we limit ourselves to consider two recent examples of analytical problems on semantic trajectories, while we refer the reader to Zheng (2015) for a survey:

Frequent sequential pattern mining The general goal is to respond to queries such as: *Where do people usually go to relax after work?* (Zhang et al. 2014; Choi et al. 2017). A specific instance of the problem considers trajectories structured as sequence of *events* annotated with categorized POIs (e.g., restaurant) (Zhang et al. 2014). In more detail, the mining task is to extract from trajectories, sequences of groups of POIs satisfying a number of constraints, in particular, the members of the groups shall be semantically homogeneous and spatially close; and the groups shall be visited in a specific order and satisfying a maximum transition time. The sequences matching those conditions are the sequential patterns. . The ultimate goal is to find those sequential patterns that are frequent in the trajectory dataset.

Topical trajectory pattern mining (Kim et al. 2015) The problem is related to the analysis of large collections of geo-tagged messages posted through micro-blogging services such as Twitter. The input dataset consists of *geo-tagged message trajectories*, namely trajectories reporting the series of time-stamped and geo-referenced messages posted by users. In more detail, a trajectory consists of units representing events, where the annotation consists of a coordinated point (x,y), i.e. the user position, and a set of words taken from a given vocabulary, and extracted from the actual text message. The mining task is to identify the geographic locations, where geo-tagged messages are posted with the same topic (*semantic regions*), and the transitions between semantic regions.

8.5 Final Remarks and Research Directions

In summary, the paradigm of semantic trajectories, encompasses a large variety of models defined at different levels of abstraction and for different purpose. Semantic trajectories are, thus, characterized by a great structural heterogeneity. Overcoming such fragmentation is a challenge. Another major goal is to provide workable solutions allowing for an effective utilization of large datasets of semantic trajectories. Although a few systems are available, e.g. Valdés and Güting (2019), Gryllakis et al. (2018), and Mello et al. (2019), the engineering efforts are still limited.

Concerning the structural properties of trajectories, the 'semantics' of the movement is often expressed by textual or spatial keywords embedded as annotations. While the integration of text, space and time raises interesting research issues, dealing with keywords only might be not sufficient to fulfill the requirements posed by modern applications, e.g. IoT. In essence, what is needed is the capability of describing the context evolution through multiple dimensions. The recent database model of multi-attribute trajectory (Valdés and Güting 2019) is in line with that view, similarly the notion of multi-aspect trajectory in Mello et al. (2019). Several issues, however, remain to be addressed, including: the specification of a unifying framework enabling the coherent representation of both discrete, stepwise and continuous movement; the specification of query processing mechanisms over multi-dimensional trajectories, e.g. Xu et al. (2019); the efficient processing of complex operations, such as semantic trajectories join. All these functionalities are instrumental in supporting trajectory data analytics and, more in general, the research and technological stream of *mobility data science*. From a data representation perspective, a challenging direction is towards *representation learning* methods to support complex tasks such as trajectory similarity detection (Li et al. 2018), applied to semantic trajectories. In this perspective, the availability of a solid and flexible semantic trajectory representation framework paves the way to novel research opportunities on fine-grained behavior analysis and prediction.

Acknowledgments This work is partially supported by the Italian government via the NG-UWB project (MIUR PRIN 2017).

References

Alvares LO, Bogorny V, Kuijpers B, de Macedo J, Moelans B, Vaisman A (2007) A model for enriching trajectories with semantic geographical information. In: Proceedings of ACM GIS

Bogorny V, Renso C, de Aquino AR, de Siqueira FL, Alvares LO (2014) CONSTAnT – a conceptual data model for semantic trajectories of moving objects. Trans GIS 18(1):66–88

Choi D-W, Pei J, Heinis T (2017) Efficient mining of regional movement patterns in semantic trajectories. Proc VLDB Endow 10(13):2073–2084

Damiani ML, Güting RH (2014) Semantic trajectories and beyond. In: Proceedings of IEEE MDM

Damiani ML, Issa H, Güting RL, Valdés F (2015) Symbolic trajectories and application challenges. SIGSPATIAL Special 7(1):51–58

Dey AK (2001) Understanding and using context. Pers Ubiquit Comput 5:4–7

Giannotti F, Pedreschi D (2008) Mobility, data mining and privacy: geographic knowledge discovery, 1 edn. Springer Publishing Company, Incorporated, Berlin/Heidelberg

Gryllakis F, Pelekis N, Doulkeridis C, Sideridis S, Theodoridis Y (2018) Spatio-temporal-keyword pattern queries over semantic trajectories with hermes@neo4j. In: Proceedings of EDBT

Güting RH, Böhlen MH, Erwig M, Jensen CS, Lorentzos NA, Schneider M, Vazirgiannis M (2000) A foundation for representing and querying moving objects. ACM Trans Database Syst 25:1–42

Güting RH, Valdés F, Damiani ML (2015) Symbolic trajectories. ACM Trans Spat Algorithms Syst 1(2):7

Issa H, Damiani ML (2016) Efficient access to temporally overlaying spatial and textual trajectories. In: IEEE MDM

Kim Y, Han J, Yuan C (2015) Toptrac: topical trajectory pattern mining. In: Proceedings of KDD

Li X, Zhao K, Cong G, Jensen C, Wei W (2018) Deep representation learning for trajectory similarity computation. In: Proceedings of ICDE

Liu J, Wolfson O, Yin H (2006) Extracting semantic location from outdoor positioning systems. In: Proceedings of the 7th international conference on mobile data management, p 73

Mello RS, Bogorny V, Alvares LO, Zambom Santana L, Ferrero C, Frozza A, Schreiner GA, Renso C (2019) Master: a multiple aspect view on trajectories. Trans GIS 23(4):805–822

OGC (2019) Moving features encoding part 1: Xml core. http://docs.opengeospatial.org/is/18-075/18-075.html

Parent C, Spaccapietra S, Renso C, Andrienko G, Andrienko N, Bogorny V, Damiani ML, Gkoulalas-Divanis A, Macedo J, Pelekis N, Theodoridis Y, Yan Z (2013) Semantic trajectories modeling and analysis. ACM Comput Surv 45(4), Article 42, 32 pp

Spaccapietra S, Parent C, Damiani ML, de Macêdo JAF, Porto F, Vangenot C (2008) A conceptual view on trajectories. Data Knowl Eng 65(1):126–146

Valdés F, Güting RH (2019) A framework for efficient multi-attribute movement data analysis. VLDB J 28(4):427–449

Valdés F, Damiani ML, Güting RH (2013) Symbolic trajectories in secondo: pattern matching and rewriting. In: DASFAA
Xu J, Bao Z, Lu H (2019) Continuous range queries over multi-attribute trajectories,. In: IEEE international conference on data engineering (ICDE)
Yan Z, Chakraborty D, Parent C, Spaccapietra S, Aberer K (2013) Semantic trajectories: mobility data computation and annotation. ACM Trans Intell Syst Technol 4:1–38
Zhang C, Han J, Shou L, Lu J, La Porta T (2014) Splitter: Mining fine-grained sequential patterns in semantic trajectories. Proc VLDB Endow 7(9):769–780
Zheng Y (2015) Trajectory data mining: an overview. ACM Trans Intell Syst Technol 6(3):1–41
Zheng Y, Zhou X (2011) Computing with spatial trajectories. Springer Publishing Company, Incorporated, New York
Zheng Y, Wang L, Zhang R, Xie X, Ma W (2008) Geolife: managing and understanding your past life over maps. In: Proceedings of IEEE MDM
Zheng K, Zheng B, Xu J, Liu G, Liu A, Li Z (2016) Popularity-aware spatial keyword search on activity trajectories. World Wide Web 20(4):749–773

Chapter 9
Multi-attribute Trajectory Data Management

Jianqiu Xu

9.1 Introduction

Trajectory data, keeping track of historical movements of moving objects such as vehicles and ships, is becoming ubiquitous due to the widespread use of GPS devices. Such data that records geographical locations changing over time is of crucial importance for emerging applications, e.g., route recommendation (Chen et al. 2010; Tong et al. 2017, 2018), tracking (Lange et al. 2011), monitoring (Yao et al. 2014), to name but a few.

Despite tremendous efforts made on studying trajectory databases, proposals in the literature mainly deal with *standard trajectories* (Tzoumas et al. 2009; Long et al. 2013; Zheng et al. 2013b), i.e., a sequence of time-stamped geo-locations. The majority of queries are limited to the spatio-temporal evaluation such as range queries (Wang and Zimmermann 2011), nearest neighbors (Güting et al. 2010b) and convoys (Jeung et al. 2008). In the real world, typical moving objects such as vehicles and persons are associated with pieces of descriptive information. The database system should represent moving objects by considering several aspects and allow users to query objects with extensive knowledge to better understand the movement and users' behavior. As a fundamental step towards that, a new form of trajectories is investigated called *multi-attribute trajectories*, each of which consists of a standard trajectory and a set of attribute values. Modeling and representing standard trajectories has been well established (Güting and Schneider 2005), while attributes have various semantics according to applications. The combination allows users to issue queries with both spatio-temporal and attribute predicates.

J. Xu (✉)
Nanjing University of Aeronautics and Astronautics, Nanjing, China
e-mail: jianqiu@nuaa.edu.cn

M. Werner, Y.-Y. Chiang (eds.), *Handbook of Big Geospatial Data*,
https://doi.org/10.1007/978-3-030-55462-0_9

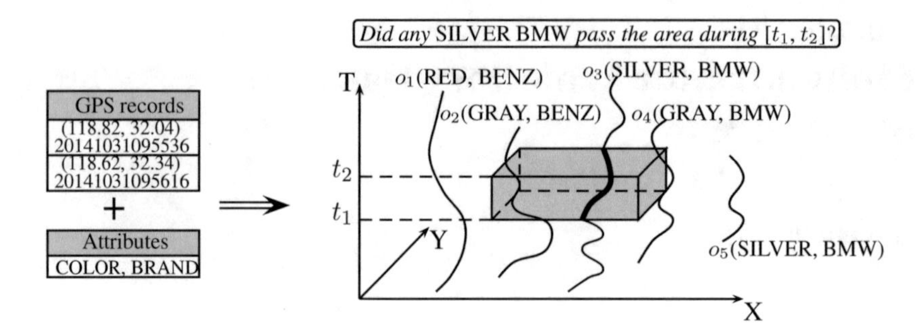

Fig. 9.1 Querying multi-attribute trajectories

Consider a database storing vehicle trips in a city. Each trip contains a standard trajectory and two attribute values over domains COLOR = {RED, SILVER, GRAY} and BRAND = {BENZ, BMW}, respectively, as illustrated in Fig. 9.1. A query that contains a tuple of attribute values and a spatio-temporal box is issued, that is, "*Did any* SILVER BMW *pass the area during* [t_1, t_2]?". Boolean range queries are studied to report objects containing query attribute values and fulfilling the spatio-temporal condition. In the example, o_3 is returned. Although o_2 intersects the query window, it is not a SILVER BMW.

Recently, researchers have started to investigate spatio-temporal trajectories annotated with additional information, e.g., semantic trajectories (Yan et al. 2011; Parent et al. 2013; Zhang et al. 2014; Zheng et al. 2015), activities trajectories (Zheng et al. 2013a), and transportation modes (Xu and Güting 2013). In particular, a semantic trajectory is essentially an enriched version of a standard trajectory in terms of locations. Labels are attached to geo-locations to describe places that users have visited or performed activities at, e.g., *hotel*, *sport*, *restaurant*. However, semantic data is restricted to locations. This is orthogonal to multi-attribute trajectories that consider location-independent attributes. The major differences include three aspects.

- Attributes represent a range of aspects and aim to provide a full picture of moving objects, as opposed to semantics limited to locations. This will support a different (even broader) range of applications.
- Semantic locations are sparsely defined because among a person's trajectory a few locations have semantics. Attributes are location-independent and associated with the complete trajectory. They are not derived from time-stamped locations or the geographical environment. For example, a semantic trajectory is of the form $o = \langle(loc_1, t_1, \textit{coffee}), (loc_2, t_2, \textit{pizza})\rangle$, where *coffee* and *pizza* are defined at two locations. That is, there is no semantic at locations between loc_1 and loc_2. A multi-attribute trajectory is of the form $o = \langle(loc_1, t_1), (loc_2, t_2)\rangle$, (RED, BENZ), where (RED, BENZ) is associated with the complete trajectory.
- Semantic trajectories cope with similarity search or ranking queries rather than the exact match on attribute values with spatio-temporal predicts, leading to

different tasks when developing the index. Semantic trajectories are grouped in terms of locations and semantics, but attributes are not related to locations.

To efficiently process multi-attribute trajectories, an index is essentially required because a sequential scan over the database is prohibitively expensive for large datasets. Standard trajectory indexes such as TB-tree (Pfoser and Jensen 2000), SETI (Chakka et al. 2003) and TrajStore (Mauroux et al. 2010) only deal with the spatio-temporal data without managing attributes. Such a method is suboptimal because one cannot use the index to prune the search space at the attribute level. As a result, objects after performing the spatio-temporal evaluation are sequentially processed, significantly inhibiting the performance. Furthermore, the pruning technique of *min* and *max* distances[1] cannot be applied for nearest neighbor queries if attribute values are not determined. This is because objects that are close to the query may not fulfill the attribute condition and cannot be used for pruning further objects. The trajectory subset containing query attributes changes according to the query setting and cannot be pre-computed. False dismissals will occur if one performs the pruning without the awareness of attribute values.

One can employ two individual indexes (e.g., a 3-D R-tree and a B-tree) on standard trajectories and attributes, respectively. The problem is, when the query evaluates the selective predicate on both parts, an intersection will be performed on two candidate sets that are separately retrieved, which is suboptimal. Another solution is to employ an attribute index. The method first receives trajectories containing query attributes and then proceeds to processing standard trajectories. However, this method is limited in scope and inherently suffers from the performance issue. Standard trajectories will be processed by either performing a sequential scan or accessing an on-line built index. If the attribute predicate is selective, the query cost may be acceptable because a small dataset is processed. If the attribute predicate has a poor selectivity, a large number of trajectories will be returned. Both the sequential scan and building an on-line index incur high CPU and I/O costs. Furthermore, creating an index for each query at runtime causes extra storage space. This calls for a structure that is able to simultaneously manage both standard trajectories and attributes. Meanwhile, the structure should be general and flexible in order to answer queries on standard trajectories and support update-intensive applications. From a system point of view, existing techniques need to be extended or adapted to deal with coming issues rather than developing individual structures each of which only applies to one problem.

The rest of the chapter is organized as follows. Related work is analyzed in Sect. 9.2. Multi-attribute trajectories and queries are defined in Sect. 9.3. Indexing and querying multi-attribute trajectories are introduced in Sects. 9.4 and 9.5,

[1]Given three rectangles a, b, c, each contains a set of points inside. We aim to find the nearest point to a given point inside a. Let *Max*(b, a) and *Min*(c, a) denote the maximum and minimum distances between two rectangles. If $Max(b, a) \leq Min(c, a)$, then no point inside the rectangle c can be closer than a point in the rectangle b to a. As a consequence, we can omit c when searching for the nearest neighbor to a.

respectively. The system development is presented in Sect. 9.6 and the performance evaluation is reported in Sect. 9.7. Future directions are pointed out in Sect. 9.8, followed by conclusions in Sect. 9.9.

9.2 Related Work

The current state-of-the-art is classified into two parts: (i) extending the representation of standard trajectories by incorporating semantics, and (ii) indexing standard trajectories with additional data.

9.2.1 Enriching Spatio-Temporal Trajectories

Semantic trajectories Emerging applications require extensive information about trajectories such as quality and semantics (Zheng and Su 2015). *Semantic trajectories* are based on discovering meaningful knowledge from locations (Alvares et al. 2007; Yan et al. 2011; Zheng et al. 2015). Formally,

Definition 1 (Semantic trajectory) A semantic trajectory is represented by a sequence of time-stamped positions complemented with annotations, that is, $o_{\text{sem}} = \langle\, (\, loc_1, t_1, \mathcal{A}_1\,), \ldots, (\, loc_n, t_n, \mathcal{A}_n\,)\,\rangle$ in which $loc \in \mathcal{R}^2$, $t \in T$, and $\mathcal{A}$ is a set of labels (strings) describing locations.

Interesting patterns can be properly defined and extracted. For example, a so-called *fine-grained sequential pattern* reports trajectories that satisfy spatial compactness, semantic consistency and temporal continuity simultaneously (Zhang et al. 2014). Consider actions that users can take at particular places such as *sport, dining* and *entertaining*. *Activity trajectories* are defined by associating geo-spatial points with activities. A similarity search returns k trajectories whose semantics contain the query and have the shortest minimum match distance (Zheng et al. 2013a). Motivated by the fact that standard trajectories do not make much sense for humans, a *partition-and-summarization* approach is proposed to automatically generate texts to highlight the significant semantic behavior (Su et al. 2015). A good survey of semantic trajectories refers to Parent et al. (2013).

Motion modes Moving objects with transportation modes are investigated in Xu and Güting (2013) and Xu et al. (2015a,b) . A trajectory over diverse geographical spaces includes time-stamped locations and a sequence of transportation modes such as *Indoor* → *Walk* → *Car*. Queries containing transportation modes can be answered, e.g., "*who arrived at the university by taxi*".

Definition 2 (Trajectories with transportation modes) A trip with transportation modes is represented by a sequence of units, each of which defines the movement over a time interval and a certain mode. That is, each unit is of the form $u_{\text{tm}} = (\, loc_1,$

$loc_2, t_1, t_2, m)$ in which $loc_1, loc_2 = (oid, loc')$, $oid \in int$, $loc' \in \mathcal{R}^2$, $t \in T$, and $m \in$ {Indoor, Walk, Car, Bus, Metro, Bike, Taxi}.

The location representation employs a reference system in which *oid* points to a geographical object such as a road, a walking area or a bus. Then, the relative location in the geographical object is recorded. The transportation mode does not change for each piece of movements.

Symbolic trajectories The task is to deal with generic semantic information including transportation modes and users' activities (Valdés and Güting 2014; Güting et al. 2015). A generic model is proposed to capture a wide range of meanings derived from a standard trajectory. The symbolic information is computed from the movement itself or obtained from the geographical environment, and a symbolic trajectory is represented by a time-dependent label. Typical examples include names of roads, activities and transportation modes. The goal is to provide a simple and flexible model for any kind of semantic information, while geometric locations are not defined.

Definition 3 (Symbolic trajectory) A symbolic trajectory is represented as a sequence of pairs (t, l), in which t is a time interval and l is a label (short character string) describing certain aspects of a trajectory.

If transportation modes are considered, a symbolic trajectory is denoted by $o_{\mathrm{sym}} = \langle([t_1, t_2], \mathit{Walk}), ([t_2, t_3], \mathit{Bus}), ([t_3, t_4], \mathit{Metro}), ([t_4, t_5], \mathit{Walk}), ([t_5, t_6], \mathit{Indoor})\rangle$.

There are fundamental differences between those works and multi-attribute trajectories. First, multi-attribute trajectories consider attributes that are location-independent, differing from attaching location labels in semantic trajectories. Symbolic trajectories do not contain geo-locations, while multi-attribute trajectories do. Multi-attribute trajectories are defined in a broad context by annotating trajectories with domain-specific attributes such that users can issue queries combining different aspects of moving objects. Second, different queries are evaluated. Multi-attribute trajectories incorporate attributes into the evaluation for Boolean queries and search for the objects fulfilling the spatio-temporal condition during a time interval or at each time point. Previous queries deal with spatial closeness and attributes similarity instead of time-dependent distances and exact matches on attributes. Labels are sparsely defined in semantic trajectories because a few locations may contain semantics. As a result, ranking queries are primarily dealt with rather than the spatio-temporal evaluation at each time point with attributes.

Heterogeneous k-nearest neighbor queries are studied in Su et al. (2007). A moving object is represented by a location-independent attribute and a set of coordinates. By defining a function that combines the costs of distances and the location-independent attribute, the query returns objects having the k-th smallest value. Although the work considers the location-independent attribute, there are three major differences in comparison with ours. First, the data representation is limited in scope because each moving object is associated with only one attribute.

Second, they query objects based on a ranking function on distance and attribute, but our queries require exact matches on attribute, leading to different results. Third, their distance function is not time-dependent, while queries of multi-attribute trajectories support distances changing over time.

Spatial keywords Queries of spatial keywords have been extensively studied in the literature (Chen et al. 2013; Lee et al. 2015; Wang et al. 2016). The task is to support queries that take a geo-location and a set of text descriptions called *keywords* as augments and return (i) objects that are close to the query location and contain the keywords called Boolean *k*NN query (De Felipe et al. 2008), or (ii) objects with the highest ranking scores measured by a combination of distances to the query location and the text relevance to the keywords called Top-*k* NN query (Cong et al. 2009). To efficiently answer the query, a spatial index such as 2-D R-tree and a text index structure are combined. For example, the IR-tree (Cong et al. 2009) augments each node of the R-tree with a pointer to an inverted file that contains a summary of the text content of the objects in the corresponding subtree. During the query procedure, one uses the combined structure to estimate both the spatial distance and the text relevancy to prune the objects that cannot contribute to the result. However, spatial keywords focus on static geo-locations and location-dependent text descriptions, leading to different queries. Text descriptions and attributes will make different tasks when designing the index structure. The index groups close spatial objects in terms of spatial distances and location-related text relevances. It is possible to attach attributes to time-stamped locations, but each piece of trajectories will have all attributes along with the trajectory, resulting in an extremely large amount of redundant data. In fact, the key issue of boosting the index for multi-attribute trajectories is to know which objects contain particular attribute values and where the objects are located in the spatio-temporal index. Therefore, a different criterion is used to design the index.

9.2.2 Indexing Spatio-Temporal Trajectories

In the last decade, a substantial number of spatio-temporal index structures have been proposed to efficiently access trajectories. A good survey on trajectory indexing and retrieved is given in Dinh et al. (2010) and Zheng and Zhou (2011). Indices can be classified into three categories according to the environment: (1) free space (Pfoser and Jensen 2000; Tao and Papadias 2001; Chakka et al. 2003; Pelanis et al. 2006); (2) road network (Frentzos 2003; Pfoser and Jensen 2003; de Almeida and Güting 2005; Popa et al. 2011); and (3) indoor (Jensen et al. 2009; Lu et al. 2012). Several algorithms are proposed to minimize the total volume of trajectory approximations given a user-specified number of splits (Hadjieleftheriou et al. 2002). Rasetic et al. (2005) provide a better solution that splits trajectories into a number of sub-trajectories and builds an index on them to minimize the number of expected disk I/Os with respect to an average size of spatio-temporal range queries.

The method expects the query window as the input but in real applications the size of the window varies and the assumption leads to inaccurate estimations.

Indexing semantic and symbolic trajectories Recently, traditional spatio-temporal indexes have been studied to incorporate semantic information. A grid index is established to organize spatio-temporal trajectories with activities in a hierarchical manner (Zheng et al. 2013a). A similar structure is developed to incorporate both spatial and semantic information for approximate keyword search (Zheng et al. 2015). The grid is in fact a spatial index and is extended to maintain objects based on spatial and activity proximities for ranking queries. This line of work is not applicable to our problem. On the one hand, our attributes are not related to locations and therefore it does not make sense to group objects by considering both spatio-temporal data and attributes. On the other hand, our query reports trajectory objects rather than individual locations. A framework of analyzing large sets of movement data having time-dependent attributes is developed (Valdés and Güting 2017, 2019). They aim to support pattern matching queries on tuples of time-dependent values, e.g., "*return all tuples that include either a flight on Tuesday or a work in Dortmund with a later bus trip*". A new pattern language is proposed and the superiority is thoroughly analyzed in terms of flexibility and expressiveness. The corresponding matching algorithm uses a collection of different indexes and is divided into a filtering and an exact matching phase. A composite index structure for sets of tuples of time-dependent value is proposed in which a single index of a suitable type is created for each time-dependent attribute.

Indexing trajectories with keywords An index structure called IOC-Tree is proposed to answer spatial keyword range queries on trajectories (Han et al. 2015). The structure consists of an inverted index and a set of 3-D quadtrees termed *octrees*. The inverted index has two components: a search structure for all keywords and lists of references to documents containing words. One is called a *dictionary* and the other is called *inverted lists*. Each keyword is combined with one reference, that is an octree built on the keyword in the dictionary to organize relevant trajectory points. In an octree, each leaf node maintains a signature represented by a bit vector to summarize the identifications of a set of trajectories intersecting the node. The signature of a non-leaf node is achieved by performing the union on the signatures of its child nodes.

The IOC-Tree can be extended to solve our problem by setting attribute values as keywords associated with trajectory points. One can implement the inverted index as an array of attribute values and each value contains a pointer to an octree. Certain parameters are defined: the maximal depth $h = 5$ and the split threshold $\varphi = 80$. Leaf nodes that do not contain enough trajectories are merged as one node (still a leaf node). A 64-bit integer is used for the signature in each node. Each bit corresponds to a range of trajectory ids. Each octree leaf node is assigned a morton code and empty nodes (no trajectory intersects) are not materialized. Since each attribute value corresponds to an octree, we will have a set of octrees and combine the attribute value and the morton code as the key for each node. A relation stores all leaf nodes and tuples are increasingly sorted on keys in order to maintain the locality of nodes in terms of the spatio-temporal proximity. A B-tree is built on the relation.

The main difference between trajectories with keywords and multi-attribute trajectories is that keywords are location-dependent texts, but attribute values are location-independent. A keyword is relevant to one or a few location points of the trajectory, while all location points of the trajectory have the same attribute values. This results in two major changes when maintaining the IOC-Tree and performing the query, in particular, inserting trajectory points into the index. A thorough analysis and comparison is provided in the following.

(i) In the context of keywords, location points will be distributed into octrees each of which corresponds to a keyword that the trajectory point contains. Each octree stores one or a few relevant location points of the trajectory. However, for multi-attribute trajectories each octree contains all location points of the trajectory because they all have the attribute value. Consider the following two trajectories.

- given a trajectory with keywords $o_1 = \langle (loc_1, t_1, coffee), (loc_2, t_2, pizza) \rangle$, we will store (loc_1, t_1) and (loc_2, t_2) in two octrees for coffee and pizza, respectively;
- given a multi-attribute trajectory $o_2 = (\langle (loc_1, t_1), (loc_2, t_2) \rangle, (\text{GRAY}, \text{BENZ}))$, we will store both (loc_1, t_1) and (loc_2, t_2) in two octrees for GRAY and BENZ, respectively.

The IOC-Tree is efficient for processing trajectories with keywords because only relevant trajectory points are indexed. However, attribute values are not related to locations but associated with the complete trajectory. That means, each attribute value is relevant to all points of the trajectory. Then, the number of trajectory points in each octree for multi-attribute trajectories is larger than that for trajectories with keywords, as demonstrated in Table 9.1. To gain trajectories with keywords, we randomly assign two attributes as keywords to each trajectory point using the dataset BTaxi in the experiment (Sect. 9.7). During the query procedure, the numbers of processed octree leaf nodes and trajectories increase, leading to more CPU and I/O costs. The values in Table 9.1 are calculated based on the condition that the number of trajectory points is the same in both cases. In fact, such a value for trajectories with keywords is much smaller than that for multi-attribute trajectories. We will explain this at point (ii) below.

The variation in processed trajectory points also makes the signature in IOC-Tree less effective when we perform the intersection on trajectories containing different keywords. Each node in the octree maintains a signature represented by a bit vector to summarize the identifications of trajectories passing through the node. Table 9.2

Table 9.1 The average number of relevant trajectory points in an octree (BTaxi, d = 10, dom(Att) = [1, 151])

Multi-attribute trajectory	Trajectory with keywords
83,974	75,198

Table 9.2 The percentage of defined bits (64 in total) in the signature at each level in IOC-Tree ($d = 10$, [1, 151])

Att	Leaf nodes	$H = 4$	$H = 3$	$H = 2$	$H = 1$
1	45%	98%	100%	100%	100%
20	49%	96%	95%	100%	100%
50	47%	97%	100%	100%	100%
100	46%	98%	100%	100%	100%
Avg	46%	97%	99%	100%	100%

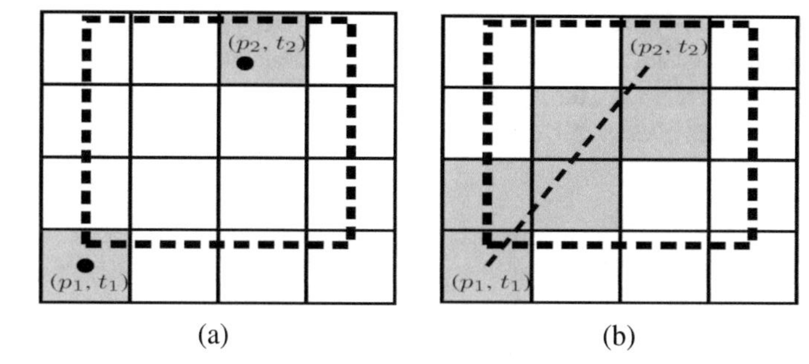

Fig. 9.2 Cells intersecting the trajectory. (**a**) Trajectories with keywords. (**b**) Multi-attribute trajectories

reports the percentage of defined bits in the vector at each level of the IOC-Tree. We can see that almost all bits are defined for signatures in non-leaf nodes, weakening the pruning ability.

(ii) Trajectory points with keywords are sparsely defined because only a few locations of the trajectory may have semantics such as *coffee* and *mall*, but attributes are associated with all locations of the trajectory. This results in different numbers of octree leaf nodes intersecting the trajectory. Still using o_1 and o_2, we assume that the space is partitioned into 2×2 cells. Figure 9.2a and b show the cells intersecting o_1 and o_2, respectively. For trajectories with keywords, each point is assigned to the cell intersecting the trajectory. Locations between two sampled points will not be addressed because keywords are not defined. For multi-attribute trajectories, attribute values are associated with the overall movement and all cells intersecting the trajectory are included. Consequently, the number of maintained trajectory IDs in the IOC-Tree is much larger than that of trajectories with keywords. Given a query window, multi-attribute trajectories process more nodes and trajectories than trajectories with keywords, increasing the query cost.

Indexing spatial objects with keywords In the field of spatial keywords search, geo-textual indexes combine spatial and text aspects such that both types of information can be utilized to prune the search space. To answer Boolean kNN queries (Wu et al. 2012), the data is partitioned into multiple indexing groups each of which shares as few attributes as possible. A hierarchical aggregate grid index called *HAGI* is developed to support heterogeneous kNN queries (Su et al. 2007). The method can be adapted to answer our queries, but it is limited as only one attribute is considered. A function is defined to combine the cost of distances and location-independent attributes, and the query returns objects having the k-th smallest function value. Each node in *HAGI* maintains *min* and *max* attribute values of all objects stored in the subtree. Although *min* and *max* values may work well for one attribute, they fail to guarantee good pruning ability for multiple attributes as *min* and *max* values are likely from different attributes. Also, the query evaluates objects based on a ranking function, whereas we require the exact match on attributes. Furthermore, the distance function is not time-dependent, whereas we deal with distances changing over time.

9.3 Problem Definition

9.3.1 Data Representation

A composite data model $\mathcal{O}(Trip; Att)$ is used to represent a multi-attribute trajectory database, in which *Trip* denotes standard trajectories and *Att* denotes multi-attributes. A standard trajectory is typically modeled by a function from time to 2D space. In a database system, a discrete model is implemented and the continuously changing locations are represented by linear functions of time, as illustrated in Fig. 9.3. That is, a trajectory is represented by a sequence of so-called *temporal units*, each of which records start and end locations during a time interval. Locations between start and end locations are estimated by interpolation. A data type called $\underline{mpoint}$ is defined (Forlizzi et al. 2000; Güting et al. 2000).

Definition 4 $D_{\underline{mpoint}} = \{< u_1, \ldots, u_n > | n \geq 1$, and $u = (loc_1, loc_2, t_1, t_2)$ where $loc_1, loc_2 \in \mathcal{R}^2, t_1, t_2 \in T\}$

Let A be the set of multiple attributes. The ith attribute and its domain are denoted by $A[i]$ and $dom(A[i])$ ($i \in 1,\ldots, |A|$), respectively. Assume that each $dom(A[i])$ is represented by a set of positive integers and a data type called $D_{\underline{att}}$ is defined for the set of attributes. For the sake of readability, the enum data type is used for attributes in the following.

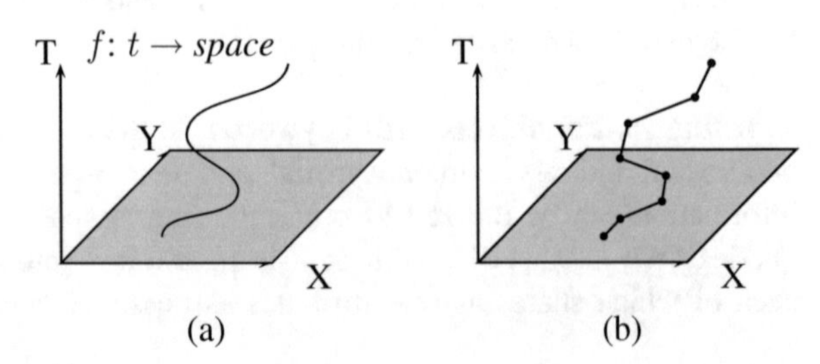

Fig. 9.3 Standard trajectory representation. (**a**) Abstract. (**b**) Discrete

Table 9.3 An integration of standard trajectories and attributes

Id: int	*Trip: mpoint*	*Att: att*
o_1	location + time	(RED, BENZ)
o_2	location + time	(GRAY, BENZ)
o_3	location + time	(SILVER, BMW)
o_4	location + time	(GRAY, BMW)
o_5	location + time	(SILVER, BMW)

Table 9.4 Summary of notations

Notation	Description
O	The set of multi-attribute trajectories
o	A multi-attribute trajectory
$\|A\|$	The number of attributes
$dom(A_i)$, $dom(A)$	The domain of A_i, the overall domain
Q_a	Query attribute expression
o_q, d	A query trajectory, the query distance
k	The number of nearest neighbors
t	A time point or interval
$T(o)$	The time period of a trajectory

Definition 5 (Multi-attribute representation) $D_{\underline{att}} = \{(a_1, \ldots, a_{|A|}) \mid a_i \in \text{dom}(A[i]), i \in \{1, \ldots, |A|\}\}$ such that

(i) $\forall i \in \{1,\ldots, |A|\}$: $dom(A[i]) \subset N^+$;
(ii) $\forall i, j \in \{1,\ldots, |A|\}$: $i \neq j \Rightarrow dom(A[i]) \cap dom(A[j]) = \varnothing$.

The data model is translated to a relation with the schema (*Id*: *int*, *Trip*: *mpoint*, *Att*: *att*) by embedding $\underline{mpoint}$ and $\underline{att}$ as relational attributes, as shown in Table 9.3. To be more specific, a relation is used to store multi-attribute trajectories by defining two attributes *Trip* and *Att*. The system manipulates multi-attribute trajectories via a relation. The advantage of using the relational interface is that (i) it allows combining heterogeneous data models, i.e., spatio-temporal and attribute; and (ii) existing operators on standard trajectories can be leveraged, benefiting the system development.

Table 9.4 summarizes notations frequently used in the chapter.

9.3.2 Queries

Attribute values are incorporated into the evaluation and the attribute expression is defined in the following.

Definition 6 (Attribute query expression) $Q_a = (a_1, \ldots, a_{|A|})$, $a_i \in dom(A_i)$ or $a_i = \perp$

Given a tuple of attribute values $Q_a = (a_1, \ldots, a_{|A|})$ and a multi-attribute trajectory $o \in \mathcal{O}$, an operator **contain**($o.Att$, Q_a) returns *True* if $\forall\ a \in Q_a$: $a \in o.Att$ or $a = \perp$.

Three types of queries are supported: (i) range queries, (ii) continuous range queries and (iii) continuous nearest neighbor queries. The range query is called RQMAT (Range Queries on Multi-attribute Trajectories) (Xu et al. 2018b). Formally,

Definition 7 (Range queries on multi-attribute trajectories (RQMAT)) Given a spatio-temporal window Q_{box} and attribute values Q_a, RQMAT returns a set of trajectories $\mathcal{O}' \subseteq \mathcal{O}$ such that $\forall\ o \in \mathcal{O}'$: (i) $o.Att$ **contains** Q_a; and (ii) $o.Trip$ **intersects** Q_{box}.

There is a variation of RQMAT that returns objects containing query attributes and keeping within a spatial range to a moving target at each query time point. The query is called CRQMAT, Continuous Range Queries on Multi-attribute Trajectories. Let $T(o)$ return the time period of an object. The function in Frentzos et al. (2007) is employed to return the time-dependent distance between two trajectories $o_1, o_2 \in \mathcal{O}$, denoted by $dist\,(o_1, o_2, T(o_1) \cap T(o_2))$.

Definition 8 (Continuous range queries on multi-attribute trajectories (CRQMAT)) Given a query trajectory o_q, a distance threshold d and an attribute predicate Q_a, CRQMAT aims to identify the result set $\mathcal{O}' \subseteq \mathcal{O}$ such that $\forall\ o' \in \mathcal{O}'$: (i) **contain**($o'.Att$, Q_a) ; (ii) $\forall\, t \in T(o_q) \cap T(o')$, $dist(\, o_q, o', t\,) \leq d$.

Consider an example in Fig. 9.4. Assume that o_3 is a special object that carries VIP passengers or sensitive materials. For security reasons, one detects whether the special object is stalked. To this end, one makes use of multiple attributes to form a semantic-richer query, e.g., *Did any GRAY BENZ always keep 50 meters to* o_3. The returned objects must satisfy the criteria: (i) time-dependent distance constraint and (ii) attribute consistency. Although o_1 is within 50 meters to o_3, it is not a GRAY BENZ and should not be returned. Note that o_4 and o_2 fulfill the condition during $[t_1, t_2]$ and $[t_2, t_3]$, respectively, but they do not fulfill the condition during the overall query time. As a result, the query reports $\{([t_1, t_2], o_4), ([t_2, t_3], o_2)\}$.

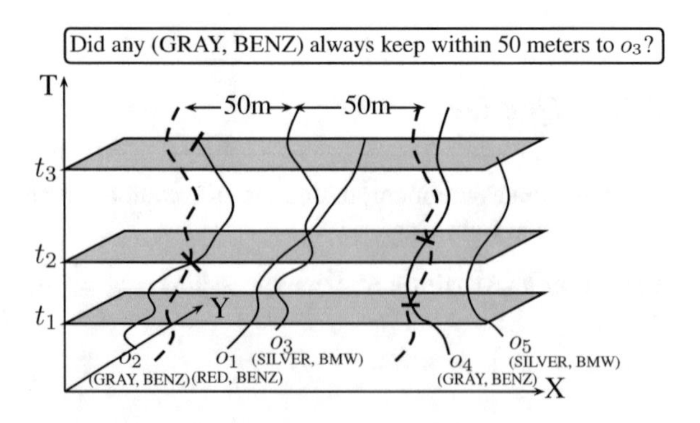

Fig. 9.4 Example of CRQMAT

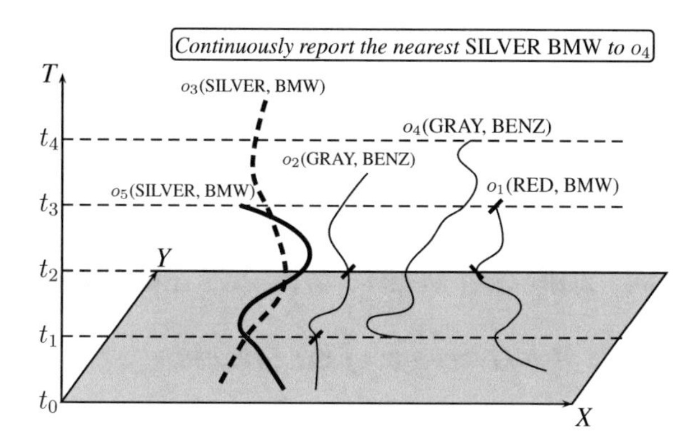

Fig. 9.5 Example of CkNN_MAT

The third type of queries is called CkNN_MAT (Continuous k Nearest Neighbor queries over Multi-attribute Trajectories) (Xu et al. 2018a). Such a query returns the objects fulfilling the condition: (i) *attribute consistency* and (ii) *time-dependent distance closeness*.

Definition 9 (Continuous k nearest neighbor queries over multi-attribute trajectories (CkNN_MAT)) Given a query standard trajectory o_q, an integer k and a set of query attributes Q_a, CkNN_MAT receives k trajectories denoted by $\mathcal{O}' \subseteq \mathcal{O}$ at each query time such that (i) $\forall\, o \in \mathcal{O}'$: **contain**($o.Att$, Q_a) returns *True*; (ii) $\nexists$ $o' \in \mathcal{O} \setminus \mathcal{O}'$: **contain**($o'.Att$, Q_a) $\wedge$ o' is closer than $\forall\, o \in O'$ to o_q.

An example is illustrated in Fig. 9.5. CkNN_MAT returns ([t_1 , t_2], o_3), ([t_2, t_3], o_5), ([t_3, t_4], o_3) indicating the key aspect that only objects fulfilling the attribute condition will be evaluated on the time-dependent closeness. Although o_1 and o_2 are closer than o_3 and o_5 to the query trajectory, they do not contain (SILVER BMW) and will not be included. Since distances between moving objects vary over time, results change at certain time points.

Generalizing query attribute expression Up to now, one assumes that the query defines a single value for each attribute. It is possible that multiple values are defined, e.g., *Continuously report the nearest* SILVER BMW or VW *to* o_4. The query expression is extended to support multiple values.

Definition 10 (An extension of query attributes) $Q_a = (X_1, \ldots, X_{|A|})$, $X_i \subseteq dom(A_i)$ or $X_i = \varnothing$

At the concept level, $Q_a = (X_1 , \ldots, X_{|A|})$ defines the component for each attribute over $\{A_1, \ldots, A_{|A|}\}$, in which X_i is a set of attribute values. The multi-value query SILVER BMW or VW is formed by Q_a = ({SILVER}, {BMW, VW}). At the implementation level, the query is defined by a relation in which a tuple supports multi-valued attributes. The operator **contain** is extended accordingly: **contain**($o.Att$, Q_a) returns *True* if $\forall\, X_i \in Q_a$: $o.Att[i] \in X_i$ or $X_i = \varnothing$.

Attribute queries with negative values are also supported, that is, $o.Att[j] \neq Q_a[j]$. Negative queries can be transformed into set queries by setting $Q_a[j]$ the values that are not equal to the query, e.g., $o.Att[j] \neq \text{RED} \Rightarrow o.Att[j] = \text{GRAY}$ or SILVER.

9.4 Indexing Multi-attribute Trajectories

9.4.1 An Overview of the Structure

To efficiently answer queries, the index should manage both spatio-temporal trajectories and attributes in order to prune the search space on both predicates. A hybrid structure is developed that consists of a 3-D R-tree and a composite structure named BAR, as shown in Fig. 9.6. The 3-D R-tree that serves as indexing standard trajectories is a height balanced tree. Each node contains an array of entries, each of which couples (i) a pointer to a subtree or an object with (ii) a rectangle that bounds data objects in the subtree. BAR is a composite structure that includes a B-tree, a relation Att_Rel and a record file RF.

The system builds BAR on top of the 3-D R-tree by extracting attribute values from multi-attribute trajectories. The structure builds the connection between attribute values and R-tree nodes and enables us to know attribute values in a subtree. For a leaf node, each entry stores a pointer to a tuple in the trajectory relation and the tuple is accessed to obtain the attribute value. For a non-leaf node, attribute values are collected by performing the union on values from child nodes. BAR maintains attribute values in an efficient way such that one is able to fast settle the R-tree nodes that (i) contain query attributes and (ii) fall into the range of the query time. Before elaborating the index structure, we first introduce pre-processing trajectories in order to have a compact dataset for building a good shape of the 3-D R-tree (nodes have similar sizes in spatio-temporal dimensions). Sections 9.4.2 and 9.4.3 present grouping small units and partitioning trips according to spatio-temporal distributions, respectively.

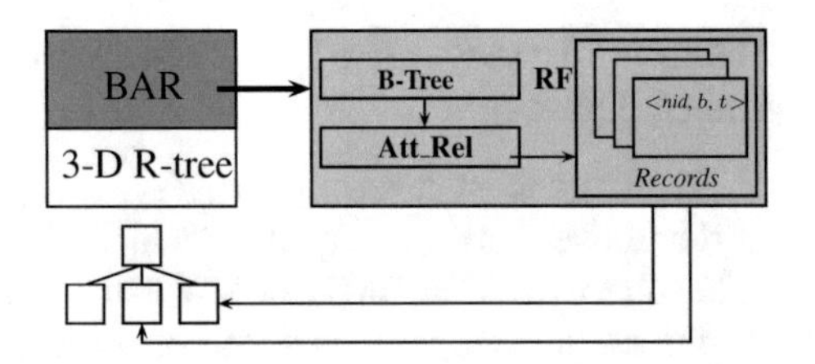

Fig. 9.6 Index architecture

9.4.2 Packing Trajectories

The R-tree is supposed to be built on sorted minimum bounding rectangles (MBRs) that approximate trajectories. By observation, raw trajectories from GPS records contain a large number of small units due to short time intervals or slow movement. In order to reduce the size of the dataset, small pieces of movements are packed to have fewer but larger units. Let u_i denote the trajectory extent in the ith dimension. The average extent over all units in the ith dimension is denoted by $\triangle_i$. Then, the deviation of a unit is given as:

$$f(u) = \sum \frac{u_i}{\Delta_i}, i \in \{d_x, d_y, d_t\} \tag{9.1}$$

A threshold *Bound* is defined to select small units. Duplicate values are removed to overcome the impact of the number of small units. The lower bound is analytically estimated.

$$Bound = \text{Avg}(\text{Unique}(\lfloor f(u) \rfloor)) \geq \text{Avg}(\lfloor f(u) \rfloor) \approx 3 \tag{9.2}$$

Let U be the set of all temporal units and the unit with the maximum deviation is

$$u^* = \arg\max_{u \in U} Unique(\lfloor f(u) \rfloor) \tag{9.3}$$

Not all values in $\{0, 1, \ldots, \lfloor f(u^*) \rfloor\}$ may be defined and thus the upper bound is

$$Bound \approx \text{Avg}(0 + \ldots + f(u^*)) \leq \frac{f(u^*)}{2} \tag{9.4}$$

The packing can be treated as building the R-tree in a different way, as demonstrated in Fig. 9.7. Small pieces of trajectories are packed to obtain large units which are taken as the input for a leaf node. The index is built by bulk load (Bercken and Seeger 2001; Bercken et al. 1997) which uses the same threshold as the standard value to group units into one leaf node, guaranteeing the spatio-temporal locality. The *Bound* is the average value over Unique($f(u)$) and thus will not result in grouping units into a large extent. During the packing procedure, neither raw units are modified/simplified nor data is lost. One does not need extra storage space and the same number of original units is maintained.

Demonstrate packing trajectories Using 500 GPS records of taxis, we calculate the unit deviation and report their values as well as *Bound* in Fig. 9.8. One can see that the majority of units have the derivation smaller than *Bound*. We pack successive small units of the trajectory as one unit such that the deviation of the unit is larger than *Bound*. The overall number of trajectory approximations (MBRs) is greatly reduced, leading to a compact dataset to build the index.

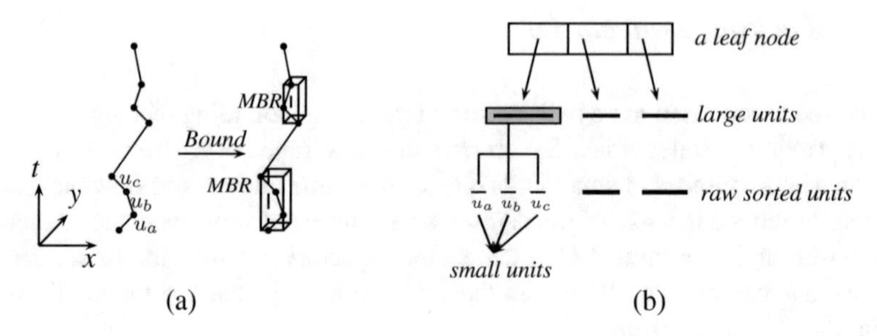

Fig. 9.7 The packing procedure. (**a**) Pack small units. (**b**) Build the index

Fig. 9.8 Effect of packing trajectories

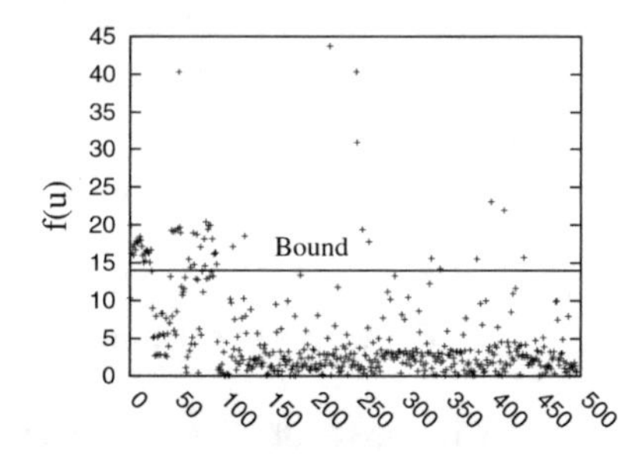

9.4.3 *Partitioning Trajectories*

Trajectories have different distributions over time and space. We would like to decompose them into pieces which have similar sizes in terms of spatial and temporal dimensions. This will benefit the index structure because spatio-temporal extents of nodes are similar, derivations among nodes are small and the area of *inactive space*[2] is reduced. The time dimension is partitioned into a set of equal-sized intervals $\{T_1,\ldots,T_K\}$ $(K > 1)$ and the 2-D space is partitioned into a set of equal-sized cells. Given a multi-attribute trajectory, its spatio-temporal trajectory is split into a set of so-called *cell trajectories*, each of which represents the movement within a cell during an interval $T_k \in \{T_1,\ldots,T_K\}$.

Definition 11 (Cell trajectory) Let $Cell(o, t)$ return the cell where o is located at a time point $t \in T(o)$. A cell trajectory $o[i]$ is a subset of $o.Trip$ such that (i) $\forall\, t_1, t_2 \in T(o[i]) : Cell(o[i], t_1) = Cell(o[i], t_2)$; (ii) $\exists\, T_k \in \{T_1,\ldots, T_K\}$: $T(o[i]) \subseteq T_k$.

We partition each $o \in \mathcal{O}$ into a set of cell trajectories in three steps: (1) $o.Trip$ is decomposed into a sequence of sub trajectories such that the time of each sub trajectory is contained by T_k; (2) For each sub trajectory, a set of cells intersecting the 2-D bounding box of the trajectory is identified, which is efficiently determined

[2]The space is contained by the node but there are few or no data objects. One can also call this *dead space* (Tao and Papadias 2001), meaning that the area will be evaluated but few objects are there or even no object exists.

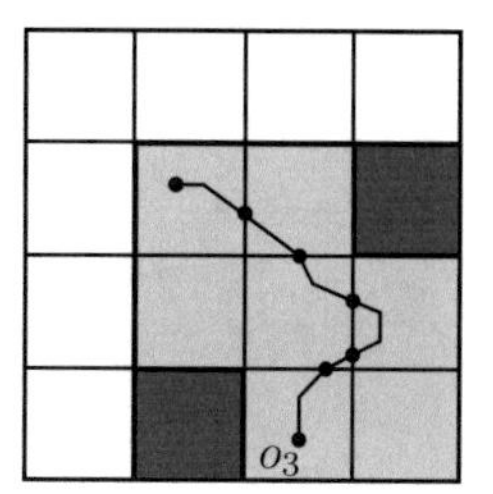

Fig. 9.9 Partitioning o_3 into cell trajectories

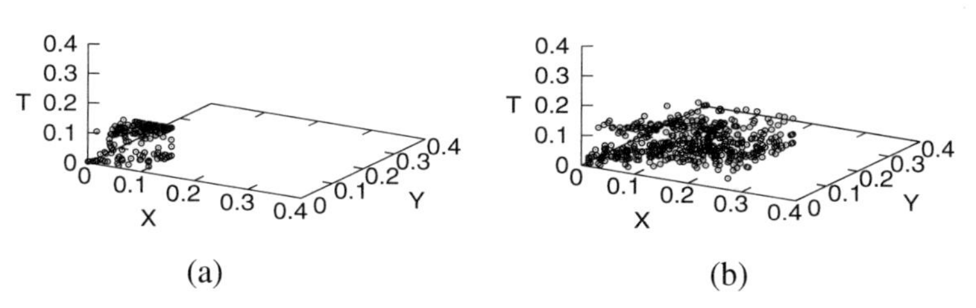

Fig. 9.10 Leaf node extents affected by partition. (**a**) Partition. (**b**) Without the partition

by finding the left-bottom and right-top cells; (3) Each sub trajectory is split into a set of cell trajectories. We may encounter the case that the object enters the cell more than once. As a consequence, there are several cell trajectories from one object located in the same cell. Assume that the 2-D space is partitioned into 4×4 cells and o_3 is contained by a time interval. The cells intersecting o_3 and o_3's cell trajectories are reported in Fig. 9.9. The index is built on cell trajectories sorted by time, cell id and 3-D bounding box following a bulk loading approach (Bercken et al. 1997).

Demonstrate partitioning the trajectories We use part of real trajectories in the experimental evaluation (66,000 taxi trajectories in Beijing) and build two indexes on trajectories with and without performing the partition, respectively. The extents on all dimensions are reported by randomly selecting 500 leaf nodes, as illustrated in Fig. 9.10. Clearly, the deviation among different dimensions after the partition is much smaller than that without the partition. This contributes to create a good R-tree.

Grid granularity Grid granularity plays a pivotal role in the index design as an arbitrary value cannot guarantee an optimal query performance. Assume the 2-D space is partitioned into $\delta \times \delta$ equal-sized cells. If we set a coarse granularity, e.g., $\delta = 1$, all trajectories are located in one cell. The index does not exhibit the spatio-temporal proximity, increasing false positives in query processing. At the opposite end, a fine granularity leads to small cells and each cell contains fewer trajectories having small extent in x and y dimensions. This is good for preserving locality. However, the finer the granularity is, more nodes are maintained. This is because the number of cell trajectories grows proportionally as a spatio-temporal trajectory is partitioned into all intersecting cells. As shown in Fig. 9.11, we will visit all cells under the setting $\delta = 2$ because they are within the d-distance to o_3. However, in

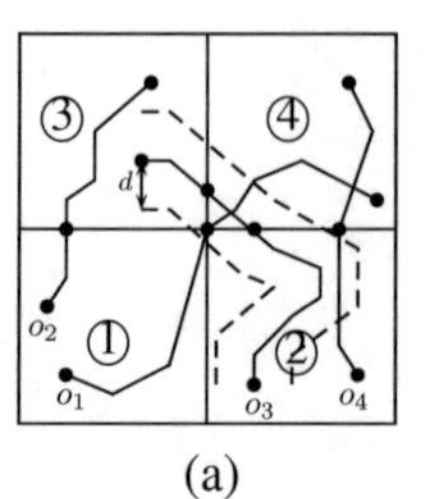

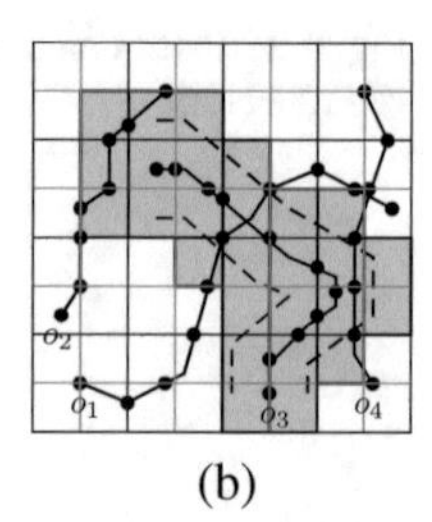

Fig. 9.11 Coarse and fine grid granularities. (**a**) $\delta = 2$. (**b**) $\delta = 8$

cells ③ and ④, cell trajectories of o_2 and o_4 do not fulfill the distance condition. Considering $\delta = 8$, although we can greatly reduce the search space (gray area), more cells are accessed and some of them do not contain trajectories.

9.4.4 BAR

The relation Att_Rel The key component in BAR is the relation Att_Rel that builds the connection between attribute values and R-tree nodes. The relation schema is defined as

Att_Rel: (A_VAL:*int*, H: *int*, RecId: *int*).

For each attribute value, the system maintains a tuple for all nodes containing the value at the same height. The nodes are stored in a record. A tuple stores the attribute value, the height, and the record identifier. The relation is created as follows. Step 1, for each $a \in dom$(A) the approach traverses the R-tree in depth-first search to collect all nodes containing a and creates an intermediate tuple for each node. One sets a as the key and records the node height. Step 2, the intermediate tuples are grouped according to the height and a record stores all nodes containing a at each height. A tuple is created to store the record id. Steps 1–2 are repeated for all $a \in dom$(A). One creates a B-tree on Att_Rel by making a key combining A_VAL and H.

A unique key is required for each attribute value. The ideal case is that attribute domains do not overlap. In practice, it may be not possible to have non-intersecting domains, but this problem can be solved. One can use a composite number to represent the attribute value. This is achieved by combining the attribute id and the value. In turn, a two-dimensional point (i, a) $(i \in [1, |A|], a \in dom(A_i))$ is formed. Then, a space-filling curve Z-order is used to map points of a two-dimensional space to one-dimensional values. This is done by interleaving the binary coordinate values, which guarantees that attribute domains do not overlap.

Record Storage The system maintains a list of items in each record. Each item is represented by a three-tuple: (*nid*, *b*, *t*), in which *nid* is the node id, *b* is a bitmap and *t* is a time interval. The bitmap represents the entries containing the attribute value in a node and *t* is the overall time of entries. The design is made based on the observation that the number of entries containing an attribute value cannot be

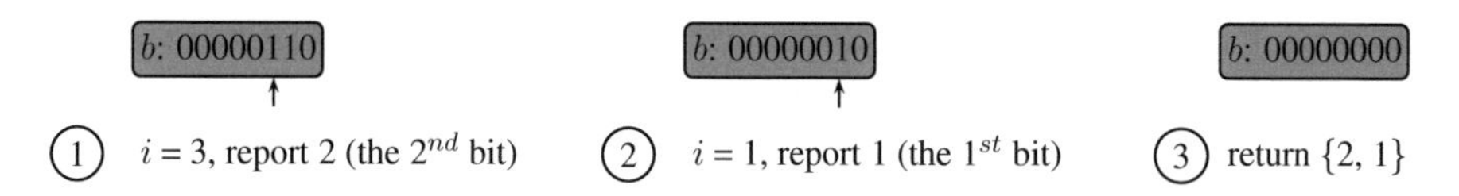

Fig. 9.12 Report defined bits

larger than the total number of entries, usually much smaller. Also, the number of entries containing an attribute value increases from leaf level to root level. This is because if a node contains the value, all its ancestor nodes will contain the value. To efficiently settle the entries fulfilling the attribute condition, the bitmap is accessed at first instead of performing a linear scan over all entries.

Let m denote the length of a collection of bit-vectors and E be the entry count of a node. A mapping between m and E is performed. There are two cases. Case (i): $m \geq E$, each bit maps to an entry. If the ith ($i \in [0, m)$) entry contains the attribute, one has $b[i] = 1$. Otherwise, $b[i] = 0$. Case (ii): $m < E$, each bit maps to a sequence of entries. The corresponding entries for the ith bit are calculated by $[i \cdot \lceil \frac{E}{m} \rceil, (i + 1) \cdot \lceil \frac{E}{m} \rceil)$. The system defines $b[i] = 1$ if one of the entries contains the attribute. The bitmap index incurs little storage overhead and is efficient for processing data in small quantity due to the speed of bit-wise operations. The length m depends on the implementation, e.g., a 32-bit integer. The bitmap fast determines qualified entries for the intersection condition of several attributes. A data type is embedded into an relation to represent the records.

Querying the bitmap In order to know the entries containing the query attribute, the method accesses the bitmap to report defined bits. Let $\mathcal{B} = < 2^0, 2^1, \ldots, 2^{m-1} >$ be a sequence of integers. Given a bitmap b, its defined bits are reported as follows: Step 1, by performing a binary search one finds the smallest $2^i \in B$ such that $2^i \geq b$. Step 2, if $2^i = b$, i is reported and the searching is terminated because the bit is already found. If $2^i > b$, the procedure updates i - 1 and $b = b - 2^{i-1}$. Then, steps 1–2 are repeated until b is equal to 0, during which bits are progressively reported from high to low positions. Figure 9.12 depicts the procedure of reporting $b = 00000110$ in Record 3.

Let P denote the set of defined bit positions, initially empty. Two indexes s and e are used to define the sth and eth integers in $\mathcal{B}$. To find the smallest i such that $2^i \geq b$, the procedure performs a binary search and terminates when either b is equal to an integer in $\mathcal{B}$ or $e = s + 1$. In the former case, all bits are found already. In the latter case, the position of the highest bit is found and put into P. To continue searching the bits, the approach updates b as well as s and e by setting $e \leftarrow s$ and $s \leftarrow 0$.

Time complexity One needs $O(\log m)$ to report the highest bit and the position is $p \in [0, m)$. To find the second highest bit, a binary search is performed, leading to $O(\log p)$. The iteration time depends on p. The smaller p is, the fewer iterations are needed. If $p \in [m/2, m)$, $\log p = \log m$ iterations are required. If $p \in [0, m/2)$,

$\log m - 1$ iterations are achieved. To report the ith highest bit, the iteration time is between $[\log m - (i - 1), \log m]$, depending on where the bit is located. To sum up,

Theorem 1 (Upper Bound)

$$O(|b| \log m) \tag{9.5}$$

Theorem 2 (Lower Bound)

$$O\left(\sum_{i=1}^{|b|}(\log m - (i-1))\right) = O(|b| \log m - (|b| - 1)|b|/2) \tag{9.6}$$

Proof One performs a binary search to look for $|b|$ defined bits in $O(m)$. In the worst case, they are the $|b|$ highest bits and each iteration needs the time $O(\log m)$, leading to $O(|b| \log m)$. An optimal case is that one needs $O(\log m - 1)$ for the second iteration when the position of the second highest bit is smaller than $\frac{m}{2}$. If the position of the second highest bit is $\leq \frac{m}{4}$, one needs $O(\log m - 2)$ for the third iteration and so on. □

9.4.5 *Updating the Index*

The database needs to keep track of the incoming data and allow querying both the historical and new data. An important task is to synchronize index structures in order to be consistent with the underlying data space. Given a set of incoming multi-attribute trajectories, inserting them into the index incurs updating two structures: (i) 3-D R-tree and (ii) BAR. In general, a new R-tree named $\mathcal{R}_u$ is created on new trajectories and BAR is built on $\mathcal{R}_u$. To distinguish between historical and new structures, the new structure is termed BAR_u. New created structures $\mathcal{R}_u$ and BAR_u are inserted into historical structures $\mathcal{R}$ and BAR, as illustrated in Fig. 9.13.

Updating 3-D R-tree The incoming trajectories are packed and a new R-tree is created by bulk load. The new R-tree is maintained by the same storage file as

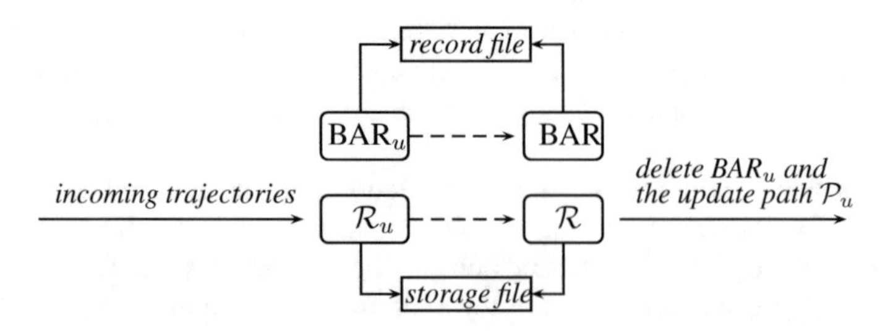

Fig. 9.13 An outline for updating

the historical R-tree in order to simplify the procedure of accessing the structure. Otherwise, one has to detect whether the accessed node belongs to the new R-tree or the historical R-tree and select the corresponding file. It is a rather complex task to maintain many storage files for frequent updates. $\mathcal{R}_u$ is inserted into R as follows. Let H_u and H denote the heights of the two R-trees, respectively. Assume that $H_u \leq H$. This is because the number of incoming trajectories for one update is usually much smaller than that of the historical data. If $H_u = H$, a new root node is created to hold root nodes of R and R_u as two entries. If $H_u < H$, the root node of R_u is inserted as an entry to an appropriate node in the target tree R whose height is equal to H_u. This is achieved by performing a top-down traversal in the target tree until a node whose height is equivalent to the new R-tree. During the traversal, the last entry of each accessed node is always chosen as the node to be processed at the next level. This is because entries are increasingly sorted by time and incoming trajectories are certainly located after existing trajectories. If the node is not full, the root of the new R-tree is inserted as an entry into the node. Otherwise, a new node is created for the R-tree.

Updating BAR We insert BAR$_u$ into BAR: step 1, for each tuple in BAR$_u$.Att_Rel, the procedure searches for the matching tuple in BAR.Att Rel and appends record items for the nodes in $\mathcal{R}_u$; step 2, record items are updated for each node appearing in $\mathcal{P}_u$.

$\mathcal{R}_u$ is inserted into $\mathcal{R}$ as a sub-tree and the nodes in $\mathcal{P}_u$ are updated in terms of (i) spatio-temporal boxes; and (ii) attribute values. For each attribute value in new trajectories, the method looks for tuples in BAR.Att_Rel having the value and the appropriate height according to $\mathcal{P}_u$. Note that the height is increasingly numbered from leaf to root level, guaranteeing that the heights of $\mathcal{R}$ and $\mathcal{R}_u$ are consistent. If the tuple is found, the record is accessed to update the item for the node. Precisely, the bitmap and the time box are updated. Later, the record is refreshed to synchronize the data.

New arrival trajectories incur an ongoing expansion of the time. The time range of the nodes in $\mathcal{P}_u$ overlaps with that of new trajectories. To enhance the update performance, record items are increasingly sorted on time and updated from the end of the list.

Definition 12 (Sorted records) $D_{\underline{rec}} = \{<\text{nid}_1, b_1, t_1\ ),\ldots,(\text{nid}_n, b_n, t_n > \mid t_1 < \ldots < t_n, \text{nid}_i \in \underline{int}, b_i \in \underline{int}, t_i \in \underline{interval}\}$

If a new root node is created, there is no matching tuple in BAR.Att_Rel. Therefore, the tuple as well as the record are created and inserted into BAR. Afterwards, BAR$_u$ and $\mathcal{P}_u$ are dropped. Let $\mathcal{O}_u$ be the set of new arrival trajectories. In order to achieve good performance for updating, a light-weight BAR named lw-BAR is proposed to reduce the I/O cost. The idea is to buffer record items in lw-BAR rather than updating BAR for each new trajectory. A relation and a B-tree make up lw-BAR and there is no record file. The relation schema is of the form

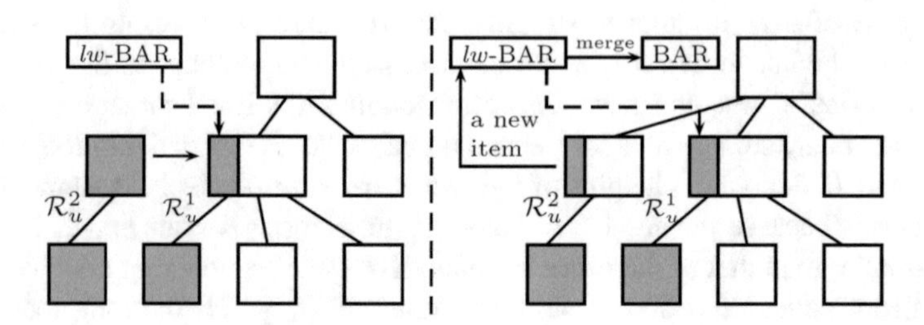

Fig. 9.14 Overflow and Update lw-BAR, BAR

lw-Att Rel:(*A_VAL*: int, *H*: int, *RecItem*: rec).

For each attribute value appearing in BAR_u, a tuple in lw-Att_Rel is maintained for the node in $\mathcal{P}_u$. In the update path, there is only one node at each height from H_u to H, and therefore the number of updated items in a record is one. The lw-BAR only stores one item in a record. To have a compact structure, the record component (managed in a record file in BAR) is merged into the relation by replacing the record id by the record item. Employing the lw-BAR, the number of I/O accesses for updating will be considerably reduced because only one tuple is processed.

To accommodate frequent updates, the record items for $\mathcal{P}_u$ have to be updated whenever $\mathcal{R}_u$ is inserted into $\mathcal{R}$. If entries in the inserted node do not overflow, only record items for historical nodes are updated. However, under a continuous updating load, frequent insertions will cause the node overflow and lead to new nodes, as illustrated in Fig. 9.14. In this scenario, the record item in lw-BAR is merged into the one in BAR and another record item in lw-BAR is created for the new node.

9.4.6 The Generality

The proposed index structure is general from three aspects: (i) packing standard trajectories, (ii) managing attribute values, and (iii) supporting a range of queries on multi-attribute trajectories and also queries on standard trajectories.

Packing. The established method produces a compact data set by reducing the number of approximations. There is no information loss and no extra storage cost. The procedure can be applied for other trajectory queries to enhance the performance.

BAR. The system is able to flexibly build the traditional trajectory index or the hybrid index, depending on whether standard trajectories or multi-attribute trajectories are processed. BAR is not tightly integrated into the spatio-temporal index and therefore can be combined with other traditional trajectory indexes, categorized into (i) R-tree based indexes, e.g., TB-tree (Pfoser and Jensen 2000), MV3R-Tree (Tao and Papadias 2001), and (ii) grid based indexes, e.g., SETI (Chakka et al. 2003). The well-established structures do not have to be modified,

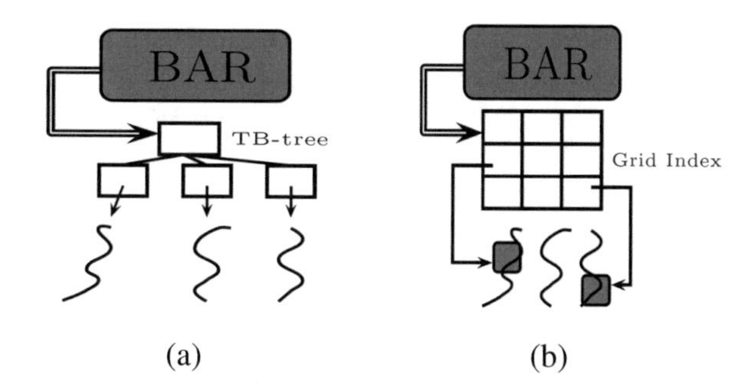

Fig. 9.15 Popularizing BAR. (**a**) TB-tree. (**b**) Grid index

benefiting the system development. Figure 9.15 reports BAR built on top of TB-tree and Grid index. One instantiates into the 3-D R-tree due to the advantage of preserving the spatio-temporal proximity and the efficiency of answering nearest neighbor queries (Güting et al. 2010b). The comparison with other trajectory indexes is as follows.

- TB-tree. The structure has the trajectory preservation property that only stores units of the same trajectory within a leaf node, resulting in a large spatial extent of leaf nodes. The spatial proximity is not preserved because segments of different trajectories that lie spatially close will be stored in different nodes. One cannot effectively prune the search space by *min* and *max* distances, resulting in poor performance for nearest neighbor queries. The STR-tree (Pfoser and Jensen 2000) introduces a parameter to balance between spatial properties and trajectory preservation, but the main concern is to handle the spatial domain and treating the temporal as a secondary issue.
- SETI. The space is divided into disjoint cells, each of which contains trajectory segments that are completely within the cell and has a temporal index (an R-tree) for objects' time intervals. The number of spatial partitions plays a crucial role in index design, but setting an optimal value is not trivial. Trajectories can be uniformly and uniformly distributed, making the performance unstable. The method focuses on the spatial proximity and has the limitation that the boundaries of the spatial dimension remain constant. The SEB-tree (Song and Roussopoulos 2003) is similar to SETI where the space is partitioned into zones, but the difference is that only the zone information is stored in the database without knowing the exact location.
- MV3R-tree. The index combines a multi-version R-tree (MVR-tree) and a small auxiliary 3-D R-tree built on the leaf nodes of the MVR-tree. The former is to process time-stamp queries and the latter is to process long interval queries. Short interval queries are processed by selecting the appropriate tree. Multi-attribute trajectories deal with interval queries and thus the structure is essentially a 3-D R-tree.

Queries. RQMAT, CRQMAT and C*k*NN_MAT can all be answered by employing the proposed index structure. One accesses BAR to find the subtrees in the spatio-temporal index fulfilling the attribute condition and then explores the spatio-temporal index. If attributes are not considered, the algorithm directly searches the spatio-temporal index without accessing BAR.

9.5 Query Algorithms

9.5.1 An Outline

The query procedure follows the *filter-and-refine* strategy. In general, one performs a traversal on the index during which objects are pruned on both spatio-temporal and attribute conditions. When the leaf level is reached, we open the node to retrieve objects from the relation. The filter step returns a set of candidate trajectories, each of which is likely to be in the result and will be iteratively evaluated, called *refinement*.

9.5.2 Processing RQMAT

The query processing runs in two steps. Step 1 accesses BAR to determine the R-tree nodes that contain query attribute values and overlap the query time. Step 2 takes the nodes returned from Step 1 as well as the spatio-temporal box to perform a breadth-first search on the R-tree, as illustrated in Fig. 9.16.

To determine the objects containing query values, the procedure accesses BAR to look for the nodes fulfilling the attribute condition. That is, the method searches for the nodes in which there are trajectories containing Q_a. For each attribute value, BAR is accessed to find the tuple and get the record. Each item in the record stores a node id and a bitmap marking the entries containing the value. The item is obtained for each attribute value and the intersection operation is performed on bitmaps to find the entries containing the query. The time dimension of the node is checked to determine whether the item (a candidate node) exists in the returned node set, denoted by N_a. If not, an item (nid, b, t) is inserted by adding a counter, initialized by 1. If yes, the counter is increased and the bitmap is updated by performing the operation AND. In the end, items in N_a that cannot contribute to the result are removed.

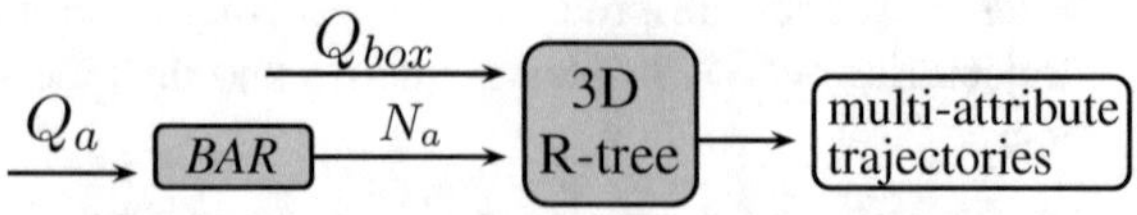

Fig. 9.16 Procedure of RQMAT

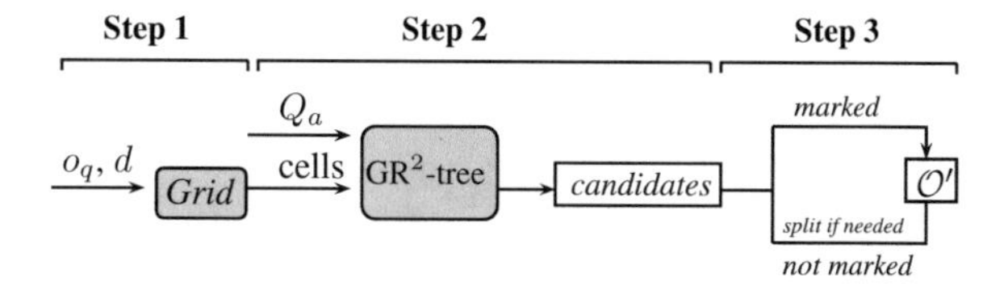

Fig. 9.17 Procedure of CRQMAT

9.5.3 *Processing CRQMAT*

The query CRQMAT is answered in three steps, as illustrated in Fig. 9.17. The index structure *GR^2-tree* includes a Grid R-tree and an attribute relation.

In Step 1, the spatio-temporal area restricted by o_q and d is established. Based on the grid partition one quickly determines the cells within the d-distance to the query. This is achieved by computing the distance between the 2-D bounding box of the query trajectory and the cell. The nodes that do not intersect the cells can be safely pruned. Usually, a cell is not always within the d-distance to the query as the location of the trajectory changes over time. Time-dependent cells are reported and maintained by a composite structure including three components: *cell tree*, *cell set* and *cell list*. The cell tree is a binary tree that records a time interval and the cells intersecting the query trajectory. The structure reports all cells within the d-distance to the query during a time interval. A cell may be valid at different time intervals. The method maintains a cell set by removing duplicate results. The cell list determines whether all trajectories in a leaf node are within the d-distance to the query. If yes, the exact distance computation can be avoided as a leaf node stores trajectories whose movements are restricted in a cell.

An example By referring to Fig. 9.18a, we enlarge the bounding box of o_3 in both x and y dimensions to find all cells within the d-distance to the query (depicted in gray). Two dashed lines are depicted to help figure out the cells. The time interval $T(o_3)$ intersects $\{T_1, T_2, T_3\}$. The cells $\{c_{5,1}, c_{5,2}, c_{6,1}, c_{6,2}, c_{7,2}\}$ are within the d-distance to the query at T_1, but they should not be considered at T_3. There are three *marked* cells $\{c_{5,5}, c_{6,3}, c_{7,4}\}$ at $T_2 \cup T_3$. Thus, the cell trajectory of o_1 in $c_{5,5}$ and the cell trajectory of o_4 in $c_{7,4}$ can be directly returned without performing the accurate distance computation (the attribute condition is not considered here). The structure of the time-dependent cells is reported in Fig. 9.18b. The cell set consists of three parts C_1, C_2 and C_3, partitioned by time intervals. The cell tree is built on cells with corresponding time intervals. Since cells $\{c_{5,5}, c_{6,3}, c_{7,4}\}$ are *marked cells*, they are put into the cell list with time intervals.

In Step 2, the procedure traverses the R-tree to return a set of candidates, each of which contains Q_a and has the distance to o_q less than d. Given an R-tree node, the algorithm determines the cells intersecting the node. The cells reported in step 1 are used to prune the node if there is no overlap between the cells intersecting the node and the cells within the d-distance to the query. When a leaf node is accessed, the

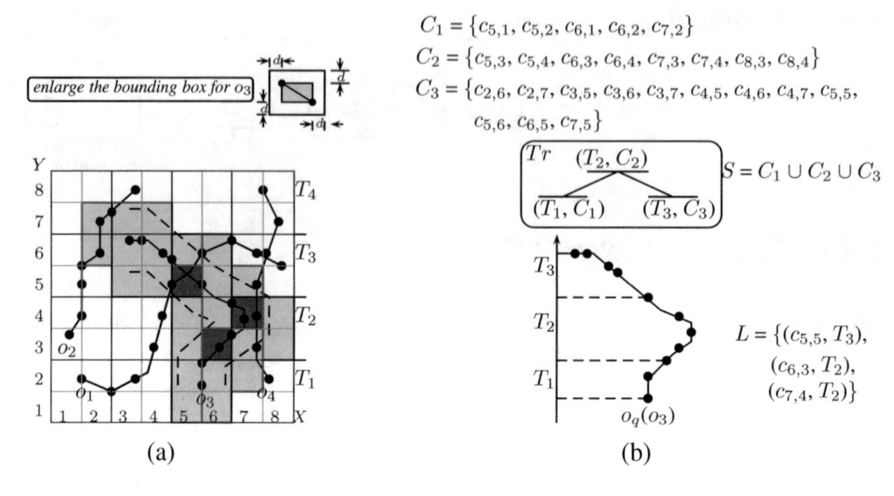

Fig. 9.18 An example of establishing qualified cells. (**a**) Cells within d to o_q. (**b**) Cell tree, cell set and cell list

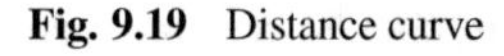

Fig. 9.19 Distance curve

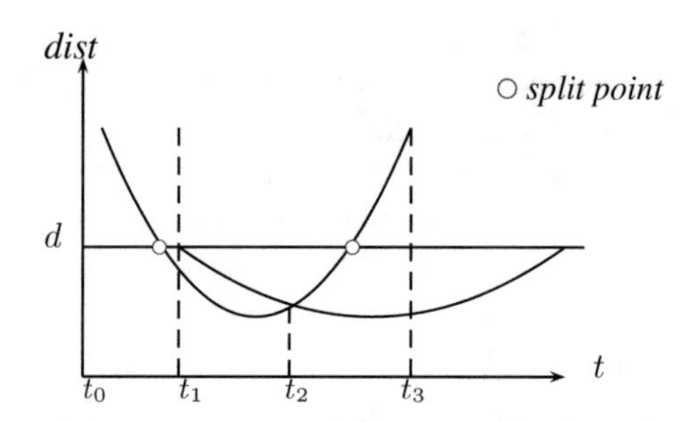

algorithm iteratively computes the exact distance between each trajectory and the query. The distance is an approximate value calculated by using minimum bounding boxes of trajectories. A candidate is marked if its maximum distance to o_q is less than d.

Step 3 iteratively checks the accurate distance. If the candidate is marked, it will be directly put into the result set. Otherwise, the actual distance is computed. A trajectory may be split because only the piece of movements fulfilling the distance condition is reported. Two trajectories are mapped into pieces with the same time. The task is to compute the intersections among a set of distance curves to determine the curves whose values to o_q are smaller than d, and return the parts corresponding to these pieces. The time-dependent distance is represented by a square root of a quadratic polynomial, as demonstrated in Fig. 9.19.

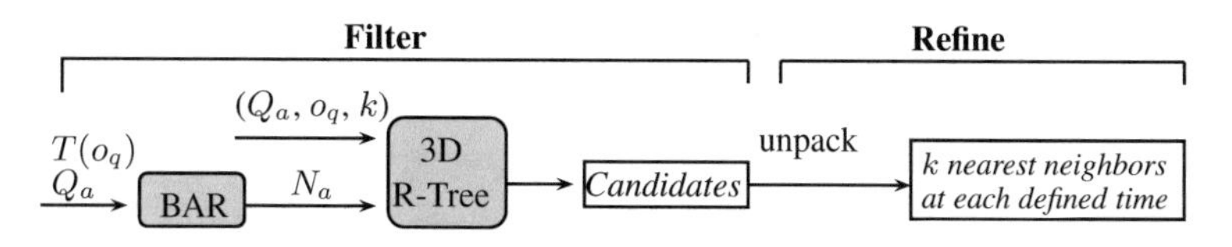

Fig. 9.20 Procedure of C*k*NN_MAT

9.5.4 Processing CkNN_MAT

In the filter step, the structure BAR is accessed to determine the R-tree nodes including query attribute values and intersecting the query time, denoted by N_a. Next, the procedure takes N_a coupled with o_q and k as input to traverse the R-Tree in breadth-first order, during which the search space is pruned by taking into account spatial and temporal parameters as well as attributes. The filter returns a set of candidates, each of which fulfills the attribute condition and approximately belongs to k nearest neighbors. In the refinement step, each candidate trajectory is unpacked to get the original temporal units and perform the exact distance computation to return k nearest neighbors at each query time. The query procedure is shown in Fig. 9.20.

Collecting R-tree nodes The R-tree nodes containing Q_a are collected level by level. For each $a \in Q_a$, the procedure starts from $h = 1$ and accesses BAR to find the records. For each item (nid, b, t) in the record, one checks whether the item identified by *nid* is already in N_a. If not, the method inserts the item into N_a by attaching a counter, initialized by 1. Such a value represents the number of query attribute values contained in the node. The extended record item is denoted by λ. If the item already exists in N_a, the counter is increased and the bitmap is updated by performing the bitwise AND. This is because a node fulfilling the condition must contain all values in Q_a.

Lemma 1 *Given an item $\lambda \in N_a$, the item is pruned if $\lambda.counter \neq |Q_a|$ or $\lambda.b = 0$.*

Proof (i) $\lambda.counter \neq |Q_a|$: Obviously, it is impossible that $\lambda.counter > |Q_a|$ as distinct values are counted. If $\lambda.counter < |Q_a|$, this means that the number of attributes contained by λ is less than $|Q_a|$ and therefore λ can be safely pruned. (ii) $\lambda.b = 0$: There is no entry in the node containing all $a \in Q_a$ and λ can be safely pruned. □

An extension: multiple values An extension is made to allow a query attribute with multiple values. A node is satisfied if it contains one of the values. The aforementioned $\lambda \in N_a$ is extended to (nid, b, t, aid) by adding the attribute id. The bitwise OR is performed on bitmaps for values from the same attribute.

Lemma 2 *Let $AttCount(Q_a)$ ($\leq |Q_a|$) return the number of query attributes. An item $\lambda \in N_a$ is pruned if $\lambda.counter \neq AttCount(Q_a)$ or $\lambda.b = 0$.*

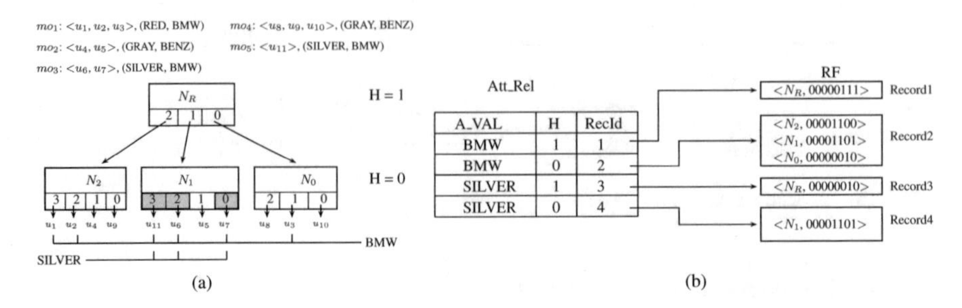

Fig. 9.21 3D R-Tree

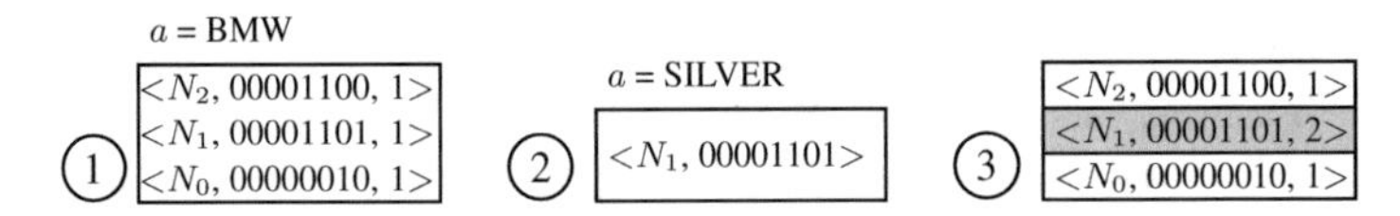

Fig. 9.22 Reporting nodes at $H = 0$

An example Using trajectories in Fig. 9.5, we report the 3D R-Tree and the structure BAR for attributes SILVER and BMW in Fig. 9.21. Consider Q_a = (SILVER, BMW). At $H = 1$, we access records 1 and 3, and have $b = 00000111$ for BMW and $b = 0000010$ for SILVER. By performing the bitwise "AND" operation, the 0th and 2nd entries are not defined and therefore N_0 and N_2 are pruned. Figure 9.22 depicts the procedure of determining nodes for Q_a = <SILVER, BMW> at $H = 0$ (t is omitted).

Reporting candidates The approach traverses the R-tree from root to leaf level and uses a list to maintain accessed nodes. For each visited node, the algorithm determines whether (i) the node fulfills the attribute condition; and (ii) objects in the subtree will contribute to the result. The query needs only k neighbors. If there are k candidates at each defined time, objects that are further than current candidates to o_q can be safely pruned. To determine whether there are enough candidates, a segment tree is maintained by storing the time interval, the distance and the number of trajectories in a node. One can do the pre-computation for each attribute value and use it for $|Q_a| = 1$. However, such a value has to be calculated on-the-fly for $|Q_a| > 1$, which is a costly procedure. The sub-tree will be traversed to count the number of trajectories containing query attributes as one needs to determine the intersection set of different query attributes.

The refinement This step includes two phases: (i) unpack each candidate to obtain temporal units; (ii) apply the slightly modified plane-sweep algorithm (Bentley and Ottmann 1979) to determine the k lowest time-dependent distance curves to report the result. Phase (ii) takes in a sequence of candidate trajectories ordered on time. The time-dependent distances to the query trajectory are computed to find k nearest objects at each time point. To achieve this, one determines pieces of movements with overlapping time and applies the distance function by employing the linear interpolation on each piece (Forlizzi et al. 2000; Frentzos et al. 2005). The method manipulates temporal units to calculate the distance. Split points between curves are found to determine the k lowest pieces of curves.

9.6 The System Development

9.6.1 The Architecture

A prototype database system is developed to efficiently manage multi-attribute trajectories including data representation, index structures, query algorithms and optimizations (Wang and Xu 2017; Xu and Güting 2017; Wei and Xu 2018). Since standard trajectories have been supported in a database system SECONDO (Güting et al. 2010a), the task is to develop modules for multi-attribute trajectories and seamlessly integrate them into the system. Key system components are shown in Fig. 9.23.

The query interface animates standard trajectories and displays multiple attributes. Not only objects whose locations changing over time are visualized, but also their time-dependent attribute values are displayed. Queries on multi-attribute trajectories include several predicates, leading to different query plans. The optimization selects the best plan according to the analytical model. Then, the corresponding algorithm is executed and the index structure is accessed. The index component is in principle made up of a 3-D R-tree and a composite structure BAR. The 3-D R-tree preserves the spatio-temporal proximity, and BAR manages attribute values. The data storage component includes several modules such as spatial and temporal data, standard trajectories, relational tables, and attributes.

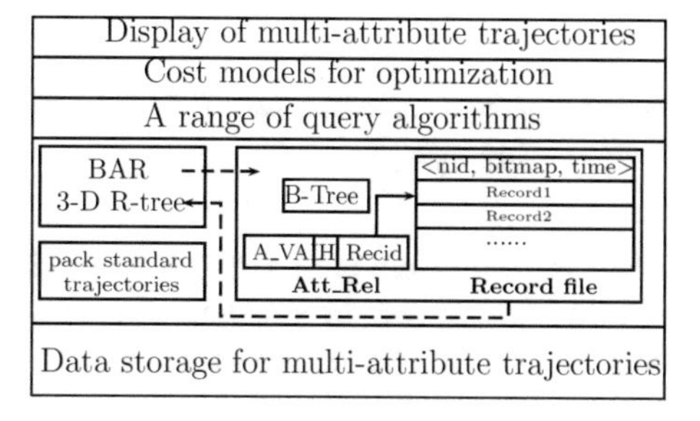

Fig. 9.23 The system architecture

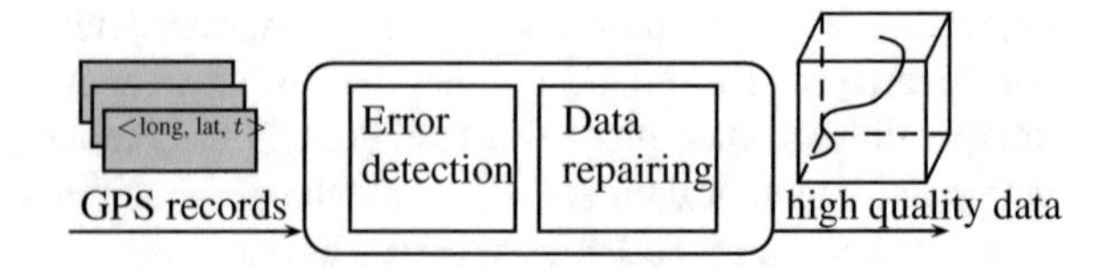

Fig. 9.24 GPS-clean workflow

9.6.2 A Tool for GPS Data Clean

Real-life data is far from being reliable enough for applications predominantly due to a large number of errors such as inaccurate measurement, noisy, distortion and outliers, mainly caused by the limitations of devices or signal loss. A data cleaning tool will serve as the pre-processing module to bring data from a messy to a neat state. The primary task is the minimization or the total removal, if possible, of GPS errors, and the repairing of trajectories after removing some sample points. The tool can be treated as an objective function *Clean*: *RawData* → *HighQualityData* and must be an integral part of a moving objects database.

The procedure consists of two steps: error detection and data repairing, as illustrated in Fig. 9.24. *Error detection* identifies incorrect data values, which can be classified into two categories: *point error* and *trajectory error*. The first is identified by checking the data item in each individual record such as time-stamp and long/lat, and the second is established by evaluating a sequence of records of the same object such as *distortion* and long time still. *Data repairing* involves updating the available data to remove any detected errors, and derives and fills in missing data from the existing data.

Refining a good clean function requires a rich set of detection rules, filtering operations, statistical analysis and missing value imputation methods. Prediction models can be built by learning features from historical data with different characters. To evaluate the data quality, we define data metrics to measure the quality and employ machine learning techniques to classify the raw mobility data.

9.6.3 The Generation of Multi-attribute Values and Query Interface

There is a number of public trajectory data but attribute values are not easy to collect. A tool is developed to generate attributes. One can flexibly scale the number of attributes and the domain of each attribute. For each attribute, the value is randomly and uniformly selected from its domain. By making use of real trajectories from a company DataTang (2018) (http://factory.datatang.com/en/) and synthetic attribute values, the chapter demonstrates queries RQMAT, CRQMAT, C*k*NN_MAT, as illustrated in Fig. 9.25. Queries are "*Find all* SILVER VWs *intersecting the query window*", "*Keep reporting all* BENZs *within 5km to the target*" and "*Continuously report the nearest* RED BMW (or SILVER VM, BLACK BENZ) *to the query*

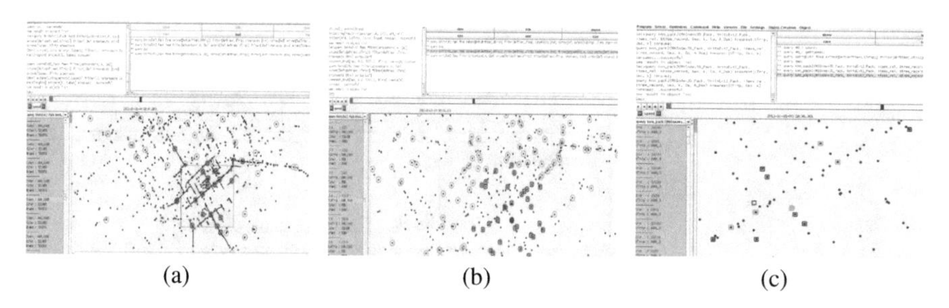

Fig. 9.25 Query and visualize multi-attribute trajectories. (**a**) RQMAT. (**b**) CRQMAT. (**c**) C*k*NN_MAT

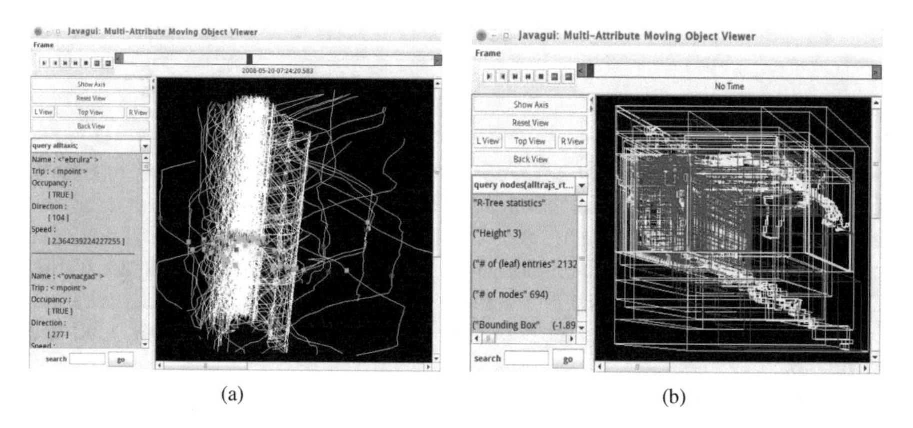

Fig. 9.26 The 3-D query interface. (**a**) Dynamic attributes. (**b**) Visualizing 3-D R-tree

trajectory". Objects in the figure are results at a time point and the interface provides the animation. Users can also define multiple values for each attribute. The query interface provides zoom in/out to let users get a closer/further view.

By making use of cab mobility traces from Piorkowski et al. (2009), the system demonstrates querying time-dependent attributes. Each taxi is associated with a flag marking whether the taxi is free or occupied. To make a good judgment about the shape of the R-tree, a tool graphically viewing the structure is needed. The developed query interface supports displaying dynamic attributes and visualizing the R-tree in a 3-D view, as illustrated in Fig. 9.26.

9.6.4 MDBF: A Tool for Monitoring Database Files

File monitoring plays an essential role in operating system that constantly watches folders and files. Actions are triggered when files are created or accessed (read-/write). A tool called MDBF (Monitoring Database Files) is developed to monitor

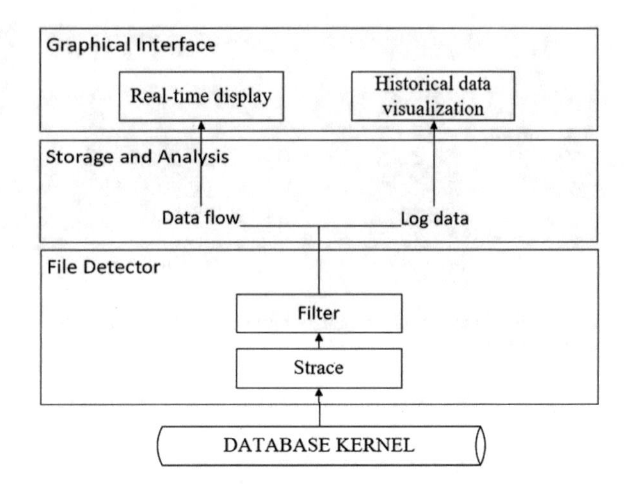

Fig. 9.27 The framework of MDBF

database files during the query execution (Wei and Xu 2018). The tool consists of three components: file detector, storage and analysis system, and graphical interface, as illustrated in Fig. 9.27. The file detector, serving as the key component, makes use of a tool Strace, which is a diagnostic, debugging and instructional userspace utility. Strace monitors and tampers with interactions between processes and Linux kernel, including system calling, signal delivering, process state changing, and read and write blocks of data.

When a query is executed, the tool detects database file operations and produces a monitoring log. The log data is automatically reported to the filter and formatted in a relational table in the system. The filtering is performed by extracting the data related to database file operations and storing action statistics. Strace captures all system calls in the query evaluation, but file operations are the main tasks. The data flow received from Strace is transformed to a certain format transferred between system modules. The information of accessing files is displayed when the querying is running and also recorded as historical data. A thorough analysis is performed on log data. MDBF monitors queries incurring database file operations. A graphical interface visualizes the access content to help understanding the query progress. Users can compare the access information of different files.

9.7 Performance Evaluation

The proposal is implemented in C/C++ and the evaluation is performed in an extensible database system SECONDO (Güting et al. 2010a). The system is a freely open source software and has an extensible architecture well-supported for spatial and spatio-temporal data management. A standard PC (Intel(R) Core(TM) i7-4770CPU, 3.4GHz, 4GB memory, 2TB hard disk) running Suse Linux 13.1 (32 bits, kernel version 3.11.6) is used.

Name	#Trips	#GPS Records
BTaxi	992,997	55,950,357

(a)

$X(Y)$	$30\% \cdot \Delta X$ (ΔY)
T	$40\% \cdot \Delta T$
$\lvert Q_a \rvert$	{1, 2, **3**, 4, 5}

(b)

Fig. 9.28 Datasets and parameters. (**a**) Standard trajectories. (**b**) Query parameters

9.7.1 Evaluation of RQMAT

We use real taxi trajectories in Beijing from DataTang (2018) (http://factory.datatang.com/en/). The statistics of standard trajectories and query parameters are reported in Fig. 9.28. Q_{box} is randomly generated with sizes $X = 30\% \cdot \Delta X$ and $Y = 30\% \cdot \Delta Y$, in which ΔX and ΔY are lengths of x and y dimensions, respectively. The time interval is 40% of the overall time. One can arbitrarily enlarge or shrink the spatio-temporal window. We did some preliminary tests and found that smaller windows may not receive any result. We focus on evaluating the performance affected by attributes and hence keep the same size for Q_{box}.

In the evaluation, CPU time and I/O accesses are used as performance metrics and the results are averaged over 20 runs. Five alternative methods are included: (i) **3D R-tree**, (ii) 3D R-tree + Attribute Set (**RAttSet** for short). The idea is similar to IR-tree (Cong et al. 2009) employed in spatial keyword querying that augments each R-tree node with a summary of keywords in the subtree, (iii) **4D R-tree**, for each multi-attribute trajectory, we distribute $(a_1, \ldots, a_{|A|})$ into $|A|$ trajectories, each of which defines the 4-D data: *location*, *time* and *a single-attribute value*. (iv) **IOC-Tree** (Han et al. 2015), the structure consists of an inverted index and a set of three-dimensional quadtrees, each of which corresponds to an attribute and stores relevant trajectory points and (v) **HAGI** (Su et al. 2007), the method employs a hierarchical aggregate grid index. The evaluation demonstrates the impact of $|Q_a|$ on the performance. Figure 9.29 shows that our method is an order of magnitude faster than other methods in most settings.

9.7.2 Evaluation of CRQMAT

We use real GPS records of Beijing taxis (DataTang (2018) http://factory.datatang.com/en/). The dataset statistics and the settings of query parameters ($|Q_a|$ and d) are reported in Table 9.5. We perform the evaluation by comparing our method named GR^2-tree (Grid R-tree with an attribute Relation) with five baseline methods in terms of scalability and efficiency: (i) **3-D R-tree**; (ii) **RIB**, we adapt the method in Wu et al. (2012). Multi-attribute trajectories are grouped on attribute values by applying Z-order to map the $|A|$-dimensional value to one-dimensional. Each R-tree node contains a pointer to an inverted bitmap that records the positions of entries

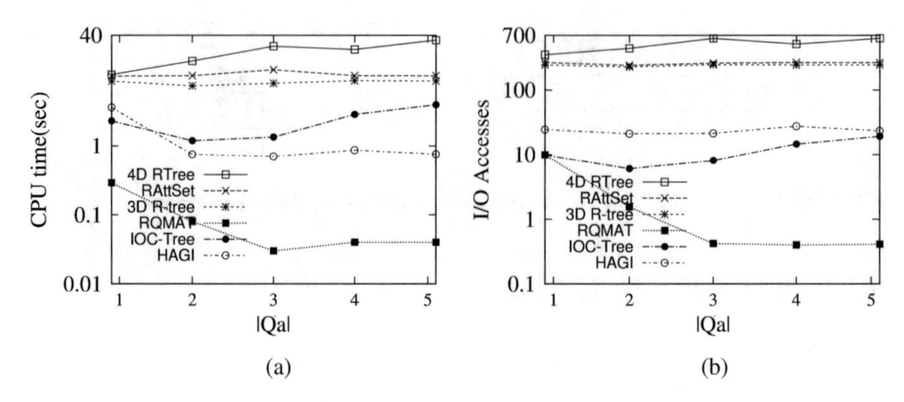

Fig. 9.29 Effect of $|Q_a|$ ($|A| = 10$). (**a**) CPU Time (sec). (**b**) I/O Accesses($\times$ 10^3)

Table 9.5 Datasets and parameter settings

Name	#GPS Records	$\|\mathcal{O}\|$	$\|A\|$	$dom(A)$	X and Y ranges
BTaxi	235634511	4220435	10	[1, 151]	[21, 119958], [0, 119653]
		Query settings			
$\|Q_a\|$: {1, 2, $\underline{3}$, 4, 5}			d (km): {1, 5, $\underline{10}$, 20, 50}		

Table 9.6 Datasets for scaling $|\mathcal{O}|$

Name	$\|\mathcal{O}\|$
BT1	533635
BT2	1009579
BT3	1424273
BT4	2757312
BT5	4220435

defining the attribute value. A relation stores the bitmaps by setting the fanout as the bit length; (iii) **4-D R-tree**; (iv) **IOC-Tree** (Han et al. 2015); (v) **HAGI** (Su et al. 2007).

Scalability. To vary the data size, different subsets of BTAXI are selected, as summarized in Table 9.6. The performance result is reported in Fig. 9.30. When the data size grows, the costs of all methods rise proportionally, but our method outperforms baseline methods by a factor of 5-50x on the largest dataset.

Varying $|Q_a|$. We perform the evaluation by varying the number of query attributes. The results, as reported in Fig. 9.31, demonstrate that our method substantially outperforms baseline methods in all settings. When $|Q_a|$ increases, the performance becomes better as the attribute predicate is more selective.

Varying the distance d. We evaluate the performance affected by d, as reported in Fig. 9.32. When d increases the performance degrades as expected due to more objects being processed. The advantage of our method is significant. When d increases, more cells will be included in the search space.

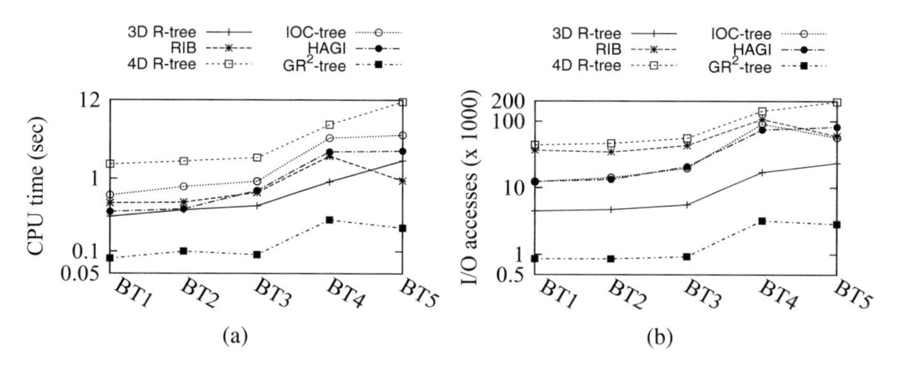

Fig. 9.30 Scaling $|\mathcal{O}|$. (**a**) CPU time. (**b**) I/O accesses

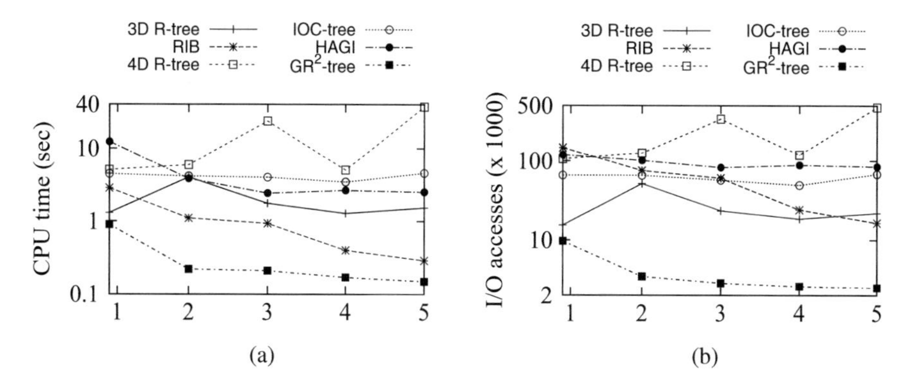

Fig. 9.31 The effect of $|Q_a|$. (**a**) CPU time. (**b**) I/O accesses

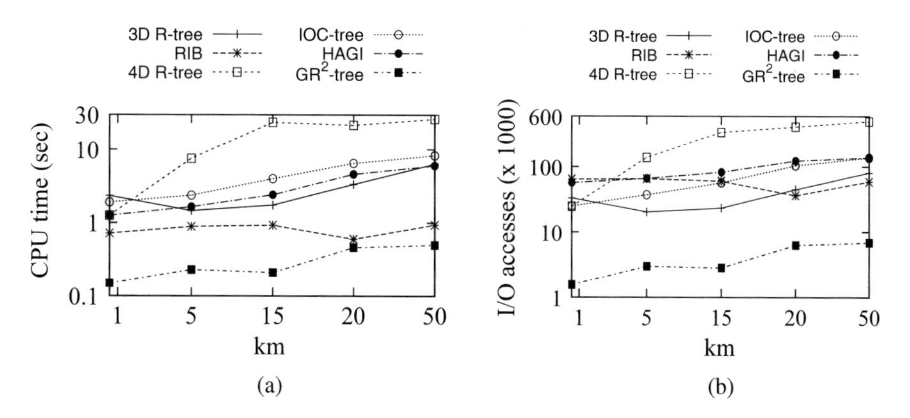

Fig. 9.32 The effect of d. (**a**) CPU time. (**b**) I/O accesses

Table 9.7 Datasets statistics and parameters

Name	$\|\mathcal{O}\|$ (million)	Temporal units # (million)	$\|A\|$	$dom(A)$
TAXI	3.0	32.5	1	[1, 4]
BUS	1.1	8.3	1	[1, 384]
MAT5	1.8	110.88	10	[1, 127]

$\|Q_a\|$	{1, 2, **3**, 4, 5}
k	{1, **5**, 10, 20, 50, 100}

9.7.3 Evaluation of CkNN_MAT

Both real and synthetic datasets are used, shown in Table 9.7. Real datasets are from a company called Datatang (2018) (http://factory.datatang.com/en/): Shanghai taxis (TAXI) and Beijing buses (BUS). TAXI includes GPS records from four taxi companies in 2014. The company id is defined as an attribute. BUS contains bus card records in 2014. Each record stores the time and bus stops where passengers gets on and off the bus. Each bus is identified by its id and bus stops are identified by the order in the route and long/lat. We build bus trips from these records. This is done by grouping records on bus id and then sorting them on time. There are 384 bus routes in total and the route id is set as an attribute. Part of the data can be found at http://dbgroup.nuaa.edu.cn/jianqiu/. Synthetic datasets are generated by utilizing a tool MWGen (Xu and Güting 2012). For each query, o_q is randomly selected over the dataset. The settings for Q_a and k are listed, in which default values are in bold. Each $a \in Q_a$ is a stochastic value from the domain.

We develop three baseline methods for performance comparison. (i) **4D R-tree**. (ii) **3D R-tree + Attribute Relation** (3D RAR). The method in Güting et al. (2010b) is extended to support proposed queries by recording the set of attributes contained by each node. During the query procedure, we first determine whether the accessed node contains Q_a. If yes, we open the node and move forward to spatial and temporal examinations. Otherwise, we prune the node. However, the R-tree does not know the number of trajectories containing Q_a for each node because queries issue different attributes. Consequently, the criterion of pruning trajectories based on distance and the number of trajectories can not be used. (iii) **3D R-tree + Inverted Bitmap** (RIB) (Wu et al. 2012). Our method is named BAR.

Scaling the number of attributes and the domain. We evaluate the scalability affected by attributes: $|A|$ and $dom(A)$. Figure 9.33 reports the settings. Figure 9.34 reports the result on scaling $|A|$. Our method achieves the best performance in all settings and RIB performs competitively to our method when $|A| = 3$. RIB manages attributes and the attribute predicate has a good selectivity when $|Q_a| = |A|$. However, when $|A|$ increases, $dom(A)$ rises proportionally, and the RIB performance degrades significantly. We analyze that Z-order values cannot well preserve the locality when the dimension becomes large, and the linear scanning method of determining entries is inferior to our bitmap querying approach. When

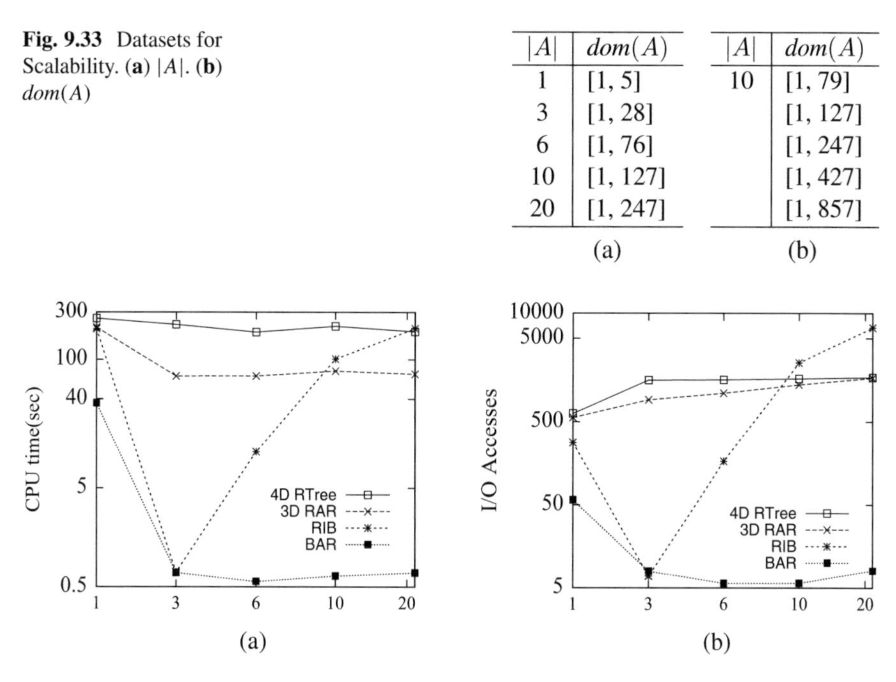

Fig. 9.33 Datasets for Scalability. (**a**) $|A|$. (**b**) $dom(A)$

$\|A\|$	$dom(A)$
1	[1, 5]
3	[1, 28]
6	[1, 76]
10	[1, 127]
20	[1, 247]

(a)

$\|A\|$	$dom(A)$
10	[1, 79]
	[1, 127]
	[1, 247]
	[1, 427]
	[1, 857]

(b)

Fig. 9.34 Scaling $|A|$. (**a**) CPU Time (s). (**b**) I/O Accesses($\times$ 10^3)

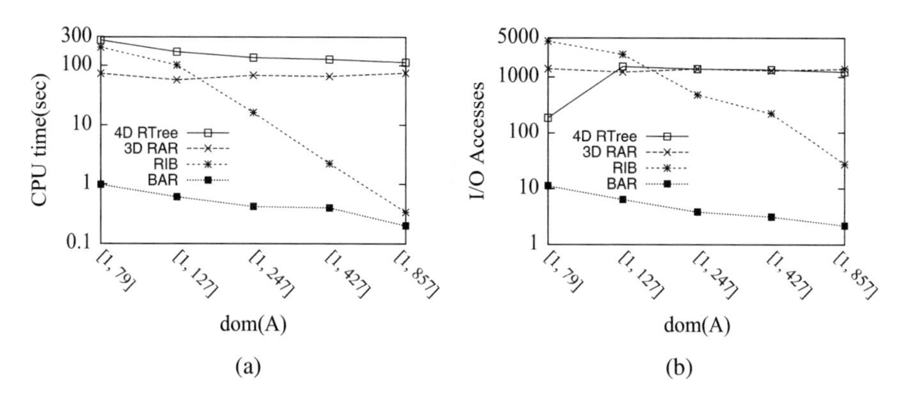

Fig. 9.35 Scaling $dom(A)$, $|A| = 10$. (**a**) CPU Time (s). (**b**) I/O Accesses($\times$ 10^3)

scaling $dom(A)$, our method also outperforms baseline methods. One can see the trend that the performance increases when $dom(A)$ becomes large as the attribute predicate is more selective. The 4D R-tree has poor performance because the dataset is enlarged when $dom(A)$ increases (Fig. 9.35).

Effect by k. We evaluate the performance effected by k and report the results in Figs. 9.36 and 9.37. TAXI and BUS contain only one attribute and therefore we

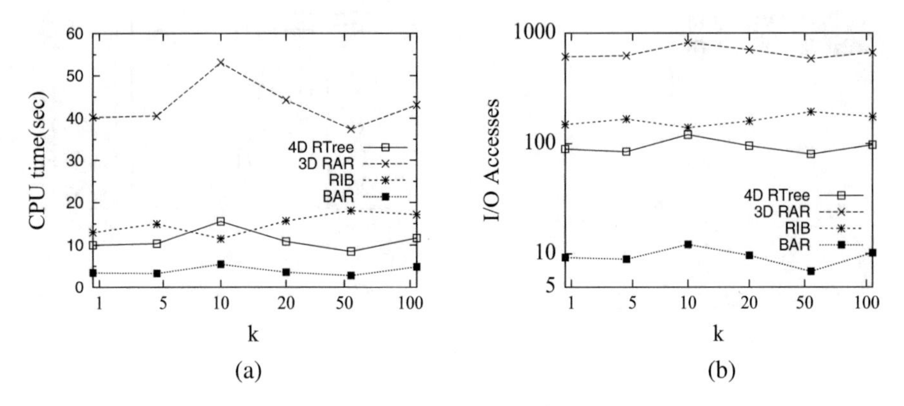

Fig. 9.36 Scaling k using TAXI. (**a**) CPU Time (s). (**b**) I/O Accesses($\times$ 10^3)

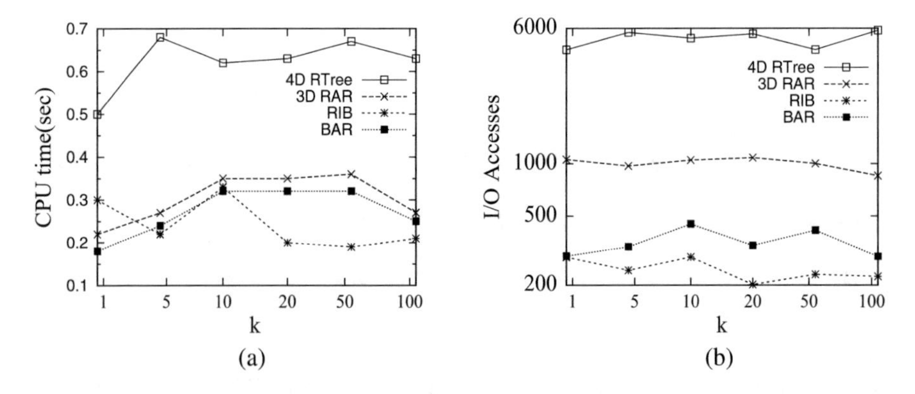

Fig. 9.37 Scaling k using BUS. (**a**) CPU Time (s). (**b**) I/O Accesses($\times$ 10^3)

set $|Q_a| = 1$ for them. All methods are not sensitive to k. Our method significantly outperforms alternative methods for TAXI. For BUS, the RIB performance is close to ours due to good attribute selectivity. BUS has $|A| = 1$ and $dom(A)$ is [1, 384]. In contrast, we have $|A| = 1$, $dom(A) = [1, 4]$ for TAXI.

The analysis. Consider the 4D R-tree. To build the index, each multi-attribute trajectory is decomposed into $|A|$ trajectories, each of which contains a single attribute. This enlarges the dataset by $|A|$ times. Also, the attribute is approximately evaluated when traversing the index. The method 3D RAR is able to select R-tree nodes containing individual attributes, but cannot determine whether data objects contain all query attributes if $|Q_a| > 1$. RIB achieves good performance when the attribute predicate is quite selective, for example, (i) $|Q_a| = |A|$, or (ii) $dom(A)$ is large and $|Q_a| = 1$. Thus, this method is limited in scope. Our method achieves the stable performance and generalizes to queries on standard trajectories, achieved by skipping the step of accessing BAR.

9.8 Future Directions

9.8.1 Data Analytics

By making use of multi-attribute trajectories, one is able to provide a fine-grained analysis on the traffic condition by discovering dense areas such as taxi pick-ups and drop-offs and crowded bus/metro stations. An optimal deployment can be made by recommending the ride-sharing and redesigning the scheduling/route. Based on the number of passengers on the bus/metro at populated places, the system can recommend people to advance or postpone their trips in a small derivation in order to avoid the peak time and reduce the waiting time.

Mining and analyzing multi-attribute trajectories is also an interesting topic. One can utilize the rich contexts from attributes to fully understand spatio-temporal trajectories and discover potential relationships and behavior. For example, in order to know whether large vehicles such as buses and trucks have a great negative impact on traffic flow, the system should consider not only the number of vehicles in crowded places but also the percentage of large vehicles. Based on that, policies can be made to improve the traffic. In some applications, attribute values change over time, e.g., taxi status (free or occupied), fuel consumption and the number of passengers in a bus. The data representation needs to be extended to support dynamic attributes. One solution is to define a moving integer/real to represent the time-dependent value. The question is how to efficiently manage several dynamic attributes in the framework and adjust the index structure correspondingly. Application queries and analysis can be performed by considering spatio-temporal parameters and dynamic attribute values to find some interesting behavior.

9.8.2 Intelligent Trajectory Data Management

The artificial intelligence community has accomplished many promising results, while the specialization to moving objects will offer new opportunities because solutions are fitted to particular properties of mobile data. Due to the increasing number of models and structures, a number of parameters are required to control the system. An automatic approach of recommending system settings is preferred that leverages past experiences of workloads. Machine learning models can be trained by transferring previous experiences to apply for a new application.

The query interface should be powerful enough to support arbitrary queries and simple enough to let users express their questions in natural language. User queries are expressed by natural language but will be translated into an executable language in the system. However, employing natural language to query the database is a non-trivial task. Most database systems are queried by structured query language which is often difficult to write for non-experts. Moving objects databases should be capable of providing a communication model that translates natural language questions into well-formed queries such that the user and the system can understand each other. There are hundreds of published papers about querying moving objects in which example queries are expressed in natural languages, but an interface translating natural languages into an SQL or SQL-like language is not available.

9.9 Conclusions

This chapter introduces multi-attribute trajectories that enrich the spatio-temporal trajectory representation. A range of new queries is studied that search for the target fulfilling both spatio-temporal and attribute conditions. A hybrid index structure is designed and updating the index is supported. Efficient query algorithms are developed with optimization strageties. A systematic design is made such that the proposed structure is also able to process standard trajectories with little effort. We develope a prototype database system for multi-attribute trajectories including data storage, access methods, index structures, data generators, monitoring tools and the query interface. The performance evaluation is conducted by using real and synthetic datasets.

Acknowledgments This work is supported by NSFC under grants 61972198, Natural Science Foundation of Jiangsu Province of China under grants BK20191273 and National Key Research and Development Plan of China (2018YFB1003902). Thanks Weiwei Wang for developing the 3-D visualization tool and Xiangyu Wei for developing the monitoring tool.

References

Alvares LO, Bogorny V, Kuijpers B, Moelans B, Fern JA, Macedo ED, Palma AT (2007) Towards semantic trajectory knowledge discovery. Data Mining and Knowl Discov

Bentley JL, Ottmann T (1979) Algorithms for reporting and counting geometric intersections. IEEE Trans Comput 28(9):643–647

Bercken J, Seeger B (2001) An evaluation of generic bulk loading techniques. In: VLDB, pp 461–470

Bercken J, Seeger B, Widmayer P (1997) A generic approach to bulk loading multidimensional index structures. In: VLDB, pp 406–415

Chakka VP, Everspaugh A, Patel JM (2003) Indexing large trajectory data sets with seti. In: CIDR

Chen Z, Tao Shen H, Zhou X, Zheng Y, Xie X (2010) Searching trajectories by locations: an efficiency study. In: SIGMOD, pp 255–266

Chen L, Cong G, Jensen CS, Wu D (2013) Spatial keyword query processing: an experimental evaluation. PVLDB 6(3):217–228

Cong G, Jensen CS, Wu D (2009) Efficient retrieval of the top-k most relevant spatial web objects. PVLDB 2(1):337–348

de Almeida VT, Güting RH (2005) Indexing the trajectories of moving objects in networks. GeoInformatica 9(1):33–60

De Felipe I, Hristidis V, Rishe N (2008) Keyword search on spatial databases. In: ICDE, pp 656–665

Dinh L, Aref WG, Mokbel MF (2010) Spatio-temporal access methods: part 2 (2003–2010). IEEE Data Eng Bull 33(2):46–55
Forlizzi L, Güting RH, Nardelli E, Schneider M (2000) A data model and data structures for moving objects databases. In: SIGMOD, pp 319–330
Frentzos E (2003) Indexing objects moving on fixed networks. In: SSTD, pp 289–305
Frentzos E, Gratsias K, Pelekis N, Theodoridis Y (2005) Nearest neighbor search on moving object trajectories. In: SSTD, pp 328–345
Frentzos E, Gratsias K, Pelekis N, Theodoridis Y (2007) Algorithms for nearest neighbor search on moving object trajectories. GeoInformatica 11(2):159–193
Güting RH, Schneider M (2005) Moving objects databases. Morgan Kaufmann, Amsterdam
Güting RH, Böhlen MH, Erwig M, Jensen CS, Lorentzos NA, Schneider M, Vazirgiannis M (2000) A foundation for representing and querying moving objects. ACM TODS 25(1):1–42
Güting RH, Behr T, Düntgen C (2010a) SECONDO: a platform for moving objects database research and for publishing and integrating research implementations. IEEE Data Eng Bull 33(2):56–63
Güting RH, Behr T, Xu J (2010b) Efficient k-nearest neighbor search on moving object trajectories. VLDB J 19(5):687–714
Güting RH, Valdés F, Damiani ML (2015) Symbolic trajectories. ACM Trans Spat Algorithms Syst 1(2):Article 7
Hadjieleftheriou M, Kollios G, Tsotras VJ, Gunopulos D (2002) Efficient indexing of spatiotemporal objects. In: EDBT, pp 251–268
Han Y, Wang L, Zhang Y, Zhang W, Lin X (2015) Spatial keyword range search on trajectories. In: DASFAA, pp 223–240
Jensen CS, Lu H, Yang B (2009) Indexing the trajectories of moving objects in symbolic indoor space. In: SSTD, pp 208–227
Jeung H, Yiu ML, Zhou X, Jensen CS, Shen HT (2008) Discovery of convoys in trajectory databases. PVLDB 1(1):1068–1080
Lange R, Dürr F, Rothermel K (2011) Efficient real-time trajectory tracking. VLDB J 20(5):671–694
Lee T, Park J, Lee S, et al (2015) Processing and optimizing main memory spatial-keyword queries. PVLDB 9(3):132–143
Long C, Wong RCW, Jagadish HV (2013) Direction-preserving trajectory simplification. PVLDB 6(10):949–960
Lu H, Cao X, Jensen CS (2012) A foundation for efficient indoor distance-aware query processing. In: ICDE, pp 438–449
Mauroux PC, Wu E, Madden S (2010) Trajstore: an adaptive storage system for very large trajectory data sets. In: ICDE, pp 109–120
Parent C, Spaccapietra S, Renso C, Andrienko GL, Andrienko NV, Bogorny V, Damiani ML, Gkoulalas-Divanis A, de Macêdo JAF, Pelekis N, Theodoridis Y, Yan Z (2013) Semantic trajectories modeling and analysis. ACM Comput Surv 45(4):42
Pelanis M, Saltenis S, Jensen CS (2006) Indexing the past, present, and anticipated future positions of moving objects. ACM TODS 31(1):255–298
Pfoser D, Jensen CS (2003) Indexing of network constrained moving objects. In: GIS, pp 25–32
Pfoser D, Jensen CS (2000) Novel approaches in query processing for moving object trajectories. In: VLDB, pp 395–406
Piorkowski M, Sarafijanovic-Djukic N, Grossglauser M CRAWDAD dataset epfl/mobility (v. 24 Feb 2009). http://crawdad.org/epfl/mobility/20090224
Popa IS, Zeitouni K, Oria V, Barth D, Vial S Indexing in-network trajectory flows. VLDB J 20(5):643–669 (2011)
Rasetic S, Sander J, Elding J, Nascimento MA (2005) A trajectory splitting model for efficient spatio-temporal indexing. In: VLDB, pp 934–945
Song Z, Roussopoulos N (2003) Seb-tree: an approach to index continuously moving objects. In: MDM, pp 340–344

Su Y, Wu Y, Chen ALP (2007) Monitoring heterogeneous nearest neighbors for moving objects considering location-independent attributes. In: DASFAA, pp 300–312
Su H, Zheng K, Zeng K, Huang J, Sadiq SW, Yuan NJ, Zhou X (2015) Making sense of trajectory data: a partition-and-summarization approach. In: ICDE, pp 963–974
Tao Y, Papadias D (2001) Mv3r-tree: a spatio-temporal access method for timestamp and interval queries. In: VLDB, pp 431–440
Tong Y, Chen Y, Zhou Z, Chen L, Wang J, Yang Q, Ye J, Lv W (2017) The simpler the better: a unified approach to predicting original taxi demands based on large-scale online platforms. In: ACM SIGKDD, pp 1653–1662
Tong Y, Zeng Y, Zhou Z, Chen L, Ye J, Xu K (2018) A unified approach to route planning for shared mobility. PVLDB 11(11):1633–1646
Tzoumas K, Yiu ML, Jensen CS (2009) Workload-aware indexing of continuously moving objects. PVLDB 2(1):1186–1197
Valdés F, Güting RH (2014) Index-supported pattern matching on symbolic trajectories. In: ACM SIGSPATIAL, pp 53–62
Valdés F, Güting RH (2017) Index-supported pattern matching on tuples of time-dependent values. GeoInformatica 21(3):429–458
Valdés F, Güting RH (2019) A framework for efficient multi-attribute movement data analysis. VLDB J 28(4):427–449
Wang H, Zimmermann R (2011) Processing of continuous location-based range queries on moving objects in road networks. IEEE Trans Knowl Data Eng 23(7):1065–1078
Wang W, Xu J (2017) A tool for 3d visualizing moving objects. In: APWeb-WAIM, pp 353–357
Wang X, Zhang Y, Zhang W, Lin X, Huang Z (2016) SKYPE: top-k spatial-keyword publish/subscribe over sliding window. PVLDB 9(7):588–599
Wei X, Xu J (2018) MDBF: a tool for monitoring database files. In: ER Workshops, pp 54–58
Wu D, Yiu ML, Cong G, Jensen CS (2012) Joint top-k spatial keyword query processing. IEEE Trans Knowl Data Eng 24(10):1889–1903
Xu J, Güting RH (2012) Mwgen: a mini world generator. In: IEEE MDM, pp 258–267
Xu J, Güting RH (2013) A generic data model for moving objects. GeoInformatica 17(1):125–172
Xu J, Güting RH (2017) Query and animate multi-attribute trajectory data. In: ACM CIKM, pp 2551–2554
Xu J, Güting RH, Qin X (2015a) Gmobench: benchmarking generic moving objects. GeoInformatica 19(2):227–276
Xu J, Güting RH, Zheng Y (2015b) The TM-RTree: an index on generic moving objects for range queries. GeoInformatica 19(3):487–524
Xu J, Güting RH, Gao Y (2018a) Continuous k nearest neighbor queries over large multi-attribute trajectories: a systematic approach. GeoInformatica 22(4):723–766
Xu J, Lu H, Güting RH (2018b) Range queries on multi-attribute trajectories. IEEE Trans Knowl Data Eng 30(6):1206–1211
Yan Z, Chakraborty D, Parent C, Spaccapietra S, Aberer K (2011) Semitri: a framework for semantic annotation of heterogeneous trajectories. In: EDBT, pp 259–270
Yao B, Xiao X, Li F, Wu Y (2014) Dynamic monitoring of optimal locations in road network databases. VLDB J 23(5):697–720
Zhang C, Han J, Shou L, Lu J, La Porta TF (2014) Splitter: mining fine-grained sequential patterns in semantic trajectories. PVLDB 7(9):769–780
Zheng K, Su H (2015) Go beyond raw trajectory data: quality and semantics. IEEE Data Eng Bull 38(2):27–34
Zheng Y, Zhou X (2011) Computing with spatial trajectories. Springer, New York
Zheng K, Shang S, Yuan NJ, Yang Y (2013a) Towards efficient search for activity trajectories. In: ICDE, pp 230–241
Zheng K, Zheng Y, Yuan NJ, Shang S (2013b) On discovery of gathering patterns from trajectories. In: ICDE, pp 242–253
Zheng B, Yuan NJ, Zheng K, Xie X, Sadiq SW, Zhou X (2015) Approximate keyword search in semantic trajectory database. In: ICDE, pp 975–986

Chapter 10
Mining Colocation from Big Geo-Spatial Event Data on GPU

Arpan Man Sainju and Zhe Jiang

10.1 Introduction

Given a set of spatial features and their instances, co-location mining aims to find subsets of features whose instances are frequently located together. Examples of colocation patterns include symbiotic relationships between species such as Nile Crocodiles and Egyptian Plover, as well as environmental factors and disease events (e.g., air pollution and lung cancer).

Societal applications: Colocation mining is important in many applications that aim to find associations between different spatial events or factors. For example, in public safety, law enforcement agencies are interested in finding relationships between different crime event types and potential crime generators. In ecology, scientists analyze common spatial footprints of various species to capture their interactions and spatial distributions. In public health, identifying relationships between human disease and potential environmental causes is an important problem. In climate science, colocation patterns help reveal relationships between the occurrence of different climate extreme events. In location based service, colocation patterns help identify travelers that share the same favourite locations to promote effective tour recommendation.

Challenges: Mining colocation patterns from big spatial event data poses several computational challenges. First, in order to evaluate if a candidate colocation pattern is prevalent, we need to generate its instances. This is computationally expensive due to checking spatial neighborhood relationships between different instances, particularly when the number of instances is large and instances are clumpy (e.g., many instances are within the same spatial neighborhoods). Second, the number

A. M. Sainju (✉) · Z. Jiang
Department of Computer Science, The University of Alabama, Tuscaloosa, AL, USA
e-mail: asainju@crimson.ua.edu; zjiang@cs.ua.edu

M. Werner, Y.-Y. Chiang (eds.), *Handbook of Big Geospatial Data*,
https://doi.org/10.1007/978-3-030-55462-0_10

of candidate colocation patterns are exponential to the number of spatial features. Evaluating a large number of candidate patterns can be computationally prohibitive. Finally, the distribution of event instances in the space may be uneven, making it hard to design parallel data structure and algorithms.

We introduce GPU colocation mining algorithms based on a grid index, including a cell-aggregate-based upper bound filter and two refinement algorithms (Sainju and Jiang 2017). Cell-aggregate-based filter is easier to implement on GPU and is also insensitive to pattern *clumpiness* (the average number of overlaying colocation instances for a given colocation instance) compared with the existing multi-resolution filter (Huang et al. 2004). We use a GPU platform due to its better energy efficiency and pricing compared to Map-reduce based clouds.

Scope and outline: We focus on spatial colocation patterns defined by the event-centric models (Huang et al. 2004). Other colocation definitions such as Voronoi diagram based focuses on addressing the problem of predefining proximity threshold among co-located instances. This is beyond the scope of this paper. We assume the underlying space is Euclidean space. In this chapter, we are only concerned with the comparison of computational performance of various colocation mining algorithms.

The outline of the chapter is as follows. Section 10.2 briefly introduces GPU computing. Section 10.3 discusses on some of the related works. Section 10.4 reviews basic concepts and the definition of the colocation mining problem. Section 10.5 introduces the GPU colocation pattern mining algorithms, and analyzes the theoretical properties of algorithm correctness and completeness. Section 10.6 summarizes our experimental evaluation of algorithms on both synthetic datasets and a real world dataset. Section 10.7 discusses memory bottleneck issues in our approach and concludes the chapter with potential future research directions.

10.2 GPU Computing

GPU refers to Graphical Processing Unit is a processor that was specialized for graphics processing but are now gaining popularity to accelerate scientific computations. In contrast to CPU, GPU contains thousands of lightweight cores, which are optimized for data-parallel tasks with simple control logic in order to maximize throughput. GPU threads are grouped into thread blocks, which can utilize the limited amount of resources called register and shared memory. The number of threads in a thread block is limited by the architecture of the GPU. GPUs have to wait for the threads to finish in the thread blocks. Therefore, the runtime of a thread block is usually the maximum runtime of threads in the same block.

10.3 Related Work

Colocation pattern mining has been studied extensively in the literature, including early work on spatial association rule mining (Koperski and Han 1995; Morimoto 2001) and colocation patterns based on event-centric model (Huang et al. 2004). Various algorithms have been proposed to efficiently identify colocation patterns, including Apriori generator and multi-resolution upper bound filter (Huang et al. 2004), partial join (Yoo et al. 2004) and joinless approach (Yoo and Shekhar 2006), iCPI tree based colocation mining algorithms (Boinski and Zakrzewicz 2012). There are also works on identifying regional (Mohan et al. 2011; Wang et al. 2013; Liu et al. 2015) or zonal (Celik et al. 2007) colocation patterns, and statistically significant colocation patterns (Barua and Sander 2011, 2014, 2017), top-K prevalent colocation patterns (Yoo and Bow 2011) or prevalent patterns without thresholding (Huang et al. 2003). Existing algorithms are mostly sequential, and can be insufficient when the number of event instances is very large (e.g., several millions). Recently, parallel colocation mining algorithms have been proposed based on the Map-reduce framework (Yoo et al. 2014) to handle a large data volume. However, these algorithms need a large number of nodes to scale up, which is economically expensive, and their reducer nodes have a bottleneck of aggregating all instances of the same colocation patterns. Another work proposes a GPU based parallel colocation mining algorithm (Andrzejewski and Boinski 2013) using iCPI tree (Andrzejewski and Boinski 2014; Yoo and Boulware 2014; Andrzejewski and Boinski 2015) and the joinless approach, but this method assumes that the number of neighbors for each instance is within a small constant (e.g., 32), and thus can be inefficient when instances are dense and unevenly distributed.

10.4 Problem Statement

10.4.1 Basic Concepts

This subsection reviews some basic concepts based on which the colocation mining problem can be defined. More details on the concepts are in Huang et al. (2004).

Spatial feature and instances: A *spatial feature* is a categorical attribute such as a crime event type (e.g., assault, drunk driving). For each spatial feature, there can be multiple *feature instances* at the same or different point locations (e.g., multiple instances of the same crime type "assault"). In the example of Fig. 10.1a, there are three spatial features (A, B and C). For spatial feature A, there are three instances (A_1, A_2, and A_3). Two feature instances are *spatial neighbors* if their spatial distance is smaller than a threshold. Two or more instances form a *clique* if every pair of instances are spatial neighbors.

Spatial colocation pattern: If the set of instances in a clique are from different feature types, then this set of instances is called a *colocation (pattern) instance*, and the corresponding set of features is a *colocation pattern*. The *cardinality* or *size* of a colocation pattern is the number of features involved. For example, in Fig. 10.1a, (A_1, B_1, C_1) is an instance of colocation pattern (A, B, C) with a size or cardinality of 3. If we put all the instances of a colocation pattern as different rows of a table, the table is called an *instance table*. For example, in Fig. 10.1b, the instance table of colocation pattern (A, B) has three row instances, as shown in the third table of the bottom panel. A spatial colocation pattern is *prevalent* (significant) if its feature instances are *frequently* located within the same neighborhood cliques. In order to quantify the prevalence or frequency, an interestingness measure called participation index has been proposed (Huang et al. 2004).

The *participation ratio* of a spatial feature within a candidate colocation pattern is the ratio of the number of unique feature instances that participate in colocation instances to the total number of feature instances. For example, in Fig. 10.1, the participation ratio of B in candidate colocation pattern $\{A, B\}$ is $\frac{2}{3}$ since only B_1 and B_2 participate into colocation instances $(\{A_1, B_1\}, \{A_3, B_2\})$. The *participation index* (PI) of a candidate colocation pattern is the minimum of participation ratios among all member features. For example, the participation index of the candidate colocation pattern $\{A, B\}$ in Fig. 10.1 is the minimum of $\frac{3}{3}$ and $\frac{2}{3}$, and is thus $\frac{2}{3}$. We use "candidate colocation patterns" to refer to those whose participation index values are undecided. PI is used as the measure of prevalence of a colocation pattern because it follows apriori property. It basically implies that if the pattern is not prevalent then any of its superset pattern will also not be prevalent which can be exploited for computational efficiency.

10.4.2 Problem Definition

We now introduce the formal definition of colocation mining problem (Huang et al. 2004).

Given:

- A set of spatial features and their instances
- Spatial neighborhood distance threshold
- Minimum threshold of participation index: θ

Find:

- All colocation patterns whose participation index are above or equal to θ

Objective:

- Minimize computational time cost

Constraint:

- Spatial neighborhood relationships are defined in Euclidean space

Figure 10.1 provides a problem example. The input data contains 12 instances of 3 spatial features A, B, and C. The neighborhood distance threshold is d. The

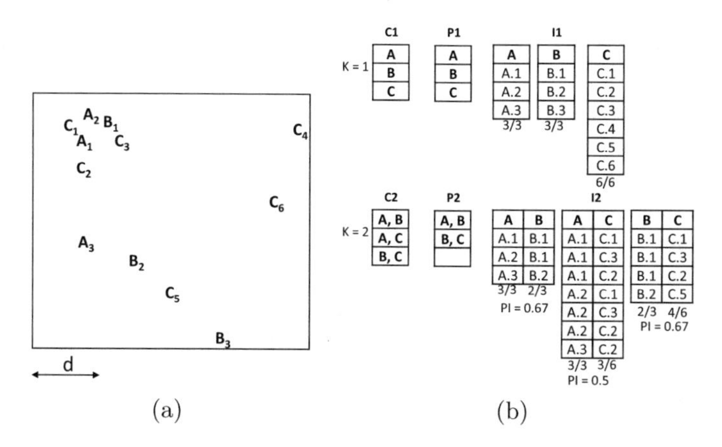

Fig. 10.1 A problem example with inputs and outputs. (**a**) Input spatial features and instances; (**b**) Candidate and prevalent colocation patterns, instance tables

prevalence threshold is 0.6. The output prevalent colocation patterns include $\{A, B\}$ (participation index 0.67) and $\{B, C\}$ (participation index 0.67). Colocation mining is similar to association rule mining in market basket analysis (Agrawal et al. 1994), but is different in that there are no given "transactions" in continuous space. Generation colocation instance tables ("transactions") is the most computationally intensive part.

Our baseline approach (Huang et al. 2004) uses the filter and refine strategy for colocation pattern mining. They proposed multiresolution filter that generates the coarse instance table for each candidate colocation pattern without performing any computationally expensive distance computation. The coarse instance table contains the entire actual as well as some false instances of the candidate colocation pattern. We can use this coarse instance table to compute the coarse participation index. Coarse participation index is the upper bound of the actual participation index for a given candidate colocation pattern. Hence, if the upper bound coarse participation index is lower than the prevalence threshold, we do not have to generate the computationally intensive instance table to calculate the actual participation index.

10.5 Approach

This section introduces our GPU colocation mining algorithm. We start with the overview of algorithm structure, and then describe the main part for parallel algorithms implemented in GPU. We prove the correctness and completeness of the algorithm.

10.5.1 Algorithm Overview

The overall structure of our algorithm is similar to the one proposed by Huang et al. in 2004 (Huang et al. 2004). We design a novel upper-bound filter based on aggregated counts of feature instances in grid cells. Compared with the multi-resolution filter (Huang et al. 2004), our upper bound filter is easier to parallelize on GPU and does not rely on the assumption that colocation instances are clumpy into a small number of cells.

The overall structure of our algorithm is shown in Algorithm 1. The algorithm identifies all prevalent colocation patterns iteratively. Candidate colocation patterns and their instance tables of cardinality $k + 1$ are generated, based on prevalent patterns and their instance tables of cardinality k. Each candidate pattern of cardinality $k + 1$ is then evaluated based on the participation index computed from its instance table. For cardinality $k = 1$, prevalent colocation patterns simply consist of the set of input features, and their instance tables are the instance lists for each feature (step 1 in Algorithm 1). Step 2 generates candidate patterns C_{k+1} of size $k+1$ based on prevalent patterns P_k using Apriori property (Agrawal et al. 1994) (i.e., a candidate pattern of size $k+1$ cannot be prevalent and thus needs not to be generated if any subset pattern of size k is not prevalent). Step 4 builds a grid index on spatial point instances with the cell size equal to the distance threshold. Step 5 counts the number of instances for each feature in every cell. This will be used in our upper bound filter. Step 7 starts the iteration. As long as the set of candidate patterns C_{k+1} is not empty, the algorithm evaluate each candidate pattern $c \in C_{k+1}$. When evaluate a candidate pattern, the algorithm first computes an upper bound of its participation index in parallel using GPU kernels based on the grid index (step 9–10). If the upper bound is below the threshold, the candidate pattern is pruned out. Otherwise, the algorithm runs into a refinement phase, generating the pattern instance table $I_{k+1}.c$ and computing the participation index PI. We design two different parallel refinement algorithms to speed up instance table generation: one using the grid to rule out unnecessary joins, the other using prefix-based hash joins (steps 13 to 16). After all prevalent patterns of cardinality $k+1$ are identified, the algorithm go to the next iteration (steps 19–22). Figure 10.1b illustrates the execution trace for $k = 1$ and $k = 2$.

10.5.2 Cell-Aggregate-Based Upper Bound Filter

The cell aggregate based upper bound filter first overlays a regular grid with its cell size equal to the distance threshold (shown in Fig. 10.2), and then computes an upper bound of participation index based on aggregated counts of feature instances in cells. Our Filter is different from the multi-resolution filter (Huang et al. 2004) in that the computation of upper bound is not based on generating coarse scale colocation instance tables. There are two main advantages of cell aggregate based

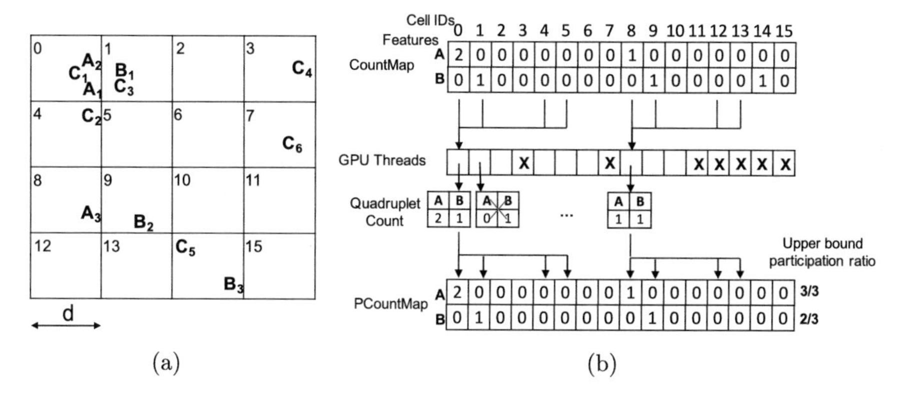

(a) (b)

Fig. 10.2 Grid-aggregate based Upper Bound Filter: (**a**) A regular grid (**b**) An execution trace of upper bound filter

filter on GPU: first, it is easily parallelizable and can leverage the large number of GPU cores; second, its performance does not rely on the assumption that pattern instances are clumpy into a small number of cells, which is required by the multi-resolution filter.

To introduce cell aggregate based filter, we define a key concept of **quadruplet**. A quadruplet of a cell is a set of four cells, including the cell itself as well as its neighbors on the right, bottom, and right bottom. For a cell that is located on the right and bottom boundary of the grid, not all four cells exist and its quadruplet is defined empty (these cells will still be covered by other quadruplets). For example, in Fig. 10.2, the quadruplet of cell 0 includes cells (0, 1, 4, 5), while the quadruplet of cell 15 is an empty set.

Based on the concept of quadruplet, we can check all potential colocation instances by examining all quadruplets. When examining a quadruplet, our filter computes the aggregated count of instances for every feature in the candidate pattern. If the aggregated count for any feature is zero, then there cannot exist colocation instances in the quadruplet. Otherwise, we pretend that all these feature instances participate into colocation pattern instances. This tends to overestimate the participating instances of a colocation pattern (an "upper bound"), but avoids expensive spatial join operations.

Algorithm 2 shows details of the cell aggregate based filter. The algorithms have three main variables, including $CountMap$, $PCountMap$, and $QuadrupletAggregate$. $CountMap$ records the true aggregated instance count for each feature in every cell. $PCountMap$ records the instance count for each feature in every cell that potentially participates in the candidate colocation pattern. $QuadrupletAggregate$ is a local array for each cell to record the aggregated count of instances within the quadruplet for each pattern feature. Specifically, steps 1 to 11 computes potential number of participating instances for each feature in each cell in parallel. A kernel thread is allocated to each cell. For a specific cell i, the kernel first gets the quadruplet (step 2). Step 3 initializes a local array $QuadrupletAggregate$

Algorithm 1 Parallel-Colocation-Miner

Input: A set of spatial features F
Input: Instances of each spatial features $I[F]$
Input: Neighborhood distance threshold d
Input: Minimum prevalence threshold θ
Output: All prevalent colocation patterns P

```
 1: Initialize P ← ∅, k ← 1, C_k ← F, P_k ← F
 2: Initialize C_{k+1} ← APRIORIGEN(P_k, k + 1)
 3: Initialize P_{k+1} ← ∅
 4: Initialize instance tables I_k (k = 1) by feature instances
 5: Overlay a regular grid with cell size d × d (total N cells)
 6: Compute CountMap[N × |F|] in one round instance scanning
 7: while |C_{k+1}| > 0 do
 8:   for each c ∈ C_{k+1} do
 9:     Initialize PCountMap[N × |c|] ← 0
10:     Upperbound =PARALLELCELLAGGREGATEFILTER(CountMap,PCountMap,c)
11:     if Upperbound ≥ θ then
12:       BitMap ← 0 //initialize bitmap for instances of each feature
13:       if Hash Join Refinement then
14:         (I_{k+1}.c, PI) ← PARALLELHASHJOINREFINE(I_k, c)
15:       else if Grid Search Refinement then
16:         (I_{k+1}.c, PI) ← PARALLELGRIDSEARCHREFINE(I_k, CInstances, c, BitMap)
17:       if PI ≥ θ then
18:         P_{k+1} = P_{k+1} ∪ c
19:   P ← P ∪ P_{k+1} //add prevalent patterns to results
20:   k ← k + 1; C_k ← C_{k+1}; P_k ← P_{k+1}, I_k ← I_{k+1} //prepare next iteration
21:   C_{k+1} ← APRIORIGEN(P_k, k + 1)
22:   P_{k+1} ← ∅
23: return P
```

with zero values. Steps 4 to 8 compute the aggregated count of instances for each pattern feature ($QuadrupletAggregate[f]$). If aggregated count of any pattern feature is zero, then there cannot be any candidate pattern instance in the quadruplet and thus the parallel kernel thread terminates (step 8). Otherwise, all feature instances in the quadruplet can potentially participate into colocation pattern instances. Steps 9 to 11 record the potential participating instances from each cell in the quadruplet. This is done by copying instance counts from $CountMap$ to $PCountMap$ for the 4 cells in the quadruplet. It is worth noting that duplicated-counting on the same cell is avoided since different GPU kernel threads may over-write the count for a cell with the same value. Finally, steps 12 to 14 compute the upper bound of participation index based on counts of potential participating instances in $PCountMap$. We use built in GPU library to compute the total counts of distinct participating instances in step 13.

Figure 10.2 provides an illustrative execution trace of Algorithm 2. Figure 10.2a shows the input spatial instances overlaid with a regular grid. The distance threshold is d. Assume that the candidate colocation pattern is (A, B). Figure 10.2b shows how the filter works. The $CountMap$ array stores the number of instances for

Algorithm 2 ParallelCellAggregateFilter

Input: $CountMap$, feature instance count in cells
Input: $PCountMap$, participating feature instance count in cells
Input: PR, participation ratio
Input: Candidate colocation pattern c
Output: Upper bound of participation index, $upperBound$

```
1:  for each cell i do in parallel
2:    QuadrupletCells = GETQUADRUPLET(cell i)
3:    QuadrupletAggregate[|c|] ← 0
4:    for each feature f ∈ c do
5:      for each cell j ∈ QuadrupletCells do
6:        QuadrupletAggregate[f] ← QuadrupletAggregate[f] + CountMap[j][f]
7:      if QuadrupletAggregate[f] == 0 then
8:        finish the parallel thread for cell i //no pattern instance in the quadruplet
9:    for each feature f ∈ c do
10:     for each cell j ∈ QuadrupletCells do
11:       PCountMap[j][f] ← CountMap[j][f] //participating instance count
12: for each feature f ∈ c do
13:   PR[f] = PARALLELSUM(CountMap[ ][f])/|I_1.f|
14: upperBound = MIN(PR)
15: return upperBound
```

feature A and B in each cell. A GPU thread is assigned to each cell to compute the counts of feature instances within its quadruplet. For example, the leftmost GPU thread is assigned to cell 0. The aggregated instance count for this quadruplet ((0, 1, 4, 5)) is shown by the leftmost $QuadrupletCount$ array, with 2 instances for A and 1 instance for B. Since instances from both features exist, the number potential participating instances in these four cells ($PCountMap$) are copied from corresponding cell values in $CountMap$, as shown by the fork branches close to the bottom. In contrast, the quadruplet of cell 1 ((1, 2, 5, 6)) does not contain instances of A, and thus cannot contain colocation pattern instances.

Lemma 1 *The participation index of a colocation pattern in the cell-aggregate-based filter is an upper bound of the true participation index value.*

Proof The proof is based on the following fact. We create an upper bound to the true number of neighboring points in neighboring cells (quadruplet) by assuming that all pairs of points of neighboring cells are within the distance threshold, which coincides with the cell size. Of course, some of them will not, but it is impossible for points not within neighboring cells to be neighboring with respect to the distance threshold. □

Theorem 1 *The cell aggregate based upper bound filter is correct and complete.*

Proof The proof is based on Lemma 1. The algorithm is complete (it does not mistakenly prune out any prevalent pattern) due to the upper bound property. The algorithm is correct since it computes the exact participation index of a candidate pattern if it passes the upper bound filter. □

10.5.3 Refinement Algorithms

The goal of the refinement phase is to generate the instance table of a candidate colocation pattern, and to compute participation index. Generating colocation instance tables is the main computational bottleneck, and thus is done in GPU. As shown in Algorithm 1, we have two options for refinement algorithms, a geometric approach based on grid search called ParallelGridSearchRefine and a combinatorics approach based on prefix-based hash join called ParallelHashJoinRefine, similar to sequential algorithms discussed in Huang et al. (2004). We now introduce the two algorithms below.

Algorithm 3 ParallelGridSearchRefine

Input: I_k, instance table of patterns of size k
Input: $CellInstances$, feature instances for each cell
Input: $BitMap$, bitmap for participating instances from different features
Output: $I_{k+1}.c$, instance table of colocation c (size $k+1$) if prevalent
Output: PI, participation index of pattern c

1: //$I_k.(c[1..k])$ is instance table of sub-pattern of c with first k features
2:
3: Initialize $I_{k+1}.c \leftarrow \emptyset$
4: **for each** row instance $rIns \in I_k.(c[1..k])$ **do in parallel**
5:   get $neighborhood$ cells of first feature instance in $rIns$
6:   **for each** cell i in neighborhood **do**
7:     **for each** instance ins of feature type $c[k+1]$ in cell i **do**
8:       **if** ins is neighbor of all feature instances in $rIns$ **then**
9:         Create new row instance of c, $rInsC =< rIns, ins >$
10:         $I_{k+1}.c = I_{k+1}.c \cup rInsC$ //add new instance into c's instance table
11:         **for each** feature $f \in c$ **do**
12:           $BitMap[f][rInsC[f]]$ = true
13: **for each** feature $f \in c$ **do**
14:   $PR[f] = \text{PARALLELSUM}(BitMap[\,][f])/|I_1.f|$
15: $PI = \text{MIN}(PR)$
16: **return** $I_{k+1}.c$, PI

Geometric approach: The geometric approach generate an instance table of a size $k+1$ pattern based on the instance table of a size k pattern. For example, when generating the instance table of pattern (A, B, C), it starts from the instance table of (A, B) and joins each row of the table with instances of the last feature type C. In order to reduce redundant computation, we utilize the grid index and only check the instances of the last feature type within neighboring cells. Algorithm 3 provides details of the grid-based refinement algorithm. Step 2 is kernel assignment. Each kernel thread first finds out all neighboring cells of the size k row instance $rIns$ (step 3). Then, for each neighboring cell, the kernel thread finds out every instance ins of the last feature type in the cell. It joins ins with size k instance $rIns$ to create a size $k+1$ pattern instance $rInsC$ if they are spatial neighbors (steps 4 to 7). The new pattern instance $rInsC$ is inserted into the final instance table $I_{k+1}.c$, and a

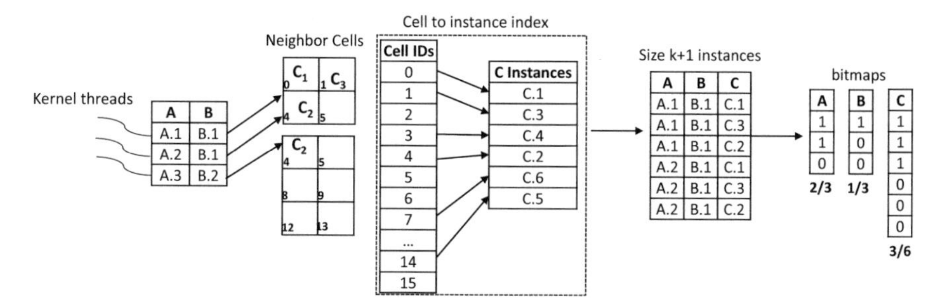

Fig. 10.3 Illustrative execution trace for grid-based refinement

bitmap is updated to mark the instances that participate into the colocation pattern (steps 8 to 10). Finally, the participation ratio and participation index are computed (steps 11 to 13).

Figure 10.3 shows an example. The input data is the same as Fig. 10.1. Assume that the candidate pattern is (A, B, C). A kernel thread is assigned to each row instance of table (A, B). Thread 1 is assigned to instance (A_1, B_1), and it scans all neighboring cells of A_1 (cells $0, 1, 4, 5$). Based on the cell to instance index, the kernel thread checks all instances of feature C (C_1, C_3, C_2) in these cells, and conducts a spatial join operation. The final output size $k + 1$ instances from this thread are (A_1, B_1, C_1), (A_1, B_1, C_3), and (A_1, B_1, C_2).

One issue in GPU implementation is that we need to allocate memory for an output instance table, and specify the specific memory location to which each kernel thread writes its results. For example, in the output instance table of pattern (A, B, C) in Fig. 10.3, the first kernel thread generates 3 row instances, so the second kernel thread has to start with the 4th row when writing its instances. It is hard to predetermine the total required memory and enforcing memory coalesce when threads are writing results. Thus, we use a two-run strategy in which in the first run we can calculate the exact size of output instance table as well as slot counts of the number of row instances generated by each kernel thread. In the second run, we allocate memory for output instance table, and use the slot counts to guide which row a kernel thread needs to start from when writing results. Similar to the grid-based refinement, we use two-run strategy to allocate memory and enforce memory coalesce.

Prefix-based hash Join based refinement: Another option is to generate size $k + 1$ instance table by a combinatorics approach. For example, when generating instance table of pattern (A, B, C), we can join rows in instance tables of (A, B) and (A, C). The join condition is that the first k instances from the two tables should be the same, and the last instances from two tables should be spatial neighbors. For example, when joining a row (A_1, B_1) with another row (A_1, C_1), we check that the first instance is the same (A_1), and the last instances B_1 and C_1 are spatial neighbors. So these two rows are joined to form a new row instance (A_1, B_1, C_1). In sequential implementations (Huang et al. 2004), the join process can be done

Algorithm 4 ParallelHashJoinRefine

Input: I_k, instance table of patterns of size k
Input: $BitMap$, bitmap for instances of different features
Output: $I_{k+1}.c$, instance table of colocation c if prevalent
Output: PI, participation index of pattern c

1: //$I_k.c1$ and $I_k.c2$ instance tables of $c1 = c[1..k]$ and $c2 = c[1..k-1, k+1]$
2: **for each** row instance $rIns1$ in $I_k.c1$ **do in parallel**
3: **for each** row instance $rIns2$ in $I_k.c2$ starting with $rIns1[1]$ **do**
4: **if** $rIns1$ and $rIns2$ forms an instance of c **then**
5: Create new instance $rInsC$ by merging $rIns1$ and $rIns2$
6: $I_{k+1}.c \leftarrow I_{k+1}.c \cup rInsC$
7: **for each** feature $f \in c$ **do**
8: $BitMap[f].[rInsC[f]] = \text{true}$
9: **for each** feature $f \in c$ **do**
10: $PR[f] = \text{PARALLELSUM}(BitMap[\,][f])/|I_1.f|$
11: $PI = \text{MIN}(PR)$
12: **return** $I_{k+1}.c$, PI

efficiently through sort-merge join. However, for GPU algorithm, sort merge is difficult due to the order, dependency and multi-attribute keys. We choose to use hash join instead. A prefix-based hash index is built on the second table based on instances of the first spatial feature. Details are shown in Algorithm 4. A kernel thread is allocated to each row in the first size k instance table $I_k.c1$ (step 2). The kernel thread then scans all rows in the second size k instance table $I_k.c2$ that has the same first feature instance. For example, if the row in the first table is (A_1, B_1), then the thread only scans rows starting with A_1 in the instance table of (A, C). If the two rows satisfy the join condition (sharing the same first k instances, and having last instances as neighbors), a size $k + 1$ instance is created and inserted into output size $k + 1$ table (steps 5 to 8). Finally, the participation index is computed (steps 9 to 11). It is worth noting that when generating instance tables of size $k = 2$ patterns, we use the grid-based method since hash-index cannot be created in that case.

An illustrative execution trace is shown in Fig. 10.4. The raw input data is still the same. Each kernel thread is allocated to a row in instance table (A, B). For example, thread 1 works on pattern instance (A_1, B_1), and scans instance table (A, C). Based on the hash index on A instances, the thread only needs to check (A_1, C_1), (A_1, C_3)

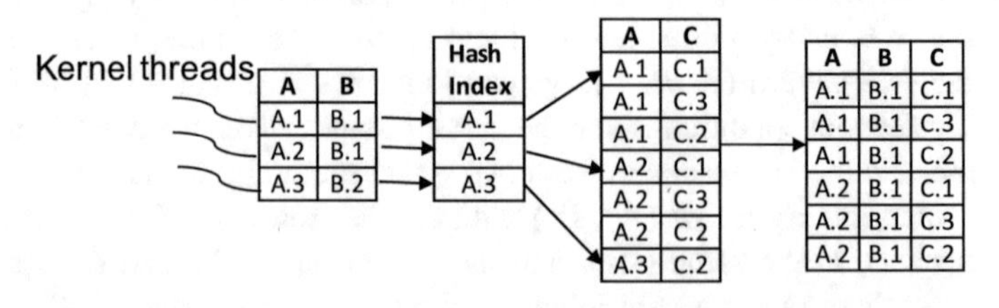

Fig. 10.4 Illustrative execution trace for hash-join-based refinement

and (A_1, C_2). It turns out that B_1 is a neighbor for all C_1, C_2 and C_3. So these instances are inserted to the final output instance table (A, B, C).

10.6 Evaluation

The goals of our evaluation are to:

- Evaluate the speedup of GPU colocation algorithms against a CPU algorithm.
- Compare cell-aggregate-based filter with multi-resolution filter on GPU.
- Compare grid-based refinement with hash-join-based refinement on GPU.
- Test the sensitivity of GPU algorithms to different factors.

Experiment Setup: As shown in Fig. 10.5, we implemented four GPU colocation mining algorithms with two filter options (**M** for multi-resolution filter and **C** for cell-aggregate based filter) and two refinement options (**G** for grid-based and **H** for hash-join based). We also implemented a CPU colocation mining algorithms by Huang et al. (2004) (multi-resolution filter, grid-based instance table generation for size $k = 2$, and sort-merge based instance table generation for size $k > 2$). We only compared computational performance since all methods produce the same patterns. For each experiment, we measured the time cost of one run for CPU algorithm, and averaged time cost of 10 runs for GPU algorithms. Algorithms were implemented in C++ and CUDA, and run on a Dell workstation with Intel(R) Xeon(R) CPU E5-2687w v4 @ 3.00 GHz, 64 GB main memory, and a Nvidia Quadro K6000 GPU with 2880 cores and 12 GB memory.

Dataset description: The real dataset contains 13 crime types and 165,000 crime event instances from Seattle in 2012 (City of Seattle 2012). The synthetic data is generated similarly to Huang et al. (2004). Figure 10.5b, c provide illustrative examples. We first chose a study area size of 10,000 × 10,000, a neighborhood

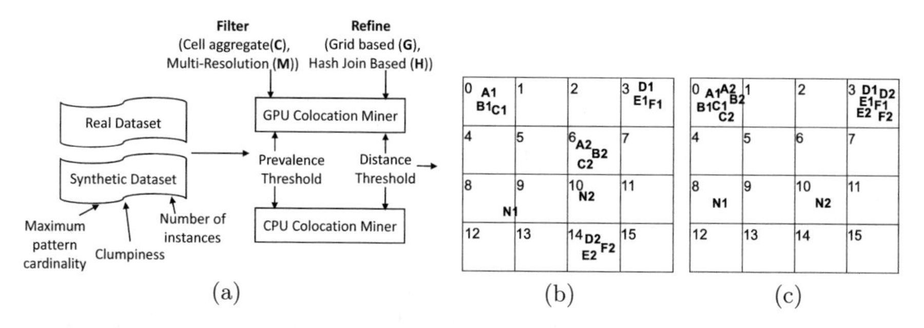

Fig. 10.5 Experiment Setup (**a**) Experiment design with different candidate approaches; (**b**) An example of synthetic dataset generated with 2 maximal patterns (A, B, C) and (D, E, F), each pattern with 2 instances with a clumpiness of 1, 2 noise instances N_1 and N_2; (**c**) Another synthetic dataset similar to (**b**) but with a clumpiness of 2

distance threshold (also the size of a grid cell) of 10, a maximal pattern cardinality of 5, and the number of maximal colocation patterns as 2. The total number of features was 12 (5×2 plus 2 additional noise features). We then generated a *number of instances* for each maximal colocation pattern. Their locations were randomly distributed to different cells according to the **clumpiness** (i.e., the number of overlaying colocation instances within the same neighborhood, higher clumpiness means larger instance tables). In our experiments, we varied the number of instances and clumpiness to test sensitivity.

Evaluation metric: We used the speedup of the GPU algorithms over the CPU algorithm on computational time.

10.6.1 Results on Synthetic Data

10.6.1.1 Effect of the Number of Instances

We conducted this experiment with two different parameter settings. For both settings, the minimum participation index threshold was 0.5. In the first setting, we set the clumpiness to 1 (very low clumpiness), and varied the number of feature instances as 250,000, 500,000, 1,000,000, 1,500,000 and 2,000,000. Results are summarized in Fig. 10.6a. GPU algorithms in the plot are based on grid-based filtering. We can see that the speedup of both GPU algorithms increases with the number of feature instances. The grid-based refinement gradually becomes superior over the hash join based refinement in GPU algorithms as the number of instances increases. The reason can be that the cell-instance index in grid-based refinement is done once and for all, while the prefix-based hash index in hash-join based refinement needs to be created repeatedly for each new instance table. The comparison of two approaches with 250,000 instances (the first two points in the curve) may be less conclusive since the running time for both approaches is too small (far below one second).

In the second setting, we set the clumpiness value as 20, and varied the number of feature instances as 50,000, 100,000, 150,000, 200,000, and 250,000. The number of feature instances were set smaller in this setting due to the fact that given the same number of feature instances, a higher clumpiness value results in far more colocation pattern instances (see Fig. 10.5b versus c) but we only have limited memory. The results are summarized in Fig. 10.6b. We can see that the grid based refinement is persistently better than hash-join based refinement (around 30 versus 5). The reason is that when the clumpiness is high, there are a large number of pattern instances being formed combinatorially. Many of them share the same prefix (i.e., first feature instance). Thus, each GPU kernel thread in prefix-based hash-join refinement was loaded with heavy computation when doing the join operation, impacting the parallel performance. In contrast, in the grid-based refinement, each GPU kernel thread only scans a limited number of instances within neighboring cells.

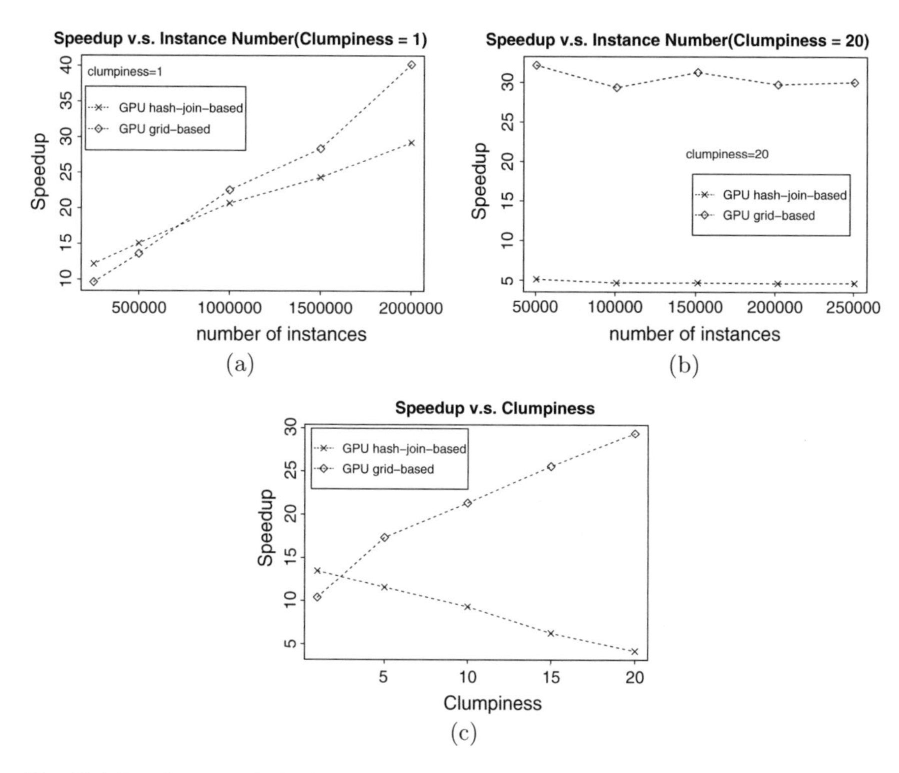

Fig. 10.6 Results on synthetic datasets: (**a**) effect of the number of instances with clumpiness as 1 (**b**) effect of the number of instances with clumpiness as 20 (**c**) effect of clumpiness with the number of instances as 250k

10.6.1.2 Effect of Clumpiness

We set the number of instances to 250k, and the prevalence threshold to 0.5. Grid-based filtering was used for GPU algorithms. We varied the clumpiness value as 1, 5, 10, 15, and 20. Results in Fig. 10.6c confirm with our analysis above that when clumpiness is higher, the performance of grid-based refinement gets better while the performance of hash-join based refinement gets worse.

10.6.1.3 Comparison on Filter and Refinement

We also compared the computational time of filter and refinement phases of GPU algorithms in the above experiments for the cases with the largest number of instances. Details are summarized in Table 10.1. When clumpiness is 1, the grid-based filter is much faster than the multi-resolution filter in GPU algorithms (0.2 s versus 0.8 s), making the overall GPU speedup better (37.9 and 28.4 times versus 17.9 times). The reason is that a low clumpiness significantly impacts the multi-

Table 10.1 Comparison of filter and refinement on synthetic data (time in secs)

Clumpiness	Approaches	Filter time	Refine time	Total time	Speedup
1	CPU baseline	15.3	18.8	34.1	-
	GPU-Filter:M, Refine:H	0.8	1.1	1.9	17.9x
	GPU-Filter:C, Refine:G	**0.2**	**0.7**	**0.9**	**37.9x**
	GPU-Filter:C, Refine:H	**0.2**	**1.0**	**1.2**	**28.4x**
20	CPU Baseline	0.9	407.5	408.4	-
	GPU-Filter:M, Refine:H	**0.1**	97.3	97.4	4.2x
	GPU-Filter:C, Refine:G	**0.1**	**13.8**	**13.9**	**29.4x**
	GPU-Filter:C, Refine:H	**0.1**	96.9	97	4.2x

resolution filter (coarse scale instance tables cannot be much smaller than true instance tables), while the grid-based filter was less sensitive to clumpiness (more robust) since the time cost of grid-based filtering does not depend on instance distribution. When clumpiness is 20, the refinement phase becomes the bottleneck. The grid-based refinement has a significantly higher speedup than the hash-join refinement (29.4 times versus 4.2 times).

10.6.2 Results on Real World Dataset

10.6.2.1 Effect of Minimum Participation Index Threshold

We fixed the distance threshold as 10 m and varied the prevalence thresholds from 0.3 to 0.9 (we did not chose thresholds lower than 0.3, because there would be too many instance tables exceeding our memory capacity). The clumpiness of the real dataset was high due to a large density of points. Results are summarized in Fig. 10.7. As we can see, as the prevalence threshold gets higher, the pruning ratio (candidate patterns being pruned out) gets improved (Fig. 10.7a). The GPU algorithm with grid based refinement is much better than the GPU algorithm with hash join based refinement. This is consistent with the results on synthetic datasets when the clumpiness is high.

10.6.2.2 Comparison of Filter and Refinement

We also compared the detailed computational time in the filter and refinement phases. The distance threshold was 10 m, and the prevalent threshold was 0.3. Results are summarized in Table 10.2. Due to a high clumpiness, the refinement phase is the bottleneck, and the grid-based refinement is better than the hash-join based refinement (63.2 times overall speedup versus 12.7 times overall speedup).

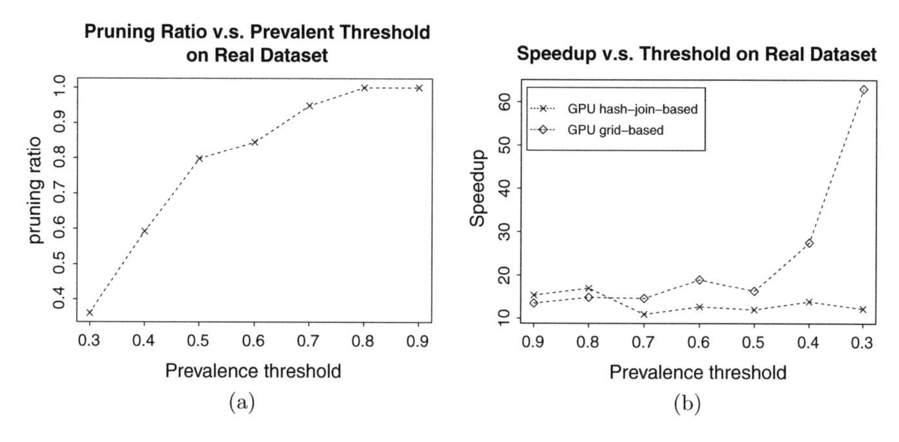

Fig. 10.7 Results on real world dataset: (**a**) pruning ratio versus prevalence thresholds (**b**) speedup versus prevalence thresholds

Table 10.2 Comparison of filter and refinement on real dataset (time in secs)

Approaches	Filter time	Refine time	Total time	Speedup
CPU baseline	6.5	340.9	347.4	-
GPU-Filter:M, Refine:H	0.2	26.	26.8	13x
GPU-Filter:C, Refine:G	**0.5**	**5.0**	**5.5**	**63.2x**
GPU-Filter:C, Refine:H	**0.5**	26.9	27.4	12.7x

10.7 Discussion and Conclusion

Results show that GPU algorithms are promising for the colocation mining problem. One limitation of the GPU algorithm is memory bottleneck. Our algorithm generates instance tables of candidate colocation patterns in GPU. When spatial points are dense and the number of points is large (e.g., millions), such instance tables can reach gigabytes in size. In case that the GPU global memory is insufficient for a very large instance table, we can slice it into smaller pieces, compute one piece each time, and transfer results to the host memory. We need to store all relevant instance tables of cardinality k in host memory when computing instance tables of cardinality $k + 1$. Thus, the host memory size can also be a bottleneck. This may be less a concern in future when the main memory price gets lower. Our recent work (Sainju et al. 2018) focuses on addressing those memory management issues.

This chapter investigates GPU-based parallel colocation mining algorithms. We introduced a cell-aggregate-based upper bound filter, which is easier to implement on GPU and less sensitive to data clumpiness compared with the multi-resolution filter. We also design two GPU refinement algorithms, based on grid-based search and prefix based hash-join. We provide theoretical analysis on the correctness and completeness of the algorithms. Experimental results on both real world data and synthetic data on various parameter settings show that the GPU algorithms are promising.

In future work, we will explore further refinements on GPU implementation to achieve higher speedup, e.g., avoid redundant distance computation in instance table generation. We will also explore other computational pruning methods.

References

Agrawal R, Srikant R et al (1994) Fast algorithms for mining association rules. In: Proceedings of 20th International Conference on Very Large Data Bases, VLDB, vol 1215, pp 487–499

Andrzejewski W, Boinski P (2013) GPU-accelerated collocation pattern discovery. In: East European Conference on Advances in Databases and Information Systems. Springer, pp 302–315

Andrzejewski W, Boinski P (2014) A parallel algorithm for building ICPI-trees. In: East European Conference on Advances in Databases and Information Systems. Springer, pp 276–289

Andrzejewski W, Boinski P (2015) Parallel GPU-based plane-sweep algorithm for construction of ICPI-trees. J Database Manag (JDM) 26(3):1–20

Barua S, Sander J (2011) SSCP: mining statistically significant co-location patterns. In: International Symposium on Spatial and Temporal Databases. Springer, pp 2–20

Barua S, Sander J (2014) Mining statistically significant co-location and segregation patterns. IEEE Trans Knowl Data Eng 26(5):1185–1199

Barua S, Sander J (2017) Statistically significant co-location pattern mining. In: Shekhar S, Xiong H, Zhou X (eds) Encyclopedia of GIS. Springer, Cham, pp. 2204–2212. https://doi.org/10.1007/978-3-319-17885-1_1552. ISBN: 978-3-319-17885-1

Boinski P, Zakrzewicz M (2012) Collocation pattern mining in a limited memory environment using materialized ICPI-tree. In: International Conference on Data Warehousing and Knowledge Discovery. Springer, pp 279–290

Celik M, Kang JM, Shekhar S (2007) Zonal co-location pattern discovery with dynamic parameters. In: Data Mining, 2007. ICDM 2007. Seventh IEEE International Conference on. IEEE, pp 433–438

City of Seattle, Department of Information Technology, Seattle Police Department Seattle Police Department 911 Incident Response (2012) https://data.seattle.gov/Public-Safety/Seattle-Police-Department-911-Incident-Response/3k2p-39jp

Huang Y, Xiong H, Shekhar S, Pei J (2003) Mining confident co-location rules without a support threshold. In: Proceedings of the 2003 ACM Symposium on Applied Computing. ACM, pp 497–501

Huang Y, Shekhar S, Xiong H (2004) Discovering colocation patterns from spatial data sets: a general approach. IEEE Trans Knowl Data Eng 16(12):1472–1485

Koperski K, Han J (1995) Discovery of spatial association rules in geographic information databases. In: International Symposium on Spatial Databases. Springer, pp 47–66

Liu B, Chen L, Liu C, Zhang C, Qiu W (2015) RCP mining: towards the summarization of spatial co-location patterns. In: International Symposium on Spatial and Temporal Databases. Springer, pp 451–469

Mohan P, Shekhar S, Shine JA, Rogers JP, Jiang Z, Wayant N (2011) A neighborhood graph based approach to regional co-location pattern discovery: a summary of results. In: Proceedings of the 19th ACM SIGSPATIAL International Conference on Advances in Geographic Information Systems. ACM, pp 122–132

Morimoto Y (2001) Mining frequent neighboring class sets in spatial databases. In: Proceedings of the Seventh ACM SIGKDD International Conference on Knowledge Discovery and Data Mining. ACM, pp 353–358

Sainju AM, Jiang Z (2017) Grid-based colocation mining algorithms on GPU for big spatial event data: a summary of results. In: International Symposium on Spatial and Temporal Databases. Springer, pp 263–280

Sainju AM, Aghajarian D, Jiang Z, Prasad SK (2018) Parallel grid-based colocation mining algorithms on GPUS for big spatial event data. IEEE Trans Big Data 6(1):107–118

Wang S, Huang Y, Wang XS (2013) Regional co-locations of arbitrary shapes. In: International Symposium on Spatial and Temporal Databases. Springer, pp 19–37

Yoo JS, Boulware D (2014) Incremental and parallel spatial association mining. In: Big Data (Big Data), 2014 IEEE International Conference on. IEEE, pp 75–76

Yoo JS, Bow M (2011) Mining top-k closed co-location patterns. In: Spatial Data Mining and Geographical Knowledge Services (ICSDM), 2011 IEEE International Conference on. IEEE, pp 100–105

Yoo JS, Shekhar S (2006) A joinless approach for mining spatial colocation patterns. IEEE Trans Knowl Data Eng 18(10):1323–1337

Yoo JS, Shekhar S, Smith J, Kumquat JP (2004) A partial join approach for mining co-location patterns. In: Proceedings of the 12th Annual ACM International Workshop on Geographic Information Systems. ACM, pp 241–249

Yoo JS, Boulware D, Kimmey D (2014) A parallel spatial co-location mining algorithm based on mapreduce. In: Big Data (BigData Congress), 2014 IEEE International Congress on. IEEE, pp 25–31

Chapter 11
Automatic Urban Road Network Extraction From Massive GPS Trajectories of Taxis

Song Gao, Mingxiao Li, Jinmeng Rao, Gengchen Mai, Timothy Prestby, Joseph Marks, and Yingjie Hu

11.1 Introduction

Increasingly, mobile Internet, ubiquitous sensors (Hancke and Hancke Jr. 2013) and growing Volunteering Geographic Information (VGI) (Goodchild 2007) altogether boost the construction of transportation information infrastructures (Shaw 2010). Urban road networks, as the important carriers of transportation in cities, provide basic support for human & goods transportation and various Location-Based Services (LBS) such as vehicle route planning and navigation, which are the keys to smart transportation. However, how to build and update the road network in a rapid and cost-effective way still remains to be a challenging problem.

Traditional methods such as field surveying and map digitalization are usually costly and cannot produce up-to-date urban road networks in time (Tao 2000). With the rapid development of information and communication technologies (ICTs) and positioning technologies such as the Global Positioning System (GPS), huge amounts of vehicle movement trajectory data have been accumulated (Liu et al.

S. Gao (✉) · M. Li · J. Rao · T. Prestby · J. Marks
GeoDS Lab, Department of Geography, University of Wisconsin-Madison, Madison, WI, USA
e-mail: song.gao@wisc.edu

M. Li
GeoDS Lab, Department of Geography, University of Wisconsin-Madison, Madison, WI, USA

State Key Laboratory of Resources and Environmental Information System, Institute of Geographic Sciences and Natural Resources Research, Chinese Academy of Sciences, Beijing, China

G. Mai
STKO Lab, Department of Geography, University of California, Santa Barbara, CA, USA

Y. Hu
GeoAI Lab, Department of Geography, University at Buffalo, Buffalo, NY, USA

M. Werner, Y.-Y. Chiang (eds.), *Handbook of Big Geospatial Data*,
https://doi.org/10.1007/978-3-030-55462-0_11

2012). By leveraging these massive GPS trajectories, automatic construction and updates to road networks can be achieved in near real-time (Li et al. 2012). Although the idea for the extraction of urban road networks from GPS trajectory data is intuitive, there is still a considerable gap between the raw GPS traces and the road network structure. On the one hand, the GPS trajectories have non-negligible errors due to the inherent noise in GPS, which makes it difficult to distinguish two closely located road segments in some cases (Cao and Krumm 2009). On the other hand, complex urban environment such as "urban canyon" often leads to the deterioration in the GPS precision, making the trajectories less accurate to represent the road segments. In addition, the diversity and complexity of the road network structure in some places (e.g., roundabouts, parking lots) also cause challenges for road network extraction from raw GPS trajectories. Besides the GPS precision and the complex road network structure, the preprocessing of trajectory data is another challenge. A large proportion of the GPS data on straight road segments are redundant since fewer points are already enough to reconstruct the linear road segments, whereas the curved roads required more points. Also, when there is traffic congestion, more redundant data (e.g., stay points) will be produced (Zheng 2015). These situations may cause high computational costs and limit the road network extraction efficiency. In addition, trajectory outliers (i.e., anomalies) will make it non-trivial to reconstruct the road network and need to be addressed during the preprocessing step (Zheng 2015; Wang et al. 2019).

To this end, in this chapter, we focus on the trajectory sampling, compression, and clustering techniques to update road geometry information with regard to the spatial coverage and topological connectivity. We conduct a literature review in the following section and then propose a geospatial-big-data-driven framework to achieve an automatic road network extraction. Specifically, we first introduce a trajectory compression approach to reduce redundant trajectory data and avoid the unnecessary computational cost. Then we present an anisotropic density-based trajectory clustering with noise (ADCN) algorithm (Mai et al. 2018) for identifying the trajectory points on the road segments, and finally, a kernel density estimation and vectorization approach is utilized for road network extraction.

11.2 Literature Review

Existing literature on road network extraction methods can be classified into two categories: density-based approaches and cluster-based approaches.

11.2.1 Density-Based Approaches

The first category mainly relies on density estimation and raster processing techniques. It converts trajectory data into raster data based on density and extracts the road network using morphological methods (Davies et al. 2006; Wu et al. 2007; Shi et al. 2009; Zhao et al. 2011; Biagioni and Eriksson 2012; Jiang et al. 2012; Wang et al. 2015b; Kuntzsch et al. 2016; Tang et al. 2017). For example, Davies et al. (2006) first generated a 2D histogram based on the GPS trajectories, then applied a global density threshold on the cells to find potential road areas, and finally computed road centerlines based on the Voronoi graph. Shi et al. (2009) converted vehicle GPS trajectories into a road network bitmap, then computed the road network skeleton on the bitmap, and finally extracted the vector road network map data from the skeleton. Biagioni and Eriksson (2012) generated a road network skeleton based on a kernel-density method and use a map-matching method to achieve topology reconstruction. Kuntzsch et al. (2016) formulated an explicit intersection model which integrated consistency measurements with the raw trajectory data to better perform geometry and topology reconstruction of the network; Tang et al. (2017) employed Delaunay triangulation with the trajectory stream fusion to improve the map generation accuracy. However, the difference in trajectory density has a great influence on the extraction effect, which could make these methods unreliable in cases with heterogenous trajectory density.

11.2.2 Cluster-Based Approaches

The second category adopted clustering methods to generate road networks (Edelkamp and Schrödl 2003; Lee et al. 2007; Worrall et al. 2007; Wu et al. 2013; Cao and Krumm 2009; Wang et al. 2015a; Aronov et al. 2016; Stanojevic et al. 2018). Trajectory clustering is usually used to find representative trajectories shared by different objects such as individuals or vehicles (Zheng 2015). For example, Gaffney and Smyth (1999) and Cadez et al. (2000) used a regression mixture model and an Expectation-Maximization (EM) model to cluster trajectories according to the overall distance between two trajectories. Lee et al. (2007) proposed TRACLUS, a modified density-based trajectory clustering algorithm for grouping close trajectory line segments into clusters, which is based on the original point-based DBSCAN algorithm. Li et al. (2010) further introduced an incremental clustering algorithm that reduces the computational and storage cost. In practice, trajectory clustering can be naturally used for road network extraction. Edelkamp and Schrödl (2003) applied the K-means algorithm to cluster the trajectories and then fit the road centerline with the spline curve. This approach is suitable for data with small density difference, low noise, and high frequency sampling. Worrall et al. (2007) used clustering to extract the skeleton points of the road network and used the least squares regression method to connect the skeleton points to generate the road network. Cao and Krumm (2009) classified the GPS traces using simulations of physical forces among the traces, and then merged the classified traces into a graph representation of the road network structure. Wang et al. (2015a) determined a proper circle boundary to cluster trajectory data into intersections and used the core points to build the road networks. Stanojevic et al. (2018) formulated

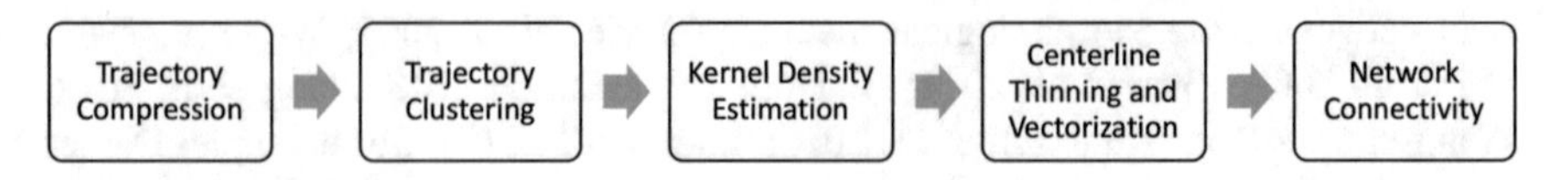

Fig. 11.1 The trajectory big data processing workflow in this study

the road network generation task as a network alignment optimization task and proposed an offline algorithm that clustered GPS points for graph construction as well as an online algorithm that can create and update the road network. However, the trajectory points along the road networks include linear features with a continuously changing density which makes current clustering methods tend to either create an increasing number of small clusters or add noise points into large clusters. Therefore, incorporating directional information into clustering methods has become an efficient way to cluster anisotropic distributed points and enhance extraction performance (Mai et al. 2018).

11.3 Methodology

In this section, we present the details of our road map generation method using GPS trajectories. As shown in Fig. 11.1, the trajectory big data processing workflow can be divided into the following steps. First, we utilized a trajectory compression approach to simplify the trajectory data and reduce unnecessary computational cost. Second, we applied the Anisotropic Density-based Clustering with Noise (ADCN) algorithm (Mai et al. 2018) to identify the trajectory points that were along the road networks with high confidence. Third, a kernel density estimation (KDE) approach was used to generate a continuous surface. Fourth, high-density areas were selected as candidates to further extract the road centerlines using thinning and vectorization operators. Finally, the extracted road network connectivity should be evaluated.

11.3.1 Trajectory Compression

With the rapid development of ICT technologies and positioning devices, the spatiotemporal resolution of trajectory data has unprecedentedly increased. However, a large amount of trajectory data lead to high computational costs, and a large storage space is necessary to support the data processing and management. With regard to road network extraction, it is straightforward to reconstruct the road segments as long as the trajectory points at the intersections can be obtained and then connected. Thus, a trajectory compression method was applied to simplify the trajectory data points and improve extraction efficiency.

The main purpose of trajectory compression in road network extraction is to maintain the shape of the trajectories. Several popular algorithms for line simplification include Douglas-Peucker algorithm (Douglas and Peucker 1973), Reumann-Witkam algorithm (Reumann and Witkam 1974), Lang simplification (Lang 1969), Opheim simplification (Opheim 1982), etc. The evaluation and comparison of these line simplification algorithms for vector generalization were conducted by Shi and Cheung (2006). McMaster (1989) proposed a conceptual model including a sequential set of five procedures for processing linear data with focuses on *geometric simplification* and *smoothing*. In our method, we applied the widely used Douglas-Peucker simplification algorithm for trajectory simplification.[1] A graphic illustration of the Douglas-Peucker algorithm is shown in Fig. 11.2. A sequence of points (P_1, P_2, P_3, P_4, P_5, P_6) represent a trajectory in Fig. 11.2a. To simplify the trajectory, we first mark the first point P_1 and the last point P_6 as endpoints and added them to the reserved point set. Then, the point that is furthest from the line segment with the endpoints is found, and the distance between the point and the line segment is calculated. If the distance is larger than a compression threshold α, the point is marked as an endpoint and added to the reserved point set (e.g. P_4 in in Fig. 11.2b and P_5 in Fig. 11.2c). Otherwise, the points between the endpoints would be discarded (e.g. P_2 and P_3 in Fig. 11.2c). The same process is iteratively performed until all the trajectory points are marked as an endpoint or discarded (Fig. 11.2d). Finally, the points in the reserved point set are sorted according to the original trajectory sequence and linked to generate the compressed trajectory.

11.3.2 Identification of the Trajectory Points Along the Road

Density-based clustering algorithms such as DBSCAN have been widely used for spatial knowledge discovery such as the detection of urban areas of interest (Hu et al. 2015) and vague cognitive regions (Gao et al. 2017). The DBSCAN algorithm offers several key advantages compared with other clustering algorithms such as K-Means. DBSCAN can discover clusters with arbitrary shapes, are robust to noise, and do not require prior knowledge (Ester et al. 1996). However, the trajectory points demonstrate clear anisotropic spatial processes, which makes these methods tend to either create an increasing number of small clusters or add noise points into large clusters. Therefore, in this section, we apply a novel anisotropic density-based clustering algorithm with noise (ADCN) (Mai et al. 2018) to cluster anisotropic points for identifying the trajectory points along the roads. The codes for implementing the ADCN algorithm are available in both Javascript[2] and Python[3] programming languages.

[1] Python code for the Douglas-Peucker algorithm https://pypi.org/project/simplification/.

[2] Javascript implementation of the ADCN algorithm at: https://github.com/gengchenmai/adcn.

[3] Python implementation of the ADCN algorithm at: https://github.com/gissong/ADCN.

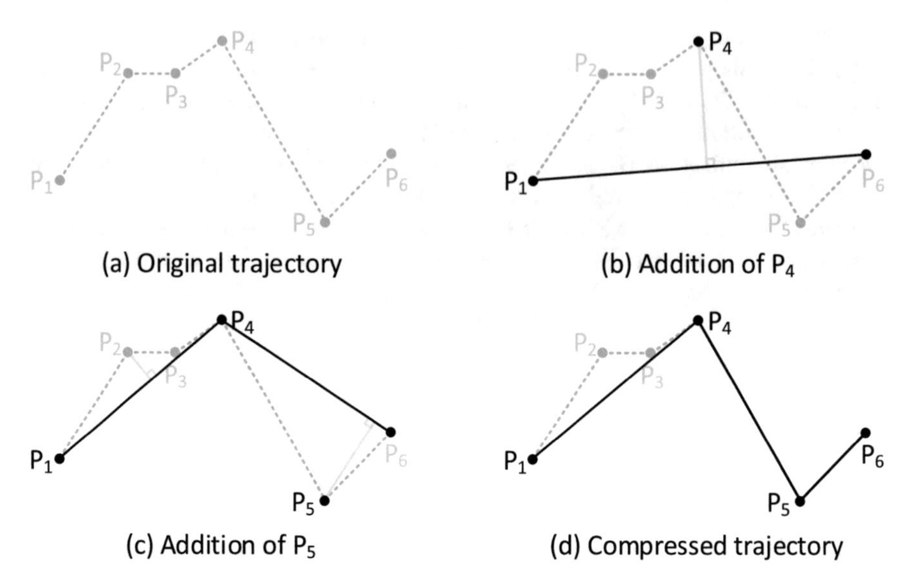

Fig. 11.2 Illustration of the Douglas-Peucker trajectory compression algorithm

11.3.2.1 Density-Based Spatial Clustering of Applications with Noise (DBSCAN)

The used ADCN algorithm was modified based on the DBSCAN algorithm. Thus, before we describe the ADCN algorithm in detail, we first introduce some fundamental concepts of the DBSCAN algorithm (Ester et al. 1996). The key idea of the DBSCAN algorithm is that: given a set of points, it groups nearby points (i.e., the points in high-density areas) together and marks the points in low-density areas as outliers. In order to group the points based on the density, the *Eps-neighborhood* of a point is defined (see Definition 1).

Definition 1 (Eps-neighborhood of a point) The *Eps-neighborhood* $N_{Eps}(p_i)$ of point p_i in a dataset D is defined as all the points within the scan circle centered at p_i with a radius *Eps*, which is expressed as follows:

$$N_{Eps}(p_i) = \{p_j(x_j, y_j) \in D | dist(p_i, p_j) \leq Eps\}$$

where $dist(p_i, p_j)$ is the distance between point p_i and point p_j.

There are two kinds of points in a cluster: *core points* (i.e., points inside of the cluster) and *border points* (i.e., points on the border of the cluster). One intuition is that for each point of a cluster, an Eps-neighborhood should contain at least a minimum number of points (*MinPts*). However, an Eps-neighborhood of a border point usually contains much fewer points than an Eps-neighborhood of a core point, and it is hard to choose the representative *MinPts* for all points. Thus, DBSCAN introduces three basic concepts: directly density-reachable, density-reachable and density-connected (see Definitions 2, 3, and 4).

Definition 2 (directly density-reachable) A point p is *directly density-reachable* from a point q wrt. *Eps* and *MinPts* if:

1. p $\in N_{Eps}$(q) and
2. $|N_{Eps}(q)| \geq MinPts$ (core point condition).

Definition 3 (density-reachable) A point p is *density-reachable* from a point q wrt. *Eps* and *MinPts* if there is a chain of points $p_1, \ldots, p_n, p_1 = q, p_n = p$ such that p_{i+1} is directly density-reachable from p_i.

Definition 4 (density-connected) A point p is *density-connected* to a point q wrt. *Eps* and *MinPts* if there is a point O such that p and q are density-reachable from O wrt. *Eps* and *MinPts*.

Figure 11.3 illustrates core points, border points, density-reachability, and density-connectivity in DBSCAN. These definitions can be used to further define the density-based notion of a cluster or a noise in DBSCAN. Specifically, the core points and border points are grouped into clustered points while the points that do not belong to any cluster are the noise points (as shown in Fig. 11.4).

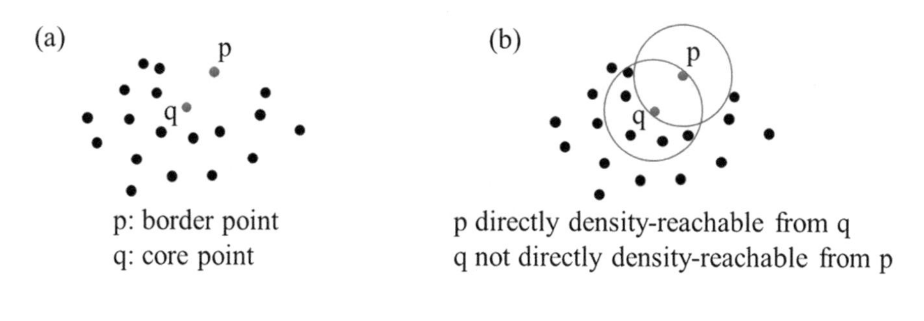

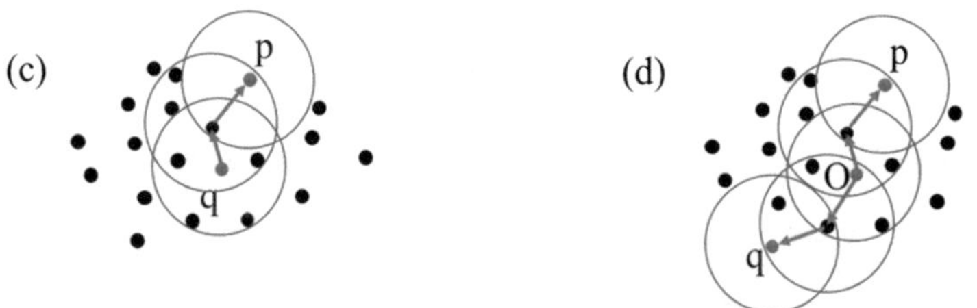

Fig. 11.3 Illustration of core points, border points, density-reachability and density-connectivity in the DBSCAN algorithm (Ester et al. 1996)

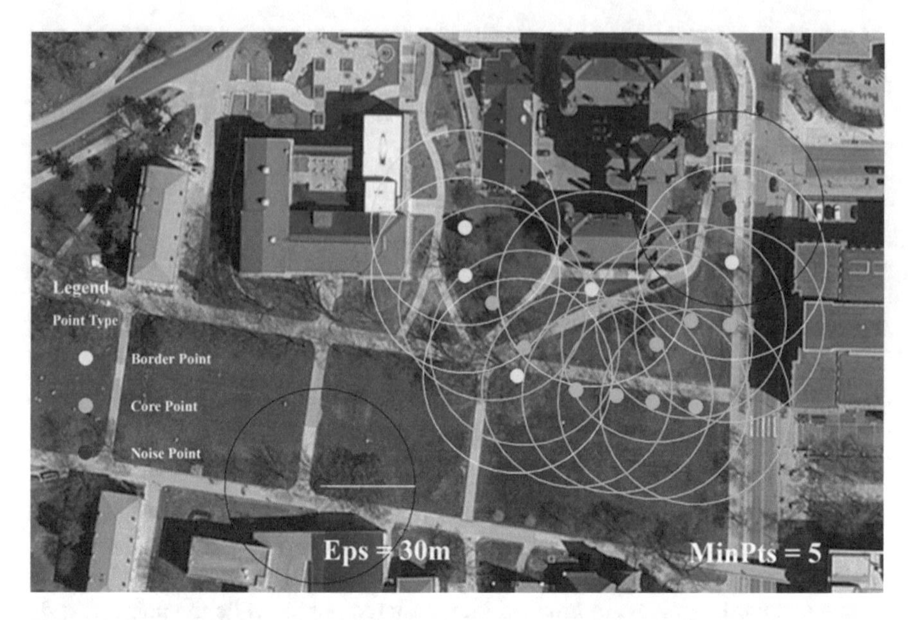

Fig. 11.4 Illustration of core points, border points, and noise points in the DBSCAN algorithm

11.3.2.2 Anisotropic Perspective on Local Point Density

One key consideration in the ADCN algorithm is the anisotropic perspective on local point density in different directions. Without predefined direction information from spatial point datasets, one has to compute the local major direction for each point based on the spatial distribution of neighboring points. The standard deviation ellipse (SDE) (Yuill 1971) is a suitable method to get the major direction of a point set. Given n points, the SDE constructs an ellipse to represent the orientation and arrangement of these points. The center of this ellipse is defined as the geometric center of these n points and is calculated as:

$$\overline{X} = \sum_{i=1}^{n} x_i \Big/ n, \quad \overline{Y} = \sum_{i=1}^{n} y_i \Big/ n$$

The coordinates (x_i,y_i) of each point are normalized to the deviation from the mean center point:

$$\widetilde{x_i} = x_i - \overline{X}, \quad \widetilde{y_i} = y_i - \overline{Y}$$

Thus, the semi-major axes σ_x and the semi-minor axes σ_y of SDE are calculated as:

$$\sigma_x = \sqrt{\sum_{i=1}^{n} \left(\widetilde{x_i} \cos\theta + \widetilde{y_i} \sin\theta\right)^2 \Big/ n}, \quad \sigma_y = \sqrt{\sum_{i=1}^{n} \left(\widetilde{y_i} \cos\theta - \widetilde{x_i} \sin\theta\right)^2 \Big/ n}$$

where θ is the rotation angle and the SDE will be further used as the search polygon for clustering neighboring points (Mai et al. 2018).

11.3.2.3 Anisotropic Density-Based Clusters with Noise (ADCN) Algorithm

In order to introduce an anisotropic perspective to density-based clustering algorithms, we have to revise some definitions. First, the original *Eps-neighborhood* of a point in a dataset D is defined by DBSCAN, as given in Definition 1. Such a scan circle results in an isotropic perspective on clustering. However, an anisotropic assumption will be more appropriate for trajectory points along the roads. Intuitively, in order to introduce the anisotropic perspective into DBSCAN, we can employ a scan ellipse instead of a circle to define the Eps-neighborhood of each point. Before that, we defined a set of points around a point to derive the scan ellipse;

Definition 5 (Search-neighborhood of a point) The k_{th} nearest neighbor $KNN(p_i)$ of point p_i. Here $k = MinPts$ and $KNN(p_i)$ does not include p_i itself.

After determining the search-neighborhood of a point, it is possible to define the *Eps-ellipse-neighborhood region* (see Definition 6) and the *Eps-ellipse-neighborhood* (see Definition 7) of each point.

Definition 6 (Eps-ellipse-neighborhood region of a point) An ellipse ER_i is called an Eps-ellipse-neighborhood region of a point p_i if:

1. Ellipse ER_i is centered at point p_i.
2. Ellipse ER_i is scaled from the standard deviation ellipse SDE_i computed from the search-neighborhood $S(p)_i$ of point p_i.
3. $\frac{\sigma'_{max}}{\sigma'_{min}} = \frac{\sigma_{max}}{\sigma_{min}}$ where σ'_{max}, σ'_{min}, σ_{max}, σ_{min} are the length of the semi-long and semi-short axes of ellipse ER_i and ellipse SDE_i.
4. $Area(ER_i) = \pi Eps^2$.

According to Definition 6, the Eps-ellipse-neighborhood region of a point is computed based on the search-neighborhood of a point. Each point should have a unique *MinPts*, as long as the search-neighborhood of the current point has at least two points for the computation of the standard deviation ellipse.

Definition 7 (Eps-ellipse-neighborhood of a point) An Eps-ellipse-neighborhood $EN_{Eps}(p_i)$ of point p_i is defined as all the points inside the ellipse ER_i, which can be expressed as:

$$EN_{Eps}(p_i) = \left\{ p_j(x_j, y_j) \in D | \frac{\left((y_j - y_i)\sin\theta_{max} + (x_j - x_i)\cos\theta_{max}\right)^2}{a^2} + \frac{\left((y_j - y_i)\cos\theta_{max} - (x_j - x_i)\sin\theta_{max}\right)^2}{b^2} \le 1 \right\}$$

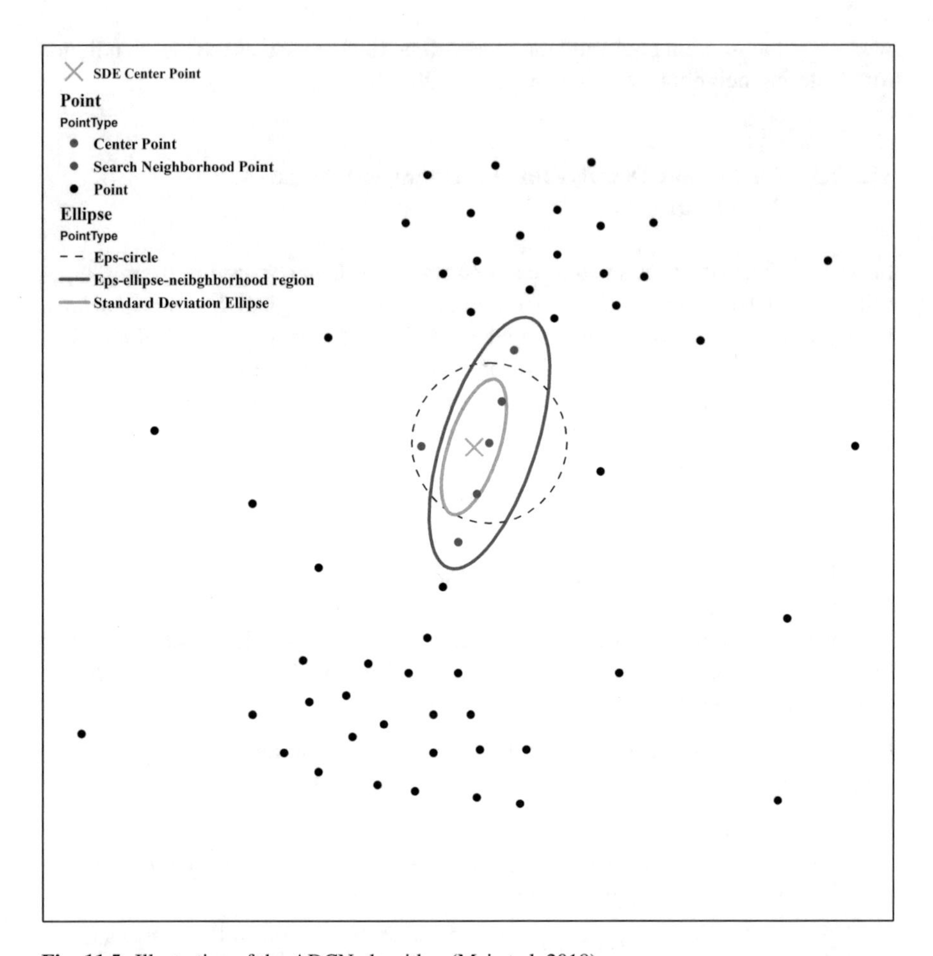

Fig. 11.5 Illustration of the ADCN algorithm (Mai et al. 2018)

Equipped with Definitions 1, 5, 6, and 7, we can introduce the anisotropic perspective to density-based clustering algorithms. The definitions of directly density reachable, density reachable, cluster, and noise in ADCN are similar to DBSCAN, which will not be repeated here. Figure 11.5 illustrates the related definitions for ADCN. The red point in the figure represents the current center point. The blue points indicate the search-neighborhood of the corresponding center point according to Definition 5. The green ellipse and the green cross stand for the standard deviation ellipse constructed from the corresponding search-neighborhood and the center point. The red ellipse is the scale-transformed Eps-ellipse-neighborhood region according to Definition 6, whereas the dashed-line circle indicates a traditional scan circle in DBSCAN. As can be seen in Fig. 11.5, ADCN could exclude the point to the left of the linear bridge pattern in the clustering process, whereas DBSCAN still includes it.

11.3.2.4 ADCN Algorithm in Road Network Extraction

The abovementioned ADCN algorithm takes the same parameters (*MinPts* and *Eps*) as the DBSCAN algorithm which must be decided before clustering. This is for good reasons, as the proper selection of DBSCAN parameters has been well studied, and ADCN can easily replace DBSCAN without any changes to established workflows.

The ADCN method starts with an arbitrary point p_i in a point dataset D and discovers all the core points which are density reachable from point p_i along the major direction. The result of the points in the clusters will be regarded as the trajectory points on the road. In order to take care of situations where all points of the search-neighborhood $S(p_i)$ of point p_i are strictly on the same line, the short axis of the Eps-ellipse-neighborhood region ER_i becomes zero, and its long axis becomes infinity. This means that $EN_{Eps}(p_i)$ is reduced to a straight line. The process of constructing the Eps-ellipse-neighborhood $EN_{Eps}(p_i)$ of point p_i becomes a point-on-line query. Furthermore, the ADCN method uses a k_{th} nearest neighborhood of point p_i as the search-neighborhood. Here, the center point p_i will not be included in its k_{th} nearest neighborhood. The runtimes of ADCN are heavily dominated by the search-neighborhood query which is executed on each point. Hence, the time complexities of ADCN, DBSCAN, and OPTICS are $O(n^2)$ without a spatial index and $O(n \log n)$otherwise (Kolatch 2001; Mai et al. 2018).

11.3.3 Road Network Generation

11.3.3.1 Road Density Surface Generation

After identifying the compressed and clustered GPS trajectory points on the roads, the kernel density estimation (KDE) is introduced to fit a smooth surface. To fit the surface, we used 30 m*30 m square grid cells to divide the space and calculated the trajectory point density of each cell. Let $d_1, d_2, \ldots, d_n$ be a given set of trajectory point densities. The kernel density estimator is defined as:

$$\hat{f}_{(d)} = \frac{1}{nh^d} \sum_{i=1}^{n} K\left(\frac{d - d_i}{h}\right)$$

where n is the number of density sets, h denotes the bandwidth parameter, and K is a kernel function. The kernel surface value is highest at the location of the center and decreases with increasing distance from the center until reaching zero at the search radius. The density at each grid cell is calculated by adding the weighted values under the kernel surface where it overlays the raster cell center. The kernel function used in this work is based on the quadratic kernel function (Silverman 1986) described as follows:

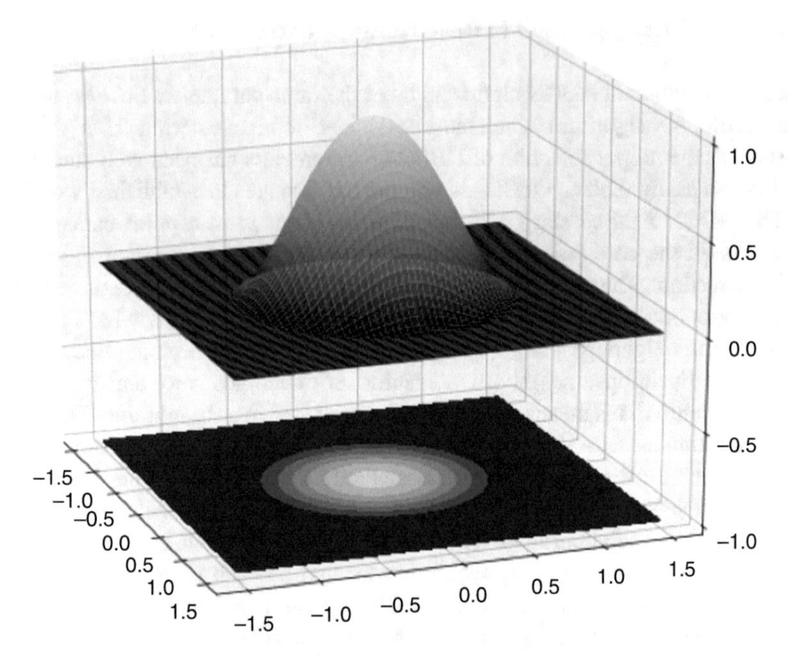

Fig. 11.6 A 3D Visualization of the selected quadratic kernel function. The vertical axis represents the derived K function value

$$K_{(d)} = \begin{cases} 3\pi^{-1}\left(1 - d^T d\right)^2 & if\ d^T d < 1 \\ 0 & otherwise \end{cases}$$

Figure 11.6 shows a 3D plot of the selected quadratic kernel function. Note that there are two reasons we choose the quadratic kernel function. First, the quadratic kernel function's density estimates have higher differentiability properties. Second, it can be calculated more quickly than the Gaussian kernel, which is suitable for massive-scale trajectory points with fine spatial resolution.

11.3.3.2 Collapse Surface to Centerline

After the previous steps, the areas whose density is above a threshold β are selected as the candidates. In this step, the main purpose is to extract the centerline of the candidate areas while keeping the road network topology. The thinning operator proposed by Yuan et al. (2012) was performed to remove certain grid cells from the candidate areas. For a given grid cell in the candidate area, whether it should be removed depends on its 8-neighboring cells. This method first divided the binary image (i.e., is a road pixel or not) into two disjointed subfields in a checkerboard pattern. Then, iterations were performed to remove redundant neighboring cells. The algorithm ensured that the connectivity of the cells was preserved when a cell was deleted. A combination of the thinning operator and the raster-to-vector operation converted the KDE surface of compressed trajectories into road centerlines. The quality measures of the extracted road networks including accuracy (correctness), coverage completeness, redundancy, and connectivity will be discussed with case study experiments in Sect. 11.4.2.

11.4 Case Study

11.4.1 Data

We applied the above introduced trajectory compression and clustering workflow in the DiDi research open data "November 2016, Chengdu City Second Ring Road Regional Trajectory Data Set"[4] to extract the road information. Figure 11.7 shows the KDE visualization of over 1 million e-hailing trip origins and destinations and part of the extracted road geometry density map from 181,172 trip trajectories in one day. The date range of this dataset is from November 1 to November 30, 2016 and the temporal sampling resolution is about 2~5 seconds. The original data is about 50 GB, and about 10 GB after compression.

11.4.2 Experiment

11.4.2.1 Evaluation Metrics

In order to determine the effectiveness of our algorithm that generates big-data-driven road networks, we performed vector-based quality measures (Wiedemann et al. 1998). These quality measures were done by comparing a reference road network data layer to our extracted road network data layers (in Fig. 11.8). To begin, the road networks were divided into small pieces of the same length. Next, a buffer was created as a zone with a consistent width that encircles each line segment for a given analysis. To determine if the extracted network roads match the reference roads, we constructed a buffer around the reference road data and determined if the portion of extracted roads inside of the buffer met our requirements. This is the *correctness*. Similarly, we constructed a buffer around the extracted road data and evaluated the portion of reference roads inside the buffer to further analyze the results. This is called the *completeness*. Together, *correctness* and *completeness* constitute a comprehensive metric known as *quality*. Another evaluation metric known as *redundancy* determines the degree of overlap in the road extraction methods.

[4]https://gaia.didichuxing.com.

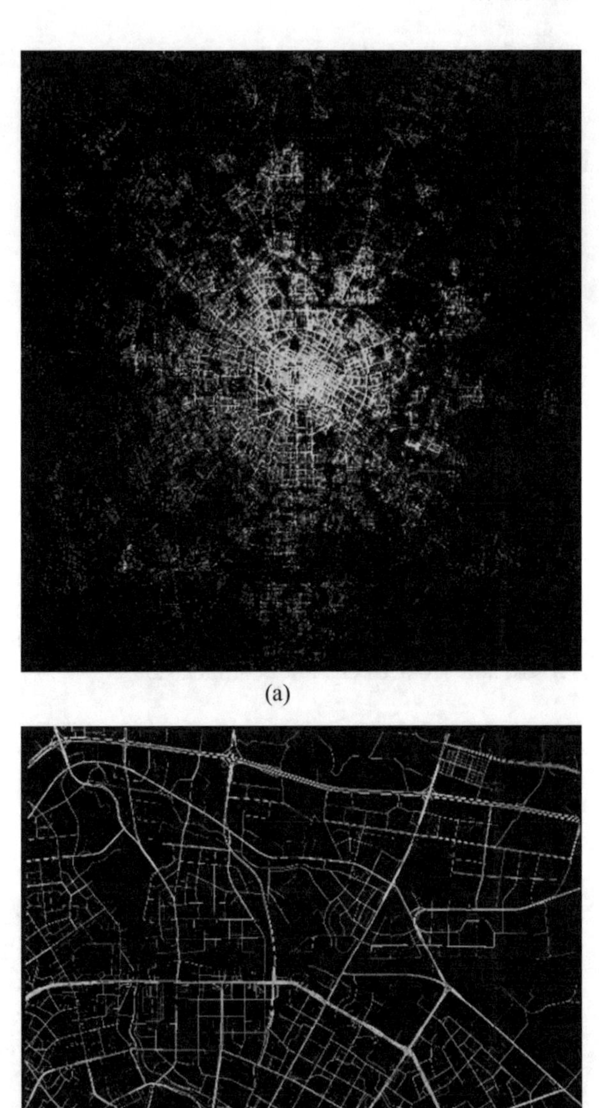

Fig. 11.7 A map visualization of the case-study DiDi GPS data: (**a**) over 1 million trip origins and destinations per day; (**b**) part of the extracted road geometry density from 181,172 trip trajectories

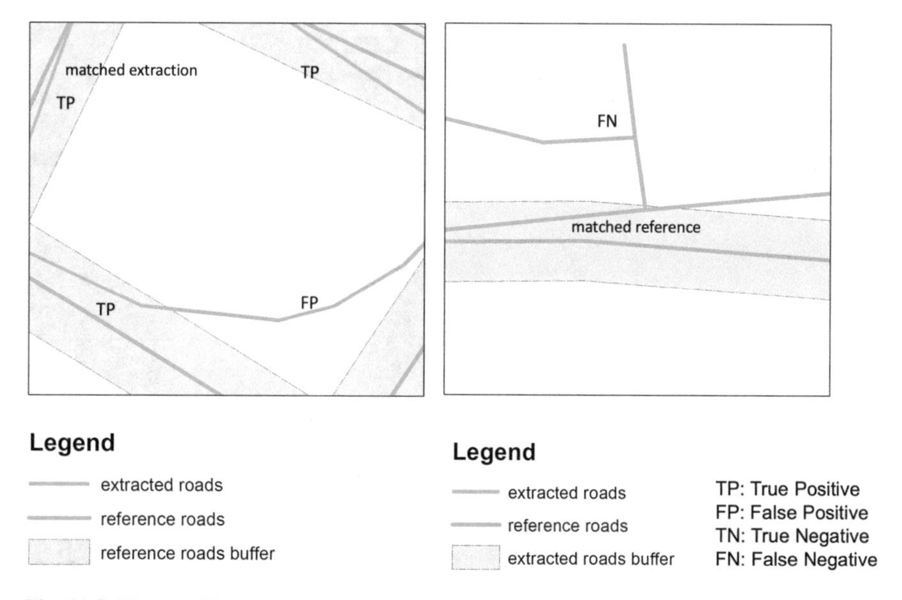

Fig. 11.8 The matching principle between extracted roads and reference roads using buffers

- Completeness

  The completeness constitutes the percentage of reference roads delineated by the extracted road network. That is, the percentage of the reference road data located within the buffer encircling the extracted roads. The closer completeness is to 1, the better the performance of the algorithm is. Let L_{mr} be the length of matched reference roads and L_r be the length of all reference roads.

$$Completeness = \frac{L_{mr}}{L_r}$$

- Correctness

  The correctness signifies the percentage of extracted road data that actually corresponds to the reference road. In other words, it is the percentage of the extracted road network that falls within the buffer surrounding the reference roads. The closer correctness is to 1, the better. Let L_{me} be the length of matched extraction roads and L_e be the length of all extracted roads.

$$Correctness = \frac{L_{me}}{L_e}$$

- Quality

  The quality summarizes the road extraction method by factoring both completeness and correctness into one value. The closer quality is to 1, the better. Let L_{ur} be the length of unmatched reference roads.

$$Quality = \frac{L_{me}}{L_e + L_{ur}}$$

- Redundancy

  The redundancy measures the percentage of the matched extraction road network which overlaps itself. The closer redundancy is to 0, the better.

$$Redundancy = \frac{L_{me} - L_{mr}}{L_{me}}$$

11.4.2.2 Results

In order to evaluate the extracted road network quality, we downloaded the OpenStreetMap (OSM) data as the reference road network layer. Moreover, we also compared our approach to the baseline approach that is directly based on the KDE surface of raw trajectory GPS points and the vectorization operation. The one-day outcomes of the correctness, completeness, quality, and redundancy using our proposed method with different buffer radiuses (10 m, 20 m, and 40 m) are shown in Table 11.1 and the results using the baseline approach are shown in Table 11.2. One can conclude that both methods achieved good performance in extracting roads which fall within the vicinity of the reference roads. Overall, a larger buffer distance gets better results but it may encounter the problem of large redundancy for nearby roads. However, the methods scored poorly in completeness as an inadequate amount of the reference roads were situated closely to the extracted roads. As seen in Fig. 11.9, many of the reference roads were simply not extracted at all which made these unsuccessfully extracted reference roads isolated from the extracted network. With no extracted roads close by, the completeness value falls. As a whole, the quality metric scored poorly due to the shortcoming of many reference road segments not being extracted at all. We would expect the completeness score

Table 11.1 The extracted road quality evaluation results using our proposed approach

Buffer distance	Completeness	Correctness	Quality	Redundancy
40 m	0.603534	0.995467	0.430439	1.00737
20 m	0.502849	0.983940	0.371855	0.692080
10 m	0.330325	0.854347	0.265552	0.280146

Table 11.2 The extracted road quality evaluation results using the baseline approach

Buffer distance	Completeness	Correctness	Quality	Redundancy
40 m	0.597233	0.986477	0.429040	0.952992
20 m	0.469571	0.925763	0.341474	0.636230
10 m	0.302925	0.768156	0.236453	0.272123

Fig. 11.9 A visual comparison of the extracted road network and the OSM road reference layer

could increase with larger spatial coverage of the datasets especially in those (non-major) tributary roads. The redundancy value scored the worst since the length of the matched extraction was much larger than the length of matched reference. Again, the failure to extract many reference road segments resulted in a low evaluation metric. However, Tables 11.1 and 11.2 showed that our approach scored best in completeness, correctness, and quality whereas the baseline approach scored best in redundancy only. Our approach outperformed the baseline approach with a small margin as it compressed clusters and could limit noise points outside the road network.

In addition, in order to determine the connectivity of the extracted road network, we applied the shortest-path based approach using the average path length similarity (APLS) metric to evaluate the extracted road network quality. The average path length was calculated for both the proposal layer (the extracted roads) and the ground-truth reference layer (the downloaded OSM road data) using the Dijkstra's shortest path algorithm (Dijkstra 1959). Then, we examined the similarity between the average path lengths to formulate an overall score. The code to run such an analysis was made available by the CosmiQ Works.[5] This measurement was performed on both approaches.

To perform the measurement, the road layers were converted to graphs, on which nodes are placed at intersections, endpoints, and midpoints. The shortest path was then calculated from each node to each other node for each graph. The differences in path length were used to calculate a metric. In order to quantify these differences in distance between the proposal and ground truth graphs, the APLS metric was computed, which sums the differences in optimal path lengths between nodes in the ground truth graph and the proposal graph. Missing paths are given the maximum proportional distance of 1.0.

Due to the nature of this metric, any missing nodes with high centrality will be penalized much more heavily than those with low centrality as high betweenness centrality roads account for larger traffic flow (Gao et al. 2013). It is important to consider how the nodes are generated and which nodes are important to the results when considering larger graphs. Accordingly, it may be necessary to exclude or alter the generation of nodes if calculating every possible path becomes infeasible—which was the case for our road extraction result.

Figure 11.10 shows the proposal graph for our road centerline extraction with nodes for intersections in sky-blue, as well as a buffer for the visible ground truth graph in yellow.

The graph was then run through multiple iterations of the APLS measurement code in order to determine a semi-optimal distance between the generated midpoints in order to approximate the best result. As shown in Table 11.3, it was determined that a midpoint distance of 300 meters gave the best suitable result with the highest APLS score.

A histogram (in Fig. 11.11) of the results was also created for the path differences when comparing the ground truth graph to the proposal graph, as well as when comparing the path differences from the proposal graph to the ground truth graph. The connectivity results were not good given the aforementioned low coverage completeness of the extracted road networks.

The same process was repeated to calculate the APLS metric for the baseline approach. Also, a distance of 300 meters between midpoints gave the best result for the total score (as shown in Table 11.4). However, the total APLS score for the baseline approach was lower than that of our proposed centerline road extraction approach, concluding that our approach provided slightly better connectivity in its resulting road network.

[5]https://github.com/CosmiQ/apls.

Fig. 11.10 The extracted road graph with ground-truth reference graph (yellow)

Table 11.3 Connectivity evaluation results with different distance settings using our proposed approach (The bold number highlights the best score)

Distance between midpoints (in meters)	Ground truth nodes snapped to proposal graph	Proposal nodes snapped onto ground truth graph	Total score
100	0.032315	0.275312	0.057841
150	0.032516	0.271229	0.058071
200	0.033787	0.287925	0.060478
250	0.034572	0.280139	0.061549
300	0.036189	0.281458	**0.064132**
400	0.034554	0.276131	0.061422
450	0.033946	0.277987	0.060503
500	0.034808	0.271562	0.061707
550	0.034434	0.265163	0.060953
600	0.034333	0.269076	0.060897
700	0.034470	0.259051	0.060844
1000	0.029633	0.243184	0.052829

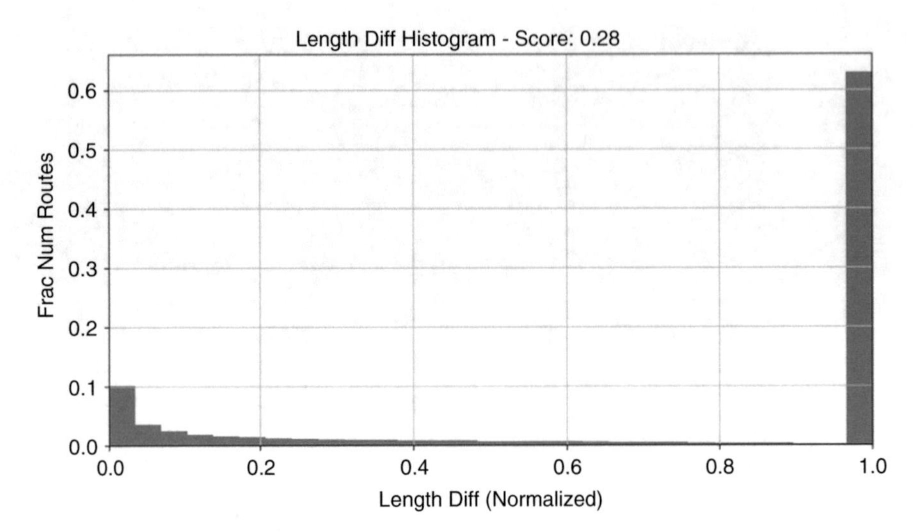

Fig. 11.11 A histogram showing the path differences from proposal (our approach) to ground truth road network

Table 11.4 Connectivity evaluation results with different distances settings using the baseline approach (The bold number highlights the best score)

Distance between midpoints (in meters)	Ground truth nodes snapped onto proposal graph	Proposal nodes snapped onto ground truth graph	Total score
250	0.030692	0.286749	0.055450
300	0.032063	0.289250	**0.057727**
328	0.031514	0.283418	0.056722
350	0.031102	0.293708	0.056248
400	0.031444	0.284132	0.056621
450	0.030946	0.285617	0.055842
500	0.031358	0.276147	0.056321
1000	0.029496	0.242499	0.052595
2000	0.029709	0.238963	0.052849

11.5 Conclusion and Future Work

In this chapter, we present a data-driven approach to extracting road centerline geometry information using large-scale GPS trajectory data. The introduced road extraction framework utilizes trajectory compression (Douglas-Peucker algorithm), an anisotropic density-based trajectory clustering (ADCN algorithm), a kernel density estimation, and vectorization approach. Compared with remote sensing-based approach, the ride-hailing service GPS trajectory data has a higher spatiotemporal resolution but a smaller geographical coverage. A case study using the DiDi open trajectory dataset in Chengdu, China demonstrates the effectiveness of our proposed approach for extracting road networks. It performed well with regard to the correctness of extracted road networks. However, the connectivity quality is bad due to the large incomplete coverage of the tributary roads. Future work needs to further improve the completeness. In addition, road networks include not only geometry information but also attributes (e.g., number of lanes, one-way restriction, speed limit). Such information may also be extracted from large-scale trajectory datasets with additional attributes (e.g., direction, speed), which require more attention in the road network generation pipeline.

Acknowledgments We thank the support of the DiDi Chuxing GAIA Open Dataset Initiative. Dr. Song Gao would like to thank the support for this research provided by the University of Wisconsin - Madison Office of the Vice Chancellor for Research and Graduate Education with funding from the Wisconsin Alumni Research Foundation (WARF).

References

Aronov B, Driemel A, Kreveld MV, Löffler M, Staals F (2016) Segmentation of trajectories on nonmonotone criteria. ACM Trans Algorithms (TALG) 12(2):26

Biagioni J, Eriksson J (2012, November) Map inference in the face of noise and disparity. In: Proceedings of the 20th international conference on advances in geographic information systems. ACM, pp 79–88

Cadez I, Gaffney S, Smyth P (2000, March) A general probabilistic framework for clustering individuals. In: Proc of ACM SIGKDD, vol 2000

Cao L, Krumm J (2009, November) From GPS traces to a routable road map. In: Proceedings of the 17th ACM SIGSPATIAL international conference on advances in geographic information systems. ACM, pp 3–12

Davies JJ, Beresford AR, Hopper A (2006) Scalable, distributed, real-time map generation. IEEE Pervasive Comput 5(4):47–54

Dijkstra EW (1959) A note on two problems in connexion with graphs. Numer Math 1(1):269–271

Douglas DH, Peucker TK (1973) Algorithms for the reduction of the number of points required to represent a digitized line or its caricature. Cartographica Int J Geogr Inf Geovis 10(2):112–122

Edelkamp S, Schrödl S (2003) Route planning and map inference with global positioning traces. In: Computer science in perspective. Springer, Berlin, Heidelberg, pp 128–151

Ester M, Kriegel HP, Sander J, Xu X (1996) A density-based algorithm for discovering clusters in large spatial databases with noise. KDD 96(34):226–231

Gaffney S, Smyth P (1999) Trajectory clustering with mixtures of regression models. KDD 99:63–72

Gao S, Janowicz K, Montello DR, Hu Y, Yang JA, McKenzie G, Ju Y, Gong L, Adams B, Yan B (2017) A data-synthesis-driven method for detecting and extracting vague cognitive regions. Int J Geogr Inf Sci 31(6):1245–1271

Gao S, Wang Y, Gao Y, Liu Y (2013) Understanding urban traffic-flow characteristics: a rethinking of betweenness centrality. Environment and Planning B: Planning and Design. 40(1):135–53

Goodchild MF (2007) Citizens as sensors: web 2.0 and the volunteering of geographic information. GeoFocus Revista Internacional de Ciencia y Tecnología de la Información Geográfica 7:8–10

Hancke GP, Hancke GP Jr (2013) The role of advanced sensing in smart cities. Sensors 13(1):393–425

Hu Y, Gao S, Janowicz K, Yu B, Li W, Prasad S (2015) Extracting and understanding urban areas of interest using geotagged photos. Comput Environ Urban Syst 54:240–254

Jiang Y, Li X, Li XJ et al (2012) Geometrical characteristics extraction and accuracy analysis of road network based on vehicle trajectory data[J]. J Geo-Inf Sci 14(2):165–170
Kolatch E (2001) Clustering algorithms for spatial databases: a survey, pp 1–22
Kuntzsch C, Sester M, Brenner C (2016) Generative models for road network reconstruction. Int J Geogr Inf Sci 30(5):1012–1039
Lang T (1969) Rules for the robot draughtsmen. Geogr Mag 42(1):50–51
Lee JG, Han J, Whang KY (2007, June) Trajectory clustering: a partition-and-group framework. In: Proceedings of the 2007 ACM SIGMOD international conference on Management of data. ACM, pp 593–604
Li Z, Lee JG, Li X, Han J (2010) Incremental clustering for trajectories. In: International conference on database systems for advanced applications. Springer, Berlin, Heidelberg, pp 32–46
Li J, Qin Q, Xie C, Zhao Y (2012) Integrated use of spatial and semantic relationships for extracting road networks from floating car data. Int J Appl Earth Obs Geoinf 19:238–247
Liu Y, Wang F, Xiao Y, Gao S (2012) Urban land uses and traffic 'source-sink areas': evidence from GPS-enabled taxi data in Shanghai. Landsc Urban Plan 106(1):73–87
Mai G, Janowicz K, Hu Y, Gao S (2018) ADCN: an anisotropic density-based clustering algorithm for discovering spatial point patterns with noise. Trans GIS 22(1):348–369
McMaster RB (1989) The integration of simplification and smoothing algorithms in line generalization. Cartographica Int J Geogr Inf Geovis 26(1):101–121
Opheim H (1982) Fast data reduction of a digitized curve. GeoProcessing 2:33–40
Reumann K, Witkam APM (1974) Optimizing curve segmentation in computer graphics. International Comp Sympo 1973. Amsterdam, pp 467–472
Shaw SL (2010) Geographic information systems for transportation: from a static past to a dynamic future. Ann GIS 16(3):129–140
Shi W, Cheung C (2006) Performance evaluation of line simplification algorithms for vector generalization. Cartogr J 43(1):27–44
Shi W, Shen S, Liu Y (2009, October) Automatic generation of road network map from massive GPS, vehicle trajectories. In: 2009 12th international IEEE conference on intelligent transportation systems. IEEE, pp 1–6
Silverman BW (1986) Density estimation for statistics and data analysis. Chapman & Hall, London
Stanojevic R, Abbar S, Thirumuruganathan S, Chawla S, Filali F, Aleimat A (2018, May) Robust road map inference through network alignment of trajectories. In: Proceedings of the 2018 SIAM International conference on data mining. Society for Industrial and Applied Mathematics, pp 135–143
Tang L, Ren C, Liu Z, Li Q (2017) A road map refinement method using Delaunay triangulation for big trace data. ISPRS Int J Geo Inf 6(2):45
Tao CV (2000) Mobile mapping technology for road network data acquisition. J Geospat Eng 2(2):1–14
Wang J, Rui X, Song X, Tan X, Wang C, Raghavan V (2015a) A novel approach for generating routable road maps from vehicle GPS traces. Int J Geogr Inf Sci 29(1):69–91
Wang S, Wang Y, Li Y (2015b, November) Efficient map reconstruction and augmentation via topological methods. In: Proceedings of the 23rd SIGSPATIAL international conference on advances in geographic information systems, pp 1–10
Wang H, Li Y, Liu G, Wen X, Qie X (2019) Accurate detection of road network anomaly by understanding Crowd's driving strategies from human mobility. ACM Trans Spatial Algorithms Sys (TSAS) 5(2):11
Wiedemann C, Heipke C, Mayer H, Jamet O (1998) Empirical evaluation of automatically extracted road axes. In: Empirical evaluation techniques in computer vision, vol 12. IEEE Computer Society Press, Los Alamitos, pp 172–187
Worrall S, Nebot E (2007) Automated process for generating digitised maps through GPS data compression. In: Australasian conference on robotics and automation, vol 6. ACRA, Brisbane
Wu C, Ayers PD, Anderson AB (2007) Validating a GIS-based multi-criteria method for potential road identification. J Terramech 44(3):255–263

Wu J, Zhu Y, Ku T, Wang L (2013) Detecting road intersections from coarse-gained GPS traces based on clustering. JCP 8(11):2959–2965

Yuan NJ, Zheng Y, Xie X (2012) Segmentation of urban areas using road networks. MSR-TR-2012–65, Tech Rep

Zhao Y, Liu J, Chen R, Li J, Xie C, Niu W, Geng D, Qin Q (2011, July). A new method of road network updating based on floating car data. In: 2011 IEEE international geoscience and remote sensing symposium. IEEE, pp 1878–1881

Zheng Y (2015) Trajectory data mining: an overview. ACM Trans Intell Sys Technol (TIST) 6(3):29

Yuill R. S (1971) The standard deviational ellipse; an updated tool for spatial description. Geografiska Annaler: Series B, Human Geography. 53(1):28–39

Chapter 12
Exploratory Analysis of Massive Movement Data

Anita Graser, Melitta Dragaschnig, and Hannes Koller

12.1 Introduction

Movement of people and goods is related to many of the most pressing issues we are facing today. Emissions from the transport sector, for example, contribute significantly to climate change (IEA 2018) and the number of road traffic deaths worldwide keeps rising (WHO 2018). Movement data enables planners and policy-makers to make more informed decisions. Traditionally, movement data was often limited to information about flows between origins and destinations (OD flows). In contrast, modern data sources provide increasingly detailed episodic or quasi-continuous movement data (Andrienko et al. 2013) on a bigger scale. These massive movement data sources cover areas including, for example, movement ecology (e.g., by understanding migration patterns or monitoring species distribution) (Brodie et al. 2018; Demšar et al. 2015), transport (e.g., detecting travel behavior or monitoring traffic quality) (An et al. 2018; Batran et al. 2018), safety (e.g., monitoring suspicious behavior) (Lei 2016), and health (e.g., recognizing physical activity) (Fillekes et al. 2019).

Data analysts face the challenge of how to extract relevant information from these data sets. Key factors that contribute to this challenge are the wide range of applications and methods in current movement data analysis research, as well as the rapidly expanding and often complex datasets that span different spatial and

A. Graser (✉)
AIT Austrian Institute of Technology, Vienna, Austria

University of Salzburg, Salzburg, Austria
e-mail: anita.graser@ait.ac.at

M. Dragaschnig · H. Koller
AIT Austrian Institute of Technology, Vienna, Austria
e-mail: melitta.dragaschnig@ait.ac.at; hannes.koller@ait.ac.at

M. Werner, Y.-Y. Chiang (eds.), *Handbook of Big Geospatial Data*,
https://doi.org/10.1007/978-3-030-55462-0_12

temporal scales (Long et al. 2018). Exploratory data analysis and corresponding interactive visualizations that combine the powers of computational tools with human visual sense-making are essential tools to gain an understanding of these massive datasets (Li et al. 2016).

Beyond the above mentioned application domains, research fields that deal with the analysis of massive movement data, and therefore stand to profit from scalable exploratory analysis, include: geography (Lovelace et al. 2016), cartography (Robinson et al. 2017), GIScience (Demšar and Virrantaus 2010; Shi et al. 2017; Graser and Widhalm 2018), computer graphics (Willems et al. 2009; Scheepens et al. 2011), and pattern analysis (Zheng 2015; Nikitopoulos et al. 2018).

This chapter does not attempt to provide an exhaustive overview of the general field of computational movement analysis, which is also known as trajectory data mining. For those discussions, readers are referred to Gudmundsson et al. (2011) and Zheng (2015). Instead, this chapter specifically deals with issues arising in the exploratory analysis of massive movement data and developments leveraging big data technologies which have not yet received much attention in movement analysis within GIScience (Long et al. 2018).

The remainder of this chapter is structured as follows: Section 12.2 provides an overview of the varying characteristics exhibited by different movement data sources. Section 12.3 introduces the core concepts of exploratory data analysis for movement data. Section 12.4 takes a closer look at key exploratory analysis tasks dealing with massive movement data and related challenges. Section 12.5 discusses privacy challenges of massive movement data. Finally, Sect. 12.6 provides recommendations for the exploratory analysis of massive movement data before Sect. 12.7 concludes this chapter.

12.2 Movement Data Characteristics & Their Relation to Big Data Vs

Before the rise of tracking technology, low-tech data sources, such as questionnaires and statistical surveys provided some movement data to derive basic information about mobility patterns. While these smaller data sources pose their own analysis challenges, this chapter focuses on movement data from tracking systems that provide larger volumes of data.

The heterogeneity of movement data sources makes it difficult to define what constitutes *Big Data* in the context of movement data. The terms "big" or "massive" movement data have been used since at least 2007 (Andrienko and Andrienko 2007; Ma et al. 2009; Gao 2015; Zheng 2015; Dodge et al. 2016; Vahedian et al. 2017; Chen et al. 2017a) but no unified definition has been established. A practical approach is to define that big data is "when it does not fit on one machine" (Schutt and O'Neil 2013, p.24). But this definition, of course, makes big data a moving target that depends on context.

While “Big Data” may be considered as much a marketing buzzword as a technical term, it is still worthwhile to look at massive movement data through the lens of big data and its defining Vs because they provide a common reference framework that ties movement data in with the wider field of big data sources. The three core Vs: Variety, Velocity, and Volume of massive movement data are discussed in the following subsections. Challenges of Validity and Veracity are discussed in Sect. 12.4.4.1.

Background Information: Big Data Vs

Big data is commonly defined as high-variety, high-velocity and/or high-volume. These three characteristics: **Variety** (diversity of data types and data sources), **Velocity** (speed with which the data is generated, processed, and analyzed), and **Volume** (amount of data, dataset size) constitute the core 3-V model of big data.

Additional Vs have been proposed to further describe the big data phenomenon, including, for example: **Validity** (guarantee of the data quality) or, alternatively, **Veracity** (authenticity and credibility of the data), as well as **Value** (added value for society, research, or business).

12.2.1 Variety

Movement data are highly heterogeneous. Datasets vary with respect to, for example, spatial and temporal resolution, spatial dimensions, movement constraints, collection models, tracking system, data size, and privacy constraints.

Some systems that collect movement data have been built for the **purpose** of tracking movement while, in other systems, tracking is a side product. For example, people can be tracked based on their personal devices, such as mobile phones. Mobile phone-based data collection includes systems that use dedicated smart phone apps to track users (including but not limited to social media apps), as well as the mobile phone network itself, which needs to track users to function correctly and – in the process – generates so-called call detail records (CDR). In the field of vehicle tracking, mandatory systems exist for marine vessels (Automatic identification system (AIS)) and aircraft (Automatic dependent surveillance-broadcast (ADS-B)). Commercial road-bound fleets may be managed by collecting and analysing floating car data (FCD). Table 12.1 provides an overview of the different dimensions of movement data and the corresponding range of characteristics in the context of GIScience, which are discussed in the remainder of this section.

The **spatial resolution** of movement data ranges from sub-meter accuracy in trajectories extracted from video material, to meter accuracy in GPS tracks, and finally to kilometer accuracy based on mobile phone cell towers in CDR. Furthermore, generalized and aggregated data sources may only provide information on the level of administrative boundaries, such as ZIP codes, towns, states, or even countries.

Table 12.1 Dimensions of movement data

Spatial resolution	Fine resolution/small location error, e.g. video trajectories	Coarse resolution/large location errors, e.g. CDR
Spatial dimensions	2 dimensional, e.g. AIS	3 dimensional, e.g. ADS-B, animal tracks (birds, whales,...)
Temporal resolution	Frequent/quasi-continuous, e.g. 1 Hz GPS tracks from smart phone apps	Sparse/episodic, e.g. social network data, CDR
Sampling interval	Regular in space or time, e.g. many GPS loggers	Irregular, e.g. AIS, social network data, CDR, acoustic telemetry
Representation	Polylines	Continuous curves
Constraints	Network-constrained, e.g. FCD	Open space, e.g. AIS, ADS-B, animal tracks
Collection models	Lagrangian approach, e.g. GPS or video trajectories	Checkpoint-based or Eulerian approach, e.g. WLAN hot spots or acoustic telemetry
Tracking system	Cooperative, e.g. AIS, FCD, ADS-B, acoustic telemetry listening to acoustic tags	Uncooperative, e.g. video or remote sensing tracking, acoustic telemetry listening to natural sounds
Privacy	Personal/individual-related, e.g. mobile phone data	Impersonal, e.g. AIS, ADS-B
Data size	Small data sets, e.g. tracks from animal collars	Big data sets/streams, e.g. social media data

While many data source only include two spatial dimensions, some sources, such as ADS-B and some animal trackers for monitoring of birds or marine life provide three-dimensional data.

The **temporal resolution** of movement data ranges from high-frequency sampling at regular intervals (1 Hz or higher) in some tracking applications to irregular sparse sampling where hours or days can pass before a moving object creates a new data record in CDR or social media data. Generalized or aggregated statistical data, such as travel demand matrices, mostly summarize movement over longer periods of time.

The most common trajectory **representation** are polylines, that is, straight lines between consecutive locations. However, particularly in robotics, aviation, and autonomous vehicle operations, continuous curve representations are used as well.

Movement is generally **constrained** by geographic context. Network-constrained movement in particular is restricted to a network of paths (most commonly street networks) which affect how data sources, such as FCD, are processed and analyzed. On the other hand, open space or non-network-constrained phenomena (Miller and Bridwell 2009), such as movement of animals through terrain, flying drones in the air, or ships on the open seas require different analysis approaches.

Collection models commonly fit into two categories: in the Lagrangian approach (Dodge et al. 2016), movement information is referenced to the moving object, for example, by fitting a GPS device to the object, or by video-tracking the object as it moves. In the checkpoint-based (Tao 2016) or Eulerian approach (Dodge et al. 2016), movement information is referenced to fixed locations, such as Bluetooth beacons, WLAN hot spots, or mobile phone cell towers.

Tracking systems can be cooperative or uncooperative. In cooperative systems, such as AIS, FCD, or ADS-B, moving objects actively provide their location information. On the other hand, uncooperative systems, such as video and remote sensing tracking do not require the moving objects' cooperation.

Finally, **privacy** concerns are important to consider in the context of personal data, such as CDR or social media data. Similarly, movement data of endangered species can be critical and may be restricted to avoid poaching. On the other hand, AIS and ADS-B data are broadcast actively to increase safety on the sea and in the air. (See Sect. 12.4.4.1 for more about the privacy challenge in massive movement data.)

12.2.2 Velocity & Volume

Volume or size of movement datasets is primarily a result of spatial and temporal resolution, the number of tracked objects, as well as the duration for which movement has been recorded. Furthermore, some data sources provide auxiliary information beyond the basic object identifier, time and location. For example, AIS messages include vessel type, size, status (such as anchored or sailing), destination port, or type of cargo.

Velocity does not just describe how quickly data is generated but can also refer to how quickly it has to be analyzed in order to be useful, for example, to optimize logistics processes or to avoid dangerous situations. Processing and storage approaches therefore have to be able to handle a certain velocity.

While some large movement datasets stem from temporally limited observations (such as data collection campaigns using dedicated trackers or apps), many sources (such as AIS, ADS-B, mobile phone networks, or social media) provide continuous streams of data at varying velocities. The following examples illustrate how Velocity and Volume can be measured as number of records per time interval:

- **Maritime transport:** 10 million AIS records per day (local coverage focusing on Denmark) (Graser and Widhalm 2018)
- **Social media:** 10 million geotagged Tweets per day (based on estimates by Leetaru et al. (2013) that 2% of all tweets include location information) (Li et al. 2017)
- **Mobile phones:** 50 million CDR per workday (40 for each one of 1.25 million subscribers of a single provider) (Zhao et al. 2016)

It is futile to attempt to define a specific quantitative threshold that delimits regular from massive movement data irrespective of a specific application context or analysis task because the limits of existing tools for storing, processing, and visualizing movement data vary. In Sect. 12.4 we therefore look into key exploratory movement analysis tasks and compare the literature on traditional GISystems and novel distributed computing systems to provide reference points for the definition of big or massive movement data. First, however, it is necessary to establish the core concepts of exploratory movement data analysis.

12.3 Exploratory Data Analysis (EDA)

Humans tend to struggle with making sense of even moderate amounts of data that is presented in tables or forms. It is therefore unsurprising that visualization and visual analytics are frequently mentioned as potential additional Vs of big data (Li et al. 2016). Visualizations are commonly understood as important tools for communicating ideas, hypotheses, and analysis results. However, in the case of big data, the role of visualization in data exploration is even more important (Li et al. 2016).

Interactive and exploratory visualization environments combine the powers of computational tools with human visual sensemaking and help analysts better understand what information their data contains (Gudmundsson et al. 2011; Cook et al. 2012). However, visualizations can be hard to scale because large input datasets mean that there are many things that can or need to be displayed (Li et al. 2016). Visualizations of movement data quickly run into issues with over-plotting and clutter that can obscure important patterns (Janetzko et al. 2013) and traditional map-based overviews routinely reach their limits with big movement data (Robinson et al. 2017). Therefore, the exploratory analysis of massive movement data is particularly challenging.

Exploratory data analysis (EDA) (Tukey 1977) aims to analyze data sets by summarizing their main characteristics, often using visualizations, to determine what information the data contains. The objectives of EDA are: to suggest hypotheses about phenomena observed in the data and their causes, to assess assumptions, to select appropriate statistical tools and techniques, and – if necessary – to provide a basis for further data collection.

Computational movement analysis commonly involves exploratory data analysis and visualization to inspect data, summarize patterns, and formulate hypotheses and research questions (Dodge 2019). Methods used include movement data aggregation, grouping or dividing movement data, transformations of space and time, and trajectory clustering (Gudmundsson et al. 2011). EDA also helps with performing sanity checks, to find out if and where data is missing or if there are outliers (Schutt and O'Neil 2013, p.36). Furthermore, EDA is also used to debug the data collection or tracking process because patterns found in the data can be

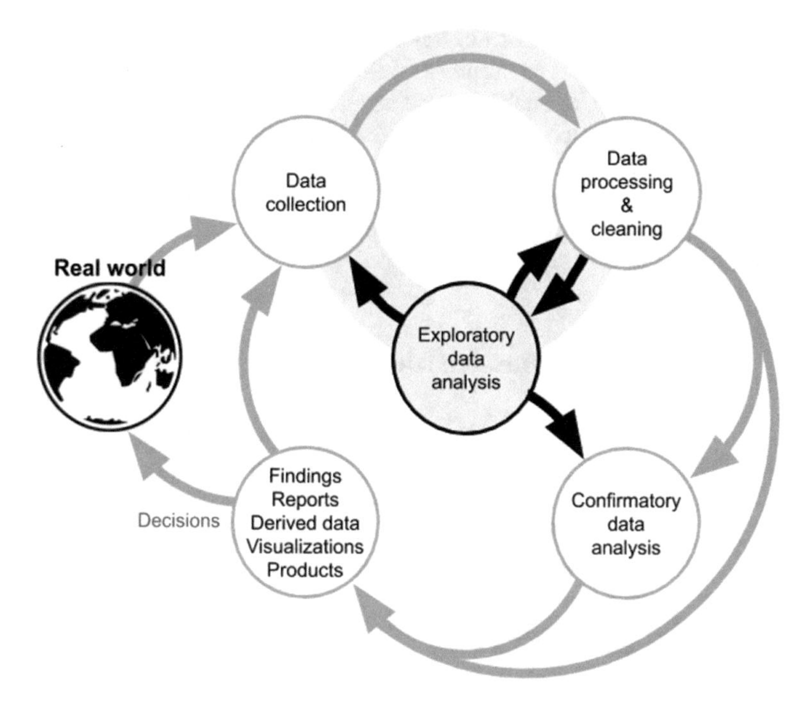

Fig. 12.1 EDA within the broader data science framework (Graser 2020)

the result of issues with the tracking process that need to be fixed. As illustrated by Fig. 12.1, this leads to feedback loops where EDA results inform data collection, processing and cleaning, as well as, eventual confirmatory data analysis.

Conceptually, the general EDA framework laid out by Andrienko and Andrienko (2006, p.158) distinguishes between tasks that deal with individual data elements, so-called elementary tasks, on the one hand and tasks that deal with the whole dataset or subsets, so-called synoptic tasks, on the other hand. With respect to exploratory movement data analysis, elementary spatial events (corresponding to individual records in a movement data source) consists of time, space, and thematic attributes (Andrienko et al. 2013, p.49). Movement trajectories consist of a sequence of these elementary spatial events. Tasks on the trajectory level therefore can be considered synoptic since they deal with the whole dataset or subsets rather than individual elements. However, for many tasks, it may be more suitable to think of trajectories as elementary units and consider only tasks on sets of trajectories as synoptic.

An alternative conceptual approach is to categorize exploratory movement data analysis tasks according to their foci (Andrienko et al. 2013): focusing either on moving objects (also called movers), spatial events, space, or time. Analyses of movers include spatial summarization and clustering of trajectories, visualization of positional attributes, as well as encounters between groups of movers and other relations between movers and their context. To analyze events in movement data, they first have to be extracted from trajectories, usually followed by a clustering step. Again, relations between events, as well as with trajectories and their context are explored. Similarly, to analyze space using movement data, places of interest are extracted from trajectories. This extraction can be based on events, time series, or individual movers. Analyses look into relations or links between places and the properties of these links, or frequent sequences. Finally, analyses focusing on time include clustering of times by similarity of spatial situations or the event extraction from spatial situations.

12.4 EDA Tasks for Massive Movement Data

This section takes a closer look at key EDA tasks and challenges dealing with massive movement data, as listed in Table 12.2. The provided examples are taken from both the literature on traditional GISystems, as well as work on cloud computing big data stacks to gain a better understanding of what kind of problems have been approached with what kind of system.

Background Information: Big Data Technologies

Traditional GISystems run on a single machine. These systems may support multi-threading to make use of multiple processor cores but the computations are not distributed over multiple different machines. In contrast, big data architectures combine the resources of many networked off-the-shelf computers in order to created a powerful environment for executing parallel programs on top of a scalable, distributed, fault-tolerant file system. (Graphics processing units (GPUs), which are commonly used in other areas of big data processing, so far do not play a prominent role in the literature on exploratory movement data analysis.)

The most commonly referred to big data programming models are MapReduce and Apache Spark.[1] **MapReduce** provides a mechanism for processing big data sets with parallel, distributed algorithms on a cluster (Lämmel 2008). In general,

Table 12.2 EDA tasks for massive movement data and related challenges

Task	Challenges
T1 – Spatio-temporal lookup	C1 – Trajectory indexing C2 – Spatio-temporal visualizations
T2 – Similar trajectory search & join	C3 – Creating & segmenting trajectories C4 – Moving object identifiers
T3 – Summarization	C5 – Representativeness & bias
T4 – Extracting events & places	C6 – Data quality
T5 – Detecting outliers	C7 – Performance assessments

[1]https://spark.apache.org

papers refer to the open source Apache Hadoop implementation of MapReduce. **Apache Spark** has been developed in response to limitations in MapReduce, which restrict the dataflow of distributed programs to reading the input data from disk, then mapping a function across the data and reducing the results before finally, storing the reduction results on disk again. In contrast, Spark does not need to constantly read from and write to disk. Spark adopts a higher-level programming paradigm that facilitates the implementation of iterative algorithms on distributed datasets. It includes libraries and high-level functions for interactive and exploratory data analysis and runs on top of existing Hadoop infrastructure.

The literature presented in this section focuses mostly on the last ten years, going back to the beginnings of distributed processing with MapReduce around 2008 (Lämmel 2008) to enable a fair comparison based on a similar state of the art.

Of course, the complexity of the analysis as well as the available hardware resources have to be factored in when comparing results. An individual computation node setup may range from consumer hardware with 4 GB RAM (Xu et al. 2018) to a large server with 256 GB RAM or more (Patroumpas et al. 2017). Papers focusing on algorithmic improvements (Xie et al. 2016; Zhang et al. 2017) often report only relative improvements or rough run times plotted in graphs on a logarithmic scale. It is also not uncommon for papers to omit processing time information (Batran et al. 2018; An et al. 2018; Wu et al. 2017).

12.4.1 Task 1: Spatio-Temporal Lookup or Range Queries

Lookup tasks are elementary EDA tasks for finding data components that correspond to given values. In massive movement data analysis, lookup tasks may refer to spatio-temporal as well as attribute values of trajectories. For example, a lookup task could return moving objects of a certain object type or within a certain spatio-temporal extent. Attribute lookup tasks are no different for movement data than for other types of data. Spatio-temporal lookup tasks however are more specific. Efficient lookups are essential to support interactive filtering in exploratory analysis tools.

Spatio-temporal lookups may be applied on the level of individual spatio-temporal events, as well as on the trajectory level where more complex queries can be formulated. For example, trajectory-based spatio-temporal range queries can be used to filter trajectories based on their origin or destination. Table 12.3 lists publications dealing with spatio-temporal lookup tasks, also commonly referred to as spatio-temporal range (STR) queries. Papers using big data technology (BDT) are marked with a check mark.

The run time of spatio-temporal lookup tasks depends on a variety of factors beyond the size of the movement dataset, hardware specifications, and the performance of the presented system. Additional factors include: spatial and temporal index granularity (which in turn affects the size of the index (Xu et al. 2018)),

Table 12.3 Spatio-temporal lookup tasks (spatio-temporal range queries): dataset sizes (in million points), and corresponding run times; for distributed computing systems, the number of computing nodes is provided in brackets (ordered by year of publication)

BDT	Paper	Data type	Size	Technology	Run time
✓	Ma et al. (2009)	Generated using Brinkhoff (2002)	1,013	MapReduce (8)	300 s
✓	Xie et al. (2016)	Generated using Brinkhoff (2002)	104	(a) Elite	(a) 10 s[a]
				(b) Spatial Hadoop (27)	(b) 10,000 s[a]
✓	Fecher and Whitby (2017)	NYC taxi ODs	1,300	GeoWave & EMR[a] (20)	<30 s
✓	Zhang et al. (2017)	Beijing taxi FCD	2,500	TrajSpark (12)	0.2–2 s[a]
–	Xu et al. (2018)	Shanghai taxi FCD	420	TripCube	54 s
–	Graser (2018)	Geolife GPS tracks	25	PostGIS	<1 s

[a] Estimated from graphs with log scale

skewedness of the data distribution (queries are faster in sparse areas (Yu and Sarwat 2017)), and size of the spatio-temporal query range (Zhang et al. 2017). The run times provided in Table 12.3 therefore only serve as a very rough performance indicator.

12.4.1.1 Challenge 1: Trajectory Indexing

In conventional applications that run on a single machine, established solutions, such as the open source library libspatialindex[2] can be used to quickly and efficiently implement spatio-temporal indexing. However, in distributed environments, different technologies and approaches are necessary to ensure efficiency and consistency. State-of-the-art big geospatial data tools, such as GeoMesa[3] and GeoWave[4], solve the problem of efficiently storing and accessing large amounts of spatio-temporal data in distributed systems. Space-filling curves are used to make multi-dimensional data sortable in one dimension (Fox et al. 2013; Hughes et al. 2015). This makes it possible to store spatio-temporal data in state-of-the-art big data stores, such as, Apache Accumulo[5] and Hive[6] (Hughes et al. 2015).

The spatio-temporal indexing approaches implemented in spatial big data tools, such as GeoMesa and GeoWave, are currently not optimized for trajectories. However, the body of computer science literature on current developments in efficient indexing schemes for movement data is extensive (e.g. Zhang et al. 2017;

[2]https://libspatialindex.org/

[3]https://www.geomesa.org

[4]http://locationtech.github.io/geowave/

[5]https://accumulo.apache.org

[6]https://hive.apache.org

Xu et al. 2018) which underlines the considerable research interest in this topic. A thorough review of these developments, however, is beyond the scope of this chapter.

12.4.1.2 Challenge 2: Spatio-Temporal Visualizations of Massive Movement Data

The visualization of massive movement data is both a conceptual, as well as a technical challenge. Recent work in visual analytics (Andrienko et al. 2017) and cartography (Robinson et al. 2017), as well as a general lack of established tools confirm as much. Part of the challenge lies in its interdisciplinary nature. While novel distributed trajectory processing tools, such as TrajSpark (Zhang et al. 2017), are developed for efficient computations, they do not deal with visualization. Apache Sedona (formerly known as GeoSpark),[7] which is more generic, just added basic visualization capabilities in version 1.2 (published in March 2019).

More advanced visualization options require integration into dedicated visualization or GIS tools. For example, GeoMesa and GeoWave make it possible to publish large distributed datasets as OGC compliant Web Map Services (WMS) and Web Feature Services (WFS) using GeoServer. However, this approach suffers from the fact that established GIS data models are based on the OGC Simple Features specification (OGC 2011) which is not time-aware (Graser 2019). This means that any optimized trajectory data types have to be broken down into Simple Features where time is not a meaningful dimension but just an attribute like any other.

12.4.2 Task 2: Similar Trajectory Search and Join

Similar trajectory search and join tasks aim to find pairs or groups of trajectories that are similar in one way or another. Trajectory similarity can be based purely on geometry (spatial distance between trajectory lines, for example, using Hausdorff (Hausdorff 1914) or Fréchet distance (Fréchet 1906)) but can also be measured using spatio-temporal distance measures (Tampakis et al. 2019) or include attributes or semantic information, such as activities (Wang and Belhassena 2017).

Due to the necessary pairwise comparisons of trajectories, this task is very computationally expensive. This is also underlined by the dominance of big data technology literature in Table 12.4. Performance of similar trajectory searches and joins can be improved through appropriate indexing techniques. However, as shown, for example, by Xie et al. (2017), building indexes for similar trajectory search can be considerably more expensive than executing search tasks once these indexes are built. Particularly for online processing of streaming data, the costs of indexing and querying therefore have to be weighed carefully.

[7] https://sedona.apache.org

Table 12.4 Similar trajectory search and join tasks: dataset sizes (in million points), and corresponding run times; for distributed computing systems the number of computing nodes is provided in brackets (ordered by year of publication)

BDT	Paper	Data type	Size	Technology	Run time
✓	Xie et al. (2017)	OSM tracks: (a) indexing, (b) querying	1,060	Spark (16)	(a) 1000 s[a] (b) 10 s[a]
✓	Wang and Belhassena (2017)	Geolife GPS tracks	25	Spark (4)	~18 s
✓	Shang et al. (2018)	GPS trajectories	141	SparkSQL (64)	100 ms[a]
		GPS trajectory join	66		500–3,000 s[a]
✓	Tampakis et al. (2019)	AIS distributed subtrajectory join	1,500	MapReduce & Hadoop (49)	n/a

[a] Estimated from graphs with log scale

12.4.2.1 Challenge 3: Building and Segmenting Trajectories

Many data sources provide continuous moving object data that contains no explicit starts and ends of trips and stays. Correctly segmenting a trajectory is highly dependent on the context and often requires expert domain knowledge (Gudmundsson et al. 2011). As a result, a variety of different segmentation approaches exist in the literature and we refer the readers to Edelhoff et al. (2016) for an overview. A common segmentation approach is to split the continuous movement data at stays. A stay is defined as a part of the movement track where the moving object stays within an area with a small radius for a certain period of time. Movement tracks may also be split at temporal or spatial gaps, at spatial or temporal events, in regular temporal cycles, or by thematic attributes (Andrienko et al. 2013).

Trajectory **simplification** is commonly performed in parallel or directly after segmentation. Trajectory simplification creates a close approximation of the original trajectory, often by using a subsampling technique. According to Gudmundsson et al. (2011), a common approach is to use algorithms based on the Douglas-Peucker algorithm (Douglas and Peucker 1973), such as Cao et al. (2006) or Gudmundsson et al. (2009). Usually, the loss of information during subsampling is minimized with respect to a distance measure, generally Euclidean distance but in principle any tolerance criterion can be used (Gudmundsson et al. 2011).

Trajectory creation, segmentation, and simplification require that the movement data is analyzed in chronological order. While this task is straightforward in traditional data processing, it poses a challenge in big data environments, where movement data of an individual object is usually massive, unsorted and physically distributed over many computation nodes.

Background Information: Distributed data processing

Distributed processing approaches are well suited for dealing with unsorted data where random order processing is acceptable. The quintessential textbook example for MapReduce is word counting, that is computing the word frequency histogram for a (huge) set of text data. In a distributed environment, this problem can easily be solved by calculating intermediate results on an arbitrary number of computation nodes, and then merging these intermediate results in a final aggregation step. Word counting lends itself to such a clean and scalable solution because the order in which the input data is processed is irrelevant for obtaining the correct result. In contrast, EDA on massive movement data often does not allow for unsorted data processing: trajectory building and segmentation is but one example for algorithms that require the data to be chronologically sorted in order to produce correct results.

Building trajectories with the current core Spark libraries[8] can be problematic. The Spark programming model adds high-level functions for grouping and aggregating records but these concepts are meant to deal with unsorted data and are not well suited to processing trajectory data, where most operations only produce correct results if applied to an entire, time-sorted group of position records. With the default Spark programming model, an aggregator therefore needs to collect and sort the entire trajectory in the main memory of a single processing node. When dealing with large datasets (such as multiple years of AIS records), using Spark's groupByKey() functionality and then sorting each group during the aggregation phase can thus easily lead to out-of-memory errors.

Third-party libraries such as spark-sorted (Tresata 2019) provide groupSort(), the functionality required to group, sort and iteratively process massive data sets. It never materializes the group for a given key in memory, but instead offers iterator-based streaming of the sorted data. This functionality helps to efficiently solve the two main tasks in this challenge: firstly, trajectories can be created in a clean and straightforward manner. Because the groupSort() function traverses the movement data of one vehicle iteratively, this approach synergizes well with an object-oriented builder pattern: the trajectory can be aggregated one position at a time. During this iteration, resampling, compression and data-imputation operations can be applied on-the-fly, as needed. Secondly, trajectory segmentation can be added to this approach quite naturally: whenever a new position is processed, the trajectory can be checked for a segmentation criterion (e.g. temporal gap detection). If segmentation is indicated, the current trajectory can be finalized and a new one can be started at the current position. Using spark-sorted is more efficient from a memory-utilization perspective and can in practice mean the difference between a query succeeding in mere minutes or failing completely. When combined with

[8]Spark 2.2.0, at the time of writing

suitable analysis algorithms, it enables processing of trajectories of arbitrary size within the Spark framework.

However, spark-sorted is no silver bullet and it is worth mentioning some limitations and drawbacks: firstly, it makes it harder to implement algorithms which require context (such as the previous *n* positions of the trajectory). This context must be provided by manually adding an appropriate caching mechanism to the aggregator. Special care must be taken when implementing the caching strategy since too liberal caching will again cause memory problems. Also, algorithms which need the entire trajectory in memory (such as the classic Douglas–Peucker algorithm) do not profit from spark-sorted. Secondly, for some algorithms a sorted forward-iterator may not be a convenient data access mode. It can thus introduce additional code complexity, especially when combined with the caching requirements mentioned above. Thirdly, while this approach improves memory utilization, it has no effect on the fundamental algorithmic complexity of the underlying algorithms. Runtime performance problems related to overall algorithmic complexity thus can not be solved with this technique. Therefore, in some cases, a different algorithmic approach might be needed to facilitate scalability in a distributed environment.

12.4.2.2 Challenge 4: Moving Object Identifiers

Reliable moving object identifiers are essential for reconstructing trajectories since they are necessary to identify consecutive locations of the same moving object. However, these are not guaranteed in all datasets. Common issues include the reassignment of mover identifiers or distinct moving objects that share the same identifier (Andrienko et al. 2013, p.344 / 349). Reassignments of mover identifiers can be introduced into a dataset on purpose (for example, as a step towards privacy protection) but reassignments can also be undocumented and may happen unintentionally. Exploratory analysis should enable the identification of undocumented reassignments in massive movement data.

Distinct moving objects sharing the same identifier can result in zigzag trajectories if both objects move at the same time but in different locations. This problem affects, for example, AIS data where this issue can be addressed by including vessel name or geographic information to attempt to identify vessels with ambiguous identifiers (Aronsen and Landmark 2016; Wang and Wu 2017). While zigzag trajectories are reasonably straightforward to spot in trajectory visualizations of small datasets, cluttered visualizations of massive movement data can hide these issues. Exploratory analysis should help determine whether a dataset contains instances where different movers share the same identifier.

12.4.3 Task 3: Density Mapping and Other Grid-Based Summarizations

Density maps are the most common example of summarization tasks that provide overviews of massive movement data in EDA. Approaches range from simple density maps that only provide counts (Shelmerdine 2015; Aronsen and Landmark 2016) (Fig. 12.2 left), to approaches that also model mean direction (Brillinger et al. 2004) or mean velocity (An et al. 2018), to more complex Gaussian mixture models (GMM) of velocity and direction (Graser and Widhalm 2018) (Fig. 12.2 right).

Besides classic two dimensional grids, some approaches employ spatio-temporal density volumes (Demšar and Virrantaus 2010). Another design choice is whether to compute density from points or from lines (Shelmerdine 2015). Besides computing densities, Nikitopoulos et al. (2018) also perform statistical Getis-Ord tests to identify significant spatio-temporal hotspots. Table 12.5 lists publications dealing with density maps and other grid-based summarization tasks.

The processing requirements of these spatial or spatio-temporal grid-based summaries increase as the corresponding grid resolution increases and cells get smaller. For example, Demšar and Virrantaus (2010)'s spatio-temporal density volume contains 1.5 million cells (or voxels), while An et al. (2018)'s analysis covers only 2,500 cells (1,680 of which contain data).

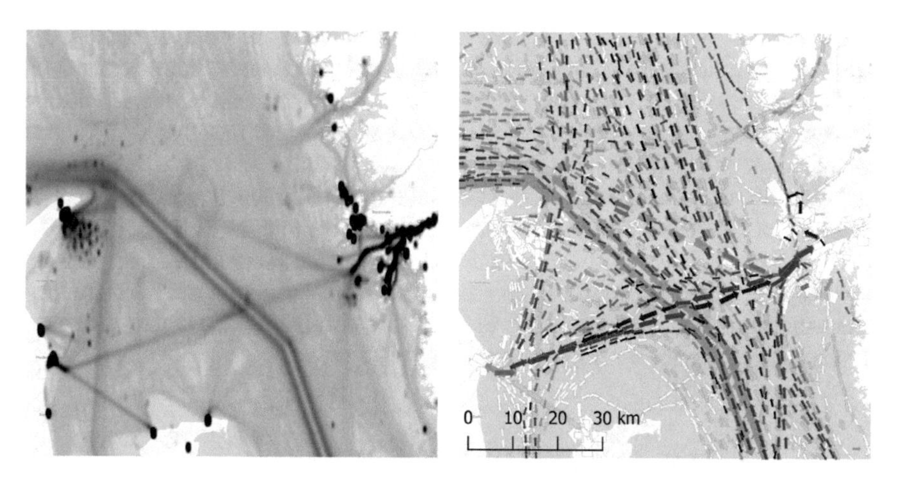

Fig. 12.2 Left: Density map of vessel data from AIS for July 2017 highlighting areas with numerous records such as harbors and anchoring areas; Right: corresponding movement model with mean direction and velocity where darker arrows signify faster movement (Graser and Widhalm 2018)

Table 12.5 Grid-based summarization tasks: dataset sizes (in million points), and corresponding run times; for distributed computing systems the number of computing nodes is provided in brackets (ordered by year of publication)

BDT	Paper	Data type	Size	Technology	Run time
–	Demšar and Virrantaus (2010)	AIS (grid: 1.5M voxels)	0.01	ArcGIS	1.5 h
–	Shelmerdine (2015)	AIS (grid: n/a)	0.06	ArcGIS	1 day [a]
–	Aronsen and Landmark (2016)	AIS (sparse global grid, cell size: 185.2 × 185.2 m)	1,500	PostGIS & GDAL	1 week
–	An et al. (2018)	Taxi FCD (counts & mean velocity, grid: 2,500 cells)	506	Oracle 11 g & ArcGIS 10.0	n/a
–	Graser and Widhalm (2018)	AIS (counts & GMMs of velocity/direction, grid: 370,000 cells)	560	Java & PostGIS	hours
✓	Nikitopoulos et al. (2018)	AIS (varying voxel sizes)	1,900	Spark (10)	17 min
✓	Graser et al. (2020)	AIS (counts & GMMs of velocity/direction, grid: 0.1° × 0.1°)	3,900	Spark & GeoMesa (8)	41 min

[a] Including AIS track generation for line-based density

In contrast to the traditional tools listed in Table 12.5, state of the art big geospatial data tools, such as GeoMesa, can provide interactive density maps on the fly.[9] This means that cell size, grid extent, and input data can be adapted during the exploration process. These tools take advantage of the fact that density grid computations can be readily distributed over multiple nodes: each node can compute a density map of its local data before finally adding the distributed results together. A downside of GeoMesa's current implementation for exploratory analysis is that the resulting density maps cannot be normalized to a specific value. Instead, the density process automatically determines the maximum density value for each density map computation individually and scales the results accordingly. This makes it hard to compare density maps for different times or locations.

12.4.3.1 Challenge 5: Representativeness & Bias

Available massive movement data is not necessarily representative of the whole population. For example, Zhao et al. (2016) caution that "How frequently one uses mobile phone to contact others and when and where those communications occur largely determine the representativeness of CDRs to reflect true mobility

[9]https://www.geomesa.org/documentation/user/process.html#density-process

characteristics." Even in more homogeneous groups, such as users of sports tracking apps, there are bias issues related to participation inequality (Oksanen et al. 2015).

Certain groups are systematically underrepresented in some data sources. While mobile phone penetration rates, for example, in the U.K. range from 93 to 99% of adults with little variation amongst age groups (Lovelace et al. 2016), Twitter, as an example of social media data, is biased towards a young (Li et al. 2013) and male demographic (Mislove et al. 2011). Besides age and gender, Taubenböck et al. (2018) find that "the economic divide influences digital participation in public life" and this in turn affect how less affluent groups living under precarious conditions are represented in digital data.

The interpretation of density maps and other summarizations in exploratory analysis therefore has to account for issues of representativeness and bias to avoid drawing wrong conclusions. For example, Oksanen et al. (2015) describe density maps that counteract participation inequality by taking user diversity into account while still preserving user privacy.

12.4.4 Task 4: Extracting Events & Places

Event or place extraction in EDA is an approach to summarize the information contained in a movement dataset. Events, places, or other significant points include, for example: stays, speed changes, turns, observation gaps, speeding, delays, or rendezvous. Events may be of direct interest if detected online from streaming data (Patroumpas et al. 2017) but they can also be used to, for example, split continuous movement tracks into individual trajectories, generalize movement (Andrienko and Andrienko 2011), or extract routing graphs from movement data (Dobrkovic et al. 2018). Beyond individual events and places, exploratory analysis of travel sequences can help reveal human mobility patterns (Zheng et al. 2009; Widhalm et al. 2015) or flow strength (Lovelace et al. 2016). Table 12.6 lists publications dealing with the extraction of events and places.

The online event detection by Patroumpas et al. (2017) illustrates the challenge of defining the border between regular and big data. By using servers with 48 GB RAM for simple event detection (implemented in C++) and 256 GB for complex event recognition (implemented in Prolog), they manage to fit processing of the AIS stream for their area of interest into main memory and thus can avoid distributed processing.

Simple events can be detected from trajectories of individual movers. Therefore, their computation can readily be parallelized by assigning each thread a subset of all monitored movers (Patroumpas et al. 2017). **Complex events** involve pairs of trajectories and therefore it is more difficult to parallelize or distribute their computation. For example, potential rendezvous or package picking events require that all combinations of movers within an area are checked (Patroumpas et al. 2017). Simple spatial partitioning of the analysis area would lead to unbalanced processing load due to skewed spatial data distribution. Furthermore, the analysis may fail

Table 12.6 Extraction of events and places: dataset sizes (in million points), and corresponding run times; for distributed computing systems the number of computing nodes is provided in brackets (ordered by year of publication)

BDT	Paper	Data type	Size	Technology	Run time
–	Zheng et al. (2009)	GPS tracks (locations & travel sequences)	5	n/a	n/a
–	Andrienko and Andrienko (2011)	FCD (significant points & aggregation)	2	n/a	n/a
–	Widhalm et al. (2015)	CDR (urban mobility patterns)	n/a	n/a	n/a
–	Patroumpas et al. (2017)	AIS (online events detection)	0.6/min	C++	<13 s
–	Dobrkovic et al. (2018)	AIS (directed routing graph)	0.6	n/a	1.7 min
–	Batran et al. (2018)	CDR (ODs)	393	n/a	n/a
✓	Ranjit et al. (2018)	FCD (ODs)	2,200	HDFS & Hive (n/a)	n/a
✓	Cao and Wachowicz (2019)	Bus FCD (stops)	14	MapReduce (n/a)	1,500 s

to identify events along the borders of analysis areas if no overlap is considered because one mover in a rendezvous event can be located in one spatial partition while the second mover can be located in another partition.

12.4.4.1 Challenge 6: Data Quality or Veracity

Challenge 4 and 5 already dealt with aspects of data quality, such as the reliability of moving object identifiers and the overall representativeness of a dataset. However, there are a additional data quality issues related to massive movement datasets that affect exploratory analyses, such as the extraction of events and places.

In cooperative tracking systems, such as AIS and FCD, participants may be less cooperative than expected. For example, vessels may turn off their AIS transponder while performing illegal activities. A more sophisticated approach to fool tracking systems is **GPS spoofing**, where location data is falsified to hide an object's true movements.

Sparse and/or coarse data sources, such as CDR and Twitter, present considerable challenges for the reliable extraction of events and places. For example, a single tweet at a new location provides no information about the duration of a potential stay at this location. The tweet could have been sent while travelling through a place rather than during a stay. Similarly, a CDR could have been generated because the mobile phone moved from one location to another and thus switched over to another cell tower. However, such handovers can also be caused by, e.g. load balancing measures implemented by the provider network, or dynamic changes in signal strength. This phenomenon, often called oscillation, causes additional CDR to be logged, however those records do not signal any actual change of the phone owner's location and have to be detected and dealt with accordingly (Wang and Chen 2018). Bigger is therefore not always better. For example, Lovelace et al. (2016) found that smaller survey data provided more realistic flow information for retail use cases than big Twitter data since "even after 'data mining' processes to refine the low grade 'ore' of the Twitter data, the data set seemed to have relatively low levels of veracity compared with the other data sets." Even relatively dense GPS-based movement data sources can provide challenges for event extraction. For example, Patroumpas et al. (2017) describe out-of-sequence positions, off-course positions, and zig-zag movement in AIS. All of these would lead to erroneous turn and speed events, if not handled properly.

Another aspect of data quality is related to the **accuracy of position measurements and veracity of quality indicators**. In general, all position measurements suffer from a certain degree of uncertainty. This measurement uncertainty should be taken into account and modelled (Kuijpers and Othman 2010; Zheng 2015). Many location sources provide an estimate for their reliability. GPS quantifies the positional measurement precision with a well-defined dilution of precision (Langley and others 1999). Similar quality indicators exist for many other data sources, for example, accuracy values reported by the location providers in Android smartphones. However, compared to GPS, the reliability of these indicators is low. For example, Seyyedhasani et al. (2016) report that the actual location error of the Google Fused location service on Android was much higher than was to be expected based on the reported accuracy. For critical applications it is suggested to verify the accuracy information provided by these providers. Widhalm et al. (2015), who worked with datasets provided by US and Austrian mobile carriers, report that "mobile carriers usually do not disclose the organization of their networks, and in practice the spatial range of network cells varies and is difficult to quantify". Furthermore, Seyyedhasani et al. (2016) caution that the algorithms behind such location measurements usually are not only proprietary, but also not standardized and thus might be subject to change at any time.

Data quality aspects also affects processes for **merging different data sources** that contain information about the same moving objects. For example, on Android smartphones, movement from multiple location providers (such as GPS and network location) can be recorded simultaneously. While GPS would generally be preferred, it is typically unavailable while travelling underground or indoors. In such situations, it can be necessary to impute the missing positioning data with measurements from less reliable sources (such as the network position). A standard solution for merging such data sources is a Kalman filter (Kalman 1960; Fritsche et al. 2009; Goh et al. 2013). As mentioned before, special care has to be taken when the positioning accuracy is used as an input to data fusion methods since some data sources report far too optimistic accuracy estimates.

12.4.5 *Task 5: Detection of Outliers and Anomalies*

This task deals with identification and handling of unusual observations and patterns in the data. These patterns are often referred to as outliers or anomalies. An outlier can generally be defined as "an observation which deviates so much from other observations as to arouse suspicions that it was generated by a different mechanism" (Hawkins 1980, p.1). Similarly, an anomaly is a pattern that does not conform to an expected behavior (Chandola et al. 2009). The distinction is subtle, and the terms are often used interchangeably in the literature (Chandola et al. 2009). Table 12.7 lists a selection of publications dealing with outlier detection in movement data sets.

While outlier or anomaly detection does not directly serve the primary EDA goal of summarizing the main dataset characteristics, it is a valuable tool to challenge assumptions about the data and to develop hypotheses. For example, "an uncommon trajectory or gathering of people in a specific area might correspond to a special event such as a festival, traffic accident or natural disaster" (Witayangkurn et al. 2013). Types of movement anomalies include:

- **Anomalous records:** unusual location, timing, speed, direction, or thematic attribute data. For example, in GPS tracking data, this can be a vehicle moving in an unusual direction (Graser and Widhalm 2018). In checkpoint-based data, such as CDR, an anomalous record can be marked by an unlikely sequences of network antennas.
- **Anomalous (sub)trajectories:** unusual trajectories or trajectory parts, for example, if a mover takes a previously unseen route to get from A to B. (Wang et al. 2014).
- **Anomalous events:** the trajectories of individual movers are unremarkable but their combined spatio-temporal pattern is unusual, for example, if movers gather at an event location (Wachowicz and Liu 2016). Anomalous events may also be detected by combining multiple data sources (Zheng et al. 2015).

To enable anomaly detection in large datasets, it is necessary to create a reference model of normal movement. The types of movement models can be broadly categorized into (Riveiro et al. 2018):

- **Grid-based movement models** commonly provide information about how likely it is to observe a moving object at a certain location, moving at a given speed, or into a given direction. Depending on the implementation, movement in a grid cell can be modelled, for example, using Gaussian Mixture Models (GMM) or Kernel Density Estimation (KDE) (Laxhammar et al. 2009; Graser and Widhalm 2018). In traditional GISystems, grid-based approaches are limited due to the "computational burden when increasing the scale, as well as the need for a priori selection of the optimal cell size" (Pallotta et al. 2013). However, grid-based movement data models where individual cells can be computed independent of each other, are a good match for distributed computing environments (Graser et al. 2020).

Table 12.7 Outlier detection tasks: dataset sizes (in million points), and corresponding run times; for distributed computing systems the number of computing nodes is provided in brackets (ordered by year of publication)

BDT	Paper	Data type	Size	Technology	Run time
✓	Witayangkurn et al. (2013)	Cellphone GPS, WiFi & tower locations (anomalous record detection using HMM)	9.2	Hive (4)	>=14 h
✓	Wang et al. (2014)	AIS (anomalous route detection using DBSCAN_SD)	1	MapReduce (20)	n/a
–	Zheng et al. (2015)	FCD, bike rentals & NYC 311 reported incidents (Multiple-Source Latent-Topic model, Likelihood Ratio Test)	165, 8, 0.2	C++	<30 min
–	Wachowicz and Liu (2016)	Twitter & CDR (outliers based on social & geographic space)	0.03 & 1,440	ArcGIS[a] & Python	n/a
✓	Chen et al. (2017b)	FCD (anomalous trajectory detection using modified line segment Hausdorff distance)	350	Spark (3)	500 s
–	Graser and Widhalm (2018)	AIS (anomalous record detection using quad trees & GMM)	560	Java	n/a
✓	Filipiak et al. (2018)	AIS (loitering detection using avg. speed and SD)	313	Spark (1)[b]	600 s
–	Wang et al. (2018)	FCD (Anomalous route detection using improved edit distance)	n/a[c]	n/a	n/a

[a] To avoid memory issues, the CDR data was split into parts with 10 million records each
[b] One-node pseudo-cluster on virtual machine
[c] 20,000 taxis for one year with reporting interval <60 s

- **Graph-based models** allow for a more compact representation of movement, particularly if the movement is constrained by regulations or geographic context. However, constructing a movement graph from significant locations, such as, for example, turning points, is a topic of ongoing research (Pallotta et al. 2013). The most common approach for clustering significant locations is density-based clustering using DBSCAN.
- **Trajectory or vector-based models** that use similarity metrics between trajectories (Laxhammar and Falkman 2011; Wang et al. 2018) are closely related to the topic of similar trajectory search and join (previously discussed in Sect. 12.4.2).

Background Information: Distributed density-based clustering

While DBSCAN is computationally expensive with a complexity of $O(n^2)$ (or $O(n \log n)$ if spatial indexes are used (Sidibé and Shu 2017)), there are distributed variants, such as P-DBSCAN a parallel DBSCAN implementation (Chen et al. 2010), MR-DBSCAN for MapReduce (He et al. 2014) or DBSCAN for Spark (Cordova and Moh 2015). Furthermore, variants like DBSCAN_SD also consider speed and direction (Wang et al. 2014).

12.4.5.1 Challenge 7: Anomaly Detection Performance

Anomaly detection is a challenging task since "in many domains normal behavior keeps evolving and a current notion of normal behavior might not be sufficiently representative in the future" Chandola et al. (2009). Approaches that consider the evolving nature of the definition of normalcy include, for example, Graser and Widhalm (2018) who propose an algorithm that "can continuously update the model using potentially endless streams of new input data".

Another challenge described by Chandola et al. (2009) stresses that the "availability of labeled data for training/validation of models used by anomaly detection techniques is usually a major issue. [...] Labeling is often done manually by a human expert and hence requires substantial effort to obtain the labeled training data set." Depending on the availability of labeled data sets, anomaly detection methods may fall into one of three categories:

- **Supervised anomaly detection** techniques which require extensive training data sets. These data sets must represent all types of behaviors, including abnormal instances (which are often difficult to obtain).
- **Semi-supervised anomaly detection** tries to mitigate the training data problem as it requires labeling of training data only for the normal class. Abnormality is detected by comparing a record to the statistical distribution learned from the training data.
- **Unsupervised anomaly detection** methods do not require any training data. They operate under the assumption that the majority of data points represent normalcy, and try to find data points that least fit with the remaining data. Consequently, "if this assumption is not true then such techniques suffer from high false alarm rate" (Chandola et al. 2009).

12.5 Privacy

Personal movement data can reveal the places a person visits. These include home and work locations, as well as other places that may be sensitive. For example,

Krumm (2007) presents an experiment to identify home locations and identities of volunteers. Outside of experimental settings, Tockar (2014) demonstrates how people can be stalked based on seemingly anonymized New York City taxi data and that it is straightforward to find pick-up locations of clients that are dropped off near a certain gentlemen's club at night. Using the same data, Douriez et al. (2016) show how taxi driver information can be reconstructed, thus potentially revealing their income, and conclude that "unless the utility of the data set is significantly compromised, it will not be possible to maintain the privacy of taxi medallion owners and drivers". Beyond human mobility, movement data of endangered species can be critical as well. Particularly data on species that are threatened by poaching requires protection (Miller et al. 2019).

An extensive overview of privacy protection strategies for movement data is provided by Duckham and Kulik (2006). They distinguish regulatory approaches and privacy policies (which set rules but cannot enforce privacy) versus anonymity approaches (including pseudonymity, spatial cloaking using k-anonymity, and zero-knowledge proofs) and obfuscation (including inaccuracy, imprecision, and vagueness).

Massive movement data collected by cloud-based services are of particular concern for user privacy. In these systems, data from client apps is sent to a cloud server where movement models are constructed and merged with other users' data. For example, Andrienko et al. (2013, p367) mention the issue of being able to find a person's home and work location from Twitter data. They therefore suggest the concept of a "trusted data transformer" that preprocesses personal data, for example, by applying k-anonymity approaches, before others get to use it.

12.5.1 k-Anonymity

k-Anonymity (Sweeney 2002) is a well-known data protection model. "A release of information provides k-anonymity protection if the information for each person contained in the release cannot be distinguished from at least k-1 individuals whose information also appears in the release. The parameter k determines the level of privacy protection." (Andrienko et al. 2013, p367f)

One approach to ensure privacy protection are aggregations using space tessellations. To ensure k-anonymity, each spatial compartment needs to contain visits of at least k different movers (Andrienko et al. 2013, p368). In areas with more movers, these compartments can be smaller than in areas with fewer movers. For example, Oksanen et al. (2015) present a method for deriving privacy-preserving heat-maps from sports tracking data. Similarly, if event clustering approaches are used, k-anonymity requires that each cluster must contain events from at least k different movers (Andrienko et al. 2013, p368).

However, as discussed by Duckham and Kulik (2006), the need for trusted transformers or brokers is just one of the disadvantages of anonymity-based approaches. They also present "a barrier to authentication and personalization, which are required for a range of applications." They can render data useless for mover-oriented tasks that deal with reconstruction of individual movement behaviors (Andrienko et al. 2013, p368). Furthermore, k-anonymity has shown to be vulnerable to attackers who possess prior knowledge (Machanavajjhala et al. 2007). This shortcoming is addressed by the l-diversity concept by Machanavajjhala et al. (2007). However, ElSalamouny and Gambs (2016) argue that "it is not possible for a privacy definition to guarantee the indistinguishability of the user's location under any prior knowledge while allowing a reasonable utility at the same time."

12.5.2 *Differential Privacy*

According to ElSalamouny and Gambs (2016), the core idea of differential privacy (Dwork 2006) is that "the presence (or absence) of an individual in the dataset should have a negligible impact on the probability of each output of a computation (e.g., a statistical query)". ElSalamouny and Gambs (2016) adapt differential privacy to location based services. They view the real location of the user as the secret to be protected. Consequently, a mechanism processing locations needs to ensure that every two adjacent locations are indistinguishable to some extent. This prevents an attacker observing the output of the mechanism from distinguishing the real location of a user from others situated within a certain distance.

12.5.3 *Privacy by Design*

Regardless of the reliability of a trusted data broker, the mere existence of individual level data on the cloud (or any other system that may be compromised) always presents a data protection issue. Therefore, an alternative approach is described by Meier (2017) who proposes to construct the user movement model on the client-side and to share only an abstracted generalised version of the model with the cloud. The added privacy protection has to be weighed against challenges regarding initial model training and limited processing capabilities on the client-side. Furthermore, changes to the movement model that require additional data would be problematic since – in this approach – there is no more detailed archived data to use.

To the best of our knowledge, no visual analytics tools so far fully implement privacy by design and there is need for an underlying fundamental framework defining which kinds of privacy-protecting transformations to apply depending on the properties of the available data and the types of information that needs to be extracted (Andrienko et al. 2013, p369).

12.6 Recommended EDA Workflow for Massive Movement Data

Building on the previously discussed EDA concepts and identified challenges, we propose a workflow for the exploratory analysis of massive movement data. Our proposed workflow is based on two core ideas: firstly, efficient exploration of movement data requires a dual approach: combining point representation of individual records and line representation of trajectories. Secondly, in order to detect patterns in massive data, aggregates and summarizations are required that communicate movement characteristics.

The structure of our proposed workflow is summarized in Fig. 12.3. It is based on EDA goals and starts with establishing an overview of the data, considering context, extracting trajectories and events, and finally exploring patterns and outliers. The following subsections cover the individual steps of the workflow and describe the corresponding questions and assumptions about the data that should be checked by the analyst.

12.6.1 Establishing an Overview

The first step in our proposed EDA workflow can be performed directly on raw input data since it does not require temporally ordered data. It is therefore suitable as a first exploratory step when dealing with new data. The following questions and assumptions should be checked:

- **Q1.1 Geographic extent**: Is the geographical extent as expected and are there holes in the spatial coverage?
- **Q1.2 Temporal extent**: Is the temporal extent as expected and are there holes in the temporal coverage?
- **Q1.3 Spatio-temporal gaps**: Does the geographic extent vary over time or do holes appear during certain times? For example, tracking data of migratory animals is expected to exhibit seasonal changes. Such changes in vehicle tracking systems however may indicate issues with data collection.

These questions can be approached using, for example, density maps, time series plots, and animated density maps (for example, using GeoMesa's built-in density function which can be published as GeoServer WMS and animated with QGIS TimeManager) or space-time density volumes (Demšar and Virrantaus 2010). When

Fig. 12.3 Recommended EDA workflow for massive movement data

doing so, it is important to consider the analysis scale since data gaps may only be visible at certain scales. For example, if there is a systematic issue with the data collection and data is missing for a specific hour of the day, aggregated daily statistics will fail to detect this gap.

Common technical issues: Time zones

It is generally recommended to use one single, standardised time representation (preferably UTC) internally, and convert all timestamps to this representation at the earliest possible moment. This approach ensures that potential timestamp issues are identified and resolved at an early stage and is geared towards a common requirement of movement data analysis: sorting items by absolute moments in time.

12.6.2 Putting Movement Records in Context

The second exploration step puts movement records in their temporal and geographic context. The exploration includes information based on consecutive movement data records, such as time between records (sampling intervals), speed, and direction. Therefore, this step requires temporally ordered data. The following questions and assumptions should be checked:

- **Q2.1 Sampling intervals**: Is the data sampled at regular or irregular intervals?
- **Q2.2 Speed values**: Are there any unrealistic movements? For example: Does the data contain unattainable speeds?
- **Q2.3 Movement patterns**: Are there any patterns in movement direction or speed?
- **Q2.4 Temporal context**: Does the movement make sense in its temporal context? For example: Do nocturnal animal tracks show movement at night?
- **Q2.5 Geographic context**: Does the movement make sense in its geographic context? For example: Do vessels follow traffic separation schemes defined in maritime maps? Are there any ship trajectories crossing land?

Movement patterns can be approached using appropriate summarizations. For example, movement prototypes (Graser and Widhalm 2018; Graser et al. 2020) provide information about the distribution of movement speed and direction.

12.6.3 Extracting Trajectories & Events

The third exploration step looks at individual trajectories. It therefore requires that the continuous tracks are split into individual trajectories. Analysis results depend on how the continuous streams are divided into trajectories and/or events (as discussed in the segmentation challenge in Sect. 12.4.2.1). The following questions and assumptions should be checked:

- **Q3.1 Object identifiers**: Does the dataset contain stable object identifiers?
- **Q3.2 Trajectory lines**: Do the trajectory lines look plausible or are there indications of out of sequence positions or other unrealistic location jumps?
- **Q3.3 Home/depot locations**: Do day trajectories start and end at the same home (for human and animal movement) or depot (for logistics applications) location?

Splitting tracks in regular temporal cycles (creating for example, daily trajectories) or according to attribute changes (for example, when the moving object's status changes) is straightforward. However, most meaningful trajectory analyses require trajectories between consecutive stay locations. Therefore, appropriate stay detection methods that are tailored to the dataset characteristics have to be applied.

12.6.4 Exploring Patterns in Trajectory and Event Data

The fourth exploration step looks at the set of all or subsets of the trajectories and events extracted from the movement data. This step involves many of the computationally more expensive operations, such as trajectory similarity computations (discussed in Sect. 12.4.2). The following questions and assumptions should be checked:

- **Q4.1 Origins/destinations**: Are there any common trip origins or destinations? For example: Where do customers of a certain retail store live?
- **Q4.2 Similar trajectories**: Are there repeating similar trajectories?
- **Q4.3 Pairwise patterns**: Are there frequent encounters or are movers avoiding one another?
- **Q4.4 Group patterns**: Are there any patterns involving groups of movers? For example: Are there any discernible flocking or dispersal patterns?
- **Q4.5 Home ranges**: Are individual home ranges overlapping, by how much, and where/when?

Interpretation of event and trajectory patterns that goes beyond basic observations generally requires domain knowledge and therefore discussions with domain experts. It is not unusual at this point in the EDA workflow to perform multiple iterations of step 3 and 4 to evaluate different trajectory segmentation and event extraction approaches.

12.6.5 Analyzing Outliers

The fifth and final exploration steps looks at potential outliers and how they may challenge preconceived assumptions about the dataset characteristics. This step in particular regularly requires domain knowledge that exceeds the information that can be discovered from the movement data itself. The following questions and assumptions should be checked:

- **Q5.1 Anomalous movement**: Are there any unusual movements? For example: a tanker vessel travelling through a region where no tankers are usually encountered.
- **Q5.2 Anomalous spatio-temporal regions/events**: Are there any unusual spatio-temporal regions or events? For example: unusual gatherings in a certain area and time.

As indicated by the data science framework in Fig. 12.1, results from any one of the steps in this EDA workflow may trigger a feedback loop that can lead to changes in data collection, processing, cleaning, exploratory or confirmatory data analysis. Especially if data analysts are not domain experts themselves, visual tools are an important asset in communicating with other experts in interdisciplinary teams.

12.7 Conclusions

In this chapter, we have discussed exploratory data analysis (EDA) for massive movement datasets. After an introduction to EDA in general and EDA for movement data, we laid out common exploratory analysis tasks for massive movement data and provided examples from the classic GIS literature as well as examples using big data technology.

We have discussed the various challenges, including indexing, visualization, segmentation, data quality, veracity, bias, and particularly, privacy. In addition to these conceptual challenges, we observe a general lack of established EDA tools for movement data. The heterogeneity of movement data sources and application domains makes it difficult to develop general purpose EDA tools.

In light of these challenges, we proposed a five-step EDA workflow for massive movement data. This workflow can serve researchers and analysts who are confronted with new datasets in getting to know their data in a structured way. The technical implementation of the proposed workflow steps, however, is up to the individual analyst. While this chapter provides some general pointers to tools within the Hadoop and Apache Spark ecosystem, these are by no means the only option.

References

An S, Yang H, Wang J (2018) Revealing recurrent urban congestion evolution patterns with taxi trajectories. ISPRS Int J Geo-Inf 7(4):128. https://doi.org/10.3390/ijgi7040128, https://www.mdpi.com/2220-9964/7/4/128

Andrienko N, Andrienko G (2006) Exploratory analysis of spatial and temporal data: a systematic approach. Springer Science & Business Media, google-Books-ID: Oqq7oP31EycC

Andrienko N, Andrienko G (2007) Designing visual analytics methods for massive collections of movement data. Cartographica Int J Geogr Inf Geovis 42(2):117–138. https://doi.org/10.3138/carto.42.2.117

Andrienko N, Andrienko G (2011) Spatial generalization and aggregation of massive movement data. IEEE Trans Vis Comput Graph 17(2):205–219. https://doi.org/10.1109/TVCG.2010.44

Andrienko G, Andrienko N, Bak P, Keim D, Wrobel S (2013) Visual analytics of movement. Springer Science & Business Media, Berlin/Heidelberg

Andrienko G, Andrienko N, Chen W, Maciejewski R, Zhao Y (2017) Visual analytics of mobility and transportation: state of the art and further research directions. IEEE Trans Intell Transp Syst 18(8):2232–2249. https://doi.org/10.1109/TITS.2017.2683539

Aronsen M, Landmark K (2016) Density mapping of ship traffic. FFI-RAPPORT 16/02061. https://www.ffi.no/no/Rapporter/16-02061.pdf

Batran M, Mejia MG, Kanasugi H, Sekimoto Y, Shibasaki R (2018) Inferencing human spatiotemporal mobility in greater maputo via mobile phone big data mining. ISPRS Int J Geo-Inf 7(7):259. https://doi.org/10.3390/ijgi7070259, https://www.mdpi.com/2220-9964/7/7/259

Brillinger DR, Preisler HK, Ager AA, Kie JG (2004) An exploratory data analysis (EDA) of the paths of moving animals. J Stat Plann Inference 122(1–2):43–63. https://doi.org/10.1016/j.jspi.2003.06.016, https://linkinghub.elsevier.com/retrieve/pii/S0378375803002404

Brinkhoff T (2002) A framework for generating network-based moving objects. GeoInformatica 6(2):153–180. https://doi.org/10.1023/A:1015231126594, https://doi.org/10.1023/A:1015231126594

Brodie S, Lédée EJI, Heupel MR, Babcock RC, Campbell HA, Gledhill DC, Hoenner X, Huveneers C, Jaine FRA, Simpfendorfer CA, Taylor MD, Udyawer V, Harcourt RG (2018) Continental-scale animal tracking reveals functional movement classes across marine taxa. Sci Rep 8(1):3717. https://doi.org/10.1038/s41598-018-21988-5, https://www.nature.com/articles/s41598-018-21988-5

Cao H, Wachowicz M (2019) The design of an IoT-GIS platform for performing automated analytical tasks. Comput Environ Urban Syst 74:23–40. https://doi.org/10.1016/j.compenvurbsys.2018.11.004, http://www.sciencedirect.com/science/article/pii/S0198971518303284

Cao H, Wolfson O, Trajcevski G (2006) Spatio-temporal data reduction with deterministic error bounds. VLDB J 15(3):211–228. https://doi.org/10.1007/s00778-005-0163-7, http://link.springer.com/10.1007/s00778-005-0163-7

Chandola V, Banerjee A, Kumar V (2009) Anomaly detection: a survey. ACM Comput Surv 41(3):1–58. https://doi.org/10.1145/1541880.1541882, http://portal.acm.org/citation.cfm?doid=1541880.1541882

Chen M, Gao X, Li H (2010) Parallel DBSCAN with priority R-tree. In: 2010 2nd IEEE International Conference on Information Management and Engineering. IEEE, Chengdu, pp 508–511. https://doi.org/10.1109/ICIME.2010.5477926, http://ieeexplore.ieee.org/document/5477926/

Chen C, Boedihardjo AP, Jenkins BS, Ellison CL, Lin J, Senin P, Oates T (2017a) STAVIS 2.0: mining spatial trajectories via motifs. In: International Symposium on Spatial and Temporal Databases. Springer, pp 433–439. https://doi.org/10.1007/978-3-319-64367-0_30

Chen Y, Yu J, Gao Y (2017b) Detecting trajectory outliers based on spark. In: 2017 25th International Conference on Geoinformatics. IEEE, Buffalo, pp 1–5. https://doi.org/10.1109/GEOINFORMATICS.2017.8090919, http://ieeexplore.ieee.org/document/8090919/

Cook K, Grinstein G, Whiting M, Cooper M, Havig P, Liggett K, Nebesh B, Paul CL (2012) VAST challenge 2012: visual analytics for big data. In: 2012 IEEE Conference on Visual Analytics Science and Technology (VAST). IEEE, Seattle, pp 251–255. https://doi.org/10.1109/VAST.2012.6400529, http://ieeexplore.ieee.org/document/6400529/

Cordova I, Moh TS (2015) Dbscan on resilient distributed datasets. In: High Performance Computing & Simulation (HPCS), 2015 International Conference on. IEEE, pp 531–540. https://doi.org/10.1109/HPCSim.2015.7237086

Demšar U, Virrantaus K (2010) Space–time density of trajectories: exploring spatio-temporal patterns in movement data. Int J Geograph Inf Sci 24(10):1527–1542. https://doi.org/10.1080/13658816.2010.511223

Demšar U, Buchin K, Cagnacci F, Safi K, Speckmann B, Van de Weghe N, Weiskopf D, Weibel R (2015) Analysis and visualisation of movement: an interdisciplinary review. Mov Ecol 3(1):5. https://doi.org/10.1186/s40462-015-0032-y

Dobrkovic A, Iacob ME, van Hillegersberg J (2018) Maritime pattern extraction and route reconstruction from incomplete AIS data. Int J Data Sci Anal 5, 111–136. https://doi.org/10.1007/s41060-017-0092-8

Dodge S (2019) A data science framework for movement. Geograph Anal p gean.12212. https://doi.org/10.1111/gean.12212, https://onlinelibrary.wiley.com/doi/abs/10.1111/gean.12212

Dodge S, Weibel R, Ahearn SC, Buchin M, Miller JA (2016) Analysis of movement data. Int J Geograph Inf Sci 30(5):825–834. https://doi.org/10.1080/13658816.2015.1132424

Douglas DH, Peucker TK (1973) Algorithms for the reduction of the number of points required to represent a digitized line or its caricature. Cartograph Int J Geograph Inf Geovisual 10(2):112–122. https://doi.org/10.3138/FM57-6770-U75U-7727, https://utpjournals.press/doi/10.3138/FM57-6770-U75U-7727

Douriez M, Doraiswamy H, Freire J, Silva CT (2016) Anonymizing NYC taxi data: does it matter? In: 2016 IEEE International Conference on Data Science and Advanced Analytics (DSAA). IEEE, Montreal, pp 140–148. https://doi.org/10.1109/DSAA.2016.21, http://ieeexplore.ieee.org/document/7796899/

Duckham M, Kulik L (2006) Location privacy and location-aware computing. In: Drummond, J et al. (eds) Dynamic & mobile GIS: Investigating change in space and time. CRC Press: Boca Raton, p 34–51

Dwork C (2006) Differential privacy. In: Proceedings of the 33rd International Conference on Automata, Languages and Programming (ICALP'06). Springer, pp 1–12

Edelhoff H, Signer J, Balkenhol N (2016) Path segmentation for beginners: an overview of current methods for detecting changes in animal movement patterns. Movement Ecology 4(1):21. https://doi.org/10.1186/s40462-016-0086-5, http://movementecologyjournal.biomedcentral.com/articles/10.1186/s40462-016-0086-5

ElSalamouny E, Gambs S (2016) Differential privacy models for location-based services. Trans Data Priv 9(1):15–48. https://hal.inria.fr/hal-01418136

Fecher R, Whitby M (2017) Optimizing spatiotemporal analysis using multidimensional indexing with GeoWave. Free and Open Source Software for Geospatial (FOSS4G) Conference Proceedings 17(1). https://doi.org/10.7275/R5639MXD, https://scholarworks.umass.edu/foss4g/vol17/iss1/12

Filipiak D, Strózyna M, Wecel K, Abramowicz W (2018) Anomaly detection in the maritime domain: comparison of traditional and big data approach. https://www.sto.nato.int/publications/STO%20Meeting%20Proceedings/STO-MP-IST-160/MP-IST-160-S2-5.pdf

Fillekes MP, Röcke C, Katana M, Weibel R (2019) Self-reported versus GPS-derived indicators of daily mobility in a sample of healthy older adults. Soc Sci Med 220:193–202. https://doi.org/10.1016/j.socscimed.2018.11.010, http://www.sciencedirect.com/science/article/pii/S0277953618306440

Fox A, Eichelberger C, Hughes J, Lyon S (2013) Spatio-temporal indexing in non-relational distributed databases. In: 2013 IEEE International Conference on Big Data, pp 291–299. https://doi.org/10.1109/BigData.2013.6691586

Fréchet MM (1906) Sur quelques points du calcul fonctionnel. Rendiconti del Circolo Matematico di Palermo (1884–1940) 22(1):1–72

Fritsche C, Klein A, Wurtz D (2009) Hybrid GPS/GSM localization of mobile terminals using the extended Kalman filter. In: 2009 6th Workshop on Positioning, Navigation and Communication. IEEE, Hannover, pp 189–194. https://doi.org/10.1109/WPNC.2009.4907826, http://ieeexplore.ieee.org/document/4907826/

Gao S (2015) Spatio-temporal analytics for exploring human mobility patterns and urban dynamics in the mobile age. Spatial Cogn Comput 15(2):86–114. https://doi.org/10.1080/13875868.2014.984300

Goh ST, Abdelkhalik O, Zekavat SAR (2013) A weighted measurement fusion Kalman filter implementation for UAV navigation. Aerosp Sci Technol 28(1):315–323. https://doi.org/10.1016/j.ast.2012.11.012, https://linkinghub.elsevier.com/retrieve/pii/S1270963812001976

Graser A (2018) Evaluating spatio-temporal data models for trajectories in PostGIS databases. GI_Forum – J Geograph Inf Sci 2018(1):16–33. https://doi.org/10.1553/giscience2018_01_s16

Graser A (2019) MovingPandas: efficient structures for movement data in Python. GI_Forum – J Geograph Inf Sci 2019(1):54–68. https://doi.org/10.1553/giscience2019_01_s54

Graser A (2020) Data science workflow framework. https://doi.org/10.6084/m9.figshare.11638368.v1, https://figshare.com/articles/Data_Science_Workflow_Framework/11638368

Graser A, Widhalm P, Dragaschnig M (2020) The M^{3}massive movement model: a distributed incrementally updatable solution for big movement data exploration. Int J Geogr Inf Sci 34(12):2517–2540. https://doi.org/10.1080/13658816.2020.1776293

Graser A, Widhalm P (2018) Modelling massive AIS streams with quad trees and gaussian mixtures. In: Mansourian A et al (ed) Geospatial technologies for all, Lund. https://agile-online.org/index.php/conference/proceedings/proceedings-2018

Gudmundsson J, Katajainen J, Merrick D, Ong C, Wolle T (2009) Compressing spatio-temporal trajectories. Comput Geom 42(9):825–841. https://doi.org/10.1016/j.comgeo.2009.02.002, https://linkinghub.elsevier.com/retrieve/pii/S0925772109000248

Gudmundsson J, Laube P, Wolle T (2011) Computational movement analysis. In: Kresse W, Danko DM (eds) Springer handbook of geographic information. Springer, Berlin/Heidelberg, pp 423–438. https://doi.org/10.1007/978-3-540-72680-7_22

Hausdorff F (1914) Grundzüge der Mengenlehre

Hawkins DM (1980) Identification of outliers, vol 11. Springer, Dordrecht

He Y, Tan H, Luo W, Feng S, Fan J (2014) MR-DBSCAN: a scalable MapReduce-based DBSCAN algorithm for heavily skewed data. Front Comput Sci 8(1):83–99. https://doi.org/10.1007/s11704-013-3158-3

Hughes JN, Annex A, Eichelberger CN, Fox A, Hulbert A, Ronquest M (2015) GeoMesa: a distributed architecture for spatio-temporal fusion. Baltimore, p 94730F. https://doi.org/10.1117/12.2177233, http://proceedings.spiedigitallibrary.org/proceeding.aspx?doi=10.1117/12.2177233

IEA (2018) CO2 emissions from fuel combustion 2018 highlights. Technical report. https://webstore.iea.org/co2-emissions-from-fuel-combustion-2018-highlights

Janetzko H, Jäckle D, Deussen O, Keim DA (2013) Visual abstraction of complex motion patterns. San Francisco, p 90170J. https://doi.org/10.1117/12.2035959, http://proceedings.spiedigitallibrary.org/proceeding.aspx?doi=10.1117/12.2035959

Kalman RE (1960) A new approach to linear filtering and prediction problems. Trans ASME–J Basic Eng 82(Series D):35–45

Krumm J (2007) Inference attacks on location tracks. In: International Conference on Pervasive Computing. Springer, pp 127–143

Kuijpers B, Othman W (2010) Trajectory databases: data models, uncertainty and complete query languages. J Comput Syst Sci 76(7):538–560. https://doi.org/10.1016/j.jcss.2009.10.002, https://linkinghub.elsevier.com/retrieve/pii/S0022000009000919

Lämmel R (2008) Google's MapReduce programming model – revisited. Sci Comput Programm 70(1):1–30. https://doi.org/10.1016/j.scico.2007.07.001, http://www.sciencedirect.com/science/article/pii/S0167642307001281

Langley RB, others (1999) Dilution of precision. GPS World 10(5):52–59

Laxhammar R, Falkman G (2011) Sequential conformal anomaly detection in trajectories based on hausdorff distance. In: Information Fusion (FUSION), 2011 Proceedings of the 14th International Conference on. IEEE, pp 1–8

Laxhammar R, Falkman G, Sviestins E (2009) Anomaly detection in sea traffic-a comparison of the gaussian mixture model and the kernel density estimator. In: Information Fusion, 2009. FUSION'09. 12th International Conference on. IEEE, pp 756–763

Leetaru K, Wang S, Cao G, Padmanabhan A, Shook E (2013) Mapping the global Twitter heartbeat: the geography of Twitter. First Monday 18(5). https://doi.org/10.5210/fm.v18i5.4366, https://journals.uic.edu/ojs/index.php/fm/article/view/4366

Lei PR (2016) A framework for anomaly detection in maritime trajectory behavior. Knowl Inf Syst 47(1):189–214. https://doi.org/10.1007/s10115-015-0845-4

Li L, Goodchild MF, Xu B (2013) Spatial, temporal, and socioeconomic patterns in the use of Twitter and Flickr. Cartogr Geogr Inf Sci 40(2):61–77. https://doi.org/10.1080/15230406.2013.777139

Li S, Dragicevic S, Castro FA, Sester M, Winter S, Coltekin A, Pettit C, Jiang B, Haworth J, Stein A, Cheng T (2016) Geospatial big data handling theory and methods: a review and research challenges. ISPRS J Photogramm Remote Sens 115:119–133. https://doi.org/10.1016/j.isprsjprs.2015.10.012, http://www.sciencedirect.com/science/article/pii/S0924271615002439

Li Y, Li Q, Shan J (2017) Discover patterns and mobility of twitter users—a study of four US college cities. ISPRS Int J Geo-Inf 6(2):42. https://doi.org/10.3390/ijgi6020042, https://www.mdpi.com/2220-9964/6/2/42

Long JA, Weibel R, Dodge S, Laube P (2018) Moving ahead with computational movement analysis. Int J Geograph Inf Sci 32(7):1275–1281. https://doi.org/10.1080/13658816.2018.1442974

Lovelace R, Birkin M, Cross P, Clarke M (2016) From big noise to big data: toward the verification of large data sets for understanding regional retail flows. Geograph Anal 48(1):59–81. https://doi.org/10.1111/gean.12081, https://onlinelibrary.wiley.com/doi/abs/10.1111/gean.12081

Ma Q, Yang B, Qian W, Zhou A (2009) Query processing of massive trajectory data based on Mapreduce. In: Proceedings of the First International Workshop on Cloud Data Management, CloudDB'09. ACM, New York, pp 9–16. https://doi.org/10.1145/1651263.1651266, http://doi.acm.org/10.1145/1651263.1651266, event-place: Hong Kong

Machanavajjhala A, Kifer D, Gehrke J, Venkitasubramaniam M (2007) *L* -diversity: privacy beyond *k* -anonymity. ACM Trans Knowl Discovery Data 1(1):3–es. https://doi.org/10.1145/1217299.1217302, http://portal.acm.org/citation.cfm?doid=1217299.1217302

Meier S (2017) Personal big data: a privacy-centred selective cloud computing approach to progressive user modelling on mobile devices. Ph.D. Thesis, Universität Potsdam, Mathematisch-Naturwissenschaftliche Fakultät

Miller HJ, Bridwell SA (2009) A field-based theory for time geography. Ann Assoc Am Geograph 99(1):49–75. https://doi.org/10.1080/00045600802471049

Miller HJ, Dodge S, Miller J, Bohrer G (2019) Towards an integrated science of movement: converging research on animal movement ecology and human mobility science. Int J Geograph Inf Sci 33(5):855–876. https://doi.org/10.1080/13658816.2018.1564317, https://www.tandfonline.com/doi/full/10.1080/13658816.2018.1564317

Mislove A, Lehmann S, Ahn YY, Onnela JP, Rosenquist JN (2011) Understanding the demographics of twitter users. In: Fifth International AAAI Conference on Weblogs and Social Media

Nikitopoulos P, Paraskevopoulos A, Doulkeridis C, Pelekis N, Theodoridis Y (2018) Hot spot analysis over big trajectory data. In: 2018 IEEE International Conference on Big Data (Big Data), pp 761–770. https://doi.org/10.1109/BigData.2018.8622376

OGC (2011) OpenGIS® implementation standard for geographic information – simple feature access – part 1: common architecture. https://www.opengeospatial.org/standards/sfa

Oksanen J, Bergman C, Sainio J, Westerholm J (2015) Methods for deriving and calibrating privacy-preserving heat maps from mobile sports tracking application data. J Transp Geograp 48:135–144. https://doi.org/10.1016/j.jtrangeo.2015.09.001, https://linkinghub.elsevier.com/retrieve/pii/S0966692315001647

Pallotta G, Vespe M, Bryan K (2013) Vessel pattern knowledge discovery from AIS data: a framework for anomaly detection and route prediction. Entropy 15(6):2218–2245. https://doi.org/10.3390/e15062218

Patroumpas K, Alevizos E, Artikis A, Vodas M, Pelekis N, Theodoridis Y (2017) Online event recognition from moving vessel trajectories. GeoInformatica 21(2):389–427. https://doi.org/10.1007/s10707-016-0266-x

Ranjit S, Witayangkurn A, Nagai M, Shibasaki R (2018) Agent-based modeling of taxi behavior simulation with probe vehicle data. ISPRS Int J Geo-Inf 7(5):177. https://doi.org/10.3390/ijgi7050177, https://www.mdpi.com/2220-9964/7/5/177

Riveiro M, Pallotta G, Vespe M (2018) Maritime anomaly detection: a review. Wiley Interdiscip Rev Data Min Knowl Disc 8(5):e1266. https://doi.org/10.1002/widm.1266, http://doi.wiley.com/10.1002/widm.1266

Robinson AC, Demšar U, Moore AB, Buckley A, Jiang B, Field K, Kraak MJ, Camboim SP, Sluter CR (2017) Geospatial big data and cartography: research challenges and opportunities for making maps that matter. Int J Cartograph 1–29. http://www.tandfonline.com/doi/abs/10.1080/23729333.2016.1278151

Scheepens R, Willems N, Wetering HVD, Wijk JJV (2011) Interactive visualization of multivariate trajectory data with density maps. In: 2011 IEEE Pacific Visualization Symposium, pp 147–154. https://doi.org/10.1109/PACIFICVIS.2011.5742384

Schutt C, O'Neil R (2013) Doing data science. O'Reilly. http://shop.oreilly.com/product/0636920028529.do

Seyyedhasani H, Dvorak JS, Sama MP, Stombaugh TS (2016) Mobile device-based location services accuracy. Appl Eng Agric 32(5):539–547. https://doi.org/10.13031/aea.32.11351

Shang Z, Li G, Bao Z (2018) DITA: distributed in-memory trajectory analytics. In: Proceedings of the 2018 International Conference on Management of Data, SIGMOD'18. ACM, New York, pp 725–740. https://doi.org/10.1145/3183713.3183743, http://doi.acm.org/10.1145/3183713.3183743, event-place: Houston

Shelmerdine RL (2015) Teasing out the detail: how our understanding of marine AIS data can better inform industries, developments, and planning. Mar Policy 54:17–25. https://doi.org/10.1016/j.marpol.2014.12.010, http://www.sciencedirect.com/science/article/pii/S0308597X14003479

Shi X, Yu Z, Fang Q, Zhou Q (2017) A visual analysis approach for inferring personal job and housing locations based on public bicycle data. ISPRS Int J Geo-Inf 6(7):205. https://doi.org/10.3390/ijgi6070205, https://www.mdpi.com/2220-9964/6/7/205

Sidibé A, Shu G (2017) Study of automatic anomalous behaviour detection techniques for maritime vessels. J Navig 70(4):847–858. https://doi.org/10.1017/S0373463317000066

Sweeney L (2002) k-anonymity: a model for protecting privacy. Int J Uncertain Fuzziness Knowl Based Syst 10(05):557–570. https://doi.org/10.1142/S0218488502001648, https://www.worldscientific.com/doi/abs/10.1142/S0218488502001648

Tampakis P, Doulkeridis C, Pelekis N, Theodoridis Y (2019) Distributed subtrajectory join on massive datasets. arXiv:190307748 [cs] http://arxiv.org/abs/1903.07748, arXiv: 1903.07748

Tao Y (2016) Data modeling for checkpoint-based movement data. In: GIScience 2016 Workshop on Analysis of Movement Data (AMD'16), 27 September 2016, Montreal

Taubenböck H, Staab J, Zhu X, Geiß C, Dech S, Wurm M (2018) Are the poor digitally left behind? Indications of urban divides based on remote sensing and twitter data. ISPRS Int J Geo-Inf 7(8):304. https://doi.org/10.3390/ijgi7080304, http://www.mdpi.com/2220-9964/7/8/304

Tockar A (2014) Riding with the stars: passenger privacy in the NYC Taxicab dataset. https://research.neustar.biz/2014/09/15/riding-with-the-stars-passenger-privacy-in-the-nyc-taxicab-dataset/

Tresata (2019) Secondary sort and streaming reduce for Apache Spark: tresata/spark-sorted. https://github.com/tresata/spark-sorted, original-date: 2015-03-06T16:04:27Z

Tukey JW (1977) Exploratory data analysis. Addison-Wesley, Reading/Menlo Park/London/Amsterdam

Vahedian A, Zhou X, Tong L, Li Y, Luo J (2017) Forecasting gathering events through continuous destination prediction on big trajectory data. In: Proceedings of the 25th ACM SIGSPATIAL International Conference on Advances in Geographic Information Systems. ACM, p 34. https://doi.org/10.1145/3139958.3140008

Wachowicz M, Liu T (2016) Finding spatial outliers in collective mobility patterns coupled with social ties. Int J Geograph Inf Sci 30(9):1806–1831. https://doi.org/10.1080/13658816.2016.1144887, http://www.tandfonline.com/doi/full/10.1080/13658816.2016.1144887

Wang H, Belhassena A (2017) Parallel trajectory search based on distributed index. Inf Sci 388–389:62–83. https://doi.org/10.1016/j.ins.2017.01.016, http://www.sciencedirect.com/science/article/pii/S0020025517300178

Wang F, Chen C (2018) On data processing required to derive mobility patterns from passively-generated mobile phone data. Transp Res Part C Emerg Technol 87:58–74. https://doi.org/10.1016/j.trc.2017.12.003, https://linkinghub.elsevier.com/retrieve/pii/S0968090X17303637

Wang Y, Wu L (2017) A functional model of AIS data fusion. In: International Conference on Intelligence Science. Springer, pp 191–199. https://doi.org/10.1007/978-3-319-68121-4_20

Wang X, Liu X, Liu B, Souza ENd, Matwin S (2014) Vessel route anomaly detection with Hadoop MapReduce. In: 2014 IEEE International Conference on Big Data (Big Data), pp 25–30. https://doi.org/10.1109/BigData.2014.7004464

Wang Y, Qin K, Chen Y, Zhao P (2018) Detecting anomalous trajectories and behavior patterns using hierarchical clustering from taxi GPS data. ISPRS Int J Geo-Inf 7(1):25. https://doi.org/10.3390/ijgi7010025, http://www.mdpi.com/2220-9964/7/1/25

WHO (2018) Global status report on road safety 2018. Technical report, World Health Organization, Geneva. https://apps.who.int/iris/bitstream/handle/10665/276462/9789241565684-eng.pdf

Widhalm P, Yang Y, Ulm M, Athavale S, González MC (2015) Discovering urban activity patterns in cell phone data. Transportation 42(4):597–623. https://doi.org/10.1007/s11116-015-9598-x, http://link.springer.com/10.1007/s11116-015-9598-x

Willems N, Wetering HVD, Wijk JJV (2009) Visualization of vessel movements. Comput Graph Forum 28(3):959–966. https://doi.org/10.1111/j.1467-8659.2009.01440.x, https://onlinelibrary.wiley.com/doi/abs/10.1111/j.1467-8659.2009.01440.x

Witayangkurn A, Horanont T, Sekimoto Y, Shibasaki R (2013) Anomalous event detection on large-scale GPS data from mobile phones using hidden markov model and cloud platform. In: Proceedings of the 2013 ACM Conference on Pervasive and Ubiquitous Computing Adjunct Publication – UbiComp'13 adjunct. ACM Press, Zurich, pp 1219–1228. https://doi.org/10.1145/2494091.2497352, http://dl.acm.org/citation.cfm?doid=2494091.2497352

Wu L, Hu S, Yin L, Wang Y, Chen Z, Guo M, Chen H, Xie Z (2017) Optimizing cruising routes for taxi drivers using a spatio-temporal trajectory model. ISPRS Int J Geo-Inf 6(11):373. https://doi.org/10.3390/ijgi6110373, https://www.mdpi.com/2220-9964/6/11/373

Xie X, Mei B, Chen J, Du X, Jensen CS (2016) Elite: an elastic infrastructure for big spatiotemporal trajectories. VLDB J 25(4):473–493. https://doi.org/10.1007/s00778-016-0425-6

Xie D, Li F, Phillips JM (2017) Distributed trajectory similarity search. Proc VLDB Endow 10(11):1478–1489. https://doi.org/10.14778/3137628.3137655

Xu T, Zhang X, Claramunt C, Li X (2018) TripCube: a trip-oriented vehicle trajectory data indexing structure. Comput Environ Urban Syst 67:21–28. https://doi.org/10.1016/j.compenvurbsys.2017.08.005, http://www.sciencedirect.com/science/article/pii/S0198971516303921

Yu J, Sarwat M (2017) Indexing the pickup and drop-off locations of NYC taxi trips in PostgreSQL – lessons from the road. In: Gertz M, Renz M, Zhou X, Hoel E, Ku WS, Voisard A, Zhang C, Chen H, Tang L, Huang Y, Lu CT, Ravada S (eds) Advances in spatial and temporal databases. Lecture notes in computer science. Springer International Publishing, Cham, pp 145–162. https://doi.org/10.1007/978-3-319-64367-0_8

Zhang Z, Jin C, Mao J, Yang X, Zhou A (2017) TrajSpark: a scalable and efficient in-memory management system for big trajectory data. In: Chen L, Jensen CS, Shahabi C, Yang X, Lian X (eds) Web and big data. Lecture notes in computer science. Springer International Publishing, Cham, pp 11–26. https://doi.org/10.1007/978-3-319-63579-8_2

Zhao Z, Shaw SL, Xu Y, Lu F, Chen J, Yin L (2016) Understanding the bias of call detail records in human mobility research. Int J Geograph Inf Sci 30(9):1738–1762. https://doi.org/10.1080/13658816.2015.1137298

Zheng Y (2015) Trajectory data mining: an overview. ACM Trans Intell Syst Technol 6(3):1–41. https://doi.org/10.1145/2743025, http://dl.acm.org/citation.cfm?doid=2764959.2743025

Zheng Y, Zhang L, Xie X, Ma WY (2009) Mining interesting locations and travel sequences from GPS trajectories. In: Proceedings of the 18th International Conference on World wide web. ACM, pp 791–800. https://doi.org/10.1145/1526709.1526816

Zheng Y, Zhang H, Yu Y (2015) Detecting collective anomalies from multiple spatio-temporal datasets across different domains. In: Proceedings of the 23rd SIGSPATIAL International Conference on Advances in Geographic Information Systems – GIS'15. ACM Press, Bellevue, pp 1–10. https://doi.org/10.1145/2820783.2820813, http://dl.acm.org/citation.cfm?doid=2820783.2820813

Part III
Statistics, Uncertainty and Data Quality

Chapter 13
Spatio-Temporal Data Quality: Experience from Provision of DOT Traveler Information

Douglas Galarus, Ian Turnbull, Sean Campbell, Jeremiah Pearce, and Leann Koon

13.1 Introduction

The motivation for this chapter stems from a 15-year collaboration between the California Department of Transportation (Caltrans) and the Western Transportation Institute, and now Utah State University. Ian Turnbull, recently retired from Caltrans, set the standard of "accurate, timely and reliable" for all projects he was involved with, particularly those in remote, rural areas of California, and related collaborations. Ian's team, now headed by Jeremiah Pearce, is responsible for the operation of a vast field sensor network and the provision of weather sensor data, CCTV images, chain control messages, changeable message sign (CMS) messages, and just about every other type of traveler information related to Caltrans District 2, particularly in remote, rural areas of Northeastern California. In turn, Sean Campbell, also from Caltrans, makes this and similar data from the rest of the state available to third-party providers of traveler information and other related applications. The experience gained in working with Ian, Sean, Jeremiah and others on the provision of traveler information and the development of systems for DOT

D. Galarus (✉)
Department of Computer Science, Montana Technological University Butte, Butte, MT, USA
e-mail: dgalarus@mtech.edu

I. Turnbull · J. Pearce
Office of ITS Engineering and Support, Caltrans District 2, Redding, CA, USA

S. Campbell
ITS Special Projects Branch, Caltrans Division of Research, Innovation, and Systems Information, Redding, CA, USA

L. Koon
Western Transportation Institute, Montana State University, Bozeman, MT, USA

M. Werner, Y.-Y. Chiang (eds.), *Handbook of Big Geospatial Data*,
https://doi.org/10.1007/978-3-030-55462-0_13

operations and maintenance personal has been invaluable for recognizing the need for and the challenges of providing quality spatio-temporal data.

The transportation industry has shifted from building more infrastructure to "smarter" operation of existing roadways so-as to better manage challenges including congestion, inclement weather, maintenance, increased traffic volumes, etc. This shift has been accompanied by a rapid advance in technologies and an increased demand for more high quality, real-time traveler information (Margiotta 2002; Schuman 2001). This need will be magnified with the plan for massive Federal infrastructure investment by the current administration and compounded by advancement in sensing technology associated with the Internet of Things as well as the push for connected and autonomous vehicles. The standard of "accurate, timely and reliable" remains elusive in this field despite best intentions, and surely is problematic in other areas that treat data quality as a pre-processing activity.

Data quality for traveler information has generally been handled on an ad-hoc basis, with little or no provision for error notification other than sometimes through user-reporting of observed errors. The quality of data - for example, whether it is accurate, timely, and reliable - is a crucial consideration for the provision and use of traveler information. When drivers access traveler information that is up to date, correct, and accessible every time they need it, they will use it to make travel decisions which ultimately impact traffic management effectiveness (Robinson et al. 2012). However, if travelers access traveler information and see old camera images, obviously incorrect weather conditions (e.g., warm temperatures when it is cold everywhere else in near proximity), or misspellings on sign messages that change the meaning of the message, then users are less likely to make travel decisions based on the traveler information and they will not trust it. This can significantly diminish the effectiveness of traffic management efforts. Even worse, if drivers (or automated systems) use incorrect information to make travel-related decisions, more serious consequences such as death may follow.

Our project team conducted related work that was sponsored by Caltrans under the auspices of the Western States Rural Transportation Consortium as a technology incubator project (Galarus et al. 2018). The goal of that project was to analyze and document existing system best practices for data quality for the aggregation and dissemination of state department of transportation traveler information. The research team conducted a survey of DOT practitioners in western states, as well as a literature review on data quality within the transportation field. "Best practices" were documented. Recommendations and next steps were formulated based on applicability to Caltrans traveler information data and processes.

Neither the survey of DOT practitioners nor the literature review identified a comprehensive, well-defined plan for unified, multi-dimensional approaches to quality assurance of traveler information. However, all DOT practitioners that were surveyed as well as the literature reviewed relative to data quality in transportation indicated that quality data was important for safe, efficient operation of the transportation system, including provision of traveler information. This observation is especially valid given the current environment that is increasingly focused on performance measurements, accountability and "smarter" operation of roadways, and is rapidly progressing with connected and autonomous vehicle initiatives.

Recommendations were made considering Caltrans' traveler information data and processes. It was recommended that relevant quality metrics and requirements should be clearly defined. This includes how to determine that requirements are being met with quality data. Common statewide standards for data quality, performance, maintenance, and calibration should be defined and established using an engineering approach. These standards should be tied to all specific uses of the data.

In this chapter, we present information from our experience with the challenges of data quality for traveler information:

- Concrete examples of spatio-temporal data quality problems.
- Spatio-temporal data quality attributes.
- State of the practice methods for assessing spatio-temporal data quality.
- Enhanced methods for assessing spatio-temporal data quality, including identification of bad metadata.
- Motivation for further research and development in this area.

While motivated by our transportation applications, these topics should be of general interest and use to practitioners and researchers alike.

13.2 Example Data Quality Problems

The process of providing accurate, timely, and reliable traveler information that effectively impacts traffic management, safety, and operations, is complex and rife with challenges. One of the core problems is determining which data quality descriptors to apply and how they should be used relative to traveler information, e.g. determining what to measure and how to measure it. In September 2000, ITS America and the U.S. Department of Transportation established guidelines for traveler information data collection and quality. In the introduction to the report, the authors commented that the early vision for traveler information was simple – data would be collected by public agencies and disseminated to various devices and media outlets. However, they stated that, "In hindsight, it is clear that the difficulty of collecting good complete and timely data, transforming data into information, then packaging, marketing and communicating that information to people and devices was underestimated" (I. A. A. C. Steering Committee 2000). In the era of big data and rapidly changing technology, this statement still holds true, possibly more so now than ever.

Data quality is a crucial consideration in the process of aggregating and disseminating meaningful traveler information. Potential issues and problems with traveler information data include but are not limited to:

- incorrect, erroneous or missing sensor data (temperature, precipitation, surface condition, etc.);
- bad meta data such as incorrect locations and timestamps;
- old, frozen, partial, poorly lit, poorly positioned or unavailable CCTV images;
- device settings visible to the public;
- etc.

In November 2010, a final rule was published establishing the Real-Time System Management Information Program (23 CFR 511) in accordance with Section 1201 of the Safe, Accountable, Flexible, Efficient Transportation Equity Act: A Legacy for Users (SAFETEA-LU) (USDOT n.d.). "The Real-Time System Management Information Program (was) to provide the capability to monitor in real-time the traffic and travel conditions of the major highways across the U.S. and provide a means of sharing these data with state and local governments and with the traveling public" (FHWA, Federal Highway Administration 2015). It provided a foundation for basic traveler information and data exchange formats and established minimum requirements for real-time traffic and road condition information for construction activities, road or lane blocking incidents, road weather observations, travel times, information accuracy, and information availability. It also specified that a real-time information program was to be 85 percent accurate at a minimum or have a maximum error rate of 15 percent (U. S. D. of T. F. H. Administration). But, no methods for measuring accuracy or other quality metrics were included. Nor did the program define metrics for specific elements such as RWIS or CMS. Instead, individual states were given the "flexibility to use methods appropriate to systems and processes used to acquire information and data" (Trachy et al. 2016).

In our survey of DOT practitioners, one state DOT representative commented relative to quality dimensions and meeting the RTSMIP requirements, "This has been problematic - how to measure accuracy of ... e.g., incident data ... we're reporting what we know but how do we know if that's all." In other words, the requirements can be met, but the quality of the information can't be quantified or defined. i.e., the requirements can be met with garbage data. As a result, DOTs find themselves challenged in assessing the quality of their traveler information data. Following are several examples of problematic traveler information from the One-Stop-Shop for Rural Traveler Information (http://oss.weathershare.org/).

In Fig. 13.1, an Arizona DOT CMS message, shows how a single missing letter dramatically changes the meaning of a message. Figure 13.2 shows a Caltrans CCTV image including configuration settings. The positioning of the camera makes it difficult to determine what else is being shown. Figure 13.3 shows an incorrect pavement surface temperature from an Oregon Department of Transportation RWIS, indicating freezing conditions when freezing is not present. Figure 13.4 shows a CMS message from Caltrans District 1 that is verified by Fig. 13.5, which shows a camera image of the CMS displaying this message. Caltrans District 1 has deployed cameras in proximity to CMS to help verify that messages are correctly displayed on CMS. This example is shown not because camera verification of sign messages is a preferred approach, but to demonstrate the concern from DOT staff that messages

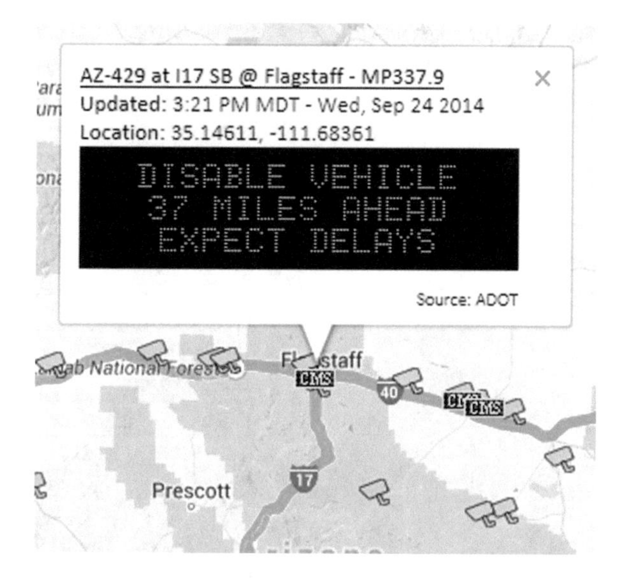

Fig. 13.1 Arizona DOT CMS message with a typo

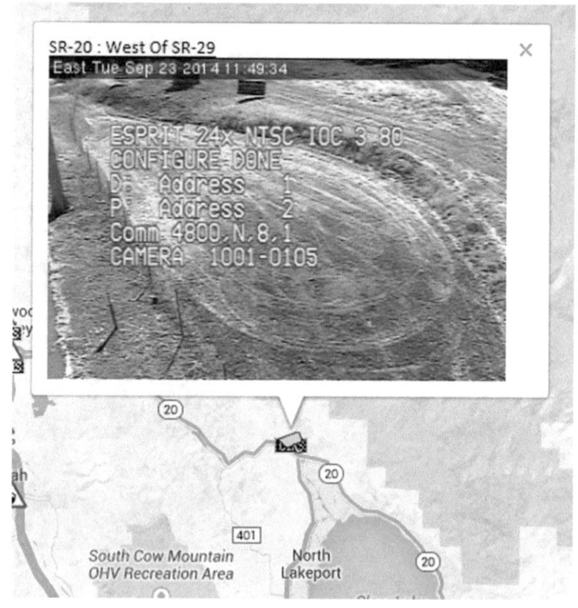

Fig. 13.2 Caltrans CCTV image with camera settings

may not be properly displayed on the physical signs. In Fig. 13.6, a Montana DOT camera is incorrectly located approximately 20 miles south of its actual location along Interstate 90. As a result, automated placement of these cameras on map-based traveler information systems is problematic, and detection likely requires user recognition of the problem.

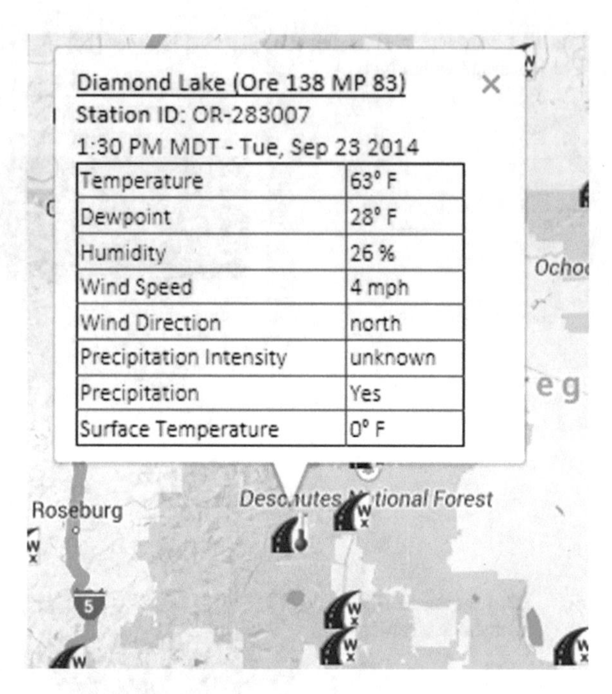

Fig. 13.3 Incorrect ODOT RWIS surface temperature

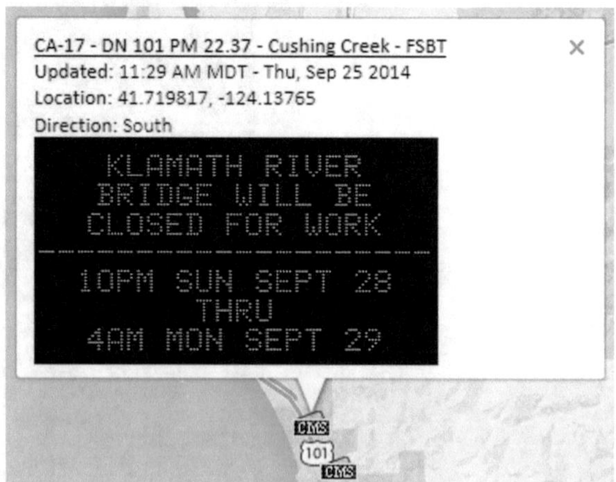

Fig. 13.4 Caltrans District 1 CMS message

Taken individually, these errors may seem anecdotal. But many errors are present at any time, indicating that there is a systemic problem that affects all state DOTs, with numerous points of failure.

Fig. 13.5 Caltrans District 1 CCTV image to confirm CMS message

Fig. 13.6 Bad location for Montana DOT CCTV image

13.3 Data Quality Attributes

Data quality from the perspective of the consumer is presented subjectively by Wang and Strong (1996) as a comprehensive framework of data quality attributes. They describe the accuracy dimension as having the following attributes: "data are certified error-free, accurate, correct, flawless, reliable, errors can be easily identified, the integrity of the data, and precise." Timeliness has a single entry in its attribute list: age of data. Reliability is mentioned throughout their paper, but it is neither given a definition nor attributes. They present many other data quality attributes and specifically mention precision, currency, completeness, relevancy, accessibility and interoperability in the context of information systems.

Batini et al. (2009) present a more recent survey and summary of data quality dimensions, and point out varying definitions for dimensions such as timeliness and completeness. They define several types of accuracy including "closeness of value ... to the elements of the corresponding definition domain." They include timeliness with currency and volatility as time-related dimensions and assert that "there is no agreement on the abstract definition of time-related (data quality) dimensions; typically, currency and timeliness are often used to refer to the same concept." They too mention reliability but provide no definition. Completeness and consistency are also mentioned as popular measures.

Luebbers et al. (2003) develop data mining tools to assist in data quality assessment, and present a definition for data auditing that includes measurement and improvement of data quality. Bisdikian et al. (2007) present overlap and differences between "Quality of Data" and "Quality of Information" (QoI). While these papers are useful in general terms, they do not include specific, comprehensive measures that can be applied to the spatio-temporal data quality challenges that we have encountered.

Devillers and Jeansoulin (2010) provide a comprehensive review of spatial data quality, including treatment of temporal aspects, and distinguish between internal and external quality. Internal quality includes dimensions such as accuracy, completeness and consistency, while external quality is defined as fitness for use or purpose. They also cite and expand on prior work from Bédard and Vallière (1995), which presented six characteristics of external quality for geospatial databases: definition, coverage, lineage, precision, legitimacy, and accessibility. They distinguish between spatial, temporal and thematic accuracy.

Work from Shi et al. in (2003a) is relevant because it presents sources of uncertainty in spatial-data mining, and these sources can also be viewed as sources of data quality problems. While these papers provide general guidance to us, they do not provide specific measures and algorithms that address our spatio-temporal situation.

In Aggarwal (2013), Aggarwal defines Data Cleaning as "Given a data set, remove discordants from it. Correct any errors in the data if possible.". Data cleaning is presented by Sathe et al. in the context of pull-based and push-based data acquisition in Sathe et al. (2013), along with a model-based approach to outlier/anomaly detection. Ives et al. (1999) present an adaptive query system for systems integrating overlapping data sources, including query optimization, while Sofra et al. (2008) investigate the trade-offs between accuracy and timeliness of information acquired in a data aggregation network. Also from the networking domain, the work presented by Charbiwala et al. in (2009) focuses on rate control guided by Quality of Information (QoI) measures. They indicate that such efforts are highly application-dependent. Another network-related publication, this one by Hermans et al., (2009) presents four components of data quality: accuracy, consistency, timeliness and completeness. Timeliness is expressed principally as a network phenomenon. Fugini et al. (2002) define completeness, currency, internal consistency, timeliness, importance, source reliability and confidentiality for cooperative web information systems. These definitions are general and useful conceptually but require further definition specific to spatio-temporal data to be of direct help to us.

Klein et al. present work in relation to the transfer and management challenges related to the inclusion of quality control information in data streams and develop optimal, quality-based load-shedding for data streams in (Klein and Lehner 2009a, b; Klein 2007; Klein et al. 2007; Klein and Hackenbroich n.d.). Specific measures presented include accuracy, confidence, completeness, data volume and timeliness, and all are presented in relation to sensor data streams. A missing component in these works relative to ours is an accounting for the spatial aspect.

For Klein et al., data is considered and managed as individual streams. In our work, it is important to not only consider data streams from individual sites and sensors but the collective of all sites and sensors and their interrelationships. We are generally working with batches of data that cover relatively short amounts of time. Data from one site may be in error while data from another nearby site may be good. Specific measures are presented by Klein et al. and are of use as examples, while some such as completeness have apparent short-comings for our application which we subsequently address. In several applications, we wish to evaluate overlapping data providers, and there is no direct mechanism to do so here. In subsequent work, Klein et al. (Jerzak et al. 2011; O'donovan et al. 2013) incorporate their data quality measures into a larger middleware architecture named GINSENG, intended for performance monitoring and control of sensor networks. The specific measures used are like those presented by Klein in prior work.

Quality of Service (QoS) is used by Tatbul for load-shedding in (2002) while noting that conflicting objectives are common. Similar work is presented in the context of operator scheduling by Carney et al. (Carney et al. 2003). Mokbel et al. (2004) present load-shedding for spatio-temporal data streams, but they do not specifically address quality control measures. Other work regarding load-shedding for data streams can be found in Babcock et al. (Babcock et al. 2004, 2007), Nehme and Rundensteiner (2007), and Tatbul et al. (2003, 2007; Tatbul and Zdonik 2006). Of these, (Nehme and Rundensteiner 2007) from Nehme and Rundenste appears most relevant due to its spatio-temporal setting, presenting a clustering approach to load-shedding in which moving objects that are similar in terms of location, time, direction and speed are clustered, and data from individual members of the cluster can be dropped with the representatives of the cluster summarizing them.

Jeung et al. (2010) present an automated metadata generation approach that includes a probabilistic measure of data quality. In Hossain et al. (2011), Hossain et al. dynamically assess three quality attributes for the detection and identification of human presence in multimedia monitoring systems, whereas Rodríguez and Riveill (2010) present data quality in relation to e-Health monitoring systems. Crowd-sourced citizen science as described by Kelling et al. in (2015) and volunteered geographic information efforts from Goodchild and Li (2012), Barron et al. (2014), and Ballatore and Zipf (2015) overlap with data quality research for obvious reasons. When the public assists in collecting data, the benefits of public collection must be weighed against the potential for poor quality submissions. These efforts do indicate that the benefits of public participation outweigh the drawbacks while leaving open paths for further research.

In Kelling et al. (2015), Kelling et al. tackle the problem of quality with analysis both of data submission and subsequent observer variation. Goodchild et al. call upon existing data quality standards such as the US Spatial Data Transfer Standard from the USGS (USGS) and the Content Standard for Digital GeoSpatial Metadata (F. (Federal G. D. Committee)) while demonstrating the open-ended nature of quality assurance for volunteered geographic information. Barron et al. (2014) reference the ISO 19113 (2002), ISO 19114 (2003) and ISO 19157 (2013) standards while pointing out that data quality for volunteered geographic information projects such as OpenStreetMap (OSM) (2013) depends on the user's purpose. In turn they present a framework tailored to "fitness for purpose" with six different categories of purpose and 25 measures within those categories, all specific to OSM. Ballatore and Zipf (2015) investigate "conceptual quality" using OSM, and indicate wider applicability than that of Baron et al. (2014). While these sources demonstrate ongoing interest and need for related research, none of these approaches directly addresses quality control for spatio-temporal data for the consumer/aggregator situation.

Accuracy, precision, error and uncertainty relative to location information are addressed in general terms by Goodchild and Gopal (1989), although the specifics of identifying such problems in data sets like ours are lacking. In Shi et al. (2003b), Shi et al. describe errors in position, in which location is distorted by some vector. They further describe the situation in which such errors vary smoothly in space, preserving continuity and spatial autocorrelation. They also describe absolute versus relative errors in location. This is useful for both location and timestamp errors. A distinction is made by Shi et al. (2003b) between attribute accuracy and thematic accuracy. Data quality as spatial metadata and use of the quality information is also addressed.

13.4 Data Quality Assessment Methods

The weather and road-weather communities employ detailed accuracy checks for individual observations. The Oklahoma Mesonet uses the Barnes Spatial Test (Barnes 1964), a variation of Inverse Distance Weighting (IDW) (see Shepard (1968)). Further examples include Shafer et al. (2000), and the Federal Highway Administration's Clarus project, as described by Limber et al. in (2010). MesoWest (U. of Utah) uses multivariate linear regression to assess data quality for air temperature, as described by Splitt and Horel in (Splitt and Michael n.d.; U. of Utah). MADIS (NOAA) implements multi-level, rule-based quality control checks including a neighbor check using Optimal Interpolation/kriging (NOAA; NOAA; Belousov et al. 1972).

These approaches (IDW, Linear Regression, kriging) can be used to check individual observations for deviation from predicted and flag individual observations as erroneous or questionable if the deviation is large. If interpolated values are erroneous, then the quality assessment will be wrong too. If metadata such as location or timestamps associated with a site is erroneous, then the quality control assessment will be invalid because of comparison with the wrong data from the wrong sites. None of these approaches identify incorrect location metadata and only one provider, Mesowest, attempts to identify bad timestamps. Their approach only identifies one of the most obvious timestamp-related problems – timestamps that cannot possibly be correct because they occur in the future relative to collection time.

Beyond bad metadata, sites that are chronically bad are identified at best by rudimentary means. MADIS' statistical spatial consistency check flags a current observation as failing if 75% of the observations for the site/sensor have failed individually in the prior week. This check will discontinue flagging observations as bad if the failure rate for other checks drops beneath 25% in subsequent weekly statistics. While this check does give an overall, general indication of site/sensor health, it is possible that there is a problem with a site while observations from the site still pass quality control enough proportion of time to go unnoticed.

Many spatial approaches use interpolation to model data for quality assessment, so it is useful to examine work that compares and enhances traditional interpolation methods. Zimmerman et al. (1999) use artificial surfaces and sampling techniques as well as noise level and strength of correlation to compare Ordinary and Universal kriging and IDW. They found that the kriging methods outperformed IDW across all variations they examined. Lu and Wong (2008) found instances in which kriging performed worse than their modified version of IDW, where they vary the exponent depending on the neighborhood. They indicate that kriging would be favored in situations for which a variogram accurately reflects the spatial structure. Mueller et al. (2004) show similar results, saying that IDW is a better choice than ordinary kriging in the absence of semi-variograms to indicate spatial structure.

In prior work, we proposed a modification of IDW that used a data-based distance rather than geographic distance to assess observation quality (Galarus et al. 2012; Galarus and Angryk 2013). That work focused on the use of robust methods to associate sites for assessment of individual observations. In Galarus and Angryk (2016a, b, 2018) we extended the mappings to better account for spatio-temporal variation and observation time differences when assessing observations. In Galarus and Angryk (2014, 2016c) we developed quality measures that extended beyond sites, to help evaluate overall spatial and temporal coverage of a region.

Shepard's method/Inverse Distance Weighting is widely applied, including applications which involve outlier detection and mitigation. Xie et al. (2004) applied it to surface reconstruction, in which they detect outliers using distance from fitted surfaces. Others extend the method in different ways including added dimensions, particularly time. Li et al. extend IDW in (Li et al. 2016) to include the time dimension in their application involving estimated exposure to fine particulate matter. Grieser warns of problems with arbitrarily large weights when sites are near in analyzing monthly rain gauge observations (Grieser 2015), and mitigates the problem in a manner that Shepard originally used by defining a neighborhood for which included points are averaged with identical weights in place of the large, inverse distance weights.

Kriging and Optimal Interpolation were developed separately and simultaneously as spatial best linear unbiased predictors (blups) that are equivalent for practical purposes. L. S. Gandin, a meteorologist, developed and published optimal interpolation in the Soviet Union in 1963. Georges Matheron, a French geologist and mathematician, developed and published kriging in 1962, named for a South African mining engineer, Danie Krige, who partially developed the technique in 1951 and later in 1962. For further information, refer to Cressie (1990).

Kriging is easily impacted by multiple data quality dimensions and its applicability is hindered by data quality problems. Kriging will down-weight observations that are clustered in direction, as indicated by Wackernagel (2013). This may be beneficial. However, a near observation can shadow far observations in the same direction, causing them to have small or even negative weights. This is problematic in the case that the near observation is bad.

Kriging is used to estimate values at locations for which measurements are unknown using observations from known locations. Covariance is typically estimated. This estimate usually takes the form of a function of distance alone and is determined by the data. A principal critique of kriging is that while it does produce optimal results when the covariance structure is known, the motivation for using kriging is questionable when the covariance structure must be estimated. Handcock and Stein (1993) make such an argument. Another critique is that kriging will yield a model that matches data input, giving the (false) impression that the model is perfect, as stated by Hunter et al. (2009).

Unfortunately, none of these approaches directly addresses outlier and anomaly detection for spatio-temporal data in a robust and comprehensive manner that meets our needs. None identify bad sites and metadata in a comprehensive manner. The methods used by the weather data providers appear to be state of the art for assessment of accuracy.

13.5 Enhanced Methods

13.5.1 General Definitions

An individual site refers to a fixed-location facility that houses one or multiple sensors that measure conditions. A measurement and associated metadata are referred to as an observation. The set of all sites, represented by S, is the set of sites for which observations are available for a time period and geographic area of interest.

An observation, obs, is represented as a 4-tuple, $obs = \langle s, t, l, v \rangle = \langle obs_s, obs_t, obs_l, obs_v \rangle$ consisting of the site/sensor s, timestamp t, location l (spatial coordinates), and an observed value v. We investigate observations from a single sensor type, so we assume that s identifies both the site and sensor. The set of all observations, represented by O, consists of observations from sites in S over a time-period of interest.

Ground-truth is the exact value of the condition that a given sensor is intended to measure at a given location and time. Ground-truth will rarely be known because of sensor error, estimation error, and high human costs, among other reasons. Human cost is a huge challenge, with agencies struggling to accurately inventory assets and technicians unable to service and maintain all equipment, including situations where they may not even be able to find the equipment.

We wish to evaluate observations to determine if they are erroneous. To do so, we compare observations to estimates of ground-truth. For our purposes, these estimates will be determined via interpolation, which is commonly used in the GIS community, as well as in the weather and road-weather communities.

13.5.2 General Approach

We measure outlyingness as the absolute deviation between an observed value and ground truth. In principal, this is our measure of accuracy. Ground truth may not be known, so we estimate outlyingness as the absolute deviation between an observation and modeled ground truth corresponding to the observed value in time and location. Given the degree of outlyingess (exact or estimated), we identify outliers using a threshold. If the degree of outlyingness for an observation meets or exceeds the threshold, then we flag the observation as an outlier. Otherwise, we flag it as an inlier. The degree of outlyingness is more informative than an outlier/inlier label.

Our process is consistent with general model-based approaches for outlier detection found in Han et al. (2011), Tan, Steinbach and Kumar (2006) and Aggarwal (2013), and follows the general data-mining framework of Train, Test and Evaluate. Our process is outlined in Fig. 13.7. Notice that quality assessment is not relegated to pre-processing but is instead part of a necessarily iterative process.

13.5.3 Interpolation to Model Ground Truth

Inverse Distance Weighting (IDW) (refer to Shepard (1968)) estimates ground truth as the weighted average of observation values using (geographic) distance from the site for which an observation is to be estimated as the weight, raised to some

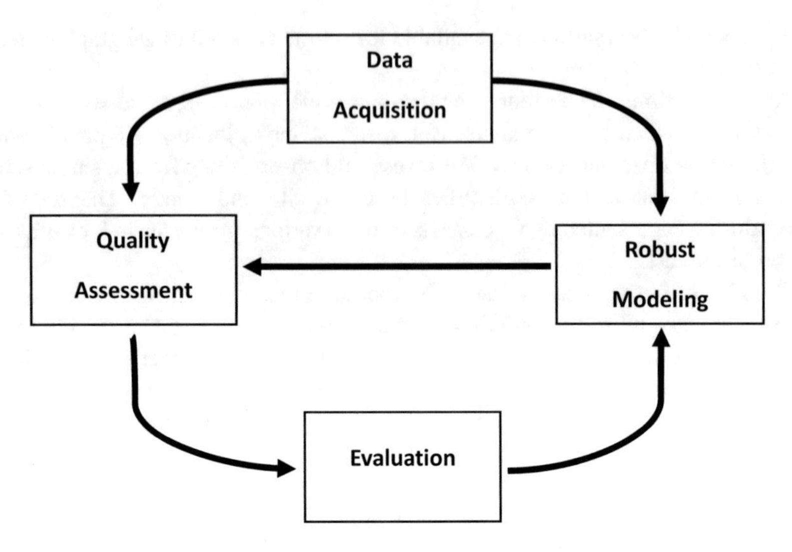

Fig. 13.7 Our general process

exponent h. If ground truth is known, a suitable exponent h can be determined to minimize error. Isaaks and Srivastava (1989) indicate that if $h = 0$, then the estimate becomes a simple average of all observations, and for large values of h, the estimate tends to the nearest neighboring observation(s). This simple version of IDW does not account for time, so it is assumed that observations fall in temporal proximity.

Least Squares Regression (LSR) estimates observed values using the coordinates of the sites. We only use x-y coordinates in our experiments for LSR. There could be benefit in using elevation and other variables including time. However, doing so compounds problems related to bad metadata such as incorrect locations, bad timestamps and inaccurate elevations.

Universal kriging (kriging with a trend) (UK) estimates observed values using the covariance between sites, the coordinates of the sites, and the observed values. In our experiments, we used a Gaussian covariance function of distance and estimated the related parameters to minimize error relative to ground-truth for our training data using data from the present time window. Refer to Huijbregts and Matheron (1971) for further information on Universal kriging. We implemented a fitter/solver for the estimation of the covariance function parameters using the Gnu Scientific Library (GSL) non-linear optimization code (Galassi et al. n.d.). Refer to Bohling (n.d.) for additional covariance functions.

These methods can be applied using a restricted radius or a bounding box to alleviate computational challenges and to focus on local trends. Other interpolators could be applied in a similar manner. There are obvious risks in using interpolators. Outliers and erroneous values will have an adverse impact on interpolation, causing poor estimates. Lack of data in proximity to a point to be estimated can also result in a poor estimate. For these reasons, we developed our own robust interpolator in prior work.

13.5.4 Our SMART Approach

In prior work, we developed a representative approach for data quality assessment of site-based, spatio-temporal data using what we call Simple Mappings for Approximation and Regression of Time series (SMART) (Klein 2007; Klein et al. 2007; Klein and Hackenbroich n.d.; Jerzak et al. 2011; O'donovan et al. 2013; Tatbul 2002; Carney et al. 2003). We used the SMART mappings to identify bad (inaccurate) observations and "bad" sites/sensors, so that they can be excluded from display and computation, and to subsequently estimate (interpolate) ground truth.

First, we compute site-to-site mappings. Let an observation be represented as $obs = \{(t, v) : t = time, v = value\}$, pairing the value with the reported time. Let obs_i be the set of observations from site i and obs_j be the set of observations from site j. For a given time radius r we pair the observations from sites i and j as $obs_{pairsi,j} = \{(x, y) : (t_1, x) \in obs_i, (t_2, y) \in obs_j, |t_2 - t_1| \leq r\}$. See Fig. 13.7 for an example pairing of an observation from one site to observations from another site.

We then define a site-to-site mapping l as a linear function of the x-coordinate (the observed value from site i) of the paired observations $obs\ \_\ pairs_{i,j}$: $l_{i,j}(x) = a + bx$. We determine this function to minimize the squared error between the values of the function and the y-coordinates (the observed values from site j) for the paired observations. See Fig. 13.8 for an example site-to-site mapping.

We next define a quadratic estimate q of the squared error of the linear mapping relative to the time offset between the paired observations. We expect an increased squared error for increased time differences. This model estimates the squared error and accounts for time offsets between observations. Our method does not require a complex, data-specific covariance model. See Fig. 13.9 for an example quadratic error mapping (Fig. 13.10).

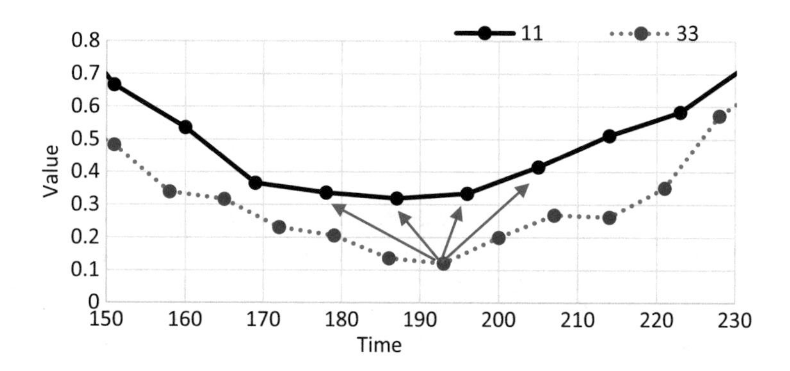

Fig. 13.8 An observation paired with observations from another site

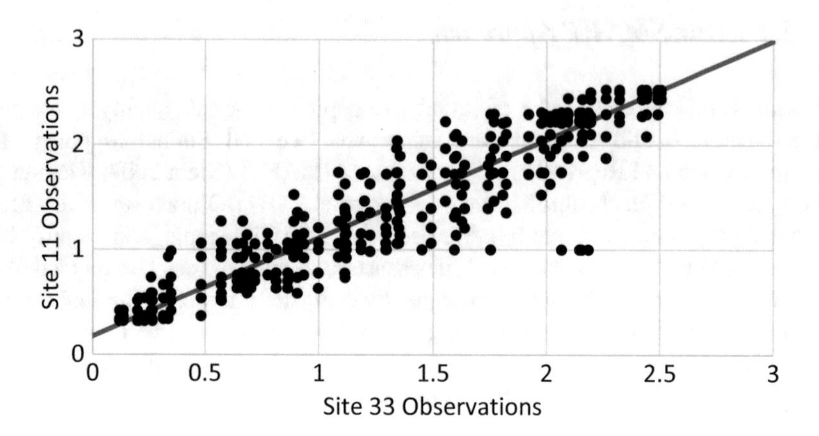

Fig. 13.9 Example site-to-site mapping between two sites

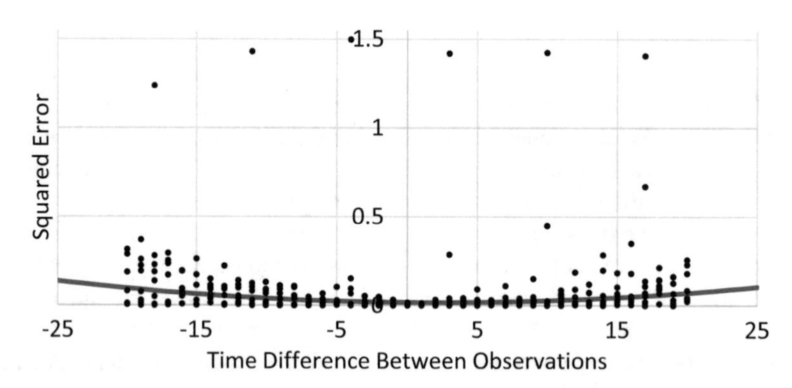

Fig. 13.10 Example quadratic error mapping

$$sq_{err_{pairs\,i,j}} = \left\{ \begin{array}{c} \left(\Delta t, \left(y - \left(l_{i,j}(x)\right)\right)^2\right) : (t_1, x) \in obs_i, \\ (t_2, y) \in obs_j, \Delta t = |t_2 - t_1| \leq r \end{array} \right\}$$

$$q_{i,j}\,(\Delta t) = a + b\,(\Delta t) + c(\Delta t)^2$$

These simple mappings are the core elements of our approach, and we must overcome the potential impact of the erroneous data in determining them. Least squares regression suffers from sensitivity to outliers. We use the method from Rousseeuw and Van Driessen to perform Least Trimmed Squares Regression (Rousseeuw and Van Driessen 2006). Least Trimmed Squares determines the least squares fit to a subset of the original data by iteratively removing data furthest from the fit. Before applying least trimmed squares to determine the linear mapping, we select the percentage of data that will be trimmed. We can interpret the trim percentage either as our willingness to accept bad data in our models or our estimation of how much data is bad. We used a trim percentage of 0.1 throughout.

For the quadratic error mappings, we experienced problems with local minima when attempting quadratic least trimmed squares. Instead we group data into intervals, determine the trimmed mean for each group, and then compute the least squares quadratic fit for the (time difference, trimmed mean) pairs.

We then check the coefficients and derived measures of the linear and quadratic mappings for outlying values relative to all other mappings. If we find outlying values, we flag the mapping as unusable. For instance, if the axis of symmetry of the quadratic error mapping is an outlier relative to that for another pairing, then there may be problem with the timestamps of at least one of the two sites.

We also investigated derived values for the quadratic error mappings and found a relationship between these values and timestamps associated with observations from a site. In turn, we found strong evidence that timestamps are incorrect for many sites. The impact of bad metadata such as location and timestamps on interpolation can be significant, adding error to estimates of ground truth and reducing our ability to characterize bad observation data (Fig. 13.11).

We next employ our SMART Interpolator. The SMART interpolator uses the site-to-site mappings. Formally: Let S be the set of all sites. Let $s \in S$ be a site for which we are evaluating observations. Let $\langle s_1, \ldots, s_n | s_i \in S, s_i \neq s \rangle$ be the set of sites other than site s. We want to estimate $obs_s(t_s)$, the value of the observation at site s at time t_s using the most recent observations from the other sites relative to time t: (t_i, v_i).

Our SMART interpolator is like IDW, but using our quadratic error estimates instead of distance given the time lag between observations and using our SMART

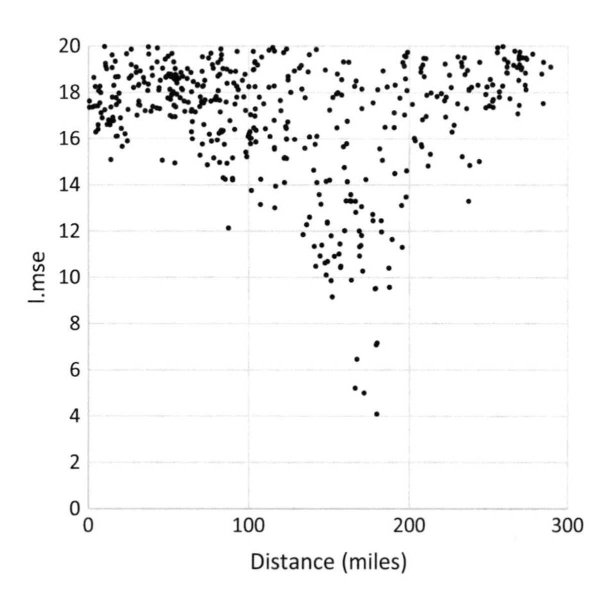

Fig. 13.11 Distance versus l.mse for GISC1 (Gibson near Castella)

linear mappings to yield estimated ground truth producing an estimate. Neither distance nor direction are directly used. The linear mappings and quadratic error estimates account for similarity between sites. No attempt is made to down-weight clustered sites, although there may be benefit in doing so.

$$\mathrm{SMART\_estimate_s}\,(\mathrm{t_s}) \approx \frac{\sum_{i=1}^{n}\left(\frac{1}{q_{s,s_i}(t_s - t_i)}\right)^{g} l_{s,s_i}(v_i)}{\sum_{i=1}^{n}\left(\frac{1}{q_{s,s_i}(t_s - t_i)}\right)^{g}}$$

We determine the exponent g by minimizing error relative to ground truth, if available, or estimated ground truth. Prior to computing the weighted estimate, we examine the weights and, if necessary, "re-balance" to reduce the potential influence of single sites on the outcome. We found it useful to restrict the maximum relative weight a site can be given to 0.25 to reduce the risk that a bad value from one site will overly influence the resulting average. Rather than take a simple weighted average, we use a trimmed mean to further reduce the influence of outliers.

Algorithm 1. SMART_ESTIMATE(s,S,t)
Input: S is the set or a subset of all sites, $s \in S$ is a site for which we are evaluating values/observations, $\langle s_1, \ldots, s_n | \, s_i \in S, \, s_i \neq s \rangle$ is the set of sites other than site s, t is the time for which the prediction will be made.
Constants: $maxweight = 0.25 \in (0, 1]$, $trimpct = 0.1 \in (0, 1]$, $minvalidqfit = 0.0001 \in (0, 1]$ $maxvalidqfit = 0.5 \in (0, 1]$, $iterations_{max} = 100 \in \mathbb{N}$
Output: The estimate.
Algorithm:

sumweights=0
weightedsum=0
for $i = 1$ ***to*** n
  if *VALID_MAPPING*(s_i,s) ***then***
    let $(t_{si}, v_{si}) = MOST\_RECENT\_OBS(s_i, t)$
    $x_i = l_{si,s}(v_{si})$
    $\Delta t = t - t_{si}$
    $qfitval = q_{si,s}(\Delta t)$
    if ($qfitval > minvalidqfit$) ***and*** ($qfitval < maxvalidqfit$) ***then***
      $weight = 1/qfitval$
      $w_i = weight$
    else
      $w_i = 0$
  else
      $w_i = 0$
      $x_i = 0$
NORMALIZE_WEIGHTS(w)
BALANCE_WEIGHTS(w,maxweight)
$predicted = W\_TRIMMED\_MEAN(x, w, trimpct, iterations_{max})$
return predicted

VALID_MAPPING($s_{i,s}$) indicates if the SMART mapping between si and s is valid. A mapping will be invalid if it is undefined or if the associated coefficients or derived values are outliers. *NORMALIZE_WEIGHTS(W)* normalizes weights to sum to 1. Weights must be non-negative. *BALANCE_WEIGHTS(w, maxweight)* reduces any weights that exceed the maximum specified and redistributes the excess weight proportionally to remaining elements. Iteration may be necessary if a redistributed weight exceeds the maximum specified.

13.5.5 Artificial Data Set

As we show subsequently, evaluation with real data for which data quality measures are inaccurate or unavailable is challenging. For this reason, we developed a weather-like artificial data set representing temperature as approximate fractal surfaces using the method of Successive Random Addition. For further information on Successive Random Addition, refer to Voss (1985), Feder (2013), and Barnsley et al. (2011). Fractional Brownian processes were used by Goodchild and Gopal to generate random fields representing mean annual temperature and annual precipitation to investigate error in (Goodchild and Gopal 1989). We generated a surface and multiple weather patterns using a simulated flow. We generated time series of "ground truth" data by combining the surface data with the weather data, a periodic effect, a north-south effect, and a diurnal effect. We selected 250 "sites" using random uniform x-y coordinates. For each site we assigned a reporting pattern with a random frequency and offset. We added errors to the observations from 25 sites via: random noise added to ground truth (NOISE), rounding of ground truth (ROUNDING), replacement of ground truth with a constant value (CONSTANT), replacement with random bad values with varying probabilities (RANDOMBAD), or negation of ground truth. The remaining 225 sites were left error-free. See Fig. 13.12 for example artificial data from multiple sites, including several apparent errors.

13.5.6 Evaluation

We evaluated the performance of the various interpolators including our SMART Method in-depth, in terms of computation and ability to identify bad data. We compared our SMART method, Inverse Distance Weighting (IDW), Least Squares Regression (LSQ), Universal Kriging (UK) and Ordinary Kriging (OK). We measured performance and scalability using run-time in milliseconds. We measured accuracy using mean-squared-error between estimated and known ground-truth. We compared means using t-tests when multiple runs were available. We used Area Under the ROC Curve (AUROC) analysis to evaluate accuracy of outlier classification given varying "threshold" values for outlier/inlier determination.

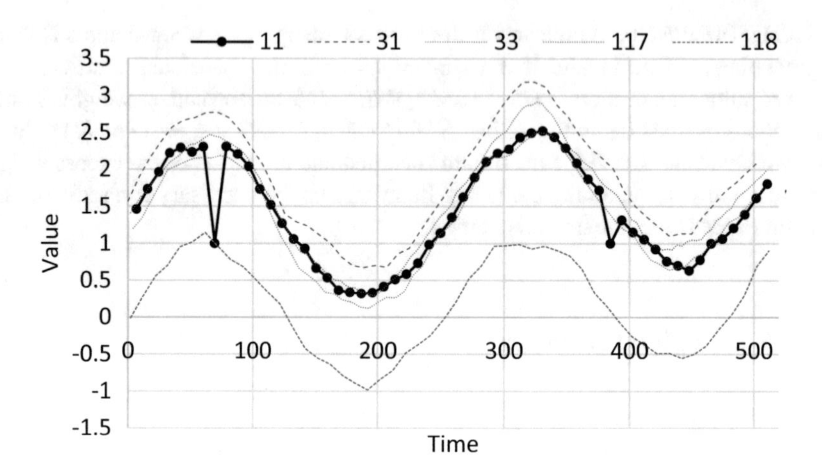

Fig. 13.12 Example artificial data

We analyzed our artificial data set, MADIS air temperature for Northern California from December 2015, MADIS air temperature for Montana from January 2017, and Average Daily USGS Streamflow for Montana from 2015, 2016, 2017.

13.5.7 Evaluation Using an Artificial Data Set

We performed an in-depth comparison of the various algorithms using our artificial data set. We enhanced the standard algorithms by randomly choosing neighboring sites using set inclusion percentages (0.1, 0.2, 0.3, ..., 0.9, 1.0). For instance, a 0.9 inclusion percentage corresponds to selecting neighboring sites individually with 0.9 inclusion/0.1 exclusion probability. We varied the radius (50, 75, 100, ..., 175, 200) over which sites were included relative to the location of the site whose observation we were testing. We repeated this procedure 10 times for each parameter combination (inclusion percent and radius) and used the median of the resulting estimates as the estimate for that parameter combination. By randomly holding out sites, bad data will be held out in some of the resulting combinations. By taking the median of the results, we eliminate the extreme estimates, particularly those impacted by bad data, and determine a robust estimate. We iterated through the observations in order by time and estimated ground truth for each observation as if computing in real time as the observations become known. Only observations that occurred at the same time as or prior to each observation were used for prediction, simulating real-time operation of the system. We averaged the MSE and run time for each configuration (inclusion radius and inclusion percent).

For Inverse Distance Weighting, the Mean-Squared-Error was least for larger radii and smaller inclusion percentages. This result indicates both the benefit of holding out potentially bad data and including a wide radius of values. The latter is somewhat surprising, indicating that data other than just the closest is of benefit to the estimate. An inclusion percentage of 1 corresponds to use of IDW without random exclusion, and full inclusion out-performed inclusion percentages of 0.5, 0.6, 0.7, 0.8 and 0.9 for inclusion radii of 150 units or greater, showing that the balance between more data versus exclusion of erroneous data tilted in favor of more data. Using similar analysis, we identified the inclusion percentages and radii for the optimal versions of the other methods for our artificial data set: IDW (0.1, 175), LSQ (0.2, 200), UK (0.2, 200). Using these parameter settings, we conducted a statistical analysis of run time and accuracy for the individual runs of the various methods.

The average run times were 4022 ms for IDW, 6337 ms for SMART, 6550 ms for LSQ and 16079 ms for UK. Corresponding p-values for t-tests (one-tail, unequal variance) comparing the mean run times by method were all less than $3 \times 10\text{-}05$. Based on these statistics, we concluded the following ordering from least to greatest of run-time required for the methods and associated parameters: IDW, SMART, LSQ, UK.

The accuracy (MSE) for the methods was 0.1026 for SMART, 0.2759 for IDW, 0.7943 for UK and 0.8035 for LSQ. Corresponding p-values for t-tests (one-tail, unequal variance) comparing the mean-squared-errors by method were all effectively 0 except for the comparison between LSQ and UK, which yielded a p-value of 0.3520. Based on these statistics and t-tests, we concluded the following ordering of accuracy (MSE): SMART is best, IDW is second best, and LSQ/UK are tied for worst.

We measured the ability of each method to distinguish increasing percentages of the bad data from good data using an AUROC analysis. True outliers were defined as data that differs from ground-truth – i.e., data that was modified to be erroneous. Predicted outliers were data that differed from estimated ground truth by a given threshold. We varied thresholds for outlier/inlier cutoffs and compared results with the actual labels identifying whether the data was truly an outlier or inlier. See Fig. 13.13. The AUROC (area under the ROC curve) values were 0.827 for the SMART method, 0.74 for UK, 0.739 for LSQ and 0.708 for IDW. The AUROC values show better discriminative power for the SMART method versus the other methods. No method will be perfect in identifying all errors. Some errors are small and impossible to distinguish from interpolation error. Known ground truth and known error from ground truth yields perfect labels. As we will see in subsequent analysis, bad labels will yield questionable results using ROC curve analysis.

The overall amount of preprocessing time required to determine the linear mappings and quadratic error functions for our SMART Method was comparable to run time required for Universal kriging. This was encouraging. Generation of the mappings will be done as an offline, batch process, so the observed time required is still within reason to help facilitate the faster and more accurate, online process. Additional benefits such as identification of bad sites and bad metadata come from these mappings, further justifying the effort required. Optimizations can reduce the

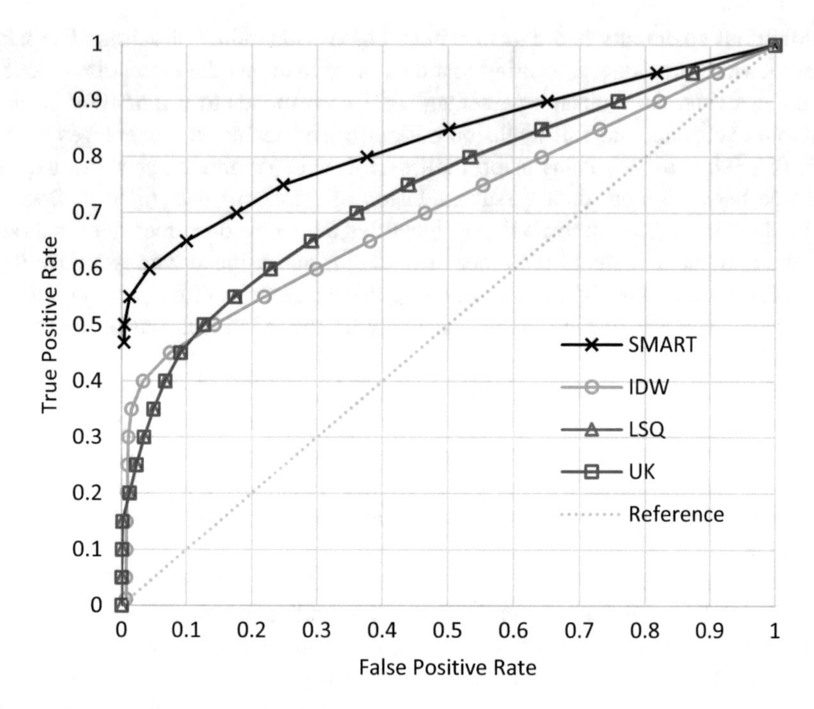

Fig. 13.13 ROC curves by method for artificial data set

overall time needed to compute the mappings. The benefits and potential to improve the run time outweigh the amount of required preprocessing time.

13.5.8 December 2015 MADIS California Data

We analyzed Northern California December 2015 ambient air temperature data from 888 sites in the MADIS Mesonet subset. We excluded observations that failed the MADIS Level 1 Quality Control Check. This range check restricts observations in degrees Fahrenheit to the interval [−60 °F, 130 °F]. Many values that failed this check fall far outside the range and can have a dramatic impact on the interpolation methods. Our SMART method performs very well in the presence of extreme bad data, and it would have easily out-performed the other methods in the presence of the range-check failed data. There were over 2 million observations. MADIS flagged 73.5% of these observations as "verified"/V, slightly less than 4% as "questioned"/Q, and 22.5% as "screened"/S, indicating that it had passed the MADIS Level 1 and Level 2 quality checks, but that the Level 3 quality checks had not been applied.

Verified (V) observations from the first week in December 2015 were used to train all methods, including our SMART method. In the absence of range-failed data, the "enhanced" (iterated subset) versions of the other algorithms showed little improvement in accuracy while consuming excessive computation time, particularly "enhanced" Universal kriging. In some cases, it would have taken days to compute results. Because of this, we used the methods directly, without enhancement. We also tested Ordinary kriging (refer to Bailey and Gatrell (1995) for further information). We do not know "ground truth" for this data and the verified data is the closest to ground truth. We trained all methods on this data to minimize mean-squared-error of predicted versus actual. We used a 50-mile inclusion radius due to the density of sites to avoid excessive computation time for the kriging approaches. The SMART mapping coefficients and derived values were examined for outliers, and ranges were determined for valid mappings. If any coefficient or derived value for a given SMART mapping fell outside these ranges, then the SMART mapping was considered bad, and that mapping was not used for predictions.

The SMART method produced significantly better results than all other methods for the training data in terms of estimation of ground truth measured by mean-squared-error. Only the verified (V) data was used in this comparison since it best approximates ground truth. The MSE values were 2.8 for SMART, 7.6 for IDW, 16.5 for UK, 17.1 For LSQ, and 18.5 for OK.

Run times were recorded for multiple training sessions for the methods. Only one run time was used for Least Squares (LSQ) because no parameters were trained for that method. There was a statistically significant difference between the run times for the methods and the order from least to greatest for average run times was: LSQ (19842), IDW (23932), SMART (113865), OK (352905) and UK (363005).

Testing. Testing was conducted using all data from the entire month of December 2015, minus the range-check-failed data, with results grouped by week. We computed the mean-squared-error for the verified (V) data since it best represents ground truth, but all observations were used in making estimates. The testing results indicate the robustness of methods in the presence of bad data. The SMART method significantly out-performed all other methods, with MSE values for SMART ranging from 2.75 to 6.63 and MSE values for the other methods ranging from 7.69 to 17.93.

We conducted an AUROC analysis to compare classification ability of the methods based on the MADIS quality control flags. We considered the following flags from MADIS to be good/inlier data: V/verified, S/screened, good. The Q/questioned, was treated as bad/outlier data. Recall that we excluded the observations having a QC flag of X, those that failed the range test, from our evaluation. Even if we accept the MADIS quality control flags as being correct, and we do not, this approach is problematic. The S flag corresponds to data for which not all the QC checks have been run. While this data had not failed any quality control checks that have been applied, it possibly would have failed the higher-level checks. The AUROC values were 0.7906 for IDW, 0.7578 for LSQ, 0.7317 for SMART, 0.6458 for OK and 0.6062 for UK. While these AUROC values seem reasonable, they are affected by incorrect outlier/inlier labels, and our SMART method suffers the greatest impact because the distance-based methods approximate the MADIS Level 3 quality control check. OK and UK fall short because they fail to make predictions for many observations.

13.5.9 December 2017 MADIS Montana Data

We investigated ambient air temperature for 497 sites from Western Montana/Northern Idaho from the MADIS Mesonet and the MADIS HFMetar subset in January 2017. We added the HFMetar data set to account for aviation AWOS/ASOS sites that included in the Mesonet data set. The bounding box was comparable in size to the one used for Northern California, although the density of sites is less. We excluded observations that failed the MADIS Level 1 Quality Control Check. All total there were over 1 million observations. MADIS flagged 71.2% of these observations as "verified"/V; 10.3% of as "screened"/S, indicating that they had passed the MADIS Level 1 and Level 2 quality checks, but that the Level 3 quality checks had not been applied; and a relatively large 18.5% of the data was labeled as "questioned"/Q. This is over four times the percentage of questioned data as there was for the California data set.

Verified (V) observations from the first week in January 2017 were used to train all methods, including our SMART method. We used a 100-mile inclusion radius due to a low density of the Montana/Idaho sites. The SMART mapping coefficients and derived values were examined for outliers, and bad mappings were identified as any mapping associated with such values. The quality of the mappings was noticeably less than that for the Northern California data set. We found problems with many of the timestamps in this data set. Recognizing that much of the Idaho data comes from the Pacific Time Zone while the Montana data comes from the Mountain Time Zone, there appeared to be many sites for which the conversion to UTC time was not consistent. The Northern California data all falls within Pacific Time, and we did not see this problem in that data set. In terms of mean-squared-error, the SMART method produced significantly better results than each of the other methods for the training data. The p-values were effectively zero for all comparisons. The MSE values 8.1513 for SMART versus 16.7217 for IDW and 29.8039 for LSQ, and 10.6726 for SMART versus 47.0028 for OK and 33.9863 for UK. Ordinary kriging and Universal kriging failed to produce estimates in nearly two-thirds of the tests, likely due to singular matrices. Methods were compared pairwise only in instances when both produced estimates.

There was a statistically significant difference between the run times for the methods and the order from least to greatest run time was: LSQ (6009), IDW (6803), SMART (33356), OK (103851) and UK (107540).

Testing was conducted using data from the remainder of January 2017. All data was used for this test except for the observations that failed the MADIS Level 1 range test. The SMART method significantly out-performed all other methods in pairwise comparisons with MSE values ranging from 7.69 to 31.51. The MSE values for the other methods ranged from 16.26 to 68.82.

We conducted an AUROC analysis to test classification ability based on the MADIS quality control flags in the same way as described for the Northern California data set in the previous section. As noted in that section, many of the MADIS QC flags are incorrect. In terms of Area Under the ROC curve, LSQ, IDW and SMART were comparable, with LSQ finishing ahead. These AUROC values are less than those for the Northern California data set at least in part because they are all adversely affected by incorrect outlier/inlier labels. The AUROC values were 0.69 for LSQ, 0.6697 for IDW, 0.6393 for SMART, 0.5432 for OK and 0.5476 for UK.

This data set includes a large percentage of observations (18.5%) that are flagged as "questionable" by MADIS. These were considered "bad"/outliers for the purposes of our analysis. It also includes a large percentage (10.3%) that are flagged as "screened" by MADIS, indicating that not all QC checks have been conducted. These are considered "good"/inliers for our analysis. There were many observations flagged as "questionable"/outliers in the HFMetar subset that should have been flagged as "good"/inliers. This data alone accounts for most of the questionable data in the data set. Aviation weather sites are well-maintained and regularly calibrated, so it is hard to believe that these sites would produce data that is entirely bad. We checked this data against predicted values and it was very close, so it is unclear why the data was labeled as questionable.

Numerous sites were flagged by our SMART method as "bad" and all observations from those sites were labeled as bad. MADIS flagged some observations from these sites as good when they were close to predicted values. In some cases, this may have been reasonable, but in others it was a random occurrence. There were some sites that produced bad data for the training period but then produced good data for at least a portion of the test period. One could argue that for such sites all associated observations should be questioned. If a site was identified as bad by the SMART method, then the V and S observations would adversely impact the SMART method in the AUROC analysis. The chance situations in which the other methods came close to the "good" values and far from the "bad" values improved their performance.

13.5.10 December 2015–2017 USGS Streamflow Data

Mean daily streamflow (ft^3/sec) was downloaded for all sites in Montana from the USGS (USGS) for every day from January 1st, 2015 through April 24th, 2017. There were 145 sites having data than spanned this period, and these sites were analyzed. This data set is far different from the air temperature data used for prior analysis. Since daily averages were used, there is no visible diurnal effect. There is a seasonal effect which varies with elevation and location relative to watersheds. Due to the dramatic fluctuations that occur in this data during times of peak runoff, the base-10 logarithm of the data was used for analysis.

This data set includes quality flags. Daily values are flagged as "A", approved for publication, and "P", provisional and subject to revision. Values may further be flagged as "e" for estimated. Values transition from provisional to approved after more extensive testing is conducted, so provisional values aren't necessarily bad. These flags were of limited use to us and we did not use them for analysis. We treated the data as being all good and subsequently introduced errors into some of the observations, making them known bad. There were 122,380 total observations.

All data from 2015 was used to train all methods, including our SMART method. We assume this data, which was mostly "approved", to be ground truth. We trained over this data to minimize mean-squared-error of predicted versus actual. We used a 200-mile inclusion radius. The SMART mapping coefficients and derived values were examined for outliers. If any coefficient or derived value for a given SMART mapping was an outlier, then the SMART mapping was considered bad, and it wasn't used for predictions. In terms of mean-squared-error, the SMART method produced significantly better results than the other methods for the training data with p-values of zero. The MSE was 0.0174 for SMART, 0.8751 for IDW, 0.9611 for LWQ, 0.9431 for OK and 0.9617 for UK.

Run times were recorded for multiple training sessions. Only one run time was used for Least Squares (LSQ) because no parameters are trained for that method. There was a statistically significant difference between the run times for the methods except between LSQ (686) and IDW (742), and the order from least to greatest was: LSQ (686)/IDW (742), SMART (2628), OK (69667) and UK (72356).

Testing was conducted using the 2016–2017 data, grouped by year. The SMART method significantly out-performed all other methods in terms of mean-squared-error. The MSE for the SMART method ranged from 0.3805 to .0438. The MSE for the other methods ranged from 0.8175 to 0.9968.

Further testing was conducted using the 2016–2017 data, with errors introduced into 10% of the observations. A random normal value with mean zero and standard deviation one was added to each observation in the 10% group. The mean-squared-error was computed relative to the known, original observations which represent ground truth, and all observations (including bad observations) were used in making estimates. The testing results help to measure the robustness of methods in the presence of bad data. The SMART method again significantly out-performed all other methods in terms of mean-squared-error, with MSE ranging from 0.0406 to 0.0463, while the MSE for the other methods ranged from 0.8242 to 1.0002.

We conducted an AUROC analysis to test the methods on classification ability based on whether observations had been altered to be erroneous by our process of randomly selecting 10% of the observations and adding a normal random variable with mean 0 and standard deviation 1 to those observations. The altered observations were labeled "bad"/outlier and the unaltered observations were labeled as "good"/inlier. Our SMART method performed far better than all the other methods, achieving an AUROC value of 0.8722. The AUROC values for the other methods were 0.6241 for IDW, 0.6136 for OK, 0.6046 for UK and 0.6031 for LSQ.

13.5.11 Evaluation Summary

For all four data sets and for every training and testing instance compared, our SMART method performed significantly better in terms of accuracy (mean-squared error) than all other methods. Its computational performance was competitive even though no effort was made to optimize it. For the two MADIS data sets, its performance for AUROC analysis of classification and discrimination capability showed it to be competitive with the best of the other methods. This comparison and evaluation made use of MADIS data quality labels for which we have found numerous problems. As such, all methods underperformed, and the SMART method was penalized most by mislabeling. For the other two data sets (artificial and USGS) in which ground truth is known or assumed and errors were introduced relative to ground truth, the SMART method outperformed the other methods by a wide margin. This further supports our assertions regarding the impact of bad labels on the MADIS data, and the need for better methods and benchmark data sets for data quality assessment.

Ordinary kriging and Universal kriging both failed to produce estimates for many observations, likely due to singular matrices. They were not competitive in terms of run time and their accuracy was no better than the other methods. Universal kriging and least squares regression are prone to very large errors if the predicted surface slopes in an extreme manner.

Our SMART method identifies "bad sites", sites that chronically produce bad data, and does not use data from these sites in estimating ground truth for other sites. Data from these "bad sites" is labeled as all bad. The SMART method falls short in cases where a site exhibits chronic behavior during training but recovers to produce good data during a testing period.

The USGS streamflow data exhibits correlation between sites, but the correlation corresponds to sites close to each other and in the same river/stream. Correlation will not necessarily be high for sites that are close but in different rivers. For rivers that have dams and other features that may influence streamflow in unusual ways, sensors will be correlated on each side of such features, but not as much on opposite sites, and certainly not as much with sites on rivers that do not have similar features.

The SMART method identifies like sites, yielding better correlations. Inverse Distance Weighting and Least Squares Regression did not perform well. And, the kriging methods will not perform well if a stationary, isotropic covariance function is used. Such an assumption is typical, and we used this assumption in determining the covariance matrices for the kriging tests.

13.6 Further Research and Development Topics

While our SMART method out-performed the other methods in nearly all instances, it was not our intent to present it as the "best" method. Instead, we present it as representative of the type of approach needed to overcome challenges of spatio-temporal data quality assessment. It makes no assumption of isotropic covariance and does not require the determination of a specific covariance function. While it requires preprocessing time, it is suitable for near-real-time, online use. It accounts for disparate reporting times and frequencies of sites. It not only helps to identify “bad data”, but it also works well in the presence of bad data. It helps to identify erroneous observations, “bad sites”, and bad metadata. It uses multiple, robust methods to mitigate the impact of bad data on its estimates. Other methods such as Least Squares Regression and the various kriging approaches could (and should) be modified in a similar manner to produce better, more robust results. Further, it is important to recognize the impact of bad data quality labels on evaluation. It is necessary to develop and use benchmark datasets with known, correct data quality labels.

In our research, we have investigated relatively simple situations and data sets involving ambient air temperature. We set out to improve our assessment of the accuracy of individual observations and found that location metadata was inaccurate. When we tried to assess the accuracy of location metadata we found that timestamp metadata was incorrect. The assessment of spatio-temporal data quality is a difficult challenge. Connected and autonomous vehicles present further challenges. We have used the site-based nature of the data to help identify mislocated sites, bad timestamps and bad sites in general. Mobile data presents further challenges.

We intend to expand our work to further examine other measures including wind and precipitation as well as CCTV camera images. Departments of Transportation use CCTV camera images to verify road weather conditions reported by sensors. These images also suffer from poor data quality. Further research is needed to develop methods for detecting bad CCTV image data and for using CCTV image data to confirm sensor conditions and vice-versa. And, we intend to further develop benchmark datasets with known, good data quality labels. Indeed, Ian’s standard of “Accurate, Timely and Reliable” is elusive. But, we believe it is attainable and efforts must continue to achieve it.

Bibliography

Margiotta R (2002) State of the practice for traffic data quality. In: Traffic data quality Wkshp

Schuman R (2001) Summary of transportation operations data issues. In: National Summit on Transportation OperationsNational Steering Committee on Transportation Operations, Federal Highway Administration, American Association of State Highway and Transportation Officials, American Public Transportation Association, Intelligent Transportation Society of America, Instit

Robinson E, Jacobs T, Frankle K, Serulle N, Pack M (2012) Deployment, use, and effect of real-time traveler information systems

Galarus D, Koon L, Turnbull I, Campbell S, Pearce J (2018) An analysis of best practices for DOT traveler information data quality. In: The intelligent transportation Society of America 2018 annual meeting

I. A. A. C. Steering Committee (2000) ATIS-related data collection and data quality. Closing the data gap: guidelines for quality advanced traveler information system (ATIS) data. Washington, DC

USDOT. Real-time system management information program, 23 CFR Part 511

FHWA, Federal Highway Administration (2015) Real-time system management infor-mation program. [Online]. Available: http://www.ops.fhwa.dot.gov/1201/. Accessed: 20 Oct 2015
U. S. D. of T. F. H. Administration. Real-time system management information program [1201] fact sheet. [Online]. Available: http://ops.fhwa.dot.gov/1201/factsheet/
Trachy L, Via R, Cowherd S (2016) SAFETEA-LU Section 1201 real-time system management information. Introductory MPO collaborative meetings. Operations Division, Virginia Department of Transportation, 2013. Presentation. [Online]. Available: http://www.mwcog.org/uploads/committee-documents/a11aXldc20130807135940.pdf
Wang RY, Strong DM (1996) Beyond accuracy: what data quality means to data consumers. J Manag Inf Syst 12(4):5–33
Batini C, Cappiello C, Francalanci C, Maurino A (2009) Methodologies for data quality assessment and improvement. ACM Comput Surv 41(3):16
Luebbers D, Grimmer U, Jarke M (2003) Systematic development of data mining-based data quality tools. Proc 29th Int Conf. Very large data bases 29:548–559
Bisdikian C, Damarla R, Pham T, Thomas V (2007) Quality of information in sensor networks. In: 1st annual conference of ITA (ACITA'07)
Devillers R, Jeansoulin R (2010) Fundamentals of spatial data quality, chapter 2. In: Spatial data quality: concepts. Newport Beach, Wiley-ISTE
Bédard Y, Vallière D (1995) Qualité des données à référence spatiale dans un contexte gouvernemental. Université Laval, Quebec
Shi W, Wang S, Li D, Wang X (2003a) Uncertainty-based spatial data mining. Proc Asia GIS Assoc. Wuhan, China, pp 124–135
Aggarwal CC (2013) Outlier analysis. Springer Publishing Company, Incorporated. Springer, New York, NY
Sathe S, Papaioannou TG, Jeung H, Aberer K (2013) A survey of model-based sensor data acquisition and management. In: Managing and mining sensor data. Springer, Boston, pp 9–50
Ives ZG, Florescu D, Friedman M, Levy A, Weld DS (1999) An adaptive query execution system for data integration. ACM SIGMOD Rec 28(2):299–310
Sofra N, He T, Zerfos P, Ko BJ, Lee KW, Leung KK (2008) Accuracy analysis of data aggregation for network monitoring. MILCOM 2008 - 2008 IEEE Mil Commun Conf, pp 1–7
Charbiwala ZM, Zahedi S, Kim Y, Cho YH, Srivastava MB (2009) Toward quality of information aware rate control for sensor networks. In: Fourth international workshop on feedback control implemenation and design in computing systems and networks
Hermans F, Dziengel N, Schiller J (2009) Quality estimation based data fusion in wireless sensor networks. MASS'09. IEEE 6th Int Conf Mob Adhoc Sens Syst, pp 1068–1070
Fugini M, Mecella M, Plebani P, Pernici B, Scannapieco M (2002) Data quality in cooperative web information systems. Personal communication. citeseer. ist. psu. edu/fugini02data. html. [Online]. Available: http://citeseerx.ist.psu.edu/viewdoc/download?doi=10.1.1.18.9821&rep=rep1&type=pdf. Accessed: 26 Dec 2015
Klein A, Lehner W (2009a) Representing data quality in sensor data streaming environments. J Data Inf Qual 1(2):1–28
Klein A, Lehner W (2009b) How to optimize the quality of sensor data streams. Proc 2009 Fourth Int Multi-Conference Comput Glob Inf Technol, pp 13–19
Klein A (2007) Incorporating quality aspects in sensor data streams. Proc {ACM} first {Ph.D.} Work. {CIKM}, pp 77–84
Klein A, Do HH, Hackenbroich G, Karnstedt M, Lehner W (2007) Representing data quality for streaming and static data. Proc Int Conf Data Eng, pp 3–10
Klein A, Hackenbroich G (n.d.) How to screen a data stream. [Online]. Available: http://mitiq.mit.edu/ICIQ/Documents/IQConference 2009/Papers/3-A.pdf. Accessed: 26 Dec 2015
Jerzak Z, Klein A, Hackenbroich G (2011) GINSENG data processing framework. In: Reasoning in event-based distributed systems. Springer, Berlin, Heidelberg, pp 125–150
O'donovan T et al (2013) The GINSENG system for wireless monitoring and control: design and deployment experiences. ACM Trans Sens Networks 10(1):4

Tatbul N (2002) Qos-driven load shedding on data streams. XML-Based Data Manag. Multimed. Eng Work, pp 566–576
Carney D, Çetintemel U, Rasin A, Zdonik S, Cherniack M, Stonebraker M (2003) Operator scheduling in a data stream manager. VLDB 29:838–849
Mokbel MF, Xiong X, Aref WG, Hambrusch SE, Prabhakar S, Hammad MA (2004) PLACE: a query processor for handling real-time spatio-temporal data streams. Proc Thirtieth Int Conf. Very large data bases 30:1377–1380
Babcock B, Datar M, Motwani R (2004) Load shedding for aggregation queries over data streams. Proc Int Conf Data Eng 20:350–361
Babcock B, Datar M, Motwani R (2007) Load shedding in data stream systems. In: Data Streams. Springer, Boston, pp 127–147
Nehme RV, Rundensteiner EA (2007) ClusterSheddy: load shedding using moving clusters over spatio-temporal data streams. In: Advanced databases concepts for system applications. Springer, Berlin, Heidelberg, pp 637–651
Tatbul N, Çetintemel U, Zdonik S, Cherniack M, Stonebraker M (2003) Load shedding in a data stream manager. Proceeding VLDB '03 Proc 29th Int Conf Very large data bases, pp 309–320
Tatbul N, Çetintemel U, Zdonik S (2007) Staying fit: efficient load shedding techniques for distributed stream processing. Proc 33rd Int Conf very large data bases, pp 159–170
Tatbul N, Zdonik S (2006) Window-aware load shedding for aggregation queries over data streams. Proc 32nd Int Conf Very large data bases 6:799–810
Jeung H et al (2010) Effective metadata management in federated sensor networks. SUTC 2010 - 2010 IEEE Int Conf Sens Networks, Ubiquitous, Trust Comput UMC 2010 - 2010 IEEE Int Work Ubiquitous Mob Comput, pp 107–114
Hossain MA, Atrey PK, El Saddik A (2011) Modeling and assessing quality of information in multisensor multimedia monitoring systems. ACM Trans Multimed Comput Commun Appl 7(1):1–30
Rodríguez CCG, Riveill M (2010) e-Health monitoring applications: what about data quality? [Online]. Available: http://ceur-ws.org/Vol-729/paper2.pdf
Kelling S, Fink D, La Sorte FA, Johnston A, Bruns NE, Hochachka WM (2015) Taking a 'Big Data' approach to data quality in a citizen science project. Ambio 44(4):601–611
Goodchild MF, Li L (2012) Assuring the quality of volunteered geographic information. Spat Stat 1:110–120
Barron C, Neis P, Zipf A (2014) A comprehensive framework for intrinsic OpenStreetMap quality analysis. Trans GIS 18(6):877–895
Ballatore A, Zipf A (2015) A conceptual quality framework for volunteered geographic information. In: Spatial information theory. Springer, Cham, pp 89–107
USGS. Spatial Data Transfer Standard (SDTS). [Online]. Available: http://mcmcweb.er.usgs.gov/sdts/. Accessed: 28 Dec 2015
F. (Federal G. D. Committee). Content standard for digital geospatial metadata. [Online]. Available: http://www.fgdc.gov/metadata/csdgm/. Accessed: 28 Dec 2015
ISO 19113:2002 (2002) Geographic information-Quality principles
ISO 19114:2003 (2003) Geographic information – Quality evaluation procedures
ISO 19157:2013 (2013) Geographic information – Data quality
OpenStreetMap Foundation (2013) OpenStreetMap. Open Database License (ODbL)
Goodchild MF, Gopal S (1989) The accuracy of spatial databases. CRC Press, Boca Raton
Shi W, Fisher P, Goodchild MF (2003b) Spatial data quality. CRC Press, Boca Raton
Barnes SL (1964) A technique for maximizing details in numerical weather map analysis. J Appl Meteorol 3(4):396–409
Shepard D (1968) A two-dimensional interpolation function for irregularly-spaced data. 23rd ACM Natl Conf, pp 517–524
Shafer MA, Fiebrich CA, Arndt DS, Fredrickson SE, Hughes TW (2000) Quality assurance procedures in the Oklahoma Mesonetwork. J Atmos Oceanic Tech 17(4):474–494
Limber M, Drobot S, Fowler T (2010) Clarus quality checking algorithm documentation report
U. of Utah. MesoWest data. [Online]. Available: http://mesowest.utah.edu/. Accessed: 26 Dec 2015

Splitt J, Michael E (n.d.) Horel, Use of multivariate linear regression for meteorological data analysis and quality assessment in complex terrain. [Online]. Available: http://mesowest.utah.edu/html/help/regress.html. Accessed: 26 Dec 2015

U. of Utah. MesoWest quality control flags help page. [Online]. Available: http://mesowest.utah.edu/html/help/key.html. Accessed: 26 Dec 2015

NOAA. Meteorological assimilation data ingest system (MADIS). [Online]. Available: http://madis.noaa.gov/. Accessed: 26 Dec 2015

NOAA. MADIS meteorological surface quality control. [Online]. Available: https://madis.ncep.noaa.gov/madis_sfc_qc.shtml. Accessed: 26 Dec 2015

NOAA. MADIS quality control. [Online]. Available: http://madis.noaa.gov/madis_qc.html. Accessed: 26 Dec 2015

Belousov SL, Gandin LS, Mashkovich SA (1972) Computer processing of current meteorological data, translated from russian to english by atmospheric environment service. Nurklik, Meteorol Transl 18:227

Zimmerman D, Pavlik C, Ruggles A, Armstrong MP (1999) An experimental comparison of ordinary and universal kriging and inverse distance weighting. Math Geol 31(4):375–390

Lu GY, Wong DW (2008) An adaptive inverse-distance weighting spatial interpolation technique. Comput Geosci 34(9):1044–1055

Mueller TG, Pusuluri NB, Mathias KK, Cornelius PL, Barnhisel RI, Shearer SA (2004) Map quality for ordinary kriging and inverse distance weighted interpolation. Soil Sci Soc Am J 68(6):2042

Galarus DE, Angryk RA, Sheppard JW (2012) Automated weather sensor quality control. FLAIRS Conf, pp 388–393

Galarus DE, Angryk RA (2013) Mining robust neighborhoods for quality control of sensor data. Proc 4th ACM SIGSPATIAL Int Work GeoStreaming (IWGS '13), pp 86–95

Galarus DE, Angryk RA (2016a) A SMART approach to quality assessment of site-based Spatio-temporal data. In: Proceedings of the 24th ACM SIGSPATIAL international conference on advances in geographic information systems (GIS '16)

Galarus DE, Angryk RA (2016b) The SMART approach to comprehensive quality assessment of site-based spatial-temporal data. In: 2016 IEEE international conference on big data (big data), pp 2636–2645

Galarus DE, Angryk RA (2018) Beyond accuracy - a SMART approach to site-based spatio-temporal data quality assessment. (Accepted) Intell Data Anal 22(1):21–43

Galarus DE, Angryk RA (2014) Quality control from the perspective of the real-time spatial-temporal data aggregator and (re)distributor. In: Proceedings of the 22nd ACM SIGSPATIAL international conference on advances in geographic information systems (SIGSPATIAL '14), pp 389–392

Galarus DE, Angryk RA (2016c) Spatio-temporal quality control: implications and applications for data consumers and aggregators. Open Geospatial Data Softw Stand 1(1):1

Xie H, McDonnell KT, Qin H (2004) Surface reconstruction of noisy and defective data sets. In: Proceedings of the conference on Visualization'04, pp 259–266

Li L, Zhou X, Kalo M, Piltner R (2016) Spatiotemporal interpolation methods for the application of estimating population exposure to fine particulate matter in the contiguous US and a Real-Time web application. Int J Environ Res Public Health 13(8):749

Grieser J (2015) Interpolation of global monthly rain gauge observations for climate change analysis. J Appl Meteorol Climatol 54(7):1449–1464

Cressie N (1990) The origins of kriging. Math Geol 22(3):239–252

Wackernagel H (2013) Multivariate geostatistics: an introduction with applications. Springer Science & Business Media. Springer-Verlag, Berlin, Heidelberg

Handcock MS, Stein ML (1993) A Bayesian analysis of kriging. Technometrics 35(4):403–410

Hunter GJ, Bregt AK, Heuvelink GBM, De Bruin S, Virrantaus K (2009) Spatial data quality: problems and prospects. In: Research trends in geographic information science. Springer, Berlin, Heidelberg, pp 101–121

Han J, Pei J, Kamber M (2011) Data mining: concepts and techniques. Elsevier, Burlington

Tan P-N, Steinbach M, Kumar V (2006) Introduction to data mining. Pearson Education, Inc., Boston

Isaaks EH, Srivastava RM (1989) An introduction to applied geostatistics. Oxford University Press, New York

Huijbregts C, Matheron G (1971) Universal kriging (an optimal method for estimating and contouring in trend surface analysis). In: Proceedings of ninth international symposium on techniques for decision-making in the mineral industry

Galassi M et al (2009) GNU scientific library reference manual, 3rd edn. Free Software Foundation, Network Theory Ltd., United Kingdom

Bohling G. Introduction to geostatistics and variogram analysis. [Online]. Available: http://people.ku.edu/~gbohling/cpe940/Variograms.pdf

Rousseeuw PJ, Van Driessen K (2006) Computing LTS regression for large data sets. Data Min Knowl Discov 12(1):29–45

Voss RF (1985) Random fractal forgeries. In: Fundamental algorithms for computer graphics. Springer, Springer-Verlag, Berlin, Heidelberg pp 805–835

Feder J (2013) Fractals. Springer Science & Business Media New York

Barnsley MF et al (2011) The science of fractal images. Springer Publishing Company, Incorporated. Springer-Verlag, New York

Bailey TC, Gatrell AC (1995) Interactive spatial data analysis, vol 413. Longman Scientific & Technical, Essex

USGS. USGS water data for the nation. [Online]. Available: https://waterdata.usgs.gov/nwis/. Accessed: 28 Apr 2017

Chapter 14
Uncertain Spatial Data Management: An Overview

Andreas Züfle

14.1 Introduction

Due to the proliferation of handheld GPS enabled devices, spatial and spatio-temporal data is generated, stored, and published by billions of users in a plethora of applications. By mining this data, and thus turning it into actionable information, The McKinsey Global Institute projects a "$600 billion potential annual consumer surplus from using personal location data globally".

As the volume, variety and velocity of spatial data has increased sharply over the last decades, uncertainty has increased as well. Until the early twenty-first century, spatial data available for geographic information science (GIS) was mainly collected, curated, standardized (Fegeas et al. 1992), and published by authoritative sources such as the United States Geological Survey (USGS) (United States Geological Survey). Now, data used for spatial data mining is often obtained from sources of volunteered geographic information (VGI) (Sui et al. 2012; Open Street Map). Consequentially, our ability to unearth valuable knowledge from large sets of such spatial data is often impaired by the uncertainty of the data which geography has been named the "the Achilles heel of GIS" (Goodchild 1998) for many reasons:

– Imprecision is caused by physical limitations of sensing devices and connection errors, for instance in geographic information system using cell-phone GPS (Couclelis 2003),

A. Züfle (✉)
Department of Geography and Geoinformation Science, George Mason University, Fairfax, VA, USA
e-mail: azufle@gmu.edu

M. Werner, Y.-Y. Chiang (eds.), *Handbook of Big Geospatial Data*,
https://doi.org/10.1007/978-3-030-55462-0_14

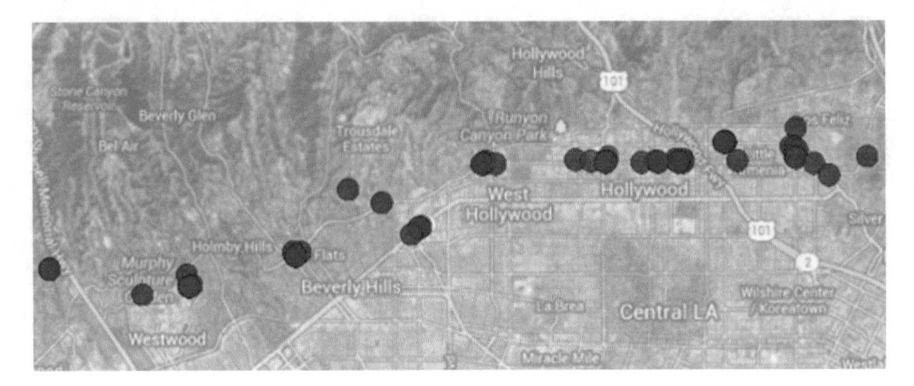

Fig. 14.1 Locations of a user of a Location-based social network (Gowalla) over a day

- Data records may be obsolete. In geo-social networks and microblogging platforms such as Twitter, users may update their location infrequently, yielding uncertain location information in-between data records (Kumar et al. 2014),
- Data can be obtained from unreliable sources, such as volunteered geographic information like data in Open-Street-Map (Open Street Map), where data is obtained from individual users, which may incur inaccurate or plain wrong data, deliberately or due to human error (Grira et al. 2010),
- Data sets pertaining to specific questions may be too small to answer questions reliably. Proper statistical inference is required to draw significant conclusions from the data and to avoid basing decisions upon spurious mining results (Hsu 1996; Casella and Berger 2002).

To illustrate uncertainty in spatial and spatio-temporal data, Fig. 14.1 shows a typical one-day "trajectory" of a prolific user in the location-based social network Gowalla (data taken from Cho et al. 2011). While a trajectory is usually defined as a function that continuously maps time to locations, we see that in this case, we can only observe the user at discrete times, having hours in-between subsequent location updates. Where was the user located in-between these updates? Should we use dead reckoning techniques to interpolate the locations or should be assume that the user stays at a location until next update? Also, users may spoof their location (Zhao and Sui 2017), either to protect their privacy or to gain advantages within the location-based social network. Given this uncertainty, how certain can we be about the location of the user at a given time t? And how does the uncertainty increase as location updates become more sparse and obsolete? The goal of this chapter is to provide a comprehensive overview of models and techniques to deal with uncertainty. To handle uncertainty, we must first remind ourselves that a database models an aspect of the real world, the universe of discourse. Information observed and stored in a database may deviate from the real-world. For reliable decision making, we need to quantify the uncertainty of attribute values stored in the database and consider potentially missing objects that may change mining results.

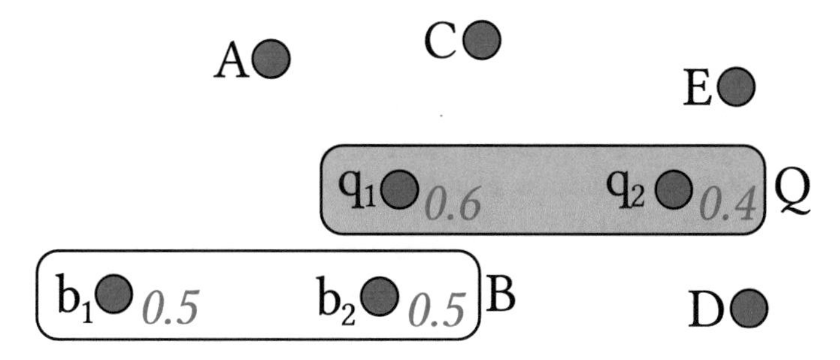

Fig. 14.2 Exemplary uncertain database

Example 1 As a running example used through this chapter, consider Fig. 14.2 which shows a toy uncertain spatial database. In this example, two objects, Q and B have uncertain locations, indicated by alternative locations $\{q_1, q_2\}$ of Q and alternative locations $\{b_1, b_2\}$ of B. In this book chapter, we will survey methods to answer questions such as "What object is closest to Q?", or "What is the probability of B to be one of the two-nearest neighbors of Q?"

To answer such queries, we first need a crisp definition of what it means for an uncertain object to be a (probabilistic) nearest neighbor of a query object and how the probability of such an event is defined. This chapter gives a widely used interpretation of uncertain databases using *Possible Worlds Semantics*. This interpretation allows to answer arbitrary queries on uncertain data, but at a computational cost exponential in the number of uncertain objects. For efficient processing, this chapter defines a paradigm of querying uncertain data that allows to efficiently answer many spatial queries on uncertain spatial data.

This chapter gives a survey on the field of modeling, managing, and querying uncertain spatial data. Parts of this section have been presented in the form of presentation slides at recent conference tutorials at VLDB 2010 (Renz et al. 2010), ICDE 2014 (Cheng et al. 2014), ICDE 2017 (Züfle et al. 2017), and MDM 2020 (Züfle et al. 2020). This section is subdivided to give a survey of definitions, notions and techniques used in the field of querying and mining uncertain spatio-temporal data.

- Section 14.2 presents a survey of state-of-the-art *data representations models* used in the field of uncertain data management. This section explain discrete and continuous models for uncertain objects.
- To interpret queries on a database of uncertain objects, well-defined semantics of uncertain database are required. For this purpose, Sect. 14.3 introduces the *possible world semantics* for uncertain data.
- To run queries on uncertain spatial data, existing systems for uncertain spatial database management are surveyed in Sect. 14.4.
- Given an uncertain database, the result of a probabilistic query can be interpreted in two ways as elaborated in Sect. 14.5. This distinction between different

probabilistic result semantics is not made explicitly in any related work, but is required to gain a deep understanding of problems in the field of querying uncertain spatial data and their complexity.
- Section 14.6 gives an overview over *probabilistic query predicates*. A probabilistic query predicate defines the requirements for the probability of a candidate result to be returned as a query result.
- Section 14.7 introduces a novel paradigm for uncertain data to efficiently answer any kind of query using possible world semantics. This *Paradigm of Equivalent Worlds* generalizes existing solutions by identifying requirements a query must satisfy in order to have a polynomial solution.
- Section 14.8 presents efficient solutions for the problem of computing range queries on uncertain spatial databases. For this purpose, the paradigm of equivalent worlds is leveraged to compute the distribution of the sum of a Poisson-binomial distributed random variable, a problem that is paramount for many spatial queries on uncertain data.
- Section 14.9 gives an overview of specific research problems using uncertain spatial and spatio-temporal data, and surveys state-of-the-art solutions.
- Finally, Sect. 14.10 concludes this book chapter and sketches future research directions that can be opened by leveraging the Paradigm of Equivalent Worlds to new applications and query types.

14.2 Discrete and Continuous Models for Uncertain Data

An object is uncertain if at least one attribute of o is uncertain. The uncertainty of an attribute can be captured in a discrete or continuous way. A discrete model uses a probability mass function (pmf) to describe the location of an uncertain object. In essence, such a model describes an uncertain object by a finite number of alternative instances, each with an associated probability (Kriegel et al. 2007; Pei et al. 2008), as shown in Fig. 14.3a. In contrast, a continuous model uses a continuous probability density function (pdf), like Gaussian, uniform, Zipfian, or

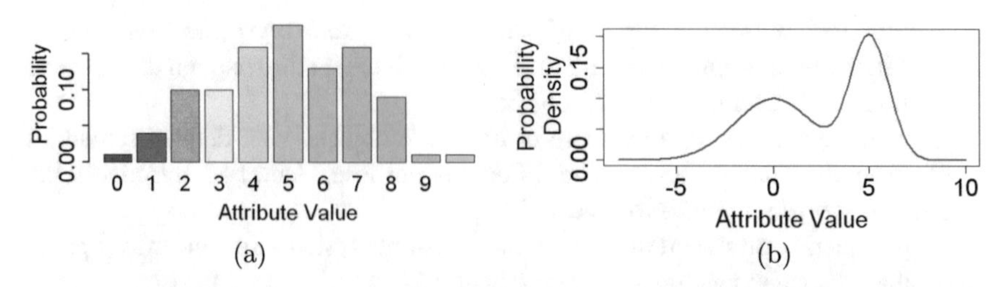

Fig. 14.3 Models for uncertain attributes. (**a**) Discrete probability mass function. (**b**) Continuous prob. density function

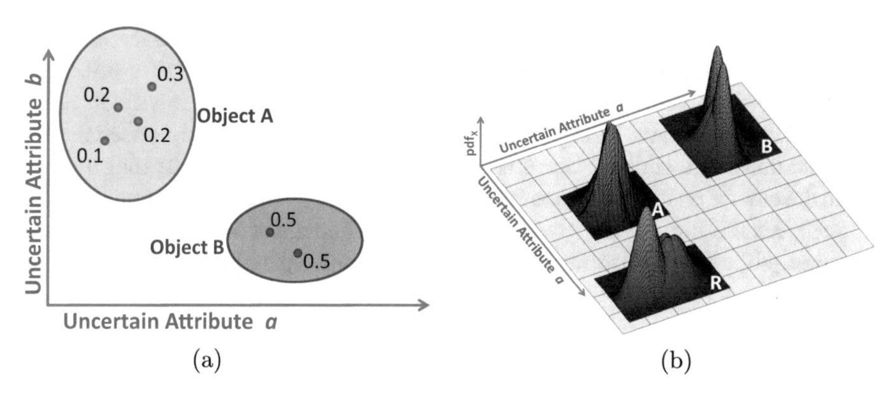

Fig. 14.4 Uncertain objects. (**a**) Discrete case. (**b**) Continuous case

a mixture model, as depicted in Fig. 14.3b, to represent object locations over the space. Thus, in a continuous model, the number of possible attribute values is uncountably infinite. In order to estimate the probability that an uncertain attribute value is within an interval, integration of its pdf over this interval is required (Tao et al. 2005). The random variables corresponding to each uncertain attribute of an object o can be arbitrarily correlated.

To capture positional uncertainty, such models can be applied by treating longitude and latitude (and optionally elevation) as two (three) uncertain attributes. In the case of discrete positional uncertainty, the position of an object A is given by a discrete set $a_1, \ldots, a_m$ of $m \in \mathbb{N}$ possible alternatives in space, as exemplarily depicted in Fig. 14.4a for two uncertain objects A and B. Each alternative a_i is associated with a probability value $p(a_i)$, which may for example be derived from empirical information about the turn probabilities of intersection in an underlying road network. In a nutshell, the position A is a random variable, defined by a probability mass function pdf_A that maps each alternative position a_i to its corresponding probability $p(a_i)$, and that maps all other positions in space to a zero probability. An important property of uncertain spatial databases is the inherent correlation of spatial attributes. In the example shown in Fig. 14.4a it can be observed that the uncertain attributes a and b are highly correlated: given the value of one attribute, the other attribute is certain, as there is no two alternatives of objects A and B having identical attribute values in either attribute.

Clearly, it must hold that the sum of probabilities of all alternatives must sum to at most one:

$$\sum_{i=1}^{m} p(a_i) \leq 1$$

In the case where $\sum_{i=1}^{m} p(a_i) \leq 1$ object A has a non-zero probability of $1 - \sum_{i=1}^{m} p(a_i) \geq 0$ to not exist at all. This case is called *existential uncertainty*, and A is denoted as *existentially uncertain* (Yiu et al. 2009). If the total number of possible instances m is greater than one, A is denoted as *attribute uncertain*. In the context of uncertain spatial data, attribute uncertainty is also referred to as *positional uncertainty* or *location uncertainty*. An object can be both existentially uncertain and attribute uncertain. In Fig. 14.4a, object A is both existentially uncertain and attribute uncertain, while object B is attribute uncertain but does exist for certain.

In the case of continuous uncertainty, the number of possible alternative positions of an object A is infinite, and given by the non-zero domain of the probability density function pdf_x. The probability of A to occur in some spatial region r is given by integration

$$\int_r pdf_A(x)dx.$$

Since arbitrary pdfs may be represented by an uncountably infinite large number of $(position, probability)$ pairs, such pdfs may require infinite space to represent. For this reason, assumptions on the shape of a pdf are made in practice. All continuous models for positionally uncertain data therefore use parametric pdfs, such as Gaussian, uniform, Zipfian, mixture models, or parametric spline representations. For illustration purpose, Fig. 14.4b depicts three uncertain objects modelled by a mixture of gaussian pdfs. Similar to the discrete case, the constraint

$$\int_{\mathbb{R}^d} pdf_A(x)dx \leq 1$$

must be satisfied, where $\mathbb{R}^d$ is a d dimensional vector space. In the case of spatial data, d usually equals two or three. The notion of existentially and attribute uncertain objects is defined analogous to the discrete case.

The following section reviews related work and state-of-the-art on the field of modeling uncertain data.

14.2.1 Existing Models for Uncertain Data

This section gives a brief survey on existing models for uncertain spatial data used in the database community. Many of the presented models have been developed to model uncertainty in relational data, but can be easily adapted to model uncertain spatial data. Since one of the main challenges of modeling uncertain data is to capture correlation between uncertain objects, this section will elaborate details on how state-of-the-art approaches tackles this challenge. Both discrete and continuous models are presented.

14.2.2 Discrete Models

In addition to reviewing related work defining discrete uncertainty models, the aim of this section is to put these papers into context of Sect. 14.2. In particular, models which are special cases or equivalent to the model presented in Sect. 14.2 will be identified, and proper mappings to Sect. 14.2 will be given.

Independent Tuple Model. Initial models have been proposed simultaneously and independently in Fuhr and Rölleke (1997b) and Zimányi (1997). These works assume a relational model in which each tuple is associated with a probability describing its existential uncertainty. All tuples are considered independent from each other. This simple model can be seen as a special case of the model presented in Sect. 14.2, where only existential uncertain but no attribute uncertainty is modelled.

Block-Independent Disjoint Tuples Model and X-Tuple model A more recent and the currently most prominent approach to model discrete uncertainty is the block-independent disjoint tuples model (Dalvi et al. 2009), which can capture mutual exclusion between tuples in uncertain relational databases. A probabilistic database is called block independent-disjoint if the set of all possible tuples can be partitioned into blocks such that tuples from the same block are disjoint events, and tuples from distinct blocks are independent. A commonly used example of a block-independent disjoint tuples model is the *Uncertainty-Lineage Database Model* (Benjelloun et al. 2006; Sarma et al. 2006; Soliman et al. 2007; Yi et al. 2008a,b), also called *X-Relation Model* or simply *X-Tuple Model* that has been developed for relational data. In this model, a probabilistic database is a finite set of *probabilistic tables*. A probabilistic table T contains a set of (uncertain) tuples, where each tuple $t \in T$ is associated with a membership probability value $Pr(t) > 0$. A *generation rule* R on a table T specifies a set of mutually exclusive tuples in the form of $R : t_{r_1} \oplus \ldots \oplus t_{r_m}$ where $t_{r_i} \in T(1 \leq i \leq m)$ and $P(R) := \sum_{i=1}^{m} t_{r_i} \leq 1$. The rule R constrains that, among all tuples $t_{r_1}, \ldots, t_{r_m}$ involved in the rule, at most one tuple can appear in a possible world. The case where $P(R) < 1$ the probability $1 - P(R)$ corresponds to the probability that no tuple contained in rule R exists. It is assumed that for any two rules R_1 and R_2 it holds that R_1 and R_2 do not share any common tuples, i.e., $R_1 \cap R_2 = \emptyset$. In this model, a possible world w is a subset of T such that for each generation rule R, w contains exactly one tuple involved in R if $P(R) = 1$, or w contains 0 or 1 tuple involved in R if $Pr(R) < 1$.

This model can be translated to a discrete model for uncertain spatial data as discussed in Sect. 14.2 by interpreting the set T as the set of all possible locations of all objects, and interpreting each rule R as an uncertain spatial object having alternatives t_{r_i}. The constraint that no two rules may share any common tuples translates into the assumption of mutually independent spatial objects. Finally, the case $P(R) < 1$ corresponds to the case of existential uncertainty (see Sect. 14.2).

A similar block-independent disjoint tuples model is called *p-or-set* (Re et al. 2006) and can be translated to the model described in Sect. 14.2 analogously. In Antova et al. (2008a), another model for uncertainty in relational databases has been proposed that allows to represent attribute values by sets of possible values instead of single deterministic values. This work extends relational algebra by an operator for computing possible results. A normalized representation of uncertain attributes, which essentially splits each uncertain attribute into a single relation, a so-called U-relation, allows to efficiently answer projection-selection-join queries. The main drawback of this model is that it is not possible to compute probabilities of the returned possible results. Sen and Deshpande (2007) propose a model based on a probabilistic graphical model, for explicitly modeling correlations among tuples in a probabilistic database. Strategies for executing SQL queries over such data have been developed in this work. The main drawback of using the proposed graphical model is its complexity, which grows exponential in the number of mutually correlated tuples. This is a general drawback for graphical models such as Bayesian networks and graphical Markov models, where even a *factorized representation* may fail to reduce the complexity sufficiently: The idea of a factorized representation is to identify conditional independencies. For example, if a random variable C depends on random variables A and B, then the distribution of C has to be given relative to all combination of realizations of A and B. If however, C is conditionally independent of A, i.e., B depends on A, C depends on B, and C only transitively depends on A, then it is sufficient to store the distribution of C relative only to the realizations of B. Nevertheless, if for a given graphical model a random variable depends on more than a hand-full of other random variables, then the corresponding model will become infeasible.

And/Xor Tree Model. A very recent work by Li and Deshpande (2009) extends the block-independent disjoint tuples model by adding support for mutual co-existence. Two events satisfy the mutual co-existence correlation if in any possible world, either both happen or neither occurs. This work allows both mutual exclusiveness and mutual co-existence to be specified in a hierarchical manner. The resulting tree structure is called an *and/xor tree*. While theoretically highly relevant, the and/xor tree model becomes impracticable in large database having non-trivial object dependencies, as it grows exponentially in the number of database objects.

If not stated otherwise, this chapter will apply the block-independent disjoint tuples model as model of choice for discrete uncertain data.

14.2.3 Continuous Models

In general, similarity search methods based on continuous models involve expensive integrations of the PDFs, hence special approximation and indexing techniques for efficient query processing are typically employed (Cheng et al. 2004b; Tao et al. 2005). In order to increase quality of approximations, and in order to reduce the computational complexity, a number of models have been proposed making assumptions on the shape of object PDFs. Such assumptions can often be made in applications where the uncertain values follow a specific parametric distribution, e.g. a uniform distribution (Cheng et al. 2003, 2008) or a Gaussian distribution

(Cheng et al. 2008; Deshpande et al. 2004; Patroumpas et al. 2012). Multiple such distributions can be mixed to obtain a mixture model (Tran et al. 2010; Böhm et al. 2006). To approximate arbitrary PDFs, Li and Deshpande (2010a) proposes to use polynomial spline approximations.

14.3 Possible World Semantics

In an uncertain spatial database $\mathcal{D} = \{U_1, \ldots, U_N\}$, the location of an object is a random variable. Consequently, if there is at least one uncertain object, the data stored in the database becomes a random variable. To interpret, that is, to define the semantics of a database that is, in itself, a random variable, the concept of *possible worlds* is described in this section.

Definition 1 (Possible World Semantics) A possible world $w = \{u_1^{a_1}, \ldots, u_N^{a_N}\}$ is a set of instances containing at most one instance $u_i^{a_i} \in U_i$ from each object $U_i \in \mathcal{D}$. The set of all possible worlds is denoted as $\mathcal{W}$. The total probability of an uncertain world $P(w \in \mathcal{W})$ is derived from the chain rule of conditional probabilities:

$$P(w) := P(\bigwedge_{u_i^{a_i} \in w} U_i = u_i^{a_i}) = \prod_{i=1}^{N} P(u_i^{a_i} | \bigwedge_{j<i} u_j^{a_j}). \tag{14.1}$$

By definition, all worlds w having a zero probability $P(w) = 0$ are excluded from the set of possible worlds $\mathcal{W}$. Equation 14.1 can be used if conditional probabilities of the position of objects given the position of other objects are known, e.g. by a given graphical model such as a Bayesian network or a Markov model. In many applications where independence between object locations can be assumed, as well as in applications where only the marginal probabilities $P(u_i^{a_i})$ are known, and thus independence has to be assumed due to lack of better knowledge of a dependency model, the above equation simplifies to

$$P(w) = \prod_{i=1}^{N} P(u_i^{a_i}). \tag{14.2}$$

Example 2 As an example, consider Fig. 14.5 where a database consisting of three uncertain objects $\mathcal{D} = \{U_1, U_2, U_3\}$ is depicted. Objects $U_1 = \{u_1^1, u_1^2\}$ and $U_2 = \{u_2^1, u_2^2\}$ each have two possible instances, while object $U_3 = \{u_3^1, u_3^2, u_3^3\}$ has three possible instances. The probabilities of these instances is given as $P(u_1^1) = P(u_1^2) = 0.5$, $P(u_2^1) = 0.7$, $P(u_2^2) = 0.2$, $P(u_3^1) = 0.5$, $P(u_3^2) = 0.3$, $P(u_3^3) = 0.2$. Note that object U_2 is the only object having existential uncertainty: With a probability of $1 - 0.7 - 0.2 = 0.1$ object U_2 does not exist at all. Assuming independence between spatial objects, the probability for the possible world where

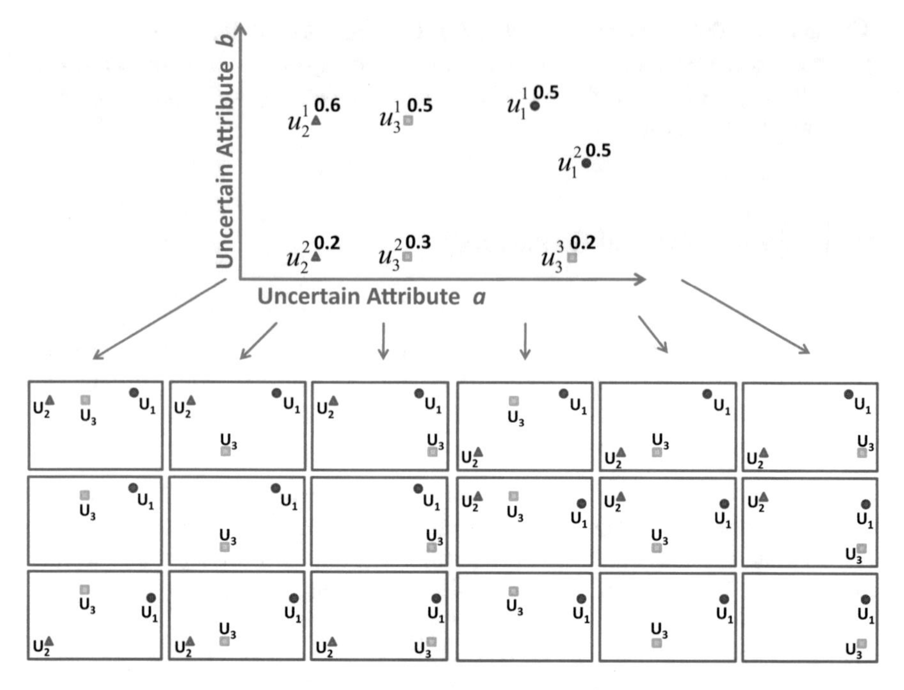

Fig. 14.5 An uncertain database and all of its possible worlds

Table 14.1 Possible worlds corresponding to Fig. 14.5

World	Probability	World	Probability
$\{u_1^1, u_2^1, u_3^1\}$	$0.5 \cdot 0.7 \cdot 0.5 = 0.175$	$\{u_1^2, u_2^1, u_3^1\}$	$0.5 \cdot 0.7 \cdot 0.5 = 0.175$
$\{u_1^1, u_2^1, u_3^2\}$	$0.5 \cdot 0.7 \cdot 0.3 = 0.105$	$\{u_1^2, u_2^1, u_3^2\}$	$0.5 \cdot 0.7 \cdot 0.3 = 0.105$
$\{u_1^1, u_2^1, u_3^3\}$	$0.5 \cdot 0.7 \cdot 0.2 = 0.07$	$\{u_1^2, u_2^1, u_3^3\}$	$0.5 \cdot 0.7 \cdot 0.2 = 0.07$
$\{u_1^1, u_2^2, u_3^1\}$	$0.5 \cdot 0.2 \cdot 0.5 = 0.05$	$\{u_1^2, u_2^2, u_3^1\}$	$0.5 \cdot 0.2 \cdot 0.5 = 0.05$
$\{u_1^1, u_2^2, u_3^2\}$	$0.5 \cdot 0.2 \cdot 0.3 = 0.03$	$\{u_1^2, u_2^2, u_3^2\}$	$0.5 \cdot 0.2 \cdot 0.3 = 0.03$
$\{u_1^1, u_2^2, u_3^3\}$	$0.5 \cdot 0.2 \cdot 0.2 = 0.02$	$\{u_1^2, u_2^2, u_3^3\}$	$0.5 \cdot 0.2 \cdot 0.2 = 0.02$
$\{u_1^1, u_3^1\}$	$0.5 \cdot 0.1 \cdot 0.5 = 0.025$	$\{u_1^2, u_3^1\}$	$0.5 \cdot 0.1 \cdot 0.5 = 0.025$
$\{u_1^1 u_3^2\}$	$0.5 \cdot 0.1 \cdot 0.3 = 0.015$	$\{u_1^2, u_3^2\}$	$0.5 \cdot 0.1 \cdot 0.3 = 0.015$
$\{u_1^1, u_3^3\}$	$0.5 \cdot 0.1 \cdot 0.2 = 0.01$	$\{u_1^2, u_3^3\}$	$0.5 \cdot 0.1 \cdot 0.2 = 0.01$

$U_1 = u_1^1$, $U_2 = u_2^1$ and $U_3 = u_3^1$ is given by applying Equation 14.2 to obtain the product $0.5 \cdot 0.7 \cdot 0.5 = 0.175$. All possible worlds spanned by $\mathcal{D}$ are depicted in Fig. 14.5. The probability of each possible world is shown in Table 14.1, including possible worlds where U_2 does not exist.

Recall that a predicate can evaluate to either true or false on a crisp (non-uncertain) database. An exemplary predicate is *There are at least five database objects in a 500 m range of the location "Theresienwiese, Munich"*. To evaluate a predicate ϕ on an uncertain database using possible world semantics, the query predicate is evaluated on each possible world. The probability that the query predicate evaluates to true is defined as the sum of probabilities of all worlds where ϕ is satisfied, formally:

Definition 2 Let $\mathcal{D}$ be an uncertain spatial database inducing the set of possible worlds $\mathcal{W}$, let ϕ be some query predicate, and let

$$\mathcal{I}(\phi, w \in \mathcal{W}) := P(\phi(\mathcal{D})|\mathcal{D} = w) \in \{0, 1\}$$

be the indicator function that returns one if world w satisfies ϕ and zero otherwise. The marginal probability $P(\phi(\mathcal{D}))$ of the event $\phi(\mathcal{D})$ that predicate ϕ holds in $\mathcal{D}$ is defined as follows using the theorem of total probability (Zwillinger and Kokoska 2000):

$$P(\phi(\mathcal{D})) = \sum_{w \in \mathcal{W}} \mathcal{I}(\phi, w) \cdot P(w) \tag{14.3}$$

The main challenge of analyzing uncertain data is to efficiently and effectively deal with the large number of possible worlds induced by an uncertain database $\mathcal{D}$. In the case of continuous uncertain data, the number of possible worlds is uncountably infinite and expensive integration operations or numerical approximation are required for most spatial database queries and spatial data mining tasks. Even in the case of discrete uncertainty, the number of possible worlds grows exponentially in the number of objects: in the worst case, any combination of alternatives of objects may have a non-zero probability, as shown exemplary in Fig. 14.5. This large number of possible worlds makes efficient query processing and data mining an extremely challenging problem. In particular, any problem that requires an enumeration of all possible worlds is #P-hard.[1] In particular, a number of probabilistic problems have been proven to be in #P (Valiant 1979). Following this argumentation, general query processing in the case of discrete data using object independence has proven to be a #P-hard problem (Dalvi and Suciu 2004) in the context of relational data. The spatial case is a specialization of the relation case, but clearly, the spatial case is in #P as well, which becomes evident by construction of a query having an exponentially large result, such as the query that returns all possible worlds. Consequently, there can be no universal solution that allows to answer *any* query in polynomial time. This implies that querying processing on models that are generalizations of the discrete case with object independence, e.g.,

[1]#P is the set of counting problems associated with decision problems in the class NP. Thus, for any NP-complete decision problem which asks if there exists a solution to a problem, the corresponding #P problem asks for the number of such solutions.

models using continuous distribution, or models that relax the object independence assumption, must also be a #P hard problem. The result of Dalvi and Suciu (2004) implies that there exists query predicates, for which no polynomial time solution can be given. Yet, this result does not outrule the existence of query predicates that can be answered efficiently. For example the (trivial) query that always returns the empty set of objects can be efficiently answered on uncertain spatial databases.

14.4 Existing Uncertain Spatial Database Management Systems

Recently developed systems provide support for spatio-temporal data in big data systems (Akdogan et al. 2010; Aji et al. 2013; Lu et al. 2012; Wang et al. 2010; Zhang et al. 2012). Such systems exhibit high scalability for batch-processing jobs (Apache; Dean and Ghemawat 2008), but do not provide efficient solutions to handle uncertain data and to assess the reliability of results. The vivid field of managing, querying, and mining uncertain data has received tremendous attention from the database, data mining, and spatial data science communities. Recent books (Aggarwal 2010) and survey papers (Aggarwal and Philip 2008; Wang et al. 2013; Li et al. 2018) provide an overview of the flurry of research papers that have appeared in these fields.

The problem of managing uncertain data has been well-studied by the database research community in the past. While the traditional database literature (Cavallo and Pittarelli 1987; Barbará et al. 1992; Bacchus et al. 1996; Lakshmanan et al. 1997; Fuhr and Rölleke 1997a) has studied the problem of managing uncertain data, this research field has seen a recent revival, due to modern techniques for collecting inherently uncertain data. Most prominent concepts for probabilistic data management are MayBMS (Antova et al. 2008b), MystiQ (Boulos et al. 2005), Trio (Agrawal et al. 2006), and BayesStore (Wang et al. 2008). These uncertain database management systems (UDBMS) provide solutions to cope with uncertain relational data, allowing to efficiently answer traditional queries that select subsets of data based on predicates or join different datasets based on conditions. Extensions to the UDBMS also allow answering of important classes of spatial queries such as top-k and distance-ranking queries (Hua et al. 2008; Cormode et al. 2009a; Li et al. 2009a; Bernecker et al. 2010; Li and Deshpande 2010b). While these existing UDBMS provide probabilistic guarantees for their query results, they offer no support for data mining tasks. A likely reason for this gap is the theoretic result of Dalvi and Suciu (2007) which shows that the problem of answering complex queries is #P-hard in the number of database objects. To illustrate this theoretic result, imagine running a simple range query with an arbitrary query point on a database having N objects each having an arbitrary non-zero probability of being in that range. Further, assume stochastic independence between these objects. In that case, any of the 2^N combinations of result objects becomes a possible result and must be returned.

Nevertheless, a number of polynomial time solutions have been proposed in the literature for various spatial query types such as nearest neighbor queries (Cheng et al. 2004a, 2008; Kriegel et al. 2007; Iijima and Ishikawa 2009), k-nearest neighbor queries (Beskales et al. 2008; Ljosa and Singh 2007; Li et al. 2009b; Cheng et al. 2009) and (similarity-) ranking queries (Bernecker et al. 2008; Cormode et al. 2009b; Li et al. 2009b; Soliman and Ilyas 2009). On first glance, these findings may look contradicting (unless $P = NP$), providing polynomial-time solution to a #P-hard problem. On closer look, it shows that different related work use different semantics to interpret a result. Aforementioned related works that provide polynomial time solutions for spatial queries on uncertain data make a simplifying assumption: Rather than computing the probability for each possible result, they compute the probability of each *object* to be part of the result. This reduces the number of probabilities that have to be reported, in the worst-case, from a number exponential in the number of database objects, to a linear number. Re-using the example of a range query on an uncertain database, it is possible to compute the probability that a single object is within the query range independent from all other objects.

Unfortunately, this simplification also yields a loss of information, as it is not possible to infer the probability of query results given only probabilities of individual objects. Let us revisit the running example from introduction, which is duplicate in Fig. 14.6 for convenience. This example will illustrate how such an object-based approach, which computes object-individual probabilities, rather than the probabilities of result sets, may yield misleading results.

Example 3 Assume that the task is to simply find the probabilistic two nearest neighbors (2NN) of uncertain object Q. Objects Q and B have two alternative positions each, yielding a total of four possible worlds. For example, in one possible world, where Q has location q_1 and B has location b_1, the two nearest neighbors of Q are A and C. This possible world has a probability of $0.6 \cdot 0.5 = 0.3$, obtained by assuming stochastic independence between objects. Following object-based result semantics, we can obtain probabilities of 0.3, 0.3, 0.6, 0.4, 0.4 for objects A, B, C, D, and E to be the 2NNs of Q, respectively. However, this result hides any dependence between these result objects, such as objects A and B are mutually exclusive, while D and E are mutually inclusive.

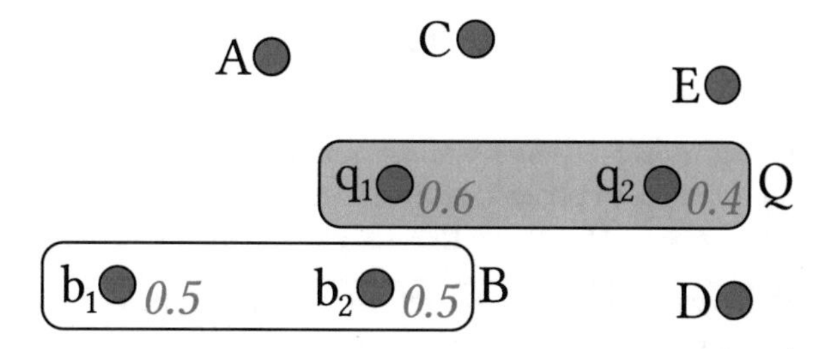

Fig. 14.6 The exemplary uncertain database from Fig. 14.2

Towards approximate solutions, the Monte-Carlo DB (MCDB) system (Jampani et al. 2008) has been proposed, which samples possible worlds from the database, executes the query predicate on each sampled world. MCDB estimates the probability of each object to be part of the result set. However, this approach of assigning a result probability to each object, as illustrate in the example above, cannot be extended to assess the probability of result sets. The problem is that the number of possible result sets may be exponentially large. To aggregate possible worlds into groups of mutually similar worlds (having similar results), an approach has been proposed for clustering of uncertain data (Züfle et al. 2014; Schubert et al. 2015) and more recently for general query processing on spatial data (Schmid and Züfle 2019). Revisiting the example of Fig. 14.2, this approach reports the results of a probabilistic query 2NN query as $\{A, C\}$, $\{B, C\}$, $\{D, E\}$, having respective probabilities of 0.3, 0.3, and 0.4. However, this approach (Schmid and Züfle 2019) can only be applied to spatial queries that return result sets, thus cannot be applied to more complex spatial queries or data mining tasks. To further elaborate the difference between solutions that compute the probability of each object to be part of the result, and solutions that compute the probability of each result, the following section will further survey the two different "Probabilistic Result Semantics": Object-based and Result-based.

14.5 Probabilistic Result Semantics

Recall that a spatial similarity query always requires a query object q and, informally speaking, returns objects to the user that are similar to q. In the case of uncertain data, there exists two fundamental semantics to describe the result of such a probabilistic spatial similarity query. These different result semantics will be denoted as *object based result semantics* and the *result based result semantics*. Informally, the former semantics return possible *result objects* and their probability of being part of the result, while the later semantics return possible results, which consist of a single object, of a set of objects or of a sorted list of objects depending on the query predicate, and their probability of being the result as a whole.

14.5.1 Object Based Probabilistic Result Semantics

Using *object based probabilistic result semantics*, a probabilistic spatial query returns a set of objects, each associated with a probability describing the individual likelihood of this object to satisfy the spatial query predicate.

Definition 3 (Object Based Result Semantics) Let $\mathcal{D}$ be an uncertain spatial database, let q be a query object and let ϕ denote a spatial query predicate. Under object based (OB) probabilistic result semantics, the result of a probabilistic spatial

ϕ query is a set $\phi_{OB}(q, \mathcal{D}) = \{(o \in \mathcal{D}, P(o \in \phi_{OB}(q, \mathcal{D})))\}$ of pairs. Each pair consists of a result object o and its probability $P(o \in \phi_{OB}(q, \mathcal{D}))$ to satisfy ϕ. Applying possible world semantics (cf. Definition 1) to compute the probability $P(o \in \phi_{OB}(q, \mathcal{D}))$ yields

$$P(o \in \phi_{OB}(q, \mathcal{D})) = \sum_{w \in \mathcal{W}, o \in \phi(q,w)} P(w), \tag{14.4}$$

where $\phi(q, w)$ is the deterministic result of a spatial ϕ query having query object q applied to the deterministic database defined by world w.

Formally, the result of a probabilistic spatial query under object based result semantics is a function

$$\phi_{OB}(q, \mathcal{D}) : \mathcal{D} \to [0, 1]$$

$$o \mapsto P(o \in \phi_{OB}(q, \mathcal{D})).$$

mapping each object o in $\mathcal{D}$ (the results) to a probability value.

Example 4 Figure 14.7 depicts a database containing objects $\mathcal{D} = \{A, B, C\}$. Objects A and B have two alternative locations each, while the position of C is known for certain. The locations and the probabilities of all alternatives are also depicted in Fig. 14.7. This leads to a total number of four possible worlds. For example, in world w_1 where $A = a_1$, $B = b_1$ and $C_1 = c_1$, object A is closest to q, followed by objects B and C. Assuming inter-object independence, the probability of this world is given by the product of individual instance probabilities $P(w_1) = P(a_1) \cdot P(b_1) \cdot P(c_1) = 0.04$. The ranking of each possible world and the corresponding probability is also depicted in Fig. 14.7. For a probabilistic $2NN$ query for the depicted query object q, the object based result semantic computes the probability of each object to be in the two-nearest neighbor set of q. For object A, the probability $P(A)$ of this event equals 0.1, since there exists exactly two possible

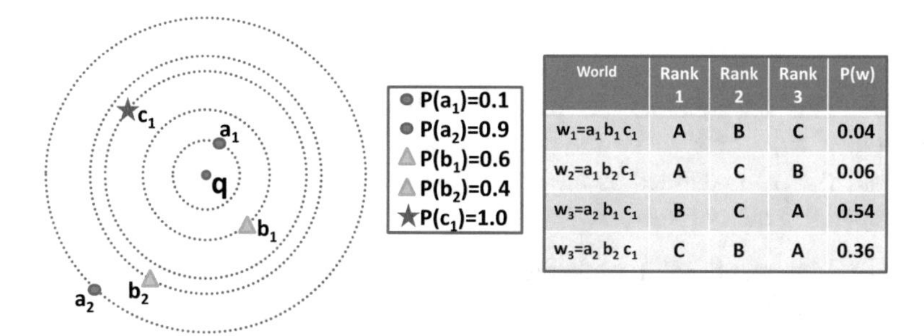

World	Rank 1	Rank 2	Rank 3	P(w)
$w_1=a_1\,b_1\,c_1$	A	B	C	0.04
$w_2=a_1\,b_2\,c_1$	A	C	B	0.06
$w_3=a_2\,b_1\,c_1$	B	C	A	0.54
$w_3=a_2\,b_2\,c_1$	C	B	A	0.36

Fig. 14.7 Example Database showing possible positions of uncertain objects and their corresponding probabilities

worlds w_1 and w_2 with a total probability of $0.04 + 0.06 = 0.1$ in which A is on rank one or on rank two, yielding a result tuple $(A, 0.1)$. The complete result of a $P2NN$ query under object based result semantics is $\{(A, 0.1), (B, 0.94), (C, 0.96)\}$. Note that in general, objects having a zero probability are included in the result. For instance, assume an additional object D such that all instances of D have a distance to q greater than the distance between q and b_2. In this case, the pair $(D, 0)$ would be part of the result.

The result of a query under object based probabilistic result semantics contains one result tuple for every single database object, even if the probability of the corresponding object to be a result is very low or zero. In many applications, such results may be meaningless. Therefore, the size of the result set can be reduced by using a probabilistic query predicate as explained later in Sect. 14.6. A computational problem is the computation of the probability $P(o \in \mathcal{D})$ of an object o to satisfy the spatial query predicate. In the example, this probability was derived by iterating over the set of all possible worlds $w_1, \ldots, w_4$. Since this set grows exponentially in the number of objects, such an approach is not viable in practice. Therefore, efficient techniques to compute the probability values $P(o)$ are required. A general paradigm to develop algorithms that avoid an explicit enumeration of all possible worlds is presented in Sect. 14.7.

14.5.2 Result Based Probabilistic Result Semantics

In the case of result based result semantics, possible result sets of a probabilistic spatial query are returned, each associated with the probability of this result.

Definition 4 (Result Based Result Semantics) Let $\mathcal{D}$ be an uncertain spatial database, let q be a query object and let ϕ denote a spatial query predicate. Under result based (RB) result semantics, the result of a probabilistic spatial ϕ query is a set

$$\phi_{RB}(q, \mathcal{D}) = \{(r, P(r))|r \subseteq \mathcal{D}, P(r) = \sum_{w \in \mathcal{W}, \phi(q,w)=r} P(w)\}$$

of pairs. This set contains one pair for each result $r \subseteq \mathcal{D}$ associated with the probability $P(r)$ of r to be the result. Following possible world semantics, the probability $P(r)$ is defined as the sum of probabilities of all worlds $w \in \mathcal{W}$ such that a spatial ϕ query returns r.

Formally, the result of a probabilistic spatial query under result based result semantics is a function

$$\phi_{RB}(q, \mathcal{D}) : \overline{\mathcal{P}}(\mathcal{D}) \rightarrow [0, 1]$$

$$r \mapsto P(r).$$

mapping a elements of the power set $\overline{\mathcal{P}}(\mathcal{D})$ (the results) to probability values.

Example 5 For a probabilistic $2NN$ query for the depicted query object q, result based result semantics require to compute the probability of each subset of $\{A, B, C\}$ to be in the two-nearest neighbor set of q. For the set $\{B, C\}$, the probability of this event is 0.90, since there is two possible worlds w_3 and w_4 with a total probability of $0.54 + 0.36 = 0.9$ in which B and C are both contained in the $2NN$ set of q. Note that in worlds w_3 and w_4 objects B and C appear in different ranking positions. This fact is ignored by a kNN query, as the results are returned unsorted. In this example, the complete result of a $P2NN$ query under object based result semantics is $\{(\{A, B, C\}, 0), (\{A, B\}, 0.04), (\{A, C\}, 0.06), (\{B, C\}, 0.90), (\{A\}, 0), (\{B\}, 0), (\{C\}, 0), (\{\emptyset\}, 0)\}$.

Clearly, the result of a query using result based result semantics can be used to derive the result of an identical query using object based result semantics. For instance, the result of Example 5 implies that the probability of object A to be a $2NN$ of q is 0.10, since there exists two possible results using result based result semantics, namely $(\{A, B\}, 0.04)$ and $(\{A, C\}, 0.06)$ having a total probability of $0.04 + 0.06 = 0.1$, which matches the result of Example 4.

Lemma 1 *Let q be the query point of a probabilistic spatial ϕ query. It holds that the result of this query using object based result semantics $\phi_{OB}(q, \mathcal{D})$ is functionally dependent of the result of this query using result based result semantics. The set $PS\phi Q_{OB}(q, \mathcal{D})$ can be computed given only the set $PS\phi Q_{RB}(q, \mathcal{D})$ as follows:*

$$PS\phi Q_{OB}(q, \mathcal{D}) = \{(o, P(o)) | o \in \mathcal{D} \wedge P(o) = \sum_{(r, P(r)) \in PS\phi Q_{RB}(q, \mathcal{D}), o \in r} P(r)\}$$

Proof Let $\mathcal{W}$ denote the set of possible worlds of $\mathcal{D}$, and let $p(w \in \mathcal{W})$ denote the probability of a possible world. Furthermore, let

$$w_{S \subseteq \mathcal{D}} := \{w \in \mathcal{W} | \phi(q, w) = S\}$$

denote the set of possible worlds such that $\phi(q, w) = S$, i.e., such that the predicate that a ϕ query using query object q returns set S holds. In each world w, query q returns exactly one deterministic result $PS\phi Q_{RB}(q, w)$. Thus, the sets $w_{S \subseteq \mathcal{D}}$ represent a complete and disjunctive partition of $\mathcal{W}$, i.e., it holds that

$$\mathcal{W} = \bigcup_{S \subseteq \mathcal{D}} w_S \tag{14.5}$$

and

$$\forall R, S \in \overline{\mathcal{P}}(\mathcal{D}) : R \neq S \Rightarrow w_R \bigcap w_S = \emptyset. \tag{14.6}$$

Using Equations 14.5 and 14.6, we can rewrite Equation 14.4

$$P(o \in \phi_{OB}(q, \mathcal{D})) = \sum_{w \in \mathcal{W}, o \in \phi(q,w)} P(w)$$

as

$$P(o \in \phi_{OB}(q, \mathcal{D})) = \sum_{S \in \overline{\mathcal{P}}(\mathcal{D})} \sum_{w \in w_S, o \in \phi(q,w)} P(w).$$

By definition, query q returns the same result for each world in $w \in w_S$. This result contains object o if $o \in S$. Thus we can rewrite the above equation as

$$P(o) = \sum_{S \in \overline{\mathcal{P}}(\mathcal{D}), o \in S} P(S).$$

The probabilities $P(S)$ are given by function $PS\phi Q_{RB}(q, \mathcal{D})$. □

In the above proof, we have performed a linear-time reduction of the problem of answering probabilistic spatial queries using object based result semantics to the problem of answering probabilistic spatial queries using result based result semantics. Thus, we have shown that, except for a linear factor (which can be neglected for most probabilistic spatial query types, since most algorithm run in no better than log-linear time), the problem of answering a probabilistic spatial query using result based result semantics is at least as hard as answering a probabilistic spatial query using object based semantics.

To summarize this section, we have learned about two different semantics to interpret the result of a spatial query on uncertain data: Object Based and Result Based. Understanding the difference of both result semantics is paramount to understand the landscape of existing research: in some related publication the problem of answering some probabilistic query may be proven to be in $\#P$, while another publication gives a solution that lies in P-TIME for the same spatial query predicate and the same probabilistic query predicate. In such cases, different result semantics may explain these results without assuming $P = NP$.

14.6 Probabilistic Query Predicates

Generally, in an uncertain database, the question whether an object satisfies a given query predicate ϕ, such as being in a specified range or being a kNN of a query object, cannot be answered deterministically due to uncertainty of object locations. Due to this uncertainty, the predicate that an object satisfies ϕ is a random variable, having some (possibly zero, possibly one) probability. A probabilistic query predicate quantifies the minimal probability required for a result to qualify as a result that is sufficiently significant to be returned to the user. This section formally define probabilistic query predicate for general query predicates. The following definition are made for uncertain data in general, but can be applied analogously for uncertain spatial data.

A *probabilistic query* can be defined without any probabilistic query predicate. In this case, all objects, and their respective probabilities are returned.

Definition 5 (Probabilistic Query) Let $\mathcal{D}$ be an uncertain database, let q be a query point and let ϕ be a query predicate. A *probabilistic query* $\phi(q, \mathcal{D})$ returns all database objects $o \in \mathcal{D}$ together with their respective probability $P(o \in \phi(q, \mathcal{D}))$ that o satisfies ϕ.

$$\phi(q, \mathcal{D}) = \{(o \in \mathcal{D}, P(o \in \phi(q, \mathcal{D})))\} \tag{14.7}$$

The term *probabilistic* query is simply derived from the fact that unlike a traditional query, a probabilistic query result has probability values associated with each result. The main challenge of answering a probabilistic query, is to compute the probability $P(o \in \phi(q, \mathcal{D}))$ for each object. Using possible world semantics, a probabilistic query can be answered by evaluating the query predicate for each object and each possible world, i.e.,

$$P(o \in \phi(q, \mathcal{D})) := \sum_{w \in \mathcal{W} find(\phi, w) \cdot P(w)} .$$

But clearly, it is necessary to avoid the combinatorial growth that would be induced by this “naive” evaluation method.

Example 6 For example, consider the query *“Return all friends of user q having a spatial distance of less than 100m to q”* depicted in Fig. 14.8. Thus, the predicate ϕ is a 100 m-range predicate using query point q. We can deterministically tell that friend A must be within $\epsilon = 100$ m Euclidean distance of q, while friends E and F cannot possibly be in range. The pairs $(A, 1)$, $(E, 0)$ and $(F, 0)$ are added to the result. For friends B, C and D, this predicate cannot be answered deterministically. Here, friend B has some possible positions located inside the 100 m range of q, while other possible positions are outside this range. The two locations inside q's range have a probability of 0.1 and 0.2, respectively, thus the total probability of object B to satisfy the query predicate is $0.1 + 0.2 = 0.3$. The pair $(B, 0.3)$ is thus added to the result. The pairs $(C, 0.2)$ and $(D, 0.9)$ complete the result 100 m-range$(q, \mathcal{D}) = \{(A, 1), (B, 0.3), (C, 0.2), (D, 0.9), (E, 0), (F, 0)\}$.

The immediate question in the above example is: “Is a probability of 0.3 sufficient to warrant returning B as a result?”. To answer this question, a probabilistic query can explicitly specify a probabilistic query predicate, to specify the requirements, in terms of probability, required for an object to qualify to be included in the result. The following subsections briefly survey the most commonly used probabilistic query predicates: probabilistic threshold queries and probabilistic Top$k$ queries.

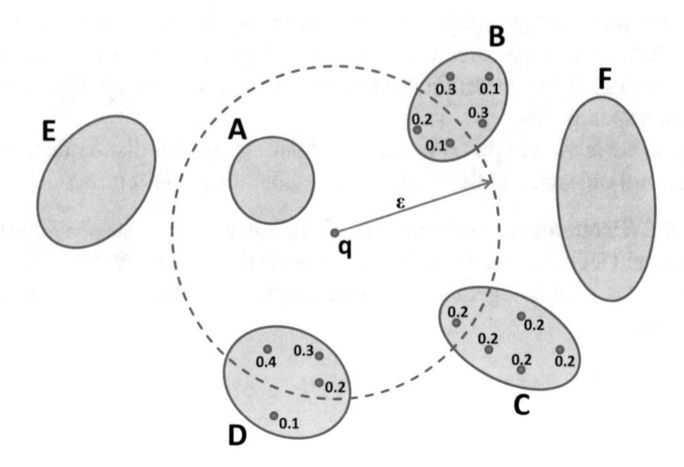

Fig. 14.8 Example of an uncertain ϵ-range query. Object A is a true hit, objects B, C and D are possible hits

14.6.1 Probabilistic Threshold Queries

This paragraph defines a probabilistic query predicate that allows to return only results that are statistically significant.

Definition 6 (Probabilistic Threshold Query(PτQ)) Let $\mathcal{D}$ be an uncertain (spatial) database, let q be a spatial query object, let $0 \leq \tau \leq 1$ be a real value and let ϕ be a spatial query predicate. A *probabilistic τ query* (PτQ) returns all objects $o \in \mathcal{D}$ such that o has a probability of at least τ to satisfy $\phi(q, \mathcal{D})$:

$$P\tau\phi(q, \mathcal{D}) := \{o \in \mathcal{D} | P(o \in \phi(q, \mathcal{D})) \geq \tau\}.$$

Example 7 In Fig. 14.8, a probabilistic threshold 100 m-range(q,$\mathcal{D}$) query with $\tau =$ 0.5 query returns the set of objects $P0.5$ 100 m-range$(q, \mathcal{D}) = \{A, D\}$, since objects A and D are the only objects such that their total probability of alternatives inside the query region is equal or greater to $\tau = 0.5$.

Semantically, a probabilistic threshold spatial query returns all results having a statistically significant probability to satisfy the query predicate. Therefore, the probabilistic threshold query serves as a statistical test of the hypothesis "o is a result" at a significance level of τ. This test uses the probability $P(o \in \phi(q, \mathcal{D}))$ as a test statistic. Efficient algorithms to compute this probability $P(o \in \phi(q, \mathcal{D}))$, for the example of kNN and similarity ranking queries will be surveyed in Sect. 14.8 similarity ranking queries and RkNN queries.

A probabilistic threshold query on uncertain spatial data is useful in applications, where the parameters of the spatial predicate τ (e.g. the range of an ϵ-range query, or the parameter k of a kNN query), as well as the probabilistic threshold τ are chosen wisely, requiring expert knowledge about the database $\mathcal{D}$. If these parameters are chosen inappropriately, no results may be returned, or the set of returned result may grow too large. For example, if τ is chosen very large, and if the database has a high grade of uncertainty, then no result may be returned at all. Analogously, if the parameter ϵ is chosen too small then no result will be returned, while a too large value of ϵ may return all objects. The special case of having $\epsilon = 0$, i.e., the case of returning all possible results (having a non-zero probability), is often used as default if no other probabilistic query predicate is specified (e.g. Soliman et al. 2007; Yi et al. 2008a). This case may be referred to as a *possibilistic query predicate*, as all possible results (regardless of their probability) are returned.

14.6.2 *Probabilistic Topk Queries*

In cases where insufficient information is given to select appropriate parameter values, the following probabilistic query predicate is defined to guarantee that only the k most significant results are returned.

Definition 7 (Probabilistic Top*k* Query (PTop*k*Q)) Let $\mathcal{D}$ be an uncertain spatial database, let q be a spatial query object, let $1 \leq k \leq |\mathcal{D}|$ be a positive integer, and let ϕ be a spatial query predicate. A *probabilistic spatial Topk query* (PTop$k$Q) returns the smallest set PTop$k\phi(q,\mathcal{D})$ of at least k objects such that

$$\forall U_i \in \text{PTop}k\phi(q, \mathcal{D}), U_j \in \mathcal{D} \backslash \text{PTop}k\phi(q, \mathcal{D}) : P(U_i \in \phi(q, \mathcal{D})) \geq P(U_j \in \phi(q, \mathcal{D}))$$

Thus, a probabilistic spatial Top$k$ query returns the k objects having the highest probability to satisfy the query predicate. Again, in case of ties, the resulting set may be greater than k.

Example 8 In Fig. 14.8, a PTop3 ϕ query using a $\phi = 100$ m-range spatial predicate returns objects $PTop3\,100\,\text{m-range}(q, \mathcal{D}) = \{A, B, D\}$, since these objects have the highest probability to satisfy the spatial predicate, i.e., have the highest probability to be located in the spatial 100 m-range.

Note, that the probabilistic Top$k$ query predicate can be combined with a kNN spatial query, i.e., with the case where $\phi = kNN$. Such a probabilistic Top$k$ jNN query returns the set of k objects having the highest probability, to be j-nearest neighbor of the query object. Clearly, k and j may have different integer values, such that differentiation is needed.

14.6.3 Discussion

In summary, a probabilistic spatial query is defined by two query predicates:

- A spatial predicate ϕ to select uncertain objects having sufficiently *high proximity* to the query object, and
- a probabilistic predicate ψ, to select uncertain objects having sufficiently *high probability* to satisfy ϕ.

It has to be mentioned, that alternatively to this definition, a single predicate can be used, that combines both spatial and probabilistic features. For example, a monotonic score function can be utilized, which combines spatial proximity and probability to return a single scalar score. An example of such a monotone score function is the expected distance function

$$E(\text{dist}(q, U \in \mathcal{D})) = \sum_{u \in U} P(u) \cdot \text{dist}(q, u),$$

where q is the query object, and $\mathcal{D}$ is an uncertain database. The expected support function is utilized by a number of related publications, such as Ljosa and Singh (2007) and Cormode et al. (2009b). Using such a monotone score function, objects with a sufficiently high score can be returned. The advantage of using such an approach, is that objects that are located very close to the query require a lower probability to be returned as a result, while objects that are located further away from the query object require a higher probability. Yet, the main problem of such a combined predicate, is that the probability of an object is treated as a simple attribute, thus losing its probabilistic semantic. Thus, the resulting score is very hard to interpret. An object that has a high score, may indeed have a very low probability to exist at all, because it is located (if it exists) very close to the query object. Consequently, the score itself no longer contains any confidence information, and thus, it is not possible to answer queries according to possible world semantics using a single aggregate, such as expected distance, only.

14.7 The Paradigm of Equivalent Worlds

In Sect. 14.3 the concept of possible world semantics has been introduced. Possible world semantics give an intuitive and mathematically sound interpretation of an uncertain spatial database. Furthermore, queries that adhere to possible world semantics return unbiased results, by evaluating the query on each possible world. Since any such approach requires to run queries on an exponential number of worlds, any naive approach is infeasible. Yet, for specific settings, such as specific result-based semantics, specific spatial query predicates and specific probabilistic query predicates, the literature has shown that it is possible to efficiently answer queries on uncertain data. While it is hardly feasible to enumerate all combinations of result-based semantics, spatial query predicates and probabilistic query predicates, this section introduces a general paradigm to find such a solution yourself. In a nutshell, the idea is to find, among the exponentially large set of possible worlds, a partitioning into a polynomially large number of subsets, which are equivalent for a given query.

14.7.1 Equivalent Worlds

The goal of this section is introduce a general paradigm to efficiently compute exact probabilities, while still adhering to possible world semantics. For this purpose, reconsider Definition 2, defining the probability that some predicate ϕ is satisfied in an uncertain database $\mathcal{D}$ as the total probability of all possible worlds satisfying ϕ. Recall Equation 14.3

$$P(\phi(\mathcal{D})) = \sum_{w \in \mathcal{W}} \mathcal{I}(\phi, w) \cdot P(w),$$

where $\mathcal{W}$ is the set of all possible worlds; $\mathcal{I}(\phi, w)$ is an indicator function that returns one if predicate ϕ holds (i.e., resolves to true) in the crisp database defined by world w and zero otherwise, and $P(w)$ is the probability of world w. To reduce the number of possible worlds that need to be considered to compute $P(\phi(\mathcal{D}))$, we first need the following definition.

Definition 8 (Class of Equivalent Worlds) Let ϕ be a query predicate and let $S \subseteq \mathcal{W}$ be a set of possible worlds such that for any two worlds $w_1, w_2 \in S$ we can guarantee that ϕ holds in world w_1 if an only if ϕ holds in world w_2, i.e.,

$$\forall w_1, w_2 \in S : \mathcal{I}(\phi, w_1) \Leftrightarrow \mathcal{I}(\phi, w_2)$$

Then set S is called a *class of worlds equivalent with respect to* ϕ. In the remainder of this chapter, if the spatial query predicate ϕ is clearly given by the context, then S will simply be denoted as a *class of equivalent worlds*. Any worlds $w_i, w_j \in S$ are denoted as *equivalent worlds*.

We now make the following observation:

Corollary 1 *Let* $S \subseteq \mathcal{W}$ *be a class of worlds equivalent with respect to* ϕ *(cf. Definition* 8*), we can rewrite Equation* 14.3 *as follows:*

$$P(\phi(\mathcal{D})) = \sum_{w \in \mathcal{W}} \mathcal{I}(\phi, w) \cdot P(w) \Leftrightarrow$$

$$P(\phi(\mathcal{D})) = \sum_{w \in \mathcal{W} \setminus S} \mathcal{I}(\phi, w) \cdot P(w) + \mathcal{I}(\phi, w \in S) \cdot \sum_{w \in S} P(w). \tag{14.8}$$

Proof Due to the assumption that for any two worlds $w_1, w_2 \in S$ it holds that ϕ holds in world w_1 if an only if ϕ holds in world w_2, we get $\mathcal{I}(\phi, w_1) = 1 \Leftrightarrow \mathcal{I}(\phi, w_2) = 1$ and $\mathcal{I}(\phi, w_1) = 0 \Leftrightarrow \mathcal{I}(\phi, w_2) = 0$ by definition of function $\mathcal{I}$. Due to this assumption, we have to consider two cases.

Case 1: $\forall w \in S : \mathcal{I}(\phi, w) = 0$
In this case, both Equations 14.3 and 14.8 can be rewritten as

$$P(\phi(\mathcal{D})) = \sum_{w \in \mathcal{W} \setminus S} \mathcal{I}(\phi, w) \cdot P(w).$$

Case 2: $\forall w \in S : \mathcal{I}(\phi, w) = 1$
In this case, both Equations 14.3 and 14.8 can be rewritten as

$$P(\phi(\mathcal{D})) = \sum_{w \in \mathcal{W} \setminus S} \mathcal{I}(\phi, w) \cdot P(w) + \sum_{w \in S} P(w)$$

□

The only difference between both cases is the additive term $\sum_{w \in S} P(w)$, which exists only in Case 2. The indicator function $\mathcal{I}(\phi, w \in S)$ ensures that this term is only added in the second case. As main purpose, Corollary 1 states that, given a set of equivalent worlds S, we only have to evaluate the indictor function $\mathcal{I}(\phi, w)$ on a single representative world $w \in S$, rather than on each world in S. This allows to reduce the number of (crisp) ϕ queries required to compute Equation 14.3 by $|S| - 1$.

Corollary 1 leads to the following Lemma.

Lemma 2 *Let $\mathcal{S}$ be a partitioning of $\mathcal{W}$ into disjoint sets such that $\bigcup_{S \in \mathcal{S}} S = \mathcal{W}$ and for all $S_1, S_2 \in \mathcal{S} : S_1 \cap S_2 = \emptyset$. Equation* 14.3 *can be rewritten as*

$$P(\phi(\mathcal{D})) = \sum_{w \in \mathcal{W}} \mathcal{I}(\phi, w) \cdot P(w) \Leftrightarrow$$

$$P(\phi(\mathcal{D})) = \sum_{S \in \mathcal{S}} \mathcal{I}(\phi, w \in S) \cdot \sum_{w \in S} P(w). \tag{14.9}$$

Proof Lemma 2 is derived by applying Corollary 1 once for each $S \in \mathcal{S}$. □

The next subsection will show how to leverage Lemma 2 to partition the set of all possible worlds into equivalence classes that are guaranteed to have the same result for a given query predicate, and how to exploit this partitioning to efficiently answer queries.

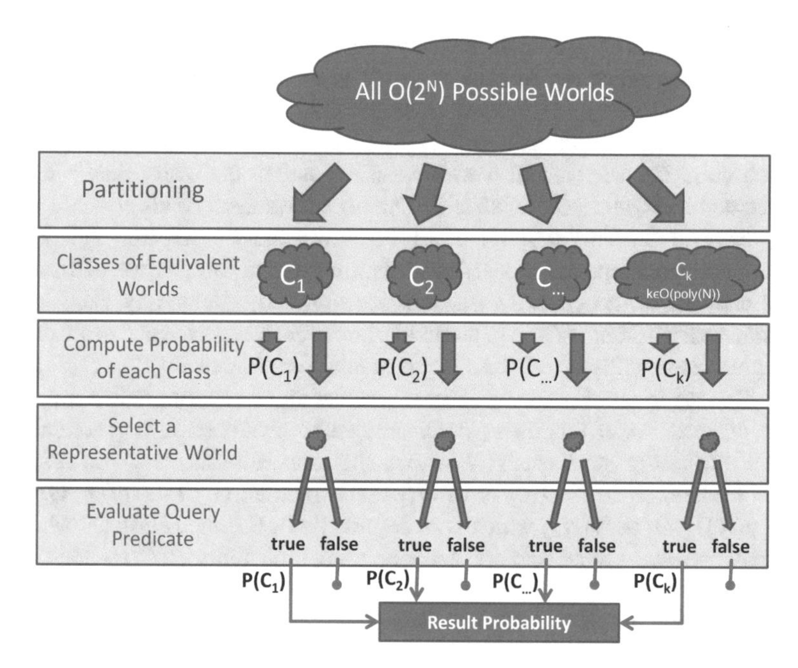

Fig. 14.9 Summary of the paradigm of equivalent worlds

14.7.2 Exploiting Equivalent Worlds for Efficient Algorithms

Given a partitioning $\mathcal{S}$ of all possible worlds, Equation 14.9 requires to perform the following two tasks. The first task requires to evaluate the indicator function $\mathcal{I}(\phi, w \in S)$ for one representative world of each partition. This can be achieved by performing a traditional (non-uncertain) ϕ query on these representatives. The final challenge is to efficiently compute the total probability $P(S) := \sum_{w \in S} P(w)$ for each equivalent class $S \in \mathcal{S}$. This computation must avoid an enumeration of all possible worlds, i.e., must be in $o(|S|)$.[2] Achieving an efficient computation is a creative task, and usually requires to exploit properties of the model (such as object independence) and properties of the spatial query predicate. The paradigm of equivalent worlds is illustrated and summarized in Fig. 14.9. In the first step, set of all possible worlds $\mathcal{W}$, which is exponential in the number N of uncertain objects, has to be partitioned into a polynomial large set of classes of equivalent worlds, such that all worlds in the same class are guaranteed to be equivalent given the query predicate ϕ. This yields a the set $\mathcal{C} = \{C_1, C_2, \ldots, C_k\}$ of classes of equivalent worlds. To allow efficient processing, this set must be polynomial in size,

[2]Note that if an exponential large set is partitioned into a polynomial number of subsets, then at least one such subset must have exponential size. This is evident considering that $O(\frac{2^n}{poly(n)}) = O(2^n)$.

since each class has to be considered individually in the following. Next, we require to compute the probability of each class C_i, without enumeration of all possible worlds contained in C_i, the number of which may still be exponential. In fact, at least one class C_i must contain $O(2^N)$ possible worlds. Next, we need to decide, for each class C_i, whether all worlds $w \in C_i$ satisfy the query predicate ϕ, or whether no world $w \in C_i$ satisfies ϕ. Due to equivalence of all possible worlds in C_i, these are the only possible cases. For some query predicates, this decision can be made using special properties of the query predicate, as we will see later in this chapter. In the general case, this decision can be made by choosing one representative world $w \in C_i$ (e.g. at random) from each class C_i, and evaluating the query predicate on this world. This yields at total run-time of $O(|\mathcal{C}|) \cdot O(\mathcal{I}(\phi, w))$, where $\mathcal{I}(\phi, w)$ is the time complexity of evaluating the query predicate ϕ on the certain database w. If this query predicate can be evaluated in polynomial time, i.e., if $O(\mathcal{I}(\phi, w)) \in O(poly(N))$, then the total run-time is in $O(poly(N))$. This is evident, since if $O(\mathcal{C})$ is in $O(poly(N))$, then $O(\mathcal{C}) \cdot O(\mathcal{I}(\phi, w))$ is in $O(poly(N)) \cdot O(poly(N))$ which is in $O(poly(N))$. For each class C_i, where the representative world satisfies ϕ, the corresponding probability $P(C_i)$ is added to the result probability.

The following lemma summarizes the assumptions that a query predicate has to satisfy in order to efficiently apply paradigm of finding equivalent worlds.

Lemma 3 *Given a query predicate ϕ and an uncertain database $\mathcal{D}$ of size $N := |DB|$, we can answer ϕ on $\mathcal{D}$ in polynomial time if the following four conditions are satisfied:*

I *A traditional ψ query on non-uncertain data can be answered in polynomial time.*
II *we can identify a partitioning $\mathcal{C}$ of $\mathcal{W}$ into classes $C \in \mathcal{C}$ of equivalent worlds (see Definition 8).*
III *The number $|\mathcal{C}|$ of classes is at most polynomial in N.*
IV *The the total probability of a class $S \in \mathcal{C}$ can be computed in at most polynomial time.*

Proof Answering a ϕ query on $\mathcal{D}$ requires to evaluate Equation 14.3 which we reformed into Equation 14.9 using Property II. This requires to iterate over all $|\mathcal{C}|$ classes of equivalent worlds in polynomial time due to Property III. For each class $C \in \mathcal{C}$, this requires to perform two tasks. The first task requires to compute the total probability of all worlds in C, and the second task requires to evaluate ϕ on a single possible world $w \in C$. The former task can be performed in polynomial time due to Property IV. The later task requires to perform a crisp ϕ query on the (crisp) world w in polynomial time due to Property I. □

14.8 Case Study: Range Queries and the Sum of Independent Bernoulli Trials

In this chapter, the paradigm of equivalent worlds will be applied to efficiently solve the problem of computing the number of uncertain objects located within a specified range.

Example 9 As an example, consider the setting depicted in Fig. 14.8. In this example, we have four objects, A, B, C, and D having probabilities of 1.0, 0.3, 0.2, and 0.9 of being located inside the query region defined by query location q and query range ϵ. Intuitively, the number of objects in this range can be anywhere between one and four, as only object A is guaranteed to be inside the range, while on B, C, and D have a chance to be inside this range among all other objects. How can we efficiently compute the distribution of this number of objects inside the query range? What is the probability of having exactly one, two, three or four object in the range? Intuitively, the number of objects corresponds depends on the result of three "coin-flips", each using a coin with a different bias of flipping heads.

Each such "coin-flip" is a Bernoulli trial, which may have a successful ("heads") of unsuccessful ("tails") outcome. In the case where all Bernoulli trials have the same probability p, the number of successful trials out of N trials is described by the well-known binomial distribution. In the case where each trial may have a different probability to succeed, the number of successful trials follows a Poisson-binomial distributions (Hoeffding et al. 1956).

Formally, let $X_1, \ldots, X_N$ be independent and not necessarily identically distributed Bernoulli trials, i.e., random variables that may only take values zero and one. Let $p_i := P(X_i = 1)$ denote the probability that random variable X_i has value one. In this section, we will show how to efficiently compute the distribution of the random variable

$$\sum_{i=1}^{N} X_i$$

without enumeration of all possible worlds. That is, for each $0 \leq k \leq N$, this section shows how to compute the probability $P(\sum_{i=1}^{N} X_i = k)$ that exactly k trials are successful.

This section shows two commonly used solutions to compute the distribution of $\sum_i X_i$ efficiently: The Poisson-binomial recurrence, and a technique based on generating functions. Both solutions have in common that they identify worlds that are equivalent to the query predicate.

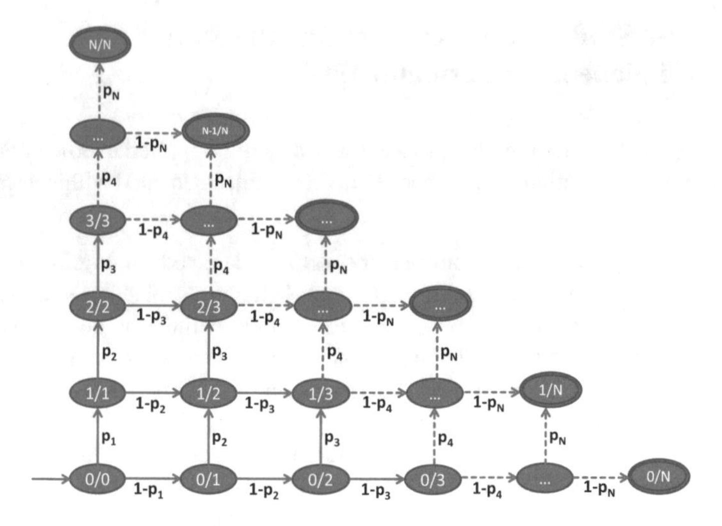

Fig. 14.10 Deterministic finite automaton corresponding to the problem of the sum of independent Bernoulli trials

14.8.1 Poisson-Binomial Recurrence

The first approach iteratively computes the distribution of the sum of the first $1 \leq k \leq N$ Bernoulli variables given the distribution of the sum of the first $k - 1$ Bernoulli variables.

To gain an intuition of how to do this efficiently, consider the deterministic finite automaton depicted in Fig. 14.10.[3] The states (i/j) of this automaton correspond to the random event that out of the first j Bernoulli trials $X_1, \ldots, X_j$, exactly i trials have been successful. Initially, zero Bernoulli trials have been performed, out of which zero (trivially) were successful. This situation is represented by the initial state (0/0) in Fig. 14.10. Evaluating the first Bernoulli trial X_1, there is two possible outcomes: The trial may be successful with a probability of p_1, leading to a state $(1/1)$ where one out of one trials have been successful. Alternatively, the trial may be unsuccessful, with a probability of $1 - p_1$, leading to a state $(0/1)$ where zero out of one trial have been successful. The second trial is then applied to both possible outcomes. If the first trial has not been successful, i.e., we are currently located in state $(0/1)$, then there is again two outcomes for the second Bernoulli trial, leading to state $(1/2)$ and $(0/2)$ with a probability of p_2 and $1 - p_2$ respectively. If currently located in state $(0/1)$, the two outcomes are state $(2/2)$ and state $(1/2)$ with the same probabilities. At this point, we have unified two different possible worlds that are

[3]Note that this automaton is deterministic, despite the process of choosing a successor node being a random event. Once the Bernoulli trial corresponding to a node has been performed, the next node will be chosen deterministically, i.e., the upper node will be chosen if the trial was successful, and the right node will be chosen otherwise. Either way, there is exactly one successor node.

equivalent with respect to $\sum_i X_i$: The world where trial one has been successful and trial two has not been successful, and the world where trial one has not been successful and trial two has been successful have been unified into state (1/2), representing both worlds. This unification was possible, since both paths leading to state (1/2) are equivalent with respect to the number of successful trials.

The three states (0/2), (1/2) and (2/2) are then subjected to the outcome of the third Bernoulli trial, leading to states (0/3), (1/3), (2/3) and (3/3). That is a total of four states for a total of $2^3 = 8$ possible worlds. In summary, the number of states in Fig. 14.10 equals $\frac{N^2}{2}$. However, it is not yet clear how to compute the probability of a state (i/j) efficiently. Naively, we have to compute the sum over all paths leading to state (i/j). For example, the probability of state (2/3) is given by $p_1 \cdot p_2 \cdot (1 - p_3) + p_1 \cdot (1 - p_2) \cdot p_3 + (1 - p_1) \cdot p_2 \cdot p_3$. This naive computation requires to enumerate all $\binom{j}{p_3}$ combinations of paths leading to state (i/j).

For an efficient computation, we make the following observation: Each state of the deterministic finite automaton depicted in Fig. 14.10 has at most two incoming edges. Thus, to compute the probability of a state (i/j), we only require the probabilities of states leading to (i/j). The states leading to state (i/j) are state $(i - 1/j - 1)$ and state $(i/j - 1)$. Given the probabilities $P(i - 1/j - 1)$ and $P(i/j - 1)$, we can compute the probability $P(i/j)$ of state (i/j) as follows:

$$P(i/j) = P(i - 1/j - 1) \cdot p_j + P(i, j - 1) \cdot (1 - p_j) \tag{14.10}$$

where

$$P(0/0) = 1 \text{ and } P(i/j) = 0 \text{ if } i > j \text{ or if } i < 0.$$

Equation 14.10 is known as the Poisson-Binomial Recurrence (To the best of our knowledge, the Poisson binomial recurrence was first introduced by Lange 1999) and can be used to compute the probabilities of states $(k/N), 0 \leq k \leq N$ which by definition, correspond to the probabilities $P(\sum_{i=1^N} X_i = k)$ that out of all N Bernoulli trials, exactly k trials are successful.

This approach follows the paradigm of equivalent worlds in each iteration k: The set of all 2^k possible worlds is partitioned into $k + 1$ equivalent sets, each corresponding to a state i/k, where $i \leq k$. Each class contains only and all of the $\binom{k}{i}$ possible worlds where exactly i Bernoulli trails succeeded. The information about the particular sequence of the successful trials, i.e., which trials were successful and which were unsuccessful is lost. This information however, is no longer necessary to compute the distribution of $\sum_{i=0}^{N} X_i$, since for this random variable, we only need to know the number of successful trials, not their sequence. This abstraction allows to remove the combinatorial aspect of the problem.

An example showcasing the Poisson binomial recurrence is given in the following.

Example 10 Let $N = 4$ and let $p_1 = 0.1$, $p_2 = 0.2$, $p_3 = 0.3$ and $p_4 = 0.4$. The corresponding DFA is depicted in Fig. 14.11. The probability of state (0/0) is

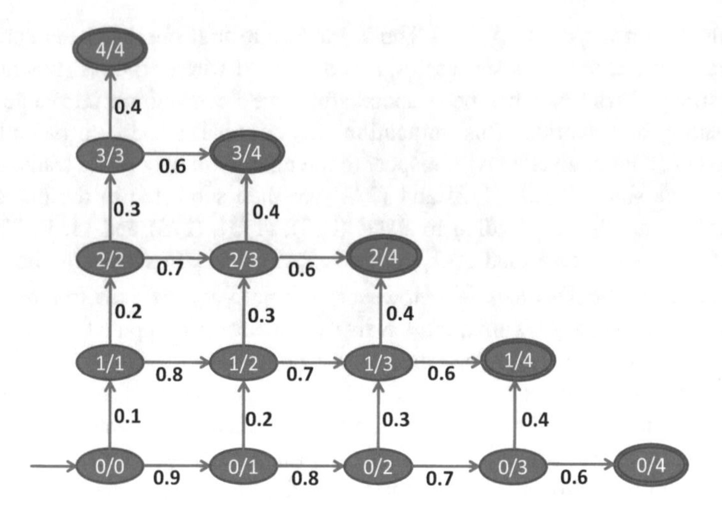

Fig. 14.11 Example deterministic finite automaton for a total of four Bernoulli random variables

explicitly set to 1.0 in Equation 14.10. To compute the probability of state (0/1), we apply Equation 14.10 to compute

$$P(0/1) = P(-1/0) \cdot p_1 + P(0/0) \cdot (1 - p_1).$$

with $P(-1/0) = 0$ and $P(0/0) = 1$ explicitly defined in Equation 14.10 this yields

$$P(0/1) = 0 \cdot p_1 + 1 \cdot (1 - p_1) = 0.9$$

Analogously, we obtain

$$P(1/1) = P(0/0) \cdot p_1 + P(1/0) \cdot (1 - p_1) = 1 \cdot p_1 = 0.1$$

Using these initial probabilities, we can continue to compute

$$P(0/2) = P(-1/1) \cdot p_2 + P(0/1) \cdot (1 - p_2) = 0 \cdot 0.2 + 0.9 \cdot 0.8 = 0.72$$

$$P(1/2) = P(0/1) \cdot p_2 + P(1/1) \cdot (1 - p_2) = 0.9 \cdot 0.2 + 0.1 \cdot 0.8 = 0.26$$

$$P(2/2) = P(1/1) \cdot p_2 + P(2/1) \cdot (1 - p_2) = 0.1 \cdot 0.2 + 0 \cdot 0.8 = 0.02$$

The probabilities $P(i/2), 0 \leq i \leq 2$ can be used to compute

$$P(0/3) = P(-1/2) \cdot p_3 + P(0/2) \cdot (1 - p_3) = 0 \cdot 0.3 + 0.72 \cdot 0.7 = 0.504$$

$$P(1/3) = P(0/2) \cdot p_3 + P(1/2) \cdot (1 - p_3) = 0.72 \cdot 0.3 + 0.26 \cdot 0.7 = 0.398$$

$$P(2/3) = P(1/2) \cdot p_3 + P(2/2) \cdot (1 - p_3) = 0.26 \cdot 0.3 + 0.02 \cdot 0.7 = 0.092$$

$$P(3/3) = P(2/2) \cdot p_3 + P(3/2) \cdot (1 - p_3) = 0.02 \cdot 0.3 + 0 \cdot 0.7 = 0.006$$

Finally, these probabilities can be used to derive the final distribution of the random variable $\sum_{i=1}^{4} X_i$:

$$P(0/4) = P(-1/3) \cdot p_4 + P(0/3) \cdot (1 - p_4) = 0 \cdot 0.4 + 0.504 \cdot 0.6 = 0.3024$$

$$P(1/4) = P(0/3) \cdot p_4 + P(1/3) \cdot (1 - p_4) = 0.504 \cdot 0.4 + 0.398 \cdot 0.6 = 0.4404$$

$$P(2/4) = P(1/3) \cdot p_4 + P(2/3) \cdot (1 - p_4) = 0.398 \cdot 0.4 + 0.092 \cdot 0.6 = 0.2144$$

$$P(3/4) = P(2/3) \cdot p_4 + P(3/3) \cdot (1 - p_4) = 0.092 \cdot 0.4 + 0.006 \cdot 0.6 = 0.0404$$

$$P(4/4) = P(3/3) \cdot p_4 + P(4/3) \cdot (1 - p_4) = 0.006 \cdot 0.4 + 0 \cdot 0.6 = 0.0024$$

These probabilities describe the PDF of $\sum_{i=1}^{4} X_i$ by definition of $P(i/j)$.

14.8.1.1 Complexity Analysis

To compute the distribution of $\sum_i X_i$ we require to compute each probability $P(i/j)$ for $0 \leq j \leq N, i \leq j$, yielding a total of $\frac{N^2}{2} \in O(N^2)$ probability computations. To compute any such probability, we have to evaluate Equation 14.10, which requires to look up four probabilities $P(i-1/j-1)$, $P(i/j-1)$, p_j and $1 - p_j$, which can be performed in constant time. This yields a total runtime complexity of $O(N^2)$. The $O(N^2)$ space complexity required to store the matrix of probabilities $P(i/j)$ for $0 \leq j \leq N, i \leq j$ can be reduced to $O(N \cdot k)$ by exploiting that in each iteration where the probabilities $P(i/k), 0 \leq i \leq k$ are computed, only the probabilities $P(i/k-1), 0 \leq i \leq k-1$ are required, and the result of previous iterations can be discarded. Thus, at most N probabilities have to be stored at a time.

14.8.2 Generating Functions

An alternative technique to compute the sum of independent Bernoulli variables is the generating functions technique. While showing the same complexity as the Poisson binomial recurrence, its advantage is its intuitiveness.

Represent each Bernoulli trial X_i by a polynomial $\text{poly}(X_i) = p_i \cdot x + (1 - p_i)$. Consider the generating function

$$\mathcal{F}^N = \prod_{i=1}^{N} \text{poly}(X_i) = \sum_{i=0}^{N} c_i x^i. \tag{14.11}$$

The coefficient c_i of x^i in the expansion of $\mathcal{F}^N$ equals the probability $P(\sum_{n=1}^{N} X_n = i)$ (Li and Deshpande 2009). For example, the monomial $0.25 \cdot x^4$ implies that with a probability of 0.25, the sum of all Bernoulli random variables equals four.

The expansion of N polynomials, each containing two monomials leads to a total of 2^N monomials, one monomial for each sequence of successful and unsuccessful Bernoulli trials, i.e., one monomial for each possible worlds. To reduce this complexity, again an iterative computation of $\mathcal{F}^N$, can be used, by exploiting that

$$\mathcal{F}^k = \mathcal{F}^{k-1} \cdot \text{poly}(X_k). \tag{14.12}$$

This rewriting of Equation 14.11 allows to inductively compute $\mathcal{F}^k$ from $\mathcal{F}^{k-1}$. The induction is started by computing the polynomial $\mathcal{F}^0$, which is the empty product which equals the neutral element of multiplication, i.e., $\mathcal{F}^0 = 1$. To understand the semantics of this polynomial, the polynomial $\mathcal{F}^0 = 1$ can be rewritten as $\mathcal{F}^0 = 1 \cdot x^0$, which we can interpret as the following tautology:"with a probability of one, the sum of all zero Bernoulli trials equals zero." After each iteration, we can unify monomials having the same exponent, leading to a total of at most $k+1$ monomials after each iteration. This unification step allows to remove the combinatorial aspect of the problem, since any monomial x^i corresponds to a class of equivalent worlds, such that this class contains only and all of the worlds where the sum $\sum_{k=1}^{N} X_k = 1$. In each iteration, the number of these classes is k and the probability of each class is given by the coefficient of x^i.

An example showcasing the generating functions technique is given in the following. This examples uses the identical Bernoulli random variables used in Example 10.

Example 11 Again, let $N = 4$ and let $p_1 = 0.1$, $p_2 = 0.2$, $p_3 = 0.3$ and $p_4 = 0.4$. We obtain the four generating polynomials $\text{poly}(X_1) = (0.1x + 0.9)$, $\text{poly}(X_2) = (0.2x + 0.8)$, $\text{poly}(X_3) = (0.3x + 0.7)$, and $\text{poly}(X_4) = (0.4x + 0.6)$. We trivially obtain $\mathcal{F}^0 = 1$. Using Equation 14.12 we get

$$\mathcal{F}^1 = \mathcal{F}^0 \cdot \text{poly}(X_1) = 1 \cdot (0.1x + 0.9) = 0.1x + 0.9.$$

Semantically, this polynomial implies that out of the first one Bernoulli variables, the probability of having a sum of one is 0.1 (according to monomial $0.1x$=$0.1x^1$), and the probability of having a sum of zero is 0.9 (according to monomial $0.9 = 0.9x^0$. Next, we compute F^2, again using Equation 14.12:

$$\mathcal{F}^2 = \mathcal{F}^1 \cdot \text{poly}(X_2) = (0.1x^1 + 0.9x^0) \cdot (0.2x^1 + 0.8x^0) =$$

$$0.02x^1x^1 + 0.08x^1x^0 + 0.18x^0x^1 + 0.72x^0x^0$$

In this expansion, the monomials have deliberately not been unified to give an intuition of how the generating function techniques is able to identify and unify equivalent worlds. In the above expansion, there is one monomial for each possible world. For example, the monomial $0.18x^0x^1$ represents the world where the first trial was unsuccessful (represented by the zero of the first exponent) and the second trial was successful (represented by the one of the second exponent). The above notation allows to identify the sequence of successful and unsuccessful Bernoulli trials, clearly leading to a total of 2^k possible worlds for $\mathcal{F}^k$. However, we know that we only need to compute the total number of successful trials, we do not need to know the sequence of successful trials. Thus, we need to treat worlds having the same number of successful Bernoulli trials equivalently, to avoid the enumeration of an exponential number of sequences. This is done implicitly by polynomial multiplication, exploiting that

$$0.02x^1x^1+0.08x^1x^0+0.18x^0x^1+0.72x^0x^0 = 0.02x^2+0.08x^1+0.18x^1+0.72x^0$$

This representation no longer allows to distinguish the sequence of successful Bernoulli trials. This loss of information is beneficial, as it allows to unify possible worlds having the same sum of Bernoulli trials.

$$0.02x^2 + 0.08x^1 + 0.18x^1 + 0.72x^0 = 0.02x^2 + 0.26x^1 + 0.72x^0$$

The remaining monomials represent an equivalence class of possible worlds. For example, monomial $0.26x^1$ represents all worlds having a total of one successful Bernoulli trials. This is evident, since the coefficient of this monomial was derived from the sum of both worlds having a total of one successful Bernoulli trials. In the next iteration, we compute:

$$\mathcal{F}^3 = \mathcal{F}^2 \cdot \text{poly}(X_3) = (0.02x^2 + 0.26x^1 + 0.72x^0) \cdot (0.3x + 0.7)$$

$$= 0.006x^2x^1 + 0.014x^2x^0 + 0.078x^1x^1 + 0.182x^1x^0 + 0.216x^0x^1 + 0.504x^0x^0$$

This polynomial represents the three classes of possible worlds in $\mathcal{F}^2$ combined with the two possible results of the third Bernoulli trial, yielding a total of 32 monomials. Unification yields

$$0.006x^2x^1 + 0.014x^2x^0 + 0.078x^1x^1 + 0.182x^1x^0 + 0.216x^0x^1 + 0.504x^0x^0 =$$

$$0.006x^3 + 0.092x^2 + 0.398x^1 + 0.504$$

The final generating function is given by

$$\mathcal{F}^4 = \mathcal{F}^3 \cdot \text{poly}(X_4) =$$

$$(0.006x^3 + 0.092x^2 + 0.398x^1 + 0.504) \cdot (0.4x + 0.6) =$$

$$.0024x^4 + .0036x^3 + .0368x^3 + .0552x^2 + .1592x^2 + .2388x^1 + .2016x^1 + .3024x^0$$

$$= 0.0024x^4 + 0.0404x^3 + 0.2144x^2 + 0.4404x + 0.3024$$

This polynomial describes the PDF of $\sum_{i=1}^{4} X_i$, since each monomial $c_i x^i$ implies that the probability, that out of all four Bernoulli trials, the total number of successful events equals i, is c_i. Thus, we get $P(\sum_{i=1}^{4} X_i = 0) = 0.0024$, $P(\sum_{i=1}^{4} X_i = 1) = 0.0404$, $P(\sum_{i=1}^{4} X_i = 2) = 0.2144$, $P(\sum_{i=1}^{4} X_i = 3) = 0.4404$ and $P(\sum_{i=1}^{4} X_i = 4) = 0.3024$. Note that this result equals the result we obtained by using the Poisson binomial recurrence in the previous section.

14.8.2.1 Complexity Analysis

The generating function technique requires a total of N iterations. In each iteration $1 \leq k \leq N$, a polynomial of degree k, and thus of maximum length $k + 1$, is multiplied with a polynomial of degree 1, thus having a length of 2. This requires to compute a total of $(k + 1) \cdot 2$ monomials in each iteration, each requiring a scalar multiplication. Thus leads to a total time complexity of $\sum_{i=1}^{N} 2k + 2 \in O(N^2)$ for the polynomial expansions. Unification of a polynomial of length k can be done in $O(k)$ time, exploiting that the polynomials are sorted by the exponent after expansion. Unification at each iteration leads to a $O(n^2)$ complexity for the unification step. This results in a total complexity of $O(n^2)$, similar to the Poisson binomial recurrence approach.

An advantage of the generating function approach is that this naive polynomial multiplication can be accelerated using Discrete Fourier Transform (DFT). This technique allows to reduce to total complexity of computing the sum of N Bernoulli random variables to $O(N\log^2 N)$ (Li et al. 2011). This acceleration is achieved by exploiting that DFT allows to expand two polynomials of size k in $O(k\log k)$ time. Equi-sized polynomials are obtained in the approach of Li et al. (2011), by using a divide and conquer approach, that iteratively divides the set of N Bernoulli trials into two equi-sized sets. Their recursive algorithm then combines these results by performing a polynomial multiplication of the generating polynomials of each set. More details of this algorithm can be found in Li et al. (2011).

14.9 Advanced Techniques for Managing Uncertain Spatial Data

The Paradigm of Equivalent worlds has been successfully applied to efficiently support many spatial query predicates and spatial data mining tasks. These more advanced techniques are out of scope of this book chapter, but the techniques presented in this chapter should help the interested reader to dive deeper into understanding state-of-the-art solutions, and to help the reader to contribute to this field. An overview of research directions on uncertain spatial is provided in Table 14.2.

Efficient solutions on uncertain data have been presented for (1)-nearest neighbor (1NN) queries (Cheng et al. 2004a, 2008; Kriegel et al. 2007; Iijima and Ishikawa 2009; Zhang et al. 2013; Niedermayer et al. 2013a; Schmid et al. 2017). The case of $1NN$ is special, as for $1NN$ the cases of object-based and result-based probabilistic result semantics are equivalent: Since a $1NN$ query only results a single result object. Thus, the probability of any object to be part of the result is equal the probability of this object to be the (whole) result. For k Nearest Neighbor queries, this is not the case, as initially motivated in Fig. 14.2. For object-based result semantics (as explained in Sect. 14.5), polynomial time solutions leveraging the paradigm of equivalent worlds have been proposed (Bernecker et al. 2011a). For result-based result semantics, where each of the (potentially exponential many in k)

Table 14.2 Advanced topics in querying and mining uncertain spatial data

Topic	Related work
Nearest neighbor query processing	Cheng et al. (2004a, 2008), Kriegel et al. (2007), Iijima and Ishikawa (2009), Zhang et al. (2013), Niedermayer et al. (2013a), and Schmid et al. (2017)
k-nearest neighbor (kNN) query processing	Kolahdouzan and Shahabi (2004), Beskales et al. (2008), Cheng et al. (2009), and Bernecker et al. (2011a)
Top-k query processing	Re et al. (2007), Soliman et al. (2007), and Yi et al. (2008b)
Ranking of uncertain spatial data	Lian and Chen (2008b, 2009b), Bernecker et al. (2008, 2010, 2012), Cormode et al. (2009b), Soliman and Ilyas (2009), Li et al. (2009b), Dai et al. (2005), and Hua et al. (2008)
Reverse kNN query processing	Lian and Chen (2009a), Cheema et al. (2010), Bernecker et al. (2011b), and Emrich et al. (2014)
Skyline query processing	Pei et al. (2007), Lian and Chen (2008a), Vu and Zheng (2013), Ding et al. (2014), and Yang et al. (2018)
Indexing uncertain spatial data	Zhang et al. (2009), Emrich et al. (2012a), and Agarwal et al. (2009)
Maximum range-sum query processing	Agarwal et al. (2018), Nakayama et al. (2017), and Liu et al. (2019)
Querying uncertain trajectory data	Emrich et al. (2012b), Niedermayer et al. (2013b), and Zheng et al. (2011)
Clustering uncertain spatial data	Schubert et al. (2015), Züfle et al. (2014), Ngai et al. (2006), and Kriegel and Pfeifle (2005)
Frequent itemset and colocation mining	Bernecker et al. (2009, 2012, 2013) and Wang et al. (2011, 2012)

results is associated with a probability, solutions have been presented in Beskales et al. (2008) and Cheng et al. (2009).

A related problem is Top-k query processing which returns the k best result objects for a given score function (Re et al. 2007; Soliman et al. 2007; Yi et al. 2008b). While these solution are not proposed in the context of spatial or spatio-temporal data, they are mentioned here as they can be applied to spatial data. For example, if the score function is defined as the distance to query object, this problem becomes equivalent to kNN. Solutions for result-based probabilistic result semantics are proposed in Soliman et al. (2007) and Re et al. (2007) and for object-based result semantics in Yi et al. (2008b).

Another problem generalization are ranking queries, which return the Top-k result ordered by score. For uncertain data using object-based result semantics, this yields a probabilistic mapping of each database mapping to each rank for the case of object-based result semantics. For example, it may return that object o_1 has a 80% probability to be Rank 1, and a 20% probability to be Rank 2. In the case of result-based probabilistic result semantics, each possible ranking of objects is mapped to a probability, for example, the ranking $[o_1, o_3, o_2]$ may have a 10% probability. Solutions for the result-based probabilistic result semantic case have been proposed in Soliman and Ilyas (2009) having exponential run-time due to the hard nature of this problem. For the case of object-based probabilistic result semantics, first solutions having exponential run-time were proposed (Bernecker et al. 2008; Lian and Chen 2008b). Applying the paradigm of equivalent worlds, a number of solutions have been proposed concurrently and independently to achieve polynomial run-time (linear in the number of database objects times the number of ranks). The generating functions technique (as explained in Sect. 14.8) was proposed for this purpose by Li et al. (2009b). An equivalent approach using a technique called Poisson-Binomial Recurrence was simultaneously proposed by Bernecker et al. (2010) and Hua et al. (2008). A comparison of the generating functions technique and the Poisson Binomial Recurrence, along with a proof of equivalence, can be found in Züfle (2013). Other works shown in Table 14.2 include solutions for the case of existential uncertainty (Dai et al. 2005), inverse ranking (Lian and Chen 2009b), and spatially extended objects (Bernecker et al. 2012), and the computation of the expected rank of an object. Cormode et al. (2009b). Solution for indexing of uncertain spatial (Agarwal et al. 2009; Chen et al. 2017) and spatio-temporal (Zhang et al. 2009; Emrich et al. 2012a) data have been proposed to speed up various of the previously mentioned query types.

The problem of finding reverse k nearest neighbors (RkNNs) have been studied for spatial data (Lian and Chen 2009a; Cheema et al. 2010; Bernecker et al. 2011b) and spatio-temporal data (Emrich et al. 2014). Solutions for skyline queries on uncertain data have been proposed in Pei et al. (2007), Lian and Chen (2008a), Vu and Zheng (2013), Ding et al. (2014), and Yang et al. (2018). More recently, the problem of answering Maximum Range-Sum Queries has been studied for uncertain data (Agarwal et al. 2018; Nakayama et al. 2017; Liu et al. 2019).

Solutions tailored towards uncertain spatio-temporal trajectories, in which the exact location of an object at each point in time is a random variable have been proposed (Emrich et al. 2012b; Niedermayer et al. 2013b; Zheng et al. 2011). In this work, the challenge is to leverage stochastic processes that consider temporal dependencies. Such dependencies describe that the location of an object at a time t depends on its location at time $t - 1$.

Solutions for clustering uncertain data have been proposed (Schubert et al. 2015; Züfle et al. 2014; Ngai et al. 2006; Kriegel and Pfeifle 2005). The challenge of clustering uncertain data is that the membership likelihood of on uncertain object to a cluster depends on other objects, making it hard to identify groups of worlds that are guaranteed to yields the same clustering result.

Finally, solutions for frequent itemset mining have been proposed for uncertain data (Bernecker et al. 2009, 2012, 2013; Wang et al. 2012). While frequent itemset mining is not a spatial problem, it has applications in spatial co-location mining (Wang et al. 2011; Chan et al. 2019).

Yet, many other spatial query predicates, as well as other probabilistic query predicates using different probabilistic result semantics are still open to study. The authors hopes that this chapter provides interested scholars with a starting point to fully understand preliminaries and assumptions made by existing work, as well as a general paradigm to develop efficient solutions for future work leveraging the Paradigm of Equivalent Worlds presented herein.

14.10 Summary

This chapter provided an overview of uncertain spatial data models and the concept of possible world semantics to interpret queries on these models. To understand the landscape of existing query processing algorithms on uncertain data, this chapter further surveyed different probabilistic result semantics and different probabilistic query predicates. To give the interested reader a start into this field, this chapter presented a general paradigm to efficiently query uncertain data based on the Paradigm of Equivalent Worlds, which aims at finding possible worlds that are guaranteed to have the same query result. As a case-study to apply this paradigm, this chapter provided solutions to efficiently compute range queries on uncertain data using an efficient recursion approach, as well as leveraging the concept of generating functions.

Given this survey on modeling and querying uncertain spatial data, this chapter further provided a brief (and not exhaustive) overview of some research directions on uncertain spatial data. Many queries on uncertain data have already been solved efficiently, but many new challenges arise. For instance, only limited work has focused on streaming uncertain data, that is, handling uncertain data that changes rapidly. Another mostly open research direction is uncertain data processing in resources-limited scenarios such as edge computing. The author hopes that readers will find this overview useful to help readers understanding existing solutions and support readers towards adding their own research to this field.

References

Agarwal PK, Cheng S-W, Tao Y, Yi K (2009) Indexing uncertain data. In: Proceedings of the Twenty-Eighth ACM SIGMOD-SIGACT-SIGART Symposium on Principles of Database Systems, pp 137–146

Agarwal PK, Kumar N, Sintos S, Suri S (2018) Range-max queries on uncertain data. J Comput Syst Sci 94:118–134

Aggarwal CC (2010) Managing and mining uncertain data, vol 35. Springer Science & Business Media, New York

Aggarwal CC, Philip SY (2008) A survey of uncertain data algorithms and applications. IEEE Trans Knowl Data Eng 21(5):609–623

Agrawal P, Benjelloun O, Sarma AD, Hayworth C, Nabar S, Sugihara T, Widom J (2006) Trio: a system for data, uncertainty, and lineage. In: Proceedings of VLDB 2006 (Demonstration Description)

Aji A, Wang F, Vo H, Lee R, Liu Q, Zhang X, Saltz J (2013) Hadoop-GIS: a high performance spatial data warehousing system over MapReduce. Proc VLDB Endowment 6(11):1009–1020

Akdogan A, Demiryurek U, Banaei-Kashani F, Shahabi C (2010) Voronoi-based geospatial query processing with MapReduce. In: 2010 IEEE Second International Conference on Cloud Computing Technology and Science. IEEE, pp 9–16

Antova L, Jansen T, Koch C, Olteanu D (2008a) Fast and simple relational processing of uncertain data. In: Proceedings of the 24th International Conference on Data Engineering (ICDE), Cancun, pp 983–992

Antova L, Jansen T, Koch C, Olteanu D (2008b) Fast and simple relational processing of uncertain data. In: 2008 IEEE 24th International Conference on Data Engineering. IEEE, pp 983–992

Apache. Hadoop. http://hadoop.apache.org/. Accessed 02/03/2021

Bacchus F, Grove AJ, Halpern JY, Koller D (1996) From statistical knowledge bases to degrees of belief. Artif Intell 87(1):75–143

Barbará D, Garcia-Molina H, Porter D (1992) The management of probabilistic data. IEEE Trans Knowl Data Eng 4(5):487–502

Benjelloun O, Sarma AD, Halevy AY, Widom J (2006) ULDBs: databases with uncertainty and lineage. In: Proceedings of the 32nd International Conference on Very Large Data Bases (VLDB), Seoul, pp 953–964

Bernecker T, Kriegel H-P, Renz M (2008) ProUD: probabilistic ranking in uncertain databases. In: Proceedings of the 20th International Conference on Scientific and Statistical Database Management (SSDBM), Hong Kong, pp 558–565

Bernecker T, Kriegel H-P, Renz M, Verhein F, Zuefle A (2009) Probabilistic frequent itemset mining in uncertain databases. In: Proceedings of the 15th ACM SIGKDD International Conference on Knowledge Discovery and Data Mining. ACM, pp 119–128

Bernecker T, Kriegel H-P, Mamoulis N, Renz M, Zuefle A (2010) Scalable probabilistic similarity ranking in uncertain databases. IEEE Trans Knowl Data Eng 22(9):1234–1246

Bernecker T, Emrich T, Kriegel H-P, Mamoulis N, Renz M, Züfle A (2011a) A novel probabilistic pruning approach to speed up similarity queries in uncertain databases. In: 2011 IEEE 27th International Conference on Data Engineering. IEEE, pp 339–350

Bernecker T, Emrich T, Kriegel H-P, Renz M, Zankl S, Züfle A (2011b) Efficient probabilistic reverse nearest neighbor query processing on uncertain data. Proc VLDB Endowment 4(10):669–680

Bernecker T, Emrich T, Kriegel H-P, Renz M, Züfle A (2012) Probabilistic ranking in fuzzy object databases. In: Proceedings of the 21st ACM International Conference on Information and Knowledge Management, pp 2647–2650

Bernecker T, Cheng R, Cheung DW, Kriegel H-P, Lee SD, Renz M, Verhein F, Wang L, Zuefle A (2013) Model-based probabilistic frequent itemset mining. Knowl Inf Syst 37(1):181–217

Beskales G, Soliman M, Ilyas I (2008) Efficient search for the top-k probable nearest neighbors in uncertain databases. PVLDB 1:326–339

Böhm C, Pryakhin A, Schubert M (2006) The Gauss-tree: efficient object identification of probabilistic feature vectors. In: Proceedings of the 22nd International Conference on Data Engineering (ICDE), Atlanta, p 9
Boulos J, Dalvi N, Mandhani B, Mathur S, Re C, Suciu D (2005) MYSTIQ: a system for finding more answers by using probabilities. In: Proceedings of the 2005 ACM SIGMOD International Conference on Management of Data. ACM, pp 891–893
Casella G, Berger RL (2002) Statistical inference, vol 2. Duxbury, Pacific Grove
Cavallo R, Pittarelli M (1987) The theory of probabilistic databases. In: VLDB, vol 87, pp 1–4
Chan HK-H, Long C, Yan D, Wong RC-W (2019) Fraction-score: a new support measure for co-location pattern mining. In: 2019 IEEE 35th International Conference on Data Engineering (ICDE). IEEE, pp 1514–1525
Cheema MA, Lin X, Wang W, Zhang W, Pei J (2010) Probabilistic reverse nearest neighbor queries on uncertain data. IEEE Trans Knowl Data Eng 22(4):550–564
Chen L, Gao Y, Zhong A, Jensen CS, Chen G, Zheng B (2017) Indexing metric uncertain data for range queries and range joins. VLDB J 26(4):585–610
Cheng R, Kalashnikov DV, Prabhakar S (2003) Evaluating probabilistic queries over imprecise data. In: Proceedings of the ACM International Conference on Management of Data (SIGMOD), San Diego, pp 551–562
Cheng R, Kalashnikov DV, Prabhakar S (2004a) Querying imprecise data in moving object environments. IEEE Trans Knowl Data Eng 16(9):1112–1127
Cheng R, Xia Y, Prabhakar S, Shah R, Vitter J (2004b) Efficient indexing methods for probabilistic threshold queries over uncertain data. In: Proceedings of the 30th International Conference on Very Large Data Bases (VLDB), Toronto, pp 876–887
Cheng R, Chen J, Mokbel MF, Chow C-Y (2008) Probabilistic verifiers: evaluating constrained nearest-neighbor queries over uncertain data. In: Proceedings of the 24th International Conference on Data Engineering (ICDE), Cancun, pp 973–982
Cheng R, Chen L, Chen J, Xie X (2009) Evaluating probability threshold k-nearest-neighbor queries over uncertain data. In: Proceedings of the 13th International Conference on Extending Database Technology (EDBT), Saint-Petersburg, pp 672–683
Cheng R, Emrich T, Kriegel H-P, Mamoulis N, Renz M, Trajcevski G, Züfle A (2014) Managing uncertainty in spatial and spatio-temporal data. In: 2014 IEEE 30th International Conference on Data Engineering. IEEE, pp 1302–1305
Cho E, Myers SA, Leskovec J (2011) Friendship and mobility: user movement in location-based social networks. In: Proceedings of the 17th ACM SIGKDD International Conference on Knowledge Discovery and Data Mining. ACM, pp 1082–1090
Cormode G, Li F, Yi K (2009a) Semantics of ranking queries for probabilistic data and expected ranks. In: 2009 IEEE 25th International Conference on Data Engineering. IEEE, pp 305–316
Cormode G, Li F, Yi K (2009b) Semantics of ranking queries for probabilistic data and expected results. In: Proceedings of the 25th International Conference on Data Engineering (ICDE), Shanghai, pp 305–316
Couclelis H (2003) The certainty of uncertainty: GIS and the limits of geographic knowledge. Trans GIS 7(2):165–175
Dai X, Yiu ML, Mamoulis N, Tao Y, Vaitis M (2005) Probabilistic spatial queries on existentially uncertain data. In: International Symposium on Spatial and Temporal Databases. Springer, pp 400–417
Dalvi NN, Suciu D (2004) Efficient query evaluation on probabilistic databases. In: Proceedings of the 30th International Conference on Very Large Data Bases (VLDB), Toronto, pp 864–875
Dalvi N, Suciu D (2007) Efficient query evaluation on probabilistic databases. VLDB J 16(4):523–544
Dalvi NN, Ré C, Suciu D (2009) Probabilistic databases: diamonds in the dirt. Commun ACM 52(7):86–94
Dean J, Ghemawat S (2008) MapReduce: simplified data processing on large clusters. Commun ACM 51(1):107–113

Deshpande A, Guestrin C, Madden S, Hellerstein JM, Hong W (2004) Model-driven data acquisition in sensor networks. In: Proceedings of the 30th International Conference on Very Large Data Bases (VLDB), Toronto, pp 588–599
Ding X, Jin H, Xu H, Song W (2014) Probabilistic skyline queries over uncertain moving objects. Comput Inform 32(5):987–1012
Emrich T, Kriegel H-P, Mamoulis N, Renz M, Züfle A (2012a) Indexing uncertain spatio-temporal data. In: Proceedings of the 21st ACM International Conference on Information and Knowledge Management. ACM, pp 395–404
Emrich T, Kriegel H-P, Mamoulis N, Renz M, Züfle A (2012b) Querying uncertain spatio-temporal data. In: IEEE 28th International Conference on Data Engineering (ICDE). IEEE, pp 354–365
Emrich T, Kriegel H-P, Mamoulis N, Niedermayer J, Renz M, Züfle A (2014) Reverse-nearest neighbor queries on uncertain moving object trajectories. In: International Conference on Database Systems for Advanced Applications. Springer, pp 92–107
Fegeas RG, Cascio JL, Lazar RA (1992) An overview of FIPS 173, the spatial data transfer standard. Cartograph Geograph Inf Syst 19(5):278–293
Fuhr N, Rölleke T (1997a) A probabilistic relational algebra for the integration of information retrieval and database systems. ACM Trans Inf Syst TOIS) 15(1):32–66
Fuhr N, Rölleke T (1997b) A probabilistic relational algebra for the integration of information retrieval and database systems. ACM Trans Inf Syst 15(1):32–66
Goodchild MF (1998) Uncertainty: the achilles heel of GIS. Geo Inf Syst 8(11):50–52
Grira J, Bédard Y, Roche S (2010) Spatial data uncertainty in the VGI world: going from consumer to producer. Geomatica 64(1):61–72
Hoeffding W et al (1956) On the distribution of the number of successes in independent trials. Ann Math Stat 27(3):713–721
Hsu J (1996) Multiple comparisons: theory and methods. Chapman and Hall/CRC, London
Hua M, Pei J, Zhang W, Lin X (2008) Ranking queries on uncertain data: a probabilistic threshold approach. In: Proceedings of the 2008 ACM SIGMOD International Conference on Management of Data, pp 673–686
Iijima Y, Ishikawa Y (2009) Finding probabilistic nearest neighbors for query objects with imprecise locations. In: Proceedings of the 10th International Conference on Mobile Data Management (MDM), Taipei, pp 52–61
Jampani R, Xu F, Wu M, Perez LL, Jermaine C, Haas PJ (2008) MCDB: a Monte Carlo approach to managing uncertain data. In: Proceedings of the 2008 ACM SIGMOD International Conference on Management of Data. ACM, pp 687–700
Kolahdouzan M, Shahabi C (2004) Voronoi-based k nearest neighbor search for spatial network databases. In: Proceedings of the Thirtieth International Conference on Very Large Data Bases, vol 30. VLDB Endowment, pp 840–851
Kriegel H-P, Pfeifle M (2005) Density-based clustering of uncertain data. In: Proceedings of the Eleventh ACM SIGKDD International Conference on Knowledge Discovery in Data Mining, pp 672–677
Kriegel H-P, Kunath P, Renz M (2007) Probabilistic nearest-neighbor query on uncertain objects. In: Proceedings of the 12th International Conference on Database Systems for Advanced Applications (DASFAA), Bangkok, pp 337–348
Kumar S, Morstatter F, Liu H (2014) Twitter data analytics. Springer, New York
Lakshmanan LV, Leone N, Ross R, Subrahmanian VS (1997) ProbView: a flexible probabilistic database system. ACM Trans Database Syst (TODS) 22(3):419–469
Lange K (1999) Numerical analysis for statisticians. In: Statistics and Computing
Li J, Deshpande A (2009) Consensus answers for queries over probabilistic databases. In: Symposium on Principles of Database Systems (PODS), Providence, pp 259–268
Li J, Deshpande A (2010a) Ranking continuous probabilistic datasets. In: Proceedings of the 36nd International Conference on Very Large Data Bases (VLDB), Singapore 3(1):638–649
Li J, Deshpande A (2010b) Ranking continuous probabilistic datasets. Proc VLDB Endowment 3(1–2):638–649

Li J, Saha B, Deshpande A (2009a) A unified approach to ranking in probabilistic databases. Proc VLDB Endowment 2(1):502–513
Li J, Saha B, Deshpande A (2009b) A unified approach to ranking in probabilistic databases. In: Proceedings of the 35nd International Conference on Very Large Data Bases (VLDB), Lyon 2(1):502–513
Li J, Saha B, Deshpande A (2011) A unified approach to ranking in probabilistic databases. VLDB J 20(2):249–275
Li L, Wang H, Li J, Gao H (2018) A survey of uncertain data management. Front Comput Sci 9:1–29
Lian X, Chen L (2008a) Monochromatic and bichromatic reverse skyline search over uncertain databases. In: Proceedings of the 2008 ACM SIGMOD International Conference on Management of Data, pp 213–226
Lian X, Chen L (2008b) Probabilistic ranked queries in uncertain databases. In: Proceedings of the 12th International Conference on Extending Database Technology (EDBT), Nantes, pp 511–522
Lian X, Chen L (2009a) Efficient processing of probabilistic reverse nearest neighbor queries over uncertain data. VLDB J 18(3):787–808
Lian X, Chen L (2009b) Probabilistic inverse ranking queries over uncertain data. In: Proceedings of the 14th International Conference on Database Systems for Advanced Applications (DASFAA), Brisbane, pp 35–50
Liu Q, Lian X, Chen L (2019) Probabilistic maximum range-sum queries on spatial database. In: Proceedings of the 27th ACM SIGSPATIAL International Conference on Advances in Geographic Information Systems, pp 159–168
Ljosa V, Singh AK (2007) APLA: indexing arbitrary probability distributions. In: Proceedings of the 23rd International Conference on Data Engineering (ICDE), Istanbul, pp 946–955
Lu W, Shen Y, Chen S, Ooi BC (2012) Efficient processing of k nearest neighbor joins using MapReduce. Proc VLDB Endowment 5(10):1016–1027
Nakayama Y, Amagata D, Hara T (2017) Probabilistic MaxRS queries on uncertain data. In: International Conference on Database and Expert Systems Applications. Springer, pp 111–119
Ngai WK, Kao B, Chui CK, Cheng R, Chau M, Yip KY (2006) Efficient clustering of uncertain data. In: Sixth International Conference on Data Mining (ICDM'06). IEEE, pp 436–445
Niedermayer J, Züfle A, Emrich T, Renz M, Mamoulis N, Chen L, Kriegel H-P (2013a) Probabilistic nearest neighbor queries on uncertain moving object trajectories. Proc VLDB Endowment 7(3):205–216
Niedermayer J, Züfle A, Emrich T, Renz M, Mamoulis N, Chen L, Kriegel H-P (2013b) Similarity search on uncertain spatio-temporal data. In: International Conference on Similarity Search and Applications. Springer, pp 43–49
Open Street Map. http://www.openstreetmap.org. Accessed 02/03/2021
Patroumpas K, Papamichalis M, Sellis TK (2012) Probabilistic range monitoring of streaming uncertain positions in geosocial networks. In: Proceedings of the 22nd International Conference on Scientific and Statistical Database Management (SSDBM), Crete, pp 20–37
Pei J, Jiang B, Lin X, Yuan Y (2007) Probabilistic skylines on uncertain data. In: Proceedings of the 33rd International Conference on Very Large Data Bases. Citeseer, pp 15–26
Pei J, Hua M, Tao Y, Lin X (2008) Query answering techniques on uncertain and probabilistic data: tutorial summary. In: Proceedings of the ACM International Conference on Management of Data (SIGMOD), Vancouver, pp 1357–1364
Re C, Dalvi NN, Suciu D (2006) Query evaluation on probabilistic databases. IEEE Data Eng Bull 29(1):25–31
Re C, Dalvi N, Suciu D (2007) Efficient top-k query evaluation on probalistic databases. In: Proceedings of the 23rd International Conference on Data Engineering (ICDE), Istanbul, pp 886–895
Renz M, Cheng R, Kriegel H-P, Züfle A, Bernecker T (2010) Similarity search and mining in uncertain databases. In: Proceedings of the 36nd International Conference on Very Large Data Bases (VLDB), Singapore 3(2):1653–1654

Sarma AD, Benjelloun O, Halevy AY, Widom J (2006) Working models for uncertain data. In: Proceedings of the 22nd International Conference on Data Engineering (ICDE), Atlanta, p 7
Schmid KA, Züfle A (2019) Representative query answers on uncertain data. In: Proceedings of the 16th International Symposium on Spatial and Temporal Databases, pp 140–149
Schubert E, Koos A, Emrich T, Züfle A, Schmid KA, Zimek A (2015) A framework for clustering uncertain data. Proc VLDB Endowment 8(12):1976–1979
Schmid KA, Zufle A, Emrich T, Renz M, Cheng R (2017) Uncertain voronoi cell computation based on space decomposition. Geoinformatica 21(4):797–827
Sen P, Deshpande A (2007) Representing and querying correlated tuples in probabilistic databases. In: Proceedings of the 23rd International Conference on Data Engineering (ICDE), Istanbul, pp 596–605
Soliman M, Ilyas I (2009) Ranking with uncertain scores. In: Proceedings of the 25th International Conference on Data Engineering (ICDE), Shanghai, pp 317–328
Soliman MA, Ilyas IF, Chang KC-C (2007) Top-k query processing in uncertain databases. In: Proceedings of the 23rd International Conference on Data Engineering (ICDE), Istanbul, pp 896–905
Sui D, Elwood S, Goodchild M (2012) Crowdsourcing geographic knowledge: volunteered geographic information (VGI) in theory and practice. Springer Science & Business Media, Dordrecht
Tao Y, Cheng R, Xiao X, Ngai WK, Kao B, Prabhakar S (2005) Indexing multi-dimensional uncertain data with arbitrary probability density functions. In: Proceedings of the 31st International Conference on Very Large Data Bases (VLDB), Trondheim, pp 922–933
Tran TT, Peng L, Li B, Diao Y, Liu A (2010) PODS: a new model and processing algorithms for uncertain data streams. In: Proceedings of the ACM International Conference on Management of Data (SIGMOD), Indianapolis, pp 159–170
United States Geological Survey. USGS science data catalog. https://data.usgs.gov/datacatalog/. Accessed 02/03/2021
Valiant L (1979) The complexity of enumeration and reliability problems. SIAM J Comput 8:410–421
Vu K, Zheng R (2013) Efficient algorithms for spatial skyline query with uncertainty. In: Proceedings of the 21st ACM SIGSPATIAL International Conference on Advances in Geographic Information Systems, pp 412–415
Wang DZ, Michelakis E, Garofalakis M, Hellerstein JM (2008) BAYESSTORE: managing large, uncertain data repositories with probabilistic graphical models. Proc VLDB Endowment 1(1):340–351
Wang K, Han J, Tu B, Dai J, Zhou W, Song X (2010) Accelerating spatial data processing with MapReduce. In: IEEE 16th International Conference on Parallel and Distributed Systems. IEEE, pp 229–236
Wang L, Wu P, Chen H (2011) Finding probabilistic prevalent colocations in spatially uncertain data sets. IEEE Trans Knowl Data Eng 25(4):790–804
Wang L, Cheung DW-L, Cheng R, Lee SD, Yang XS (2012) Efficient mining of frequent item sets on large uncertain databases. IEEE Trans Knowl Data Eng 24(12):2170–2183
Wang Y, Li X, Li X, Wang Y (2013) A survey of queries over uncertain data. Knowl Inf Syst 37(3):485–530
Yang Z, Li K, Zhou X, Mei J, Gao Y (2018) Top k probabilistic skyline queries on uncertain data. Neurocomputing 317:1–14
Yi K, Li F, Kollios G, Srivastava D (2008a) Efficient processing of top-k queries in uncertain databases. In: Proceedings of the 24th International Conference on Data Engineering (ICDE), Cancun, pp 1406–1408
Yi K, Li F, Kollios G, Srivastava D (2008b) Efficient processing of top-k queries in uncertain databases with x-relations. IEEE Trans Knowl Data Eng 20(12):1669–1682
Yiu ML, Mamoulis N, Dai X, Tao Y, Vaitis M (2009) Efficient evaluation of probabilistic advanced spatial queries on existentially uncertain data. Knowl Data Eng IEEE Trans 21(1):108–122

Zhang M, Chen S, Jensen CS, Ooi BC, Zhang Z (2009) Effectively indexing uncertain moving objects for predictive queries. Proc VLDB Endowment 2(1):1198–1209

Zhang C, Li F, Jestes J (2012) Efficient parallel kNN joins for large data in MapReduce. In: Proceedings of the 15th International Conference on Extending Database Technology. ACM, pp 38–49

Zhang P, Cheng R, Mamoulis N, Renz M, Züfle A, Tang Y, Emrich T (2013) Voronoi-based nearest neighbor search for multi-dimensional uncertain databases. In: 2013 IEEE 29th International Conference on Data Engineering (ICDE). IEEE, pp 158–169

Zhao B, Sui DZ (2017) True lies in geospatial big data: detecting location spoofing in social media. Ann GIS 23(1):1–14

Zheng K, Trajcevski G, Zhou X, Scheuermann P (2011) Probabilistic range queries for uncertain trajectories on road networks. In: Proceedings of the 14th International Conference on Extending Database Technology, pp 283–294

Zimányi E (1997) Query evaluation in probabilistic relational databases. Theor Comput Sci 171(1–2):179–219

Züfle A (2013) Similarity search and mining in uncertain spatial and spatio-temporal tatabases. Ph.D. thesis, Ludwig-Maximilians University Munich

Züfle A, Emrich T, Schmid KA, Mamoulis N, Zimek A, Renz M (2014) Representative clustering of uncertain data. In: Proceedings of the 20th ACM SIGKDD International Conference on Knowledge Discovery and Data Mining. ACM, pp 243–252

Züfle A, Trajcevski G, Pfoser D, Renz M, Rice MT, Leslie T, Delamater P, Emrich T (2017) Handling uncertainty in geo-spatial data. In: 33rd International Conference on Data Engineering (ICDE). IEEE, pp 1467–1470

Züfle A, Trajcevski G, Pfoser D, Joon-Seok K (2020) Managing uncertainty in evolving geo-spatial data. In 2020 21st IEEE International Conference on Mobile Data Management (MDM). IEEE, pp. 5–8.

Zwillinger D, Kokoska S (2000) CRC standard probability and statistics tables and formulae. CRC Press, Boca Raton

Chapter 15
Spatial Statistics, or How to Extract Knowledge from Data

Anna Antoniuk, Miryam S. Merk, and Philipp Otto

15.1 Introduction

"Everything is related to everything else, but near things are more related than distant things" according to the first law of geography by Tobler (1970). That is, observations of a random process appearing close together in space are more similar than observations that are more distant. In statistics, this is known as dependence or, more precisely, spatial dependence. To avoid producing artifacts when analyzing spatial data, it is important that the model accounts for any dependence in the data. How to account for spatial dependence in different modeling approaches is the subject of this paper. More precisely, we focus on so-called geostatistical models and spatial econometric models. Both of them grasp spatial dependence in different ways, leading to different model properties. However, if all parameters are chosen in a suitable manner, both approaches lead to the same conclusions and results.

The major aim of statistical modeling is to gain knowledge from data and to find statistical evidence for certain phenomena, which could be expected according to theoretical models on the underlying process. Such models could be, for instance, physical models about the ocean currents to describe displacements of sand or economical models assuming a certain behavior of individuals and postulating, for example, tax rates of municipalities. Contrary to methods in artificial intelligence like deep learning or artificial neural networks, which are mostly used for pattern

A. Antoniuk
European University Viadrina, Frankfurt (Oder), Germany

M. S. Merk
University of Göttingen, Göttingen, Germany

P. Otto (✉)
Leibniz University Hannover, Hanover, Germany
e-mail: otto@ikg.uni-hannover.de

M. Werner, Y.-Y. Chiang (eds.), *Handbook of Big Geospatial Data*,
https://doi.org/10.1007/978-3-030-55462-0_15

recognition, prediction, and classification, the focus of this paper is on models that allow for understanding the driving factors of the data. Thus, on one hand, the model should be simple enough to understand its structure and draw certain conclusions on the effect of these driving factors. On the other hand, the model should have a certain flexibility and complexity to accurately account for spatial dependence and to reproduce the data.

However, in the case of a large number of observations ($\geq$10,000) or big geospatial data, the estimation of the model parameters might be computationally challenging for both geostatistical and econometric approaches. Hence, the dimensionality of the so-called spatial covariance matrix regarding geostatistical models or the spatial weight matrix regarding spatial econometric models must be reduced. In this overview, we will address the dimensionality reduction by discussing selected approaches.

In the next section, we summarize different types of spatial stochastic processes and data, which are illustrated using simple examples. Further, the geostatistics section focuses on the estimation of spatial dependence using covariograms and geostatistical modeling approaches. In Sect. 15.4, we provide an overview of spatial autoregressive (SAR) models with simultaneous spatial dependencies originating in spatial econometrics. Finally, we utilize both approaches to model the atmospheric concentration of $PM_{2.5}$ (particles with diameters less than 2.5 μm). Section 15.6 concludes the paper.

15.2 Spatial Data

Let $\{Y(s) : s \in D\}$ be a univariate stochastic process at known locations s, where D is a subset of the d-dimensional real numbers $\mathbb{R}^d$. That is, stochastic processes are (random) observations drawn from data-generating processes having a certain order in a predefined space. Certainly, a d-dimensional rectangle of positive volume must exist in D (see, e.g., Cressie 1993; Cressie and Wikle 2011). Moreover, s may vary discretely or continuously over D. For instance, if the stochastic process lies in a one-dimensional space (i.e., $d = 1$), the process could be a time series, like daily prices/returns of financial assets or the acceleration of a car measured each second, millisecond, and so on. In Fig. 15.1, two examples of time series data with a one-dimensional support D are depicted. More precisely, the stock prices and returns of Apple Inc. are plotted on the left-hand side, and the simulated acceleration of a car is shown in the right-hand plot.

If s represents a location in the d-dimensional space with $d > 1$, the stochastic process is called a spatial or geospatial process. In the simplest case, $d = 2$, the process lies on a plane, like the surface of a certain workpiece (e.g., when inspecting the quality of produced metal sheets). Locally, the surface of the Earth could also be considered as a plane. Of course, the Earth is a sphere, or rather an ellipsoid, such that spherical spaces $\mathbb{S}^d = \{s \in \mathbb{R}^{d+1} : ||s||_2 = 1\}$ (i.e., unit spheres) should be used instead of projections into Euclidean space $\mathbb{R}^2$. Note that map projections

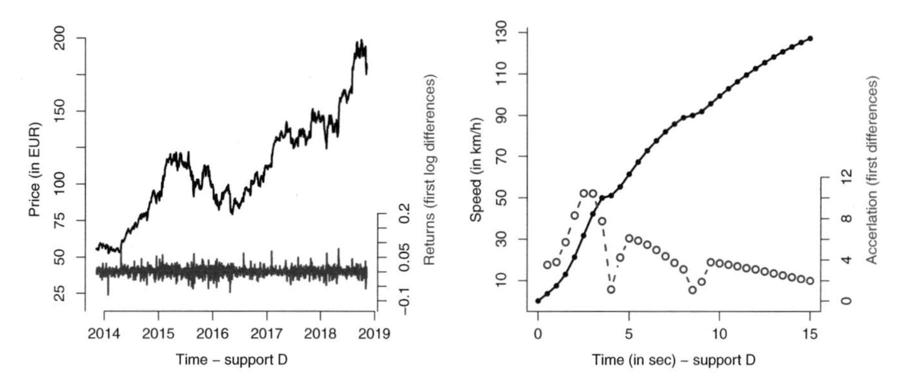

Fig. 15.1 Two time series examples. Left: Stock prices (black, left axis) and logarithmic returns (blue, right axis) of Apple Inc. Right: Simulated speed (black, dotted, left axis) and acceleration of a car (blue, dashed, right axis)

are more suitable for local processes being close to the equator (see also Banerjee 2005).

Moreover, one might distinguish between spatial point processes, geostatistical processes and spatial lattice processes, which might be observed on regular or irregular grids. In the case of spatial point processes, data are observed at a random set of locations, where the location of the measurements is of primary interest. For example, in analyzing the performance of professional dart players, the location of the arrows hitting the board is of interest. Another example from the field of astrostatistics would be the location of galaxies in space. The second case covers, for example, data measured at several measurement stations, like ground temperature or precipitation measurement stations. More precisely, D would be a subset of the two-dimensional real numbers $\mathbb{R}^2$ in these cases. That is, we observe realizations of a random process at certain spatial locations/points. Thus, these processes are also called marked point processes. In the upper left plot of Fig. 15.2, an example of such a marked point process or geostatistical process is shown. In particular, we depict copper concentrations in the topsoil along the river Meuse (see Pebesma 2004; Burrough et al. 2015). The data are measured at several points in space, or more precisely, at coordinates. If the spatial locations are not "randomly" distributed across space but lie on a regular or irregular grid, the process would be called the lattice process. Typical examples of such data are images ($d = 2$), sequences of images ($d = 3$), or raster data ($d = 2$). Images typically result from aggregating observations from point processes into grid cells, e.g., by averaging all observation of one cell. One example of such data is depicted in the upper right corner of Fig. 15.2. It visualizes the percentage of underweight children (under the age of 5) in the years 1990 to 2002 (Center for International Earth Science Information Network 2005). In contrast to the previous example, the data are not measured at several coordinates, but in larger grid cells of size $2.5° \times 2.5°$, resulting in 4×4 raster cells per $10°$ in longitude and latitude. Moreover, D might be a discrete set of locations

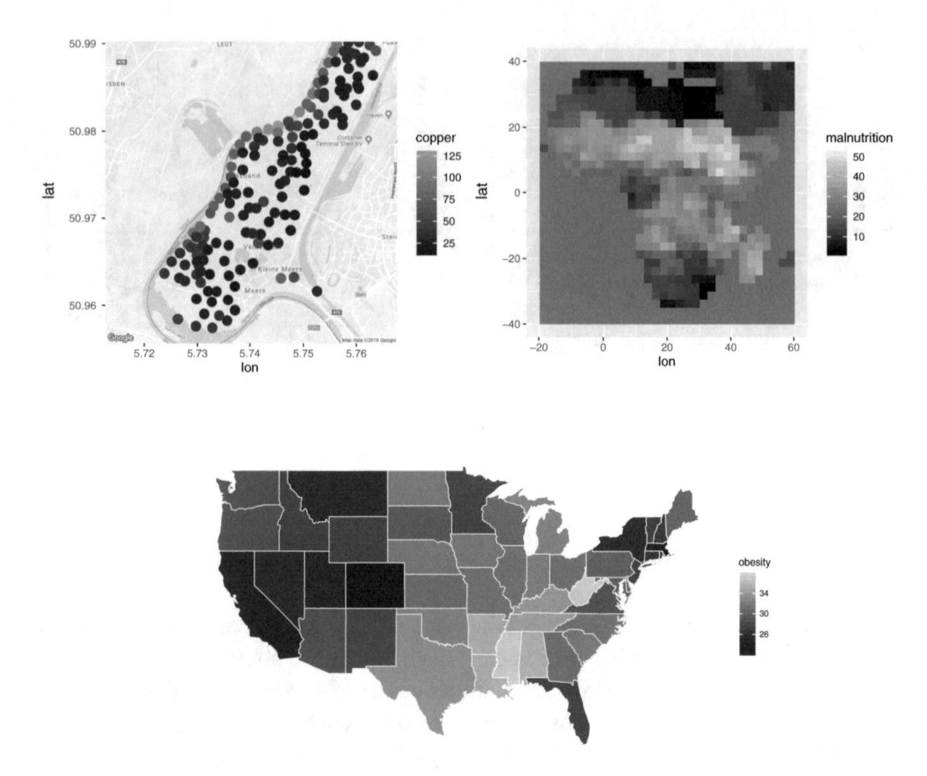

Fig. 15.2 Three examples of spatial processes. Top left: Geostatistical process of the copper concentration along the river Meuse. Top right: Percentage of malnutrition in Africa on a regular spatial raster/lattice. Bottom: Percentage of obesity in all US states (except Alaska and Hawaii) as an example of data on irregular polygons

$\{s_1, \ldots, s_n\}$ representing irregular polygons, like municipalities, counties, or states. Typical applications of such processes can be found in econometrics, epidemiology, or disease modeling. An example of this kind of process is presented in the bottom plot of Fig. 15.2. In particular, the percentage of adults suffering from obesity is shown for each state (Centers for Disease Control and Prevention 2017).

15.3 Geostatistical Models

In general, assessing the effect of observable or latent variables and gaining knowledge about the spatial dependence structure may be of interest. For that reason, geostatistical or spatial econometric models can be applied. Such models could be of a simple linear form but also of a highly complex and nonlinear form. Initially, in this section, we focus on geostatistics, while the focus of Sect. 15.4 is on spatial econometric models. We refer to Cressie (1993) and Cressie and

Wikle (2011) for an exhaustive and detailed overview of models for spatial and spatiotemporal data from the viewpoint of geostatistics.

Consider that the **random process** $\{Y(s)\}$ is drawn from a distribution $F_{Y(s_1),\dots,Y(s_n)}$, which is defined as

$$F_{Y(s_1),\dots,Y(s_n)}(y_1,\dots,y_n) = P(Y(s_1) \leq y_1,\dots,Y(s_n) \leq y_n),$$

where P is a probability measure. The idea of any statistical model is to approximate the true distribution function F using a sample of observations. Furthermore, this approximation allows us to draw conclusions on the true distribution. In particular, certain measures describing characteristics or properties of a distribution function F, like the expectation of F or (co-)variance of F, etc., are easier to interpret than the full distribution function. Consider the example of malnutrition in Fig. 15.2; using a statistical model, one might answer questions like "Does the percentage of underweight children in central Africa (expectation of F) deviate from 25% (target/reference value)?", "Adjusting for a set of predictive variables (covariates), like poverty or employment rates, could we expect that there are fewer underweight children (expectation of F) in central Africa than in southern Africa (reference value is zero considering differences of the expectations)?", or "Do the covariates describe the malnutrition equally well ((co-)variance of F) in all parts of Africa?" Moreover, certain regularity assumptions are often made on the distribution F to reduce the modeling complexity.

The random process is called **first-order stationary** if

$$E(Y(s_i)) = \mu \qquad \text{for all } i.$$

That is, the mean does not depend on the location. However, the focus should not only be on the mean of the distribution but also on the variance and covariances describing the dependence in space. More precisely, the spatial dependence is characterized by the covariances

$$\text{Cov}(Y(s_i), Y(s_j)) \qquad \text{for all i and j},$$

where $\text{Cov}(Y(s_i), Y(s_i)) = \text{Var}(Y(s_i))$. Often it is also assumed that

$$\text{Cov}(Y(s_i), Y(s_j)) = C(s_i - s_j)\,,$$

that is, the covariance is a function C of the difference between s_i and s_j. If these two assumptions are fulfilled, the process is called weakly or **second-order stationary**. This means that the expectation of the random process is the same in all locations, and the dependence between the random variables does not depend on the location but only on the difference between the two locations. We might also consider the special case in which the dependence is the same for all directions (depending only on the distance between two locations but not their orientation to each other). Thus,

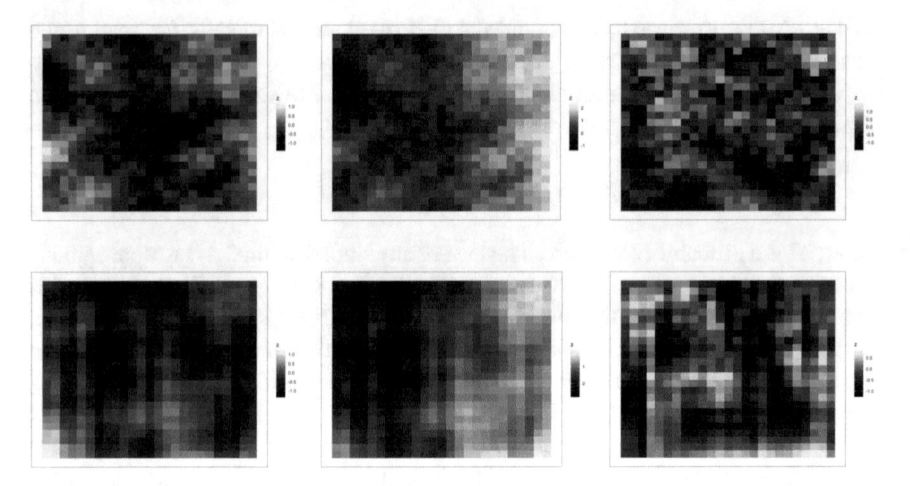

Fig. 15.3 Simulated isotropic (1st row) and anisotropic (2nd row) random processes that are second-order stationary (1st column), not mean-stationary (2nd column) and not covariance-stationary (3rd column)

if there is a norm $||\cdot||$, such that $C(\cdot)$ is only a function of $||s_i - s_j||$, the process (or rather $C(\cdot)$) is called **isotropic**.

Figure 15.3 illustrates six different random processes to visualize these properties. In the first row, isotropic processes are depicted, while the second row shows anisotropic processes with stronger north-south than east-west dependencies. In addition, we distinguish between three cases regarding first or second-order stationarity. The first column depicts second-order stationary processes, while the second and third columns reflect cases where either the mean or the covariance is not constant across space. More precisely, the mean value increases from west to east in the second column (i.e., $E(Y(s_i)) = \mu(s_i)$ and the colors are brighter in the eastern parts) and the covariance grows stronger from north to south in the third column (i.e., $\text{Cov}(Y(s_i), Y(s_j)) = C(s_i - s_j, s_i, s_j)$ and the clusters get more pronounced in the south).

15.3.1 Covariogram Estimation

If the (geo-)physical drivers of spatial dependence are known, the covariance function C might also be known. However, in most real cases, the covariance structure is unknown. Hence, C must be estimated. One straightforward estimation strategy is based on the generalized method of moments (GMM), originally proposed by Hansen (1982). That is, the covariance $\text{Cov}(Y(s_i), Y(s_j))$ between two observations at s_i and s_j is estimated by the respective sample covariance. Examining

$$\mathrm{Var}(Y(s_i) - Y(s_j)) = \mathrm{Var}(Y(s_i)) + \mathrm{Var}(Y(s_j)) - 2\mathrm{Cov}(Y(s_i), Y(s_j)),$$

we see that there is a relation between $C(s_i - s_j)$ and $\mathrm{Var}(Y(s_i) - Y(s_j))$. Thus, in geostatistics, one typically estimates $\mathrm{Var}(Y(s_i) - Y(s_j))$ instead of $\mathrm{Cov}(Y(s_i), Y(s_j))$. Moreover, suppose that

$$\mathrm{Var}(Y(s_i) - Y(s_j)) = 2\gamma(s_i - s_j) \text{ for all } i \neq j, \text{ and } s_i, s_j \in D.$$

The function γ is called a **semivariogram** (because 2γ is a variogram). Assuming isotropy (i.e., γ is only a function of the distance $||s_i - s_j||$), the locations can be grouped in classes with equal distances from each other. Particularly, for spatial data on regular grids, this approach leads to almost equally sized groups of more than two locations. More formally, an estimator of γ is given by

$$2\hat{\gamma}(h) = \frac{1}{|\Xi(h)|} \sum_{\Xi(h)} (y(s_i) - y(s_j))^2, \tag{15.1}$$

where $\Xi(h)$ is the set of all observations with distance h, that is,

$$\Xi(h) = \{(s_i, s_j) : ||s_i - s_j|| = h; i, j = 1, \ldots, n\}. \tag{15.2}$$

Moreover, the observations are denoted by $y(s_1), \ldots, y(s_n)$. However, data are typically not "evenly" distributed across space and, therefore, the distance between the points is irregular and one might not even have one pair of equal distances. In these cases, one would group the locations into bins of distances from $h - l$ to $h + l$, that is,

$$\Xi(h) = \{(s_i, s_j) : h - l < ||s_i - s_j|| \leq h + l; i, j = 1, \ldots, n\}. \tag{15.3}$$

Meaning, we allow for a certain tolerance l. However, this also implies that the estimated variogram is smoothed to some extent.

Obviously, $\gamma(h) = \gamma(-h)$ and $\gamma(0) = 0$, but $\gamma(h)$ approaches a certain constant c_0 as h approaches zero. This constant is the so-called **nugget effect**. For modeling spatial dependence, certain parametric models are often estimated for γ, like the exponential, spherical, or Matérn model.

In Fig. 15.4, we illustrate the abovementioned properties using a simulated geostatistical process on a regular grid. More precisely, the process is weakly stationary and isotropic (i.e., mean and variance are constant across space, and the covariance between two observations only depends on the distance between them). In the side view on the left-hand plot, one may see that the mean and variance do not vary across space in terms of having a trend, for instance. That is, the observed/simulated values visualized by the color of the locations are evenly spread around zero. However, we observe a strong spatial dependence from the overhead view. More precisely, the red/blue locations representing positive/negative

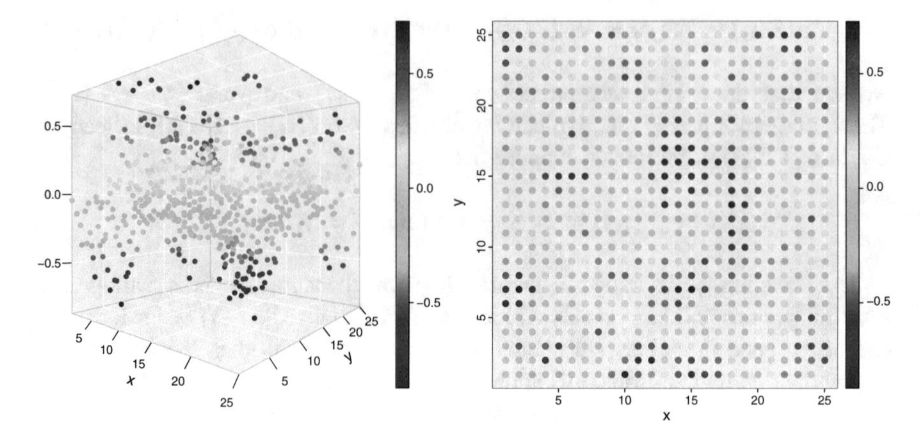

Fig. 15.4 Simulated random processes with an exponential covariance function

observations are located close to each other. These so-called spatial clusters indicate a positive spatial dependence. In most real situations, the spatial dependence would be positive (see, e.g., Cressie 1993).

The illustrated process can be reproduced using the R package MASS. First, a $d \times d$ regular grid is created that represents the spatial region of interest.

```
R> require("MASS")
R> d <- 25
R> n <- d^2
R> s <- array(c(rep (1:d, d), sort(rep (1:d, d))), dim = c(n, 2))
```

The next step involves determining the covariance matrix. In the example, we employ an exponential covariance function $C(||s_i - s_j||) = ae^{b||s_i - s_j||}$ with $a = 0.3$ and $b = -0.1$.

```
R> a     <- 0.3
R> b     <- -0.1
R> Sigma <- array(, dim = c(n, n))
R> for(i in 1:n){
+    for(j in 1:n){
+      Sigma[i, j] <- a * exp(b * sqrt(sum((s[i, ] - s[j, ])^2)))
+    }
+ }
```

Assuming that the expected value in each location is equal to 0, the random process can be simulated from a multivariate normal distribution with $\mu = 0$ and Σ resulting from C.

```
R> mu <- rep(0, n)
R> z  <- mvrnorm(1, mu = mu, Sigma = Sigma)
```

To visualize the simulated process as 3D and 2D images, the functions scatter3D() and scatter2D() from the plot3D package can be used, respectively.

```
R> require("plot3D")
R> scatter3D(x = s[, 1], y = s[, 2], z = z, phi = 0,
+   bty ="g", lwd = 6, pch = 19, ticktype = "detailed")
R> scatter2D(x = s[, 1], y = s[, 2], colvar = z, type = "p",
+   bty = "g", lwd = 10)
```

15.3.2 Modeling Approaches

In most cases, researchers are interested in the mean behavior of the random process (i.e., in the expectation $E((Y(s_1), \ldots, Y(s_n))'))$ and the effect of potential covariates on the mean. Let $Y = (Y(s_1), \ldots, Y(s_n))'$ and consider a general regression model of the mean

$$E(Y) = X\beta + f(Z), \tag{15.4}$$

where X is a $n \times (p + 1)$ matrix of potential covariates and regressors, and β is a $(p + 1)$-dimensional vector of regression coefficients. Note that the one column of X, mostly the first column, is assumed to be a vector of ones representing the intercept. Hence, the model includes p deterministic regressors. In the case of stochastic regressors, X could be replaced by $E(X)$. Furthermore, the function $f : \mathbb{R}^n \times \mathbb{R}^l \to \mathbb{R}^n$ allows for a wide range of potential models, such as spatial and spatiotemporal models or hierarchical regression models. The support of the function is a set of l covariates observed at all n locations (i.e., Z is an $n \times l$ matrix).

Generally, there are two leading ways of estimating the parameters of such models. First, they can be estimated by the maximum-likelihood approach. However, this approach requires the computation of the determinant of a large dimensional matrix. Thus, in most cases, only the expected likelihood is maximized by an expectation-maximization (EM) algorithm, which is computationally more efficient. Below, we go into further detail on this approach. Second, Bayesian approaches are often used for parameter estimation, particularly, the integrated nested Laplace approximation (INLA).

Without going into theoretical details, we sketch the abovementioned estimation methods and particularly their computational implementation. For the sake of simplicity, we focus only on the purely spatial setting, but not on the spatiotemporal case. Thus, consider that f covers a latent spatial random effect without any further regressors. In this case, f is given by

$$f(Z) = K, \tag{15.5}$$

where K is an n-dimensional random vector. More precisely, suppose that K follows a multivariate Gaussian distribution with zero mean and a weakly stationary and isotropic covariance matrix $\Sigma = (\sigma_{ij})_{i,j=1,\ldots,n}$. Hence, the (i, j)-th element σ_{ij} is given by

$$\sigma_{ij} = C(||s_i - s_j||),$$

as in the above definition. The covariance function C can be any valid covariance function, which might also depend on further parameters. For instance, C could be an exponential covariance function, Matérn covariance function, or a spherical covariance function. Many covariance functions/models have been proposed in the literature for different kinds of spatial data, such as space-time data, stationary and nonstationary data, or isotropic and anisotropic data (e.g., Matérn 1960; Cressie and Huang 1999; Gneiting 2002; Fuentes 2002; Apanasovich and Genton 2010). Note that all interpretations based on such models must be done conditioned on the specification of the assumed covariance model. Similar issues also occur in spatial econometrics, which we describe in more detail in the ensuing section. It is important to note that not all covariance functions are well-defined if the data are observed on a sphere, such as the surface of the Earth. Moreover, using Euclidean distances on map projections or chordal distances instead of geodetic distances, meaning distances on the sphere, may cause severe anomalies, especially for large distances (see Banerjee 2005; Porcu et al. 2016).

This model is a special case of the spatiotemporal model implemented in the `D-STEM` software by Finazzi and Fasso (2014) or Cameletti (2015). While Finazzi and Fasso (2014) provided an overview of the MATLAB software `D-STEM`, Cameletti (2015) implemented the same model in an R package. In particular, `D-STEM` can be used for spatiotemporal predictions and kriging, using hierarchical models. The model parameters can be estimated by the maximum-likelihood approach using an EM algorithm. In addition, the MATLAB software allows for distributed computing on several processing units and incorporates several techniques for reducing the dimensionality of the spatial covariance matrix Σ, such that complex spatiotemporal models can be estimated in a reasonable amount of time and memory even for big geospatial data.

From a Bayesian perspective, spatial and spatiotemporal models can efficiently be estimated using the INLA approach proposed by Rue et al. (2009). The INLA is an alternative to the computationally intensive Markov chain Monte Carlo (MCMC) simulations. The idea of this Bayesian approach is that the process of the parameter estimation is not deterministic but is being achieved through an approximation (see Rue et al. 2009). In the first step of such a method, the prior distribution must be assumed. This prior distribution is further supplemented with the new information coming from the observed data to transform it into the posterior distribution, which is more suitable to describe the data. More precisely, INLA uses a basis representation, which is similar to the fixed-rank kriging, which we describe below. However, the number of basis functions is larger compared to the fixed-rank kriging, yielding a high-dimensional basis representation (see Lindgren et al. 2015; Blangiardo and Cameletti 2015).

Given the general model of the mean in (15.4), we define the parameter vector, which should be estimated, as $\theta = \{\beta, f\}$, where f is some function or a set of functions that depicts additional influence on the observed variable, such as spatial dependencies. Generally, the Bayesian method aims to compute the posterior marginals given by

$$\begin{aligned} p(x_i|y) &= \int p(x_i|\theta, y)p(\theta|y)d\theta, \\ p(\theta_j|y) &= \int p(\theta|y)d\theta_{-j}, \end{aligned} \tag{15.6}$$

where $p(\cdot|\cdot)$ stands for conditional density, which should be approximated.

To estimate the posteriors, the INLA approach incorporating three steps can be used. First, the marginal θ should be estimated by applying the Laplace approximation. Based on the results of this operation, the approximation of x_i can be done in the next step. Usually, this is also based on the Laplace approximation. However, a simplified Laplace or Gaussian approximation can also be used at this point. After $\tilde{p}(x_i|\theta, y)$ is determined, the numerical integration is used to determine the posterior $\tilde{p}(x_i|y)$. Then, we obtain the following marginal posterior

$$\tilde{p}(x_i|y) \approx \sum_{k=1}^{K} \tilde{p}(x_i|\theta_k, y)\tilde{p}(\theta_k|y)\Delta_k, \tag{15.7}$$

where θ_k stands for the set of the weighted points, and Δ_k represents their weights. For a detailed explanation of this approximation method, we refer to Rue et al. (2009).

The INLA approach is computationally implemented in the R package `INLA`, which can directly be downloaded on the INLA project website. Moreover, this approach allows for modeling spatiotemporal processes, as done by Cameletti et al. (2013) for modeling particulate matter concentrations. With the function `inla()`, one can conduct the estimation of the model parameters, obtain fitted values of the observed variable, and gain information criteria. The details of the implementation of this package and the related examples can be found, for instance, in Martins et al. (2013), but we also sketch it for our empirical example below.

15.3.3 Dimensionality Reduction of the Spatial Covariance Matrix

There are several ways to reduce the dimension of the spatial covariance matrix Σ, which could be very large for spatial data (see, e.g., Banerjee et al. 2008). For instance, global raster data with a resolution of 2.5 arc-minutes, approx. 5 km, would result in more than 37 million spatial locations (i.e., Σ would have over 1.3 quadrillion or $1.3 \cdot 10^{15}$ entries). As mentioned at the very beginning of the paper, the first law of geography by Tobler (1970) says that "everything is related to everything else, but near things are more related than distant things." Thus, the idea would be to restrict the spatial dependence to only close neighbors (i.e., nearby locations). Consequently, Σ would have a sparse matrix. Although no exact definition of sparsity exists, a matrix is typically considered sparse if more than 50% are zero entries. Restricting the spatial dependence to neighbors lying within a certain distance is known as covariance tapering (cf. Furrer et al. 2006). By multiplying the covariance matrix Σ element-wise by another valid covariance matrix Σ_θ, the resulting direct product is a valid covariance matrix. This is ensured by Shur's theorem (i.e., the direct product of two positive definite matrices is positive definite). The matrix Σ_θ is a covariance matrix that has zero elements for all pairs of locations with a distance greater than θ.

However, long-range spatial dependence is ignored by this approach. Thus, another approach for reducing the dimensionality is the so-called fixed-rank kriging. For this approach, the spatial process is approximated by a random-effects model, including r spline basis functions. As long as r is smaller than the number of locations n, the precision matrix Σ^{-1}, which is, e.g., needed for kriging, can be efficiently computed. More precisely, the Sherman-Morrison-Woodbury representation of the inverse only requires the inverse of the r-dimensional covariance matrix of the random-effects model to be computed. Contrary to the previous approach, this allows modeling long-range spatial dependence. Furthermore, full-scale approximation can be regarded as a combination of both approaches. Vetter et al. (2014) compared all three approaches with respect to their computational costs and accuracy of the approximation.

15.4 Spatial Regression Models

In spatial regression models, interactions may arise between observations collected from points or regions in space. These spatial interactions are essentially characterized in two different ways, namely, in the form of **spatially autoregressive** dependent variables or **spatially autocorrelated** residuals. For the former, observations on the response variable $Y(s_i)$ in location s_i may depend on observations in other locations s_j (e.g., real estate agents may base their pricing on neighboring properties with comparable amenities). An alternative motivation is latent unobservable characteristics exhibiting spatial patterns (e.g., power plants or other combustion sources will adversely affect real estate prices in different adjacent districts). Thus, observations on the response variable are likely to resemble neighboring observations, which can be used as substitutes for latent influences.

In general, endogenous spatial interaction effects arise when the i-th observation on the dependent variable affects the j-th observation, and vice versa. These simultaneous autoregressive dependencies can be expressed as an extension of the regular linear regression model and are commonly referred to as the spatial lag or simultaneous autoregressive model, that is,

$$Y = \rho WY + X\beta + \epsilon \tag{15.8}$$

or the following reduced form

$$Y = (I - \rho W)^{-1} X\beta + (I - \rho W)^{-1}\epsilon,$$

where I is the identity matrix, ϵ is an n-dimensional vector of residuals, and it is typically assumed that $\epsilon \sim \mathcal{N}(0, \sigma^2 I)$. The spatial dependence structure is characterized by the $n \times n$ spatial weight matrix W, and ρ reflects the strength of spatial dependence implied by W (see Ord 1975). More precisely, w_{ij} reflects whether and to what extent the i-th location is affected by the j-th location (i.e., how $Y(s_j)$ affects $Y(s_i)$). The specification of spatial weighting matrices is one of the key issues in spatial econometric research and is described in more detail in the following Sect. 15.4.1.

If $E(\epsilon) = 0$, the expectation of the spatial autoregressive process is given by

$$E(Y) = (I - \rho W)^{-1} X\beta, \tag{15.9}$$

where $(I - \rho W)^{-1}$ induces a "multiplier" effect (see LeSage and Pace 2009). Intuitively, the individual units not only depend on their direct neighbors (their first-order neighbors) but also on their neighbors' neighbors (second-order neighbors) and so forth. These higher-order **spillover effects** typically diminish quickly with increasing order.

In contrast to spatial lags, modeling spatial autocorrelation in the disturbances is not primarily aimed at assessing the nature of spatial dependence or spillover effects. In fact, spatial error models do not differ from conventional nonspatial models in terms of their expectations (see LeSage and Pace 2009). Consider the following random process with spatially autocorrelated residuals e

$$\begin{aligned} Y &= X\beta + e \\ e &= \lambda W e + \epsilon, \end{aligned} \tag{15.10}$$

or

$$Y = X\beta + (I - \lambda W)^{-1}\epsilon,$$

where λ reflects the strength of that spatial autocorrelation, conditional on the selection of W. The expectation of a random process with spatially autocorrelated residuals corresponds to a regular linear regression model, namely,

$$E(Y) = X\beta. \tag{15.11}$$

However, classical estimation procedures, like ordinary least squares, do not lead to consistent estimates in the presence of spatially correlated errors. Thus, for these kinds of models, the spatial dependence appears in the error variance-covariance matrix, that is,

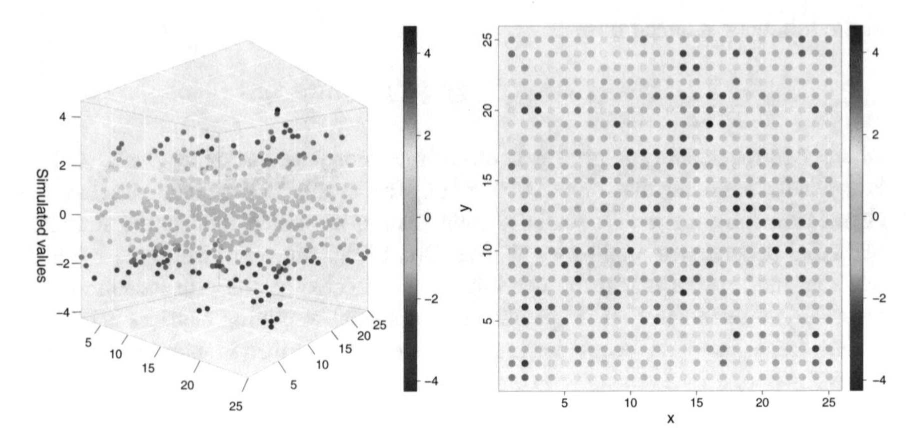

Fig. 15.5 Simulated random process with spatial autoregressive dependencies, where W is a Rook's matrix and $\rho = 0.8$

$$E(ee') = \sigma^2(I - \lambda W)^{-1}(I - \lambda W')^{-1}. \tag{15.12}$$

Figure 15.5 depicts a simulated spatial process of the following form

$$Y = (I - \rho W)^{-1}\epsilon, \tag{15.13}$$

where $\epsilon \sim \mathcal{N}(0, \sigma^2 I)$ and $E(Y) = 0$ due to the lack of regressors. Thus, there is no difference between the spatial lag and error models if the explanatory variables are omitted from the regression model. It is further assumed that the spatial dependence arises from a Rook's matrix (i.e., each location equally depends on its four nearest neighbors to the north, south, east, and west) and the strength of spatial dependence is $\rho = 0.8$. Since the upper bound for the strength in the context of normalized matrices is $\rho < 1$, a value of 0.8 reflects fairly strong spatial dependencies. These dependencies arise in the form of clusters of similar observations on the response variable Y.

In order to simulate the shown process the R package `spatialreg` is used. First, the parameters of the $d \times d$ regular spatial grid are specified as follows

```
R> require("spatialreg")
R> d <- 25
R> n <- d^2
```

In the next step, the spatial weight matrix can be determined with the functions implemented in `spatialreg`.

```
R> nb <- cell2nb(d, d, type = "rook")
R> W  <- nb2mat(nb)
```

Based on the specification of W and $\rho = 0.8$, the spatial autoregressive process can be simulated from a normal distribution by

```
R> rho <- 0.8
R> eps <- rnorm(n, mean = 0, sd = 1)
R> Y   <- solve(diag(n) - rho * W) %*% eps
```

To visualize the process, we again use the following functions

```
R> require("plot3D")
R> scatter3D(x = s[, 1], y = s[, 2], z = Y, phi = 0, lwd = 6,
+   bty ="g", pch=19, ticktype="detailed")
R> scatter2D(x = s[, 1], y = s[, 2], colvar = Y, type = "p",
+   lwd = 10, bty = "g")
```

In addition to spatial lag and error models, alternative specifications have been suggested, where variations of the response variable may be induced by spatially correlated explanatory variables WX, such as the spatial Durbin model based on both a spatially lagged dependent variable and spatially lagged explanatory variables. Moreover, both spatially autoregressive dependent variables and residuals may be included, yielding the spatially autoregressive combined model. Finally, the Manski model takes all potential sources of spatial dependence into account, that is,

$$Y = \rho WY + X\beta + WX\gamma + \alpha\iota + (I - \lambda W)^{-1}\epsilon, \tag{15.14}$$

where γ is a p-dimensional vector that captures local spillover effects, and $\alpha\iota$ is a constant vector. Thus, X is a $n \times p$ matrix comprising the p exogenous regressors. The individual spatial weighting matrices might be different but are often assumed to be equal and standardized for reasons of simplicity. Manski (1993) and Elhorst (2010) provided a comprehensive overview of different spatial dependence models and their combination and selection.

The point of departure to investigate whether a variable is spatially autocorrelated is Moran's I statistic (see Moran 1950), which is one of the suitable measures for spatial dependence in spatial econometrics. In addition, more specific tests may be employed to identify the type of spatial dependence inherent in the data. More precisely, three asymptotic procedures are available in spatial econometric research, namely, the likelihood ratio test, Wald test, and Lagrange multiplier test (see, e.g., Anselin 1988; Anselin et al. 1996). The R packages `spatialreg` and `spdep` (see Bivand et al. 2013) provide functions for specification testing, such as `lm.LMtests()` for (robust) testing in the presence of spatial lags vs. spatial errors and `LR.sarlm()` for comparing the likelihood values of different nested model specifications. The function `moran.test()` calculates Moran's I statistic for residuals from a linear regression model.

15.4.1 Specification of Spatial Weighting Matrices

Modeling spatial interactions between n cross-sectional units or locations theoretically requires the examination of $n^2 - n$ unknown spatial connections. This results in the incidental parameter problem that occurs when the number of observations

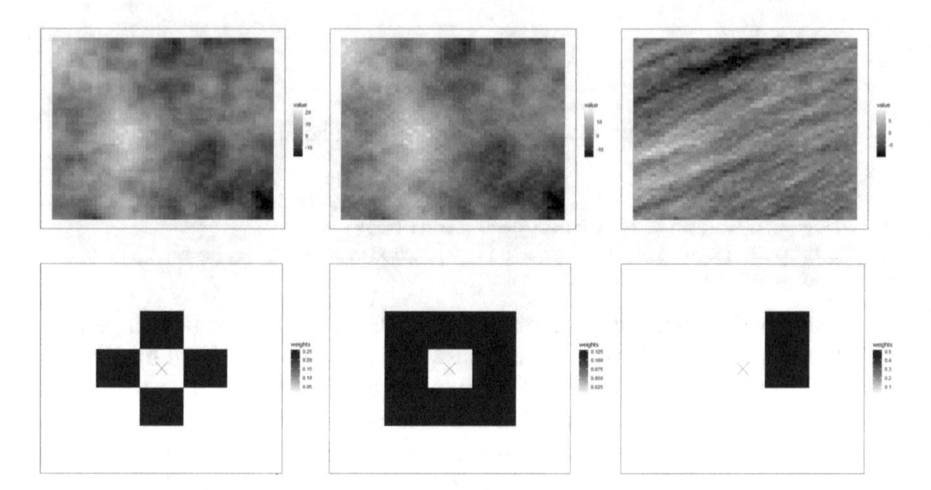

Fig. 15.6 Simulated random process with spatial autoregressive dependencies, where W is a Rook's matrix (1st column), Queen's matrix (2nd column) and anisotropic matrix with east, north-east dependencies (3rd column)

is smaller than the number of parameters. In spatial econometric research, it is common practice to replace the unknown spatial interrelations with a linear combination ρW, where W denotes the prespecified spatial weight matrix that captures how each spatial unit is related to all other units of the sample. Then, $W = (w_{ij})_{i,j=1,\ldots,n}$ is assumed to be a deterministic spatial weight matrix, and ρ is a scalar parameter that reflects the strength of spatial dependence implied by W. In the context of georeferenced data, the spatial weights are typically based on some measure of geographical proximity (e.g., p-order binary contiguity, q nearest neighbors or distance decays; Anselin 1988). Figure 15.6 illustrates three different spatial dependence structures (i.e., the commonly used Rook's and Queen's matrix and an anisotropic weighting scheme).

In addition, further assumptions concerning the construction of W are usually required (i.e., all diagonal elements of W are zero, $(I - \rho W)$ is nonsingular, and all row and column sums of W are uniformly bounded in absolute value; see, e.g., Kelejian and Prucha 1999; Elhorst 2010). The estimation of the unknown parameter ρ depends on the specification of the spatial weight matrix. Interpretations of the estimated strength of spatial dependence $\hat{\rho}$ should always be carried out with respect to the selection of W, as for geostatistical models, where a suitable covariance model is assumed. In particular, the feasible parameter space is determined by the eigenvalues of W (i.e., $\rho \in (\nu_{\min}^{-1}, \nu_{\max}^{-1})$, where $\nu_{\max}$ and $\nu_{\min}$ are the maximum and minimum real characteristic roots, respectively; see, e.g., LeSage and Pace 2009). It is common practice to row-normalize spatial weights such that each matrix row sums to unity and, hence, the principal eigenvalue of that normalized matrix is 1. Therefore, row-normalization imposes restrictions on the corresponding parameter space and facilitates inferences on parameter estimates. While row-normalization is straightforward in the case of binary weights, it is less suitable for inverse distance matrices because it may distort the proportions between rows and the meaning in terms of geographic distances (Anselin 1988; Elhorst 2001; Kelejian and Prucha 2010).

However, the deterministic specification of spatial weighting matrices has provoked some criticism (see, e.g., Corrado and Fingleton 2012). In particular, misspecifications of the spatial weights may yield biased parameter estimates (see, e.g., Stakhovych and Bijmolt 2009; Smith 2009; Lee and Yu 2012). As a result, a variety of alternative approaches for estimating the spatial weight matrix has been suggested. Pinkse et al. (2002) proposed a semiparametric estimator based on distance measures. Bhattacharjee and Jensen-Butler (2013) estimated the spatial weight matrix from the spatial autocovariance matrix for spatial panel models under the assumption of symmetry. Moreover, sparsity has been considered the identifying assumption to estimate the spatial weights from spatial or spatiotemporal data (see, e.g., Zhu et al. 2010; Ahrens and Bhattacharjee 2015; Bailey et al. 2016; Otto and Steinert 2018; Merk and Otto 2020). In the context of spatiotemporal data, time-varying spatial weighting matrices may be considered (Lee and Yu 2012; Qu et al. 2017; Billé et al. 2019; Merk and Otto 2019).

From a computational perspective, handling $n \times n$ weighting matrices can be challenging for large sample sizes n, which is relevant, especially in the context of big data. In addition to the construction of the weight matrix, further operations are usually required, such as inversion and calculation of eigenvalues or determinants. Imposing structural assumptions about W, like sparsity, symmetry, or a triangular form, contributes to reducing the computing time for these operations. Commonly used specifications, such as p-order binary contiguity or q nearest neighbor weights satisfy the assumption of sparsity because the proportion of nonzero spatial weights is small and remains constant even as the sample size increases.

15.4.2 Inferences on Parameter Estimates

For spatial error models, the interpretation of the regression parameter estimates $\hat{\beta}$ is analogous to simple linear regression models without spatial dependencies. More precisely, the first partial derivative, reflecting the effect of the p-th explanatory variable on Y, is given by the following

$$\frac{\partial E(Y)}{\partial X_p} = \beta_p. \tag{15.15}$$

However, for spatial autoregressive models with spatially lagged response variables, the interpretation of the model coefficients is quite different from linear regression and spatial error models. In particular, the partial derivatives of Y with respect to the p-th explanatory variable are given by

$$\frac{\partial E(Y)}{\partial X_p} = (I - \rho W)^{-1}\beta_p, \tag{15.16}$$

which yields an $n \times n$ matrix for each explanatory variable instead of a single scalar. Thus, changes in the p-th explanatory variable in location j can affect the response variables in other locations (i.e., $\partial E(Y_i)/\partial X_{pj} \neq 0,\ i \neq j$). LeSage and Pace (2009) referred to these cross-partial derivatives as indirect effects in contrast to direct effects $\partial E(Y_i)/\partial X_{pi}$ that reflect the effect of changes in the same location. Summary scalar measures, such as the average diagonal and cumulative off-diagonal elements over all observations, may be taken to interpret the direct and indirect effects, respectively. The differences between the estimated parameter estimates $\hat{\beta}$ and corresponding average direct effects reflect feedback effects that pass through neighboring units back to the region where the change originally occurred (cf. LeSage and Pace (2009), p. 71). Since the spatial multiplier effect $(I - \rho W)^{-1}$ is identical for all exogenous variables, no distinction exists regarding the importance of spatial spillover effects. In particular, Elhorst (2010) pointed out that the ratio between direct and indirect effects is the same for all exogenous regressors, which is not a likely scenario in empirical applications.

Anselin (2003) referred to these spillover effects as global effects because they are transferred to all locations regardless of whether they are actually connected through W. Local effects, on the other hand, only transmit changes in explanatory variables to locations that are connected to each other. Local spillovers are typically considered by including spatially lagged explanatory variables. Spatial spillover and feedback effects are precluded by spatial error models, which can be considered as the main motivation to distinguish between lag and error models.

15.4.3 Estimation Procedures

Regarding the data-generating process in (15.8), it is apparent that the spatial lag WY is correlated with the residuals ϵ. Namely,

$$E((WY)'\epsilon) = \sigma^2 W(I - \rho W)^{-1} \neq 0. \tag{15.17}$$

Ordinary least squares estimators are inconsistent in the context of simultaneity that occurs because the endogenous variable appears on both the left- and right-hand sides of the regression model. Consequently, alternative estimation procedures, such as the ML approach (Ord 1975; Anselin 1988; Lee 2004), Bayesian MCMC (LeSage 1997; LeSage and Pace 2009), generalized two-stage least squares (Kelejian and Prucha 1998), or generalized method of moments (Kelejian and Prucha 1999), have been proposed to circumvent endogeneity issues. The latter method has been primarily applied to estimate the spatial error autocorrelation and will therefore not be considered in further detail.

Estimating spatial models using the ML approach has received much attention in spatial econometrics. It typically requires distributional assumptions regarding the regression residuals (i.e., $\epsilon \sim \mathcal{N}(0, \sigma^2 I)$; see also Lee 2004 regarding the properties of these estimators if the errors are not normal). Thus, the joint likelihood results from the multivariate normal distribution for Y and the Jacobian term $|I - \rho W|$, which must be included because of the transformation rule of random variables. Bayesian methods additionally incorporate model uncertainty by including parameter distributions to derive posterior distributions. Calculating $\ln |I - \rho W|$ and the matrix inverse $(I - \rho W)^{-1}$ can be very computationally demanding, especially in the context of big data and therefore represents a major disadvantage of the ML and Bayesian MCMC estimation. However, computational aspects of the ML and MCMC have received considerable attention and improvement, especially regarding the computation of log determinants (Pace and Barry 1997; Barry and Pace 1999; Smirnov and Anselin 2001; LeSage and Pace 2007). For example, Ord (1975) proposed calculating the determinant based on eigenvalues ν_i of W (i.e., $\ln |I - \rho W| = \sum_{i=1}^{n} \ln(1 - \rho \nu_i)$). Other approaches have been suggested, such as the Monte Carlo approximation (Martin 1993; Barry and Pace 1999) or the Chebyshev log-determinant approximation proposed by Pace and LeSage (2004). LeSage and Pace (2009) provided very detailed guidance on estimating models using the ML and MCMC and explicitly elaborated on computational aspects. Since specification testing is often based on comparing log-likelihood values, the ML and Bayesian methods are also very suitable for model selection.

To control for the endogeneity inherent in spatial autoregressive models, instrumental variables or two-stage least squares can be used. The right-hand side endogenous variable (i.e., observations on the response variable of neighboring locations) is replaced with predicted values. These predictions are obtained from an instrumental variable regression, where the instruments are composed of exogenous regressors X and their spatial lags and higher-order spatial lags $(WX, W^2X, \ldots)$, as suggested by Kelejian and Prucha (1998).

For the estimation of these models, the R package `spatialreg` provides several functions. The ML estimation of the spatial lag, error, and combined model that consists of both spatially autoregressive lags and errors can be carried out by the functions `lagsarlm()`, `errorsarlm()`, and `sacsarlm()`, respectively. The function `stsls()` fits a spatial lag model (SLM) based on generalized two-stage least squares. The spatial autoregressive parameter in an error model can be estimated using the `GMerrorsar()` function.

Up to now, we have focused on modeling spatial dependence in the expectation and conditional expectations. As a side note, however, it is important to mention that spatial dependence might also occur in higher-order moments. For instance, Otto et al. (2018) introduced an autoregressive model for conditional heteroscedasticity (i.e., the conditional variance in a certain location depends on the variance in its neighboring locations). Using this approach, spatial dependence in local risks and (model) uncertainties can be described. For these models, the parameters can be estimated using the ML approach, which is computationally implemented in the `spGARCH` package in R (see Otto 2019 for more details).

15.5 Case Study

In the following empirical analysis, we consider regular lattice data with a grid-cell resolution of $0.1°$ that captures $PM_{2.5}$ concentrations in $\mu g/m^3$ for Colorado in 2000. In total, $n = 2730$ observations cover a land surface area from 37° to 41° north and 102° to 109° west. The regressors comprise NO_X concentrations (X_1), temperature (X_2), population density (X_3), and elevation (X_4). Both $PM_{2.5}$ concentrations and the exogenous regressors are obtained from the NASA Socioeconomic Data and Applications Center.

In addition, we provide source code that may be used for the analysis of spatial data with the `spatialreg` and `INLA` packages. In particular, we assess the effect of the four exogenous variables, which we refer to as `X1`, `X2`, `X3` and `X4` in the source code. Hence, the model formula is given by

```
R> f <- Y ~ X1 + X2 + X3 + X4
```

including an intercept that can be dropped by adding `-1`.

We consider the two predominant spatial econometric models described in Sect. 15.4: the spatial lag (SAR) and error model (SEM). Both model specifications are estimated using the ML approach. For the geostatistical analysis, the INLA approach is used to estimate two models, namely the spatial lag (SLM) and the stochastic partial differential equations (SPDE) model described in Sect. 15.3.2. For computational simplicity, we consider a sparse spatial weighting matrix for the SAR, SEM, and SLM. More precisely, it is assumed that each location is affected by its eight nearest neighbors that it either shares a common edge or vertex with, which is commonly known as the Queen's matrix. The corresponding list of neighbors for a $d \times d$ lattice, which can be used to construct the spatial weights matrix is generated as follows

```
R> require("spatialreg")
R> nblist <- cell2nb(nrow = d, ncol = d, type = "queen")
R> Wlist  <- nb2listw(nblist)
R> W      <- listw2mat(Wlist, style = "W")
```

where the option `style = "W"` creates row-normalized matrices such that the strength of spatial dependence is constant across all matrix rows and locations. As mentioned in Sect. 15.4.3, the following functions can be used for the ML estimation of the econometric spatial lag and error model

```
R> lag.ml <- lagsarlm(f, data = data, listw = mat2listw(W))
R> err.ml <- errorsarlm(f, data = data, listw = mat2listw(W))
```

where `data` is a data.frame object consisting of the response variable `data$Y`, regressors `data$X1`, `data$X2`, `data$X3`, `data$X4`, longitudes `data$long` and latitudes `data$lat`. The `summary()` command

```
R> summary(lag.ml)
```

returns the following summary measures such as parameter estimates, standard errors and several test statistics

```
Call:lagsarlm(formula = Y ~ X1 + X2 + X3 + X4,
listw = mat2listw(W))

Residuals:
       Min          1Q      Median          3Q         Max
-2.3724439 -0.1287108  0.0017116  0.1334192  1.7599943

Type: lag
Coefficients: (asymptotic standard errors)
              Estimate Std. Error z value  Pr(>|z|)
(Intercept)  0.1434069  0.0210097  6.8257 8.747e-12
X1          -0.0287564  0.0070383 -4.0857 4.395e-05
X2           0.0691952  0.0085066  8.1343 4.441e-16
X3           0.0810635  0.0064164 12.6338 < 2.2e-16
X4           0.0049193  0.0071976  0.6835    0.4943

Rho: 0.97731, LR test value: 9036.9, p-value: < 2.22e-16
Asymptotic standard error: 0.0032237
    z-value: 303.16, p-value: < 2.22e-16
Wald statistic: 91909, p-value: < 2.22e-16

Log likelihood: -647.5127 for lag model
ML residual variance (sigma squared): 0.073284, (sigma: 0.27071)
Number of observations: 2730
Number of parameters estimated: 7
AIC: 1309, (AIC for lm: 10344)
LM test for residual autocorrelation
test value: 108.12, p-value: < 2.22e-16
```

Analogously,

```
R> summary(err.ml)
```

returns the summary for the SEM. For more detailed instructions on implementing the `spatialreg` package, we refer to Bivand and Piras (2015).

Alternatively, the INLA approach can be used to estimate the spatial lag model. The following initial commands must be carried out in advance of the actual estimation

```
R> require("INLA")
R> data$ind <- 1:n
R> mmatrix  <- model.matrix(f, data)
R> Q        <- Diagonal(n = ncol(mmatrix), 0.0001)
```

where the `mmatrix` is the design or model matrix consisting of the exogenous regressors and an intercept, and `Q` is the precision matrix for the resulting coefficients (i.e., α and β). Moreover, the feasible parameter space for the scalar autoregressive coefficient ρ is obtained by

```
R> rho.min <- 1/max(eigen(W, only.values = TRUE)$values)
R> rho.max <- 1/min(eigen(W, only.values = TRUE)$values)
```

as discussed in Sect. 15.4.1. Further, prior distributions for the precision parameter (i.e., the reciprocal residual variance) and spatial autoregressive coefficient can be selected. The list of priors can be retrieved by the command `names(inla.models()$prior)` or, alternatively, the default specification

```
R> hyper <- list(
+ prec = list(prior = "loggamma", param = c(0.01, 0.01)),
+ rho  = list(initial = 0, prior = "logitbeta", param = c(1, 1)))
```

can be used.

Finally, the `inla()` function estimates the spatial lag model as follows

```
R> slm.ml <- inla(Y ~ -1 + f(ind, model = "slm",
+                      args.slm = list(rho.min = rho.min,
+                                      rho.max = rho.max,
+                                      W       = W,
+                                      X       = mmatrix,
+                                      Q.beta  = Q,
+                                      hyper   = hyper),
+                 data = data)
```

and the estimated parameters – as part of the random effects – can be returned by the command `slm.ml$summary.random$ind[n+1:ncol(mmatrix),]`. For more detailed theoretical and practical instructions on the INLA estimation of econometric models, we refer to Bivand et al. (2014) or Gomez-Rubio et al. (2017).

One of the modeling approaches, which was proposed by Lindgren et al. (2011) and is also widely used due to its effectiveness, is based on stochastic partial differential equations (SPDE). More precisely, a continuous field with the observed data should be approximated using a Gaussian Markov random field (see Rue and Tjelmeland 2002). For the INLA-SPDE approach, we assume that the observed variable in location i follows the normal distribution. Hence, the SPDE model can be specified as

$$
\begin{aligned}
Y &\sim \mathcal{N}(A\eta, \sigma_e^2 I_n) \\
\eta &= 1\alpha + X'\beta + \tilde{A}\tilde{\xi},
\end{aligned}
\tag{15.18}
$$

where the vector $\tilde{\xi}$ represents the Gaussian Markov random field, matrix A is a specific matrix resulting from the triangulation of the spatial field, and σ_e^2 stands for the variance of residuals, which are assumed to be independent in all locations (see Blangiardo et al. (2013) for further details). In this case, the estimation function changes as follows

```
R> spde.ml    <- inla(Y ~ -1 + Intercept + X1 + X2 + X3 + X4 +
+                          f(field, model = spde),
+    data              = inla.stack.data(stack, spde = spde),
+    family            = "gaussian",
+    control.predictor = list(A = inla.stack.A(stack),
+                             compute = FALSE),
+    control.compute   = list(cpo = FALSE, dic = TRUE),
+    keep              = FALSE,
+    verbose           = TRUE)
```

Table 15.1 Parameter estimates and metrics of prediction errors for ML and INLA

	ML-SAR	ML-SEM	INLA-SLM	INLA-SPDE
ρ	0.9773 (0.0032)		0.9999 (0.0000)	
λ		0.9916 (0.0023)		
α	0.1434 (0.0210)	6.3262 (0.6127)	0.0004 (0.0052)	6.4870 (1.2941)
β_1	-0.0288 (0.0070)	0.0426 (0.0277)	-0.0368 (0.0070)	0.0419 (0.0265)
β_2	0.0692 (0.0085)	0.1230 (0.0170)	0.0029 (0.0066)	0.1263 (0.0165)
β_3	0.0811 (0.0064)	0.1029 (0.0087)	0.0831 (0.0064)	0.1311 (0.0091)
β_4	0.0049 (0.0072)	0.0101 (0.0185)	0.0001 (0.0071)	0.0160 (0.0186)
RMSE	0.2707	0.2702	0.2691	0.1561
MAE	0.1850	0.1823	0.1833	0.1029

Standard errors in parentheses

where `stack` is a specific object comprising all data and the triangulation `A` of the spatial field, which might be a continuous field as well. In order to print the results of the estimation procedure, the summary function can be used as explained above

```
R> summary(spde.ml)
```

Table 15.1 reports the estimation results for both econometric and geostatistical model specifications obtained using the `spatialreg` and `INLA` packages, respectively.

First, inferences on parameter estimates in models with spatial dependencies and the comparison of geostatistical INLA and econometric ML approaches should be handled carefully. For example, as pointed out in Sect. 15.4.2, econometric models may implicitly incorporate spatial multiplier effects, which must be considered when interpreting regression coefficients. Hence, the size of the parameter coefficients may vary between estimation approaches. However, comparable conclusions can be drawn regarding the sign of the coefficients. The SAR and SLM models yield similar parameter estimates regarding the strength of the spatial dependencies ρ and effect of exogenous regressors captured by β. The SEM and SPDE models also produce similar parameter estimates. In particular, all exogenous variables exhibit a positive (but, as in the case of elevation level X_4, potentially insignificant) influence on the response variable. The SAR, SEM, and SLM, where the strength of spatial dependence is implied by W, are explicitly modeled to indicate strong spatial dependencies. Thus, $PM_{2.5}$ concentrations exhibit a strong positive spatial dependence, possibly resulting in clusters of high/low observations.

Moreover, both approaches, namely econometric ML estimation and INLA, are compared based on the root mean squared error (RMSE) and mean absolute deviation (MAE) to evaluate the fit of the respective model to the data. The SAR model apparently yields the highest prediction error in terms of both RMSE and MAE. However, the difference between the SAR, SEM, and SLM models is negligible. The SPDE model produces by far the lowest prediction errors and therefore appears most suitable regarding the prediction accuracy of $PM_{2.5}$ concentrations, given the data. However, the prediction accuracy and parameter estimates of the SAR, SEM, and SLM depend strongly on the specification of the spatial weight matrix. Thus, alternative specifications of W may yield very different estimation results and predictions, which represents a major disadvantage of models, where the weight matrix must be specified in advance.

15.6 Conclusion

The overview provided in this paper summarizes selected statistical models for spatial and spatiotemporal data, which are suitable for big data. We distinguish between models in the field of geostatistics and spatial econometrics. Although the latter models are not tied to applications in economics, they are typically called econometric models. To combine both fields, we applied several models and estimation techniques to an empirical dataset in environmetrics, namely $PM_{2.5}$ concentrations.

Regarding geostatistics, spatial dependence is usually modeled via spatial covariance models. These models, which may depend on additional parameters, rebuild the covariance structure of the data or the error process of a linear or nonlinear model. To gain insight into the spatial covariance, the so-called (spatial) covariogram can be computed. For large data sets, however, the covariance matrix can become very large, such that conventional estimation methods could become infeasible in a reasonable amount of computing time and using a reasonable amount of memory. Thus, we summarized selected Bayesian and frequentist estimation methods and procedures to reduce the complexity of spatial covariance matrices.

Moreover, we explain models emerging in the field of spatial econometrics. Instead of modeling spatial dependence in the covariance matrix directly, these models include so-called spatial weighting matrices to incorporate an autoregressive dependence on the spatially adjacent observations. Classical econometric models typically require the prior specification of the spatial weighting matrix. Although these approaches are close to geostatistical approaches, they lead to a slightly different covariance structure. To estimate the parameters of such models, the computation of the inverse and determinant of a high dimensional matrix is required. This matrix involves the spatial weight matrix. We sketch how this can efficiently be done or avoided when the number of observations is large. Moreover, approaches to estimating the entire weight matrix are briefly mentioned.

However, in this paper, we only focused on large spatial and spatiotemporal data (i.e., a large number of observations), but big data commonly also covers more complex data structures. These have not been addressed in this overview. In particular, functional data in space and time, such as coastal profiles or concentration profiles of environmental pollutants or network data, are interesting emerging fields where geospatial, statistical models for big data are needed. More precisely, the structure of a network can be interpreted in the same manner as spatial proximity. Thus, there is a strong relationship between the network and spatial models.

15.7 Further Reading

For a deeper view into spatial statistics and spatial regression models, we recommend the following text books. Cressie (1993) provides a thorough overview on the statistical analysis of spatial data. In particular, all important concepts in geostatistics are explained, like stationarity or isotropy of spatial processes, as well as important modeling and estimation procedures. When analyzing spatiotemporal data, one might additionally have a look into the textbook of Cressie and Wikle (2011) extending the above mentioned textbook by discussing temporal effects in spatial (or rather spatiotemporal) data. Regarding spatial regression models, a classical textbook has been written by LeSage (2008). The focus of this book is on models for analyzing economic data. Finally, we would like to recommend Banerjee et al. (2014) for further reading. In particular, they provide a comprehensive introduction to the INLA approach for spatial data.

References

Ahrens A, Bhattacharjee A (2015) Two-step lasso estimation of the spatial weights matrix. Econometrics 3(1):128–155

Anselin L (1988) Spatial econometrics: methods and models, vol 1. Kluwer Academic Publishers, Dodrecht

Anselin L (2003) Spatial externalities, spatial multipliers, and spatial econometrics. Int Reg Sci Rev 26(2):153–166

Anselin L, Bera AK, Florax R, Yoon MJ (1996) Simple diagnostic tests for spatial dependence. Reg Sci Urban Econ 26(1):77–104

Apanasovich TV, Genton MG (2010) Cross-covariance functions for multivariate random fields based on latent dimensions. Biometrika 97(1):15–30

Bailey N, Holly S, Pesaran MH (2016) A two-stage approach to spatio-temporal analysis with strong and weak cross-sectional dependence. J Appl Econ 31(1):249–280

Banerjee S (2005) On geodetic distance computations in spatial modeling. Biometrics 61(2):617–625

Banerjee S, Gelfand AE, Finley AO, Sang H (2008) Gaussian predictive process models for large spatial data sets. J R Stat Soc Series B Stat Methodol 70(4):825–848

Banerjee S, Carlin BP, Gelfand AE (2014) Hierarchical modeling and analysis for spatial data. CRC Press, Boca Raton

Barry RP, Pace RK (1999) Monte Carlo estimates of the log determinant of large sparse matrices. Linear Algebra Appl 289(1–3):41–54
Bhattacharjee A, Jensen-Butler C (2013) Estimation of the spatial weights matrix under structural constraints. Reg Sci Urban Econ 43(4):617–634
Billé AG, Blasques F, Catania L (2019) Dynamic spatial autoregressive models with time-varying spatial weighting matrices. Available at SSRN 3241470
Bivand R, Piras G (2015) Comparing implementations of estimation methods for spatial econometrics. J Stat Softw 63(18):1–36
Bivand RS, Pebesma E, Gomez-Rubio V (2013) Applied spatial data analysis with R, 2nd edn. Springer, New York
Bivand RS, Gómez-Rubio V, Rue H (2014) Approximate bayesian inference for spatial econometrics models. Spat Stat 9:146–165
Blangiardo M, Cameletti M (2015) Spatial and spatio-temporal bayesian models with R-INLA. Wiley, Chichester
Blangiardo M, Cameletti M, Baio G, Rue H (2013) Spatial and spatio-temporal models with R-INLA. Spat Spatio-temporal Epidemiol 4:33–49
Burrough PA, McDonnell R, McDonnell RA, Lloyd CD (2015) Principles of geographical information systems. Oxford University Press, Oxford
Cameletti M (2015) Stem: spatio-temporal EM. R package version 1.0
Cameletti M, Lindgren F, Simpson D, Rue H (2013) Spatio-temporal modeling of particulate matter concentration through the spde approach. AStA Adv Stat Anal 97(2):109–131
Center for International Earth Science Information Network (2005) Poverty mapping project: global subnational prevalence of child malnutrition
Centers for Disease Control and Prevention (2017) Division of nutrition, physical activity, and obesity. Data, trend and maps. https://www.cdc.gov/nccdphp/dnpao/data-trends-maps/index.html
Corrado L, Fingleton B (2012) Where is the economics in spatial econometrics? J Reg Sci 52(2):210–239
Cressie N (1993) Statistics for spatial data. Wiley, New York
Cressie N, Huang HC (1999) Classes of nonseparable, spatio-temporal stationary covariance functions. J Am Stat Assoc 94(448):1330–1339
Cressie N, Wikle CK (2011) Statistics for spatio-temporal data. Wiley, New York
Elhorst JP (2001) Dynamic models in space and time. Geograph Anal 33(2):119–140
Elhorst JP (2010) Applied spatial econometrics: raising the bar. Spat Econ Anal 5(1):9–28
Finazzi F, Fasso A (2014) D-STEM: a software for the analysis and mapping of environmental space-time variables. J Stat Softw 62(6):1–29
Fuentes M (2002) Spectral methods for nonstationary spatial processes. Biometrika 89(1):197–210
Furrer R, Genton MG, Nychka D (2006) Covariance tapering for interpolation of large spatial datasets. J Comput Graph Stat 15(3):502–523
Gneiting T (2002) Nonseparable, stationary covariance functions for space–time data. J Am Stat Assoc 97(458):590–600
Gomez-Rubio V, Bivand RS, Rue H (2017) Estimating spatial econometrics models with integrated nested laplace approximation. arXiv preprint arXiv:170301273
Hansen LP (1982) Large sample properties of generalized method of moments estimators. Econ J Econ Soc 50:1029–1054
Kelejian HH, Prucha IR (1998) A generalized spatial two-stage least squares procedure for estimating a spatial autoregressive model with autorgegressive disturbance. J Real Estate Fin Econ 17(1):99–121
Kelejian HH, Prucha IR (1999) A generalized moments estimator for the autoregressive parameter in a spatial model. Int Econ Rev 40:509–533
Kelejian HH, Prucha IR (2010) Specification and estimation of spatial autoregressive models with autoregressive and heteroskedastic disturbances. J Econ 157(1):53–67

Lee LF (2004) Asymptotic distributions of quasi-maximum likelihood estimators for spatial autoregressive models. Econometrica 72(6):1899–1925
Lee LF, Yu J (2012) QML estimation of spatial dynamic panel data models with time varying spatial weights matrices. Spat Econ Anal 7(1):31–74
LeSage JP (1997) Bayesian estimation of spatial autoregressive models. Int Reg Sci Rev 20(1–2):113–129
LeSage JP (2008) An introduction to spatial econometrics. Revue d'économie industrielle (3):19–44
LeSage JP, Pace RK (2007) A matrix exponential spatial specification. J Econ 140(1):190–214
LeSage J, Pace RK (2009) Introduction to spatial econometrics. Chapman &Hall/CRC, Boca Raton
Lindgren F, Rue H, Lindström J (2011) An explicit link between gaussian fields and Gaussian Markov random fields: the stochastic partial differential equation approach. J R Stat Soc Series B Stat Methodol 73(4):423–498
Lindgren F, Rue H et al (2015) Bayesian spatial modelling with R-INLA. J Stat Softw 63(19):1–25
Manski CF (1993) Identification of endogenous social effects: the reflection problem. Rev Econ Stud 60(3):531–542
Martin RJ (1993) Approximations to the determinant term in Gaussian maximum likelihood estimation of some spatial models. Commun Stat Theory Methods 22(1):189–205
Martins TG, Simpson D, Lindgren F, Rue H (2013) Bayesian computing with INLA: new features. Comput Stat Data Anal 67:68–83
Matérn B (1960) Spatial variation: meddelanden fran statens skogsforskningsinstitut. Lect Notes Stat 36:21
Merk MS, Otto P (2019) Estimation of anisotropic, time-varying spatial spillovers of fine particulate matter due to wind direction. Geograph Anal
Merk MS, Otto P (2020) Estimation of the spatial weighting matrix for regular lattice data–an adaptive lasso approach with cross-sectional resampling. arXiv:200101532
Moran PAP (1950) Notes on continuous stochastic phenomena. Biometrika 37:17–23
Ord K (1975) Estimation methods for models of spatial interaction. J Am Stat Assoc 70(349):120–126
Otto P (2019) spGARCH: an R-package for spatial and spatiotemporal ARCH models. R J 11(2):401–420
Otto P, Steinert R (2018) Estimation of the spatial weighting matrix for spatiotemporal data under the presence of structural breaks. arXiv:181006940
Otto P, Schmid W, Garthoff R (2018) Generalised spatial and spatiotemporal autoregressive conditional heteroscedasticity. Spat Stat 26:125–145
Pace RK, Barry R (1997) Quick computation of spatial autoregressive estimators. Geograph Anal 29(3):232–247
Pace RK, LeSage JP (2004) Chebyshev approximation of log-determinants of spatial weight matrices. Comput Stat Data Anal 45(2):179–196
Pebesma EJ (2004) Multivariable geostatistics in S: the gstat package. Comput Geosci 30:683–691
Pinkse J, Slade ME, Brett C (2002) Spatial price competition: a semiparametric approach. Econometrica 70(3):1111–1153
Porcu E, Bevilacqua M, Genton MG (2016) Spatio-temporal covariance and cross-covariance functions of the great circle distance on a sphere. J Am Stat Assoc 111(514):888–898
Qu X, Lee Lf, Yu J (2017) QML estimation of spatial dynamic panel data models with endogenous time varying spatial weights matrices. J Econ 197(2):173–201
Rue H, Tjelmeland H (2002) Fitting Gaussian Markov random fields to Gaussian fields. Scand J Stat 29(1):31–49
Rue H, Martino S, Chopin N (2009) Approximate bayesian inference for latent Gaussian models by using integrated nested Laplace approximations. J R Stat Soc Ser B 71(2):319–392
Smirnov O, Anselin L (2001) Fast maximum likelihood estimation of very large spatial autoregressive models: a characteristic polynomial approach. Comput Stat Data Anal 35(3):301–319

Smith TE (2009) Estimation bias in spatial models with strongly connected weight matrices. Geograph Anal 41(3):307–332

Stakhovych S, Bijmolt TH (2009) Specification of spatial models: a simulation study on weights matrices. Papers Reg Sci 88(2):389–408

Tobler WR (1970) A computer movie simulating urban growth in the detroit region. Econ Geograph 46(sup1):234–240

Vetter P, Schmid W, Schwarze R (2014) Efficient approximation of the spatial covariance function for large datasets – analysis of atmospheric CO2 concentrations. J Environ Stat 6(3):1–36

Zhu J, Huang HC, Reyes PE (2010) On selection of spatial linear models for lattice data. J R Stat Soc Series B Stat Methodol 72(3):389–402

Part IV
Information Retrieval from Multimedia Spatial Datasets

Chapter 16
A Survey of Textual Data & Geospatial Technology

Jochen L. Leidner

16.1 Introduction

The geographic realm can be viewed as a three-dimensional space projected onto the ellipsoid that represents planet Earth. For navigation purposes, this space has been projected down to two dimensions to create maps for centuries, and human communications and actions have been made more precise by using a grid of coordinates, latitude and longitude, to uniquely and exactly identify any point location on our planet of origin. But latitude/longitude pairs are not the first or only way to communication about locations: human communication has used *language* to *name* and *describe* places and how to get there, before a grid coordinate system was conceived, and referring by name ("New York") or description ("the green hill") remain more popular usage for human-to-human communication than grid references: people name the most relevant locations they inhabit by assigning words to them (*toponyms*) by convention, and then use these to collaborate (e.g. to instruct another human how to reach a place using navigation instructions).

In this chapter, we discuss how these two ways, the numeric, precise but less human-friendly way to reference locations can be linked with our primary means of communication, languages like English and others, through automatic means, and we explore what application uses are enabled now this is possible.

The remainder of this chapter is structured as follows. Section 16.2 disects the notion of location from a different perspectives and poses a list of research questions that we may ask when looking at the domain where geographic space and textual data intersect. Section 16.3 describes some data structures for spatial indexing, which permit fast computational operations. Section 16.4 describes

J. L. Leidner (✉)
Department of Computer Science, University of Sheffield, Sheffield, UK

Polygon Analytics Ltd, Edinburgh, UK

M. Werner, Y.-Y. Chiang (eds.), *Handbook of Big Geospatial Data*,
https://doi.org/10.1007/978-3-030-55462-0_16

address geocoding, a common way to link places identified by postal addresses to geographic space. Section 16.5 describes other ways to analyze and link linguistic signs (typically expressed as pieces of text) to geo-coordinates. Section 16.6 is concerned what we can do with spatial meta-data. Section 16.7 describes the task of searching for text when a geographic dimension is involved. Section 16.8 discusses geofencing, a notification technology that alerts users or software application once an object enters a location Sect. 16.9 describes a number of applications which have been proposed that are enabled by georeferenced text. Section 16.10 summarizes this chapter and concludes with some outlook and pointers for further reading.

16.2 Research Questions & Different Notions of "Where"

Location processing is about asking the question "where?", and one may legitimately ask the follow-on question *where* the information about the "where" may potentially come from. The following notions of "where" are conceivable today: we may obtain location data in a way that is already structured (digital and ready for computational downstream processing). In this structured case, (1) a piece of information pertaining to a location may be obtained via a *sensor*. For instance, a GPS chip in a mobile, tablet computer or car may provide location information based on triangulation across multiple satellites. Or an aeroplane's location information can be obtained by radar sensors: an airport control tower uses two radar systems, the so-called primary and secondary radar to locate it. (2) a location-bearing piece of information is available in the meta-data that describes a piece of content (text, image, movie or other). Alternatively, by processing unstructured (textual) information we can (3) extract location one or more mentions from text documents, potentially disambiguate them and map to spatial footprints in gazetteers (Hill 2006). In the latter case, we may distinguish between (3a) coarse-grained processing, i.e. a whole document will be mapped to one crude spatial footprint e.g. a country-level bounding rectangle based on the most commonly referred country. Alternatively, (3b) more fine-grained processing may resolve, using linguistic and spatial knowledge, each mention to a spatial footprint, taking into account its spatial and linguistic *context*.

In doing so, we can ask a number of research questions:

- Where is the action in the document happening ("where" is the document "about", so to speak)?
- What is the geographic footprint that best represents a document? What polygon or other spatial footprint covers the geographic space talked about in a document?
- "Where" (what place) is this document relevant "for"? We normally ask "who" is a document relevant to, but we may instead ask for a geographic space such that people inhabiting that space find the document relevant. An example would be a village downstream a dam, the inhabitants of which would be expected to find a document about the dam leaking relevant, even if it is many kilometers away.

- Where was this document written? Can we determine the location of the author at the time of writing? This is a question the answer of which is relevant where no explicit location information is given, e.g. in a diary or logbook entry.
- When does a toponym *not* denote a place? For example, in the sentence "Last week, the U.S. agreed to begin bilateral trade talks with Japan.", the phrase "the U.S." denotes a geo-political entity (GPE)—i.e., the U.S. government—rather than "just" a geographic location (LOC), a case of *metonymy*.
- Where was the user when a document was written and where are the author and the user of a document now? Does a document reveal any insights about based on the evidence that they accessed a given document?
- "Where" (which places) does the user know about (how much)? What does a document reveal about the geographic knowledge of its author? Typically, less familiar places are explicitly introduced by providing more detail. This question is also about authors' expectation to their audiences.
- When can we safely assume the current location of the user has a bearing on his or her search intent?
- Does the current search need of the user have a geographic element?

These and potentially many other questions can be asked, so clearly, the "where" question has dimension beyond just aggregating toponyms in a document collection if we are capable of taking into account spatial background knowledge and linguistic context.

16.3 Spatial Indexing

16.3.1 Spatial Data Structures

A set of data structures have been conceived to access spatial data efficiently (Samet 1989, 1990, 2006). The intuition behind *Rectangle trees* or *R-trees* is that nearby objects can be grouped together and be represented by a minimum bounding rectangle in the next higher tree level (Guttman 1984a,b). Since all children's objects lie within this bounding rectangle, a query that does not cut across the parent bounding box transitively cannot intersect with its children. R-trees are balanced, maintain their data in pages, and can be used well for persistent storage. Search in an r-tree is a top-down tree traversal, where at each non-leaf node, its bounding box is used to decide whether or not to search inside a subtree. While R-trees do not exhibit good worst-case performance, they mostly perform in practice. Data in an R-tree permits the efficient search for neighbors within a given distance r, and the k nearest neighbors can efficiently be computed with a spatial join operation. *Quad trees* (Finkel and Bentley 1974) are tree data structures in which each internal node has exactly four children, arranged into 2 quadrants, which are recursively made up of four quadtrees. Leaf nodes store spatial payload information. One parameter along which quadtrees can vary is that the shape of the four subdivided regions may

(a)	`6gkzwgjzn820`	represents the coord.	`-25.382708`	and `-49.265506`
(b)	`6gkzwgjz`		`-25.383`	and `-49.266`
(c)	`6gkzmg1w`		`-25.427`	and `-49.315`

Fig. 16.1 Geohash example from http://geohash.org (cited 2008-03-04)

Table 16.1 Some spatially-enabled database management systems

RDBMS name	Relational	NoSQL	in-RAM	Spatial	Key/Value
Oracle Database Server Enterprise Ed. with Spatial option (commercial)	Yes	No	No	Yes	No
PostgreSQL with PostGIS (free)	Yes	No	No	Yes	No
MongoDB (free)	No	Yes	No	Yes	No
REDIS (free)	No	No	Yes	Yes	Yes

be square or rectangular, or may indeed have arbitrary shapes. Each cell (sub-tree or bucket) has a maximum capacity, and when it is reached, it splits. Other spatial data structures include simple grids/bitmaps, Z-order curves, octrees, UB-trees, variants of R-trees (R+ trees, R* trees, Hilbert R-trees), X-trees, kd-trees and m-trees.

A *Geo-hash* is a sequence of characters or a number that encodes a latitude longitude pair for fast look-up and proximity comparisons. The code is designed so that the hash code calculated from a pair has two properties: a substring of a string represents a location that is geographically contained in the geographic space it is contained in (Fig. 16.1a verus b) and prefix-sharing strings are near each other geographically (Fig. 16.1b verus c).

16.3.2 Spatially Enabled Database Management Systems

Table 16.1 shows some available spatially-enabled database management systems. Oracle Database Server (Enterprise Edition) has a priced Spatial add-on option sometimes known as “Oracle Spatial” or “Spatial option”, which includes spatial datatypes like polygons (`SDO_GEOMETRY`) datatype. The free, open-source database PostgreSQL has a spatial counterpart called PostGIS, which likewise has native geometric types, spatial indexing and retrieval side by side of traditional relational functionality. Oracele and PostGIS are both Relational Database Management Systems (RDBMS). In contrast, MongoDB is a so-called “NoSQL” (schema-free) database management system, which can index semi-structured documents in JSON format, and it also provides spatial indexing and querying in two or three dimensions. REDIS is a simple in-RAM key/value store which has no SQL-like

```
CREATE TABLE businesses (
    id        NUMBER,
    poi_name  VARCHAR2(64),
    location  SDO_GEOMETRY
);
```

Fig. 16.2 A business and its location in Oracle Spatial's data definition language

```
INSERT (poi_name, location) INTO businesses VALUES (
    'PIZZA EXPRESS',
    SDO_GEOMETRY(
        2001, -- SDO_GTYPE attribute: 2-dim.
        NULL,
        SDO_POINT_TYPE(-0.132687, 51.514311, NULL), -- lon/lat/alt
                                -- PizzaExpress Jazz Club, Soho, London
        NULL,
        NULL
    )
);
```

Fig. 16.3 Creating a new business entry in Oracle Spatial's query Language

query language but it is capable of indexing points, and it can retrieve all points within a certain radius.

Other spatially-enabled database management systems include AllegroGraph, Caliper, Geocouch for CouchDB, GeoMesa, H2, IBM DB2 Spatial Extender, Linter SQL Server, Microsoft SQL Server, MySQL, OpenLink Virtuoso, RethinkDB, SAP HANA, Smallworld VMDS, Boeing Spatial Query Server for Sybase, SpatiaLite for SQLite, Tarantool, Teradata Geospatial and Vertica Place. Rigaux et al. (2002) is a standard textbook on spatial databases (Figs. 16.2 and 16.3).

16.4 Address Geocoding

The task of address geocoding takes as input a postal address in textual form, typically stored in a relational database, and maps it to a centroid (given as latitude and longitude coordinate pair) or a polygon that describes the geographic feature's spatial extent (e.g., a building). For example, the postal street address

```
John Smith
7 Cosin Court
Cambridge CB2 1QU,
United Kingdom.
```

can be mapped to

```
{  lat: 52.20114,
   lon: 0.120340 }
```

In the USA, street address geocoding is much facilitated by a public domain data set provided by the U.S. Census, namely TIGER+4 (at the time of writing in version 2018)[1] In other countries, address geocoding typically often have to rely on commercial providers. Figure 16.4 shows how a spreadsheet table containing postal addresses is batch-geocoded using ESRI ArcMap.[2]

Rhind (1999, 2001) are the authoritative sources for international address formats and managing of global address data, which varies significantly in form and regarding the degree of standardization. Goldberg (2013) describes geocoding in connection with its use to support location-based services. Li et al. (2014)

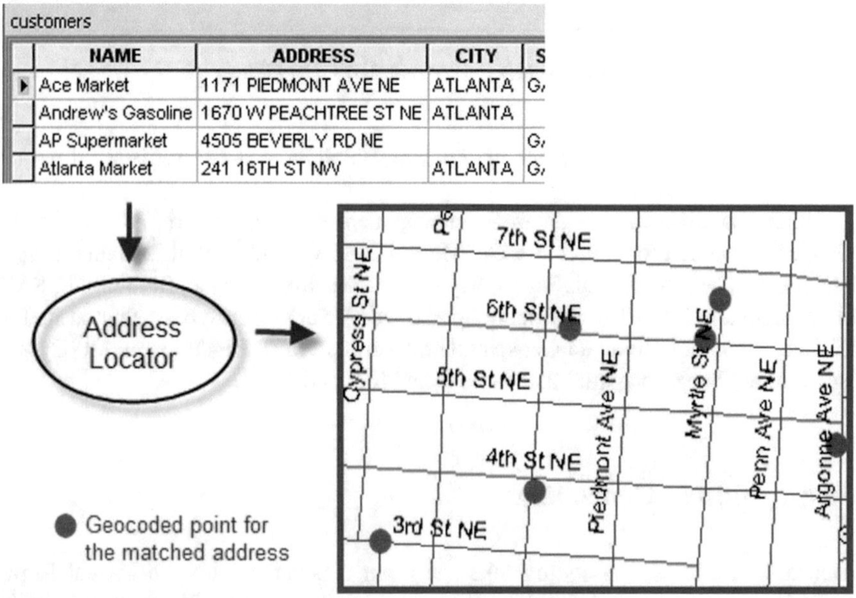

customers

NAME	ADDRESS	CITY	S
Ace Market	1171 PIEDMONT AVE NE	ATLANTA	G/
Andrew's Gasoline	1670 W PEACHTREE ST NE	ATLANTA	
AP Supermarket	4505 BEVERLY RD NE		G/
Atlanta Market	241 16TH ST NW	ATLANTA	G/

Fig. 16.4 ESRI geocoder in ArcMap applied to a table of postal addresses. (Source: ESRI)

[1]https://www.census.gov/cgi-bin/geo/shapefiles/index.php (cited 2019-01-23.)

[2]http://desktop.arcgis.com/en/arcmap/latest/manage-data/geocoding/\discretionary-geocoding-a-table-of-addresses-about.htm (cited 2018-01-15).

describe how machine learning based on Hidden Markov Models (HMMs, a limited window sequence prediction model) can be applied to the problem of address parsing using a distributed approach. Wang et al. (2016) use a similar approach but based on Conditional Random Fields (CRFs), a more general model than HMMs, and stochastic regular grammars. These works are concerned with identifying addresses in text and breaking them down into their constituent parts (street, house number, postcode/zip code etc.). Surprisingly little published work is available on algorithms, data structures and experimental evaluations of methods for the second step, the matching of these parts to gazeteers holiding the geocoordinates at the street and building levels.

16.5 Geoparsing and Spatial Resolution

16.5.1 Toponym Resolution

The term "toponym resolution" was first introduced by Leidner (2007) to denote the resolution of mentions of names for populated places to spatial footprints. It is a special case of (geo-)spatial resolution of expressions in general, which we discuss in the next sub-section. Toponym resolution is about the interface between language and the world, as proper names are said to *refer* to things in the world (i.e., outside of the linguistic system). The task of toponym resolution is made difficult because to achieve comprehensiveness, we would need to build up a complete system of place names and their footprints, so-called *gazetteers*. Frequently, names get re-used to refer to places, for example in the process of migration, either with or without modification (York→New York, Boston→Boston[3]); this is known as geo/geo ambiguity (Fig. 16.5). Words can also be ambigous between common English noun, verbs, prepositions etc. and toponyms, which is known as geo/non-geo ambiguity.[4]

For example, *Arrow, Beer, Box, Cargo, Cotton, Crackpot, Crow, Eagle, Whale* and *Wool* are all English place names as well as ordinary nounds from common English lexicon. *Send* and *Settle* are English towns and also verbs.[5]

Whereas early approaches toward georeferencing text targeted the document level (Amitay et al. 2004; Zong et al. 2005), more recently a wide range of methods

---

[3]Boston is a port town in Lincolnshire, England, GB located approx. 160 km north of London. Boston is also a city in the Commonwealth of Massachusetts in the United States of America.

[4]Metonymic use of place names (Markert and Nissim 2002; Nissim and Markert 2003), for example place-for-government metonymy as in *The United States condemned the attack strongly.* can be seen as a special case of geo/non-geo ambiguity that looks like geo/geo ambiguity at the surface.

[5]http://mentalfloss.com/article/88110/ham-sandwich-40-odd-british-place-names (cited 2019-01-29). They are capitalized in headlines or in sentence-initial position, were capitalized writing does not help the computer with disambiguating them.

geo/geo ambiguity	Cambridge (MA, USA)	⟷	Cambridge (England, GB)
	Paris (TX, USA)	⟷	Paris (France)
geo/non-geo ambiguity	Of (Turkey)	⟷	Of (preposition)
	March (England, GB)	⟷	March (month)

Fig. 16.5 Examples of Geo/Geo and Geo/Non-Geo ambiguity in English

are more granular and provide disambiguation and resolution at the toponym mention level (Leidner et al. 2003; Pouliquen et al. 2006; Leidner 2008; Speriosu and Baldridge 2013).

Generally speaking, in order to resolve a toponym in context, a number of sub-problems must be addressed:

1. A sub-sequence of characters in a text document must be recognized as containing a toponym (e.g. "New York" in "The company is based in New York, NY, USA").
2. The identified sequence must be mapped to one or more records in a database that contains information about its *feature type* and *spatial footprint*. The identified sequence must be classified (e.g. "New York" must be looked up as either the name of a city or the name of a U.S. state, and for each alternative a spatial representation e.g. a centroid given as a latitude/longitude number pair must be retrieved.
3. All alternative readings must be disambiguated, using a set of knowledge sources, which may comprise geographic knowledge (e.g. proximity information) or linguistics knowledge (e.g. context rules of the type 'if "New York" is followed by "NY" then the former toponym refers to the city and the second toponym refers to the U.S. state.)'

In general, we can distinguish a number of different methods for the toponym resolution task: first, *heuristic methods* rely on rules created by a system's developers, which are informed by human intuitions on how geospatial knowledge as well as linguistic knowledge can help disambiguate toponym mentions in context. *Probabilistic methods* rely on the statistical distribution of places, their names and co-occurrences thereof, and this group can be further divided into methods for *supervised learning* and *unsupervised learning* (also *clustering*). Leidner et al. (2003) and Leidner (2007), inspired by the proposal by Gardent and Webber (2001) to use "minimality" in different areas of reasoning, proposed an algorithm to disambiguate topomyms by assigning those candidate locations as part of the selected interpretation that jointly minimize the area of the smallest polygon that contains all centroids of the candidate locations.

Leidner (2007) surveys a dozen linguistic and geospatial heuristics that have been used to disambiguate place names, and population statistics ("prefer the interpretation that is inhabited by more people") and local patterns ("interpret 'X, Y' so as to mean that X is inside of/part of Y"; for instance, with "Cambridge, England", select the Cambridge interpretation that is part of England) are clearly

No.	Description
/H1/	'Contained-in' qualifier following
/H2/	Superordinate mention
/H3/	Largest population
/H4/	One referent per discourse
/H5/	Geometric minimality
/H6/	Singleton capitals
/H7/	Ignore small places
/H8/	Focus on geographic area
/H9/	Distance to unambiguous textual neighbours
/H10/	Discard off-threshold
/H11/	Frequency weighting
/H12/	Prefer higher-level referents
/H13/	Feature type disambiguators
/H14/	Textual-spatial correlation
/H15/	Default referent
/H16/	Preference order

Fig. 16.6 Assorted heuristics in toponym resolution strategies (Leidner 2007)

among the most commonly used pieces of evidence; Fig. 16.6 shows a list of sixteen common heuristics (Leidner 2007, p. 102–111).

Batista et al. 2012 propose a heuristic approach to toponym resolution where toponyms from Portuguese-language news are disambiguated against an ontology by using semantic similarity metrics. They compare Resnik's, Jiang and Conrath's Jiang and Conrath (1997) and Lin's semantic similarity metrics and the frequency of mention of a concept and its descendants (IC), formally defined as $IC(c) = -\log \frac{f(c)}{\max f(c)}$, to find that Conrath and Lin's performs best. It is defined as $Jiang - Conrath(c_1, c_2) = 1 - (IC(c_1) + IC(c_2) - 2IC_{MICA}(c_1, c_2))$, $Amc(c)$ denotes the ancestors of c in the ontology. Their method is greedy and processes toponyms in a pairwise fashion from left to right as they occur in the text; they also assume the one referent per discourse heuristic. While the authors give runtime information, no evaluation in terms of precision or recall are given; instead, they report an average distance measure (defined as the arithmetic mean of three ontology path distance measures) between automatically resolved and human-annotated ground truth in a corpus of annotated news called Geo-Chave-PT. Chen et al. (2018) compare sixteen unsupervised clustering methods for toponym resolution and introduce DensityK, their own method, which is inspired by Riepley's K function (related to divergence from randomness) and the density-based, spatial DBSCAN clustering algorithm. In the authors' evaluation on 1,000 toponyms comprising two subsets (University of

Melbourne campus descriptions and Wikipedia+Blogs) their DensityK performed best at $P \approx 83\%$.

Recent progress in natural language processing includes the use of *word embeddings*, low-dimensional and vectorial representations of words and their meaning defined by the context of use, starting with *word2vec* Mikolov et al. (2013). In the context of spatial text processing, hybrid textual/geospatial embedding models have been devices that include textual context and location information (Xie et al. 2016; Kejriwal and Szekely 2017).

16.5.2 Geospatial Expression Resolution

Compared to the number of proposed method for the resolution of individual toponyms, there is less published work on the problem of resolving *geographic expressions*. The may contain zero or more toponyms embedded in ordinary linguistic phrases such as noun phrases (NPs) and other signs denoting descriptions that can be used to refer to locations:

(1) *30 min north of Paris by car*
(2) *Clapham, a district south-west London lying mostly within the London Borough of Lambeth*
(3) *near Bruntsfield Links*
(4) *approximately seven kilometers from the German border*
(5) *halfway between Glasgow and Edinburgh*

While the understanding of the cognitive processing of spatial language is still at an early stage (e.g. Coventry and Olivier (eds.) (2002)), symbolic and statistical techniques for practical processing that work, but that cannot claim to model how the brain works are now available.

Piton and Maurel (2001) focus on topomym recognition for French using finite state transducers, and in doing so, they model the relation between multiple toponyms within the same document (e.g. demonyms like *French* ~ *France*). Bilhaut et al. (2003) describe a rule-based, compositional system for processing French toponyms and spatial expressions describing places in France using a unification-based grammar approach. Nagel (2008) describes a system for the analysis of geo-spatial (locative) expressions in the German language using shallow local grammars.

16.6 Content Enrichment with Geospatial Metadata

Once a centroid or a point part of a polygon have been identified given as lat/lon coordinates, it can be represented for purposes of data interchange in one of many different formats:

```
<?xml version="1.0" encoding="UTF-8"?>
<kml xmlns="http://www.opengis.net/kml/2.2">
  <Placemark>
    <name>Pizza Express (Soho)</name>
    <description>
       POI place mark attached; intelligently places
       itself at the height of the respective ground level.
     </description>
    <Point>
      <coordinates>-0.132687,51.514311,0</coordinates>
    </Point>
  </Placemark>
  </kml>
```

Fig. 16.7 Example point of interest in KML

- The *Keyhole Markup Language* (KML)[6] is an XML application part of Google Earth (originally called Keyhole), which has become one quasi-standard for sharing Point of Interest (POI) data on the Web, and it permits a Google Earth visualization directly from a Web browser (see Fig. 16.7 for an example). KMZ is a compressed form of KML. The advantage of XML-based formats is that many libraries exist to read and validate XML document instances.
- *GeoRSS* is a variant of the Real Simple Syndication (RSS) format, a Web standard for sharing newsfeeds. GeoRSS is a variant of RSS that incorporates geographic metadata encoded in the *Geography Markup Language* (GML).
- The *GPS EXchange Format* (GPX) has its origins as an exchange format of GPS trackers and navigation aids.
- The *GeoJSON* format is a variant of the Javascript Object Notation (JSON), a lightweight data exchange meta-format based on potentially hierarchically-nested attribute-value pairs. GeoJSON is less verbose than KML, but does not support any validation (JSON is schema-free).
- The Text Encoding Initiative (TEI) has a `<geo>` tag, which can model lat/lon coordinates embedded in running text (Fig. 16.10).[7]
- Figure 16.8 shows an example of a *Point of Interest* (POI) marked up using the *Resource Description Framework* (RDF) standard used by "Semantic Web" efforts to promote data interoperability on the Web and elsewhere.

[6]https://developers.google.com/kml/documentation/ (cited 2019-01-26).

[7]http://www.tei-c.org/release/doc/tei-p5-doc/en/html/ref-geo.html
TEI Specification, Section 13.3.4.1 Varieties of Location (cited 2019-01-28).

```
<!-- RDF geo meta-data: Berlin (Mitte), Germany -->
<rdf:RDF xmlns:rdf="http://www.w3.org/1999/02/22-rdf-syntax-ns#"
         xmlns:geo="http://www.w3.org/2003/01/geo/wgs84_pos#">
  <geo:Point>
    <geo:lat>52.531677</geo:lat>
    <geo:long>13.381777</geo:long>
  </geo:Point>
</rdf:RDF>
```

Fig. 16.8 Geographic meta-data in the semantic web standard RDF

```
<!-- GeoURL format -->
<meta name="ICBM" content="50.167958, -97.133185">

<!-- Geo Tag format -->
<meta name="geo.position"  content="50.167958;-97.133185">
<meta name="geo.placename" content="Rockwood Rural Municipality,
                                    Manitoba, Canada">
<meta name="geo.region"    content="ca-mb">
```

Fig. 16.9 GeoURL and GeoTag meta-data tags in HTML headers on the world wide web (WWW)

- The GeoURL method of putting meta-data in the `<head>` part of HTML pages on the Web (Fig. 16.9).[8]

Note that from the above formats only the last one is able to annotate text documents *inline* with geo-coordinates at the individual toponym mention level (see Appendix "Curating Gold Standard Data for Evaluation and Training" for more work in that direction); the other forms of meta-data must be separately kept, and they describe a document as a whole.

Luo et al. (2011) is a survey that covers how to annotate multi-medial content with geographic meta-data; at the time of writing it has become quite rare that missing geo-metadata has to be derived from textual sources using toponym resolution because many devices contain GPS sensors (Fig. 16.10).

[8]https://web.archive.org/web/20080515145826/http://geourl.org/add.html (cite 2019-01-29).

```
<place xml:id="pl-c-H" type="county">
 <placeName>Herefordshire</placeName>
 <listPlace type="villages">
  <place xml:id="pl-v-AD">
   <placeName>Abbey Dore</placeName>
   <location>
    <geo>51.969604 -2.893146</geo>
   </location>
  </place>
  <place xml:id="pl-v-AB">
   <placeName>Acton Beauchamp</placeName>
  </place>
<!-- ... -->
 </listPlace>
 <listPlace type="towns">
  <place xml:id="pl-t-H">
   <placeName>Hereford</placeName>
  </place>
  <place xml:id="pl-t-L">
   <placeName>Leominster</placeName>
  </place>
<!-- ...  -->
 </listPlace>
 </place>
```

Fig. 16.10 The geo element of the text encoding initiative (TEI)

The GeoXml library[9] is an example software component by Microsoft and part of the Bing Web controls framework that can process KML/KMZ, GeoRSS, GML via GeoRSS and GPX. GeoTools[10] is an open source library for Java that includes geographic format conversion (see Fig. 16.11). The *GDAL/OGR Geospatial Data Abstraction Library*[11] is an open source C++ library for geometric and geographic computing.

[9]https://docs.microsoft.com/en-us/bingmaps/v8-web-control/modules/geoxml-module/ (cited 2019-01-26).

[10]http://www.geotools.org (cited 2019-01-26).

[11]https://www.gdal.org (cited 2019-01-26).

```
Point point = new Point(51.514311, -0.132687);
GeometryJSON g = new GeometryJSON();
g.writePoint(point, "point.json"));

Point point2 = g.readPoint("point.json");
```

Fig. 16.11 Example conversion to and from GeoJSON mit GeoTools in Java

16.7 Hybrid Textual/Spatial Document Retrieval

In ordinary document retrieval (search), a set of pre-indexed documents are ranked from most relevant to least relevant, given a user's query. There are many methods for doing this, for explanatory purposes, we can look at the popular Vector Space Model (VSM) in this section (Salton et al. 1975). The query is transformed from a sequence of space-separated word tokens (terms) into a sparse numeric vector where each component contains the frequency of the k-th term in the lexicon. During indexing time, each document was already transformed in the same way, from a representation as a sequence of space-separated word tokens into a numeric vector of term frequencies. Simplifying somewhat, the cosine of the angle between the query vector and any document vector can be ordered from smallest angle to largest angle. The smaller the angle between query vector and a document vector, the more similar the sets of terms they represent, and the idea is that the more similar the document to the query, the more relevant that document is for the query, too.

Now to bring in a geographic aspect into the ranking process, we can either create an integrated ranking score that also takes location into account, or we can use an off-the-shelf relevance score $s(q, d)$ and combine it with a *geo-relevance score* $g(q, d)$.

There are multiple ways to integrate textual relevance and geographic relevance The first option is to integrate both using linear interpolation:

$$r(d, q) = \lambda \cdot s(q, d) + (1 - \lambda) \cdot g(q, d) \tag{16.1}$$

where r is the overall relevance ranking function and λ is a weight that determines how much influence should be given to the text-based relevancy (larger values for λ) and how much influence should be given for the geospatial dimension (smaller λ).

How a good geographic scoring function looks like is still an open research question (Leidner 2006b), but one intuition is that it should be designed so that values of g are higher for query-document pairs where the spatial footprint associated by the query's information need overlaps mover with the spatial footprint of a document. Figure 16.12 shows some example queries that have a geospatial dimension as well as some candidate documents, some of which match the queries' information needs. Query (a) matches (i) and (ii). A purely text-based search method would be able to match (a) and (i), but would miss the connection between (a)

Queries:

(a)	shark attacks in Australia or California	$\{x \mid x \in Australia \lor x \in California\}$
(b)	bomb explosion in Basque country	$\{x \mid x \in Basque\ country\}$
(c)	Nigeria elections	$\{x \mid x \in Nigeria\}$

Documents:

(i) Australia has already seen 56 shark fatalities this year.
(ii) Yesterday, two tourists died in a shark attack near Sydney.
(iii) The election of a new Nigerian president will be a challenge.
(iv) The device exploded in Spain after an anonymous call.

Fig. 16.12 Some queries with a geographic information need and candidate documents

and (ii) because the term "Sydney" mentioned in the document is not mentioned in the query; in contrast, a geographically-aware search engine has spatial footprints associated with documents and queries, so it is able to match the centroid of Sydney as located inside the polygon or bounding box for Australia, therefore giving is a higher geo-score g.

The second way to integrate textual relevance and geo-relevance is by using *geo-filtering* (Leidner 2006b):

$$r(d, q) = \begin{cases} s(d, q) & \text{GEO-FILTER(d, q)} \\ 0 & \text{otherwise} \end{cases} \tag{16.2}$$

In Leidner (2006b), three geo-filters are proposed: ANY-INSIDE: this filter lets a document pass if at least one toponym's spatial footprint is contained in the query's spatial footprint(s); MOST-INSIDE: this filter lets a document pass if most toponyms' spatial footprints are contained in the query's spatial footprint(s); and ALL-INSIDE: this (most aggressive) filter lets a document pass only if all toponyms' spatial footprints are contained in the query's spatial footprint(s).

Clough et al. (2006) and Purves et al. (2018) are good introductions to the topic of *geographic information retrieval* (GIR), and the theses (Andogah 2010) and the monograph (Sallaberry 2013) provide more detailed accounts.

16.8 Geofencing

Document retrieval as described in the previous section is a task initiated by a user's query ("pull" by the user). Sometimes, we want to initiate some action by a system based on some previously defined condition. *Geofencing* is the task of ongoing monitoring whether a (set of) candidate points (typically locations of a moving object) fall inside a geographic footprint or not. A geographic footprint demarcating an area of interest can be given as a (centroid; radius) pair or as a polygon

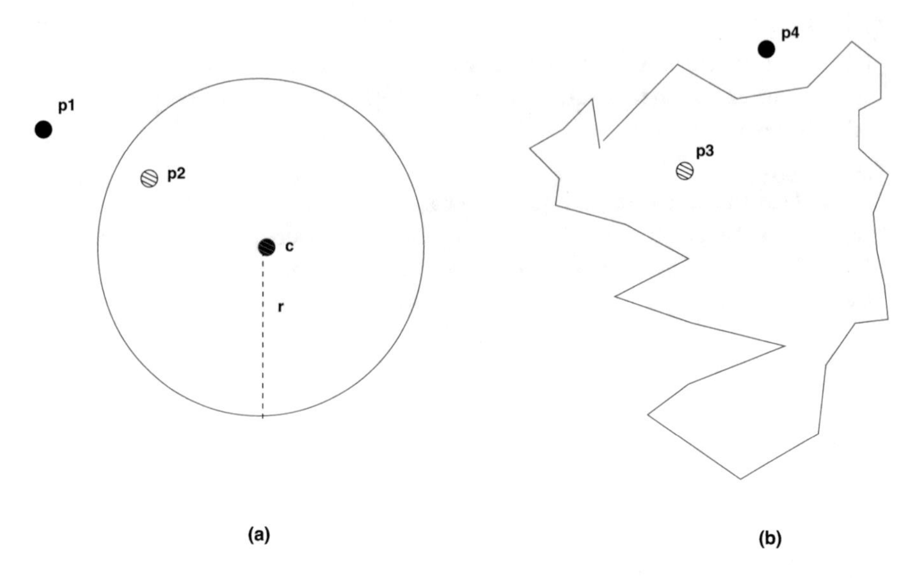

Fig. 16.13 Two types of geo-spatial footprints for geo-fencing

(Fig. 16.13). So geofencing means defining a spatial footprint (the *geofence*) and associating with it an action such that whenever an object (e.g. a car or a user carrying a mobile phone) enters the spatial footprint, the action (e.g. send a message to the user) is carried out. Geofencing is also associated with "push" delivery, a mode of notification that potentially interrupts other activities of the user to make him or her aware of a event connected with his or her current location.

Figure 16.13a on the left shows a spatial footprint in the form of a circle, given as a pair (c, r) of a centroid c and a radius r. Candidate point p_1 is clearly outside, so no monitoring event is raised, whereas point p_2 is inside the circle ($p_2 < r$), and this triggers an event of notifying the application that uses the geofence.[12] Figure 16.13b on the right shows another spatial footprint, this time given as a polygon. Point p_3 is located inside (thus triggering an event) and point p_4 is located outside of the polygon.

The spatial footprint may be given as a rectangular bounding box, a circle defined by centroid and radius or a polygon comprising a set of points if more complex shapes must be taken into consideration. Conceptually, at alerting time the system must be able to check quickly whether any previously defined geofences, so bespoke spatial data structures can be devised and studied that permit fast processing (e.g. of point-in-polygon checks).

An example is a store that may advertise special offers to passers-by that get close to the store if they have installed a certain application on their mobile device.

[12] A "push" notification can also be emulated by repeat pull requests (polling).

The spatial footprints can be stored using intensional descriptions as geometric objects or they can be translated to a bitmap, which has all bits set that correspond to "inside" locations. The former is slower but the latter is more memory-intensive.

Li et al. (2014) describe an approach for dynamic geo-fencing, which was one of the winning entries in the ACM SigSpatial GIS Cup 2013,[13] a shared task, and which exploits the trajectory of moving object to make the point-in-polygon test more efficient ("boundary partitioning"). They also facilitate the processing by reducing the complexity of the polygons that describe the geofence's boundary (*boundary simplification*). Polygons checks are often preceeded by simpler Minimum Bounding Rectangle (MBR) checks to speed up retrieval from spatial data structures like R-trees and quad-trees.

16.9 Applications

In this section, we cover a set of example applications, which represent typical uses of geospatial technologies. Our selection should not be seen as comprehensive.

16.9.1 Location Search

A mobile user is a human, changes location and carries a mobile computation device (e.g. a mobile phone or tablet). While traveling, an information need such as wanting to find out where the nearest pizza restaurant is located, May lead to a query such as "pizza" can be issued (using voice or as typed text). This really is the textual part of the underlying full query "pizza near me", where the "near me" part is implicit context and resolves to the user's current location. The actual information need itself is typically expressed using language (text, voice), whereas the "near me" part is realized by a background computation that involves obtaining the user's location via a GPS receiver built into the mobile device and making a spatial query against a database that holds a set of restaurants, including their pre-calculated geo-coordinates, held in a spatial data structure like an $R*$ tree described earlier. This is possible because an offline process has already geocoded the restaurant addresses ahead of time. The nearest restaurant, then, is the one that has the smallest distance to the user's location among all candidates.

Figure 16.14 is a screen capture from Google Maps showing a search query for a cafe in Brooklyn, at the time of writing the most popular local search site on the Web. Other local search offerings include Apple Maps, Bing, Citymapper, Yelp Foursquare, Yellow Pages Super Pages, Yellow Book or Local.com.

[13]http://dmlab.cs.umn.edu/GISCUP2013/ (cited 2018-09-13).

Fig. 16.14 A local search on Google maps (maps.google.com, cited 2019-01-23)

16.9.2 Crime Mapping, Hotspot Analysis and Forecasting

Like every event has a location, every crime has one. *Crime mapping* is the process of marking on a geographic map a set of historic crime locations with the aim to carry out more effective policing going forward (Maltz et al. 1990; Leitner 2013). Often crimes cluster in certain areas, and identifying and dealing with these is known as *hotspot analysis*. Hotspots can inform increased policing activity from intensified patrols to the establishment of new police wards. One step further goes crime forecasting, which aims to anticipate potential crimes with the objective of preventing them from becoming actual crimes. The special report (Harries 1999; Gonzales et al. 2005) describe some crime mapping techniques and existing systems from a U.S. perspective, and Hill and Paynich (2014) is a textbook by practitioners. Mohler et al. (2015) is a study that analyzes *predictive* modelling of law enforcement. *Geographic profiling* (Canter 2003) aims to analyze geospatial patterns of one particular criminal at large to establish behavioral patterns.

16.9.3 Political Anaysis and Intelligence Applications

The political sciences (geopolitical analysis) and national intelligence services can benefit from georeferenced data in ways similar to law enforcement applications mentioned above. For example, Duru (2018) present and analyse a map covering

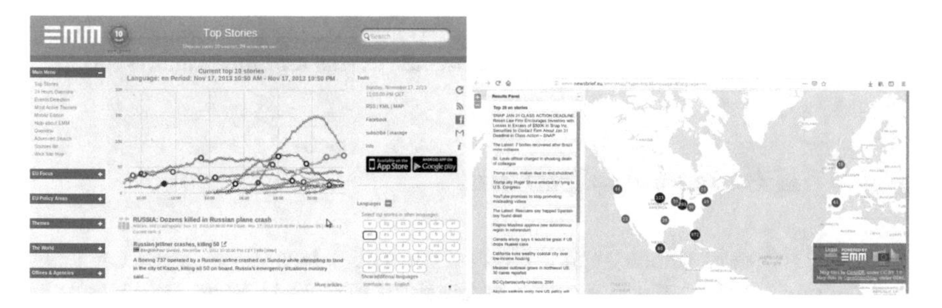

Fig. 16.15 The European media monitor (emm.newsbrief.eu, cited 2019-01-26)

terrorist attacks in Turkey derived from the *Global Terrorism Database (GTD)*[14] (LaFree and Dugan 2007). GDELT Leetaru and Schrodt (2013) is another dataset of incidents, built to enable the qualitative and quantitative study of political events across the globe. From an ethics point of view, this is a dual purpose technology: segmentation analysis of the electorate can also be used to *influence* elections, as the case of the now-defunct company Cambridge Analytica demonstrated.[15] Political actors can group sub-population by criteria including personality and location, and then target them with messaging that they will be more likely to respond to, including lies to generate anger. The role of geolocation in such an immoral activity, which is however technically very similar to targeting individuals with personalized online ads to influence their buying behavior, is paramount.

The European Media Monitor (EMM) is a suite of applications developed at the European Commission's Joint Research Centre (JRC) for decision support of policy makers (Pouliquen et al. 2006, 2004; Steinberger et al. 2013). It aims to extract named entities and recognizes events including resolving location names to geo-coordinates for mapping in order to track disease outbreaks, political crises or terrorist trends. Figure 16.15 shows some example screen captures that demonstrate its geospatial mapping and social network analysis capabilities. Petroni et al. (2018) present an event extraction system, which comprises two independent sub-systems, one that extracts events from news and another one that extracts events from social media (Twitter). It is trainable and initially covers a set of human-made and natural disaster event types (including fires, terrorist attacks) and is able to perform *event co-reference* across the two different text types (meeting the challenge that news stories are much longer and more formal than Twitter posts, whereas Twitter posts have more varied/creative forms of expression, are very short, and arrive at big daily

[14]https://www.start.umd.edu/gtd/ (cited 2019-01-20).

[15]At the time of writing, the detailed nature of the connection between Cambridge Analytica and the election of Donald Trump to the office of U.S. president as well as alleged voter manipulation in the context of Britains referendum about the departure from the European Union ("Brexit") is still ongoing.

volume and velocity namely 500 millions/day). Resolving place names to lat/lon coordinates permits the mapping of the extracted events for human analysis.

16.9.4 Healthcare Applications

Arguably the most famous application of geospatial location analysis happened in London in 1854: after a massive outbreak of cholera in Soho (120 deaths in just 3 days), John Snow[16] mapped the locations of the lethalities, and he found they clustered around a pump in Broad Street. Once that pump was closed, which ended the epedemic and perhaps began the geographical analysis of disease data. Much more recently, Rushton et al. (2008) discuss the use of address geocoding in cancer research. Fiscella and Fremont (2006) combined geocoding and name ethnicity analysis as two indirect methods to assess disparities in care. Dredze et al. (2013) introduce Carmen, a software component for georeferencing Twitter micro-blog posts ("tweets"), which uses location meta-data communicated by the sending device in the form of coordinates or textual form, locations mentioned in the user profile or toponym resolution applied to the message's content. The system was applied to the healthcare domain (influenza surveillance), and is easily applicable to other domains.

16.9.5 Location-Based Services and Location-Aware Advertising

Location Based Services (LBS) are generally speaking all kinds of services that take into account the location of their users (Schiller and Voisard 2004). For example, location-aware B2C (business to consumers) advertising services offer customers of a particular business some special offers once they are near that business' premises (*geomarketing*, see Cliquet (2006) and Faber and Prestin (2012), for instance). For example, an advertisement for a nearby restaurant is perhaps more likely to receive a visit from someone who is alerted to its existance and is 100 m away compared to someone who is 500 km away. LBS also include location-aware social media e.g. Foursquare,[17] a mobile application that lets its users "check in", i.e. indicate they arrived in a particular place such as a restaurant or bar, and they can let they friends know they are now there, in case the friends are also nearby and would like to meet up. Geospatial analysis can also be used in real estate planning (B2B, business to business); for example, the valuation of residential or commercial properties can

[16] Snow, John (1854), "Epidemiological Society" *Medical Times and Gazette* **9** (16 December 1854): 629. http://johnsnow.matrix.msu.edu/work.php?id=15-78-A8 (cited 2019-01-25).

[17] https://foursquare.com (cited 2019-01-26).

be modeled taking into account evidence from the vicinity: for instance, an airport 15 min drive away could be very positive for a business and very negative for a residential property (*geospatial analytics*). Hu et al. (2018) use toponym resolution to harvesti local place names from geotagged housing advertisements Supermarket chains may use Geographic Information Systems (GIS) to conduct branch location optimization, i.e. to find the best location for a new supermarket, based on a set of criteria such as availability of parking space, other nearby amenities that already attract drive-through traffic (e.g. next to a gas station, accessible via a highway).

16.9.6 Other Applications

Numerous other application of geospatially-connected textual data are conceivable and have been proposed. In conditions, where collections of biological speciments (animals or plants) need to be cataloged, the location where an item was found is often described using place names or spatial expressions. Automated methods like toponym resolution can be used to transform specimen or herbarium collections into georeferenced collections (Beaman and Conn 2003; Guralnick et al. 2006; Bloom et al. 2017) that can be mapped. The emerging discipline of *Digital Humanities* (Gregory and Geddes 2014), i.e. the application of information technology to scholarship in history, sociology, or literature offers additional application potential; for example, DeLozier et al. (2016) annotate the location of places in literary works so the spatial trajectory of characters can be investigated.

16.10 Summary, Conclusion and Future Work

In this Chapter, we have reviewed a set of technologies and methods for interlinking the geospatial realm with the realm of textual data. We showed how geocoding does for postal addresses what toponym resolution does for place names and spatial grounding in general does for geographic expressions, namely link text spans to the geographic locations they refer to. We reviewed how spatial data-structures can be used independently or together with full-text indices to retrieve information (e.g. documents given a query comprising a text element and a location element) in a geo-aware way after text collections have become georeferenced. We described how geofences can alert applications or users of objects entering a location, and we portrayed several applications: location-based mobile search, crime mapping, hotspot analysis and crime forecasting, political analysis and intelligence applications, location-aware advertising, real estate spatial analytics, among others.

In conclusion, the georeferencing of text offers a broad range of valuable applications, in particular given the high degree of mobility of modern humans and the wide-spread availability of GPS-enabled devices such as mobile phones and car navigation systems, we can reasonably expect that there will be even more activity in geospatial technologies in the near future.

In the future, smart cities, self-driving cars, mobile and ambient communication devices may become more tightly integrated, so it will be more easily possible for a software application to identify a user's location. To exploit this knowledge, i.e. to turn location knowledge into situational awareness, systems need to be mindful of privacy considerations, learn what a place means for the user's information needs, and improve the quality and coverage of georeferenced material available to the user informed by his or her whereabouts.

Appendix: Ancillary Tasks

Augmenting Gazetteers via Web Mining

Gazetteers comprises names of places and associated information about feature type and geographic footprint (Hill 2006); often, they are incomplete and suffer from quality issues. Leidner (2007) reports that the USGS carry out many manual corrections each month. Uryupina (2003) report an early attempt to gather place name information automatically from the Web. Official (government-maintained) toponyms are often already contained in many gazetteers, and governments increasingly make their data available as part of linked open data initiatives to stimulate innovation in the area of spatial technology; however, non-official, locally-used names such as nick-names are harder to obtain. These can be useful for "smart cities" and disaster response applications. Hu et al. (2018) mined Craiglist, a classified advertisement Web site, by analyzing housing advertisements. They apply named entity taggers followed by a frequency-weighted form of scale-structured identification clustering (first proposed by Ratttenbury et al. (2007) for extracting place semantics from Flickr data). The yield of their method can be measured by looking at how the Web-mined place names can be added to gazetteers to augment them. For some cities and gazeteers, up to 330 additional placenames were mined; on the other hand, in all six large cities evaluated only a few entries mined were not already part of the part-curated, part-crowdsourced Foursquare gazetteer, a proprietary resource. Because the original Craiglist houssing advertisements were geocoded, approximate footprints can also be obtained by the method.

Curating Gold Standard Data for Evaluation and Training

Leidner (2006a, 2007) describes TAME, the Web-based Toponym Annotation Mark-Up Editor, which was used by subjects to annotate the TR-CoNLL and TR-MUC gold data corpora for toponym resolution. It reads documents in which named entities are already tagged (in the CoNLL BIO format, named after shared tasks regularly conducted at the Conference on Natural Language Learning(CoNLL) conference) and converts them to TRML (using a command of his TextGIS suite,[18] conll2trmlpl), an XML annotation specifically designed for toponym resolution, toponym tokens are then annotated with candidate interpretations from a gazetteer (using another TextGIS command, trgaz). TAME then permits interactive Web editing by associating TRML files with a CSS style for rendering them dynamically as XHTML. In this process, the human annotator selects human-friendly generated path descriptions (e.g. Cambridge > Cambridgeshire > England > GB > Europe) and each choice is represented by setting the "selected" XML attribute to "yes" for the toponym's mention in the XML document instance.

Wallgrün et al. (2017) describe GeoCorpora, an environment to manually annotate Twitter micro-blog posts ("tweets") with geographic centroid meta-data, and such a corpus (of natural and human-made disaster tweets) that is the output of using the tool, which can be downloaded.[19] The GeoCorpora annotation tool looks up tweet sub-strings in an (Apache Solr) index of the GeoNames gazetteer, whereupon a set of human subjects (experts or anonymous Amazon Mechanical Turk workers) choose the most likely intended interpretation on an interactive map.

Bibliography

Abernathy D (2017) Using geodata & geolocation in the social sciences: mapping Our connected world. Sage, London

Ahern S, Naaman M, Nair R, Yang JH-I (2007) World explorer: visualizing aggregate data from unstructured text in geo-referenced collections. In: Proceedings of the 7th ACM/IEEE-CS Joint Conference on Digital Libraries, JCDL'07. ACM, New York, pp 1–10

Akdag F, Eick CF (2016) Interestingness hotspot discovery in spatial datasets using a graph-based approach. In Perzner P (ed) LNAI, Volume 9729 of MLDM 2016, Chur. Springer International, pp 530–544

Buscaldi D (2010) Toponym disambiguation in information retrieval. Ph.D. thesis, Universidad Politécnica de Valencia, Valencia

Cardoso N (2011) Evaluating geographic information retrieval. SIGSPATIAL Spec 3(2):46–53

Christen P, Belacic D (2005) Automated probabilistic address standardisation and verification. In: Australasian Data Mining Conference, AusDM 2005

Ferdous MS, Chowdhury S, Jose JM (2017) Geo-tagging news stories using contextual modelling. Int J Inf Retr Res 7(4):50–70

Garzon SR, Elbehery M, Deva B, Küpper A (2016) Reliable geofencing: assisted configuration of proactive location-based services. In: 2016 IEEE International Conference on Mobile Services (MS), pp 204–207

Gey F, Larson R, Sanderson M, Joho H, Clough P, Petras V (2006) GeoCLEF: the CLEF 2005 cross-language geographic information retrieval track overview. In: Proceedings of the 6th International Conference on Cross-Language Evalution Forum: Accessing Multilingual Information Repositories, CLEF'05. Springer, Heidelberg, pp 908–919

[18]A research prototype and precursor of the Polygon Analytics suite (polygonanalytics.com).

[19]Available from https://github.com/geovista/GeoCorpora (cited 2018-04-24).

Ireson N, Ciravegna F (2010) Toponym resolution in social media. In: Patel-Schneider PF, Pan Y, Hitzler P, Mika P, Zhang L, Pan JZ, Horrocks I, Glimm B (eds) The semantic web – ISWC 2010. Springer, Heidelberg, pp 370–385

Johnson K (2013) Geocoding patent data. Technical Report 2013.08.07, College of Engineeringg, University of California at Berkeley, 130 Blum Hall, 5580 Berkeley, 94720-5580

Kamalloo E, Rafiei D (2018) A coherent unsupervised model for toponym resolutions. In: The 2018 Web Conference, 23–27 Apr, Lyon, WWW 2018. IW3C2: ACM, pp 1287–1296

Karimzadeh M, Pezanowski S, MacEachren AM, Wallgrün JO (2019) GeoTxt: a scalable geoparsing system for unstructured text geolocation. Trans GIS 23:1–19

Kitchin R, Lauranlt TP, Wilson MW (2017) Understanding spatial media. Sage, London

Küpper A (2005) Location-based services: fundamentals and operation. Wiley, Chichester

LaMarca A, de Lara E (2008) Location systems: an introduction to the technology behind location awareness. Synthesis lectures on mobile and pervasive computing. Morgan Claypool, San Rafael

Lee S, Farag M, Kanan T, Fox EA (2015) Read between the lines: a machine learning approach for disambiguating the geo-location of tweets. In: Proceedings of the 15th ACM/IEEE-CS Joint Conference on Digital Libraries, JCDL'15. ACM, New York, pp 273–274

Leidner JL (2017) Georeferencing: from text to maps. In: Richardson D, Castree N, Kobayashi MFGA, Liu W, Marston RA (eds) International encyclopedia of geography: people, the earth, environment and technology, 1st edn, Volume VI (Geo-Gra). Wiley-Blackwell, Oxford, pp 2897–2906

Leidner JL, Lieberman MD (2011) Detecting geographical references in the form of place names and associated spatial natural language. SIGSPATIAL Spec 3(2):5–11

Li S, Sun W, Song R, Shan Z, Chen Z, Zhang X (2013) Quick geo-fencing using trajectory partitioning and boundary simplification. In: Proceedings of the 21st ACM SIGSPATIAL International Conference on Advances in Geographic Information Systems, SIGSPATIAL. ACM, New York, pp 580–583

Mamoulis N (2012) Spatial data management. Synthesis lectures on data management. Morgan and Claypool, San Rafael

Meier P (2015) Digital humanities: how big data is changing the face of humanitarian response. CRC Press, Boca Raton

Mostern R, Southall H (2016) Gazetteers past: placing names from antiquity to the internet. The spatial humanities. Indiana University Press, Bloomington

Müller S, Schweers S, Siegers P (2017) Geocoding and spatial linking of survey data: an introduction for social scientists. Technical Report 2017/15, GESIS – Leibniz-Institut für Sozialwissenschaften, Cologne

Müller-Budack E, Pustu-Iren K, Ewerth R (2018) Geolocation estimation of photos using a hierarchical model and scene classification. In: Ferrari V, Hebert M, Sminchisescu C, Weiss Y (eds) Computer Vision – ECCV 2018 – 15th European Conference, Munich, 8–14 Sept 2018, Proceedings, Part XII, Volume 11216 of lecture notes in computer science. Springer, Cham, pp 575–592

Nguyen D, Eisenstein J (2017) A kernel independence test for geographical language variation. Comput Linguist 43(3):567–592

Olivier P, Gapp K-P (eds) (1998) Representation and processing of spatial expressions. Lawrence Erlbaum, Mahwah

Overell SE (2009) Geographic information retrieval: classification, disambiguation and modelling. Ph.D. thesis, Imperial College London, Department of Computing, London

Paule JDG, Moshfeghi Y, Jose JM, Thakuriah PV (2017) On fine-grained geolocalisation of tweets. In: Proceedings of the International Conference on Theoretical Information Retrieval, 1–4 Oct 2017, ICTIR. Amsterdam, pp 313–316

Reclus F, Drouard K (2009) Geofencing for fleet amp; freight management. In: Ninth International Conference on Intelligent Transport Systems Telecommunications. ITST, pp 353–356

Richter L, Geiß J, Spitz A, Gertz M (2017) HeidelPlace: an exstensible framework for geoparsing, EMNLP

Ritam D, Hiware K, Ghosh A, Bihaskaran R (2018) SAVITR: a system for real-time location extraction from microblogs during emergencies

Rupp C, Rayson P, Baron A, Donaldson C, Gregory I, Hardie A, Murrieta-Flores P (2013) Customising geoparsing and georeferencing for historical texts. In: Proceedings of the 2013 IEEE International Conference on Big Data. IEEE, pp 59–62

Salfinger A, Salfinger C, Pröll B, Retschitzregger W, Schwinger W (2018) Pinpointing the eye of the hurricane – creating a gold-standard corpus for situative geo-coding of crisis tweets based on linked open data. In: LDL 2016: 5th Workshop on Linked Data in Linguistics: Managing, Building and Using Linked Language Resources. Held at the 11th Language Resources and Evaluation Conference, 7–12 May 2018, Miyazaki

Scharl A, Stern H, Weichselbraun A (2008) Annotating and visualizing location data in geospatial web applications. In: Proceedings of the First International Workshop on Location and the Web, LOCWEB'08. ACM, New York, pp 65–68

Scharl A, Tochtermann K (2007) The geospatial web: how browsers, social software and the web 2.0 are shaping the network society. Advanced information and knowledge processing. Springer, Heidelberg

Šidlauskas D, Šaltenis S, Jensen CS (2012) Parallel main-memory indexing for moving-object query and update workloads. In: Proceedings of the 2012 ACM SIGMOD International Conference on Management of Data, SIGMOD'12. ACM, New York, pp 37–48

Skoumas G, Pfoser D, Kyrillidis A (2013) On quantifying qualitative geospatial data: a probabilistic approach. In: Pfoser D, Voisard A (eds) Proceedings of the Second ACM SIGSPATIAL International Workshop on Crowdsourced and Volunteered Geographic Information, GEOCROWD'13. ACM, New York, pp 71–78

Vilain M, Hyland R, Holland R (2000) Exploiting semantic extraction for spatiotemporal indexing in GeoNODE. In: Content-Based Multimedia Information Access, Volume 2 of RIAO'00. Le Centre de Hautes Etudes Internationales d'Informatique Documentaire, Paris, pp 1440–1149

Weissenbacher D, Tahsin T, Beard R, Figaro M, Riveira R, Scotch M, Gonzalez G (2015) Knowledge-driven geospatial location resolution for phylogeographic models of virus migration

Weyand T, Kostrikov I, Philbin J (2016) PlaNet – photo geolocation with convolutional neural networks. In: European Conference on Computer Vision, ECCV'16

Wieczorek J, Guo Q, Hijmans RJ (2004) The point-radius method for georeferencing locality descriptions and calculating associated uncertainty. Int J Geograph Inf Sci 18(8):745–767

Yu Y, Tang S, Zimmermann R (2013) Edge-based locality sensitive hashing for efficient geofencing application. In: Proceedings of the 21st ACM SIGSPATIAL International Conference on Advances in Geographic Information Systems, SIGSPATIAL'13. ACM, New York, pp 576–579

Zhang L, Rushton G (2008) Optimizing the size and locations of facilities in competitive multi-site service systems. Comput Oper Res 35(2):327–338. Part special issue: location modeling dedicated to the memory of Charles S. ReVelle

Zhou T, Wei H, Zhang H, Wang Y, Zhu Y, Guan H, Chen H (2013) Point-polygon topological relationship query using hierarchical indices. In: Proceedings of the 21st ACM SIGSPATIAL International Conference on Advances in Geographic Information Systems, SIGSPATIAL'13. ACM, New York, pp 572–575

References

Amitay E, Har'El N, Sivan R, Soffer A (2004) Web-a-where: geotagging web content. In: Sanderson M, Järvelin K, Allan J, Bruza P (eds) SIGIR 2004: Proceedings of the 27th Annual International ACM SIGIR Conference on Research and Development in Information Retrieval, Sheffield, 25–29 July 2004. ACM, pp 273–280

Andogah G (2010) Geographically constrained information retrieval. Ph.D. thesis, University of Groningen, Groninigen

Batista DS, Ferreira JD, Conto FM, Silva MJ (2012) Toponym disambiguation using ontology-based semantic similarity. In: Caseil H (ed) Proceedings of PROPOR, Volume 7243 of LNAI. Springer, Heidelberg, pp 179–185

Beaman RS, Conn BJ (2003) Automated geoparsing and georeferencing of Malesian collection locality data. Telopea 10(1):43–52

Bilhaut F, Charnois T, Enjalbert P, Mathet Y (2003) Geographic reference analysis for geographic document querying. In: Proceedings of the HLT-NAACL 2003 Workshop on Analysis of Geographic References, Edmonton, Alberta, 27 May – 1 Jun 2003, HLT-NAACL'03. ACL: Association for Computational Linguistics, Stroudsburg, pp 55–62

Bloom TDS, Flower A, DeChaine EG (2017) Why georeferencing matters: introducing a practical protocol to prepare species occurrence records for spatial analysis. Ecol Evol 8:765–777

Canter D (2003) Mapping murder: the secret of geographical profiling. Virgin, London

Chen H, Vasardani M, Winter S (2018) Disambiguating fine-grained place names from descriptions by clustering. J Spat Inf Sci 17:31–62

Cliquet G (2006) Gemarketing: methods and strategies in spatial marketing. Geographical information systems series. ISTE, London

Clough PD, Joho H, Purves R (2006) Judging the spatial relevance of documents for gir. In: Lalmas M, MacFarlane A, Rüger S, Tombros A, Tsikrika T, Yavlinsky A (eds) Advances in information retrieval. Springer, Heidelberg, pp 548–552

Coventry KR, Olivier P (eds) (2002) Spatial language. Kluwer, Dordrecht

DeLozier G, Wing B, Baldridge J, Nesbit S (2016) Creating a novel geolocation corpus from historical texts. In: Proceedings of the 10th Linguistic Annotation Workshop Held in Conjunction with ACL 2016 (LAW-X 2016). Association for Computational Linguistics, Berlin, pp 188–198

Dredze M, Paul MJ, Bergsma S, Tran H (2013) Carmen: a Twitter geolocation system with applications to public health. In: Michalowski M, Michalowski W, O'Sullivan D, Wilk S (eds) Expanding the Boundaries of Health Informatics Using Artificial Intelligence: Papers from the AAAI 2013 Workshop. AAAI, pp 20–24 Technical Report WS-13-09

Duru H (2018) Spatial and temporal distribution of terrorist attacks in turkey, 1970–2016. In: Abdiraim KK (ed) Proceedings of the International Conference on Management and Social Sciences, 17–19 Nov, Istanbul, pp 220–224. Uysad

Faber R, Prestin S (2012) Social media and location-based marketing. Hanser, Munich

Finkel R, Bentley JL (1974) Quad trees: a data structure for retrieval on composite keys. Acta Inform 4(1):1–9

Fiscella K, Fremont AM (2006) Use of geocoding and surname analysis to estimate race and ethnicity. Health Serv Res 41(4 Part 1):1482–1500

Gardent C, Webber B (2001) Towards the use of automated reasoning in discourse disambiguation. J Logic Langu Inf 10(4):487–509

Goldberg DW (2013) Geocoding techniques and technologies for location-based services. In: Karimi HA (ed) Advanced location-based technologies and services. CRC Press, Boca Raton, pp 75–106

Gonzales AR, Schofield RB, Hart SV (2005) Mapping crime: understanding hotspots. Special Report NCJ 209393, U.S. Department of Justice, Office of Justice Programs, Washington, DC

Gregory IN, Geddes A (eds) (2014) Toward spatial humanities: historical GIS & spatial history. The spatial humanities. Indiana University Press, Bloomington

Guralnick RP, Wieczorek J, Beaman R, Hijmans RJ, The BioGeomancer Working Group (2006) BioGeomancer: automated georeferencing to map the world's biodiversity data. PLOS Biol 4(11):1–2

Guttman A (1984a) R-trees: a dynamic index structure for spatial searching. In: Proceedings of the 1984 ACM SIGMOD International Conference on Management of Data, SIGMOD'84. ACM, New York, pp 47–57

Guttman A (1984b) R-trees: a dynamic index structure for spatial searching. SIGMOD Rec 14(2):47–57
Harries K (1999) Mapping crime: principle and practice. Research Report NCJ 178919, U.S. Department of Justice, Office of Justice Programs, Washington, DC
Hill LL (2006) Georeferencing: the geographic associations of information. MIT Press, Cambridge, MA
Hill B, Paynich R (2014) Fundamentals of crime mapping, 2nd edn. Jones & Bartlett, Burlington
Hu Y, Mao H, McKenzie G (2018) A natural language processing and geospatial clustering framework for harvesting local place names from geotagged housing advertisements. Int J Geograph Inf Syst 22(32):1–24
Jiang JJ, Conrath DW (1997) Semantic similarity based on corpus statistics and lexical taxonomy. In: Chen K, Huang C, Sproat R (eds) Proceedings of the 10th Research on Computational Linguistics International Conference, ROCLING 1997, Taipei, Aug 1997, ROCLING, pp 19–33. The Association for Computational Linguistics and Chinese Language Processing (ACLCLP)
Kejriwal M, Szekely P (2017) Neural embeddings for populated geonames locations. In: d'Amato C, Fernández M, Tamma VAM, Lécué F, Cudré-Mauroux P, Sequeda JF, Lange C, Heflin J (eds) The Semantic Web – ISWC 2017 – 16th International Semantic Web Conference, Vienna, 21–25 Oct 2017, Proceedings, Part II, Volume 10588 of Lecture Notes in Computer Science. Springer Nature, pp 139–146
LaFree G, Dugan L (2007) Introducing the global terrorism database. Terrorism Polit Violence 19(2):181–204
Leetaru K, Schrodt PA (2013) GDELT: global data on events, location, and tone, 1979–2012. In: International Studies Association Annual Convention, vol 2, pp 4
Leidner JL (2006a) An evaluation dataset for the toponym resolution task. Comput Environ Urban Syst 30(4):400–417
Leidner JL (2006b) Re-ranking for geo-relevance with non-contextual heuristics at geoCLEF 2006. In: Proceedings of the 6th International Conference on Cross-Language Evalution Forum: Accessing Multilingual Information Repositories, CLEF'05. Springer, Cross Language Evaluation Forum
Leidner JL (2007) Toponym resolution in text: annotation, evaluation and applications of spatial grounding of place names. Ph.D. thesis, School of Informatics, University of Edinburgh, Edinburgh
Leidner JL (2008) Toponym resolution in text: annotation, evaluation and applications of spatial grounding of place names. Universal Press, Boca Raton
Leidner JL, Sinclair G, Webber B (2003) Grounding spatial named entities for information extraction and question answering. In: Kornai A, Sundheim B (eds) Proceedings of the Workshop on Analysis of Geographic References held at HLT-NAACL 2003, Edmonton, pp 31–38
Leitner M (2013) Crime modelling and mapping using geospatial technologies. Geotechnologies and the environment. Springer, New York
Li X, Kardes H, Wang X, Sun A (2014) HMM-based address parsing: efficiently parsing billions of addresses on MapReduce. In: Proceedings of the 22Nd ACM SIGSPATIAL International Conference on Advances in Geographic Information Systems, SIGSPATIAL. ACM, New York, pp 433–436
Luo J, Joshi D, Yu J, Gallagher A (2011) Geotagging in multimedia and computer vision–a survey. Multimedia Tools Appl 51(1):187–211
Maltz MD, Gordon AC, Friedman W (1990) Mapping crime in its community setting: event geography analysis. Springer, New York
Markert K, Nissim M (2002) Towards a corpus annotated for metonymies: the case of location names. In: Calzolari N, Choukri K, Maegaard B, Angel Martin Municio JM, Tapias D, Zampolli A (eds) Third International Conference on Language Resources and Evaluation, 29 May – 31 May 2002, LREC 2002. ELRA, Las Palmas, pp 1385–1392

Mikolov T, Sutskever I, Chen K, Corrado G, Dean J (2013) Distributed representations of words and phrases and their compositionality. In: Proceedings of the 26th International Conference on Neural Information Processing Systems – Volume 2, NIPS'13. Curran Associates Inc., pp 3111–3119

Mohler GO, Short MB, Malinowski S, Johnson M, Tita GE, Bertozzi AL, Brantingham PJ (2015) Randomized controlled field trials of predictive policing. J Am Stat Assoc 110(512):1399–1411

Nagel S (2008) Lokale Grammatiken zur Beschreibung von lokativen Sätzen und ihre Anwendung im Information Retrieval. Ph.D. thesis, Ludwig-Maximilians-Universität München, Munich

Nissim M, Markert K (2003) Syntactic features and word similarity for supervised metonymy resolution. In: Proceedings of the 41st Annual Meeting of the Association for Computational Linguistics. Association for Computational Linguistics, Sapporo, pp 56–63

Petroni F, Raman N, Nugent T, Nourbakhsh A, Panic Z, Shah S, Leidner JL (2018) An extensible event extraction system with cross-media event resolution. In: Guo Y, Farooq F (eds) Proceedings of the 24th ACM SIGKDD International Conference on Knowledge Discovery & Data Mining, KDD 2018, 19–23 Aug 2018. ACM, London, pp 626–635

Piton O, Maurel D (2001) "Beijing frowns and Washington pays close attention" computer processing of relations between geographical proper names in foreign affairs. In: Bouzeghoub M, Kedad Z, Métais E (eds) Proceedings of the 5th International Conference on Applications of Natural Language to Information Systems (NLDB 2000), Versailles, 28–30 June 2000, Revised Papers, Volume 1959 of Lecture notes in computer science. Springer, pp 66–78

Pouliquen B, Steinberger R, Ignat C, Groeve TD (2004) Geographical information recognition and visualization in texts written in various languages. In: Haddad H, Omicini A, Wainwright RL, Liebrock LM (eds) Proceedings of the 2004 ACM Symposium on Applied Computing (SAC), Nicosia, 14–17 Mar 2004. ACM, pp 1051–1058

Pouliquen B, Kimler M, Steinberger R, Ignat C, Oellinger T, Blackler K, Fuart F, Zaghouani W, Widiger A, Forslund A, Best C (2006) Geocoding multilingual texts: recognition, disambiguation and visualisation. In: Calzolari N, Choukri K, Gangemi A, Maegaard B, Mariani J, Odijk J, Tapias D (eds) Proceedings of the Fifth International Conference on Language Resources and Evaluation, LREC 2006, Genoa, 22–28 May 2006. European Language Resources Association (ELRA), pp 53–58

Purves RS, Clough P, Jones CB, Hall MH, Murdock V (2018) Geographic information retrieval: progress and challenges in spatial search of text. Found Trends Inf Retr 12(2–3):164–318

Ratttenbury T, Good N, Naaman M (2007) In: SIGIR'07, ACM

Rhind G (1999) Global sourcebook of address data management: a guide to address formats and data in 194 countries. Routledge, Abingdon

Rhind G (2001) Practical international data management: a guide to working with international names and addresses. Gower, Aldershot

Rigaux P, Scholl M, Voisard A (2002) Spatial databases with application to GIS. The Morgan Kaufmann series in data management systems. Morgan Kaufmann, San Francisco

Rushton G, Armstrong MP, Gittler J, Greene BR, Pavlik CE, West MM, Zimmerman DL (eds) (2008) Geooding health data: the use of geographic codes in cancer prevention and control, research, and practice. CRC Press, Boca Raton

Sallaberry C (2013) Geographical information retrieval in textual corpora. Gegraphical information systems series. Wiley, Hoboken

Salton G, Wong A, Yang CS (1975) A vector space model for automatic indexing. Commun ACM 18(11):613–620

Samet H (1989) The design and analysis of spatial data structures. Addison-Wesley, Reading (2nd corr. printing, 1990)

Samet H (1990) Applications of spatial data structures: computer graphics image processing and GIS. Addison-Wesley, Reading

Samet H (2006) Foundations of multidimensional and metric data structures. The Morgan Kaufmann series in computer graphics. Morgan Kaufmann, San Francisco

Schiller J, Voisard A (2004) Location-based services. The Morgan Kaufmann series in data management systems. Morgan Kaufmann, San Francisco

Speriosu M, Baldridge J (2013) Text-driven toponym resolution using indirect supervision. In: Proceedings of the 51st Annual Meeting of the Association for Computational Linguistics. Association for Computational Linguistics, Sofia, pp 1466–1476

Steinberger R, Pouliquen B, van der Goot E (2013) An introduction to the Europe Media Monitor family of applications. CoRR abs/1309.5290, n.p. arXiv: 1309.5290

Uryupina O (2003) Semi-supervised learning of geographical gazetteer from the internet. In: Proceedings of the HLT-NAACL 2003 Workshop on Analysis of Geographic References, Edmonton, 27 May – 1 Jun 2003, HLT-NAACL'03. ACL: Association for Computational Linguistics, Stroudsburg, pp 18–25

Wallgrün JO, Karimzadeh M, MacEachren AM, Pezanowski S (2017) GeoCorpora: building a corpus to test and train microblog geoparsers. Int J Geograph Inf Syst 13(19):1–27

Wang M, Haberland V, Yeo A, Martin AO, Howroyd J, Bishop JM (2016) A probabilistic address parser using conditional random fields and stochastic regular grammar. In: Domeniconi C, Gullo F, Bonchi F, Domingo-Ferrer J, Baeza-Yates RA, Zhou Z, Wu X (eds) IEEE International Conference on Data Mining Workshops, ICDM Workshops 2016, 12–15 Dec 2016. IEEE Computer Society, Barcelona, pp 225–232

Xie M, Yin H, Wang H, Xu F, Chen W, Wang S (2016) Learning graph-based POI embedding for location-based recommendation. In: Proceedings of the 25th ACM International on Conference on Information and Knowledge Management, CIKM'16. ACM, New York, pp 15–24

Zong W, Wu D, Sun A, Lim E-P, Goh DH-L (2005) On assigning place names to geography related web pages. In: Proceedings of the 5th ACM/IEEE-CS Joint Conference on Digital Libraries, JCDL'05. ACM, New York, pp 354–362

Chapter 17
Harnessing Heterogeneous Big Geospatial Data

Bo Yan, Gengchen Mai, Yingjie Hu, and Krzysztof Janowicz

17.1 Introduction

Among the often mentioned four characteristics, i.e., volume, variety, velocity, and veracity, of big data, variety is one of the most prominent in the geospatial domain. One grand challenge of consuming and utilizing big geospatial data is finding ways to utilize heterogeneous data, despite differences in their representations, resolution, data quality, semantics, data collection strategy, data cultures, and so forth (Janowicz 2010). For example, remote sensing images are typically collected based on a field view, while most Points-of-Interest (POI) data are constructed using an object view. The vocabularies, often called feature type ontologies, used to categorize these POI vary between a handful and more than 1000 types making their integration challenging. In terms of data formats, geospatial data can be in the forms of unstructured data, semi-structured data, and structured data. A rich volume of geospatial data (such as place names and addresses) are contained in unstructured natural language texts, such as Wikipedia, news, books, and even in social media. Structured geospatial data have a well-defined schema and are contained in geospatial databases, shapefiles, gazetteers, and knowledge graphs, e.g., so-called Linked Data. Data will also show significant variation based on whether it is collected and maintained in the form of Volunteered Geographic Information (VGI) or by an authoritative source such as a government agency. In the age of Big Data, a research project often requires the use and integration of geospatial data from different sources which may have been collected using different

B. Yan (✉) · G. Mai · K. Janowicz
University of California, Santa Barbara, CA, USA
e-mail: boyan@ucsb.edu; boyan@geog.ucsb.edu; gengchen_mai@ucsb.edu; janowicz@ucsb.edu

Y. Hu
University at Buffalo, Buffalo, NY, USA
e-mail: yhu42@buffalo.edu

M. Werner, Y.-Y. Chiang (eds.), *Handbook of Big Geospatial Data*,
https://doi.org/10.1007/978-3-030-55462-0_17

approaches. The most common source of heterogeneity, however, are differences in semantics, i.e., in the conceptualizations related to used domain vocabulary such as `River`, `Poverty`, or `Neighborhood` (Harvey et al. 1999; Frank and Raubal 1999; Bennett 2001; Kuhn et al. 2014; Scheider and Kuhn 2015) as well as cultural difference. For example, in Germany a bus stop is typically differentiated from other public spaces such as pavement because it usually has a distinct area with a roof. In some other countries, however, the concept of a bus stop may not exist at all (e.g., people in Turkey can stop the bus wherever they want to get on).

In this chapter, we review how big geospatial data can be conflated, integrated, and enriched (Kyriakidis et al. 1999; Arens et al. 1993; Samal et al. 2004; Lees and Ritman 1991; Cobb et al. 1998; Fonseca et al. 2002). These terms themselves have different definitions across and even within communities. In the context of our overview, conflation is usually the initial step which involves combining and consolidating multiple instances of the same geographic entity with various lineage. To give an intuitive example from everyday experience, in order to gain a more comprehensive understanding of a Point of Interest, such as a particular restaurant, people may check multiple sources, e.g., website listings, social media reviews from multiple vendors such as Yelp, and even images, and cross-verify, combine, and mix them to provide more accurate and complete thematic (the type of restaurant, the food they serve, and other amenities for the restaurant), temporal (hours of operation), and spatial (coordinates and neighborhood) components. After conflating all this information, the integration step comes into play by which the data are combined, e.g. in the form of layers, into a larger project. One example for this integration step are map mashups, a term first made popular in 2004 (Batty et al. 2010). Almost all web maps we use today, e.g. Google Maps, are products of integration. These maps usually contain a base map and several thematic layers (such as the POI layer, terrain layer, satellite imagery layer, traffic layer, and transit layer). These layers can be in the form of vector data, raster data, or a combination of both. Different components of the map complement each other so that users can obtain a more comprehensive view of geographic entities from different perspectives. In a general geospatial data integration workflow, the conflation stage aims to retain accurate data, reconcile conflicting data, and minimize redundant data by considering different but overlapping sources; the integration stage aims to unify different aspects of the data after the initial conflation stage. Today, geospatial data enrichment plays an increasing role as an additional step following conflation and integration. With the current development of Web-accessible knowledge graphs, the barrier for interlinking and enriching geospatial data has become less severe. Geospatial data enrichment presents a convenient way of retrieving personalized, timely, and relevant geographic information. In this stage, methods in text mining or scene classification are also frequently used depending on whether text or image sources are considered. Machine learning and data mining models are essential to the success of geospatial data enrichment. For example, in

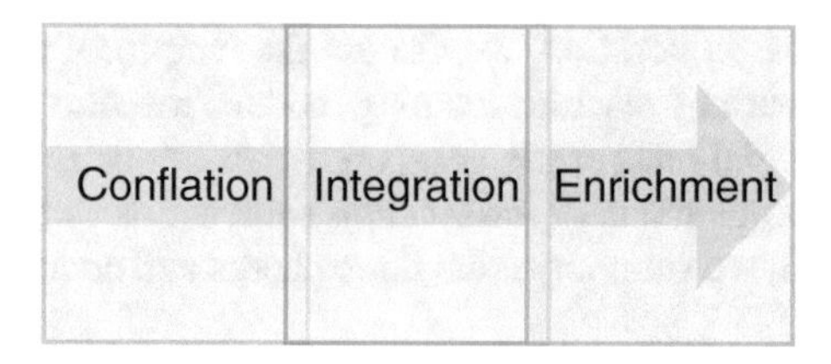

Fig. 17.1 The overlaps between the conflation process and the integration process as well as between the integration process and the enrichment process mean that there are no clear boundaries between them

semantic publishing,[1,2] in order to link different articles to the same geographic entity, preprocessing steps typically include named entity recognition, place name disambiguation, and coreference resolution. Another example is event detection.[3,4] Such enriched data includes texts, temporal information, spatial footprints, and multimedia. Besides data mining approaches in obtaining events update in a geographic context, structured and standardized data markup guidelines can also facilitate the process. In this sense, geospatial data enrichment is the highlight of the marriage of top-down theory-driven and bottom-up data-driven approaches. It is worth noting that although there seems to be a linear order to which these three phases are conducted during the whole process, they do have some overlap and are not mutually exclusive, as shown in Fig. 17.1.

The content of this chapter is organized following the thread of conflation, integration, and enrichment. In Sect. 17.2, we lay out the major obstacles in conflating geospatial data from different sources, such as discrepancies in semantics, and examine previous research studies in tackling these challenges. In Sect. 17.3, we review existing methods in spatial data integration, pointing out that the heterogeneous nature of geospatial data is, in fact, a blessing in disguise. In Sect. 17.4, we introduce a combination of top-down and bottom-up methods in enriching geospatial data and demonstrate the ways in which geospatial domain knowledge can benefit existing machine learning models. Throughout these sections, we discuss the Linked Data and geospatial semantics paradigm in which various knowledge graphs, such as DBpedia,[5] Freebase,[6] Wikidata,[7] and LinkedGeoData,[8] emerged in an attempt to faciliate the conflation, integration, and enrichment of geospatial data. While these semantically-rich datasets improve the interoperability, they also

[1] https://en.wikipedia.org/wiki/Semantic_publishing
[2] http://now.ontotext.com
[3] http://eventregistry.org
[4] https://developers.google.com/search/docs/data-types/event
[5] https://wiki.dbpedia.org/
[6] https://en.wikipedia.org/wiki/Freebase
[7] https://www.wikidata.org
[8] http://linkedgeodata.org

bring new challenges. In addition, we discuss the reciprocal relationship between geospatial data and various machine learning models: machine learning models can help integrate geospatial data, while geospatial data can be integrated into various other domains as complementary information sources for supporting cutting-edge models. In Sect. 17.5, we summarize the three phases and conclude our chapter.

17.2 Geospatial Data Conflation

In most geographic information systems and services, geospatial data can be explored from both a map-centric view and a tabular-centric view (Mai et al. 2016). Research on facilitating geospatial data conflation can be organized from these two perspectives as well.

From a map view, geospatial data most often comes in two flavors: raster data and vector data. Accordingly, studies in geospatial data conflation have considered raster and raster conflation, raster and vector conflation, and vector and vector conflation (shown in Fig. 17.2). For raster and raster conflation, Lynch and Saalfeld (1985) described it as a problem of combining two raster maps to create a third map that is better than each of the two input maps in some regards, e.g., by reducing *NoData* cells. This definition considers geospatial data conflation as *map conflation* or *map compilation*. Lupien and Moreland (1987) decomposed the task into two generic problems, namely feature alignment and feature matching. In this context, raster pixels for points, lines, and polygons on the maps are referred to as features. Since different maps may have different projections and resolutions, feature alignment is applied to transform the coordinates of one map to fit another one. A common technique called *rubber-sheeting* is utilized to solve this problem, which is a transformation technique that preserves the topology of different features on the map. Typical rubber-sheeting algorithms utilize control points, triangulation, and other computational geometry concepts to provide computationally efficient ways to transform the maps and induce coincidence between different maps (Saalfeld 1985; White Jr and Griffin 1985; Gillman 1985). Feature matching is then applied after feature alignment and the performance of feature matching depends on the strength of the feature alignment. Nearest neighbor pairings and intersection matching are commonly adopted for feature matching based on different criteria (Rosen and Saalfeld 1985). The feature alignment and feature matching processes are often done in an iterative manner to increase the matched features. In order to solve the problem of positional discrepancy and the challenge of conflation maps with different levels of detail, Liu et al. (2018) proposed a multiscale polygonal object matching approach, called the minimum bounding rectangle combinatorial optimization (MBRCO). This algorithm finds corresponding minimum bounding rectangles (MBRs) of matching pairs and aligns them to identify object-matching pairs.

For raster and vector conflation, the core idea is to find the registration between the raster map data and the vector map data. For instance, Filin and Doytsher (2000)

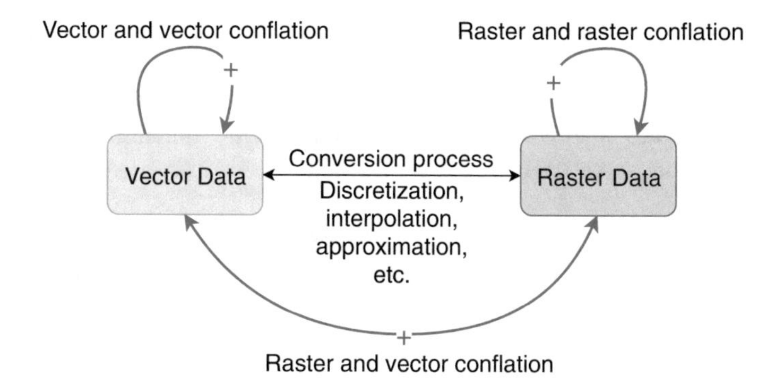

Fig. 17.2 Vector data and raster data are two commonly used types. Practically, a conversion process can be applied to switch between these two types. However, such a conversion is usually not lossless. As a result, three types of conflation, namely raster and raster conflation, vector and vector conflation, and raster and vector conflation, are studied in relevant research

utilized linear features as seed entities for registering map data by detecting the counterpart elements, establishing correspondence between matched entities, and transforming the data. Chen et al. (2004) used point pattern matching and exploited common vector datasets as 'glue' to automatically conflate street map imagery. Raster and vector data conflation is closely related to feature/object extraction from images and vector data update. As our geographic environment is constantly changing, updating vector maps automatically is of significance in order to provide the most relevant and accurate information. By successfully conflating satellite imagery with vector map data, street networks and other geographic features can be updated in a more timely manner. Baltsavias and Zhang (2005) automated the process of 3D road network reconstruction using aerial images and knowledge-based image analysis. Conflation of vector data and satellite imageries is also used for automatically geocoding satellite imageries (Hild and Fritsch 1998).

For vector and vector conflation, existing research often examines three components, namely geometric component (Beeri et al. 2004; Foley 1997), spatial relationship component (Fan et al. 2016), and attribute component (Hastings 2008; Samal et al. 2004). Many conflation models also combine all of the three components, provided that such information is available in the dataset. For example, a hierarchical rule-based approach was proposed to take into account both geometric proximity and attribute information to match features (Cobb et al. 1998). A weighted average of positional measure, shape measure, directional measure, and topological measure was proposed as the criteria for point, linear, and areal feature matching and achieved better results compared with traditional distance-based counterparts (Fan et al. 2016). Li and Goodchild (2011) developed an optimization model to improve linear feature matching that could handle one-to-one, one-to-many, and one-to-none correspondence by making use of directed Hausdorff distance (an asymmetric dissimilarity metric). Instead of using a proximity-based matching approach, Song et al. (2011) adopted a relaxation labeling approach by utilizing iterated local context updates to match the road intersections for vector road datasets. After initializing the point-to-point matching confidence matrix using the road connectivity information, in each iteration update, the relative distances between points are incorporated into the compatibility function. The proposed relaxation labeling approach yielded much better result than proximity matching approaches.

From a tabular view, geospatial data conflation can be performed based on place names, spatial and non-spatial relations, and other attributes. McKenzie et al. (2014) used a weighted multi-attribute method that considered categorical information, activities, and topic similarity to match place entries in the gazetteers of Foursquare and Yelp. To reduce redundancy in a place database, Dalvi et al. (2014) employed a language model that encapsulates domain knowledge (core words) and geographic knowledge (spatial context) to detect duplicate place entities. For example, in the place name “Fresca’s Peruvian Restaurant”, “Fresca’s” is the core word and “Peruvian Restaurant” is the description word. By accurately detecting core words and weighing them by their spatial context (such as the city or country these places are located in), their model outperformed other models. Another research area that is closely related to geospatial data conflation based on non-spatial attributes is place name disambiguation. The task is to identify the corresponding geographic entity given a place name in a text or other unstructured format. This problem is due to the one-to-many mapping between place names and geographic entities, which is also frequently encountered in geospatial data conflation. Hu et al. (2014a) used Wikipedia and enhanced Term Frequency-Inverse Document Frequency (TF-IDF) with DBpedia terms to improve place name disambiguation. Ju et al. (2016) integrated entity co-occurrence and topic modeling and outperformed benchmark systems such as DBpedia Spotlight and Open Calais in terms of F1 score and Mean Reciprocal Rank for place name disambiguation in short texts.

The recent marriage of geospatial data and the Linked Data paradigm (Kuhn et al. 2014) also increases the demand for data conflation from a tabular view. Linked Data uses a graph data model based on the Resource Description Framework (RDF) to describe both statements about the world and schema knowledge, called an ontology. RDF triples or statements consist of three parts: subject, predicate, and object. Subjects can be entities while objects can be both entities and literals. Predicates are the relationships between subjects and objects. For example, in the triple *:Santa_Barbara :isPartOf :California*, *:isPartOf* is the predicate that connects the subject *:Santa_Barbara* and the object *:California*. Geospatial Linked Data are often conflated using a tabular view. Related research has focused on reconciling data conflict, reducing data redundancy, and providing comprehensive data on both the ontology level and instance level. Geo-ontology alignment itself is a research topic that has attracted a lot of attention. Existing ontology matching or alignment systems include: Falcon (using a divide-and-conquer approach) (Hu and Qu 2008), DSSim (using an agent-based framework) (Nagy et al. 2006), RiMOM (using a dynamic multi-strategy framework) (Li et al. 2009), AgreementMaker (Cruz et al. 2009), and so on. Although most ontology alignment systems are domain-agnostic, some research specifically took a geospatial perspective. Janowicz

(2012) proposed an observation geo-ontology engineering framework that takes into account thematic, spatial, and temporal components. Zhu et al. (2016a) implemented a feature engineering approach using spatial statistics in an attempt to align three major geo-ontologies, namely DBpedia Places,[9] GeoNames,[10] and Getty Thesaurus of Geographic Names (TGN).[11] While these works emphasize aligning geographic concepts or place types from different data sources, Yan (2016) investigated how modeling bias would influence geo-ontologies and developed a data-driven method to detect these issues in order to harmonize the conflict between geo-ontology and the actual geographic entities that populate the ontology. Along the same line, Janowicz et al. (2018) discussed the issue of bias in Linked Data from data, schema, and inferential perspectives, implying potential challenges for data conflation. On the instance level, Zhu et al. (2016b) utilized spatial statistics and semantics to conflate entities in different geospatial Linked Datasets. Systems that focus particularly on the coreference resolution aspects of conflation include frameworks such as LIMES (Ngomo and Auer 2011) and SILK (Volz et al. 2009).

17.3 Geospatial Data Integration

Geospatial data integration focuses on combining data about different themes or covering different geographic areas into a unified and semantically-consistent database for various geospatial applications (Abdalla 2016). It should be differentiated from geospatial data conflation where the major goal is to reconcile the conflicts or duplications in datasets about the same theme and the same geographic areas (e.g., conflating the transportation network data from OpenStreetMap and Google Map within the same geographic area). In geospatial data integration, different datasets often provide perspectives that are complementary to each other. For example, these datasets may provide different information, such as road network and POI, about the same region. The datasets to be integrated may also focus on the same theme but are about different geographic regions, and geospatial data integration can combine them and produce a whole dataset for the entire area. An example is to integrate the temperature measurements from different sensors which monitor the temperature in neighboring counties. Since the data from different sources can have varied interpretation (e.g. some sensors measure temperature in Celsius while others measure temperature in Fahrenheit), it is important to identify and accommodate data inconsistency during the integration process to achieve semantic interoperability. In other words, we need to ensure the semantic interoperability in order to produce a correctly integrated dataset.

[9]http://mappings.dbpedia.org/server/ontology/classes/

[10]http://www.geonames.org/ontology/documentation.html

[11]http://www.getty.edu/research/tools/vocabularies/tgn/index.html

Geospatial data integration is closely related to spatial data infrastructure (SDI) (Janowicz et al. 2010) and CyberGIS (Wang 2010). One can distinguish five typical activities performed within SDIs in the context of Web-scale systems: finding, accessing, updating, processing, and visualizing geospatial data. In fact, these five activities are also among the most commonly used steps for geospatial data integration. In the following, we will discuss each of these five steps with an example of disaster mapping for Santa Barbara County after the 2017 Thomas Fire, the largest wildfire on record in California. In each step, we will emphasize the importance of semantic interoperability, and discuss how Semantic Web and Linked Data can help to ensure the propagation of semantics during the data integration process.

Discovering relevant data sources is the first step for geospatial data integration. In order to produce a disaster map for the Thomas Fire, multiple datasets for Santa Barbara county and Ventura County have to be retrieved, such as updated remote sensing images, transportation network data, wind direction, wind speed, and air pollution information from the sensor network, population data, and so on. Geographical information retrieval (GIR) systems (Jones and Purves 2008) can support such a data discovery process. Query term recommendation and query expansion techniques (Delboni et al. 2007; Mai et al. 2018) are necessary to reformulate the search query in order to find relevant datasets. In this step, semantically-similar terms like similar geographic feature types can be suggested by using either ontology-based methods like SIM-DL or machine learning based method like Place2Vec (Yan et al. 2017).

Accessing the content and metadata of the retreived datasets is the next step for geospatial data integration. It should be noted that the information required by disaster mapping or other tasks are usually from different data sources, represented in different data models, or have different internal meanings. Accordingly, it is essential to have a clear semantic interpretation of the data to achieve semantic interoperability. For example, we have two datasets about wind directions. Dataset A has the wind direction for Goleta city with a *wind blow from* conceptualization while Dataset B has the wind direction for Santa Barbara city with a *wind blow to* conceptualization. In this data accessing step, we need to have a clear understanding on the semantics of the different datasets when pass these data to the following workflow. Otherwise, such semantic inconsistency can introduce serious error in the analysis results. Semantic annotation (Janowicz et al. 2010) on the data can help to clarify the semantic inconsistency and lead to a meaningful integration result, such as a wind direction map layer of Santa Barbara County.

Registration of geospatial data is another step for integration. Data conflicts or redundancy should be removed (if they were not done in data conflation). Also some new updates that have not made to the datasets need to be added. For example, some road segments were blocked during the Thomas Fire, and such real-time road connectivity information is typically not available in the original transportation network data. In the data registration process, semantically-supported integrity check also needs to be done to preserve data quality.

Processing geospatial data is a necessary step when the initial datasets do not directly satisfy the needs of an application. Consider the example of producing a map showing the air pollution in the next 24 h of Thomas Fire. An air pollution dispersion model needs to be developed and executed based on the current wind direction, wind speed, the current air pollution distribution, the digital elevation map, and the fire locations. Let's assume that we have a geoprocessing service available for the air pollution dispersion model through an OGC Web Processing Service interface (WPS). The challenge here is not to understand the theory behind this service but to correctly interpret the intended meaning of the output of this service (Janowicz et al. 2010). Semantic inconsistency may occur when the semantics of the data in hand is not in line with the semantic definition of the input for the current service. For example, the wind speed data we have are measured in feet per second (ft/s) while the service requires the input wind speed data to be measured in miles per hour (mph). Using Semantic Web technologies to conceptualize the geoprocessing services (Scheider and Ballatore 2018) can help to clarify the semantics of each service, improve their reusability, and achieve semantic interoperability among these services.

Visualizing the integrated dataset is the last step for geospatial data integration. After we have obtained various geographic layers such as the transportation layer, the predicted air pollution layer, the fire zones layer, the POI layer, we need to combine them to produce a visualization to end users. In this step, semantics also plays an important role because the visualization need to be aware of the semantics of different geographic features in order to select the appropriate styles and symbols for each element. For example, we cannot use the blue color to represent fire zones because they may be confused with water bodies which are also colored blue on maps.

In conclusion, geospatial data integration combines heterogeneous data for addressing various spatial problems. Semantic interoperability and propagations are critical for effectively and correctly integrating geospatial datasets from different sources.

17.4 Geospatial Data Enrichment

Geospatial data enrichment aims to augment existing datasets with additional cross-domain information, typically streamed on-the-fly from an external API or knowledge graph endpoint. This can be seen as a way to contextualize data (Janowicz et al. 2019). Compared with geospatial data conflation and integration which are probably the bread and butter for harnessing heterogeneous big data, geospatial data enrichment is a relatively novel step that emerged within the last few years. However, it is playing an increasingly important role with the fast advancements in machine learning models as well as knowledge engineering in the context of global, Web-accessible knowledge graphs such as Linked Data that aims at breaking apart data silos. A simple example would be to access up-to-date demographic data from within a GIS while loading a shapefile about towns and cities. However, the term should be defined more broadly, e.g., including data about events, relevant research literature that uses the area currently loaded into a GIS as study area (Gahegan and Adams 2014; Lafia et al. 2016), the biographies of historic figures and their travels, enriching 3D models with semantic annotations from social media (Jones et al. 2014), and so on.

There are many areas that may benefit from geospatial data enrichment. One is to enrich data streams with geosocial events that happened at certain locations during a particular time period. The challenge of event detection stems from the sheer amount of streaming data and the overwhelmingly large number of noise associated with them. Weng and Lee (2011) attempted to tackle these problems by proposing a clustering algorithm with wavelet-based signals using Twitter streams and showed promising result. In order to detect important geospatial events such as earthquakes, Sakaki et al. (2010) examined Twitter streams, applied Kalman filtering and particle filtering, and developed a probabilistic spatiotemporal model to find the center and the trajectory of the event location. Pat and Kanza (2017) utilized geotagged posts in social media and developed a geosocial search system that effectively finds geospatial events. Zhu et al. (2017) developed a deep learning framework to analyze geo-tagged videos in a real-time manner in order to recognize events and activities on the map. Balduini et al. (2013) used their streaming Linked Data Framework to give city managers real-time access to event data for large-scale events and integrate the data with GIS functionality such as heatmaps.

Geospatial data enrichment is not limited to the domain of geography. The core idea of geospatial data enrichment lies in its intricate interplay between other domain areas or knowledge. In the following, we provide two examples that demonstrate the value of enriching datasets in other domains with geographic information. The first example is in scientometrics, and, in particular, spatial scientometrics which study the spatial aspect of science systems (e.g., scientific collaboration) (Frenken et al. 2009). By enriching scientometrics data with geospatial information such as the countries from which conference participants and authors came, the geographic distributions of co-authorship, the local or global scope of certain subdisciplines, and so on, researchers are able to explore spatial distributions of citations, spatial biases in collaborations, and differences between local and global citation impact. Frenken et al. (2009) also pointed out that the affiliation information that is frequently used to provide the geographic knowledge has some issues. For example, it only reflects the home institute of a visiting scholar, and the granularity of this information is very coarse. Gao et al. (2013) proposed a series of *s_indices* to evaluate the spatial impact of scientists and developed a framework that used the statistics of categorical places, spatiotemporal kernel density estimations, cartograms, distance distributions, and point-pattern analysis to identify spatiotemporal citation patterns. Hu et al. (2013, 2014b) developed several visualization components using scientometrics and geospatial Linked Data to provide analysis functions for scientific knowledge discovery from a geographic perspective.

Semantic publishing provides a second example where geospatial data enrichment can benefit the analysis of the initial data. The idea of semantic publishing is to enhance online documents with linked metadata, which facilitates machines and softwares to understand the structure and consume the information in order to provide richer content. Many of the online documents contain spatial as well as temporal information. In order to extract semantics and create structured content, methods involving geographic information retrieval are frequently utilized. For instance, place name disambiguation is used to determine the corresponding geographic entity for geographic terms in the document and coreference resolution is used to connect different surface forms of the same geographic entity. By enriching these geographic entities with semantic content, users can either follow their nose to explore the information or the system can generate analytics and graphs to summarize the geographic knowledge. OpenCalais[12] is such a system that can highlight places mentioned in a document and link them to facts from an external knowledge graph.

Geospatial data enrichment can also improve machine learning models by enriching the input training data with additional geographic information. For example, aiming to provide better embeddings for map search and location recommendation, Yan et al. (2017) devised an augmented spatial context-based algorithm that considered both local and global geographic context to learn embeddings for different place types and achieved better results based on three different evaluation schemes. Berg et al. (2014) estimated the spatiotemporal priors given locations of bird species and developed an image classifier that can greatly improve the accuracy of categorizing highly similar species of birds. Tang et al. (2015) explored different ways of encoding features extracted from the GPS information of images into Convolutional Neural Networks (CNN) and improved the mean average precision on classifying Flickr images by 7%. Along the same line, Yan et al. (2018) incorporated location Bayesian priors based on spatial contexts into the state-of-the-art CNN models, such as ResNet and DenseNet, using different approaches, such as co-occurrence models and Long Short-Term Memory (LSTM), and improved the classification of the exterior and interior images of different places collected on Google Maps, Google Street View, and Yelp by over 40% in accuracy. In addition, Mai et al. (2020) proposed a general-purpose location encoding model called Space2Vec. By combining it with the state-of-the-art image classification model, the hybrid model achieved better performances on fine-grained image recognition tasks. All these examples have shown that geospatial data and domain knowledge can further enhance machine learning models.

[12]http://www.opencalais.com/

17.5 Summary

In this chapter, we discussed three aspects of harnessing the power heterogeneous geospatial data, namely conflation, integration, and enrichment. These three parts are not mutually exclusive and can overlap. Conflation deals with reconciling data from multiple sources to resolve inconsistencies and arrive at a new dataset that is improved in terms of spatial accuracy, feature completeness, logical consistency, and so on. Data integration focuses on combining datasets in meaningful ways, e.g., as part of larger workflows or to arrive at a new, more holistic data product. In many regards conflation can be considered as a strategy of data integration (Saalfeld 1988), e.g., in the form of vector and raster conflation to correct street networks. However, conflation is just one such strategy, and, thus, we decided to address both separately here and also focus on *integration* as a driver of cross-thematic analysis. This view seems more in line with the recent thinking about heterogeneity in the context of big data. A similar concept that often occurs in discussions about the integration of geospatial data is semantic interoperability which studies how to ensure that services can exchange information meaningfully, i.e., in a way that preserves the intended interpretation of domain vocabularies. Finally, enrichment is the step of getting additional information, e.g., in the form of statements from a knowledge graph, about entities in the current dataset or project to provide additional contextual information. A typical example would be up-to-date demographics for a study area as well as events that happened in the past. This last enrichment step is part of ongoing research.

References

Abdalla R (2016) Geospatial data integration. In: Introduction to Geospatial Information and Communication Technology (GeoICT). Springer, pp 105–124

Arens Y, Chee CY, Hsu CN, Knoblock CA (1993) Retrieving and integrating data from multiple information sources. Int J Intell Coop Inf Syst 2(02):127–158

Balduini M, Della Valle E, Dell'Aglio D, Tsytsarau M, Palpanas T, Confalonieri C (2013) Social listening of city scale events using the streaming linked data framework. In: International semantic web conference. Springer, pp 1–16

Baltsavias E, Zhang C (2005) Automated updating of road databases from aerial images. Int J Appl Earth Obs Geoinf 6(3–4):199–213

Batty M, Hudson-Smith A, Milton R, Crooks A (2010) Map mashups, web 2.0 and the gis revolution. Ann of GIS 16(1):1–13

Beeri C, Kanza Y, Safra E, Sagiv Y (2004) Object fusion in geographic information systems. In: Proceedings of the thirtieth international conference on Very large data bases-Volume 30, VLDB Endowment, pp 816–827

Bennett B (2001) What is a forest? on the vagueness of certain geographic concepts. Topoi 20(2):189–201

Berg T, Liu J, Woo Lee S, Alexander ML, Jacobs DW, Belhumeur PN (2014) Birdsnap: Large-scale fine-grained visual categorization of birds. In: Proceedings of the IEEE conference on computer vision and pattern recognition, pp 2011–2018

Chen CC, Knoblock CA, Shahabi C, Chiang YY, Thakkar S (2004) Automatically and accurately conflating orthoimagery and street maps. In: Proceedings of the 12th annual ACM international workshop on Geographic information systems. ACM, pp 47–56

Cobb MA, Chung MJ, Foley III H, Petry FE, Shaw KB, Miller HV (1998) A rule-based approach for the conflation of attributed vector data. GeoInformatica 2(1):7–35

Cruz IF, Antonelli FP, Stroe C, Keles UC, Maduko A (2009) Using agreement maker to align ontologies for oaei 2009: overview, results, and outlook. In: Proceedings of the 4th international conference on ontology matching-volume 551, CEUR-WS. org, pp 135–146

Dalvi N, Olteanu M, Raghavan M, Bohannon P (2014) Deduplicating a places database. In: Proceedings of the 23rd international conference on World wide web. ACM, pp 409–418

Delboni TM, Borges KA, Laender AH, Davis Jr CA (2007) Semantic expansion of geographic web queries based on natural language positioning expressions. Trans GIS 11(3):377–397

Fan H, Yang B, Zipf A, Rousell A (2016) A polygon-based approach for matching openstreetmap road networks with regional transit authority data. Int J Geogr Inf Sci 30(4):748–764

Filin S, Doytsher Y (2000) A linear conflation approach for the integration of photogrammetric information and gis data. Int Arch Photogramm Remote Sens 33(B3/1; PART 3):282–288

Foley HA (1997) A multiple criteria based approach to performing conflation in geographical information systems. Tulane University

Fonseca FT, Egenhofer MJ, Agouris P, Câmara G (2002) Using ontologies for integrated geographic information systems. Tran GIS 6(3):231–257

Frank AU, Raubal M (1999) Formal specification of image schemata–a step towards interoperability in geographic information systems. Spat Cogn Comput 1(1):67–101

Frenken K, Hardeman S, Hoekman J (2009) Spatial scientometrics: Towards a cumulative research program. J Informet 3(3):222–232

Gahegan M, Adams B (2014) Re-envisioning data description using peirce's pragmatics. In: International conference on geographic information science. Springer, pp 142–158

Gao S, Hu Y, Janowicz K, McKenzie G (2013) A spatiotemporal scientometrics framework for exploring the citation impact of publications and scientists. In: Proceedings of the 21st ACM SIGSPATIAL international conference on advances in geographic information systems. ACM, pp 204–213

Gillman D (1985) Triangulations for rubber sheeting. In: Proceedings of 7th International symposium on computer assisted cartography (AutoCarto 7), vol 199

Harvey F, Kuhn W, Pundt H, Bishr Y, Riedemann C (1999) Semantic interoperability: a central issue for sharing geographic information. Ann Reg Sci 33(2):213–232

Hastings J (2008) Automated conflation of digital gazetteer data. Int J Geogr Inf Sci 22(10): 1109–1127

Hild H, Fritsch D (1998) Integration of vector data and satellite imagery for geocoding. Int Arch Photogramm Remote Sens 32:246–251

Hu W, Qu Y (2008) Falcon-ao: a practical ontology matching system. Web Semant Sci Serv Agents World Wide Web 6(3):237–239

Hu Y, Janowicz K, McKenzie G, Sengupta K, Hitzler P (2013) A linked-data-driven and semantically-enabled journal portal for scientometrics. In: International semantic web conference. Springer, pp 114–129

Hu Y, Janowicz K, Prasad S (2014a) Improving wikipedia-based place name disambiguation in short texts using structured data from dbpedia. In: Proceedings of the 8th workshop on geographic information retrieval. ACM, p 8

Hu Y, McKenzie G, Yang JA, Gao S, Abdalla A, Janowicz K (2014b) A linked-data-driven web portal for learning analytics: data enrichment, interactive visualization, and knowledge discovery. In: LAK Workshops

Janowicz K (2010) The role of space and time for knowledge organization on the semantic web. Semant Web 1(1, 2):25–32

Janowicz K (2012) Observation-driven geo-ontology engineering. Trans GIS 16(3):351–374

Janowicz K, Schade S, Bröring A, Keßler C, Maué P, Stasch C (2010) Semantic enablement for spatial data infrastructures. Trans GIS 14(2):111–129

Janowicz K, Yan B, Regalia B, Zhu R, Mai G (2018) Debiasing knowledge graphs: why female presidents are not like female popes. In: Proceedings of the 17th international semantic web conference

Janowicz K, Gao S, McKenzie G, Hu Y, Bhaduri B (2019) Geoai: spatially explicit artificial intelligence techniques for geographic knowledge discovery and beyond. Int J Geogr Inf Sci 0(0):1–12. https://doi.org/10.1080/13658816.2019.1684500

Jones C, Rosin P, Slade J (2014) Semantic and geometric enrichment of 3d geo-spatial models with captioned photos and labelled illustrations. In: Proceedings of the third workshop on vision and language, pp 62–67

Jones CB, Purves RS (2008) Geographical information retrieval. Int J Geogr Inf Sci 22(3):219–228

Ju Y, Adams B, Janowicz K, Hu Y, Yan B, McKenzie G (2016) Things and strings: improving place name disambiguation from short texts by combining entity co-occurrence with topic modeling. In: European knowledge acquisition workshop. Springer, pp 353–367

Kuhn W, Kauppinen T, Janowicz K (2014) Linked data-a paradigm shift for geographic information science. In: International conference on geographic information science. Springer, pp 173–186

Kyriakidis PC, Shortridge AM, Goodchild MF (1999) Geostatistics for conflation and accuracy assessment of digital elevation models. Int J Geogr Inf Sci 13(7):677–707

Lafia S, Jablonski J, Kuhn W, Cooley S, Medrano FA (2016) Spatial discovery and the research library. Trans GIS 20(3):399–412

Lees BG, Ritman K (1991) Decision-tree and rule-induction approach to integration of remotely sensed and gis data in mapping vegetation in disturbed or hilly environments. Environ Manag 15(6):823–831

Li J, Tang J, Li Y, Luo Q (2009) Rimom: a dynamic multistrategy ontology alignment framework. IEEE Trans Knowl Data Eng 21(8):1218–1232

Li L, Goodchild MF (2011) An optimisation model for linear feature matching in geographical data conflation. Int J Image Data Fusion 2(4):309–328

Liu L, Zhu X, Zhu D, Ding X (2018) M: N object matching on multiscale datasets based on mbr combinatorial optimization algorithm and spatial district. Trans GIS 22(6):1573–1595

Lupien AE, Moreland WH (1987) A general approach to map conflation. In: Proceedings of 8th international symposium on computer assisted cartography (AutoCarto 8), Citeseer, pp 630–639

Lynch MP, Saalfeld AJ (1985) Conflation: automated map compilation—a video game approach. In: Proceedings Auto-Carto, vol 7, pp 343–352

Mai G, Janowicz K, Hu Y, McKenzie G (2016) A linked data driven visual interface for the multi-perspective exploration of data across repositories. In: VOILA@ ISWC, pp 93–101

Mai G, Janowicz K, Yan B (2018) Combining text embedding and knowledge graph embedding techniques for academic search engines. In: Semdeep/NLIWoD@ ISWC, pp 77–88

Mai G, Janowicz K, Yan B, Zhu R, Cai L, Lao N (2020) Multi-scale representation learning for spatial feature distributions using grid cells. In: The eighth international conference on learning representations

McKenzie G, Janowicz K, Adams B (2014) A weighted multi-attribute method for matching user-generated points of interest. Cartogr Geogr Inf Sci 41(2):125–137

Nagy M, Vargas-Vera M, Motta E (2006) Dssim-ontology mapping with uncertainty. In: Proceedings of 1st international workshop on ontology matching, Online

Ngomo ACN, Auer S (2011) Limes-a time-efficient approach for large-scale link discovery on the web of data. In: IJCAI, pp 2312–2317

Pat B, Kanza Y (2017) Where's waldo? Geosocial search over myriad geotagged posts. In: Proceedings of the 25th ACM SIGSPATIAL international conference on advances in geographic information systems. ACM, p 37

Rosen B, Saalfeld A (1985) Match criteria for automatic alignment. In: Proceedings of 7th international symposium on computer-assisted cartography (Auto-Carto 7), pp 1–20

Saalfeld A (1985) A fast rubber-sheeting transformation using simplicial coordinates. Am Cartographer 12(2):169–173

Saalfeld A (1988) Conflation automated map compilation. Int J Geogr Inf Syst 2(3):217–228

Sakaki T, Okazaki M, Matsuo Y (2010) Earthquake shakes twitter users: real-time event detection by social sensors. In: Proceedings of the 19th international conference on World wide web. ACM, pp 851–860
Samal A, Seth S, Cueto K (2004) A feature-based approach to conflation of geospatial sources. Int J Geogr Inf Sci 18(5):459–489
Scheider S, Ballatore A (2018) Semantic typing of linked geoprocessing workflows. Int J Digital Earth 11(1):113–138
Scheider S, Kuhn W (2015) How to talk to each other via computers: semantic interoperability as conceptual imitation. In: Applications of conceptual spaces. Springer, pp 97–122
Song W, Keller JM, Haithcoat TL, Davis CH (2011) Relaxation-based point feature matching for vector map conflation. Trans GIS 15(1):43–60
Tang K, Paluri M, Fei-Fei L, Fergus R, Bourdev L (2015) Improving image classification with location context. In: Proceedings of the IEEE international conference on computer vision, pp 1008–1016
Volz J, Bizer C, Gaedke M, Kobilarov G (2009) Silk-a link discovery framework for the web of data. LDOW 538
Wang S (2010) A cybergis framework for the synthesis of cyberinfrastructure, gis, and spatial analysis. Ann Assoc Am Geogr 100(3):535–557
Weng J, Lee BS (2011) Event detection in twitter. ICWSM 11:401–408
White Jr MS, Griffin P (1985) Piecewise linear rubber-sheet map transformation. Am Cartographer 12(2):123–131
Yan B (2016) A Data-Driven Framework for Assisting Geo-Ontology Engineering Using a Discrepancy Index. University of California, Santa Barbara
Yan B, Janowicz K, Mai G, Gao S (2017) From itdl to place2vec: Reasoning about place type similarity and relatedness by learning embeddings from augmented spatial contexts. In: Proceedings of the 25th ACM SIGSPATIAL international conference on advances in geographic information systems. ACM, p 35
Yan B, Janowicz K, Mai G, Zhu R (2018) xnet+sc: Classifying places based on images by incorporating spatial contexts. In: LIPIcs-Leibniz international proceedings in informatics, Schloss Dagstuhl-Leibniz-Zentrum fuer Informatik, vol 114
Zhu R, Hu Y, Janowicz K, McKenzie G (2016a) Spatial signatures for geographic feature types: examining gazetteer ontologies using spatial statistics. Trans GIS 20(3):333–355
Zhu R, Janowicz K, Yan B, Hu Y (2016b) Which kobani? a case study on the role of spatial statistics and semantics for coreference resolution across gazetteers. In: International conference on GIScience short paper proceedings, vol 1
Zhu Y, Liu S, Newsam S (2017) Large-scale mapping of human activity using geo-tagged videos. In: Proceedings of the 25th ACM SIGSPATIAL international conference on advances in geographic information systems. ACM, p 68

Chapter 18
Big Historical Geodata for Urban and Environmental Research

Hendrik Herold

Historical geoinformation is a valuable resource for various scientific disciplines, ranging from urban and environmental research to the emerging field of digital humanities. This chapter elucidates potentials and applications of big geospatial data which it has recently become possible to automatically retrieve from historical records. Large volumes of historical textual and cartographic documents are currently being made digitally accessible by libraries and other institutions. With the help of computer vision and image analysis techniques, the hitherto only implicitly, i.e. human-readable, contained historical geoinformation can be made machine-readable and can hence be spatiotemporally analyzed and associated with current big geospatial databases filled with satellite imagery, digital maps or user-generated geocoded content. The chapter begins with an overview of existing geohistorical data sources and processing approaches and describes challenges posed by the sheer number and diversity of the sources. The main part is dedicated to potentials and applications of the derived geoinformation in the various environmental research domains, such as long-term land change monitoring, sustainability research, and Earth system modeling for studying the complex human-environment interactions between land, climate change, ecosystem and biodiversity changes during the Anthropocene.

18.1 Introduction

The term "big geospatial data" should not be used to refer only to recent geocoded information; rich digital collections of historical cartographic records also provide

H. Herold (✉)
Leibniz Institute of Ecological Urban and Regional Development, Dresden, Germany
e-mail: h.herold@ioer.de

M. Werner, Y.-Y. Chiang (eds.), *Handbook of Big Geospatial Data*,
https://doi.org/10.1007/978-3-030-55462-0_18

an enormous amount of geospatial data. For centuries, these cartographic records have been unique and "efficient storages devices" for geospatial data (Roberts 1962, p. 12). In recent years, an ever growing number of these maps have been scanned and put online by libraries and national mapping agencies. This data, however, fulfills not only the 'Volume' (1) criterion, but also all other 'V'-characteristics of big data, namely:

- Variety (2), which refers to the diversity of formats and data types, such as structured and unstructured data: old maps come in various qualities, geographical scales, scanning resolution, and as mostly un- or semi-structured image data as well as structured textual metadata,
- Veracity (3), which refers to the trustworthiness of the data, is particularly relevant in terms of the potentially uncertain geolocation and historical meaning of cartographic representations,
- Validity (4), which refers to the appropriateness for the intended use, plays a key role in terms of the correct data interpretation according to the respective abstraction level, as cartography always depicts an interpreted reality.

Parallels to 'Velocity' (5) and 'Variability' (6) of data can also be drawn. While the velocity of newly emerging data is high but not a relevant issue in comparison to sensor or video data, variability is highly relevant. The immanent variability of the mostly hand-drawn cartographic documents is comparable to that of text in handwritten books and closely related to the variety of formats, qualities and geographical scales.

All these characteristics pose specific challenges to the management, extraction, assessment and analysis of big historical geodata. The research of the past decades focused mainly on extraction and analysis. Challenges of big data have – inspired by advancements in extraction techniques – just recently addressed in studies. Both the availability of datasets and the techniques have raised awareness of the cultural and enormous scientific 'Value' (7) of the data. In the following, available data sources, time spans, and their potential applications are described.

18.2 Data Sources and Time Spans

In referring to geospatial data of the past, there are two strands to consider: textual documents and cartographic records, which are both examined in the following.

Textual documents There exist vast archives of textual documents, such as books, travel notes, itineraries, gazetteers and historical tax records. Most of them implicitly contain geographical information, such as place names, travel times, and information on land ownership. Libraries and local archives already provide millions of these documents, containing millions of potentially valuable geocoded entries. However, many collections of paper documents are still waiting to be discovered, catalogued, scanned and digitally published in the future (cf. Fig. 18.1).

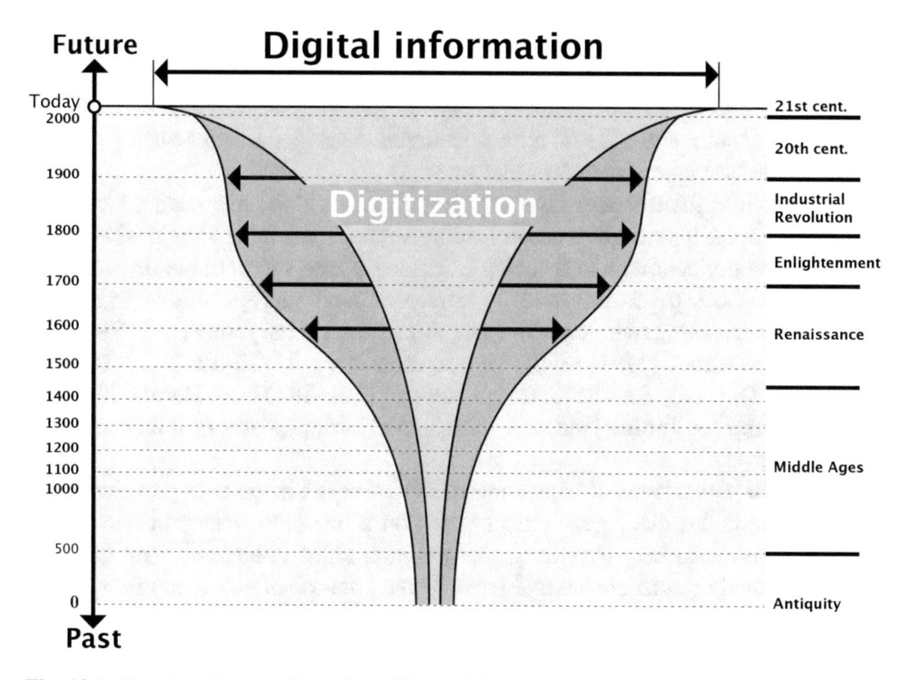

Fig. 18.1 "Le champignon informationnel" – the information mushroom, depicting schematically the relation of digital information currently and in future available for different periods of time. (Source: Modified after Kaplan (2013))

The computational analysis and interpretation of these documents is one of the main research efforts in the emerging digital humanities. A large-scale initiative addressing this challenge is the pan-European Time Machine project (Time Machine Organization 2019). This initiative also deals with the second source of geospatial data: cartographic documents.

Cartographic records A more explicit source of historical geodata consists in cartographic documents. For centuries, these records have been the most comprehensive storage and communication tools for geospatial information. In contrast to the former, they additionally represent geographical positions, relations, and dimensions of historical entities. Like textual documents, they have been collected and stored as paper documents in libraries and archives. In the last decade, large-scale digitization projects have been started. Digital libraries, national mapping agencies (e.g. the USGS), and individuals have put vast numbers of cartographic collections online.

The largest online meta-collection of historical cartographic documents is the portal "OldMapsOnline" (OldMapsOnline 2019). It provides access to many digital map repositories, such as the Harvard Library Map Collection, the ETH Library Map Collection, the National Library of Scotland, the Saxon State and University Library, the Land Survey Office of the Czech Republic, and the David Rumsey

Map Collection. A more comprehensive overview of collections is given in (Herold 2017). Here, the term old maps refers in its general meaning to all cartographic records, such as historical cadastral plans and topographic map series. While the former are usually available on a local to regional scale, the latter often provide a larger, in the best case, (inter-)national scale.

Of particular interest are those topographic maps that have been trigonometrically surveyed. The trigonometric land surveys started in 1744 with Cassini de Thury's new projection and surveying in France (Carte géométrique de la France, see Cavelti 1989, p. 2), followed by numerous land surveys all across Europe, e.g. the Austro-Hungarian surveys (e.g., Josephine military survey, 1763–1787), the Ordnance survey (1791–1850), the Saxonian survey (1780–1806), the Prussian survey (1830–1865), and the Gaussian survey (1821–1825) (cf. Herold 2017, pp. 22–23). Numerous further land surveys followed in the spheres of influence of the imperial powers.

The major advantages of trigonometrically surveyed maps over previous cartographic records are their geometrical precision at medium scale and their representational homogeneity over large areas, which allow comparative studies on a (trans-)national or even continental level. Given a temporal coverage of more than 200 years before the present, they provide an essential data source for environmental research covering the Anthropocene.

In their function as archived documents, topographic maps have preserved scale-dependent states of land surface patterns at certain points in time. With these characteristics, they provide a unique and valuable source for the reconstruction of historical land use and land cover (LULC) and its long-term changes. The land surface, in turn, is one essential component in Earth system modeling (ESM). Figure 18.2 exemplarily shows the data sources for inferring LULC data for the respective time spans.

18.3 From the Data Source to Big Geospatial Data

Scanned cartographic documents are digital images, i.e. the desired geographical information is implicitly contained and not readily available for computational approaches. This has often been confused, as the terminology of "digital" and "digitization" is ambiguous here. "Digitization" can refer to the transformation of the map into a digital document, i.e. the scanning process. In other contexts, "digitization" refers to the process of extracting features. For the latter, just like in text mining with optical character recognition (OCR), computer vision and pattern recognition algorithms need to be applied to extract machine-readable geospatial data from the scanned maps. Figure 18.3 shows the several processing steps and methods of the procedure. For relatively small study areas, the preprocessing and feature extraction can be performed manually. For the large-scale and long-term applications considered here, however, hundreds or thousands of map sheets need to be processed. Hence, the research on methods for automated feature extraction has

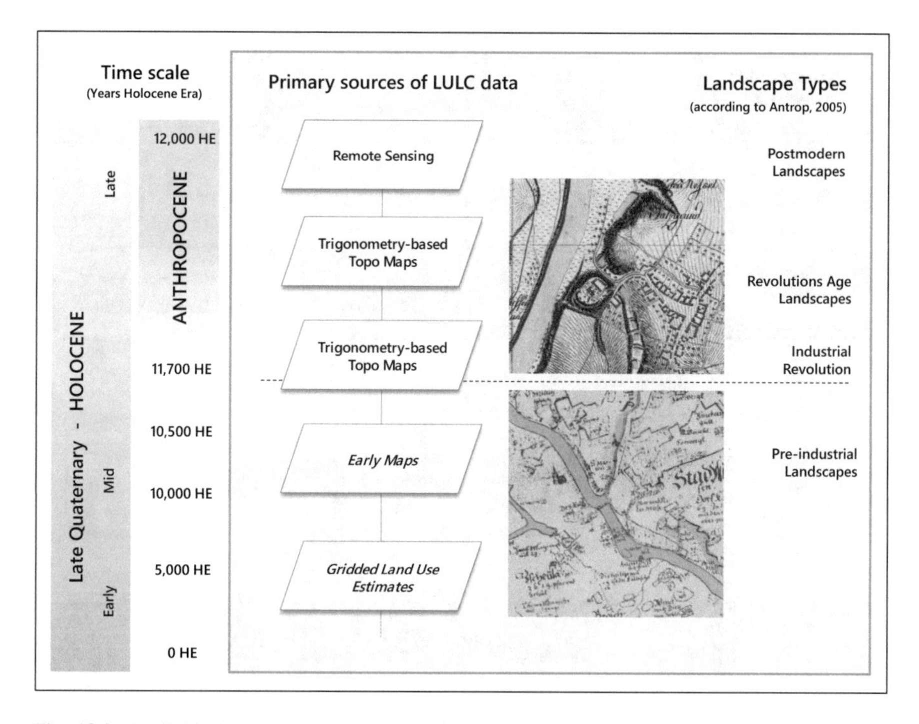

Fig. 18.2 Available historical data sources for the example of inferring long-term land change data during the Anthropocene. The land surface is one key component in Earth system modelling (ESM). (Source: Author, extended after Herold (2017), embedded sample maps © SLUB 2020)

a relatively long scientific history. A survey can be found in (Chiang et al. 2014), for instance. Some of the research efforts have led to commercial products or academic expert systems such as PROMAP (Lauterbach et al. 1992), MAGELLAN (Samet and Soffer 1998), KAMU (Frischknecht and Kanani 1998), SEMENTA (Meinel et al. 2009), and STRABO (Chiang and Knoblock 2014).

Map-extracted geodata has been applied in many studies and various scientific disciplines. An overview of studies and research contexts can be found in (Herold 2017). Besides these local and regional studies, large-scale applications on a national or continental scale have only emerged in recent years. Examples can be found in (Fuchs et al. 2015) for Europe, in (Perret et al. 2015) for France, and in (Leyk and Uhl 2018) for the US. The tendency to large-scale applications poses new challenges to the research field in terms of the big data characteristics, such as variability and veracity.

The latest advancements in computer vision and artificial intelligence, namely deep learning, enhanced data manipulation and integration techniques, will fuel this research and further boost the capabilities of geographical feature extraction algorithms. In conjunction with the simultaneously increasing number of accessible geocoded documents, this will let big historical geodata drastically grow further within the next few years.

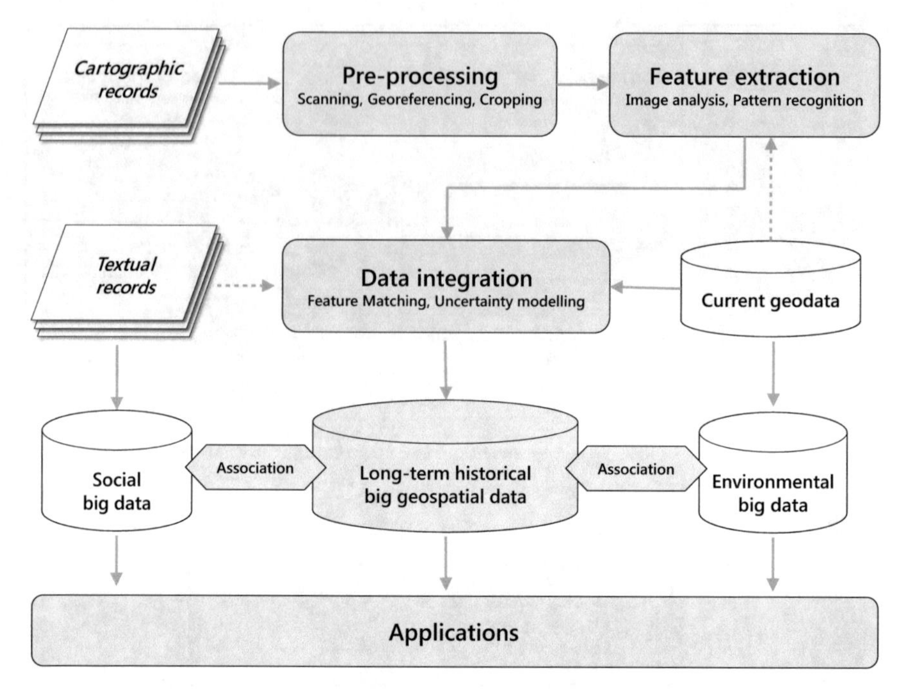

Fig. 18.3 Conceptual framework for building long-term historical geodatabases from data sources and their association with scientific data of other research domains. (Source: Author)

18.4 Potentials of Big Historical Geodata

Both the research efforts on automated feature extraction and the tedious work of manual digitization are matched by the great value of the extracted geospatial content. This section argues for the long-term data perspective in geospatial research. As discussed in the data source section, geospatial data can be retrieved from both textual and cartographic documents. While the former is especially relevant for digital humanities research, the latter rather unfolds its potential primarily within the spatial and environmental sciences. In the following, potentials and applications for both strands are given. The application fields are manifold (see Fig. 18.4), but some of them are closely linked.

18.4.1 Human-Environment Interactions

Studying and understanding the complex set of human-environment interactions is one of the key research issues in Earth system science. Climate change and the loss of ecosystems and biodiversity during the Anthropocene are among mankind's most urgent research topics and fields of action. As the human-environment interactions

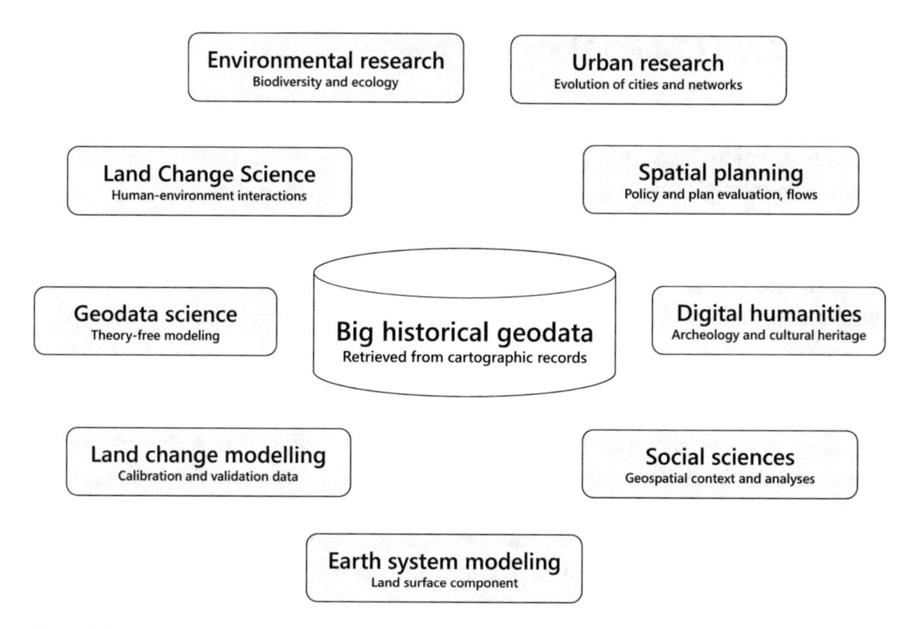

Fig. 18.4 Potentials and applications of big historical geodata in the various research domains. (Source: Author)

and system changes occur gradually – and thus mostly beyond the human horizon of perception – they can only be understood and studied in a long-term perspective.

For studying the long-term changes of the land surface during the Anthropocene, topographic map series are, as depicted in Fig. 18.2, a unique resource. The science behind studying the land surface component is referred to as Land Change Science (LCS). LCS investigates the complex dynamics of land cover and land use as a coupled human-environment system and seeks to develop new concepts and tools for improved understanding and management of land resources (cf. Turner et al. 2007). The objectives of LCS can, according to (Rindfuss et al. 2004; Verburg et al. 2004; Turner et al. 2007), be summarized as (1) improving the monitoring of land change patterns and dynamics; (2) understanding these changes as a coupled human-environment system; (3) disentangling the complex suite of biophysical and socioeconomic forces; and (4) spatially explicitly modeling land change in a manner compatible with Earth system models. Thus, long-term spatially explicit evidence is essential for perceiving, quantifying, and understanding the complex and gradually proceeding geospatial processes such as the land use and land cover change (cf. Herold 2017, p. 137). Last but not least, using the geodata for scientific time-lapse visualizations of system changes is among the most effective tools for awareness raising and problem communication.

18.4.2 Land Change Model Calibration

To understand the complex interactions of the coupled human-environment system, spatially explicit land change models are employed. Their level of application reaches from regional to global scales. Most land change models are designed for regional applications which extend from the local to the national level, with resolutions between 50 square meters and 1000 square kilometres (Verburg et al. 2006, p. 118). There exist various modeling approaches; however, one of the most crucial steps in land change modelling is the calibration and validation of the model. It is often also the computationally most expensive step in spatial modeling. In many studies both terms are used interchangeably, or only a model calibration is performed. Calibration refers to the parameter fitting, i.e., the adjustment of the model parameters such that the modeling result fits real system states at discrete points in time as well as possible. Validation, in turn, refers to the evaluation of the adjusted model against unseen data, i.e., data that was not used for calibration (Verburg et al. 2006, p. 130). That is, the calibrated model has to be tested against data for another area and/or another point in time. In the past, this validation process has often been neglected in the development and application of land change models (Wu 2002, p. 795). The issue of insufficient model validation (Wu 2002, p. 795) is in many cases primarily due to a lack of sufficient historical records, i.e., the lack of sufficient long-term spatiotemporal data (Goldstein et al. 2004, p. 128). Here the potentials of big geospatial data become obvious. The effects of long-term model calibration and validation based on historical geographical data have been investigated (cf. Fig. 18.5) and quantitatively assessed in Goldstein et al. (2004) and Akin et al. (2014).

18.4.3 Data-Driven Geoscience and Geodata Science

The traditional research approach in the geosciences is increasingly challenged by the advent of big data. In particular geographical research has shifted from a data-scarce to a data-rich environment (Miller and Goodchild 2015). While traditional space-related analytical methods are confirmatory and require the researcher to have *a priori* assumptions, the data-driven approach offers – partially at least – the possibility of hypothesis-free modeling for knowledge generation. This can be advantageous, as the classical approach may not readily discover new and unexpected patterns, trends, and relationships that can be hidden within large and diverse geographical datasets (Miller and Han 2009, p. 2). Although the data-driven approach still faces various methodological challenges, it offers the opportunity to test spatial theories against large geohistorical databases using tools such as Visual Analytics (VA), Exploratory Data Analysis (EDA), and Geospatial Knowledge Discovery (GKD, cf. Miller and Han 2009; Mennis and Guo 2009). Big historical spatiotemporal data, on the other hand, can support the knowledge generation using

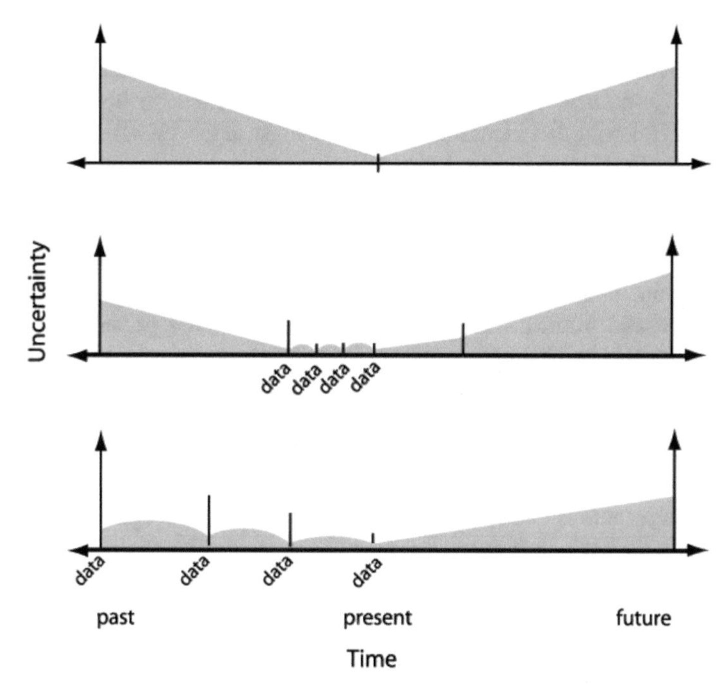

Fig. 18.5 The effects of long-term model calibration and validation using historical geospatial data. (Source: Modified after Goldstein et al. (2004, p. 129))

spatial data science and geographical knowledge discovery by providing a wider temporal scope for hypothesis generation as well as the investigation of driving forces (cf. Herold 2017). This may result in the confirmation or the rejection of long-standing spatial theories; or, it might even give rise to new theories on the spatiotemporal evolution of geographical entities.

18.4.4 Digital Humanities and Cultural Heritage

Humanities and social sciences have increasingly adopted computational approaches in the past decades. The mass digitization of books and historical documents has generated a need for analyzing this big textual data corpus. Computational models support e.g. the analysis and association of historical events and persons, i.e., building social networks of the past. As discussed above, the geographical component is mainly retrieved from textual documents. Thus, the research is mainly focused on OCR for textual documents, or, in other contexts, is focused on 3D laser scanning techniques for the digital representation of monuments and exhibition objects as part of the cultural heritage. In recent years, however, cartographic documents, such as architectural plans and topographic maps, have gained in importance. The derived geographical data can not only be used to link historical events, persons and documents to places but also to build historical city models. An approach to urban 3D reconstructions based on historical maps is described in (Herold and Hecht 2018). The reconstructed virtual city models can be of interest to scientists from the spatial sciences, architects and historians, and also to game developers who are looking for efficient ways of creating large 3D landscapes.

In the urban history domain, the model can be used to communicate the historical urban structure and land use. It may also help "historians to locate, analyze, contextualize and compare meaningful and relevant photographic sources", as applied in the project UrbanHistory4D (cf. Visual Humanities 2019). To build and interconnect many of these local or national "time machines" for spatial scientist, historians and the general public is the aim of the Time Machine project (Time Machine Organization 2019). Besides the location and dimensions of geographical entities, there also exists research interest in the geohistorical place names. Chiang (2017) gives potentials and applications of automatic text recognition of historical map labels in the social sciences.

18.4.5 Urban Research and Spatial Planning

Detailed historical data on the urban infrastructure at the level of single buildings or roads offers a wide range of applications in the interdisciplinary fields of urban research and spatial planning. These detailed and vast (up to national level) datasets provide an excellent data base for, e.g., the ex-post assessment of planning policies and the estimation and modeling of material and energy flows. Urban infrastructure is a relatively stable, long-term oriented entity and is hence also being considered as material storage. Thus, building age information, for example, can improve models to estimate the material stocks and flows as a basis for future material resource management (cf. Kleemann et al. 2017). Buildings and building construction are considered to be a major driver of global energy use. In this regard, the building age information can be a crucial factor for assessing energy consumption (e.g. Delmastro et al. 2016) as well as for developing strategies for energy optimization. Long-term data on the spatiotemporal dynamics of the urban structure also enables the assessment of spatial planning instruments as well as the ex-post study the effects of land use policies and master plans on a regional and even a national scale (e.g., Jehling et al. 2018; Xie et al. 2018). In historical demography, data on urban morphology can be used to estimate the population distribution over time. The greatest potential in urban research, however, lies in the possibility of testing hypotheses and theories (see Sect. 18.4.3) against this data on the long-term evolution of cities, urban morphologies and transportation networks.

18.5 Conclusion

There is good reason to give the term "big geospatial data" a historical dimension. This chapter has covered the sources, characteristics and time spans of big historical geodata, has depicted the path from the data source to big data, and has given potential applications across a wide range of scientific disciplines. It has shown that there already exists an enormous amount of historical geospatial data. This amount will further increase through the ongoing mass digitization efforts of libraries and other institutions as well as the advancements in artificial intelligence-based feature extraction algorithms unlocking the geospatial content that is still hidden in millions of historical textual and cartographic records. The extracted geodata offers unprecedented views and insights into the spatiotemporal evolution of land change patterns, urban structures and networks. Massive spatially explicit data could support theory building in the spatial sciences through hypothesis testing against large datasets using (geo)data science approaches. On the other hand, leaving out the historical evidence in the respective disciplines may lead to flawed assumptions or – when machine learning approaches are employed – incompletely trained computational models. The applications reach from local studies of material and energy flows to global-scale Earth system modeling studying the complex human-environment interactions between land change, ecosystems and biodiversity during the Anthropocene. In this sense, long-term historical geodata may be considered much like what ice cores are for climate research. Thus, big geospatial data of the past may in this way indirectly contribute to tackling some of humanity's greatest environmental challenges.

References

Akin A, Clarke KC, Berberoglu S (2014) The impact of historical exclusion on the calibration of the SLEUTH urban growth model. Int J Appl Earth Obs Geoinf 27(2):156–168

Antrop M (2005) Why landscapes of the past are important for the future. Landsc Urban Plan 70(1–2):21–34

Cavelti A (1989) Der Weg zur modernen Landkarte 1750–1865. Rickli + Wyss, Bern, p 43

Chiang Y-Y (2017) Unlocking textual content from historical maps – potentials & applications, trends, and outlooks. In recent trends in image processing and pattern recognition. Commun Comput Inf Sci 709:111–124

Chiang YY, Knoblock CA (2014) Recognizing text in raster maps. GeoInformatica 19(1):1–27

Chiang Y-Y, Leyk S, Knoblock CA (2014) A survey of digital map processing techniques. ACM Comput Surv 47(1):1–44

Delmastro C, Mutani G, Schranz L (2016) The evaluation of buildings energy consumption and the optimization of district heating networks: a GIS-based model. Int J Energy Environ Eng 7(3):343–351

Frischknecht S, Kanani E (1998) Automatic interpretation of scanned topographic maps: a raster-based approach. In: Graphics recognition algorithms and systems, vol 1389. Springer, Heidelberg, pp 207–220

Fuchs R, Herold M, Verburg PH, Clevers JGPW, Eberle J (2015) Gross changes in reconstructions of historic land cover/use for Europe between 1900 and 2010. Glob Chang Biol 21(1):299–313
Goldstein NC, Candau J, Clarke K (2004) Approaches to simulating the ''March of Bricks and Mortar'. Comput Environ Urban Syst 28(1–2):125–147
Herold H (2017) Geoinformation from the past – computational retrieval and retrospective monitoring of historical land use. Springer, Wiesbaden, p 192
Herold H, Hecht R (2018) 3D Reconstruction of urban history based on old maps. In: Digital research and education in architectural heritage, edited by Munster S, Friedrichs K, Niebling F, Seidel-Grzesinska, A. Springer International, Cham, pp 63–79
Jehling M, Hecht R, Herold H (2018) Assessing urban containment policies within a suburban context – an approach to enable a regional perspective. In: Land use policy, vol 77, pp 846–858
Kaplan F (2013) Lancement de la Venice Time Machine. Online document available at: https://fkaplan.wordpress.com/2013/03/14/lancement-de-la-venice-time-machine. Last accessed 2019/06/01
Kleemann F, Lederer J, Rechberger H, Fellner J (2017) GIS-based analysis of Vienna's material stock in buildings. J Ind Ecol 21:368–380
Lauterbach B, Ebi N, Besslich P (1992) PROMAP - a system for analysis of topographic maps. In: Proceedings IEEE workshop on applications of computer vision, pp 46–55
Leyk S, Uhl J (2018) HISDAC-US, historical settlement data compilation for the conterminous United States over 200 years. Sci Data 5:180175
Meinel G, Hecht R, Herold H (2009) Analyzing building stock using topographic maps and GIS. Build Res Inf 37(5–6):468–482
Mennis J, Guo D (2009) Spatial data mining and geographic knowledge discovery - an introduction. Comput Environ Urban Syst 33(6):403–408
Miller HJ, Goodchild MF (2015) Data-driven geography. GeoJournal 80:449–461
Miller HJ, Han J (2009) Geographic data mining and knowledge discovery - an overview. In: Geographic data mining and knowledge discovery. CRC Press, Taylor and Francis, Boca Raton, pp 1–26
OldMapsOnline (2019) http://www.oldmapsonline.org. Last accessed 2019/06/18
Perret J, Gribaudi M, Barthelemy M (2015) Roads and cities of 18th century France. Scientific Data 2:150048
Rindfuss RR, Walsh SJ, Turner BL, Fox J, Mishra V (2004) Developing a science of land change: challenges and methodological issues. Proc Natl Acad Sci U S A 101(39):13976–13981
Roberts JA (1962) The topographic map in a world of computers. Prof Geogr 14(6):12–13
Samet H, Soffer A (1998) Magellan: map acquisition of geographic labels by legend analysis. Int J Doc Anal Recognit 1(2):89–101
Time Machine Organization (2019) https://www.timemachine.eu/time-machine-organisation. Last accessed 2019/11/01
Turner BL, Lambin EF, Reenberg A (2007) The emergence of land change science for global environmental change and sustainability. Proc Natl Acad Sci U S A 104(52):20666–20671
Verburg PH, Schot PP, Dijst MJ, Veldkamp A (2004) Land use change modelling: current practice and research priorities. GeoJournal 61(4):309–324
Verburg PH, Kok K, Pontius R Jr, Veldkamp A (2006) Modeling land-use and land-cover change. In: Land-use and land-cover change: local processes and global impacts. Springer, Berlin, pp 117–135
Visual Humanities (2019) http://www.visualhumanities.org. Last accessed 2019/06/01
Wu F (2002) Calibration of stochastic cellular automata: the application to rural-urban land conversions. Int J Geogr Inf Sci 16(8):795–818
Xie X, Hou W, Herold H (2018) Ex post impact assessment of master plans – the case of Shenzhen in shaping a polycentric urban structure. ISPRS Int J Geo Inf 7(252):1–14

Chapter 19
Harvesting Big Geospatial Data from Natural Language Texts

Yingjie Hu and Benjamin Adams

19.1 Introduction and Motivation

Geospatial information is produced by a wide variety of data sources. In addition to commonly used datasets from agencies such as the US Geological Survey (USGS) and the US Census, geospatial information is contained in news articles (Lieberman and Samet 2011; Liu et al. 2014), encyclopedia entries (Hecht and Raubal 2008; Salvini and Fabrikant 2016), social media posts (Keßler et al. 2009b; Zhang and Gelernter 2014), historical archives (Southall 2014; DeLozier et al. 2016), housing advertisements (Madden 2017; McKenzie et al. 2018), online reviews (Cataldi et al. 2013; Wang and Zhou 2016), travel blog entries (Adams and McKenzie 2013; Ballatore and Adams 2015), and other sources. From these sources, geospatial data is embedded in natural language texts and is often presented in the form of place name mentions and place descriptions. For example, a social media post or a news article might mention multiple places through their names, or a travel blog might describe the experience of the writer at a particular place. In today's Big Data era, the *volume* and *variety* of the data from these sources are increasing at an unprecedented *velocity*, and it has become feasible to harvest big geospatial data from texts.

Why do we want to harvest geospatial data from texts? Asking this question is important, since collections of natural language text, e.g., those from social media or news articles, are often not representative of the entire population (Hecht and Stephens 2014; Malik et al. 2015; Jiang et al. 2019). There are at least three aspects in which the geospatial data harvested from texts is valuable. First, they can provide

Y. Hu (✉)
Department of Geography, University at Buffalo, Buffalo, NY, USA
e-mail: yhu42@buffalo.edu

B. Adams
Department of Computer Science and Software Engineering, University of Canterbury, Christchurch, New Zealand

M. Werner, Y.-Y. Chiang (eds.), *Handbook of Big Geospatial Data*,
https://doi.org/10.1007/978-3-030-55462-0_19

valuable human experience information, which is not available in other datasets. Travel blog entries, for example, do not simply describe where people have been but also what their *feelings* are toward these places. Such information about human experience is critical for building computational models of *places* (Goodchild 2011; Merschdorf and Blaschke 2018). Second, geospatial data harvested from some natural language texts, such as social media posts, reflect near real-time situations and are valuable for applications such as disaster response (MacEachren et al. 2011; Crooks et al. 2013; Huang and Xiao 2015). This is an important advantage compared with data from questionnaire-based surveys or face-to-face interviews which can take often months or even a few years to produce. While the geospatial data harvested from social media may not be representative, disaster response and other situation awareness applications often focus on identifying incidents, rather than, for example, whether the three people trapped in a collapsed building represent the entire population in the study area. Third, some geospatial data is only available in unstructured texts. Examples include events reported in newspapers, historical battles recorded in old archives, or business addresses contained in Web pages (Nesi et al. 2016; Hu et al. 2017; Barbaresi 2017). In these cases, harvesting geospatial data from texts is necessary for enabling advanced spatial analysis.

Harvesting geospatial data from unstructured texts has been frequently studied in geographic information retrieval (GIR) under the topic of *geoparsing* (Jones and Purves 2008; Purves et al. 2018). The goal of geoparsing is to recognize the place names, or *toponyms*, mentioned in texts, and identify the corresponding instances and the location coordinates of the recognized place names (Freire et al. 2011; Gritta et al. 2018). A software tool developed for geoparsing is called a *geoparser*, which takes unstructured natural language texts as the input, and outputs structured geographic data with the recognized place names and their location coordinates. Some geoparsers, e.g., GeoTxt (Karimzadeh et al. 2013), are published as Web services which provide easy access for general users through the Internet.

Geoparsing is typically performed in two consecutive steps: toponym recognition and toponym resolution. For the first step, the goal is to recognize place names from natural language texts without identifying the particular place instance referred by a name. For example, in the sentence, "Washington was an important stop on the rugged Southwest Trail.", the term "Washington" will be recognized as a toponym, but this step will not attempt to understand which Washington this term specifically refers to (there are more than 50 places named "Washington" in the United States). The second step, toponym resolution, aims to address the place name ambiguity and resolve the place name to its correct instance and geographic location. The toponym resolution step will (ideally) find out that the name "Washington" refers to "Washington, Arkansas" in the sentence, and will locate the place name to its corresponding spatial footprint, such as the geometric center of the city boundary. Figure 19.1 provides an overview of the two steps of geoparsing. The geospatial data harvested from natural language texts usually contain the recognized place names and their spatial footprints, such as points, lines, and polygons.

Geospatial data can also be harvested from texts that do not explicitly mention place names (Wing and Baldridge 2014). Non-spatial words, such as *beach* and

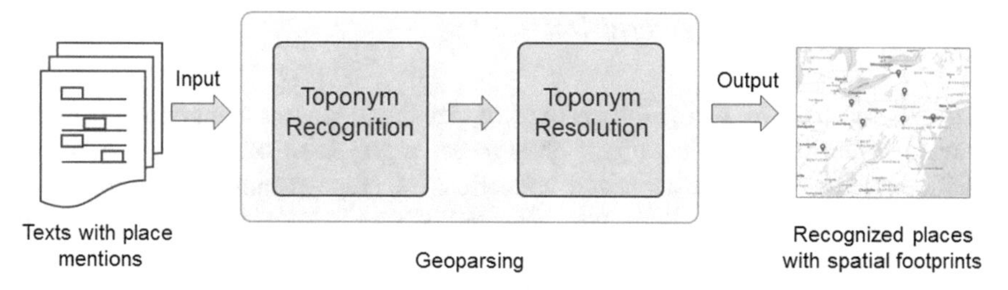

Fig. 19.1 An overview of the input, output, and the two steps of geoparsing

sunshine, can be *geo-indicative* (Adams and Janowicz 2012). That is, in the context of a textual corpus containing documents which are associated with locations on the Earth, certain words and phrases can be more or less likely to be associated with specific locations. Words with non-random spatial distributions will be most apparent in texts that describe physical environments and/or local cultural practices. Texts that are geo-referenced enable us to discover useful knowledge about places. This can be done subsequent to geoparsing as well as on texts that are already geo-referenced by the source. Examples of the latter include tweets with GPS location and travel blog entries tagged with named places (Hahmann et al. 2014; Adams and McKenzie 2013). For shorter documents it is often the case that the entire text content can be associated with one or a few toponyms. However, for longer texts the task of associating toponyms with the correct selections from the text is still an open research problem and may require more sophisticated semantic entity linking and relation extraction, reflecting a lack of easy-to-use tools in this space.

The remainder of this chapter is organized as follows. Section 19.2 reviews methods on recognizing and resolving place names from texts, and lists existing geoparsers and human-annotated corpora. Section 19.3 discusses a number of studies that have harvested big geospatial data from natural language texts for various applications. Particularly, these studies are organized into three topics: place-related studies, time-sensitive applications, and special information extraction. Finally, Sect. 19.4 presents the challenges and possible directions for the near future.

19.2 Methods and Tools

Various methods have been proposed for harvesting big geospatial data from natural language texts. In this section, we first review the existing methods for toponym recognition and resolution respectively, and then describe the existing tools for completing these two steps. We also discuss location inference from texts using language models, and such approaches are especially useful when texts do not explicitly contain toponyms.

19.2.1 Toponym Recognition

The goal of toponym recognition is to recognize the toponyms mentioned in natural language texts. One typical approach is to use a *gazetteer* which is a geographic dictionary that contains organized collections of place names, place types, and spatial footprints (Hill 2000; Janowicz and Keßler 2008). Since humans refer to places via their names while machines represent places by their coordinates, gazetteers fill the critical gap between informal human discourses and formal computer representations (Goodchild and Hill 2008; Keßler et al. 2009a). Accordingly, we can compare natural langauge texts with the entries in a gazetteer to identify the contained place names. For example, Woodruff and Plaunt (1994) used a subset of the Geographic Names Information System (GNIS) gazetteer to identify place names from textual documents related to the region of California. Amitay et al. (2004) proposed a system called *Web-a-Where* which can recognize place names from Web pages based on a gazetteer containing continents, countries, states, and cities throughout the world. While straightforward, a main disadvantage of this direct matching approach is that some place names or their vernacular versions may not be contained in a gazetteer and therefore cannot be recognized. To address this issue, methods have been proposed to enrich existing gazetteers with vernacular or vague place names. For example, Twaroch and Jones (2010) proposed a platform, called "People's Place Names" (http://www.yourplacenames.com), which encourages local people to contribute vernacular place names. Gelernter et al. (2013) developed an automatic algorithm which can add place names from OpenStreetMap and Wikimapia into a gazetteer. Jones et al. (2008) developed an approach that leverages a Web search engine to harvest entities related to a vague place name in order to construct its boundary. Geotagged photos and the associated textual tags were also used by many researchers for adding vague places into gazetteers (Grothe and Schaab 2009; Keßler et al. 2009b; Intagorn and Lerman 2011; Li and Goodchild 2012). More recently, geotagged housing posts, in which vernacular place names are often mentioned, were examined for their potential in providing local place names and enriching gazetteers (McKenzie et al. 2018; Hu et al. 2018).

Another approach for recognizing place names from texts is to use natural language processing (NLP) techniques. A key advantage of this approach is that it can be used to identify place names without relying on a gazetteer: it makes use of the words within the local context of a target word (e.g., the previous and next five words surrounding the target word) to infer whether the target word is part of a place name. One simple way to implement this idea is to define a set of grammartical rules for recognizing toponyms. For example, names in the patterns of "City of ⟨name⟩" and "⟨name⟩ Boulevard" are often place names, while those in the patterns of "Firstname ⟨name⟩" are typically not (Purves et al. 2018). Since these grammatical rules need to be defined manually, machine learning based approaches were proposed to recognize toponyms based on contextual evidence in the text. From this perspective, toponym recognition can be considered as a sub-task of named entity recognition (NER). One frequently used NER tool is Stanford NER which is based on a Conditional Random Field (CRF) sequence model (Finkel et al. 2005) and can recognize multiple types of named entities from texts, such as locations, persons, and organizations. To recognize toponyms, one can limit the identified entities to locations only. Many existing studies have included Stanford NER as part of their workflows. For example, Karimzadeh et al. (2013) developed GeoTxt in which the Stanford NER is employed for the named entity recognition step. Gelernter and Mushegian (2011) also used Stanford NER to identify location names from the tweets after the 2011 earthquake in Christchurch, Canterbury. Lieberman et al. (2010) leveraged Stanford NER to find location entities from local news articles in order to build spatial indices for textual data. In addition to Stanford NER, researchers also made use of other NER models. For example, Gelernter et al. (2013) employed OpenCalais to find building names from texts, and Hu et al. (2018) used spaCy NER as one of their four NER models to recognize place names from geotagged housing posts. Many studies also trained their own NER models for toponym recognition by leveraging a variety of evidence from the data, such as part of speech (POS) tags, left words, right words, entity relations, and other possible cues (Lieberman and Samet 2011; Inkpen et al. 2015).

19.2.2 Toponym Resolution

Once place names are recognized from texts in the first step, the second step aims to resolve these names to their corresponding geographic instances. This step is necessary because of the ambiguity existing in the semantics of place names (Leidner 2008). Amitay et al. (2004) discussed two types of ambiguities: geo/geo ambiguity, i.e., the same name, such as *London*, can refer to different geographic instances in the world; and geo/non-geo ambiguity, i.e., the same name, such as *Washington*, can refer to not only places but also persons and other types of entities. Besides, there is the issue of metonymy. For example, we may have a sentence "London voted to pass an act", in which "London" may not represent the place but the government entity, although it is not entirely unreasonable to recognize and resolve "London" to the capital of the UK in this sentence. Perhaps due to this debatable issue, many geoparsers do not directly handle metonymies. In addition, the toponyms recognized in the first step may contain false positives and false negatives. The false positives, i.e., the non-place phrases that are mistakenly recognized as toponyms, can be handled by toponym resolution methods in the process of resolving geo/non-geo ambiguity. The false negatives, i.e., the place names that are missed by the toponym recognition step, are more difficult to deal with, since most toponym resolution methods start with only the recognized toponyms rather than trying to expand the set. How to recover these false negatives could be an interesting future research topic.

A variety of methods have been developed for toponym resolution. Early approaches often make use of certain domain knowledge about places (e.g., total population) to define heuristic rules for disambiguation. A simple approach is to resolve a place name to its most prominent or *default* place instance, such as the one that has the highest population or the largest total area (these types of information are often available in gazetteers). Li et al. (2002) proposed a method for identifying the default sense of a place name based on the results returned by a search engine (Yahoo!), and their experiments showed that using the obtained default senses alone can already achieve a fair performance (i.e., resolving 78% of their ambiguous place names). Ladra et al. (2008) developed a toponym resolution Web service which combined administrative hierarchies, the populations of different places, whether a place is a capital or a main city, and some other information to perform place name disambiguation. Some other rules, such as *one referent per document* (i.e., a toponym that appears in different parts of the same document will most likely refer to the same place instance), were also developed (Leidner 2008). While hand-crafted rules can already resolve many toponyms, they can be incomplete or arbitrary: Which rules should be included and which should not? How to define the threshold for a city to be considered as a *main city*? And which rules should have higher priorities over other rules? Besides, much manual effort is needed to develop these rules.

Due to the limitations of hand-crafted rules, automatic or semi-automatic approaches are proposed for toponym resolution. Overell and Rüger (2008) proposed a co-occurrence model based on how place names occur together in Wikipedia, and then applied the co-occurrence model to disambiguate place names from texts. Buscaldi and Rosso (2008) developed a conceptual density based approach which disambiguates toponyms using an external reference corpus GeoSemCor. Lieberman and Samet (2011) proposed a multifaceted toponym recognition and resolution approach by leveraging a wide range of methods and information resources including a dictionary of entity names and cue words, statistical methods such as POS tagging and NER, and rule-based toponym refactoring. Speriosu and Baldridge (2013) trained a toponym resolver using geotagged Wikipedia articles which associates geo- and non-geo-words with toponyms, and used the trained resolver to disambiguate place names based on the words in their surrounding contexts. Santos et al. (2015) proposed a machine learning approach for place name disambiguation which combined multiple learning features such as the geospatial distances between candidates and other locations in a document and the textual context where the place references occur. Ju et al. (2016) combined entity co-occurrence and topic modeling to identify various contextual clues (i.e., related entities and topical words) to enhance place name disambiguation. There are also many other place name disambiguation studies that focused on social media data (e.g., tweets) and leveraged social media specific features, such as social interactions, location consistency of users, and metadata fields associated with tweets (Zhang and Gelernter 2014; Awamura et al. 2015; Di Rocco et al. 2016).

19.2.3 Developed Geoparsers and Tools

A number of software tools have been developed that can recognize and resolve toponyms from texts. This section provides a discussion on these tools and their advantages and limitations, with the goal of helping potential users choose the right tools for their applications. Our discussion is organized into two parts: general NER tools that can be used for identifying toponyms and specifically designed geoparsers.

General NER tools. Toponym recognition and resolution could be considered as a subtask of named entity recognition or word sense disambiguation. As a result, one way to extract place names from texts is to use existing NER tools developed from the computer science community and to keep only *locations* in the extracted entities. As discussed previously, Stanford NER is a tool that has been widely used for recognizing place names. It is based on CRF and implemented using Java (Finkel et al. 2005). While possessing the capability of recognizing toponyms not contained in gazetteers, Stanford NER does not geo-locate the identified place names to its corresponding geographic coordinates, since it is designed as a general NER tool. spaCy NER (https://spacy.io/) is an open source tool implemented in Python. Similar to Stanford NER, it can only recognize toponyms without being able to link toponyms with their coordinates. DBpedia Spotlight (Mendes et al. 2011; Daiber et al. 2013) and Open Calais (http://www.opencalais.com) are two general NER tools based on external knowledge bases (e.g., Wikipedia). A major disadvantage of them is that they can identify only those place names that are recorded in a knowledge base such as Wikipedia or a gazetteer. An advantage of DBpedia Spotlight, compared with Stanford NER, is that it links the recognized place names to the corresponding entities on DBpedia, which enables the geo-locating of these place names based on their geographic coordinates in DBpedia. Open Calais, however, does not provide such direct links for the recognized place names.

Geoparsers. There exist geoparsers specifically designed for the task of recognizing and resolving place names. Since Stanford NER already provides a strong tool for toponym recognition, many geoparsers were developed by integrating Stanford NER with a toponym resolution component. For example, Karimzadeh et al. (2013) developed *GeoTxt*, a Web-based geoparsing tool, that leverages Stanford NER for toponym recognition, and used GeoNames and a set of heuristic rules for toponym resolution. DeLozier et al. (2015) designed TopoCluster which is a geoparser that can perform geoparsing without using a gazetteer. They used Stanford NER to recognize toponyms from texts and then resolve toponyms based on the *geographic profiles* of words in the surrounding context. The geographic profile of a word is the spatial distribution of the word characterized by local spatial statistics, and (DeLozier et al. 2015) derived geographic profiles of words using a set of geotagged Wikipedia articles. Cartographic Location And Vicinity INdexer (CLAVIN) is an open-source geoparser that employs both Stanford NER and Apache OpenNLP in its different implementations for toponym recognition, and utilizes a gazetteer and fuzzy search for toponym resolution. Some geoparsers were developed using their own approaches for toponym recognition. For example, the Edinburgh Geoparser is a geoparsing system developed by the Language Technology Group at Edinburgh University (Alex et al. 2015), which used a software package developed by the same group for toponym recognition. The toponym resolution step of the Edinburgh Geoparser can be based on different gazetteers, such as GeoNames and Unlock. There are also commercial geoparsers, such as Yahoo PlaceSpotter (https://developer.yahoo.com/boss/geo/docs/PM_KeyConcepts.html) and Geoparser.io (https://geoparser.io/), which often put constrains on the number of free API calls that can be requested.

Comparing the performances of geoparsers is often challenging, largely because of a lack of openly available and human annotated corpora (Monteiro et al. 2016; Gritta et al. 2018). Some researchers have made great efforts to alleviate this dearth of open data for testing and training geoparsers. Leidner (2008) contributed TR-CoNLL which is a human annotated news corpus consisting of about 1,000 international news articles from Reuters and about 6,000 toponyms. Lieberman et al. (2010) shared a human annotated dataset called Local-Global Lexicon (LGL) corpus, which contains 588 news articles published by 78 local newspapers from highly ambiguous places, such as *Paris News* (Texas) and *Paris Beacon-News* (Illinois). Hu et al. (2014) contributed a semi-automatically annotated corpus containing textual descriptions from city websites with two highly ambiguous place names in the U.S., namely *Washington* and *Greenville*. Gritta et al. (2018) contributed WikToR which is a corpus of Wikipedia articles with ambiguous names, such as *Lima, Peru*, *Lima, Ohio*, and *Lima, Oklahoma*, automatically annotated by a Python script. Wallgrün et al. (2018) published GeoCopora, a dataset of tweets manually annotated using a crowdsourcing approach based on Amazon's Mechanical Turk and further verified by experts. In addition to contemporary corpora, some historical datasets are also made available, such as *War Of The Rebellion* by DeLozier et al. (2016). Finally, the ACE 2005 English SpatialML is an annotated news corpus shared on the Linguistic Data Consortium (Mani et al. 2008), but it charges a fee ($1,000) for non-members.

19.2.4 Location Inference from Language Modeling

While geoparsers are effective in recognizing and geo-locating toponyms mentioned in texts, there are situations when place names are not explicitly mentioned in texts. A variety of language models have been developed for geo-referencing texts using all the terms present in a document rather than toponyms only (see Purves et al. 2018, Ch. 4.6 for a comprehensive survey). Approaches vary from developing machine learning classifiers of document-level location based on word features (Wing and Baldridge 2011; Adams and Janowicz 2012) to creating more tailored linguistic models that analyze spatial language (e.g., spatial prepositions, adjectives, and reference frames) in text in order to identify locations above and beyond place names (Tenbrink and Kuhn 2011; Stock and Yousaf 2018). The former often utilize simplistic spatial models, such as regions and geodesic grids, which allows us to train predictive classifiers relatively easily on large amounts of data (Roller et al. 2012; Wing and Baldridge 2014; Han et al. 2014). When these classifiers are trained on words as features, they are usually single-language models; however, a Unicode character level classifier has been developed that is language independent (Adams and McKenzie 2018). Linguistic models, in contrast, involve formalisms of spatial language that attempt to capture the semantics of spatial relations in natural language discourse. The developed linguistic models can potentially extract spatial information that is opaque to the other methods, but also make for a more onerous task when applied to big data. For example, one can differentiate between a *locatum* (an object in space) and a *relatum* (another object that the locatum is related to), which can be used by a reader in a (geo)spatial scene to orient and locate the elements described in texts (Bateman et al. 2007). Doing so in an automated manner requires a full NLP pipeline that can identify parts-of-speech and dependencies within the texts prior to the spatial analysis (Chen and Manning 2014; Avvenuti et al. 2018). In addition, corpus linguistics research is also relevant to location inference. Lexical dialectology (the study of dialects through computational means) can be used to associate specific language features with places on the Earth, which in turn can be used to improve the models for geo-locating texts (Rahimi et al. 2017; Dunn 2018).

Unlike the geoparsing tools based on toponym resolution that were described in the previous section, location inference from language modeling is still largely done on a bespoke basis in the context of individual research projects. Among the geoparsers listed in the previous section, only TopoCluster (DeLozier et al. 2015) utilizes language modeling as a significant component in the pipeline.

19.2.5 Summary

This section discusses the main methods and tools developed for harvesting big geospatial data from natural language texts. We started from geoparsing, one major approach that collects geospatial data by recognizing and resolving toponyms mentioned in texts. The geo-located toponyms can be used as a basis for geo-locating a whole document (Monteiro et al. 2016; Melo and Martins 2017). It is necessary to differentiate geoparsing, i.e., the task of recognizing and resolving (potentially colloquial) toponyms from natural language texts, from geocoding in conventional GIS, i.e., the task of locating formatted addresses (e.g., door number with a street name) (Goldberg et al. 2008). Both are important in geographic information science. In addition to geoparsing, we also discussed the harvesting of geospatial data when toponyms are not explicitly mentioned in texts, through the use of language modeling via machine learning and linguistic approaches.

19.3 Applications of Geospatial Data Harvested from Texts

This section discusses some applications that leverage geospatial data harvested from natural language texts. We will start from understanding human experiences toward places, move to using near real-time data for situation awareness, and finally discuss extracting information about place relations in virtual or cognitive spaces.

19.3.1 Understanding Places and Human Experiences

Space and place are two related, but differently conceived concepts in academic geography. Until recently, quantitative statistical analysis of geographic information focused almost exclusively on spatial analysis, while *place* has been a rich subject of academic study in human geography. Recently with the advent of more geographic user-generated content being posted online (a.k.a. volunteered geographic information or VGI), especially on social media, *place* has become a subject of increasing interest for those doing quantitative data-driven research (Elwood et al. 2012; Sui and DeLyser 2012). In a phenomenological sense, *place* has often been described as *space* engendered with meaning through human experience (either direct or indirect) (Tuan 1977). Large amounts of unstructured observations of people's experiences in text thus provide a new window to investigate this phenomenological perspective on place, in ways that were previously restricted to smaller scaled humanisitic inquiries. Multiple kinds of textual analysis have been used on this data to provide these sorts of insights. Keyword-based, topical, sentiment, and emotion analyses all provide different ways to generalize about multiple human experiences (cf. Mei et al. (2006); Hollenstein and Purves (2010); Chon et al. (2012); Adams and McKenzie (2013); Adams (2015); Ballatore and Adams (2015); Doytsher et al. (2017)). Apart from providing better understanding of place in a generic sense, analysis of big-geo data to understand place has been used for a variety of applications, including tourism (Hao et al. 2010; Xiang et al. 2015; Rahmani et al. 2018; McKenzie and Adams 2018), urban research (Cranshaw and Yano 2010; Campagna 2014; van Weerdenburg et al. 2019), political science (Bastos et al. 2014), public health (Ghosh and Guha 2013), marketing (Caverlee et al. 2013), and sociolinguistic research (Eisenstein et al. 2010).

Another domain where place-based geospatial data harvested from texts is increasingly being used is the digital (geospatial) humanities (Bodenhamer et al. 2010). Geospatial information that is buried in massive collections in libraries and online has been seen as a goldmine for spatial historical and literary analysis (Gregory et al. 2015). Historical datasets pose unique challenges, however, as many geoparsing tools are built on gazetteers of modern place names, and therefore custom solutions are often required to automatically extract geographic information from historical texts (Rupp et al. 2013). In this context, historical gazetteers, such as Pleiades (https://pleiades.stoa.org) and World-Historical Gazetteer (http://

whgazetteer.org), have been developed to provide services for finding and using information related to ancient places. In addition to supporting direct analysis, geospatial data can be extracted from the various documents used in humanities to build spatial indices which provide an alternative way of exploring textual content from a geographic perspective (McCurley 2001; Purves et al. 2007; Adams et al. 2015).

19.3.2 Situation Awareness for Emergency Response

Emergency response applications usually need real-time data about the situations on the ground. A lot of such data comes in the form of natural language text. Examples include social media posts, short text messages, texts converted from phone calls (or voice messages), and news reports sent by the journalists at emergency scenes. After an emergency, information from different sources often flood into the emergency operations center, overwhelming first responders. Accordingly, automated methods and tools become very useful for extracting location information (e.g., who needs help at which location) from massive amounts of data.

Many studies have used geospatial data harvested from texts for emergency responses. Social media data, especially Twitter data, has been widely utilized by many researchers (Tsou 2015; Haworth and Bruce 2015). For example, De Longueville et al. (2009) investigated the spatial, temporal, and social dynamics of tweets during a major forest fire in the South of France in 2009. Crooks et al. (2013) examined the spatial and temporal characteristics of tweets after a 5.8 magnitude earthquake occurred on the East Coast of the US in 2011. Nagar et al. (2014) used daily geotagged tweets in NYC to investigate the spatiotemporal tweeting behavior related to influenza-like illness (ILI). Although a small percentage of tweets are already geotagged (about 1–2%), it is estimated that more than 10% tweets contain place references in their texts (Wallgrün et al. 2018). Thus, researchers also focused on extracting place reference information from the textual content of tweets. For example, MacEachren et al. (2011) developed SensePlace2, a visual analytics system that supports the space-time-theme exploration of Twitter data for situation awareness and crisis management. In SensePlace2, the researchers differentiated *tweets from* (i.e., geotagged location) and *tweets about* (i.e., the locations mentioned in tweet content). Gelernter and Balaji (2013) proposed an algorithm for extracting place names in various forms, such as abbreviated, misspelled, or highly localized names, from the content of tweets posted after the 2011 earthquake in Christchurch, New Zealand. Issa et al. (2017) studied the spatial diffusion of tweets about flu in four different cities using both geotagged and non-geotagged tweets. In addition to social media, news articles were also used by researchers to understand the situations related to natural hazards. For example, Wang and Stewart (2015) examined the impact of Hurricane Sandy by extracting place names, timestamps, and emergency information (e.g., power failure) from the news texts.

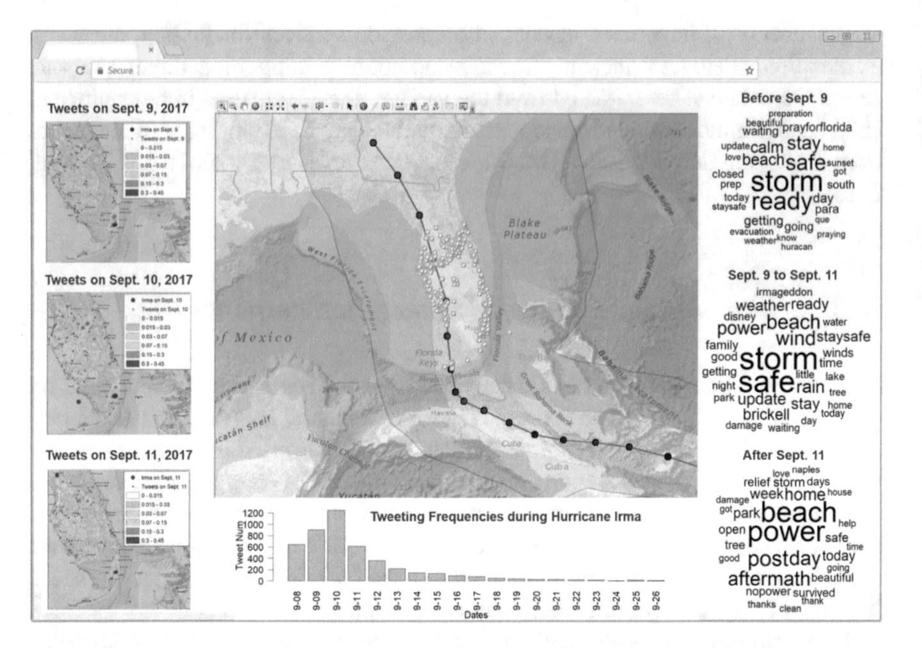

Fig. 19.2 A possible GUI of an information system for using the spatial, temporal, and textual information harvested from tweets for situation awareness using an example of Hurricane Irma

To give an intuitive idea of using social media data for situation awareness, we show a possible graphic user interface (GUI) of an information system in Figure 19.2 based on a sample of tweets collected during Hurricane Irma in September 2017. In this user interface, the main map shows the current and predicted trajectory of the hurricane and its impact area. The locations of geotagged tweets are visualized on the ground (one can also visualize the locations mentioned in the content of tweets using an approach by MacEachren et al. 2011). The bar chart at the bottom shows the tweeting intensities on different days. In the case of Hurricane Irma, most tweets were made between September 9th and 11th when Irma made Florida landfall and moved inland. On the left side of the interface, a user can pick three specific days and examine the intensities and geographic distributions of the tweets on those days. On the right side, three word clouds summarize the main topics of the tweets in three different time periods. In the case of Hurricane Irma, the tweets were summarized based on the periods of *before*, *during*, and *after* Irma. As can be seen, there were many words related to preparation and evacuation *before* the hurricane, and words about winds, rain, and trees were seen frequently *during* the event; and *after* the hurricane, the frequent words were about disaster damage and relief. Such information collected from social media and processed in a near real-time manner can help support the decision makings of emergency responders.

19.3.3 Place Relations in Virtual or Cognitive Space

Another special and valuable sort of geospatial information captured by texts is the relationships between places in virtual or cognitive space. Most traditional geographic datasets are organized based on *spatial proximity*. For example, we may have a dataset of land parcels located in the same geographic region. By contrast, texts, such as Web pages, social media posts, and news articles, can mention multiple places that are far apart and even in global scale, thereby relating these places together, often representing social, economic, and historical relationships that are non-spatially determined (Adams 2018). Place name co-occurrences, thus, are often considered as evidence for these sorts of place relations (Hecht and Raubal 2008; Twaroch et al. 2009; Ballatore et al. 2014; Liu et al. 2014; Spitz et al. 2016). Depending on application needs, different textual contexts, such as sentences, paragraphs, and even entire articles, can be used for determining place name co-occurrences. Place relations can also be established via hyperlinks, such as those in Wikipedia articles and other Web pages.

Places can be related together in texts for a variety of reasons. News articles can report different events that involve multiple places: a sports team may travel from their hometown to another city for a game; a company based in one country may establish a new branch office in another country (Toly et al. 2012; Sassen 2016); a natural disaster, such as hurricane and flooding, can impact multiple cities and towns. In addition, Wikipedia pages and online blogs can discuss the similarities and dissimilarities of two places in terms of their climates, populations, geographic locations, and other aspects. In social media posts, people can talk and compare the life styles, food, and cultures in different places. In today's digital society empowered by information and communication technologies, a majority of places are interlinked together in the virtual or cyberspace, forming place networks (Taylor and Derudder 2015; Shaw et al. 2016). As a result, big geospatial data harvested from natural language texts provide one important source for understanding the diverse and dynamic place relations in the virtual space, as well as the those perceived by people, i.e., the relations in cognitive space.

Many studies have examined place relations using different types of texts. Hecht and Moxley (2009) conducted an early study on place relations using hyperlinks in Wikipedia pages, and found that nearby places are more likely to have relations than distant ones, although places far away can still have relations. Liu et al. (2014) examined place name co-occurrences in a set of news articles, and found that place relatedness in news articles has a weaker distance decay effect compared with those derived from human movements. Zhong et al. (2017) also looked into place name co-occurrences in news articles, and concluded that places are more likely to be related if they are in the same administrative level or have a part-whole relation (e.g., Seattle is part of Washington State). Salvini and Fabrikant (2016) analyzed place name co-occurrences in Wikipedia pages and examined the semantics of place relations via the categories of Wikipedia pages. Also based on the co-occurrences of place names in Wikipedia articles, Spitz et al. (2016) constructed toponym

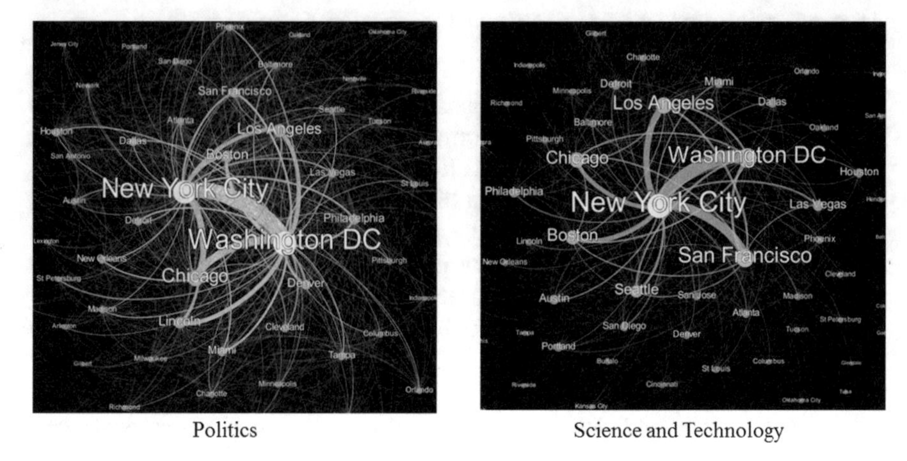

Fig. 19.3 Relations of places under different semantic topics extracted from a corpus of news articles from The Guardian

networks for place name disambiguation. Adams and Gahegan (2016) performed spatio-temporal (*chronotopic*) analysis on Wikipedia corpus by analyzing the co-occurrences of places and times in texts to understand the intrinsic relations between place, space, and time in narrative texts. Hu et al. (2017) examined place name co-occurrences in news articles, and employed a topic modeling approach to annotate the semantic topics of place relations. Figure 19.3 shows the relations of places extracted from a corpus of The Guardian newspapers under different semantic topics, as discussed in Hu et al. (2017). As can be seen, places can have different strengths of relations under different semantic topics and thus different position prominence in the place networks: Washington DC plays a much more important role under the topic of *Politics* than under the topic of *Science and Technology*; by contrast, *San Francisco* has a largely increased prominence in the network under the topic of *Science and Technology* compared with its role under the topic of *Politics*.

19.4 Summary and Future Directions

Geospatial data exist in various types of natural language texts, such as news articles, social media posts, Wikipedia pages, travel blogs, historical archives, housing advertisements, and so forth. Many of these data sources provide large amounts of data (e.g., millions or even billions of social media posts) which are constantly increasing as the time goes by. As a result, it becomes possible to harvest big geospatial data from natural language texts. Compared with the data from more conventional sources, such as the USGS and the US Census, geospatial data from texts capture valuable human experiences toward places, provide near real-time information after a disaster, and record place relations in virtual and cognitive spaces. In this chapter, we discussed the methods and tools that can be used for harvesting geospatial data from texts. Geoparsing is a major approach that can extract structured geographic information from unstructured texts by recognizing and resolving the place names mentioned in texts. When toponyms are not explicitly contained in texts, other approaches based on language modeling can help us derive geographic information from texts.

A number of research directions can be pursued in the near future. For toponym recognition, the performances of existing approaches still vary depending on the tested datasets. Advancements in deep learning, such as bidirectional recurrent neural networks, can help increase the accuracy of recognizing place names from texts. New NLP methods may also help better identify the metonymies used in the texts. For toponym resolution, most approaches currently still resolve place names only to point-based locations, and there are rivers, countries, and other geographic features whose spatial footprints can be better represented as polylines, polygons, and even polyhedras (in a 3D space). In addition, although a number of geoparsers exist, it is difficult to directly compare the performances of these geoparsers. One reason is a lack of open and annotated corpora. Although researchers have started to address this issue in recent years, it still takes a considerable amount of time and effort to implement existing baselines and run them against common datasets. Thus, a benchmarking platform, such as EUPEG (Wang and Hu 2019), could be helpful for comparing and evaluating geoparsers. From a perspective of applications, while this chapter has highlighted the use of geospatial data from texts in studies about place, digital humanities, situation awareness, and place relations, other applications are waiting to be explored and examined in the near future.

References

Adams B (2015) Finding similar places using the observation-to-generalization place model. J Geogr Syst 17(2):137–156

Adams B (2018) From spatial representation to processes, relational networks, and thematic roles in geographic information retrieval. In: Proceedings of the 12th workshop on geographic information retrieval. ACM, New York, GIR'18, pp 1:1–1:2

Adams B, Gahegan M (2016) Exploratory chronotopic data analysis. In: International conference on geographic information science. Springer, pp 243–258

Adams B, Janowicz K (2012) On the geo-indicativeness of non-georeferenced text. In: Proceedings of the international conference on web and social media (ICWSM). AAAI Press, pp 375–378

Adams B, McKenzie G (2013) Inferring thematic places from spatially referenced natural language descriptions. In: Crowdsourcing geographic knowledge. Springer, pp 201–221

Adams B, McKenzie G (2018) Crowdsourcing the character of a place: Character-level convolutional networks for multilingual geographic text classification. Trans GIS 22(2):394–408

Adams B, McKenzie G, Gahegan M (2015) Frankenplace: interactive thematic mapping for ad hoc exploratory search. In: Proceedings of the 24th international conference on world wide web, International World Wide Web conferences steering committee, pp 12–22

Alex B, Byrne K, Grover C, Tobin R (2015) Adapting the Edinburgh geoparser for historical georeferencing. Int J Humanit Arts Comput 9(1):15–35

Amitay E, Har'El N, Sivan R, Soffer A (2004) Web-a-where: geotagging web content. In: Proceedings of the 27th annual international ACM SIGIR conference on research and development in information retrieval. ACM, pp 273–280

Avvenuti M, Cresci S, Nizzoli L, Tesconi M (2018) Gsp (geo-semantic-parsing): geoparsing and geotagging with machine learning on top of linked data. In: European semantic web conference. Springer, pp 17–32

Awamura T, Aramaki E, Kawahara D, Shibata T, Kurohashi S (2015) Location name disambiguation exploiting spatial proximity and temporal consistency. SocialNLP 2015@ NAACL, pp 1–9

Ballatore A, Adams B (2015) Extracting place emotions from travel blogs. In: Proceedings of AGILE, vol 2015, pp 1–5

Ballatore A, Bertolotto M, Wilson DC (2014) An evaluative baseline for geo-semantic relatedness and similarity. GeoInformatica 18(4):747–767

Barbaresi A (2017) Towards a toolbox to map historical text collections. In: Proceedings of the 11th workshop on geographic information retrieval. ACM, p 5

Bastos MT, Recuero R, Zago G (2014) Taking tweets to the streets: a spatial analysis of the vinegar protests in Brazil. First Monday 19(3)

Bateman J, Tenbrink T, Farrar S (2007) The role of conceptual and linguistic ontologies in interpreting spatial discourse. Discourse Process 44(3):175–212

Bodenhamer DJ, Corrigan J, Harris TM (2010) The spatial humanities: GIS and the future of humanities scholarship. Indiana University Press

Buscaldi D, Rosso P (2008) A conceptual density-based approach for the disambiguation of toponyms. Int J Geogr Inf Sci 22(3):301–313

Campagna M (2014) The geographic turn in social media: opportunities for spatial planning and geodesign. In: International conference on computational science and its applications. Springer, pp 598–610

Cataldi M, Ballatore A, Tiddi I, Aufaure MA (2013) Good location, terrible food: detecting feature sentiment in user-generated reviews. Soc Netw Anal Min 3(4):1149–1163

Caverlee J, Cheng Z, Sui DZ, Kamath KY (2013) Towards geo-social intelligence: mining, analyzing, and leveraging geospatial footprints in social media. IEEE Data Eng Bull 36(3):33–41

Chen D, Manning C (2014) A fast and accurate dependency parser using neural networks. In: Proceedings of the 2014 conference on empirical methods in natural language processing (EMNLP), pp 740–750

Chon Y, Lane ND, Li F, Cha H, Zhao F (2012) Automatically characterizing places with opportunistic crowdsensing using smartphones. In: Proceedings of the 2012 ACM conference on ubiquitous computing. ACM, pp 481–490

Cranshaw J, Yano T (2010) Seeing a home away from the home: distilling proto-neighborhoods from incidental data with latent topic modeling. In: CSSWC workshop at NIPS, vol 10

Crooks A, Croitoru A, Stefanidis A, Radzikowski J (2013) # earthquake: Twitter as a distributed sensor system. Trans GIS 17(1):124–147

Daiber J, Jakob M, Hokamp C, Mendes PN (2013) Improving efficiency and accuracy in multilingual entity extraction. In: Proceedings of the 9th international conference on semantic systems. ACM, pp 121–124

De Longueville B, Smith RS, Luraschi G (2009) Omg, from here, i can see the flames!: a use case of mining location based social networks to acquire spatio-temporal data on forest fires. In: Proceedings of the 2009 international workshop on location based social networks. ACM, pp 73–80

DeLozier G, Baldridge J, London L (2015) Gazetteer-independent toponym resolution using geographic word profiles. In: Proceedings of the AAAI conference on artificial intelligence (AAAI). AAAI Press, pp 2382–2388

DeLozier G, Wing B, Baldridge J, Nesbit S (2016) Creating a novel geolocation corpus from historical texts. In: Proceedings of The 10th linguistic annotation workshop. Association for Computational Linguistics, pp 188–198

Di Rocco L, Bertolotto M, Catania B, Guerrini G, Cosso T (2016) Extracting fine-grained implicit georeferencing information from microblogs exploiting crowdsourced gazetteers and social interactions. In: AGILE international conference on geographic information science
Doytsher Y, Galon B, Kanza Y (2017) Emotion maps based on geotagged posts in the social media. In: Proceedings of the 1st ACM SIGSPATIAL workshop on geospatial humanities. ACM, pp 39–46
Dunn J (2018) Finding variants for construction-based dialectometry: a corpus-based approach to regional CxGs. Cogn Linguist 29(2):275–311
Eisenstein J, O'Connor B, Smith NA, Xing EP (2010) A latent variable model for geographic lexical variation. In: Proceedings of the 2010 conference on empirical methods in natural language processing. Association for Computational Linguistics, pp 1277–1287
Elwood S, Goodchild MF, Sui DZ (2012) Researching volunteered geographic information: spatial data, geographic research, and new social practice. Ann. Assoc Am Geogr 102(3):571–590
Finkel JR, Grenager T, Manning C (2005) Incorporating non-local information into information extraction systems by gibbs sampling. In: Proceedings of the 43rd annual meeting on association for computational linguistics. Association for Computational Linguistics, pp 363–370
Freire N, Borbinha J, Calado P, Martins B (2011) A metadata geoparsing system for place name recognition and resolution in metadata records. In: Proceedings of the 11th annual international ACM/IEEE joint conference on digital libraries. ACM, pp 339–348
Gelernter J, Balaji S (2013) An algorithm for local geoparsing of microtext. GeoInformatica 17(4):635–667
Gelernter J, Mushegian N (2011) Geo-parsing messages from microtext. Trans GIS 15(6):753–773
Gelernter J, Ganesh G, Krishnakumar H, Zhang W (2013) Automatic gazetteer enrichment with user-geocoded data. In: Proceedings of the second ACM SIGSPATIAL international workshop on crowdsourced and volunteered geographic information. ACM, pp 87–94
Ghosh D, Guha R (2013) What are we 'tweeting' about obesity? Mapping tweets with topic modeling and geographic information system. Cartogr Geogr Inf Sci 40(2):90–102
Goldberg DW, Wilson JP, Knoblock CA, Ritz B, Cockburn MG (2008) An effective and efficient approach for manually improving geocoded data. Int J Health Geogr 7(1):60
Goodchild MF (2011) Formalizing place in geographic information systems. In: Communities, neighborhoods, and health. Springer, pp 21–33
Goodchild MF, Hill LL (2008) Introduction to digital gazetteer research. Int J Geogr Inf Sci 22(10):1039–1044
Gregory I, Donaldson C, Murrieta-Flores P, Rayson P (2015) Geoparsing, gis, and textual analysis: current developments in spatial humanities research. Int J Humanit Arts Comput 9(1):1–14
Gritta M, Pilehvar MT, Limsopatham N, Collier N (2018) What?s missing in geographical parsing? Lang Resour Eval 52(2):603–623
Grothe C, Schaab J (2009) Automated footprint generation from geotags with kernel density estimation and support vector machines. Spat Cogn Comput 9(3):195–211
Hahmann S, Purves R, Burghardt D (2014) Twitter location (sometimes) matters: exploring the relationship between georeferenced tweet content and nearby feature classes. J Spat Inf Sci 2014(9):1–36
Han B, Cook P, Baldwin T (2014) Text-based twitter user geolocation prediction. J Artif Intell Res 49:451–500
Hao Q, Cai R, Wang C, Xiao R, Yang JM, Pang Y, Zhang L (2010) Equip tourists with knowledge mined from travelogues. In: Proceedings of the 19th international conference on World wide web. ACM, pp 401–410
Haworth B, Bruce E (2015) A review of volunteered geographic information for disaster management. Geogr Compass 9(5):237–250
Hecht B, Moxley E (2009) Terabytes of tobler: evaluating the first law in a massive, domain-neutral representation of world knowledge. In: International conference on spatial information theory. Springer, pp 88–105

Hecht B, Raubal M (2008) Geosr: geographically explore semantic relations in world knowledge. The European Information Society, pp 95–113

Hecht BJ, Stephens M (2014) A tale of cities: urban biases in volunteered geographic information. ICWSM 14:197–205

Hill LL (2000) Core elements of digital gazetteers: placenames, categories, and footprints. In: International conference on theory and practice of digital libraries. Springer, pp 280–290

Hollenstein L, Purves R (2010) Exploring place through user-generated content: using flickr tags to describe city cores. J Spat Inf Sci 2010(1):21–48

Hu Y, Janowicz K, Prasad S (2014) Improving wikipedia-based place name disambiguation in short texts using structured data from dbpedia. In: Proceedings of the 8th workshop on geographic information retrieval. ACM, pp 1–8

Hu Y, Ye X, Shaw SL (2017) Extracting and analyzing semantic relatedness between cities using news articles. Int J Geogr Inf Sci 31(12):2427–2451

Hu Y, Mao H, McKenzie G (2018) A natural language processing and geospatial clustering framework for harvesting local place names from geotagged housing advertisements. International Journal of Geographical Information Science, pp 1–25

Huang Q, Xiao Y (2015) Geographic situational awareness: mining tweets for disaster preparedness, emergency response, impact, and recovery. ISPRS Int J Geo Inf 4(3):1549–1568

Inkpen D, Liu J, Farzindar A, Kazemi F, Ghazi D (2015) Location detection and disambiguation from twitter messages. J Intell Inf Syst, 49(2):237–253

Intagorn S, Lerman K (2011) Learning boundaries of vague places from noisy annotations. In: Proceedings of the 19th ACM SIGSPATIAL international conference on advances in geographic information systems. ACM, pp 425–428

Issa E, Tsou MH, Nara A, Spitzberg B (2017) Understanding the spatio-temporal characteristics of twitter data with geotagged and non-geotagged content: two case studies with the topic of flu and ted (movie). Ann GIS 23(3):219–235

Janowicz K, Keßler C (2008) The role of ontology in improving gazetteer interaction. Int J Geogr Inf Sci 22(10):1129–1157

Jiang, Y, Li, Z, Ye, X (2019) Understanding demographic and socioeconomic biases of geotagged Twitter users at the county level. Cartogr Geogr Inf Sci 46(3):228–242

Jones CB, Purves RS (2008) Geographical information retrieval. Int J Geogr Inf Sci 22(3):219–228

Jones CB, Purves RS, Clough PD, Joho H (2008) Modelling vague places with knowledge from the web. Int J Geogr Inf Sci 22(10):1045–1065

Ju Y, Adams B, Janowicz K, Hu Y, Yan B, McKenzie G (2016) Things and strings: improving place name disambiguation from short texts by combining entity co-occurrence with topic modeling. In: 20th international conference on knowledge engineering and knowledge management. Springer

Karimzadeh M, Huang W, Banerjee S, Wallgrün JO, Hardisty F, Pezanowski S, Mitra P, MacEachren AM (2013) Geotxt: a web api to leverage place references in text. In: Proceedings of the 7th workshop on geographic information retrieval. ACM, pp 72–73

Keßler C, Janowicz K, Bishr M (2009a) An agenda for the next generation gazetteer: Geographic information contribution and retrieval. In: Proceedings of the 17th ACM SIGSPATIAL International Conference on Advances in Geographic Information Systems, ACM, pp 91–100

Keßler C, Maué P, Heuer J.T, Bartoschek T (2009b) December Bottom-up gazetteers: Learning from the implicit semantics of geotags. In: Proceedings of the International Conference on GeoSpatial Semantics, 83–102. Springer, Berlin, Heidelberg

Ladra S, Luaces MR, Pedreira O, Seco D (2008) A toponym resolution service following the ogc wps standard. In: International symposium on web and wireless geographical information systems. Springer, pp 75–85

Leidner JL (2008) Toponym resolution in text: annotation, evaluation and applications of spatial grounding of place names. Universal-Publishers

Li H, Srihari RK, Niu C, Li W (2002) Location normalization for information extraction. In: Proceedings of the 19th international conference on Computational linguistics-Volume 1. Association for Computational Linguistics, pp 1–7

Li L, Goodchild MF (2012) Constructing places from spatial footprints. In: Proceedings of the 1st ACM SIGSPATIAL international workshop on crowdsourced and volunteered geographic information. ACM, pp 15–21

Lieberman MD, Samet H (2011) Multifaceted toponym recognition for streaming news. In: Proceedings of the 34th international ACM SIGIR conference on research and development in information retrieval. ACM, pp 843–852

Lieberman MD, Samet H, Sankaranarayanan J (2010) Geotagging with local lexicons to build indexes for textually-specified spatial data. In: 2010 IEEE 26th international conference on data engineering (ICDE). IEEE, pp 201–212

Liu Y, Wang F, Kang C, Gao Y, Lu Y (2014) Analyzing relatedness by toponym co-occurrences on web pages. Trans GIS 18(1):89–107

MacEachren AM, Jaiswal A, Robinson AC, Pezanowski S, Savelyev A, Mitra P, Zhang X, Blanford J (2011) Senseplace2: Geotwitter analytics support for situational awareness. In: 2011 IEEE conference on visual analytics science and technology (VAST). IEEE, pp 181–190

Madden DJ (2017) Pushed off the map: toponymy and the politics of place in new york city. Urban Studies p Online First

Malik MM, Lamba H, Nakos C, Pfeffer J (2015) Population bias in geotagged tweets. People 1(3,759.710):3–759

Mani I, Hitzeman J, Richer J, Harris D (2008) ACE 2005 english spatialML annotations. Linguistic Data Consortium, Philadelphia

McCurley KS (2001) Geospatial mapping and navigation of the web. In: Proceedings of the 10th international conference on World Wide Web. ACM, pp 221–229

McKenzie G, Adams B (2018) A data-driven approach to exploring similarities of tourist attractions through online reviews. J Locat Based Serv 12(2):94–118

McKenzie G, Liu Z, Hu Y, Lee M (2018) Identifying urban neighborhood names through user-contributed online property listings. ISPRS Int J Geo Inf 7(10):388

Mei Q, Liu C, Su H, Zhai C (2006) A probabilistic approach to spatiotemporal theme pattern mining on weblogs. In: Proceedings of the 15th international conference on World Wide Web. ACM, pp 533–542

Melo F, Martins B (2017) Automated geocoding of textual documents: a survey of current approaches. Trans GIS 21(1):3–38

Mendes PN, Jakob M, García-Silva A, Bizer C (2011) Dbpedia spotlight: shedding light on the web of documents. In: Proceedings of the 7th international conference on semantic systems. ACM, pp 1–8

Merschdorf H, Blaschke T (2018) Revisiting the role of place in geographic information science. ISPRS Int J Geo Inf 7(9):364

Monteiro BR, Davis Jr CA, Fonseca F (2016) A survey on the geographic scope of textual documents. Comput Geosci 96:23–34

Nagar R, Yuan Q, Freifeld CC, Santillana M, Nojima A, Chunara R, Brownstein JS (2014) A case study of the New York City 2012–2013 influenza season with daily geocoded Twitter data from temporal and spatiotemporal perspectives. J Med Internet Res 16(10):236

Nesi P, Pantaleo G, Tenti M (2016) Geographical localization of web domains and organization addresses recognition by employing natural language processing, pattern matching and clustering. Eng Appl Artif Intell 51:202–211

Overell S, Rüger S (2008) Using co-occurrence models for placename disambiguation. Int J Geogr Inf Sci 22(3):265–287

Purves RS, Clough P, Jones CB, Arampatzis A, Bucher B, Finch D, Fu G, Joho H, Syed AK, Vaid S, et al (2007) The design and implementation of spirit: a spatially aware search engine for information retrieval on the internet. Int J Geogr Inf Sci 21(7):717–745

Purves RS, Clough P, Jones CB, Hall MH, Murdock V, et al (2018) Geographic information retrieval: Progress and challenges in spatial search of text. Found Trends® Inf Retr 12(2–3):164–318

Rahimi A, Cohn T, Baldwin T (2017) A neural model for user geolocation and lexical dialectology. In: Proceedings of the 55th annual meeting of the association for computational linguistics (Volume 2: Short Papers), vol 2, pp 209–216
Rahmani K, Gnoth J, Mather D (2018) Tourists' participation on web 2.0: A corpus linguistic analysis of experiences. J Travel Res, 57(8):108–1120
Roller S, Speriosu M, Rallapalli S, Wing B, Baldridge J (2012) Supervised text-based geolocation using language models on an adaptive grid. In: Proceedings of the 2012 joint conference on empirical methods in natural language processing and computational natural language learning. Association for Computational Linguistics, pp 1500–1510
Rupp C, Rayson P, Baron A, Donaldson C, Gregory I, Hardie A, Murrieta-Flores P (2013) Customising geoparsing and georeferencing for historical texts. In: 2013 IEEE international conference on big data. IEEE, pp 59–62
Salvini MM, Fabrikant SI (2016) Spatialization of user-generated content to uncover the multirelational world city network. Environ Plann B Plann Des 43(1):228–248
Santos J, Anastácio I, Martins B (2015) Using machine learning methods for disambiguating place references in textual documents. GeoJournal 80(3):375–392
Sassen S (2016) The global city: strategic site, new frontier. In: Managing urban futures. Routledge, pp 89–104
Shaw SL, Tsou MH, Ye X (2016) Human dynamics in the mobile and big data era. Int J Geogr Inf Sci 30(9):1687–1693
Southall H (2014) Rebuilding the great britain historical gis, part 3: integrating qualitative content for a sense of place. Hist Methods J Quantitative Interdiscip Hist 47(1):31–44
Speriosu M, Baldridge J (2013) Text-driven toponym resolution using indirect supervision. In: ACL (1), ACL, pp 1466–1476
Spitz A, Geiß J, Gertz M (2016) So far away and yet so close: augmenting toponym disambiguation and similarity with text-based networks. In: Proceedings of the third international ACM SIGMOD workshop on managing and mining enriched geo-spatial data. ACM, p 2
Stock K, Yousaf J (2018) Context-aware automated interpretation of elaborate natural language descriptions of location through learning from empirical data. Int J Geogr Inf Sci 32(6):1087–1116. https://doi.org/10.1080/13658816.2018.1432861
Sui D, DeLyser D (2012) Crossing the qualitative-quantitative chasm I: hybrid geographies, the spatial turn, and volunteered geographic information (vgi). Prog Hum Geogr 36(1):111–124
Taylor PJ, Derudder B (2015) World city network: a global urban analysis. Routledge
Tenbrink T, Kuhn W (2011) A model of spatial reference frames in language. In: Egenhofer M, Giudice N, Moratz R, Worboys M (eds) Spatial information theory. Springer, Berlin/Heidelberg, pp 371–390
Toly N, Bouteligier S, Smith G, Gibson B (2012) New maps, new questions: global cities beyond the advanced producer and financial services sector. Globalizations 9(2):289–306
Tsou MH (2015) Research challenges and opportunities in mapping social media and big data. Cartogr Geogr Inf Sci 42(sup1):70–74
Tuan YF (1977) Space and place: the perspective of experience. University of Minnesota Press
Twaroch FA, Jones CB (2010) A web platform for the evaluation of vernacular place names in automatically constructed gazetteers. In: Proceedings of the 6th workshop on geographic information retrieval. ACM, p 14
Twaroch FA, Jones CB, Abdelmoty AI (2009) Acquisition of vernacular place names from web sources. In: King I, Baeza-Yates R (eds) Weaving services and people on the World Wide Web. Springer, pp 195–214
Wallgrün JO, Karimzadeh M, MacEachren AM, Pezanowski S (2018) Geocorpora: building a corpus to test and train microblog geoparsers. Int J Geogr Inf Sci 32(1):1–29
Wang J, Hu Y (2019) Enhancing spatial and textual analysis with EUPEG: an extensible and unified platform for evaluating geoparsers. Trans GIS 23(6):1393–1419
Wang M, Zhou X (2016) Geography matters in online hotel reviews. ISPRS-international archives of the photogrammetry, Remote Sensing and Spatial Information Sciences, pp 573–576

Wang W, Stewart K (2015) Spatiotemporal and semantic information extraction from web news reports about natural hazards. Comput Environ Urban Syst 50:30–40
van Weerdenburg D, Scheider S, Adams B, Spierings B, van der Zee E (2019) Where to go and what to do: extracting leisure activity potentials from web data on urban space. Comput Environ Urban Syst 73:143–156
Wing B, Baldridge J (2014) Hierarchical discriminative classification for text-based geolocation. In: Proceedings of the 2014 conference on empirical methods in natural language processing (EMNLP), pp 336–348
Wing BP, Baldridge J (2011) Simple supervised document geolocation with geodesic grids. In: Proceedings of the 49th annual meeting of the association for computational linguistics: human language technologies-volume 1. Association for Computational Linguistics, pp 955–964
Woodruff AG, Plaunt C (1994) Gipsy: automated geographic indexing of text documents. J Am Soc Inf Sci 45(9):645–655
Xiang Z, Schwartz Z, Gerdes Jr JH, Uysal M (2015) What can big data and text analytics tell us about hotel guest experience and satisfaction? Int J Hosp Manag 44:120–130
Zhang W, Gelernter J (2014) Geocoding location expressions in twitter messages: a preference learning method. J Spat Inf Sci 2014(9):37–70
Zhong X, Liu J, Gao Y, Wu L (2017) Analysis of co-occurrence toponyms in web pages based on complex networks. Physica A Stat Mech Appl 466:462–475

Chapter 20
Automating Information Extraction from Large Historical Topographic Map Archives: New Opportunities and Challenges

Johannes H. Uhl and Weiwei Duan

20.1 Introduction

Map processing, or information extraction from map documents, is a branch of document analysis that focuses on developing methods for the extraction and the recognition of information in scanned map documents. Map processing is an interdisciplinary field that combines elements of computer vision, pattern recognition, geographic information science, cartography, and geoinformatics. The main goal of map processing is to unlock spatial (and spatio-temporal) information from scanned map documents and to provide this information in digital, machine-readable data formats to preserve the information digitally and to facilitate the use of these data for analytical purposes (Chiang et al. 2014). Earth observation via space and airborne remote sensors in a systematically operational manner has been conducted since the early 1970s. Hence, for earlier time periods, there is little digital information available about the dynamics of features at the earth surface, such as the evolution of human settlements, changes in land cover and land use, or the development of transportation networks. Thus, map processing typically focuses on the development of information extraction methods from map documents or engineering drawings created prior to the era of digital cartography.

Traditional information extraction from map documents includes the steps of recognition (i.e., identifying objects in a scanned map such as groups of contiguous pixels with homogeneous semantic meaning), and extraction as a subsequent step

J. H. Uhl (✉)
Department of Geography & Institute of Behavioral Science, University of Colorado Boulder, Boulder, CO, USA
e-mail: johannes.uhl@colorado.edu

W. Duan
Spatial Sciences Institute, University of Southern California, Los Angeles, CA, USA
e-mail: weiweidu@usc.edu

M. Werner, Y.-Y. Chiang (eds.), *Handbook of Big Geospatial Data*,
https://doi.org/10.1007/978-3-030-55462-0_20

of transferring these objects into a machine-readable format (e.g., through vectorization). The application of methods from pattern recognition or computer vision to map documents often constitutes additional challenges as compared to traditional document analysis due to low graphical quality and complex, human-made map content (e.g., overlapping cartographic symbols). Map processing typically involves an image segmentation step, and subsequent recognition and extraction procedures (Chiang et al. 2014).

Whereas image segmentation partitions a digital image into groups of homogeneous characteristics such as color or texture, the recognition of objects (i.e., obtaining the semantics for the created segments) in classical map processing include threshold-based methods (e.g., Iosifescu et al. 2016), or template matching based on (typically manually) generated templates of the cartographic symbol of interest. Template matching can be based on shape descriptors, cross-correlation measures or based on feature descriptors. Furthermore, morphological operations can be applied to filter out irrelevant map content based on a previously created segmentation. For the detection of linear features, methods such as Hough transform have been employed (Yamada et al. 1993). Optical Character Recognition (OCR) techniques have been successfully applied for text recognition in map documents (Chiang et al. 2014). Moreover, contextual reasoning methods have been employed for recognition tasks by taking into account map content in spatial proximity of an object or by considering spatial relationships to already recognized map features (e.g., Gamba and Mecocci 1999). Malerba et al. (2001) propose an inductive machine learning approach for automated map interpretation, involving user-defined spatial, topological, and semantic rules. In these approaches, once the semantic meaning of a map segment is recognized, the segment can be extracted by registering its location or shape typically in a geospatial vector data format. At this stage, common vectorization techniques are typically employed. Here, it is worth noting that approaches tackling the digitization and recognition of black-and-white documents such as cadastral maps or floorplans are typically in an inverted order, i.e., recognition takes place in a previously vectorized version of the document.

Exemplary applications of map processing techniques applied to topographic maps include the extraction of buildings (Miyoshi et al. 2004; Laycock et al. 2011; Arteaga 2013), of road networks (Chiang et al. 2011), of contour lines (Miao et al. 2016), of composite forest symbols (Leyk and Boesch 2008), the recognition of text (Chiang et al. 2014) as well as the digitization of cadastral maps (e.g., Katona and Hudra 1999). Chiang et al. (2014) and Liu et al. (2019) provide useful overviews on existing traditional methods and applications in map processing. Whereas most existing approaches are either based on map-specific thresholds and have been developed and tested on a limited number of individual maps, the lack of flexibility and a relatively high degree of user interaction required for many of those approaches did not constitute a major shortcoming.

However, three recent developments are expected to considerably change the field of map processing in the near future: (a) There is an increased availability of large map collections holding thousands of map documents as digital and georeferenced archives and thus, there is an urgent demand to reduce or even

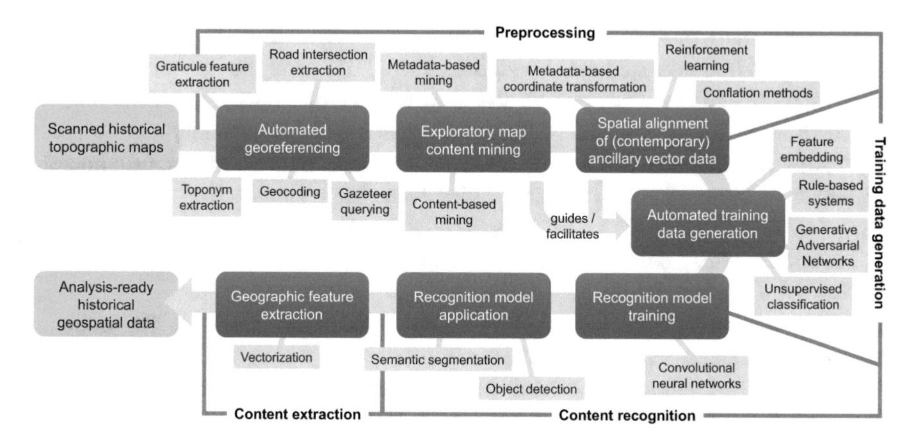

Fig. 20.1 Overview of a possible workflow for automated information extraction from historical topographic map archives, highlighting exemplary methods and approaches for each step

eliminate user interaction in information extraction from such large and potentially heterogeneous amounts of data. (b) Advances in (deep) machine learning for information extraction from geospatial data, e.g., remotely sensed earth observation data (see Ball et al. 2017 for an overview) naturally project into the idea of similar applications for scanned map documents, such as the use of Convolutional Neural Networks (CNN). (c) Increasingly public availability of geospatial data of high spatial, temporal, and semantic granularity potentially provides new sources of ancillary data to be employed in automated information extraction methods based on multi-source data integration.

These trends constitute new challenges, but also opportunities for the field of map processing and have catalyzed the development of several new approaches tackling these challenges. In this contribution, we give an overview on these developments. More specifically, we provide a brief overview on existing, and publicly available large digital map archives (Sect. 20.2), on explorative methods and innovative preprocessing techniques (Sect. 20.3), and we provide a review of recently developed advanced methods suitable for automated recognition and extraction of information large amounts of map documents (Sect. 20.4). Figure 20.1 shows a typical workflow including preprocessing, automated training data generation, recognition, and extraction steps and highlights some exemplary approaches for each step, as discussed in Sects. 20.3 and 20.4.

20.2 Digital Historical Map Archives

Recently, several efforts have been conducted in order to make large amounts of historical map documents available to the public. Some examples are the United States Geological Survey (USGS), that scanned and georeferenced approx. 200,000 topographic maps published between 1884 and 2006 at multiple cartographic scales between 1:24,000 and 1:250,000 (Fishburn et al. 2017) and the Sanborn fire insurance map collection maintained by the U.S. Library of Congress, which contains approximately 700,000 sheets of large-scale maps of approximately 12,000 cities and towns within the U.S., Canada, Mexico, and Cuba, out of which approximately 25,000 map sheets from over 3,000 cities have been published as scanned map documents (U.S. Library of Congress 2018a,b,c). Moreover, the National Library of Scotland scanned and georeferenced more than 200,000 topographic maps from the United Kingdom dating back to the 1840s and offers many of them as seamless georeferenced raster layers (National Library of Scotland 2018a,b). The national cartographic agency of Switzerland Swisstopo scanned and georeferenced approximately 52,000 printed and hand-made historical topographic maps from Switzerland dating back to 1840 (Swisstopo 2018a), available as a seamless raster layer as well (Swisstopo 2018b). David Rumsey Map Center at the Stanford University Library published more than 88,000 historical maps from different regions online (Stanford University Library – David Rumsey Map Center 2018).The web map portal Mapire (Biszak et al. 2017) holds large amounts of historical topographic and cadastral maps covering many countries in Europe of the nineteenth century, which have been reprojected into a common, modern spatial reference system and are accessible online (Mapire 2018). Other efforts include the web portals "Old Maps Online"[1] and "PAHAR".[2] Moreover, there are efforts to digitize large amounts of cartographic and other historical documents (e.g., cadastral maps) of the city of Venice (Italy) covering large periods of time (Abbott 2017). Figure 20.2 exemplarily shows a time

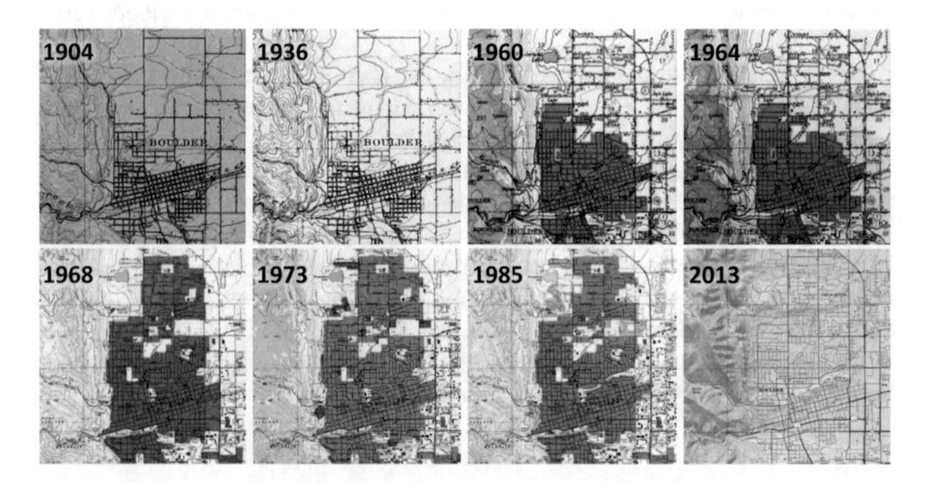

Fig. 20.2 USGS topographic maps at a cartographic scale of 1:24,000 covering Boulder, Colorado (USA), from 1904 to 2013. (Source: adapted from Uhl et al. 2018)

[1] www.oldmapsonline.org

[2] Pahar – the Mountains of Central Asia Digital Dataset: http://pahar.in

series of USGS topographic maps documenting the evolution of Boulder, Colorado (USA) between 1904 and 2013.

20.3 Preprocessing Methods

In order to extract geospatial information from large digital map archives, optimally, several conditions should be satisfied: (a) The scanned map documents are stored as georeferenced digital images, (b) the level of accuracy of the georeference allows for the spatial alignment of the map content between maps and to geospatial data from different sources, such as contemporary geospatial vector data, at sufficiently high spatial accuracy, and (c) the user (i.e., the developer of map information extraction techniques) is informed about the approximate content of the map archive (e.g., map content heterogeneity such as inconsistent symbology due to changes in cartographic design over time), and is aware of the georeference accuracy of the map documents contained in an archive. This section gives an overview on existing methods regarding the described issues.

20.3.1 Automated Georeferencing

Preprocessing steps of map processing procedures typically include the automation of the georeferencing process of scanned map documents. This is particularly important regarding the overlay of the scanned maps and other georeferenced data and for the extraction of geospatial vector data in a world coordinate system. Existing approaches for automated georeferencing can be grouped into two main categories: Metadata-based approaches and matching-based approaches.

Metadata-based approaches aim to locate features in a scanned map that allow for georeferencing the map using an ancillary metadata source. Such approaches are suitable for map collections organized systematically in quadrangles, such as the USGS topographic map archive. In such cases, map sheet corners or graticule marks correspond to apriori known coordinates that can be identified and located using morphological operations, as proposed by Herold et al. (2011) who, additionally, use a multilayer perceptron to automatically extract the map sheet number from the scanned map document. Based on the map sheet number, corner coordinates of the quadrangle corresponding to the map sheet can be obtained from a metadata source,[3] see Uhl et al. (2019). Burt et al. (2019) developed a software tool facilitating the georeferencing process of scanned USGS topographic maps by searching for perpendicular linear intersections in the scanned map document in order to locate

[3]Metadata for the USGS topographic map archive is available under https://thor-f5.er.usgs.gov/ngtoc/metadata/misc

the graticule marks. More recently, Dong et al. (2018) use a CNN to locate map corner points, using a CNN trained on image patches of map corner points generated from manually georeferenced maps. Based on the image coordinates of identified map corners and graticule marks, and the corresponding world coordinates of these locations obtained or derived from the metadata, the scanned image can be warped into the target reference system.

Matching-based approaches make use of independent, geospatial ancillary data sources and typically aim to find identical locations in the georeferenced ancillary data and the scanned map in order to obtain transformation parameters allowing for (co)registering the map document in a world coordinate system. There are two types of matching-based approaches: **Geometric matching** is based exclusively on geometric similarities between map content and ancillary data (vector or raster data) and typically make use of conflation techniques (Saalfeld 1988) or point pattern matching techniques (Li and Briggs 2006, 2012). In this context, Saeedimoghaddam and Stepinski (2020) propose and evaluate the use of deep CNNs for the detection of road intersections in historical topographic maps. The detected road intersections, as an example, can then be used for co-registering the underlying map with ancillary geospatial road network data.

Geometric matching is opposed to **semantic matching**, which involves the extraction of semantic information from the scanned map prior to the co-registration step. This semantic information is then geocoded using gazetteers or other spatial databases. For example, Weinman (2013) use a spatial toponym database (i.e., a database containing geographic names and corresponding geolocations) and extract text contained in scanned maps using OCR to georeference the maps. The extracted map text elements are matched to the toponym database, and outlier detection is used to find the matches most likely to be correct in order to establish geometric transformation rules between image and world coordinates. In Weinman (2017) this approach has been refined by employing a CNN for text recognition, by taking into account cartographic labeling styles, and by including probabilistic models to estimate the correct map projection based on toponym locations. A similar approach has been proposed by Tavakkol et al. (2019), who make use of cloud-based services for both, the text recognition step and the geocoding step. While geometric matching has widely been used in the past, semantic matching has become popular in recent years, catalyzed by the increasing availability of web-based gazetteers and geocoding infrastructures.

20.3.2 Spatial Data Alignment

It is well-known that scanned map documents are affected by several kinds of geometric distortions, introduced by deformations of the paper caused by humidity or heat, by scanner miscalibration, cartographic generalization and displacement for better map readability, and by inaccuracies in the underlying topographic measurements or the cartographic projection used (Kaim et al. 2014). For these reasons, scanned and georeferenced (historical) map documents rarely align perfectly to contemporary geospatial vector data, since transformation rules warping image to world coordinates often do not account for the inherent (and typically unknown) map distortions. However, such spatial alignment is necessary for certain automated procedures involving spatial ancillary data, typically in vector format, as described in the subsequent sections.

Thus, recently, several methods have been developed to improve spatial alignment between scanned map documents and geospatial ancillary data. Duan et al. (2017, 2020) apply systematic shifts to vector data in order to find the optimum alignment to underlying linear cartographic symbols of interest using a customized decision making approach, and Reinforcement Learning, respectively, requiring a minimum degree of user interaction. Uhl et al. (2017) use metadata available for the USGS map archive to extract the transformation parameters used to warp each individual map sheet and apply the same transformation to vector data covering the map extent in order to improve spatial alignment between objects in an individual georeferenced map and the corresponding vector data. Moreover, the automatic alignment of geospatial vector data, such as cadastral or road network data to aerial imagery is an active research field (e.g., Song et al. 2013; Ruiz-Lendínez et al. 2019), that potentially can be applied to scanned map documents as well.

20.3.3 Exploratory Methods

Exploratory methods in the context of large map document archives may be applied at two levels: (a) metadata level and (b) content level. Uhl et al. (2018) present a framework to visually-analytically explore the map archive based on metadata and low-level image descriptors: In their approach, ground control point coordinate pairs reported in USGS topographic map archive metadata are used to reconstruct and visualize the distortions applied to the scanned map image during warping, allowing for fast identification of potentially inaccurately georeferenced map sheets (Fig. 20.3).

Furthermore, Uhl et al. (2018) use ancillary geospatial data representing settlement locations to extract samples of settlement signatures across large amounts of topographic maps. For these collected samples, multidimensional feature descriptors based on color moments Huang et al. (2010) are computed and visualized in a low-dimensional space using t-SNE dimensionality reduction Maaten and Hinton (2008), as shown in Fig. 20.4.

Moreover, Zhou et al. (2018) present an approach employing CNNs for the classification of map types, which may be useful to explore the content in and across heterogeneous digital map collections.

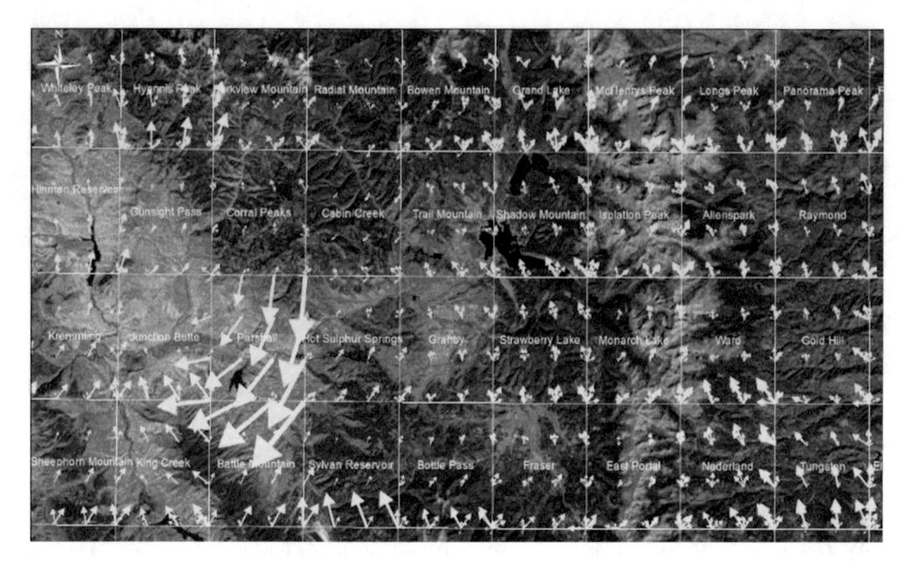

Fig. 20.3 Displacement vector field at GCP locations over multiple USGS map quadrangles of scale 1:24,000, located Northwest of Denver (Colorado), reflecting different types of distortions introduced to the map documents during the georeferencing process. (Source: Uhl et al. 2018)

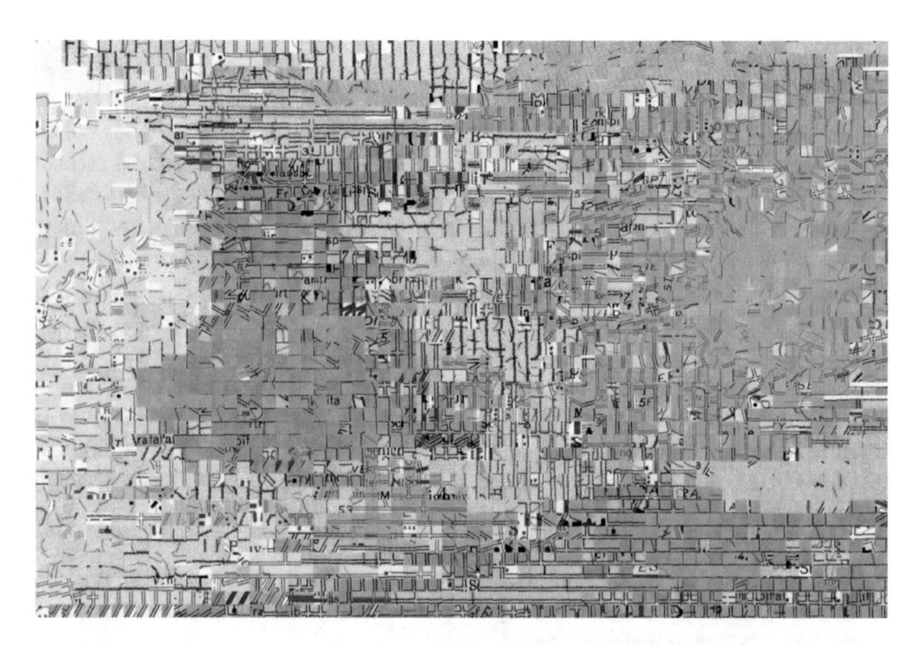

Fig. 20.4 Samples of urban settlement symbols collected across 50 maps from the USGS topographic map archive, visualized in a two-dimensional t-SNE space using low-level image descriptors. (Source: Uhl et al. 2018)

20.4 Automated Map Content Recognition and Extraction

Most existing approaches for content extraction from historical maps still require a certain degree of user interaction to ensure acceptable extraction performance for individual map sheets. To overcome this persistent limitation, several approaches have been proposed, such as the application of deep learning models in combination with automated training data generation, and the creation of benchmark datasets for pretraining and method evaluation. This section aims to give a brief overview on these advanced efforts.

20.4.1 Training Data Collection

Artificial intelligence methods such as deep convolutional neural networks typically require large amounts of training data. In order to overcome manual training data generation, **spatial ancillary data** can be employed.

The use of contemporary geospatial data as ancillary information to facilitate map processing tasks has recently received some attention (Hurni et al. 2013; Tsorlini et al. 2014; Leyk and Chiang 2016). In these approaches, ancillary geographic information that contains the feature of interest such as gazetteers or other map series for guided graphics sampling in training a recognition model (Chiang and Leyk 2015; Chiang et al. 2016; Yu et al. 2016). For example, it can be assumed that many building symbols in a historical map spatially overlap or are in close proximity to building objects in a contemporary geographic dataset. Thus, sampling nearby the contemporary building objects enables a system to collect graphic examples of building symbology in historical maps, thus creating highquality training data at image patch-level. Uhl et al. (2019) automatically collect training data using an unsupervised rule-based system involving scale-invariant feature transform (SIFT, Lowe 1999) keypoint detection to center the collected samples at the feature of interest to improve training data quality and feature extraction results and employ Local Binary Pattern (Ojala et al. 2002) feature embedding in combination with unsupervised clustering methods to automatically generate reliable training labels. In order to increase the number of training samples, Li (2019) proposes the use of Generative Adversarial Networks to generate synthetic training samples of historical maps, generated from contemporary online map services.

Moreover, **training and benchmark datasets** tailored to machine learning tasks applied to map processing have been published: Ray et al. (2018) published a dataset containing annotated text samples collected from historical maps for text detection and recognition purposes. In Zhou et al. (2018), a dataset called "deepmap" is used to train a CNN for map type classification tasks. Useful for information extraction approaches from floor plans is the LIFULL HOME's dataset containing large amounts of labelled floor plan images (Kiyota 2018).

20.4.2 Recognition and Extraction Methods

Recognition methods in map processing that are suitable to extract information from large and heterogeneous map collections mainly need to comply two criteria: (a) They need to be capable to solve complex classification problems on heterogeneous and often noisy data, and (b) they need to work in an automated manner minimizing user interaction. According to VoPham et al. (2018), information extraction from map documents is considered a key challenge in geospatial artificial intelligence (AI) applications.

Budig and Van Dijk (2015) and Budig et al. (2016a) propose the use of active learning and similar interactive concepts for more efficient recognition of cartographic symbols in historical maps, whereas Budig et al. (2016b) examine the usefulness of crowdsourcing for the same purpose. In addition to that, automated large-scale generation of training data based on ancillary geospatial data gives way to the application of deep learning techniques for recognition and detection tasks. For example, Uhl et al. (2019, 2017, 2018); Uhl (2019) collect large amounts of settlement symbol training data in order to train a CNN in a weakly supervised manner for the recognition of settlement symbols in USGS maps, based on automatically collected and labelled training data using geospatial ancillary data. Such approaches allow to combine the traditionally separated processes of segmentation and recognition into a single, integrated process: semantic segmentation. These CNN-based models assign semantic labels to each pixel of an input image and have shown impressive performance in computer vision tasks in recent years. Duan et al. (2018) and Chiang et al. (2020) test the performance of state-of-the-art CNN models (i.e., Long et al. Long et al. 2015) and DilatedNet (Yu and Koltun 2015) for the recognition of several linear geographic features from USGS historical topographic maps. The employed CNN models require pixel-level training data, which was generated automatically using contemporary geospatial vector data (i.e., railroad and waterline network data) automatically aligned to the corresponding map symbols. Heitzler and Hurni (2019, 2020) use an ensemble of U-Net CNNs (Ronneberger et al. 2015) to extract building footprints from historical topographic maps from Switzerland, based on manually digitized training data, and followed by a vectorization method that includes geometric refinement of the semantic segmentation results. Moreover, Uhl (2019) uses a trained VGGNet-16 (Simonyan and Zisserman 2014) in combination with unsupervised (color-based) image segmentation. These so-called "superpixels" obtained through color segmentation are used as analytical units to which CNN-based semantic predictions are referred (see Fig. 20.5 for an example).

Li et al. (2018) propose a framework for "topographic map understanding" that uses a CNN to (a) localize text elements in map documents, (b) employs a pretrained OCR engine for text recognition, and (c) compares the recognized text to a gazetteer containing spatially referenced geographic names in order to understand the map content. The manually collected training data to train the CNN in step (a) was augmented, e.g., by adding additional rotations. In the specific case of information

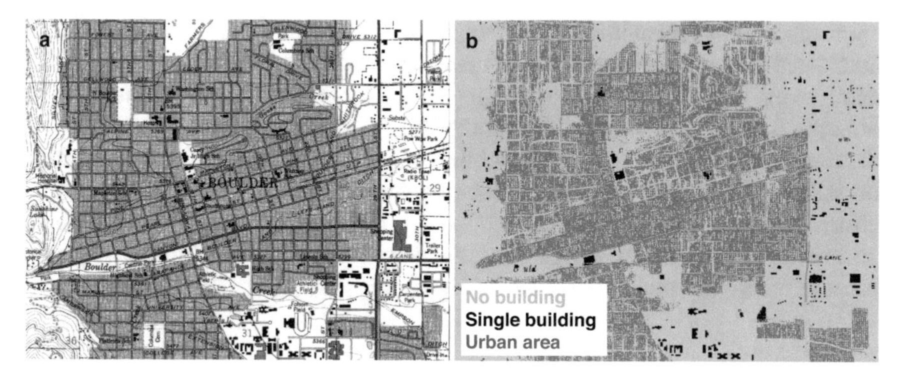

Fig. 20.5 Preliminary results for superpixel-based extraction of settlement features from historical topographic maps using VGGNet-16: (a) USGS topographic map from Boulder (Colorado) from 1960, and (b) extracted superpixels with semantic attributes inferred from the trained CNN. (Source: Uhl 2019; Chiang et al. 2020)

extraction from cadastral maps, which differ significantly from topographic maps in content and symbology, Ignjatić et al. (2018) provide an overview on deep learning methods applied to cadastral maps.

Ares Oliveira et al. (2017) use a CNN in combination with unsupervised segmentation for the extraction and recognition of cadastral parcels and handwritten digits in historical cadastral maps. Worth mentioning here is also the work of Karabork et al. (2008) that use artificial neural networks, and of Liu et al. (2017) that employ a CNN for the vectorization of cadastral maps or floor plans, respectively. Typically, the results of the discussed recognition methods are subsequently vectorized (i.e., converted into geospatial vector data) to be used in analytical environments such as Geographic Information Systems.

20.5 Conclusions and Outlook

Herein, we presented an overview on recent developments towards automated information extraction from large digital archives of historical map documents. Recent advances in deep learning, the increasing public accessibility of large amounts of scanned and often georeferenced map documents represents new opportunities for unlocking spatial-temporal information about the past and thus, preserving the information contained in such historical documents. Such extracted, analysis-ready historical datasets allow for quantitative retrospective analysis of spatio-temporal geographic processes, relevant to many scientific disciplines, but also provide baseline information for data-driven future projections of the processes of interest.

However, the described recent developments also constitute new challenges to the scientific community, such as the need for versatile and robust recognition methods, approaches for automated and reliable training data generation, and methods for automated spatial alignment between map documents and contemporary geospatial data, being able to handle the complexity of human-made map content in an effective and efficient manner. The complexity of human-made map content, and the resulting sparsity of training data often impede the straightforward application of state-of-the-art information extraction methods. Nevertheless, the presented efforts illustrate impressively the ongoing transformation of the field of map processing towards a data-driven discipline involving artificial intelligence, cloud-based services, and multi-source data integration.

References

Abbott A (2017) Nat News 546(7658):341

Ares Oliveira S, di Lenardo I, Kaplan F (2017) In: Premiere annual conference of the international alliance of digital humanities organizations (DH 2017), CONF

Arteaga MG (2013) In: Proceedings of the 1st ACM SIGSPATIAL international workshop on mapinteraction, pp 66–71

Ball JE, Anderson DT, Chan CS (2017) J Appl Remote Sens 11(4):042609

Biszak E, Biszak S, Timár G, Nagy D, Molnár G (2017) In: Proceedings of 12th ICA conference digital approaches to cartographic heritage, Venice, 26–28 Apr 2017, pp 204–208

Budig B, Van Dijk TC (2015) In: International conference on discovery science. Springer, pp 33–47

Budig B, Dijk TCV, Wolff A (2016a) ACM Trans Spatial Algorithm Syst 2(4):1

Budig B, van Dijk TC, Feitsch F, Arteaga MG (2016b) In: Proceedings of the 24th ACM SIGSPATIAL international conference on advances in geographic information systems, pp 1–4

Burt JE, White J, Allord G (2019) Quad-g: Automated georeferencing of scanned map images

Chiang Y, Leyk S (2015) In: Proceedings of the 27th international cartographic conference ICC, pp 23–28

Chiang YY, Leyk S, Knoblock CA (2011) In: International workshop on graphics recognition. Springer, pp 25–35

Chiang YY, Leyk S, Knoblock CA (2014) ACM Comput Surv 47(1):1

Chiang YY, Moghaddam S, Gupta S, Fernandes R, Knoblock CA (2014) In: Proceedings of the 22nd ACM SIGSPATIAL international conference on advances in geographic information systems, pp 581–584

Chiang YY, Leyk S, Nazari NH, Moghaddam S, Tan TX (2016) Comput Geosci 93:21

Chiang Y, Duan W, Leyk S, Uhl J, Knoblock C (2020) Using historical maps in scientific studies: challenges and best practices. Springer

Dong L, Zheng F, Chang H, Yan Q (2018) Earth Sci Inf 11(1):47

Duan W, Chiang YY, Knoblock CA, Jain V, Feldman D, Uhl JH, Leyk S (2017) In: Proceedings of the 1st workshop on artificial intelligence and deep learning for geographic knowledge discovery, pp 45–54

Duan W, Chiang Y, Knoblock CA, Leyk S, Uhl J (2018) In: Proceedings of the autocarto

Duan W, Chiang Y-Y, Leyk S, Uhl JH, Knoblock CA (2020) Automatic alignment of contemporary vector data and georeferenced historical maps using reinforcement learning. Int J Geogr Inf Sci 34(4):824–849. https://doi.org/10.1080/13658816.2019.1698742

Fishburn KA, Davis LR, Allord GJ (2017) Scanning and georeferencing historical usgs quadrangles. Technical report, US Geological Survey

Gamba P, Mecocci A (1999) Pattern Recogn Lett 20(4):355

Heitzler M, Hurni L (2019) Abstr ICA 1:1

Heitzler M, Hurni L (2020) Trans GIS
Herold H, Roehm P, Hecht R, Meinel G (2011) In: Proceedings of the ICA 25th international cartographic conference, vol 1, pp 1–5
Huang ZC, Chan PP, Ng WW, Yeung DS (2010) In: 2010 international conference on machine learning and cybernetics, vol 2. IEEE, pp 719–724
Hurni L, Lorenz C, Oleggini L (2013) In: Proceedings of the 26th international cartographic conference, Dresden, pp 25–30
Ignjatić J, Nikolić B, Rikalović A (2018) In: Computer science research notes (CSRN 2803), Proceedings of 26. international conference in central Europe on computer graphics, visualization and computer vision WSCG 2018
Iosifescu I, Tsorlini A, Hurni L (2016) e-Perimetron 11(2):57
Kaim D, Kozak J, Ostafin K, Dobosz M, Ostapowicz K, Kolecka N, Gimmi U (2014) Quaestiones Geographicae 33(3):55
Karabork H, Aktas E, et al (2008) The international archives of the photogrammetry, remote sensing and spatial information sciences, pp 1716–1723
Katona E, Hudra G (1999) In: Proceedings 10th international conference on image analysis and processing. IEEE, pp 792–797
Kiyota Y (2018) In: Proceedings of the 2018 ACM on international conference on multimedia retrieval, pp 6–6
Laycock SD, Brown PG, Laycock RG, Day AM (2011) Comput Graph 35(2):242
Leyk S, Boesch R (2008) In: 17th International research symposium on computer-based cartography (AutoCarto 2008), pp 8–11
Leyk S, Chiang YY (2016) In: Proceedings of AutoCarto, pp 100–110
Li Z (2019) In: Proceedings of the 27th ACM SIGSPATIAL international conference on advances in geographic information systems, pp 610–611
Li Y, Briggs R (2006) In: The international symposium on automated cartography (AutoCarto), Vancouver. Citeseer
Li Y, Briggs R (2012) Cartogr Geogr Inf Sci 39(4):199
Li H, Liu J, Zhou X (2018) IEEE Access 6:25363
Liu C, Wu J, Kohli P, Furukawa Y (2017) In: Proceedings of the IEEE international conference on computer vision, pp 2195–2203
Liu T, Xu P, Zhang S (2019) Neurocomputing 328:75
Long J, Shelhamer E, Darrell T (2015) In: Proceedings of the IEEE conference on computer vision and pattern recognition, pp 3431–3440
Lowe DG (1999) In: Proceedings of the seventh IEEE international conference on computer vision, vol 2. IEEE, pp 1150–1157
Maaten LVD, Hinton G (2008) J Mach Learn Res 9:2579
Malerba D, Esposito F, Lanza A, Lisi FA, Appice A (2001) Geographic data mining and knowledge discovery, pp 291–314
Mapire (2018) MAPIRE map portal. https://mapire.eu. Online; Accessed 15 Oct 2018
Miao Q, Liu T, Song J, Gong M, Yang Y (2016) IEEE Trans Geosci Remote Sens 54(11):6265
Miyoshi T, Li W, Kaneda K, Yamashita H, Nakamae E (2004) In: Proceedings of the 17th international conference on pattern recognition, ICPR 2004, vol 3. IEEE, pp 626–629
National Library of Scotland (2018a) Ordnance Survey maps. https://maps.nls.uk/os/index.html. Online; Accessed on 28 Feb 2018
National Library of Scotland (2018b) Ordnance Survey maps. http://maps.nls.uk/geo/explore. Online; Accessed on 28 Feb 2018
Ojala T, Pietikainen M, Maenpaa T (2002) IEEE Trans Pattern Anal Mach Intell 24(7):971
Ray A, Chen Z, Gafford B, Gifford N, Kumar JJ, Lamsal A, Niehus-Staab L, Weinman J, Learned-Miller E (2018) Grinnell College, Grinnell, Iowa, Technical Report
Ronneberger O, Fischer P, Brox T (2015) In: International conference on medical image computing and computer-assisted intervention. Springer, pp 234–241
Ruiz-Lendínez J, Maćkiewicz B, Motek P, Stryjakiewicz T (2019) Surv Rev 51(365):123
Saalfeld A (1988) Int J Geogr Inf Sci 2(3):217

Saeedimoghaddam M, Stepinski TF (2020) Automatic extraction of road intersection points from USGS historical map series using deep convolutional neural networks. Int J Geogr Inf Sci 34(5):947–968. https://doi.org/10.1080/13658816.2019.1696968

Simonyan K, Zisserman A (2014) arXiv preprint arXiv:1409.1556

Song W, Keller JM, Haithcoat TL, Davis CH, Hinsen JB (2013) Photogramm Eng Remote Sens 79(6):535

Stanford University Library – David Rumsey Map Center (2018) David Rumsey Map Collection. https://www.davidrumsey.com. Online; Accessed on 15 Oct 2018

Swisstopo (2018a) Historical maps of Switzerland. https://www.swisstopo.admin.ch/en/knowledge-facts/maps-and-more/historical-maps.html. Online; Accessed on 15 Oct 2018

Swisstopo (2018b) Maps of Switzerland. https://map.geo.admin.ch. Online; Accessed on 15 Oct 2018

Tavakkol S, Chiang YY, Waters T, Han F, Prasad K, Kiveris R (2019) In: Proceedings of the 3rd ACM SIGSPATIAL international workshop on AI for geographic knowledge discovery, pp 48–51

Tsorlini A, Iosifescu I, Iosifescu C, Hurni L (2014) e-Perimetron 9(4):153

Uhl JH (2019) Spatio-temporal information extraction under uncertainty using multi-source data integration and machine learning: applications to human settlement modelling. Ph.D. thesis, University of Colorado at Boulder

Uhl JH, Leyk S, Chiang YY, Duan W, Knoblock CA (2017) IET Image Process

Uhl JH, Leyk S, Chiang YY, Duan W, Knoblock CA (2018) ISPRS Int J Geo Inf 7(4):148

Uhl JH, Leyk S, Chiang YY, Duan W, Knoblock CA (2018) IET Image Process 12(11):2084

Uhl JH, Leyk S, Chiang YY, Duan W, Knoblock CA (2019) IEEE Access 8:6978–6996

U.S. Library of Congress (2018a) Sanborn fire insurance maps. http://www.loc.gov/rr/geogmap/sanborn/san6.html. Online; Accessed on 28 Feb 2018

U.S. Library of Congress (2018b) Sanborn fire insurance maps. http://www.loc.gov/rr/geogmap/sanborn/. Online; Accessed on 28 Feb 2018

U.S. Library of Congress (2018c) Sanborn fire insurance maps. https://www.loc.gov/item/prn-17-074/sanborn-fire-insurance-maps-now-online/2017-05-25/. Online; Accessed on 28 Feb 2018

VoPham T, Hart JE, Laden F, Chiang YY (2018) Environ Health 17(1):40

Weinman J (2013) In: 2013 12th international conference on document analysis and recognition. IEEE, pp 1044–1048

Weinman J (2017) In: 2017 14th IAPR international conference on document analysis and recognition (ICDAR), vol 1. IEEE, pp 957–964

Yamada H, Yamamoto K, Hosokawa K (1993) IEEE Trans Pattern Anal Mach Intell 15(4):380

Yu F, Koltun V (2015) arXiv preprint arXiv:1511.07122

Yu R, Luo Z, Chiang YY (2016) In: 2016 23rd international conference on pattern recognition (ICPR). IEEE, pp 3993–3998

Zhou X, Li W, Arundel ST, Liu J (2018) arXiv preprint arXiv:1805.10402

Part V
Governance, Infrastructures and Society

Chapter 21
The Integration of Decision Maker's Requirements to Develop a Spatial Data Warehouse

Sana Ezzedine, Sami Yassine Turki, and Sami Faiz

21.1 Introduction

Nowadays, technologies such as positioning systems and Internet make it easier to produce and access to geographic information. During the recent years, this fact led to an increasing availability of diverse, heterogeneous and distributed spatial data sources. Those sources contain information collected at different times and use different techniques to aliment spatial data warehouses (SDWs) . The specificities of the SDWs are: (1) The nature of the spatial data requires taking into account possible incompatibilities: the spatial reference (position, shape, orientation, size), the reference systems, in the measure units, the spatial uncertainty, the precision, the size, etc. (2) Other elements to be considered in a warehouse of spatial data: the topology, the spatial integrity constraints, the consistency between scales, etc.

The architecture of a SDW is divided on five layers (Rifaie et al. 2009): The first layer consists on the gathring of data sources: external and internal ones.

Internal data is captured and maintained by operational systems inside an organization and external data refers to those that originate outside an organization.

The second one is the integration of heterogeneous data using the Extract Transformation Loading (ETL).

Extraction is the process of identifying and collecting relevant data from different sources. These data will go through the transformation and the cleansing process. Transformation is the process of converting data using a set of business rules (such as aggregation functions) into consistent formats for reporting and analysis. At last the data in staging area are loaded into target repository.

S. Ezzedine (✉) · S. Y. Turki · S. Faiz
LTSIRS Laboratory, National School of Engineers, Tunis, Tunisia
e-mail: yassine.turki@isteub.rnu.tn; sami.faiz@insat.rnu.tn

M. Werner, Y.-Y. Chiang (eds.), *Handbook of Big Geospatial Data*,
https://doi.org/10.1007/978-3-030-55462-0_21

The third layer defines the structure of the multidimensional model related to a SDW; in this layer the facts that contain the business metrics (i.e. measures) and the dimensions which describe the facts and their contexts are described. The next layer is the customization layer that encourages the construction of a preaggregate data cube for the different analysis tools.

The last layer is the application layer which defines the applications used by the decision maker. Tt is based on (OLAP), Data Mining, reporting tools and other techniques.

To design a Spatial Data Warehouse (SDW), many approaches were proposed in the literature. These approaches did not propose a standard framework and did not integrate spatial and non-spatial the decision maker's (DM) requirements to design and to implement a SDW. Bimonte et al. integrated spatial information and ensured correct aggregation over spatial measures (Bimonte et al. 2008). A multidimensional analysis tool was defined by another work which modeled spatial data in a SDW (Gomez et al. 2007; Malinowski and Zimanyi 2007). Alternatively, the authors (Da Silva et al. 2007) defined a query language that allowed the use of multidimensional, spatial and topological operators such as GeoMDQL (Rivest et al. 2009). Not all these approaches defined transformations between the conceptual level of an SDW and its implementation. Moreover, they did not suggest an automatic transformation from the SDW design to the possible implementation representation. Furthermore, they did not integrate spatial requirements of a group of DMs.

To overcome these limitations, most approaches use the standard framework, the Model Driven Architecture (MDA). MDA separates independent models and platform-specific models using transformation techniques. Glorio and Trujillo (2008) proposed an alignment of multidimensional spatial model with MDA. To design an SDW, the second MDA model, the platform independent model (PIM) is used. (Glorio and Trujillo 2009) extended this approach to include spatial data in the SDW design level. They define the geographical queries of DMs independently of the implementation presentation. Mazon and Trujillo (2009) defined some spatial elements describing the top DMs goals and aims. To model both spatial and non spatial data in the SDW design, Fidalgo and Cuzzocrea (2012) used a case tool based on unified modeling language (UML) standard and propose to generate automatically the data and the analysis models. They focused on the use of transformations based on MDA.

These approaches outline some limits since they do not consider all the spatial specifications of DMs such as topological relationships, projection mode, spatial presentation and other DM's requirements. Furthermore, the different presented conceptual models in the literature are lacking the integration of all DMs' spatial requirements in the SDW design.

The present work offers a new approach that takes into account the principal DM's requirements. This approach defines multiple schemas, which are appropriate to DMs to design and implementation of the SDW.

To reach this finality, four main steps are to be followed: the first one is to identify requirements' models containing an outline of published user's profile models and spatial needs collected from Geographic metadata. The second step is to classify these models into clusters using an appropriate algorithm. The third step is to use the MDA to present requirement elements to design a SDW. The last one is to pass from the design model to the implementation model using rules.

This chapter is organized as follows: in Sect. 21.2, the authors describe the existing approaches in the literature. In Sect. 21.3, they state the main steps of the approach. In Sect. 21.3, they define the first model presenting requirements. They describe the extended clustering algorithm in Sect. 21.4. In Sect. 21.5, this work presents the conceptual model with the formal design model and the most relevant transformations between the requirements' model and the design one of the SDW. In Sect. 21.6, the authors define the implementation model. Then, they present transformations between the SDW design model and the implementation model. In addition, the experimentation is described, and the proposed approach is illustrated through a case study dealing with the design and the implementation of a SDW in a relational platform. The SDW is related to a sales manager. Finally, the authors draw their conclusion and projects on the future work.

21.2 Overview of the Existing Approaches

The methods, proposed in the literature, to design an SDW are not standard and do not take into account of all needs and profiles of the DMs.

The Table 21.1 present the approaches proposed to integrate the spatial context in a Data Warehouse (DW).

A set of solutions are presented in Table 21.2 to overcome the limits described in Table 21.1. The approaches have some disadvantages that the authors try to remedy in their proposal.

The proposed approaches do not take into account spatial and non-spatial DM's requirements of a SDW. Add to that, they do not supply a personalized SDW to every DM. Consequently, the multidimensional design model is very general and does not introduce the spatial requirements and the DM profile.

21.3 Overview of the Proposal

Several approaches provide conceptual models in order to obtain the SDW design. However, the study of the existing literature reveals that the integration of DM's requirements is not developed enough neither for descriptive nor for spatial requirements. Moreover, current methods offered to design a SDW, define a single schema for all DMs (Glorio et al. 2010). The existing systems are not able to satisfy all the DMs' contexts and the proposed SDW is not appropriate for every DM.

The authors aim consists in designing and implementing the SDW for a set of DMs. The present framework describes the spatial and non-spatial requirements. The first standard model of MDA, Geographic Computation Independent Model (GeoCIM), is used to model these requirements. Then the framework allows the classification of GeoCIMs, corresponding to requirements models and related to

Table 21.1 The integration of the spatial context in a DW

Approach and paper	Description	Advantages	Disadvantages
Toward better support for spatial decision-making: Defining the characteristics of spatial on-line analytical processing (Rivest et al. 2009).	Rivest et al. define the spatial measures and represent these measures as spatial objects or spatial metrics or topological operators.	This approach gives a precise definition of spatial measures that combines the definition of spatial objects, sptatial metrics and topological operators.	They do not define a standard notation for the design model. The transformations produce logical and physical models from the design model. In addition, the generation of models is not totally automatic. The use of the additional complex structures to support spatial data decreases the simplicity of the model.
A Metamodel for the Specification of Geographical Data Warehouses. Representing spatiality in a conceptual multidimensional model (Malinowski and Zimanyi 2004).	Malinowski et al. integrate spatial dimensions, measures and facts into a concise model supported by a diagram named MAD.	The spatial dimensions and facts in a SDW are presented by a diagram.	The personalization is lacked in the conceptual model. The absence of transformations between the spatial multidimensional model and the implementation model of a SDW.
Geocube, A multidimensional model and navigation operators handling complex measures: Application in spatial olap; (Bimonte et al. 2006a, b) Integration of geographic information into multidimensional models; (Bimonte et al. 2008)	Bimonte et al. (2006a) propose a multidimensional model named GeoCube model, which integrates the geographical information and allows their accumulation. Then they present the GeWOlap model, which supports tools for the spatial OLAP. They propose the extension of the traditional spatial dimensions to support the complexity of the navigation in a map.	An approach which integrates the space in the construction of the multidimensional model. The authors propose a model that supports the navigation in a map.	An approach, which does not present an appropriate SDW for a DM. The absence of mechanisms which introduce measures of spatiality according to the needs of a DM.

Spatial aggregation: Data model and implementation; (Gomez et al. 2007).	Gomez et al. define a formal model for the representation of the spatial data and a multidimensional tool for the analysis of the spatial data.	The use of a formal model to present and analyze spatial data in the multidimensional model of a SDW makes the approach standard.	This proposal includes the spatiality in the multidimensional model without considering DMs requirements.
Spatio semantic SOLAP approache; (Aissi et al. 2015).	Aissi et al. propose an approach, which helps the users to extract the relevant information by means of a personalized request. The approach detects implicitly the preferences and the needs for the users by using a spatial and a semantic measure.	The proposal allows users to develop a personalized SDW taking into account of his needs and profile.	The approach does not integrate the preferences of the user into the design nor into the implementation of the SDW. They gather users having only similar needs in terms of regions; they do not consider other requirements.

Table 21.2 The integration of a DM profile and his requirements in a SDW using MDA

Approach and paper	Advantages	Disadvantages
Using Web based Personalization on Spatial Data Warehouses (Glorio et al. 2010).	The approach integrates a set of DM requirements in a SDW (spatial context, profile).	The realized integration does not take into account all the spatial and non-spatial requirements.
A personalization process for spatial data warehouse development (Glorio et al. 2012).	The proposal defines both the conceptual model and the implementation one of a SDW. The models describe important spatial criteria which is the geometry of the spatial object.	The transformations between the models do not take into account all the spatial characteristics such as the topologic relations, the scale, etc. They take into account only the geometry of the spatial objects.
Conceptual Model for Spatial Data Cubes (Boulil et al. 2015).	The approach is formal and based on the UML standard.	The absence of an automatic integration of requirements and spatial users profiles into the design and the logical models of a SDW. A set of spatial requirements are lacked in this work.

different DMs, in clusters. The obtained clusters are used to align the SDW design with the standard model Geographic Platform Independent Model (GeoPIM). Then a Geographic Platform Specific Model (GeoPSM) model is derived from the GeoPIM to present the SDW implementation.

Figure 21.1 presents the different transformations between Geographic MDA models.

A set of transformations are established to derive the GeoPIM from the GeoCIM corresponding to the centroid of each cluster as shown in Fig. 21.2. For each cluster, a unique multidimensional model is adopted to design the SDW.

This work uses also the architecture MDA in order to obtain formal, automatic and understandable transformations to pass from one GeoPIM to the corresponding GeoPSM.

To conclude there are five stages:

- The GeoCIM definition: the target of this step is a requirements model that describes DM requirements and their profiles.
- The classification of GeoCIM models: this section provides k clusters from n GeoCIMs models ($k <= n$). Every cluster contains DMs with similar requirements.
- The transition from the clusters to the design model (GeoPIM): the target of this step is the development of a GeoPIM for every cluster.
- The update of the design model: The GeoPIM is enriched when a new spatial DM's requirement appears.
- The transition from the clusters to the design model (GeoPIM): this stage provides the implementation model named GeoPSM.

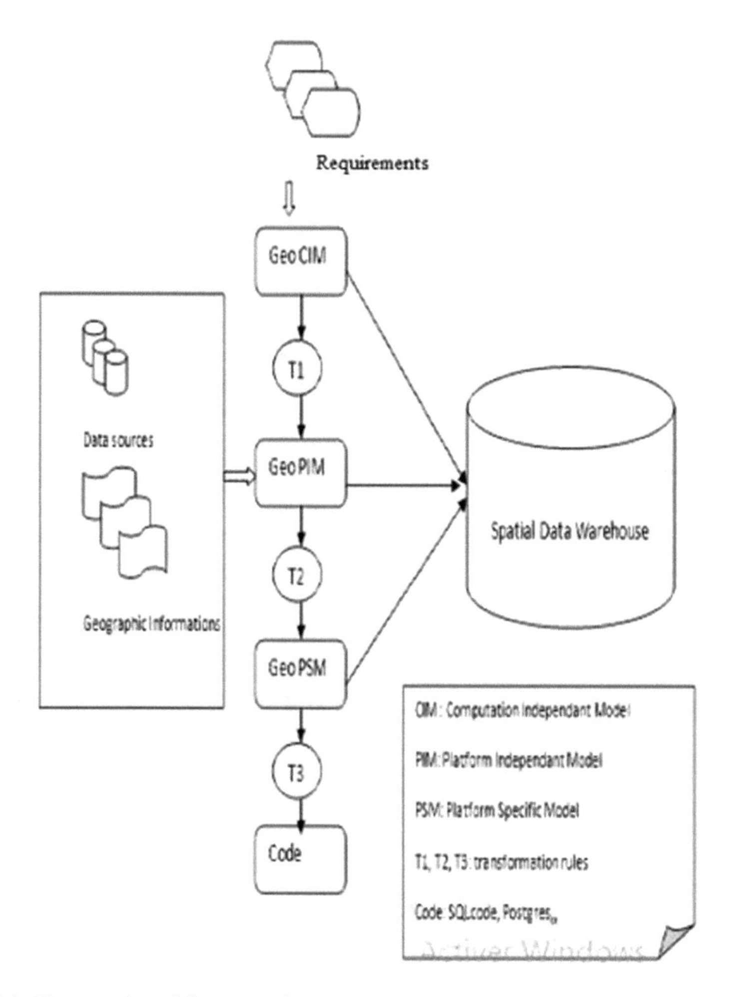

Fig. 21.1 The overview of the approach

As shown in Fig. 21.1, five GeoPIMs should be created, corresponding to the five layers of the SDW architecture (Rifaie et al. 2009). The first layer is the Geographic Source layer that consists on operational and external data source. The second one is the Geographic Integration layer that uses Extract Transformation Loading (ETL), which focuses on the integration of heterogeneous data sources. The third layer defines the structure and the multidimensional model of a SDW. It is based on facts that contain the business metrics (i.e. measures) and dimensions, which describe the facts and their contexts. The next layer is the Customization Layer that encourages the construction of a pre-aggregate data cube for the different analysis tools. The last layer is the application layer. It defines applications used by the end user. This layer is based on OLAP, Data Mining, reporting tools and other techniques.

In the present work, the authors focus on the third layer describing the multi-dimensional conceptual model. This approach allows also the transition from the

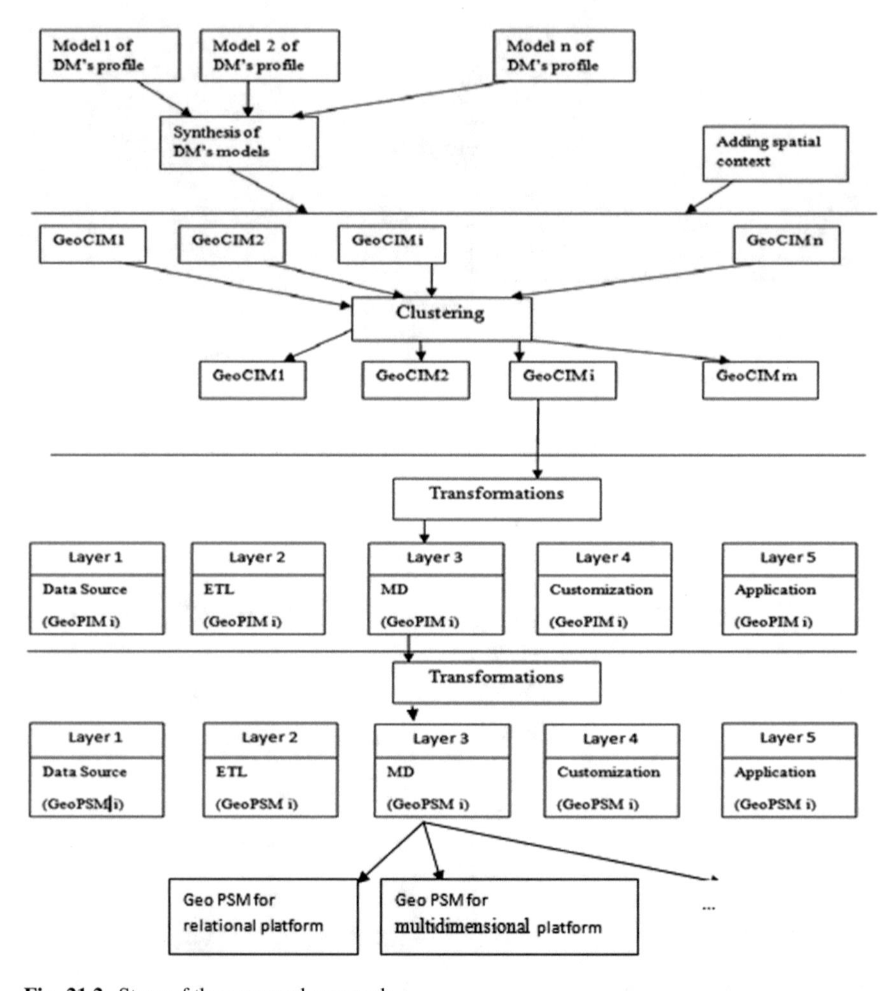

Fig. 21.2 Steps of the proposed approach

design model, GeoPIM, to different platforms. Authors choose in this work to define transformations between the GeoPIM and the relational platform, the most commonly used platform.

Table 21.3 shows the main characteristics of the proposal and its advantages compared to SDW design approaches existing in the literature. The most approaches presented in the literature lack of several points, which are required by a DM. Firstly, the approaches in the literature, define a SDW for users without considering their profiles and their spatial context. The spatial context define DM's requirements corresponding to the spatiality of the used objects like the scale of the spatial objects, the system coordinates, the projection mode, etc. Consequently, these works do not propose an appropriate SDW neither for a DM nor for a group of DMs having similar spatial and non-spatial requirements. In addition, the approaches do not integrate requirements into the implementation model. Finally, these works do not

Table 21.3 Comparison between approaches of modelling SDWs

	Approaches			
Criterion	Glorio et al. (2012)	Fidalgo and Cuzzocrea (2012)	Others approaches	This proposal
Standard framework	Yes	Yes	No	Yes
Spatial Attributes integration	Yes	Yes	Not all	Yes
Non spatial requirements	Not all	No	No	Yes
Spatial requirements	Not all	No	No	Yes
Multiple SDW conceptual models for the same query	No	No	No	Yes

offer an automatic transition from the design model to the implementation model. To resolve these limits, the authors propose a personalized SDW for a group of DMs.

Based on these Criteria, the present work offers a solution containing (1) a list of spatial and non spatial requirements integrating in requirements models (2) a classification of requirements models in clusters (3) a set of transformations to pass automatically from requirements models to the conceptual model of each cluster (4) transformations from the conceptual model to the implementation one for each cluster.

In the following sections, a detailed description of each step of the approach is presented.

21.4 GeoCIM Definition

Several context-aware approaches were proposed to model a user's profile in diverse application fields such as the search of information in digital theses (Bohé and Rumpler 2007), the bridging of the gap between the existing Internet content and today's heterogeneous computing environments (Naderi and Rumpler 2007) and the design of multi channel Web applications and the engineering design through the interactive goal programming (Kharrat et al. 2011). Recently, Glorio et al. (2012) presented a modeling approach for a personalized SDW by providing two design artifacts: a spatial aware user model, which defines the user's information needed for personalization and a set of spatial personalization rules specifying the required personalization actions. The spatial aware user model is lacked some important elements such as the presentation attributes, the semantic attributes and the spatial cover attributes. Furthermore, this framework did not allow an automatic transition from the user model to the SDW design.

The first step of the present approach is to define all spatial requirements according to every DM using a GeoCIM. The UML profile shown in Fig. 21.3 contains the necessary classes describing the main spatial and non-spatial context

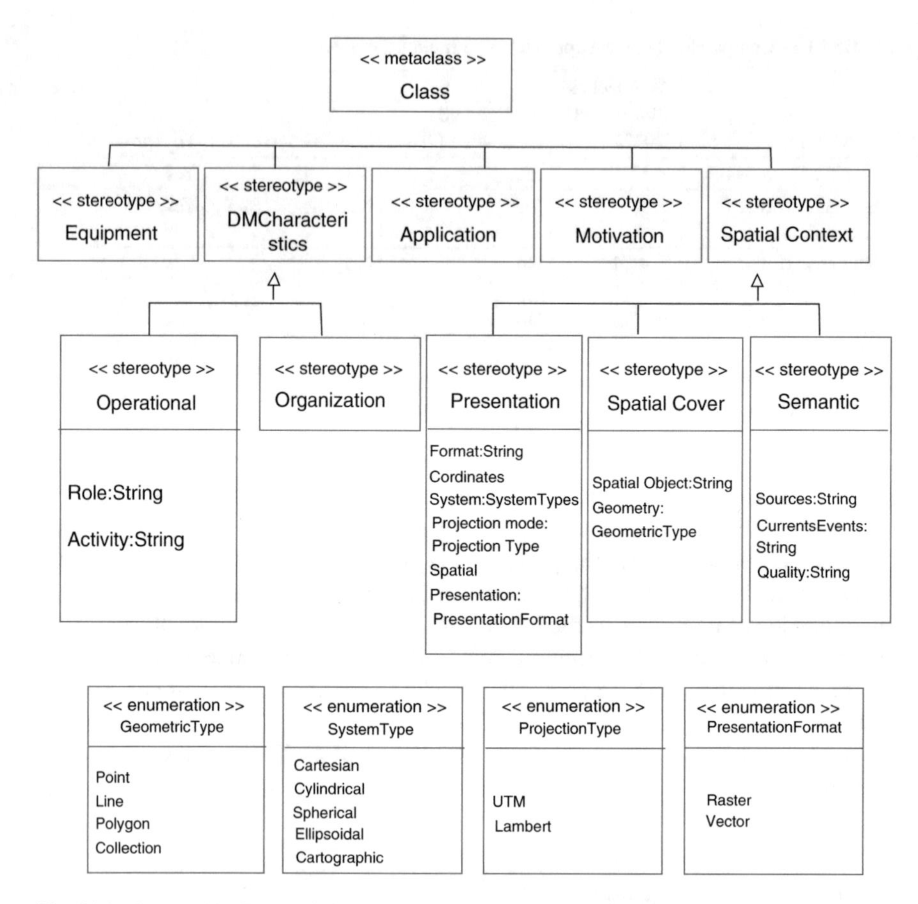

Fig. 21.3 Geographic CIM Definition

at the requirements level. Specifically, the structural properties of GeoCIM are presented by means of stereotypes. These stereotypes correspond to an inbuilt mechanism ensuring logical extensions or modifications of the meaning, the display and the syntax of a model element. Descriptive requirements are synthesized from requirements models existing in the literature presented previously. The authors keep the common descriptive elements from the studied user profiles and remove the stereotypes that are not necessary for a SDW design such as the personal context (name, language and phone) and the physical context (location and time). The description of spatial requirements is subsequently achieved by the use of geographic metadata.

The non spatial stereotypes described in this work are: (1) the 'Equipment' that describes the equipment needed to accomplish the user's needs, (2) the 'DMCharacteristics' which details the DM's organization where he belongs, his activity and his role in this organization, (3) the 'Application' which describes different tasks and goals that the DM has to achieve and (4) the 'Motivation' that describes internal and external factors to stimulate the DM's goals. The spatial context stereotypes contains: (1) the 'Presentation' stereotype describes the data formats needed (XMI, Geographic Data Bases, etc), the system coordinates (cartesian, cylindrical, etc), the projection mode (lambert, UTM) and the presentation attribute (raster, vector), (2) the 'Spatial Cover' which enumerates the spatial objects and their geometries (line, point, etc) and (3) the 'Semantic' stereotype with sources attribute to indicate the origin of data sources, the Current Events attribute which describes the date when data are taken and the quality which show the data quality.

All the geometric types (point, line, polygon and collection) are grouped in an enumeration element called GeometricTypes. The authors define also all projection types in the enumeration element called ProjectionType such as Lambert and UTM. Thus, they present all the Presentation Formats as Raster and Vector in the enumeration element PresentationFormat. In addition, they describe System Coordinates Types as Cartesian, Cylindrical, Spherical, Ellipsoidal and Cartographic.

21.5 Classification of the GeoCIMs Models

In the previous section, a GeoCIM for every DM is designed. However, the conceptual model of a SDW can be dedicated to a group of GeoCIMs related to a set of DMs. In this section, the authors group similar GeoCIMs in one cluster. Each cluster will be used for the design of a SDW.

To resolve the problem of classification, it is necessary to use a clustering algorithm. The most known ones are k-means and k-modes. K-means algorithm processes the numerical data and k-modes algorithm uses both numerical and categorical ones (Aranganayagi and Thangavel 2010) but none can process spatial data (Broda and Mazur 2012). In this work, the authors extend k-means in order to use both categorical and spatial data. The extension is made through a new similarity measure that processes numerical, categorical and spatial data.

First, the authors choose k-centroids, as shown in Algorithms 21.1 and 21.2. The k-centroids contain GeoCIMs of k DMs. The classification is not limited to spatial objects of each Geo CIM; it takes into account adjacent objects existing in other clusters. The objective is not to have a constraint-based spatial clustering (Pattabiraman and Nedunchezhian 2012), but to make it possible to gather DMs working in adjacent territories. Adjacent spatial objects as defined by Egenhofer and Franzosa (1991) are objects with external, internal, intersection, equality or inclusion relations with the required spatial objects.

Then, the authors assign every GeoCIM to the cluster having the most similar centroid. The cluster's centroid with the higher number of spatial objects and adjacent spatial objects in common with the GeoCIM is adopted. If there are non-common spatial objects, a new cluster is created containing the GeoCIM of this DM. In the case where more than one cluster is similar to the spatial objects of DM, the authors consider the similarity between the rest of DM's requirements and all requirements of the centroid before assigning the GeoCIM to the most similar cluster.

As shown in Algorithm 21.1, resemblance is a function, which counts the number of times that elements of DM's spatial cover equal elements of centroid's spatial cover. The term D in this algorithm corresponds to the density related to every DM. It contains the result of the function resemblance divided by the number of spatial objects of the DM's centroid. The assigning GeoCIMs in clusters is repeated after updating the centroid of each cluster until centroids do not change.

Algorithm 21.1 Assigning Geo CIMs into clusters **Function Assign (clusters, users)**

BEGIN

1. Repeat
2. Extract the adjacent objects of a user and add them to the spatial cover (Spatial cover requirements = spatial cover of the user + adjacent spatial cover)
3. i ← i + 1
4. Repeat
5. Resemblance (i) ← 0
6. For every spatial object_user
7. For every spatial cover centroid (i)
8. If spatial object_user = spatial cover centroid (i) then
9. Resemblance (i) ← Resemblance (i) + 1
10. END If
11. END For
12. END For
13. D (i) ← Number of object_user - Resemblance (i) / Number of object_user
14. i ← i+1
15. Until (i> number of cluster)
16. If (D=1) then
17. K ← k+1
18. Assign the user to the new cluster
19. ELSE
20. If there is a single cluster with minimal D(i) then
21. Assign the user to the cluster i
22. ELSE
23. Assign the user to the cluster having minimal D (i) and more common requirements
24. END If
25. Until all users are classified

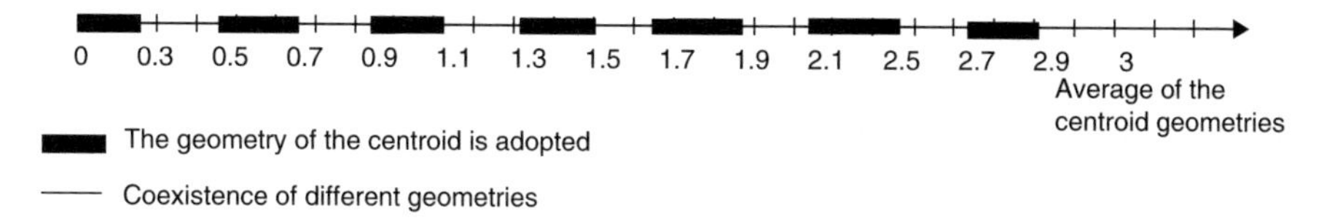

Fig. 21.4 Adopted geometries for spatial objects

The purpose of the clustering part is to group similar users i.e. users who work on neighboring spatial objects or working on the same objects but with different geometries.

When users are working on the same object but with different geometries, we propose an algorithm to present the different geometries at the centroid that will present the GeoCims group. For that we have just assigned a number to each geometry and then choose intervals for which we keep the geometry otherwise we adopt the coexistence of geometries.

To resolve the problem of the geometry to be adopted for every spatial object in the centroid, the authors assign 0 to the point geometry, 1 to the line geometry, 2 to the polygon geometry and 3 to the collection geometry. Then, they calculate the average of the different geometries of each spatial object of the centroid. To determinate these intervals, the authors take a set of geometries of spatial objects and calculate the probability to have the geometry of the centroid. If the probability exceeds 0.5 the authors adopt the geometry of the centroid. Otherwise, they adopt the coexistence of different geometries for the same spatial object.

If the average belongs to [0,0.3], [0.5,0.7], [0.9,1.1], [1.3,1.5], [1.7,1.9], [2.1,2.5], [2.5,2.7] or [2.9,3], the authors adopt the geometry of the centroid described by the continuous line in Fig. 21.3, else they adopt the coexistence of all geometries of the same spatial object of GeoCIMs existing in the same cluster (described by the dotted line in Fig. 21.4).

Outputs of this algorithm are centroids representing clusters. These centroids are used to design the SDW for each cluster.

21.6 K = Random Number of the Clusters Containing Adjacent Objects

The authors consider clusters containing spatial objects having an external connection, an internal connection, an intersection and equality with the spatial objects required

Algorithm 21.2 The extended k-means Algorithm

1. Repeat
2. Assign (users, clusters)
3. Update the centroid

4. Until centroids do not change
5. For each cluster (i) do
6. For each spatial_cover_object_centroid (i) do
7. Average = $\sum$ existing geometries of this elements in cluster (i)/number of occurrences

   Of this element in cluster (i)

8. If average [0, 0.3] or [0.5, 0.7] or [0.9, 1.1] or [1.3, 1.5] or [1.7, 1.9] or [2.1, 2.5] or [2.7, 2.9] then one geometry is presented which is the geometry of centroid's element
9. ELSE one geometry is presented which is the centroid's element
10. END If
11. END For
12. END For

The complexity of the k-means algorithm is O (KNIS), where: I is the number of iterations of the algorithm, S is the complexity of calculating the similarity, K is the number of clusters and N is the number of objects that will be classed.

21.7 From GeoCIM to GeoPIM

21.7.1 GeoPIM Definition

In this proposed approach, the GeoPIM presents the SDW design. This model describes the conceptual level of a SDW and hides all details related to a specific platform or technology that can be used later to implement the system. This work is based on the formalism defined by the PIM presented by Glorio (Glorio et al. 2010).

The most important and necessary stereotypes presented in this model are Facts (■) and Dimensions (⟀). The measure or the attribute presented in a Fact corresponds to FactAttribute (FA). With respect to dimensions, classes stereotyped as Base specify each aggregation level of a hierarchy (▰). The attributes corresponding to dimensions are stereotyped as DimensionAttribute (DA). A dimension has a description attribute stereotyped as Descriptor (D). An association between Bases is stereotyped as Rolls-up to (◎). The role R represents the direction in which the hierarchy rolls up, whereas D represents the direction in which the hierarchy drills down.

Glorio et al. (2010) introduced the spatial level (▱) as a hierarchy level with the attribute Geometry. They also introduced the spatial measure (*) to support multidimensional analysis for geometric objects. The same authors presented another element describing adjacent spatial objects. This element is stereotyped as Layer (▰).

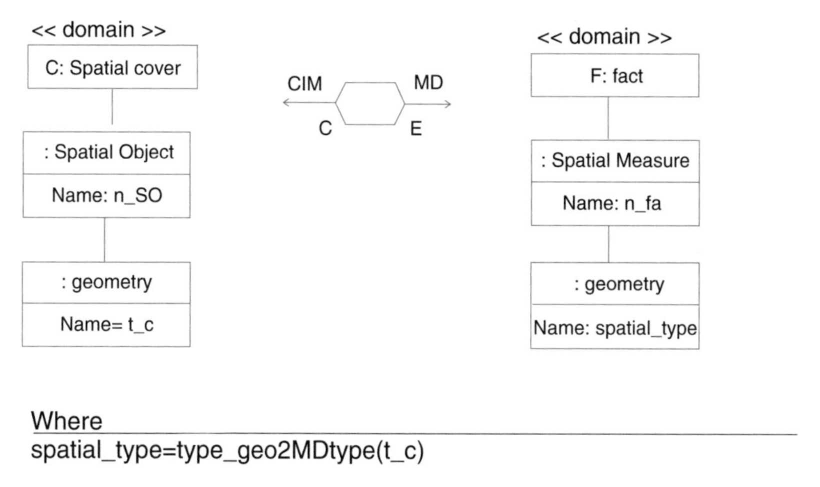

Fig. 21.5 Spatial Object 2 Spatial Measure Transformation

In this model, all the geometric primitives were grouped into an enumeration element named Geometric Types. In the present work, the authors use the same process to enumerate the spatial projection and the spatial presentation.

21.7.2 Formal Transformations from GeoCIM to GeoPIM

Defining formal transformations allows to automatically deriving every GeoPIM from the GeoCIM that represents the cluster's centroid. To perform transformations, the authors adopt the QVT with graphical notation that allows readable, understandable, adaptable and maintainable transformations.

The authors present in this section the most relevant transformations from the GeoCIM to the GeoPIM. The transformations are printed out in Figs. 21.5, 21.6, 21.7, 21.8, 21.9, 21.10, and 21.11. The authors do not provide a detailed description of each transformation, only the DMCharacteristics 2 Dimension relation is further explained. The remaining transformations are easily understood thanks to the readability of the QVT.

The QVT (Query / View / Transformation) in the MDA architecture is a standard for model transformations defined by the OMG (Object Management Group) in 2007. It is central to any proposed MDA. It defines a way to transform source models to target models.

We need QVT in our approach to pass from every element existing in the source model which is the GeoCIM or GeoPIM to the target one that can be GeoPIM or GeoPSM.

The graphical notation for the DM Characteristics 2 Dimension relation can be seen in Fig. 21.10. On the left hand side of this relation, the reader can see the source

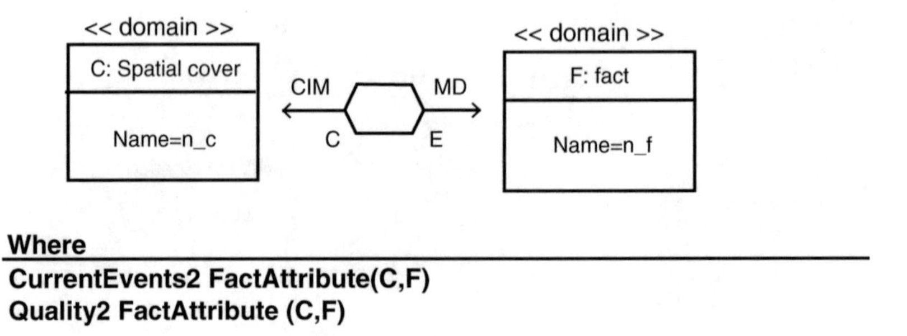

Fig. 21.6 Spatial cover 2 Fact Transformation

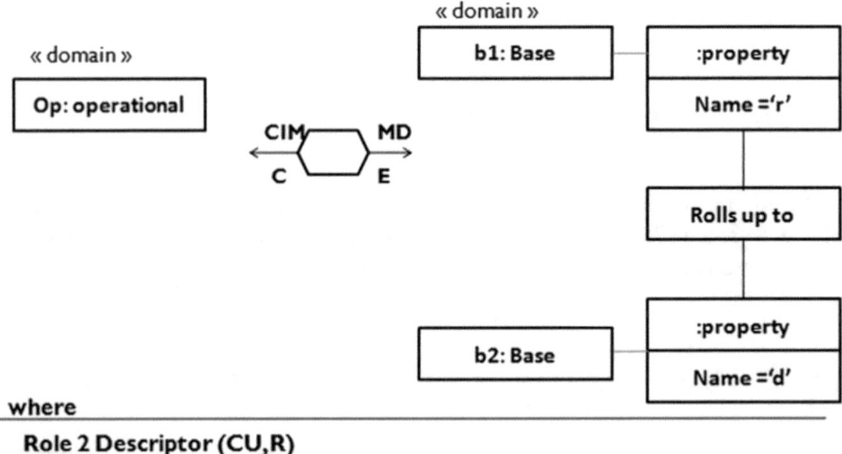

Fig. 21.7 Operational 2 Base Transformation

model, and the target model on the right hand side. The source presents the part of the GeoCIM that has to match with the part of the Geo PIM, which presents the target model. In this case, the authors use a set of elements from the UML profile that represents the DM Characteristics stereotype. The terminal level corresponds to the Dimension stereotype.

This relation determines the transformation in the following way: it is checked (C arrow) that the pattern on the left side (source model) exists in the Geo CIM. The transformation subsequently enforces (E arrow) and the following stereotypes (and their associations) are created according to the Geo PIM (MD in Fig. 21.10). A Dimension stereotype with the same name that Geo CIM component is obtained.

Once this relation is set up, the relations Organisation 2 Spatial Level, Operational 2 Base, Role 2 Descriptor and Activity 2 Descriptor must be performed (according to the where clause).

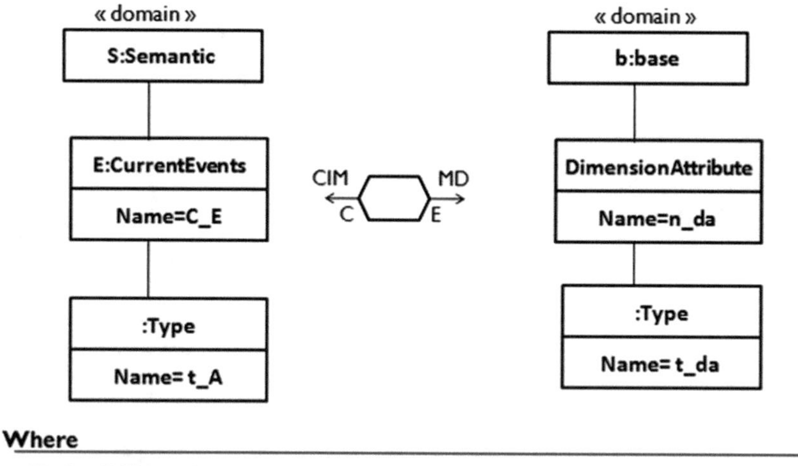

Fig. 21.8 Semantic 2 Base Transformation

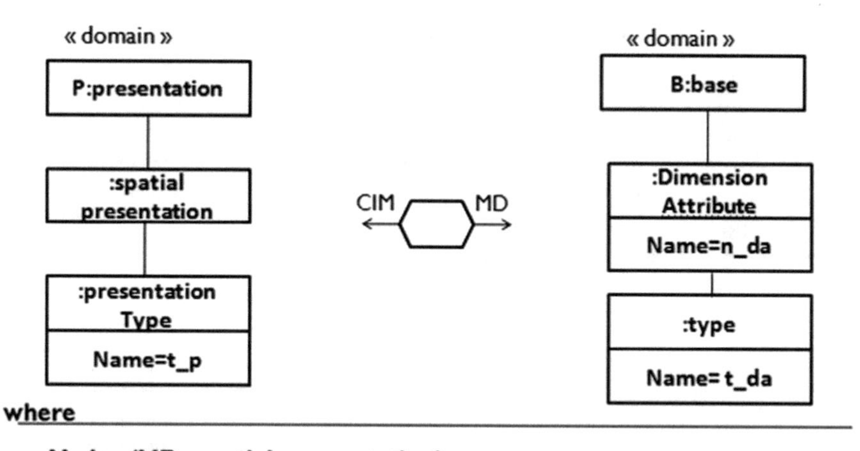

Fig. 21.9 Presentation 2 Base Transformation

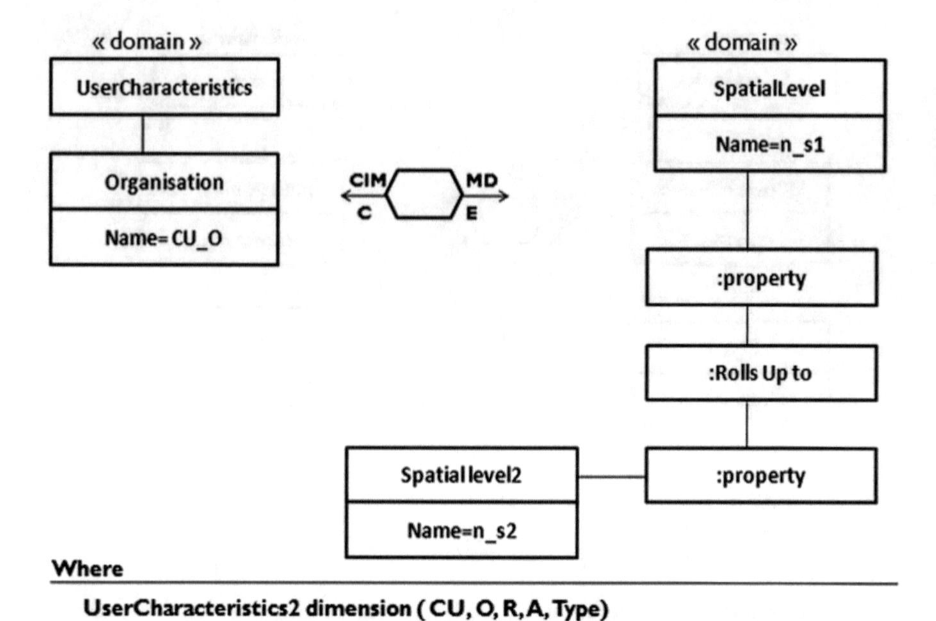

Fig. 21.10 Organization 2 Spatial Level Transformation

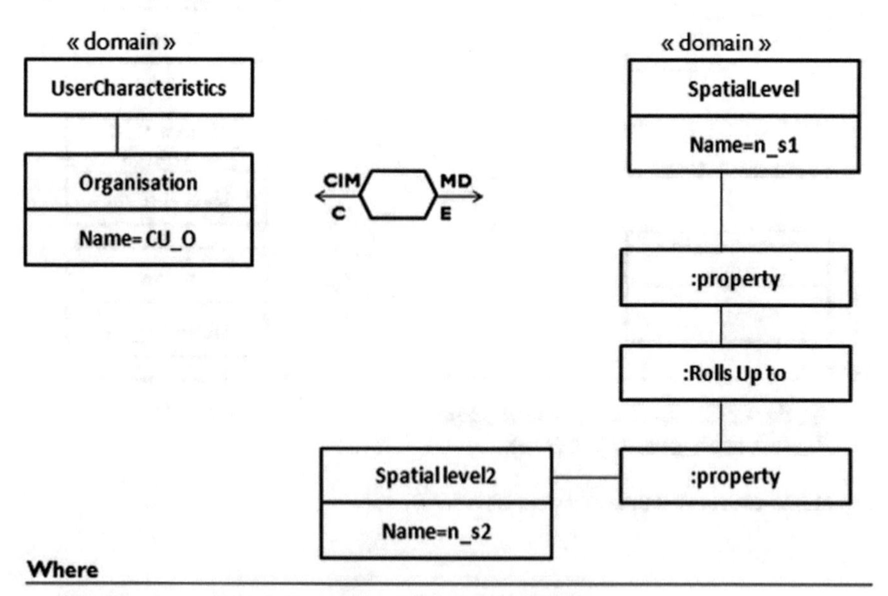

Fig. 21.11 DM Characteristics 2 Dimension Transformation

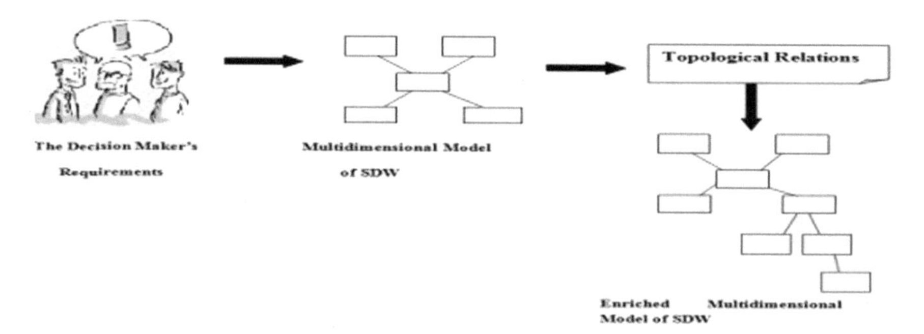

Fig. 21.12 Using Topological Relations to enrich the Multidimensional Model of SDW

Table 21.4 Spatial Hierarchy stereotype used to develop the conceptual model of a SDW

Stereotype	Description	Presentation
Spatial Hierarchy	Spatial Hierarchy present spatial Dimension Hierarchy with their attributes	SH

21.8 Using Topological Relationships to Enrich Dimension Hierarchies

The authors treat updating in terms of adding a new spatial requirement. They propose to enrich dimension hierarchies by adding new levels of aggregation in order to obtain the required hierarchies.

To accomplish this goal, they propose the use of semantic relations among spatial concepts provided by topological relationships (Egenhofer and Franzosa 1991). The initial hypothesis is that both SDWs and topological relationships present hierarchical structures: dimension hierarchies in SDWs show the relationships between value domains from different dimension attributes (levels of aggregation) (Bimonte et al. 2006a), while topological relations present hierarchical semantic relations between spatial concepts, such as adjacency or inclusion or intersection, etc. (Egenhofer and Franzosa 1991). Therefore, the present approach is based on using these topological relations to add new levels to dimension hierarchies in order to obtain the required hierarchies. Fig. 21.12 summarizes this scenario.

In this work, the authors define another stereotype based also on UML named Spatial Hierarchy, as shown in Table 21.4.

Spatial Hierarchy is added in the conceptual model of a SDW when the DM needs to take account of a new spatial requirement in the developing of the SDW.

The proposal consists of identifying topological relationships between existing dimensions in the conceptual model and the new added spatial requirements given by the DM.

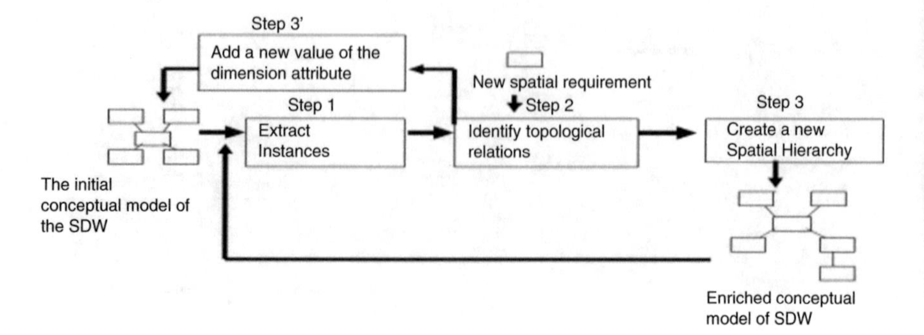

Fig. 21.13 Different Steps to update a SDW

With each identified topological relationship, the authors create a Spatial Hierarchy, which is named with the same name of the identified topological relationship and has as attributes the characteristics of the added requirement.

Following, the authors explain the main steps of the enriched design model of a SDW (an overview is shown in Fig. 21.13):

Prerequisite 1. A dimension attribute is chosen from the initial conceptual model of the SDW. The spatial hierarchy will be added starting from this attribute.

Prerequisite 2. The DM has proposed a new spatial requirement, which is in relation with instances of the dimension attribute chosen in the initial conceptual model.

Step 1. Extract different instances from the dimension attribute chosen from the initial conceptual model.

Step 2. Identify topological relationships between spatial objects recently required with spatial objects existing in the dimension attribute chosen.

Step 3. If there are relationships between the required spatial objects and the existed ones, a spatial hierarchy for every relationship is created having the same name as the topological relationship.

Step 3′. If there are no relationships between the required spatial objects and the existed ones, a new record is inserted in the selected dimension attribute without creating a new hierarchy.

In Fig. 21.12, every step of the enriched SDW's design model is illustrated. From a dimension or a dimension hierarchy in a multidimensional model, a dimension attribute is chosen. Then the topological relationships are identified between instances of the dimension attribute and the new requirement in order to create a new level of the spatial dimension hierarchy. If there are no relationships between added requirement and existed dimensions, a new record is inserted. Iterations are repeated until all required spatial objects are classified.

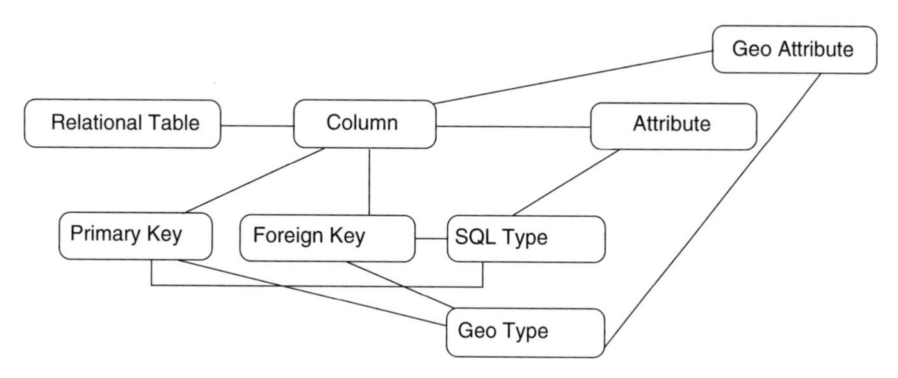

Fig. 21.14 Relational GeoPSM Definition

21.8.1 Geo SM Definition

As shown in Fig. 21.1, the authors can define a multiple GeoPSMs from a GeoPIM. It depends on the implementation platform. In this work, the authors model the GeoPSM with a geographic relational platform.

Figure 21.14 describes a GeoPSM based on the relational model. A relational model is composed of a set of relational table. A relational table has different columns, which contain attributes. Among the columns, there is a column primary key and a set of columns described the foreign keys if it is necessary. A column contains Geographic Attributes (Geo Attribute) and non-geographic attributes. Geo Attribute has a Geographic Type (GeoType). A column has a primary key and foreign key, which are defined with the SQL Type or Geo Type.

21.9 Transformations from GeoPIM to GeoPSM

Defining formal transformations allows to automatically deriving GeoPSMs from a GeoPIM. To perform transformations, the authors adopt the QVT with graphical notation. This section presents the most relevant transformations from the GeoPIM to the GeoPSM. The transformations are shown in Figs. 21.15, 21.16, 21.17, 21.18, 21.19, and 21.20. The authors provide a detailed description of Dimension 2 Relational Table relation. The remaining transformations are easily understood thanks to the readability of the QVT.

The graphical notation for the Dimension 2 Relational Table relation can be seen in Fig. 21.14. The source model is a part of a GeoPIM that has to match with the part of the GeoPSM. In this case, the authors use a set of elements from the UML profile that represents the Dimension Attribute stereotype. The terminal level corresponds to the relational table.

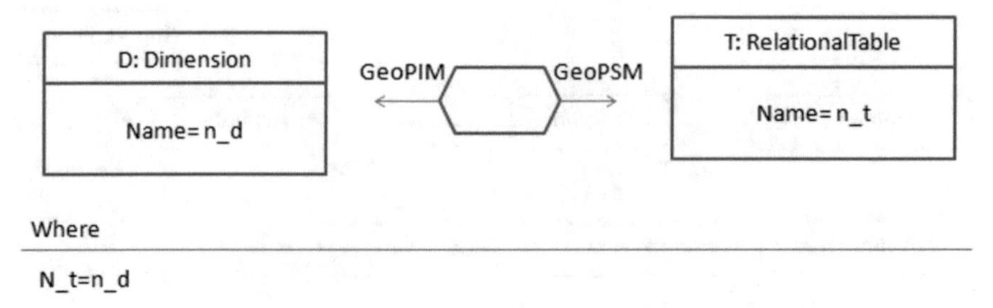

Fig. 21.15 Dimension 2 Relational Table

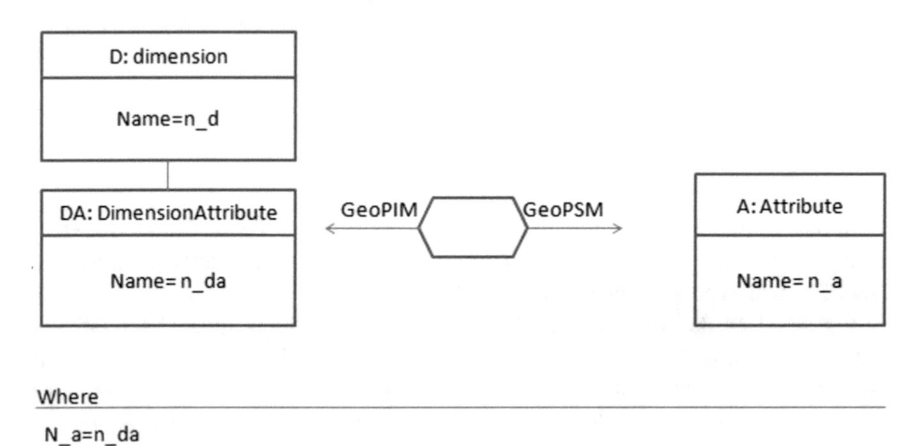

Fig. 21.16 Dimension Attribute 2 Attribute

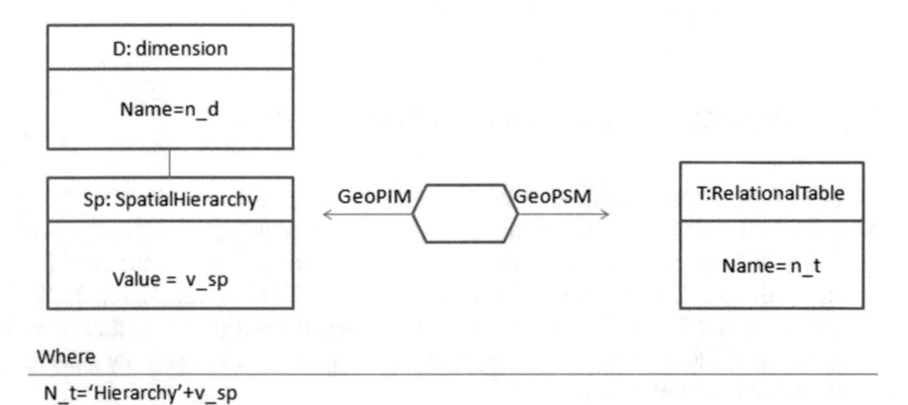

Fig. 21.17 Spatial Hierarchy 2 Relational Table

Figure 21.16 presents the transformation between DimensionAttribute which a source element and Attribute that presents a GeoPSM's element.

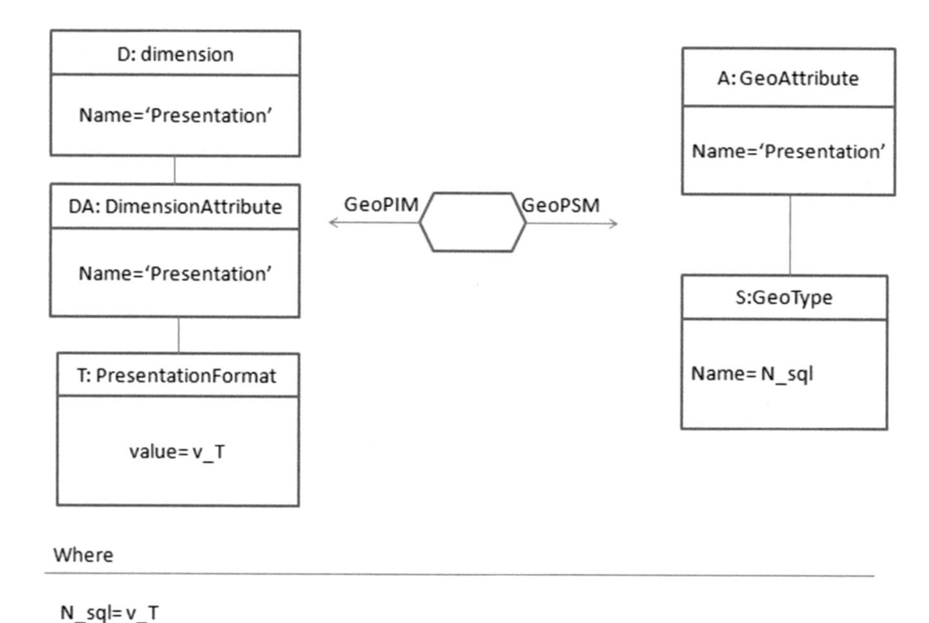

Fig. 21.18 Presentation Format 2 Geo Type

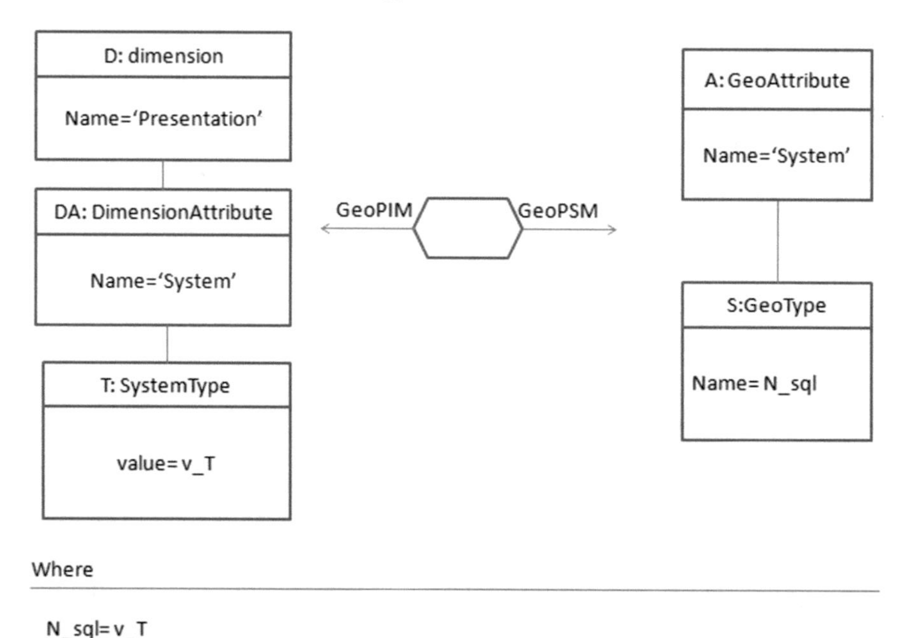

Fig. 21.19 System Type 2 Geo Type

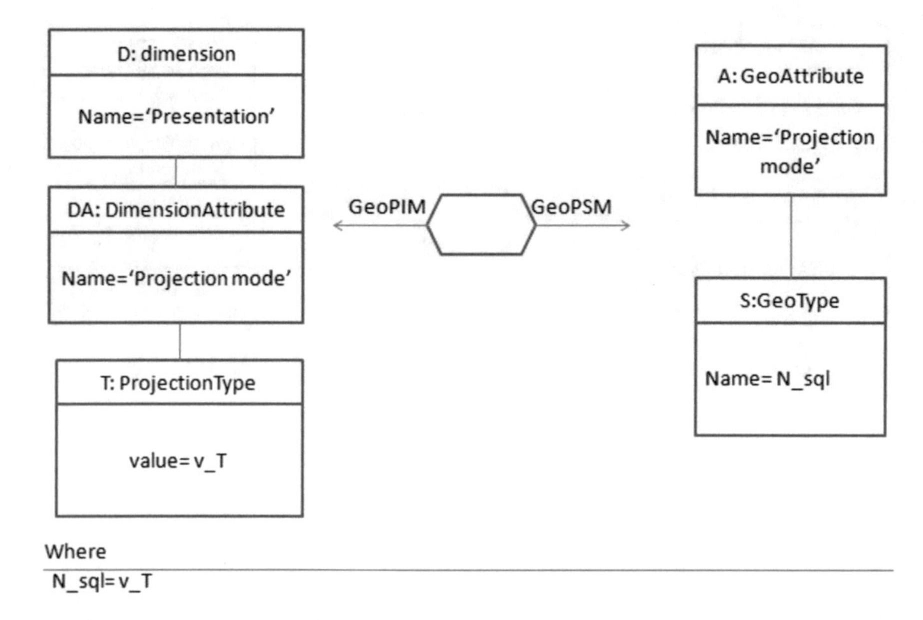

Fig. 21.20 Projection Type 2 Geo Type

Figure 21.17 describes the relation between l'élément Spatial Hierarchy de GeoPIM et Relational Table de GeoPSM. Every Spatial Hierarchy with the related attributes is transformed to a relational table.

Figure 21.18 defines a relation between Presentation Format and the related attribute value and the target element GeoType and the attribute name.

Figure 21.19 describes links System Type, value, Geo Type and name.

Figure 21.20 defines the transformation between the two parts. On the one hand the Projection Type with the attribute value and which are a GeoPIM's elements and on the other hand the GeoPSM's elements GeoType and name.

21.10 Experimentation

21.10.1 Transition from the Requirements Model to the Implementation Model of a SDW

This part summarizes the main rules of the passage between GeoCIM model and GeoPIM model described in a previous session.

- Rule 1: The transformation Spatial Cover 2 MD Fact

  The classe named Spatial Cover is transformed into the classe Fact.

```
rule SpatialCover2MDFact
{
        from
          cim: MRequirements!CouS
   to
        MD: MConceptual!Fact (
                name_f <- cim.name


                          )
        OwnedAttribute <- cim.ownedAttribute ->select (p|p.type.OclIsKindOf(Class!DataType)

   }
```

- Rule 2: The transformation Primitives2MDSpatialMeasure

  The attribute Primitive of the requirements model GeoCIM is transformed into the stereotype Spatial Measures of the design model.

```
Rule Primitives2MDSpatialMeasure

{From p : MRequirements!primitives (

To tp : MConceptual! SpatialMeasure

  (        Name <- p.name,

        SpatialType <-p.type,

)

}
```

- Rule 3: DM Characteristics 2 Dimension

```
rule DMCharacteristics2Dimension{
    from
            a: Mexigences! DMCharacteristics
          to
            out:: MConceptual!Dimension
                  (
                          name<- a.name+' DMCharacteristics''
                  )
}
```

Transition from the design model to the implementation model of a SDW

The authors describe in this part some transformations from the GeoPIM to the GeoPSM with the language Atlas Transform Language (ATL).

- Rule 1: Dimension 2 Relational Table

This rule allows passing from the design element Dimension to the element Relational Table of the implementation model.

```
rule Dimension2RelatinalTable
{
        from
         MD: MConceptual!Dimension
   to
        PSM: MImplementation! RelatinalTable (
              name_RT <- MD.name
                                   )
       OwnedAttribute <- MD.ownedAttribute ->select (p|p.type.OclIsKindOf(Class !DataType)

   }
```

- Rule 2: Dimension Attribute 2 Geo Attribute

The transformation Dimension Attriute 2 Geographic Attribute serves to transform the dimension attributes to the relational geographic attributes.

```
rule DimensionAttribute2GeoAttribute
{From p : MConceptual!DimensionAttribute (
To          tp : MImplementation!GeoAttribute
  (        Name <- p.name,
   )
}
```

- Rule3: Projection Mode 2 Geo Type

The rule transforms the attribute Projection Mode of the model GeoPIM to the item Geo type of the GeoPSM.

```
rule ProjectionMode2GeoType
{From p: MConceptual! Projection Mode (
To
        tp: MImplementation!GeoType
  (        Name <- p.value,
   )
}
```

- Rule 4: Spatial Hierarchy 2 Relational Table

Each stereotype named Spatial Hierarchy of the source model GeoPIM is transformed into the relational table in the target model GeoPSM.

```
rule SpatialHierarchy2RelationalTable
{From p: MConceptual! SpatialHierarchy (
To
        tp: MImplementation!RelationalTable
  (        Name <- p.value,
   )
}
```

- Rule 5: Base 2 Foreign Key

  A Base in the model GeoPIM is transformed to the foreign key stereotype.

```
rule Base2ForeignKey
{From p: MConceptual! Base (
 To
        tp: MImplementation! ForeignKey
 (          Name <- p.name,
   )}
```

The models GeoPIM and GeoPSM, automatically generated, are aligned to the common storage format XMI, the XML (Extensible Markup Language) Metadata Interchange.

Figures 21.21 and 21.22 show the GeoCIM and the GeoPIM in XMI format.

21.11 Case Study

In this case study, the authors apply the four steps of the proposal. First, they make three requirements models for three DMs having different spatial, descriptive needs and profiles. Then they classify them in clusters. Next, a set of QVT transformations are applied to generate the SDW design from every cluster.

Three different DMs contexts are chosen: the first DM is a sales manager who wants to analyze sales operations in stores situated 2 km around the airport (SM1). The second DM corresponds to the municipality, which aims to regulate goods transportation in streets that are 10 km around the airport. The last DM is a sales manager who wants to analyze sales in a city that contains the airport (SM2).

A GeoCIM is performed for each DM. As a result of the clustering step based on the implementation of the extended k-means algorithm, two clusters are obtained. The first cluster contains the municipality and SM1. The second cluster corresponds to SM2.

Subsequently, QVT transformations are used to move from the centroid of each cluster to a GeoPIM that model the SDW's design.

Figure 21.23 presents the GeoCIM with the geographic elements of the cluster's centroid corresponding to SM1.

Figure 21.24 presents the part of the obtained GeoPIM after applying several relations described in Figs. 21.5, 21.6, 21.7, 21.8, 21.9, 21.10, and 21.11. These transformations allow passing from Geo elements such as Application, Equipment,

```xml
<?xml version="1.0" encoding="ISO-8859-1"?>
<xmi:XMI xmi:version="2.0" xmlns:xmi="http://www.omg.org/XMI" xmlns:xsi="http://www.w3.org/2001/XMLSchema-instance" xmlns:ecore="http://www.eclipse.org/
  <ecore:EPackage name="PrimitiveTypes">
    <eClassifiers xsi:type="ecore:EDataType" name="String"/>
    <eClassifiers xsi:type="ecore:EDataType" name="Integer"/>
    <eClassifiers xsi:type="ecore:EDataType" name="Boolean"/>
  </ecore:EPackage>
  <ecore:EPackage name="MExigences">
    <eClassifiers xsi:type="ecore:EClass" name="Generalisation"/>
    <eClassifiers xsi:type="ecore:EClass" name="Classifier"/>
    <eClassifiers xsi:type="ecore:EClass" name="Materiel">
      <eStructuralFeatures xsi:type="ecore:EAttribute" name="nameMat" ordered="false" unique="false" lowerBound="1" eType="/0/String"/>
    </eClassifiers>
    <eClassifiers xsi:type="ecore:EClass" name="Application">
      <eStructuralFeatures xsi:type="ecore:EAttribute" name="description_Application" ordered="false" unique="false" lowerBound="1" eType="/0/String"/>
    </eClassifiers>
    <eClassifiers xsi:type="ecore:EClass" name="Motivation">
      <eStructuralFeatures xsi:type="ecore:EAttribute" name="description_Motivation" ordered="false" unique="false" lowerBound="1" eType="/0/String"/>
    </eClassifiers>
    <eClassifiers xsi:type="ecore:EClass" name="User"/>
    <eClassifiers xsi:type="ecore:EClass" name="Op" eSuperTypes="/1/User">
      <eStructuralFeatures xsi:type="ecore:EAttribute" name="Role" ordered="false" unique="false" lowerBound="1" eType="/0/String"/>
      <eStructuralFeatures xsi:type="ecore:EAttribute" name="Act" ordered="false" unique="false" lowerBound="1" eType="/0/String"/>
    </eClassifiers>
    <eClassifiers xsi:type="ecore:EEnum" name="Type_geometrique">
      <eLiterals name="point"/>
      <eLiterals name="ligne"/>
      <eLiterals name="polygone"/>
      <eLiterals name="relief"/>
      <eLiterals name="objet"/>
    </eClassifiers>
    <eClassifiers xsi:type="ecore:EClass" name="Orga" eSuperTypes="/1/User">
      <eStructuralFeatures xsi:type="ecore:EAttribute" name="name_Org" ordered="false" unique="false" lowerBound="1" eType="/0/String"/>
      <eStructuralFeatures xsi:type="ecore:EAttribute" name="geo" ordered="false" unique="false" lowerBound="1" eType="/1/Type_geometrique"/>
```

Fig. 21.20 GeoCIM in XMI

```xml
<?xml version="1.0" encoding="ISO-8859-1"?>
<xmi:XMI xmi:version="2.0" xmlns:xmi="http://www.omg.org/XMI" xmlns:xsi="http://www.w3.org/2001/XMLSchema-instance" xmlns:ecore="http://www.eclipse.or
  <ecore:EPackage name=" MConceptual ">
    <eClassifiers xsi:type="ecore:EClass" name="Dimension">
      <eStructuralFeatures xsi:type="ecore:EAttribute" name="nameD" ordered="false" unique="false" lowerBound="1" eType="/1/String"/>
    </eClassifiers>
    <eClassifiers xsi:type="ecore:EClass" name="Base">
      <eStructuralFeatures xsi:type="ecore:EAttribute" name="nameB" ordered="false" unique="false" lowerBound="1" eType="/1/String"/>
      <eStructuralFeatures xsi:type="ecore:EAttribute" name="Dimension" ordered="false" unique="false" lowerBound="1" eType="/1/String"/>
    </eClassifiers>
    <eClassifiers xsi:type="ecore:EClass" name="Property"/>
    <eClassifiers xsi:type="ecore:EClass" name="Descriptor" eSuperTypes="/0/Property">
      <eStructuralFeatures xsi:type="ecore:EAttribute" name="derivationRule" ordered="false" unique="false" lowerBound="1" eType="/1/String"/>
      <eStructuralFeatures xsi:type="ecore:EAttribute" name="Base" ordered="false" unique="false" lowerBound="1" eType="/1/String"/>
    </eClassifiers>
    <eClassifiers xsi:type="ecore:EEnum" name="GeometricTypes">
      <eLiterals name="point"/>
      <eLiterals name="ligne"/>
      <eLiterals name="polygone"/>
      <eLiterals name="relief"/>
      <eLiterals name="objet"/>
    </eClassifiers>
    <eClassifiers xsi:type="ecore:EEnum" name="Type_sys">
      <eLiterals name="cartesien"/>
      <eLiterals name="cylindrique"/>
      <eLiterals name="spherique"/>
    </eClassifiers>
    <eClassifiers xsi:type="ecore:EEnum" name="Type_projection">
      <eLiterals name="UTM"/>
      <eLiterals name="Lambert"/>
    </eClassifiers>
    <eClassifiers xsi:type="ecore:EEnum" name="Type_presentation">
      <eLiterals name="raster"/>
      <eLiterals name="vecteur"/>
```

Fig. 21.21 GeoPIM, automatically generated in XMI

DM Characteristics, Presentation and Semantic to a set of dimensions in Geo PIM. The transformation Operational 2 base generates the Base Operational with different Dimension Attributes.

The obtained Spatial Level Organization element is derived from the Geo PIM element named Organization. Furthermore, the transformations move from the spatial cover to Fact, which has as spatial measures the spatial objects. The authors add the layer adjacent object to the resulting Geo PIM and apply the transformation Sources 2 Spatial Level to present all details related to the sources of spatial objects.

The use of the proposed approach enables an automatic generation of the SDW design. The resulting GeoPIM integrates the entire DM's requirements.

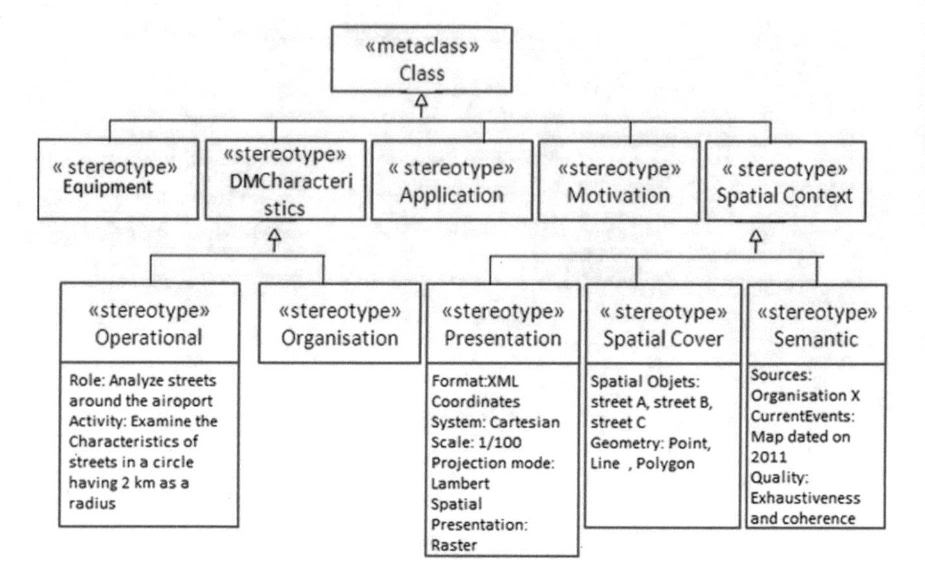

Fig. 21.22 Designed GeoCIM of the cluster containing the sales manager SM1

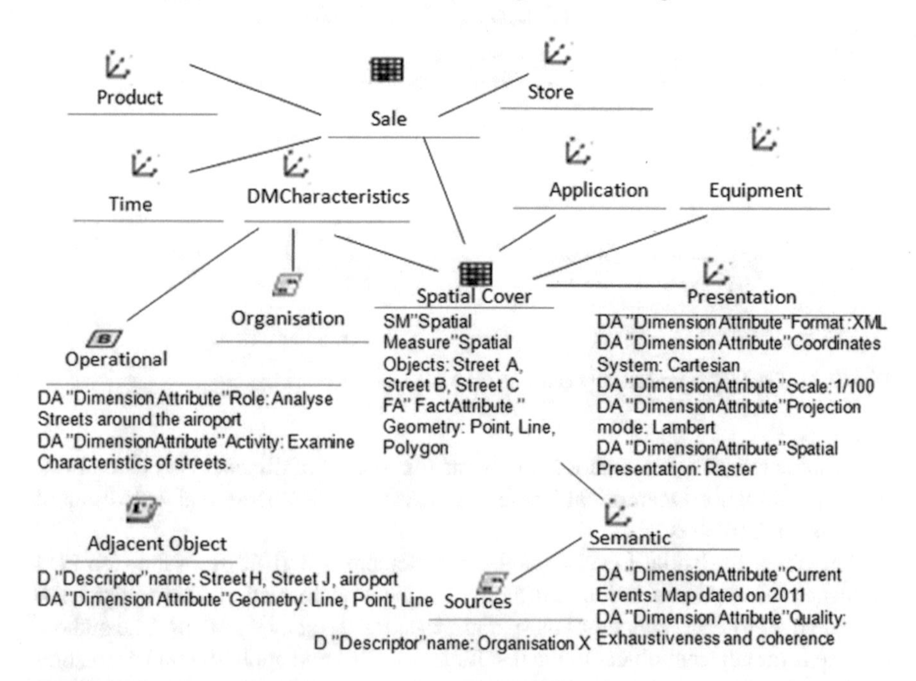

Fig. 21.23 Automatically generated GeoPIM of the cluster containing the sales manager DM

The representation of the design model of the same cluster in Glorio et al. (2012) can be performed as shown in Fig. 21.25.

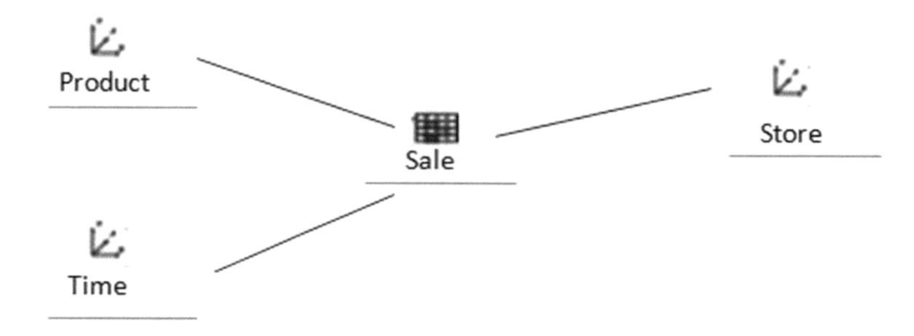

Fig. 21.24 The GeoPIM of the sales manager SM2 case before the proposed extension

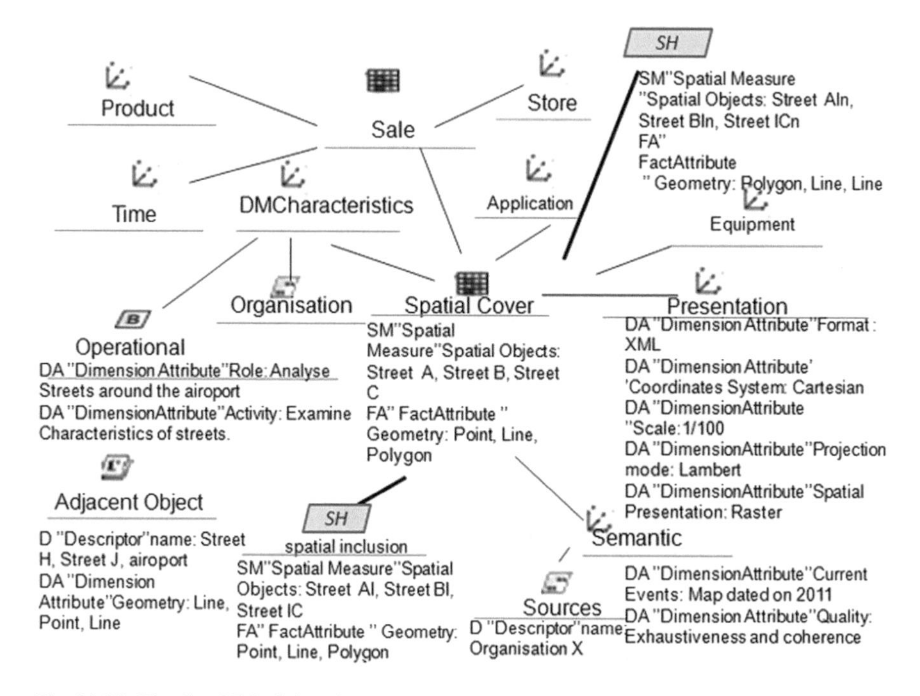

Fig. 21.25 The Geo PIM of the sales manager DM

The SDW design resulting from the application of this approach (Fig. 21.24) is more adapted to DMs than the one presented in Fig. 21.25 since in this case, DM's requirements are considered.

Figure 21.26 presents the extended GeoPIM, which integrates Spatial Hierarchies: spatial inclusion and spatial intersection.

Figures 21.27 and 21.28 present the part of the obtained GeoPSM after applying several relations.

The use of the proposed approach enables an automatic generation of the SDW implementation. The resulting GeoPSM integrates the entire DMs requirements existing in the design model.

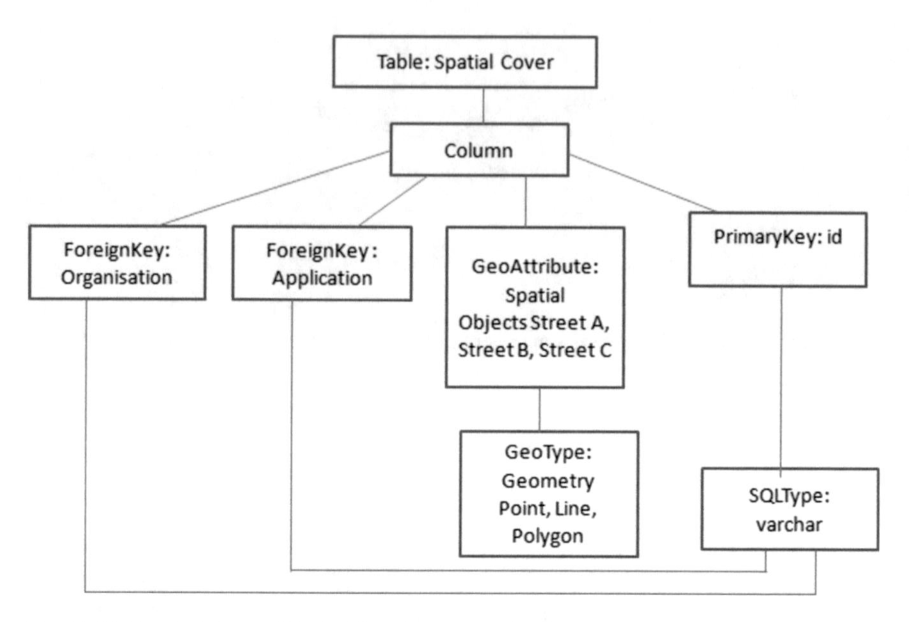

Fig. 21.26 Part 1 of the GeoPSM of the sales manager DM

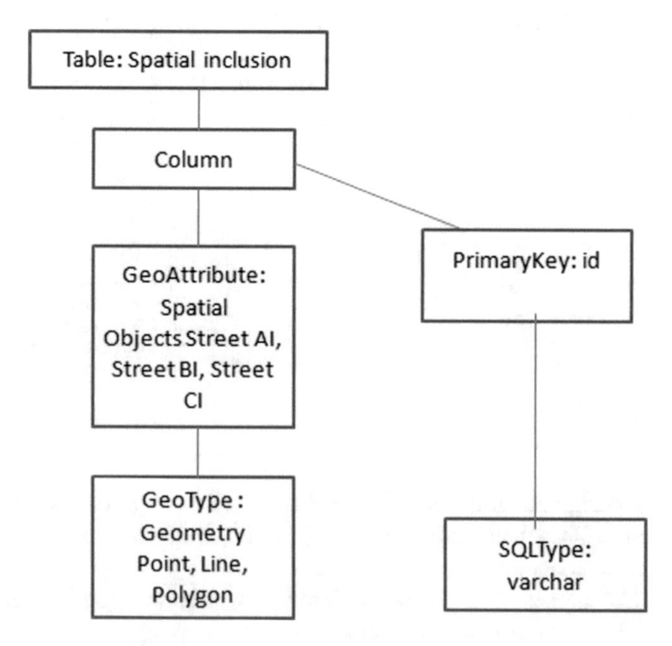

Fig. 21.27 Part 2 of the GeoPSM of the sales manager DM

21.12 Evaluation of the Proposal

To validate this approach, the authors opted for an empirical validation of the proposed approach with the valuation method.

The valuation method is defined based on the diffusion theory (Rogers 1995) which examines the rate and the motivations of the adoption of a technological innovation by a group of potential users. The diffusion theory demonstrates that the technological innovation has a chance to succeed if the community of the users appreciates its quality. This approach is also fruitful for the evaluation of a new abstract tool such as a method of design by estimating how the users' community accepts it. We choose for evaluation the Rogers's method because it describes the adoption of an innovation with five characteristics: the relative advantage, the compatibility, the complexity, the testability and the observability. These characteristics exist in the most innovative evaluation methods. We ask SDW users' students a set of questions to measure the satisfaction of the Rogers's method characteristics. Then, we calculate statistics that show the percentage of satisfaction of the SDW users' students according to every characteristic.

As indicated previously, the theory of diffusion defines five characteristics (Rogers 1995) which would determine the adoption of a new technology:

- The relative advantage: the degree in which an innovation is perceived as being better than the already existing ones. It is not necessary that this innovation possesses has more advantages than the others do but users think that it is more important and more advantageous than others.
- The compatibility: the degree of the approach's compatibility with the existing values, experiences of the users. An innovation which is incompatible with the values and the current used standards would set more time to be adopted.
- The testability: consists in the possibility of testing and modifying an innovation before using it. The opportunity to test an innovation allows trusting in the product because there is the possibility of learning now to use it.
- The observability: the results and the benefits of an innovation should be clear. When the benefits of the adoption of the innovation are clearer, users will adopt it easier.

The quality of the document presenting the approach of the design and the implementation of the SDW is evaluated by three attributes: consistency, efficiency of the examples and the clarity of the document structure.

Based on these attributes, an evaluation is done focusing on the quality of the approach. The authors choose to realize this evaluation with students and to use SDW in their projects. During a session of course, they presented them the approach proposed in detail and through examples. In return, they asked them to supply a structured feedback concerning the appreciation of the proposed approach. Because of all the known criteria indicated, the authors developed a questionnaire with two parts; a first part reserved for the evaluation of the approach and another one for the evaluation of the document.

- Part 1: evaluation of the approach

  Question 1: Do you think that the adoption of the approach can help improve the design and the implementation phases of the SDW?
  Question 2: Is the described approach compatible and coherent with the existing practices shared in your discipline specialty.
  Q3: Do you think that the approach is difficult to understand and to use?
  Q4: Do you think that the approach supplies enough elements to be tested before adoption?
  Question 5: Do you think that the results of the application of the approach proposed at the level of the design and the implementation of SDW are visible?

- Part 2: evaluation of the document

  Question 6: Do you find that the terminology used in the document is clear?
  Question 7: Do you find that the examples are useful to give you a clear idea onto the subject?
  Question 8: Do you consider that the structure and the format of the document are clear enough?

For every question, the students can choose among the following options to express their level of satisfaction: very satisfied (TS)/Satisfied (S)/unsatisfied (NS).

The authors did not define voluntarily the neutral level to incite the students to express their judgment.

- Part 3 Results

Generally, the proposed approach is considered effective and of a high quality. The authors give, in what follows, an overview on the result of the evaluation, as shown in Fig. 21.29.

(i) The criterion advantage: 45% of the "Novice" students found respectively that the approach presents an advantage in the improvement of the quality of the SDW due to a better guide and a good cover of the various aspects of the design. However, only 24% of the "Expert" students judged the "not satisfaction" work. Were unsatisfied with the work?
(ii) The compatibility criterion: 54% of the "Expert" students find that the adopted approach is not compatible with the way they are used to design and develop DW and SDW.

    The discussions with the students who followed the session of presentation of the approach revealed that if 45% of them consider that the approach is not compatible, it is because they lacked knowledge of the formalism of the MDA.
(iii) The complexity criterion: only 30% of the "Novice" students and 30% of the "Expert" students answered "Unsatisfied". This justifies the ease of use of the proposed approach what makes possible its adoption by a large number of users.

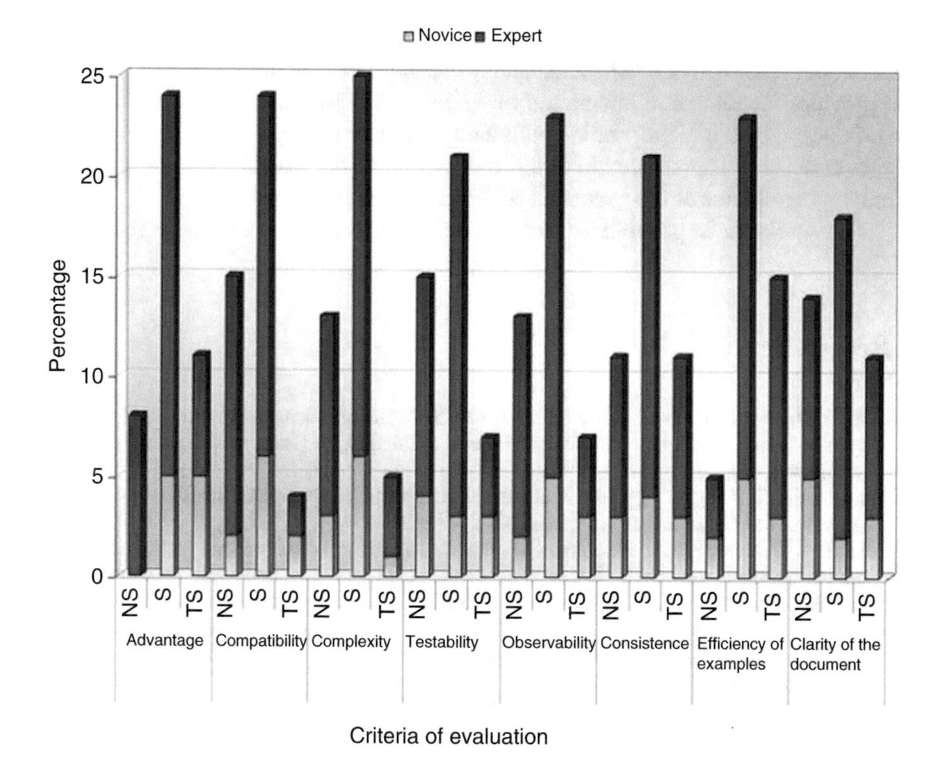

Fig. 21.28 Synopsis of the results

(iv) The testability criterion: 33% of the "Expert" students consider that the approach is "not satisfied" with this criterion; this is due to the limited number of users' profiles used in the implementation.

(v) The observability criterion: more than 60% of the "Expert" and "Novice" students covered by the questionnaire were convinced of the results and the profits of the approach.

21.13 Conclusion

An overview of the existing literature reveals that most of the SDW models lack an automatic integration of spatial and descriptive DMs requirements in the SDW design and implementation.

Basing on these limitations, the authors developed, in this chapter, an approach allowing an automatic transition from the DM's requirements to the SDW design in one hand and on other hand from the conceptual level to the implementation level by the means of model transformations.

The main contribution of this chapter is to provide a design and implementation model of an SDW that can be adequate for spatial DM's requirements. In addition, it provides an automatic integration of spatial and descriptive requirements in the SDW design and implementation without a DM intervention.

A case study is presented in order to demonstrate the feasibility of the proposal and the importance of the generated SDW implementation.

This work can be generalized to cover other platforms such as Oracle platform.

References

Aissi S, Gouider MS, Sboui T, Ben Said L (2015) Personalized recommendation of SOLAP queries: theoretical framework and experimental evaluation. In: Human-centric computing and information sciences, pp 1008–1014

Aranganayagi S, Thangavel K (2010) Extended K-modes with new weighted measure based on the domains. Int J Data Min Model Manag 2(3):288–299

Bimonte S, Tchounikine A, Bertolotto M (2008) Integration of geographic information into multidimensional models. In: International conference on computational science applications, pp 316–329

Bimonte S, Tchounikine A, Miquel M (2006a) Geocube, a multidimensional model and navigation operators handling complex measures: application in spatial OLAP. In: Advances in information systems ADVIS, pp 100–109

Bimonte S, Wehrle P, Tchounikine A, Miquel M (2006b) Gewolap: a web based spatial OLAP proposal. In: Office of technology management workshops, pp 1596–1605

Bohé SB, Rumpler B (2007) Modèle évolutif d'un profil utilisateur. In: Conférence en Recherche d'Information et Applications 2007, pp 197–210

Boulil K, Bimonte S, Pinet F (2015) Conceptual model for spatial data cubes: a UML profile and its automatic implementation. Comp Stand Interf 38:113–132

Broda B, Mazur W (2012) Evaluation of clustering algorithms for word sense disambiguation. Int J Data Anal Tech Strat 4(3):219–236

Da Silva J, Times VC, Salgado AC, Souza C, do Nascimento Fidalgo R, de Oliveira AG (2007) Querying geographical data warehouses with GeoMDQL. In: Brazilian symposium on databases, pp 223–237

Egenhofer J, Franzosa DR (1991) Point-set topological spatial relations. Int J Geograph Inf Syst 5(2):161–174

Fidalgo R, Cuzzocrea A (2012) An enhanced spatial data warehouse metamodel. In: CAiSE forum of CEUR workshop proceedings, pp 32–39

Glorio O, Trujillo J (2008) An MDA approach for the development of spatial data warehouses'. In: Data Warehouse and Knowledge discovery'08, pp 23–32

Glorio O, Trujillo J (2009) Designing Data warehouses for geographic OLAP querying by using MDA. In: International conference on computational science and its applications'09, pp 505–519

Glorio O, Mazón J, Garrigós I, Trujillo J (2010) Using web-based personalization on spatial data warehouses. In: International conference on extending database technology'10

Glorio O, Mazón J, Garrigós I, Trujillo J (2012) A personalization process for spatial data warehouse development. Decis Support Syst 52(4):884–898

Gomez L, Haesevoets S, Kuijpers B, Vaisman A (2007) Spatial aggregation: data model and implementation. In CoRR, abs/0707.4304

Kharrat A, Dhouib S, Chabchoub H, Aouni B (2011) Decision-maker's preferences modelling in the engineering design through the interactive goal-programming. Int J Data Anal Tech Strat 3(1):85–104

Malinowski E, Zimanyi E (2004) A Metamodel for the Specification of Geographical Data Warehouses. Representing spatiality in a conceptual multidimensional model. In: Proceedings of the 12th annual ACM international workshop on geographic information systems, pp 12–22

Malinowski E, Zimanyi E (2007) Implementing spatial data warehouse hierarchies in object-relational DBMSs. In: International conference of entreprise information systems, pp 186–191

Mazon J, Trujillo J (2009) A hybrid model driven development framework for the multidimensional modeling of data warehouses. Spec Int Group Manag Data 38(2):12–17

Naderi H, Rumpler B (2007) Physical document adaptation to user's context and user's profile. In: Proceedings of international conference on enterprise information systems, pp 92–97

Pattabiraman V, Nedunchezhian R (2012) A new constraint-based spatial clustering algorithm based on spatial adjacent relation for GML data. Int J Knowl Eng Data Min 2(1):14–34

Rifaie M, Kianmehr K, Alhajj R, Ridley MJ (2009) Data modeling for effective data warehouse architecture and design. Int J Inform Decis Sci 1(3):282–300

Rivest S, B'edard Y, Marchand P (2009) Toward better support for spatial decision making: defining the characteristics of spatial on-line analytical processing. In: Geomatica, pp 539–555

Rogers EM (1995) Diffusion of innovations. The Free Press, New York

Chapter 22
Smart Cities

Mayank Kejriwal

22.1 Introduction

The term 'Smart City', although far from ubiquitous, has gained increasing prominence in recent decades Hollands (2008), Nam and Pardo (2011), Neirotti et al. (2014), Su et al. (2011), especially given alarming facts about such aspects of daily urban life over the last century such as increased air pollution Mayer (1999), plastic and other kinds of waste Jambeck et al. (2015), emissions Boden et al. (2009), increasing population in several nations (Commoner 1991), global water supply (Famiglietti 2014), poverty, crime, violence (Bourguignon 2000), and strain on infrastructure. At the same time, urbanization is a trend that is ever on the rise (Vlahov and Galea 2002), with UNESCO[1] touting in one[2] of its well-cited reports that global population living in urban areas has increased from a third in 1960 to 47% in 1999, or about 2.8 billion people. Furthermore, the world's urban population is now growing by 60 million people annually, outpacing rural population growth by a factor of 3x. Urbanization in developed regions, such as Europe, North America and Japan are all more than 75%, and that of Latin America and the Caribbean has now started to approach those levels. UNESCO estimates that, by 2030, nearly 5 billion (or 61% of the projected world population of 8.1 billion) people will be living in cities.

Also illuminating is the growth of existing cities themselves (Table 22.1). UNESCO implicitly defines a *megacity* as one with 10 million or more inhabitants,

[1] United Nations Educational, Scientific and Cultural Organization: https://en.unesco.org/

[2] Accessed here: http://www.unesco.org/education/tlsf/mods/theme_c/popups/mod13t01s009.html

M. Kejriwal (✉)
Information Sciences Institute, University of Southern California, Los Angeles, CA, USA
e-mail: kejriwal@isi.edu

M. Werner, Y.-Y. Chiang (eds.), *Handbook of Big Geospatial Data*,
https://doi.org/10.1007/978-3-030-55462-0_22

Table 22.1 A brief illustration of megacity growth, location and features

	Slow-growing	Growing	Rapidly growing
Typical locations	South East Asia, Europe, and North America	South America and South East Asia	South/South East Asia and Africa
Features	Over 70& urban population, no squatter settlements	40–50% in urban population, under 20% in squatter settlements	Under 50% urban population, over 20% in squatter settlements
Examples	Tokyo, Moscow and Los Angeles	Rio de Janeiro, Beijing and Mexico City	Jakarta, Delhi and Lagos

and most of the new megacities today are in less developed regions, especially in Asia (Silver 2007; Gurjar et al. 2004), putting even more strain on resources. In 1960 there were only 2 megacities per this definition (New York and Tokyo), but by 1999, there were 17 such cities, and at the time of writing there are 31 such cities, significantly greater than the number (26) UNESCO had projected for roughly this time period (2015). The UN estimates that by 2030, this number will climb to 41.

Given these trends and the increasing emphasis on environmental issues at national and global scales, it is unsurprising that many are looking to both technology and collective community-driven initiatives to solve some of the problems mentioned earlier, in addition to (and in some cases, as alternatives to) centralized government initiatives. The pace of technological innovation, especially in the digital realm, has accelerated since the 1990s, which is also when the Kyoto protocol was enacted and signed by 192 parties, including the European Union (Grubb et al. 1997). The rise of the Internet, increased decentralization and *democratization* of communications (especially via social media and other platforms) (Welzel et al. 2018), rise of a younger generation of environmentally and socially conscious individuals and heightened populism (especially in the mid-2010s) due to pressing issues like the failure of capitalist models to tackle growing inequality (Bergmann 2018), have all contributed to grassroots-level efforts where the average citizen has greater visibility and participation than ever before (Rimmerman 2018).

Given this complex context, it is perhaps unsurprising that a unique definition of Smart City does not exist yet, and the literature on the subject is awash in synonyms, many of which conceptually overlap with each other, and some of which tend to focus more on one dimension than others. Per the overview in Cocchia (2014), examples of these roughly synonymous terms include Digital City (Aurigi 2016), Intelligent City (Hollands 2008), and Learning City (Michel 2005).

Drawing on shared characteristics between these various definitions, it is fair to synthesize the definitions as falling along one or more of three primary *dimensions*: *technological, human* and *institutional* (Cocchia 2014). These dimensions are largely self-explanatory e.g., the technological dimension is based on the use of technology and infrastructure to improve life and work within the city, the human dimension is based on people, education, learning and knowledge dissemination, while the institutional dimension is based on governance and policy. All of these dimensions are important, and continue to be studied by experts, but the scope of this chapter will be mostly limited to the technological dimension.

In the research community, academics have also started to take notice of the rise and problems associated with smart cities. For example, in their systematic literature review, Cocchia (2014) found that the number of papers about 'Smart City' and 'Digital City' increased more than 500% in just the five year period from 2007–2012. Although the review did not cover all papers that have been published in this area, and there was also likely some bias due to the term 'Smart City' not becoming popular in the earlier part of this phase, the increase is still significant. Furthermore, the number of actual smart cities is now quite impressive, especially in Europe. This makes the current time especially apt for studying smart cities and their emergence.

The rest of this chapter is structured as follows. We begin in Sect. 22.2 with a brief history and background of smart cities, including the notable events that seem to have led to an interest in forming a cohesive vision around smart cities. As stated earlier, smart cities have many synonyms in the literature, many of which overlap or otherwise focus on a subset of the relevant dimensions. In Sect. 22.3, we cover the multiple definitions of smart cities, or related synonyms, that have been proposed over the years. Sections 22.4 and 22.5 contains much of the core material of this chapter, where we detail in depth the context in which smart cities develop, and the role of technology, especially Big Data and analytics, in realizing a robust smart city vision. Section 22.6 describes examples of actual smart cities around the world. Finally, Sect. 22.7 covers some important future directions for the field, both academically and in practice, and Sect. 22.8 concludes the chapter with closing notes.

22.2 History and Background: A Brief Review

The notion of a smart city is predated by several events spanning geopolitics and information technology. We cover below some of the events that are hypothesized to have played an especially critical role.

First among these was the Kyoto protocol (Grubb et al. 1997), which sought to limit carbon dioxide emissions and safeguard the environment. The protocol was signed by all United Nations members (with the exception of the United States, Andorra, Canada, South Sudan) and by the European Union in 1997, and was entered into force after Russia ratified it in October 2004. According to the protocol, there are two commitment periods in which developed countries have to achieve *binding reductions* of greenhouse gas emissions. These periods are 2008–2012 and 2013–2020. The Kyoto Protocol has greatly influenced how we think about industrialized urban areas, and was itself, arguably, in response to rising environmental awareness and activism (Grubb et al. 1997; Weyant and Hill 1999; Victor 2011). It has not been ineffective, though the extent of the effect is debatable (Victor 2011). For example, over the last couple of decades, all state parties who signed on to the protocol have fostered initiatives to reduce carbon dioxide emissions within their own boundaries. In the context of the current topic, one can argue that the move by national and local governments to design and apply environmental policies has also driven interest about smart cities insofar as environmental issues are concerned. We note that it was during 2005 that the Kyoto Protocol *entered into force*, and although a decade has passed since then, the full impact of the protocol may only become clear in the next decade after the second of the two commitment periods comes to an end (in 2020).

On the corporate side, IBM was one of the first companies in modern history to bring attention (in 2008) to the concept of 'Smart Planet', an even grander vision than smart cities (Palmisano 2008; Harrison et al. 2010). For IBM, Smart Planet is conceived as an instrumented, interconnected and intelligent planet where Big Data could be used to 'transform enterprises and institutions through analytics, mobile technology, social business and the cloud'. During that era, IBM started a new business in this sector, powered by cloud computing (Zhu et al. 2009), and supplying to governments, solutions focused on 'smart' communications, energy and utilities, and other services. Competition soon followed, with companies such as Cisco, Siemens and Ericsson entering the space for studying and supplying solutions for new smart projects in urban areas (Höjer and Wangel 2015). It is possible that even the terminology of 'Smart City' was directly influenced by IBM's Smart Planet vision.

Although the IBM vision was US-centric, the *Covenant of Mayors* was also instituted during that time as a self-started initiative by European cities (Torres and Doubrava 2010; Christoforidis et al. 2013; Pablo-Romero et al. 2015). The idea was to reduce carbon dioxide emissions by more than 20% by 2020 through increased energy efficiency, and to increase adoption of renewable energy technology. The European Commission fostered this agreement in the frame of fulfilling *Strategy 2020*, with actions primarily focusing on clean mobility, redevelopment of (public and private) buildings and citizen awareness on energy consumption (Torres and Doubrava 2010). Signatory cities agreed to issue their own PAES (Action Plan for Sustainable Energy), a roadmap for fulfilling the objectives laid out by the Commission. Although the Covenant of Mayors was not as tech-focused or forward as the IBM Smart Planet vision, it was more synergistic with existing initiatives such as the Strategic Plan for Energy Technologies (Ruester et al. 2014), and like the Kyoto Protocol, sought to realize the Smart City vision by improving environmental sustainability, quality improvement and pollution reduction.

Slightly more comprehensive was the Europe 2020 Strategy (Lundvall and Lorenz 2011), launched by the EU in 2010 to foster investments in education, research and innovation, achieve sustainable growth in a low-carbon economy, and emphasize an inclusive agenda to ensure holistic job creation and povery reduction. The Europe 2020 Strategy was highly influential in popularizing, and spreading deep awareness and appreciation of, smart and digital city initiatives within Western Europe. In contrast, despite a strong start, the IBM vision was less successful in popularizing such a vision in the US.

Finally, and especially in the context of technology, the more diffuse 'event' of the rapid growth of the Internet, and incumbent technologies like smartphone use

(Penwarden 2014), the proliferation of easy and intuitive mobile apps (Goldsmith 2014), social media (Perrin 2015), and even the growing popularity of the sharing or 'gig' economy (Matzler et al. 2015), cannot be discounted in influencing the smart city vision, although that was not the primary goal of the Internet. In the smart and digital city literature, it is common to characterize the rise of the 'ICT (Information and Communications Technology) infrastructure' (Pickavet et al. 2008), which comprises broadband, wireless sensors, networked applications, open platforms, cloud and other similar technologies that all work in tandem to form the backbone of an 'intelligent' infrastructure (Aktan et al. 1998), (Banerjee 2009). The very concept of a digital city is predated on the feasibility of a wired city where it is possible to provide public and private services over digital platforms to create socio-economic value for city stakeholders and the larger civic society. Slowly, this vision is starting to be realized, since at the time of writing, many services centered in healthcare, energy, transportation, public safety and more controversially, voting and governance, have a digital footprint. An important topic that was not anticipated as much by early pioneers in the digital city space was *cybersecurity* (Khatoun and Zeadally 2017), and the interference of foreign governments and/or rogue parties in critical democratic processes (Norden and Vandewalker 2017).

22.3 Defining Smart Cities in Practice

The previous section suggested that there is a tension in the definition of a smart city, with Europe seeming to place a premium on environmental issues and inclusive growth versus a more tech-focused vision in the US. For this reason perhaps, the terms Smart City and Digital City (Hollands 2008; Nam and Pardo 2011; Aurigi 2016), among others (Michel 2005), have both become popular in their own right despite heavy overlap and the same overarching goals (Cocchia 2014).

In part, this terminological confusion is unsurprising because 'Smart City' is such a broad concept, including aspects of urban life ranging from planning, sustainable development, environmental issues, energy grid, water supply (Famiglietti 2014; Cash et al. 2003), to economic development and issues of income inequality (Kuznets 1955), to technological and social participation (Rimmerman 2018). Generally, when we consider the different definitions of Smart City that have been proposed in papers, whether under the banner of 'Smart' or some other moniker (e.g., 'Knowledge' or 'Digital' City) there is considerable practical overlap between the definitions, with the term primarily influenced by *application*. What is clear however is that there is no one comprehensive definition that has come to be accepted by academics, businesses and institutions. The differences between a city that is smart or non-smart are also not completely clear; one could go so far as to argue that any city with a sufficiently advanced mass transit system and uptake of technology is 'smart'. Interestingly, a similar argument and concern applies to other 'smart' technologies, including Artificial Intelligence (Davenport 2018). A

Table 22.2 Alternate definitions and characterizations of Smart Cities

Moniker [Reference]	Definition
Information city (Anthopoulos and Fitsilis 2010)	"Digital environments collecting official and unofficial information from local communities and delivering it to the public via web portals are called information cities"
Digital city (Couclelis 2004)	"The digital city is as a comprehensive, web-based representation, or reproduction, of several aspects or functions of a specific real city, open to non-experts. The digital city has several dimensions: social, cultural, political, ideological, and also theoretical"
Sustainable city (Bătăgan 2011)	"Sustainable city uses technology to reduce CO2 emissions, to produce efficient energy, to improve the buildings efficiency. Its main aim is to become a green city"
Intelligent city (Komninos 2006)	"Intelligent cities are territories with high capability for learning and innovation, which is built-in the creativity of their population, their institutions of knowledge creation, and their digital infrastructure for communication and knowledge management"
Knowledge city (Ergazakis et al. 2004)	"A Knowledge City is a city that aims at a knowledge- based development, by encouraging the continuous creation, sharing, evaluation, renewal and update of knowledge. This can be achieved through the continuous interaction between its citizens themselves and at the same time between them and other cities? citizens. The citizens? knowledge-sharing culture as well as the city?s appropriate design, IT networks and infrastructures support these interactions"

recent report by MMC Ventures,[3] for example, suggested that companies may be misleadingly reporting the use of AI in their offerings (and/or internal processes) when they are only exploiting standard digital technologies and efficiencies that academics would not call AI. Perhaps the problem then is not with the overall concept of a 'Smart City' but with the general moniker 'smart'.

In part, an important motivation behind this chapter was to try and distinguish between such cases and cities that are truly shaping the Smart City movement by adopting bold initiatives that are motivated by pressing infrastructural and social problems. In Table 22.2, we provide some alternate terms and definitions that could substitute for 'Smart City'. These are, by no means, exhaustive; some terms that we did not cover in the table include Learning City, Virtual City, Wired City and Ubiquitous City. We chose the terms in the table to illustrate the diversity of the definitions in use, but also to express the overlap. In all cases, the overarching goal is to better the lives of the citizens of the city, either by reducing environmental footprint, improving governance and strengthening social institutions, and making everyday life easier or more seamless through the power of inexpensive and scalable technology, like widespread Internet access.

[3]https://www.mmcventures.com/wp-content/uploads/2019/02/The-State-of-AI-2019-Divergence.pdf

22.4 Context Variables Affecting Smart Cities

Various factors can influence the manner in which cities choose to develop and maintain Smart City initiatives. The abundance of data has clearly had an effect on this, but even predating Big Data, *contextual factors* have always been important in focusing the resources and needs of Smart Cities. We describe four context variables in this section, inspired by prior work in the area (Neirotti et al. 2014).

22.4.1 Structural Factors

Both city size and population density can be relevant for the development patterns of Smart Cities. First, it is well known that bigger cities tend to attract more human capital and can usually rely on heavier investment in infrastructure pertaining to critical areas as electricity, water and telecommunication. Large cities also have critical masses of tech-adopting users, which tends to increase both demand and supply of new digital services. Examples include Bus Checker in London (Stone and Aravopoulou 2018), but also more advanced mass transit options in bigger cities and even ridesharing apps (Anderson 2014), which tend to offer more options in bigger, more hub-like cities. Recently, there have also been indications that ridesharing apps may start encroaching on the business model of mass transit,[4] which are often a result of centralized government planning (that has its fair share of inefficiencies) and can run into cost and time setbacks.

However, while it may seem to be a largely positive influence, large city size can also stifle Smart City innovation. Smaller cities are more ideal settings for pilot projects, as they can deal with shorter installation times when projects requiring investments in distributed infrastructures (e.g. street lighting, smart waste) are needed. Such projects are also more amenable to agile experimentation, allowing rapid pivoting and modification without causing much disruption on everyday services.

Large cities also often have a high population density, which can ease the flow of knowledge and ideas by facilitating social interactions, and via idea generation and innovation. Some other effects of higher population density are more tangential to Smart City initiatives but can make the difference between an initiative 'taking off'. For example, we mentioned earlier that larger cities have better mass transit options, which is fertile ground for testing and implementing Smart City initiatives in the digital and communications realm. Once again, we note that, over a certain threshold, both population density and size can lead to diseconomies in areas like transportation, real estate, and energy consumption, and can stifle innovation. Thus, we predict that, for bolder, riskier and potentially higher-impact ideas, smaller cities

[4]https://www.upi.com/Top_News/US/2019/01/08/After-taking-on-taxis-ride-share-services-now-challenging-public-transit-in-US/5281546857951/

will prove to be much more fertile ground than larger cities in terms of prototyping and 'failing fast and smart'. However, it is inevitable that for a model to take off and scale, adoption in large cities cannot be ignored. There is no real consensus on whether there is a single city size that can optimize for this interesting tradeoff.

22.4.2 Economic Development

A city's economy, usually measured by Gross Domestic Product (GDP), can also significantly influence the development of Smart Cities. Some of the reasons are obvious i.e. a city cannot be expected to be ready to implement Smart City initiatives till it can fulfill the basic needs of its citizens, including infrastructural needs like water, transport, sewage and welfare services. Generally speaking, both cities and countries with a higher GDP growth rate undergo a higher economic expansion, which influences the financial resources (and impetus) available for investments in new (or upgraded) educational, environmental and infrastructural initiatives. There is also a 'spiral effect': cities with a greater economic development appear more attractive to those people who wish to increase their standard of life, and consequently attract human capital; this in turn leads to the positive effects of size and demography on smart city initiatives taking off (see previous section). Human capital provides both a talent pool for corporations, but also end-users and consumers for initiatives that might otherwise not scale.

However, just like with the first context variable (structural factors), it is important not to treat a high GDP as a *prerequisite* for the development of Smart City initiatives. Some of the best Smart City initiatives may, in fact, be suitable for less developed cities since there would be greater need, and most likely, greater subsequent adoption. This should be borne in mind when considering the potential of a Smart City initiative in the context of a city: does it *really* require a sophisticated underlying infrastructure for an initial rollout? Is there a creative workaround, one that could lead to new innovation?

22.4.3 Technology

Technological development, adoption and diffusion are hard to predict, and can be susceptible to unexpected dynamics. At the very beginning, no one thought smartphones or even personal computers would have wide market traction. Yet, by employing a 'sweet spot' offering of price, design and functionality, companies have managed to turn these products into global sub-economies.

Generally, systems and organizations that have started to invest earlier in a technology trajectory are more likely also to develop and adopt emerging technologies belonging to the same trajectory. For example, a city that has historically invested heavily in smart transportation, as opposed to telecommunications, is more likely to continue doing so in the near future. In practice, it can be difficult to draw such distinctions, leading to the popular characterization that some cities can be more 'tech-leaning' than others.

Cities that are more liberal with technology are also ripe for Smart City initiatives that fall within the scope of digital, internet or telecommunication innovation. Internet access, across all incomes and social classes, can be a particularly important facilitator of smart initiatives in many urban settings. On the other hand, a limited diffusion could reflect a digital divide that hinders the achievement of a critical mass of users. This could jeopardize the development of a variety of initiatives. The budget for R&D investments in both private and public expenditure can also be a valid metric for measuring technological progress and human capital development. The density of tech companies or labs can be another metric, albeit one that is not without its controversies. The countries and cities in which these sectors are more developed are more likely to effectively deploy Smart City technologies.

22.4.4 Effective Environmentally-Progressive Governance

Environmental sustainability is an important determinant of quality of life in urban settings (Cash et al. 2003). The availability of green spaces can generate extensive socio-economic benefits, and cities with stronger environmental policies can face lower marginal costs for development of Smart City initiatives aimed at improving their environmental sustainability. Such cities can also rely on a more developed infrastructure than polluted cities with limited green areas. However, although the costs for more polluted cities to adopt smart initiatives in transportation, energy and urban planning can be higher, their relative advantage, as well as the effort spent by local policy-makers to enact initiatives aimed at mitigating pollution, can be more evident given their relevance in public opinion and in the political agenda.

Unlike technology and demographics, the effect of environmentally progressive governance is not as clear-cut or quantifiable as a context variable. However, it has generally been observed that such cities are more open to liberal, regulation-favoring initiatives that seek to curb the impact of carbon-intensive industries. It is also more difficult to study the effects of progressive governance because of definitional controversies over what qualifies as progressive or effective governance.

22.5 The Role of Data

Recent decades have witnessed a heavy proliferation of open data, including national censuses, government surveys, and statistically-oriented economic data such as collected by organizations such as the Bureau of Labor Statistics (BLS) in the United States (Goldberg and Moye 1985). These datasets are a valuable source of information about cities, countries and (both current and future) populations, since they can be used for both diagnosis and forecasting. Likewise, businesses periodically collect and analyze significant amounts of data on different segments of their value chain, including operations, markets, suppliers, distributors and customers. Limitations of many of these datasets are that they tend to be expensive to collect, usually relying on surveys and careful process monitoring, can be lagging, and are aggregated at coarse spatiotemporal scales. Any insights that are yielded by such datasets may rightfully be termed as 'small data studies' and tend to include questionnaire surveys, case studies, city audits, interviews and focus groups, and ethnographies. Arguably, much of what we have learned about cities from data is actually characterized by sparseness (Miller 2010).

Potentially, 'Big Data' (Boyd and Crawford 2012) could help transform the knowledge and governance of cities through the creation of a large and shared pool of data that seeks to provide much more sophisticated, wider-scale, finer-grained, and with powerful computation and algorithms, real-time understanding and regulation of urban settings and independent variables. For example, in the power industry, AI and Big Data could be used to understand supply and demand of power in the various states, and be used for power brokerage so that everyone benefits (Kezunovic et al. 2013; Zhou et al. 2016). With the rise of renewable energy, accurate forecasting of supply and demand is equally important. Thus, there are clear motivations in refining and adopting Big Data technology.

Although there are multiple conferences and journals on Big Data at the time of writing, a general definition still tends to be lacking (e.g., would 1 terabyte be considered Big Data today?) Generally speaking, Big Data is believed to involve at least four issues, known as the four Vs:

Volume: As is evident from the name itself, Big Data should be data that has high volume i.e. usually terabytes or even petabytes. Furthermore, not only can each data source contain a huge volume of data, but also the number of data sources, even for a single domain, could potentially be in the tens of thousands.

Velocity: The data is generally being created in near real-time, as a direct consequence of the rate at which data is being collected and continuously made available. A good example for understanding velocity of Big Data is the stock market, where there are many data sources that provide near real time, continuously changing information about stocks, including bid and ask prices, and volume of shares traded.

Variety: The data should be diverse, generally comprising a mix of structured and 'unstructured' (usually, natural language) data. Variety could also be measured in spatiotemporal terms, rather than the structure of the data.

Veracity: It should be possible to derive reasonably accurate and trustworthy insights from the data i.e. the data should not be 'falsified' or misleading at its source. This is challenging, because data sources (even in the same domain) are of widely differing qualities, with significant differences in the coverage, accuracy and timeliness of data provided.

Other than the four Vs, Big Data also tends to obey other criteria e.g., big datasets in a particular domain tend to be exhaustive in scope, striving to capture entire populations or systems rather than limited samples. As mentioned before, big datasets are also finer-grained in resolution, and hence with the potential to yield more detailed insights than small-data studies. Big datasets are flexible and scalable (with the potential to grow in size indefinitely), and finally, big datasets are usually relational i.e. contain common fields that enable deriving joint insights. For more details on characteristics of Big Data, we refer the reader to the article in Yin and Kaynak (2015), and also Dong and Srivastava (2013).

22.5.1 *Smart City and Big Data*

Sources of 'Smart City'-relevant big data can be broadly divided into three categories: **directed**, **automated** and **volunteered** (Kitchin 2014):

1. *Directed datasets* are generated by traditional forms of surveillance, examples being immigration passport control systems (e.g., at airports) that record and validate passenger details in real-time, data generated via CCTV, photographs, fingerprints and iris scans, spatial video, LiDAR, and thermal (or other electro-magnetic) scans that enable mobile and real-time mapping of two-dimensional and three-dimensional structures. It is important to note that directed datasets are generally 'domain specific' e.g., immigration passport control systems have a specific purpose and modality of data generation and processing, but even more importantly, are usually not fully automated and are augmented with some kind of human-in-the-loop check.
2. *Automated datasets*, as the name suggests are generated as an inherent, automatic function of the device or system, in contrast with directed datasets that tend to be generated by systems that have an explicit human-in-the-loop component. Examples include *capture systems*, which are used for task and performance monitoring e.g., using the outputs of a scanner at a store's check-out counter to monitor check-out operator performance, as well as collecting information on what items were purchased, and by whom (followed by data mining and data analytics techniques like itemset mining Zaki and Hsiao 2002); *digital devices*, including mobile phones that capture both background and targeted (e.g., when the user is using an app) data, digital networks that capture data on transactions and interactions; *clickstream data* that records how people navigate through a website or app; *sensed data* generated by a variety of sensors and actuators embedded into objects or environments that regularly communicate their measurements; *scanners* that scan machine-readable objects such as travel passes, passports, or barcodes on parcels that register payment and movement through a system; and devices that facilitate the Internet of Things (IoT) vision (Ashton et al. 2009), such as sensors embedded in the home or the environment.
3. Finally, *volunteered datasets* tend to be user-generated. Social media is the best example, but less well-known examples of volunteered data also include the generation and uploading of GPS traces into a public resource like OpenStreetMap to create a common, open mapping system (Haklay and Weber 2008).

In comparing these different types of datasets, we note that, while directed and volunteered data can provide useful insights into urban systems and city lives, automated data hold the most promise for scaling using computational and cloud technology that continues to grow ever-cheaper. In particular, there has been an interest in automated forms of surveillance, sensor networks and IoT devices (especially at the 'edge'), and the tracking and tracing of people and objects. Such an 'instrumented city' (Kim et al. 2017) offers the promise of an objectively measured, real-time analysis of urban life and infrastructure, especially using technologies like IoT, but can also pose threats to citizen liberties as the controversy surrounding China's 'social credit' system would indicate (Botsman 2017).

It is also important to distinguish between data that are generated in real-time and can be utilized upon generation vs. data that can be stored and analyzed post-hoc. We mentioned earlier that GPS traces can be uploaded into OpenStreetMap to help build out a public mapping resource. Social media is a resource that could be analyzed post-hoc, but that holds more promise if analyzed in real-time (e.g., to detect the emergence of a riot, terrorist act or other emergency event that requires an immediate mobilization of resources). Sensor networks generate so much real-time data that post-hoc analysis can be practically impossible; it is much more feasible to use fast signal processing algorithms to detect problems as they arise (Arasteh et al. 2016). Sensors at the edge of a network may not even have access to sufficient storage or computing resources and may have to make all decisions in real-time using minimal resources (Shi et al. 2016).

The discussion above indicates that the type of device generating the data, its placement and energy source, and its intended utility, are all factors in determining the design of algorithms that are deployed for processing the data. Even a simple fact like storage (big cities could potentially generate petabytes of data in short bursts of time) could influence algorithmic design, since it may not be possible to store data being generated continuously.

Concerning Internet of Things (IoT), what kinds of Smart City-relevant systems and applications does it entail? Based on the recent survey by Mehmood et al. Mehmood et al. (2017), we posit that IoT-based smart city applications can be categorized on the basis of network type, scalability, coverage, flexibility, heterogeneity, repeatability, and end-user involvements (Gluhak et al. 2011). Applications can be taxonomized along the lines of personal and home, utilities, mobile, and enterprises. Personal and home applications could include ubiquitous e-healthcare services to live independently via body area networks (BANs) (Chen et al. 2011), which help doctors monitor patients remotely. Utilities applications include smart grid, water network monitoring, and video-based surveillance. Mobile applications include congestion control and waste management.

IoT devices include, but are not limited to, automatic doors, lighting and heating systems, security alarms, wifi router boxes, entertainment gadgets, television recorders, and all such devices that have the ability to transfer data between each other (possibly through a mediator), leading to the emergence of *derived* (e.g., aggregated) data. Devices such as mobile phones can be traced through space by triangulation, and can record and transmit their own trails. In practice, datasets generated by IoT systems are quite diverse, usually generated by a mix of public and private agencies, and not all open. However, there is no question that, if used properly and with the right safeguards in place, the combination of such diverse, comprehensive datasets makes possible real-time analytics and adaptive forms of management and governance that were previously impossible.

22.5.2 Real-Time Data

Several city governments now use real-time analytics to manage and regulate both the functional and operational aspects of a city. A ubiquitous example relates to vehicle movement and route planning around a transportation network, where data from a network of cameras and transponders are fed back to a central control hub. The hub monitors the flow of traffic and could adjust traffic light sequences and speed limits, while automatically setting other parameters to prevent congestion while minimizing costs and delivery times (Dodge and Kitchin 2007).

Similarly, the police might monitor a suite of cameras and live incident logs in order to efficiently and reactively direct appropriate resources to particular locations. In yet another example, data relating to environmental conditions might be collated from a sensor network distributed throughout the city, for example measuring air pollution (Xiaojun et al. 2015) or providing earthquake warnings.

Many local governments use management systems to log public engagement with their services and to monitor whether staff have dealt with any issues. In nearly all cases, these are isolated systems dealing with a single issue and are controlled by a single agency. More recently there has been an attempt to draw all of these kinds of surveillance and analytics into a single hub, supplemented by broader public and open data analytics (Kitchin 2014). Some examples are noted below:

1. The Centro De Operacoes (COR)[5] in Rio de Janeiro, Brazil is a partnership between the city government and IBM. In 2010, IBM employed their first integrated operations center in Rio de Janeiro, pooling generous investments in sensor networks after signing a contract with the city. Rio had recently experienced devastating landslides that killed hundreds of people, not to mention upcoming challenges of hosting the 2016 Olympics and 2014 World Cup. The city felt that there was a need to develop an Emergency Response System, with real-time automated C2 (Command and Control) of emergency responses. The partnership between the city and IBM has led to a citywide instrumented system that draws together streams of data from 30 agencies, including traffic and public transport, emergency services and weather feeds, into a single data analytics centre. Algorithms and analytics are executed over this data for varied purposes, including for building predictive models applicable to everyday city development

[5]http://cor.rio/

and management, as well as humanitarian and disaster situations such as flooding. A virtual operations platform complements the existing technology. With the support of analytics programs, city-wide operational processes using data from any number of domains can continuously predict and react to events and trends that are affecting the city. Unfortunately, the smart city project has faced some negative press in popular media, which serves as a cautionary tale. For example, there have been concerns as to what degree city management should be delegated to private companies. Furthermore, not enough attention was paid to using resources to deal with problems such as high crime rates, social inequality and environmental degradation.

2. In the Spring of 2011, Mayor Bloomberg's office of Media and Entertainment in New York City released a roadmap for securing the future of New York City as a digital city. The report was informed by 90 days of research and over 4,000 points of engagement from residents, City employees, and technologists who shared insights and ideas. Chief among public interests were calls for expanded Internet access, a refreshed nyc.gov interface, real-time information, and more digital 311 tools. Businesses and technologists sought greater broadband connectivity, a deeper engineering employment pool, and read/write API access to City information. Finally, City employees proposed ideas for next-generation strategy, new coordination tools, and shared resources to enhance digital communications efforts. Specific elements of the roadmap included:

    a. **Access:** Ensuring that all New Yorkers can access the Internet and take advantage of public training sessions to use it effectively by supporting more vendor choices to New Yorkers, and introducing Wi-Fi in more public areas.
    b. **Open Government:** Unlocking important public information and supporting policies of Open Government, further expanding access to services, enabling innovation that improves the lives of New Yorkers, and increasing transparency and efficiency.
    c. **Engagement:** Improving digital tools including nyc.gov and 311 online to streamline service and enable citizen-centric, collaborative government, expanding social media engagement, implementing new internal coordination measures, and continuing to solicit community input.
    d. **Industry:** Continuing to support (through the New York City Economic Development Corporation) a vibrant digital media sector through a wide array of programs, including workforce development, the establishment of a new engineering institution, and a more streamlined path to do business.

3. Intelligent Nation 2015 (iN2015) was Singapore?s 10- year masterplan to help realize the potential of 'infocomm' over the next decade. iN2015 was designed to be a multi-agency effort that was the result of combining private, public and even individual inputs, with the last providing their ideas and views through focus groups and the Express IT! iN2015 Competition. The competition attracted thousands of entries from students and the general public on how they envisioned infocomm would impact the way they live, work, learn and play in 2015. In addition, hundreds of private and public sector representatives participated in

numerous discussions to come up with ideas for transforming their sectors through infocomm, and how to translate these ideas into reality. Ultimately, iN2015 was a plan to develop Singapore further as a smart city, specifically through investments in infocomm. Specific reach goals included ensuring that by the end of the effort, there were 80,000 additional jobs, 90% of homes were using broadband, and there was 100% computer ownership in homes with school-going children.

4. In 2018, Sadiq Khan, Mayor of London, launched a roadmap to better utilize tech and data in support of smart city initiatives. The roadmap was called 'Smarter London Together' and includes more than 20 initiatives designed to support the development of the next generation of smart technology and promote greater data sharing among the city?s public services. Plans include achieving full fibre connectivity for all new homes and supporting the commission of smart technology such as a 'hyper local' sensor network, which will create the world's most sophisticated air monitoring system in the UK capital. From July, 100 sensors will be attached to lamp posts and buildings in the most affected areas, alongside two dedicated Google Street View cars that will record air quality in greater detail than before. A new Connected London program is also proposed to coordinate efforts and increase connectivity in the city. Measures include expanding public Wi-Fi in streets and buildings, supporting 5G projects and promoting a new generation of smart infrastructure to help solve the city?s biggest challenges.

Advocates of such systems and efforts argue that they ultimately present a data-driven, efficient form of governance as opposed to one based on intuition and political ideology. However, there is controversy over whether such initiatives are best (i.e. most effectively) implemented by governments and also implications for privacy and incumbent issues like data theft and misuse.

22.5.3 Open Government Data

According to the OECD (Organization for Economic Cooperation and Development), Open Government Data (OGD) is a 'philosophy, and increasingly a set of policies, that promotes transparency, accountability and value creation by making government data available to all'. Public bodies produce and commission huge quantities of data and information. By making their datasets available, public institutions become more transparent and accountable to citizens. By encouraging the use, reuse and free distribution of datasets, governments promote business creation and innovative, citizen-centric services.

OGD can pose some tricky questions for governments, such as who will pay for the collection and processing of public data if it is made freely available? What are the incentives for government bodies to maintain and update their data? And what data sets should be prioritised for release in order to maximise public value? Steps are therefore needed to develop a framework for cost and benefit analysis, to collect data, and to prepare case studies demonstrating the concrete benefits – economic, social, and policy – of opening government data.

As we saw earlier with the case studies described briefly on Singapore, London and New York, OGD can be used for building up Smart City initiatives like traffic management, digitized government services and investments of Wifi and other ICT technologies to ensure that no precinct is left behind in the race to digitize.

The OECD Open Government Data project aims to progress international efforts on OGD impact assessment. The mapping of practices across countries will help establish a knowledge base on OGD policies, strategies and initiatives and support the development of a methodology to assess the impact and creation of economic, social and good governance value through OGD initiatives. In the last decade, in particular, there have been many conferences and initiatives for (1) building 'OGD cultures' especially in the Middle East and North Africa regions to combat endemic corruption; (2) developing useful indices that can allow one to compare OGD success and adoption rates in a quantitative way across countries; (3) promoting the movement, through reviews, blogs and meetings.

As a specific example of a federal initiative, data.gov is managed and hosted by the U.S. General Services Administration, Technology Transformation Service, is developed publicly on GitHub, and is powered by two open source applications, CKAN and WordPress. Data.gov follows the Project Open Data schema, a set of required fields for every data set displayed on Data.gov. Although the total number of datasets, available on the Data.gov Metrics page, can fluctuate, the range and growth has been impressive in recent years. According to official statistics, as of June 2017, there were approximately 200,000 datasets reported as the total on Data.gov, representing about 10 million data resources.

Importantly, we note that releasing data in many cases is now no longer 'voluntary'. Under the terms of the 2013 Federal Open Data Policy, newly-generated government data is required to be made available in open, machine-readable formats, while continuing to ensure privacy and security. Federal CFO-Act agencies are required to create a single agency data inventory, publish public data listings, and develop new public feedback mechanisms. Agencies are also required to identify public points of contacts for agency datasets.

22.5.4 The Semantic Web and Linked Open Data

Beyond Open Government Data, there is an entire movement in the Computer Science community that involves more intelligent sharing, modeling, publication and standardization of data. The idea behind the Semantic Web is a growing recognition for making data, not just documents, the 'first class citizen' of the Web in support of an emerging *data economy* (Berners-Lee et al. 2001). What this really means is that a *systematic* framework is desired for publishing, representing and providing *direct* access to *raw* data that currently needs to be wrapped in an HTML document before being publicly exposed on the Web. Yet, the Web was originally designed to render documents on a browser for human consumption. How can we publish raw data using such a systematic framework without re-designing the Web itself?

The Linked Data movement, a direct product of a grassroots effort called the W3C (World Wide Web Consortium) Linking Open Data (LOD) project that was founded in January 2007, emerged as a potential (albeit, not unique) solution to this problem (Bauer and Kaltenböck 2011). In the years since then, the movement has grown, and many datasets have been published using the four Linked Data principles. Some well-known Linked Open Datasets include Wikidata, DBpedia, GeoNames and OpenStreetMap (Vrandečić and Krötzsch 2014; Auer et al. 2007; Wick 2006; Haklay and Weber 2008). Some of these were not originally published as Linked Data, but at the time of writing, Linked Data versions of these datasets now exist. Of these examples, OpenStreetMap and GeoNames are particularly important examples of geospatial data that have proven to be ubiquitous in a number of smart city and digital government initiatives due to their global coverage of the planet and the obvious importance of maps and geographical entities to city planning efforts.

22.5.4.1 OpenStreetMap

OpenStreetMap[6] is built by a community of mappers that contribute and maintain data about roads, trails, railway stations, among other things, all over the world (Haklay and Weber 2008). OpenStreetMap emphasizes local knowledge. Contributors use aerial imagery, GPS devices, and low-tech field maps to verify that OSM is accurate and up to date. OpenStreetMap powers map data on thousands of web sites, mobile apps, and hardware devices, a testament to its impact. Importantly, OpenStreetMap is open data, and users are free to use it for any purpose as long as they credit OpenStreetMap and its contributors. Concerning Smart Cities, OpenStreetMap is arguably the best example of a Big Geospatial Dataset that is also open and that can be used in support of many fo the initiatives we earlier described, especially considering the prohibitive cost of licensing and using technologies like Google Maps.

22.5.4.2 GeoNames

The GeoNames geographical database[7] is available for download free of charge under a creative commons attribution license (Ahlers 2013; Wick 2006). It contains over 25 million geographical names and consists of over 11 million unique features

[6]https://www.openstreetmap.org/#map=4/38.01/-95.84

[7]https://www.geonames.org/

whereof 4.8 million populated places and 13 million alternate names. GeoNames is integrating geographical data such as names of places in various languages, elevation, population and others from various sources. All lat/long coordinates are in WGS84 (World Geodetic System 1984). Users may manually edit, correct and add new names using a user friendly wiki interface. Features for online users include searching for names using full-text search, bookmarking maps, sending maps via email, exporting names as character separated value files or png images, adding new names to the database (for registered users) and geotagging of names (for registered users).

22.6 Examples of Smart Cities

To conduct a geographic analysis (including understanding the geographic distribution) of Smart Cities, Cocchia (2014) distilled over 700 papers into 162 case studies for an empirical analysis of Digital and Smart Cities. While full details of their analysis may be found in their original paper (though some of the data will be stale since new Smart Cities are emerging every year), the highlights of the analysis were particularly informative. We list some of the critical outputs of their analysis below:

1. Europe and Asia account for more than 85% of Smart and Digital[8] Cities in their case studies, with Asia accounting for 49%, and Europe accounting for 36% of the 162 case studies. In contrast, North America only accounted for 9% of the cases. The other continents together contributed only 6%, with Middle/South America contributing the least (1%).
2. Narrowing in on Asia and Europe, the authors found that there were 'macro-clusters' of Smart Cities both in Asia and in Europe. Asia was found to exhibit greater diversion than Europe, but it is unclear if this is also the case after controlling for the size difference between the two continents. Smart Cities in Asia tend to be on the Chinese east coast, while European Smart Cities appear to be more concentrated in the North Sea Region (including countries like the Netherlands, Belgium, United Kingdom, Scandinavia) and in the Mediterranean Region (Spain, France, Italy).
3. In North America, a cluster of Smart Cities can be found near the Great Lakes Region between the United States and Canada while in Oceania and Africa, small clusters are located along the most populated and developed areas e.g., the Australian east coast and South African coast.

These analyses and the authors' data were also validated through similar findings reached by the Ericsson Report about Networked Society City Index. That report

[8]Although Cocchia (2014) use the term Smart/Digital City, we continue to use Smart City to refer to the same in this section.

also concluded that cities located in Northern Europe, North America and parts of East Asia have a longer tradition of producing and using ICT equipment, and have therefore been able to benefit from their investments over longer periods of time. Furthermore, from the literature review about city case studies, we can observe that the spread of Smart Cities in Asia, Europe and North America have some shared features insofar as Big Geospatial Data is concerned. Arguably, without the cost-effective rise of Big Data and AI/ICT technology (Boyd and Crawford 2012; Davenport 2018; Pickavet et al. 2008), such cities (and Smart City initiatives in these cities) might not have seen adequate adoption due to cost and scaling issues.

Another insightful, case study-driven analysis on fifteen cities driven by Smart City technology can be found in Angelidou (2017). We summarize the various case studies in that analysis in Table 22.3. Some of the cases, such as the initiative in Rio de Janeiro, were covered in earlier sections.

22.7 Future Directions

Over the next decade, the real-time city is likely to become a growing reality as urban administrations and municipalities seek to capitalise on novel data sources and commercial innovations that become more affordable and accurate over time and with more processing and fine-tuning (both of which scale with increased adoption). Although Big Data offers a number of opportunities, it also raises significant concerns with respect to the politics and privacy of such data, technocratic governance, technical debt and corporate influence, system vulnerabilities especially in the realm of cybersecurity, ethical issues with respect to government and corporate surveillance, among others (including how data is interpreted). Given the role that such systems are likely to play in shaping urban governance there is a pressing need to understand and regulate their functions, ethical use and limitations in governance and matters of citizenry. There has been a growing number of arguments, especially given the rise in populism around the world, that without effective controls and safeguards in place, the smart cities of the future will likely reflect narrow corporate and state visions, rather than the interests of society at large.

Some of the backlash has also come about because of economic issues, and the perception of growing inequality and homelessness, particularly in tech-heavy US cities like San Francisco and Seattle. For example, Brad Smith, President and CLO at Microsoft at the time of writing, writes in a recent post:[9]

... rapid [tech] growth creates new strains for a community's infrastructure, including its schools and transportation network. As 2018 ended, there was some well-deserved focus on another aspect – the affordability of housing.

When the housing supply fails to keep pace with added population, housing prices rise, and some people are pushed out. There is growing awareness that this

[9]https://www.linkedin.com/pulse/today-technology-top-10-tech-issues-2019-brad-smith/

Table 22.3 Cases of Smart City strategies analyzed in Angelidou (2017)

Amsterdam smart city	Partnership among businesses, authorities, research institutions, and the people of Amsterdam to reduce CO2 emissions and improve the environmental record of the city
Barcelona smart city	Strategy focusing on "international promotion", "international collaboration", and "local projects". The number of local projects is more than a hundred, and the overall strategy is structured around the collaboration among government, industry, academia, and local citizenry
Smart London plan	Smart city plan created in 2013, and revolving around seven key themes in the domains of services for citizens, citizen engagement, development of businesses, smart infrastructure, and networking among stakeholders
PlanIT valley (Portugal)	Private, planned smart city to be developed in Portugal, to showcase the "Urban Operating System" which was developed by the software company Living PlanIT. This system will accumulate information from sensors placed throughout the city, which it will then feed to the applications that monitor and control the city's systems
Stockholm smart city	Strategy whereby environmental and information technologies are tested and used extensively throughout the city's infrastructure, with the purpose of creating a flourishing ecosystem that involves the city's inhabitants, private industry, and the public sector
Cyberjaya (Malaysia)	Planned smart city that is part of a broader government policy for advancing the country's innovation and knowledge economy. The city is expected to become a global ICT hub by attracting world-class multimedia companies, professionals, and students. ICT-wise, seven flagship applications are offered to citizens and businesses
Singapore intelligent nation 2015	10-year masterplan based on innovation, integration, and internationalization. It spans the digital media and entertainment sector, education and learning, financial services, healthcare and biomedical sciences, manufacturing and logistics, tourism, hospitality and retail, land and transport, and government and society
King Abdullah Economic city (Saudi Arabia)	Planned smart city focusing on manufacturing and logistics, shipping, light and processing industry and financial services. It will be wired with high-speed broadband infrastructure and all urban operations will be managed through Integrated Operation Centers, meant to act as the "brain of the city"
Masdar city (United Arab Emirates)	Planned smart city close to Abu Dhabi, designed on the basis of sustainable urban design. Its economy revolves around clean-tech research and development, pilot projects, technology, and materials testing
Skolkovo (Russia)	Planned city to be built close to Moscow, expected to contribute to the modernization of the Russian economy. It will forge a knowledge-and-innovation ecosystem by developing collaboration channels between industry and academia in five clusters: ICTs, biomedical, energy efficiency, space and nuclear technology
Songdo International business district (South Korea)	Already developed (again planned) city which is a model of sustainable, city-scale development and innovation and aims to become a central business hub in Northeast Asia

(continued)

Table 22.3 (continued)

Chicago smart city	Strategy for leveraging technology in order to promote opportunity, inclusion, engagement, and innovation. It foresees the collaboration of the public, the private, and the "third" sector to develop the city?s infrastructure, "smart" communities, governance, civic innovation, and the technology business sector
New York digital city	Strategy for the city of New York to become "the world's most digital city", developed with the engagement of residents, city employees, and technologists. Its four core areas are Access, Open Government, Engagement, and Industry, comprising altogether forty initiatives
Rio de Janeiro smarter city	Smart city initiative that was a collaboration of the city with technology vendor IBM to become a "smarter city" for the 2014 World Cup and the 2016 Olympics. Rio is now equipped with a citywide Emergency Response System that collects sensor-and-camera-generated data that enables informed decision making in policing, traffic, and energy management
Konza technology city (Kenya)	Planned smart city close to Nairobi, designed on the basis of sustainable design principles and expected to advance technology growth in Kenya. Its economy will focus on four sectors: education, life sciences, telecom, and Information Technology Outsourcing and Business Process Outsourcing

contributes to growing homelessness, as well as up to daily four-hour commutes in key U.S. cities for teachers, nurses, first responders and many other middle-income individuals who play vital community roles. For example, since 2011, as the greater Seattle area has evolved from the Emerald City to Cloud City, median home prices have increased by more than 80 percent while median household income has risen by only 30 percent. Other tech centers confront similar trends.

What this shows, and what has become disturbingly apparent to those who live in these cities, is that far from being a panacea for many of society's problems, technology may have contributed to them with economic and political repercussions that could last a long time. Consequently, there has been an uptick in protests in some of these cities. In San Francisco, activists now regularly blockade the private buses that tech workers take to Silicon Valley, and even protest directly outside the houses of tech executives. Some of the activism is organized e.g., a group calling itself the 'Counterforce' has been taking credit for some of these actions.

In the introduction, we drew on the characterization of Cocchia (2014) to specify smart cities as being defined along technological, human and institutional dimensions. The rising activism in tech-friendly cities presents evidence that the technological dimension, by itself, is not going to lead to stable and democratic smart cities that cater to all its stakeholders. Institutional and human dimensions are equally important, and more dimensions may start to emerge as the theory on smart cities continues to evolve. We believe that this is the most pressing issue for current and future work to address, both academically and in practice.

22.8 Conclusion

Smart Cities have gained much traction (and increased scrutiny) in recent years as a vision for stimulating entrepreneurship, and supporting innovation and economic growth, and providing avenues for sustainable and efficient urban management and development. A significant tool that has been utilized in this vision is data analytics, including social media and big geospatial data, for understanding, monitoring, regulating and planning such cities (Angelidou 2017). As cities have become increasingly embedded with digital infrastructure and networked devices, the volume of data produced about them has grown exponentially, providing rich streams of information about not just cities but *citizenry*. Big Data of this nature is varied, fine-grained, indexical, dynamic and relational enabling real-time analysis of different systems and inter-connections between systems. For citizens, such data and its analysis offers insights into city life, aids everyday living and decision-making, and empowers alternative visions for city development. For governments, Big Data and integrated analysis and control centers offer more efficient and effective city management and regulation. For corporations, big data analytics offers new, long term business opportunities as key players in city governance.

In conclusion, there is the proverbial ‘something for everyone’ in realizing and implementing the Smart City vision. However, there are also tradeoffs, as we noted in the earlier section. A Smart City based on the idea of a ‘technological utopia’ ends up, in practice, in public disenchantment, protests and economic crowding-out. Some argue it has led to the rise in populism, including movements such as the anti-immigration fervor and rise of conservatism in the US, Brexit in the UK, and the rise of far-right governments across Europe. If true, then this argument illustrates the importance of taking a holistic view in implementing and designing Smart Cities, one that balances and takes into account all three aspects: technological, human and institutional.

References

Ahlers D (2013) Assessment of the accuracy of geonames gazetteer data. In: Proceedings of the 7th workshop on geographic information retrieval, pp. 74–81. ACM

Aktan A, Helmicki A, Hunt V (1998) Issues in health monitoring for intelligent infrastructure. Smart Mater Struct 7(5):674

Anderson DN (2014) “not just a taxi”? for-profit ridesharing, driver strategies, and vmt. Transportation, 41(5):1099–1117

Angelidou M (2017) The role of smart city characteristics in the plans of fifteen cities. J Urban Technol 24(4):3–28

Anthopoulos L, Fitsilis P (2010) From digital to ubiquitous cities: Defining a common architecture for urban development. In: 2010 Sixth international conference on intelligent environments, pp 301–306. IEEE

Arasteh H, Hosseinnezhad V, Loia V, Tommasetti A, Troisi O, Shafie-Khah M, Siano P (2016) Iot-based smart cities: a survey. In: 2016 IEEE 16th international conference on environment and electrical engineering (EEEIC), pp. 1–6. IEEE

Ashton K et al (2009) That "internet of things" thing. RFID J 22(7):97–114
Auer S, Bizer C, Kobilarov G, Lehmann J, Cyganiak R, Ives Z (2007) Dbpedia: a nucleus for a web of open data. In: The semantic web. Springer, pp 722–735
Aurigi A (2016) Making the digital city: the early shaping of urban internet space. Routledge
Banerjee P (2009) An intelligent it infrastructure for the future. In: 2009 IEEE 15th international symposium on high performance computer architecture. IEEE, pp 3–4
Bătăgan L (2011) Smart cities and sustainability models. Informatica Economică 15(3):80–87
Bauer F, Kaltenböck M (2011) Linked open data: the essentials. Edition mono/monochrom, Vienna, 710
Bergmann E (2018) Dissecting populism. In: Conspiracy & populism. Springer, pp 71–97
Berners-Lee T, Hendler J, Lassila O et al (2001) The semantic web. Sci Am 284(5):28–37
Boden TA, Marland G, Andres RJ (2009) Global, regional, and national fossil-fuel co2 emissions. Carbon Dioxide Information Analysis Center, Oak Ridge National Laboratory, US Department of Energy, Oak Ridge, Tenn., doi, 10
Botsman R (2017) Big data meets big brother as china moves to rate its citizens. Wired UK, 21
Bourguignon F (2000) Crime, violence and inequitable development. In: Annual world bank conference on development economics 1999, pp 199–220
Boyd D, Crawford K (2012) Critical questions for big data: provocations for a cultural, technological, and scholarly phenomenon. Inf Commun Soc 15(5):662–679
Cash DW, Clark WC, Alcock F, Dickson NM, Eckley N, Guston DH, Jäger J, Mitchell RB (2003) Knowledge systems for sustainable development. Proc Natl Acad Sci 100(14):8086–8091
Chen M, Gonzalez S, Vasilakos A, Cao H, Leung VC (2011) Body area networks: a survey. Mobile Netw Appl 16(2):171–193
Christoforidis GC, Chatzisavvas KC, Lazarou S, Parisses C (2013) Covenant of mayors initiative–public perception issues and barriers in greece. Energy Policy 60:643–655
Cocchia A (2014) Smart and digital city: a systematic literature review. In: Smart city. Springer, pp 13–43
Commoner B (1991) Rapid population growth and environmental stress. Int J Health Serv 21(2):199–227
Couclelis H (2004) The construction of the digital city. Environ Plann B Plann Des 31(1):5–19
Davenport TH (2018) The AI advantage: how to put the artificial intelligence revolution to work. MIT Press
Dodge M, Kitchin R (2007) The automatic management of drivers and driving spaces. Geoforum 38(2):264–275
Dong XL, Srivastava D (2013) Big data integration. In: 2013 IEEE 29th international conference on data engineering (ICDE). IEEE, pp 1245–1248
Ergazakis K, Metaxiotis K, Psarras J (2004) Towards knowledge cities: conceptual analysis and success stories. J Knowl Manag 8(5):5–15
Famiglietti JS (2014) The global groundwater crisis. Nat Clim Chang 4(11):945
Gluhak A, Krco S, Nati M, Pfisterer D, Mitton N, Razafindralambo T (2011) A survey on facilities for experimental internet of things research. IEEE Commun Mag 49(11):58–67
Goldberg JP, Moye WT (1985) The first hundred years of the Bureau of Labor Statistics. Number 2235. US Department of Labor
Goldsmith B (2014) The smartphone app economy and app ecosystems. In: The Routledge companion to mobile media, pp 195–204. Routledge
Grubb M, Vrolijk C, Brack D (1997) The Kyoto Protocol: a guide and assessment. Royal Institute of International Affairs Energy and Environmental Programme
Gurjar B, Van Aardenne J, Lelieveld J, Mohan M (2004) Emission estimates and trends (1990–2000) for megacity delhi and implications. Atmos Environ 38(33):5663–5681
Haklay M, Weber P (2008) Openstreetmap: user-generated street maps. IEEE Pervasive Comput 7(4):12–18
Harrison C, Eckman B, Hamilton R, Hartswick P, Kalagnanam J, Paraszczak J, Williams P (2010) Foundations for smarter cities. IBM J Res Dev 54(4):1–16

Höjer M, Wangel J (2015) Smart sustainable cities: definition and challenges. In: ICT innovations for sustainability. Springer, pp 333–349
Hollands RG (2008) Will the real smart city please stand up? Intelligent, progressive or entrepreneurial? City 12(3):303–320
Jambeck JR, Geyer R, Wilcox C, Siegler TR, Perryman M, Andrady A, Narayan R, Law KL (2015) Plastic waste inputs from land into the ocean. Science 347(6223):768–771
Kezunovic M, Xie L, Grijalva S (2013) The role of big data in improving power system operation and protection. In: 2013 IREP symposium bulk power system dynamics and control-IX optimization, security and control of the emerging power grid. IEEE, pp 1–9
Khatoun R, Zeadally S (2017) Cybersecurity and privacy solutions in smart cities. IEEE Commun Mag 55(3):51–59
Kim T-H, Ramos C, Mohammed S (2017) Smart city and iot
Kitchin R (2014) The real-time city? big data and smart urbanism. GeoJournal 79(1):1–14
Komninos N (2006) The architecture of intelligent cities. In: Conference proceedings intelligent environments, vol 6, pp 53–61. IET
Kuznets S (1955) Economic growth and income inequality. Am Econ Rev 45(1):1–28
Lundvall B-Å, Lorenz E (2011) From the lisbon strategy to europe 2020 Toward a social investment welfare state, pp 333–351
Matzler K, Veider V, Kathan W (2015) Adapting to the sharing economy. MIT
Mayer H (1999) Air pollution in cities. Atmos Environ 33(24–25):4029–4037
Mehmood Y, Ahmad F, Yaqoob I, Adnane A, Imran M, Guizani S (2017) Internet-of-things-based smart cities: recent advances and challenges. IEEE Commun Mag 55(9):16–24
Michel H (2005) e-administration, e-government, e-governance and the learning city: a typology of citizenship management using icts. Electron J e-Government 3(4):213–218
Miller HJ (2010) The data avalanche is here. Shouldn't we be digging? J Reg Sci 50(1):181–201
Nam T, Pardo TA (2011) Conceptualizing smart city with dimensions of technology, people, and institutions. In: Proceedings of the 12th annual international digital government research conference: digital government innovation in challenging times, pp 282–291. ACM
Neirotti P, De Marco A, Cagliano AC, Mangano G, Scorrano F (2014) Current trends in smart city initiatives: some stylised facts. Cities 38:25–36
Norden LD, Vandewalker I (2017) Securing elections from foreign interference. Brennan Center for Justice at New York University School of Law
Pablo-Romero MDP, Sánchez-Braza A, Manuel González-Limón J (2015) Covenant of mayors: reasons for being an environmentally and energy friendly municipality. Rev Policy Res 32(5):576–599
Palmisano SJ (2008) A smarter planet: the next leadership agenda. IBM November 6:1–8
Penwarden R (2014) The rise of the smartphone
Perrin A (2015) Social media usage: 2005–2015
Pickavet M, Vereecken W, Demeyer S, Audenaert P, Vermeulen B, Develder C, Colle D, Dhoedt B, Demeester P (2008) Worldwide energy needs for ict: the rise of power-aware networking. In: 2008 2nd international symposium on advanced networks and telecommunication systems. IEEE, pp 1–3
Rimmerman CA (2018) The new citizenship: unconventional politics, activism, and service. Routledge
Ruester S, Schwenen S, Finger M, Glachant J-M (2014) A post-2020 eu energy technology policy: Revisiting the strategic energy technology plan. Energy Policy 66:209–217
Shi W, Cao J, Zhang Q, Li Y, Xu L (2016) Edge computing: vision and challenges. IEEE Internet Things J 3(5):637–646
Silver C (2007) Planning the megacity: jakarta in the twentieth century. Routledge
Stone M, Aravopoulou E (2018) Improving journeys by opening data: the case of transport for london (tfl). Bottom Line 31(1):2–15
Su K, Li J, Fu H (2011) Smart city and the applications. In: 2011 international conference on electronics, communications and control (ICECC). IEEE, pp 1028–1031

Torres PB, Doubrava R (2010) The covenant of mayors: cities leading the fight against the climate change. In: Local governments and climate change. Springer, Dordrecht/Heidelberg/London/New York, pp 91–98

Victor DG (2011) The collapse of the Kyoto Protocol and the struggle to slow global warming. Princeton University Press

Vlahov D, Galea S (2002) Urbanization, urbanicity, and health. J Urban Health 79(1):S1–S12

Vrandečić D, Krötzsch M (2014) Wikidata: a free collaborative knowledge base

Welzel C, Haerpfer CW, Bernhagen P, Inglehart RF (2018) Democratization. Oxford University Press

Weyant J, Hill J (1999) The costs of the Kyoto Protocol: a multi-model evaluation. International Association for Energy Economics

Wick M (2006) GeoNames. GeoNames

Xiaojun C, Xianpeng L, Peng X (2015) Iot-based air pollution monitoring and forecasting system. In: 2015 international conference on computer and computational sciences (ICCCS). IEEE, pp 257–260

Yin S, Kaynak O (2015) Big data for modern industry: challenges and trends [point of view]. Proc IEEE 103(2):143–146

Zaki MJ, Hsiao C-J (2002) Charm: an efficient algorithm for closed itemset mining. In: Proceedings of the 2002 SIAM international conference on data mining. SIAM, pp 457–473

Zhou K, Fu C, Yang S (2016) Big data driven smart energy management: from big data to big insights. Renew Sust Energ Rev 56:215–225

Zhu J, Fang X, Guo Z, Niu MH, Cao F, Yue S, Liu QY (2009) Ibm cloud computing powering a smarter planet. In: IEEE international conference on cloud computing, pp 621–625. Springer

Chapter 23
The 4th Paradigm in Multiscale Data Representation: Modernizing the National Geospatial Data Infrastructure

Barbara P. Buttenfield, Lawrence V. Stanislawski, Barry J. Kronenfeld, and Ethan Shavers

23.1 Access to Nationally Managed Spatial Data in the United States

The need of citizens in any nation to access geospatial data in readily usable form is critical to societal well-being, and in the United States (US), demands for information by scientists, students, professionals and citizens continue to grow. Areas such as public health, urbanization, resource management, economic development and environmental management require a variety of data collected from many sources to identify problems, monitor trends and propose solutions. Such information needs and demands have driven the coordination of federal and regional government agencies with respective private sector participation to develop national geospatial data infrastructures in many countries.

Early spatial data infrastructures emerged in Germany, Switzerland, Canada, Australia, United Kingdom, and France, followed by national infrastructures in other nations, as for example Poland (Bialecka et al. 2018), the Netherlands, Brazil, Bolivia, Malaysia and smaller countries such as the Canary Islands. Buddhathoki et al. (2008) estimated the existence of roughly 100 national spatial data infrastructures, with more in development at regional and subnational levels. A sample

B. P. Buttenfield (✉)
University of Colorado-Boulder, Boulder, CO, USA
e-mail: babs@colorado.edu

L. V. Stanislawski · E. Shavers
U.S. Geological Survey, Center of Excellence for Geospatial Information Science, Reston, VA, USA
e-mail: lstan@usgs.gov; eshavers@usgs.gov

B. J. Kronenfeld
Eastern Illinois University, Charleston, IL, USA
e-mail: bjkronenfeld@eiu.edu

M. Werner, Y.-Y. Chiang (eds.), *Handbook of Big Geospatial Data*,
https://doi.org/10.1007/978-3-030-55462-0_23

of current spatial data infrastructure websites for various nations is at https://en.wikipedia.org/wiki/Spatial_data_infrastructure. In the European Union, a federation of countries collaborate on the Infrastructure for Spatial Information in Europe (INSPIRE) project (European Parliament and Council 2007; Craglia et al. 2014). In the United States, the National Spatial Data Infrastructure (NSDI) evolved with initial collaboration by 16 federal agencies (NRC 1993) that has since doubled in size, as described later in this chapter.

A spatial data infrastructure (SDI) differs from a database in several ways. The SDI incorporates more than a suite of data layers. It is comprised of facilities, software, systems and installations that provide necessary information integration and services for the proper functioning of a specific realm of society, such as a community of users, or an organization, city, or country (Coetzee et al. 2019). In addition to data and algorithms, an infrastructure comprises "the technology, policies, standards, human resources and related activities necessary to acquire, process, distribute, use, maintain, and preserve spatial data" (OMB 2002). Nebert (2004) defined SDI as a "framework of spatial data, metadata, tools, and a user community that are interactively connected so that spatial data can be used in an efficient and flexible way." Many SDIs have been developed for various reasons.

The U.S. Geological Survey (USGS) Topographic Mapping Program (currently the National Geospatial Program) is an example of a longstanding SDI for the United States (NRC 2012). The program began in 1884 and its general-use topographic maps are still recognized today as a signature and versatile product for viewing and evaluating the landscape, for managing natural and human resources, and for environmental stewardship. Over the years, USGS topographic maps evolved from single-version paper products to digital form called US Topo that is revised on a 3-year cycle and generated from the most current data available (Usery et al. 2009). A graphic comparison of a 1:24,000-scale scanned paper map for Fort Logan, Colorado with the current (2019) US Topo version is shown in Fig. 23.1.

Being a larger country than most, with a diverse range of geographic conditions, the United States continues development of automated workflows for multiscale mapping and feature representation, whereas smaller European countries are more advanced in automated multiscale mapping. The NSDI contains much larger data stores and a larger range of spatial resolutions than what is available in most other countries, including finer granular content in elevation and contour representations, and in hydrographic databases.

Multiscale representation strategies help to steward and sustain any national geospatial data infrastructure, especially those generated by multiple agencies with differing mandates that must align with federal, state and local policies and a variety of user needs. Such strategies are commonly subsumed under the rubric of data modeling and generalization. Generalization processing systematically modifies details in geospatial data for appropriate use in mapping and spatial analysis at reduced scales. The shift toward fully automated generalization has emerged as computing technology advances to support much larger data volumes (also referred to as "big data"), distributed data delivery, and streaming data sources that enable frequent or in some cases near-real-time data updates. At present, multiscale representation has become an umbrella term that incorporates generalization, data modeling to transform data from one scale or resolution to another, data integration to provide linkages between data versions across multiple scales, and geovisualization methods that preserve visual logic of graphical depictions across display scales.

The principles underlying multiscale representation build upon traditional objectives for generalization and include (1) retention of essential details that provide evidence of surface processes; (2) omission of details that become visually indis-

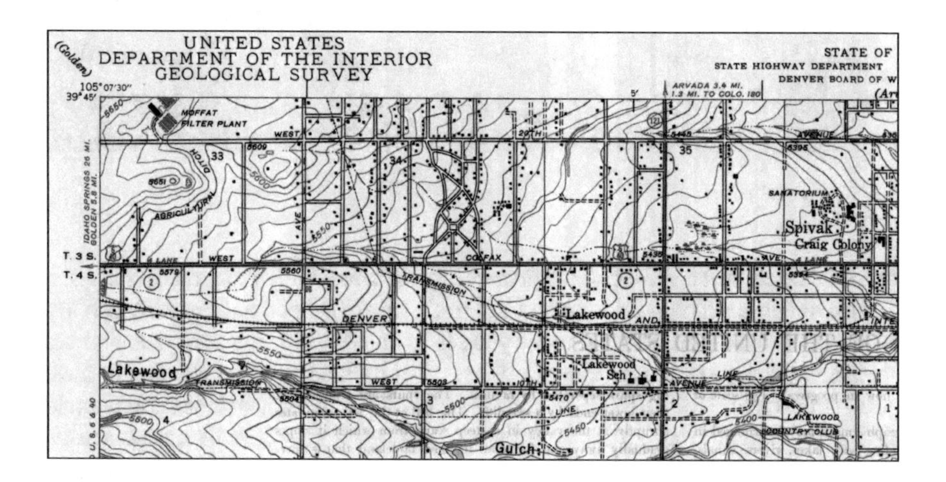

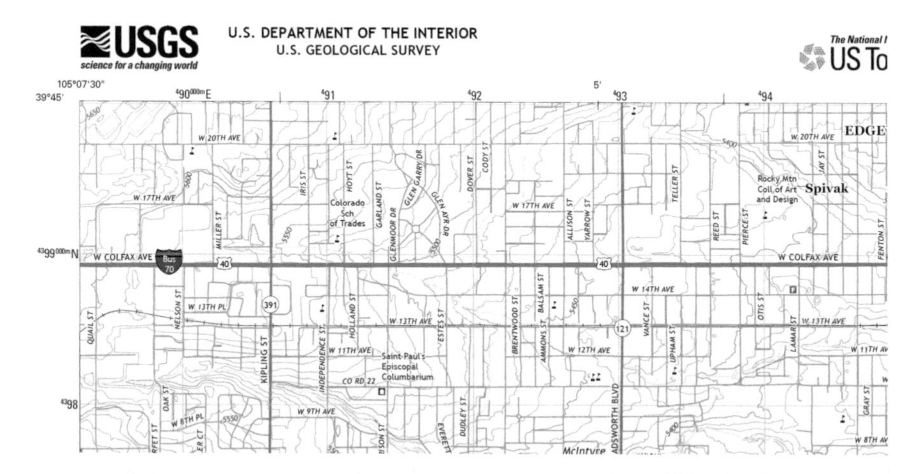

Fig. 23.1 Portions of 1:24,000-scale US topographic maps for Fort Logan Quadrangle. 1948 scanned paper edition (top), 2016 U.S. Topo digital GeoPDFderived from The National Map databases without image background (center), 2016 digital US Topo digital GeoPDF derived from The National Map datasets with National Agriculture Image Program image background (bottom)

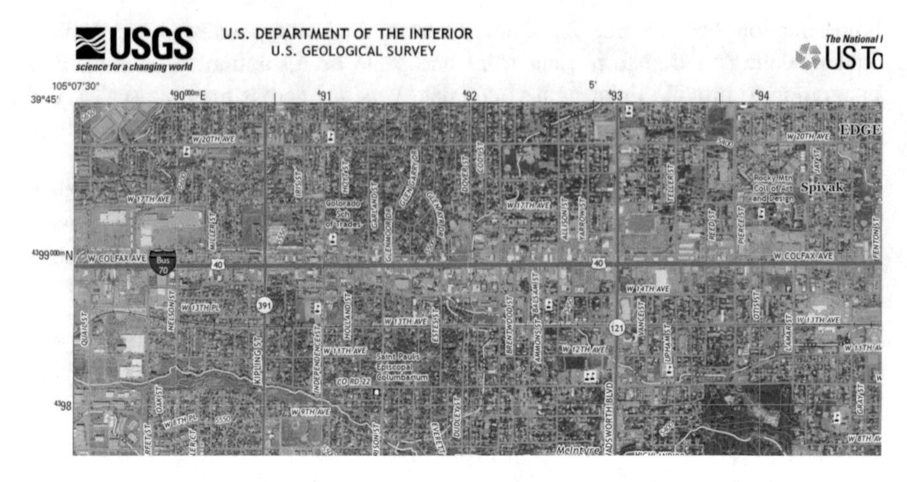

Fig. 23.1 (continued)

cernible at reduced scales; and (3) preservation of spatial relations (e.g., adjacency, displacement, topology) that support scientific investigations and modeling. Early strategies focused on basic geometry such as length and angularity; and more sophisticated approaches have evolved that tailor multiscale data modeling to local geographies (e.g., Touya 2008; Mackaness et al. 2007; Stanislawski and Buttenfield 2011; Buttenfield et al. 2011). Appropriate methods for multiscale representation can improve preservation of landscape conditions and characteristics that are important for geomorphometric analysis (Lindsay 2016; Sangireddy et al. 2017; Newman et al. 2018; Oguchi 2019). And increasingly, multiscale representation involves data integration that harmonizes information horizontally (across environmental, economic, and demographic databases) and vertically (from local to regional to national to global levels).

This chapter reviews existing approaches to multiscale representation (including data modeling, generalization and data integration, and geovisualization) in support of the NSDI, emphasizing where additional research is needed and how technological advances in data structures and algorithms, such as machine learning and deep learning for pattern recognition or feature extraction, may provide a trajectory to improve maintenance and update procedures, and to facilitate access to the NSDI for a variety of scientific, economic and societal applications. Machine learning and deep learning encompass the development, testing and examination of algorithms and statistical models that automatically effectively perform a specific task without explicit instructions, relying on pattern recognition, training algorithms, and inference. It is considered an important component of artificial intelligence. Incremental but cumulatively significant improvements in the speed, robustness and accessibility of existing data structures and algorithms (e.g., network datasets, Triangulated Irregular Networks (TINs), skeletons and related structures, numerical equation solvers) have advanced to a point where ideas that were not practical to implement in the past are now much easier to translate from theory to implementation. Best practices of data curation, processing and dissemination have also changed, marking a shift towards open and reproducible science. Tenets of the Fourth Paradigm (Hey et al. 2009) refer to all of these changes as a formal strategy to address the big data revolution and data intensive discovery.

Challenges persist. The standard "framework data sets" included in the NSDI include geodetic control, orthoimagery, terrain, cadastral information, surface hydrography, transportation and administrative boundaries. Additional data layers are NSDI components as well, such as soils, land use, land cover, geographic names, and thematic information including demographic and economic data. These data are maintained by multiple agencies and private sector stakeholders, with varying scale sensitivities, attribute aggregation levels, and accuracy standards; and this complicates efforts for horizontal and vertical data integration that are essential to analysis and mapping applications, and to the integration of multiple SDIs into an NSDI. Moreover, one of the most significant and valuable objectives of the NSDI is that data products and services for access and exchange should be made freely available both internally (among the data producing agencies) and externally to the nation as a whole, without jeopardizing individual anonymity or national security. The objective of broadest possible access, use and distribution distinguishes the NSDI from some other countries, and in some respects complicates data delivery as the NSDI evolves to accommodate changing definitions of geospatial data and services.

The question is, what would it take (in terms of data-driven science) to overcome these challenges and generate an NSDI for the United States that resides within a unified data framework, that can be accessed across a wide range of user-specified spatial resolutions, with item-level metadata, linked versions of features, and landscape descriptors that can be modified for a wide range of natural and social science applications, in an integrated suite of data models and interoperable formats? Is this unachievable in a country the size of the United States? Is it a quixotic fantasy? Or could adoption of data management and manipulation strategies such as those described in Jim Gray's Fourth Paradigm (2009) bring GIScience communities closer to meeting and overcoming this Grand Challenge?

This chapter provides a brief chronology of the United States NSDI and subsequently presents a visionary approach to modernize and integrate the NSDI that may or may not be fully operational or even achievable at present. Nonetheless, current advances in data-driven science, machine learning, and emerging statistical methods open the door to opportunities to consider a NSDI for larger nations that can be compiled, maintained, integrated and distributed in a manner analogous to those for smaller nations. The chapter will cover strategies for how this vision might be implemented, which could require advanced harmonization and statistical reasoning using ancillary layers, advanced data linking strategies, and a number of data science methods that are just beginning to emerge. The authors emphasize at the outset that solving the technical aspects of multiscale data representation cannot by itself lead to a fully integrated NSDI, as institutional mechanisms and governmental policies must be considered also. In the limited space of this chapter however, the focus will remain on multiscale representation.

23.2 Chronology and Current Status of NSDI in the United States

The US Federal Geographic Data Committee (FGDC) was established in 1990, and is an organized structure of 32 geospatial agencies, professionals and constituents at federal, state, tribal and local government levels that guides and directs geospatial activities and data sharing in the nation (https://www.fgdc.gov/organization). A primary task of the FGDC is coordination and implementation of the NSDI, which "leverages investments in people, technology, data, and procedures to create and provide the geospatial knowledge required to understand, protect, and promote our national and global interests" (FGDC 2013). In describing its vision for a spatial data infrastructure for the country, a national panel suggested that the phrase "discover and share for the long term" should be the mantra for spatial data handling (NRC 2012).

A key condition for sharing data and technology services is interoperability. Interoperability provides the ability to access, exchange and use information in a uniform and efficient manner. Priorities of access, exchange and use have guided the emergence and evolution of the NSDI since its inception with the signing of Executive Order 12906 by President William Clinton on 11 April 1994. The Order mandated FGDC to coordinate geospatial data access and sharing across the nation as a whole (The White House 1994). The Order required all US federal agencies that receive funds for the production of geospatial data to generate appropriate metadata and adopt standards for describing and distributing the created data through the NSDI. FGDC's roles were codified in federal law in the Geospatial Data Act (GDA) of 2018 (https://www.fgdc.gov/gda/geospatial-data-act-of-2018.pdf). The GDA adds more structure for accountability by federal agencies implementing the NSDI and collaborating with public and private sectors. Today, the FGDC is chaired by the Secretary of the Interior with the Deputy Director for Management, Office of Management and Budget (OMB) as vice chair. The FGDC continues to guide, steward, and coordinate the proper generation, collection, and distribution of geospatial data, metadata and services within the NSDI (https://www.fgdc.gov/gda/gda-fact-sheet-may-2019.pdf).

The national importance and value of available, interoperable data is most recently reflected in the passage of key United States laws and Administration directives designed to improve data availability, access, management processes, privacy protections, and preservation. These include the "Foundations for Evidence-Based Policymaking Act of 2018" (Evidence Act) that also includes the "Open Government Data Act", and Title III, the "Confidential Information Protection and Statistical Efficiency Act". These laws are accompanied by the Executive Order on Maintaining American Leadership in Artificial Intelligence (The White House 2019), and the "Federal Data Strategy (FDS) 2020 Action Plan" (https://strategy.data.gov/). These U.S. laws and policies are being implemented in an integrated manner with the FDS guiding the actions that establish a more unified federal data enterprise.

As directed by the GDA, the FGDC "shall lead the development and management of and operational decision making for the National Spatial Data Infrastructure strategic plan and geospatial data policy" (https://www.congress.gov/115/plaws/publ254/PLAW-115publ254.pdf, section 753). The most recent NSDI Strategic Plan (2014–2016, discussed below) focused predominantly on improving the federal geospatial data portfolio and supporting services. Currently in development, the FGDC is working with federal agencies and non-federal partners to develop a consensus-based strategic plan that supports the new laws and policies. Key elements are projected to include: strengthening the role of non-federal entities in the NSDI and increased integration of their geospatial data; continuing to improve discoverability, access and use of geospatial data and services; improving data and services interoperability through the implementation of standards; and advancing the use of the Geospatial Platform shared service.

Additionally, the Evidence Act's establishment of a Federal Chief Data Officer's (CDO) Council, and the cross-representation on the Council of FGDC members, provides a new unique opportunity to advance spatial analytics, by providing a policy body who can establish interagency processes and policies to improve alignment and integrated use of spatial and non-spatial data. As part of the FDS Action Plan, FGDC members will begin working across the spatial, statistical and other federal data communities to develop practices that enable machine readable methods to identify and relate data with spatial features to other statistical, structured and unstructured data with comparable spatial attributes or characteristics. This will broaden data use, enable innovative data analysis and applications, and expand the use of spatial analytics.

Tremendous growth in geospatial data, technologies, and industries over recent years (2010–2019) has spurred the FGDC to set priorities for the continued enhancement (i.e., development, integration and maintenance) of the NSDI. Goals laid out in the 2014–2016 NSDI strategic plan include: developing capabilities for national shared services; ensuring accountability and effective management of federal geospatial resources; and convening leadership of the national geospatial community. It is expected that enhancing the NSDI will better leverage investments in people, technology, data, and procedures for increased growth of geospatial applications and knowledge. Details of the full strategic plan can be found in FGDC (2013), but here we briefly describe objectives for the first goal. Development of capabilities for national shared services requires four foundations: a Geospatial Interoperable Reference Architecture (GIRA); a Geospatial Platform to share geospatial data, services, and applications through an internet portal; appropriate application of cloud computing; and promotion of multiagency geospatial acquisition vehicles. These four requisites are described in detail below.

23.2.1 Geospatial Interoperability Reference Architecture (GIRA)

The FGDC worked closely with partners to develop a GIRA (https://www.fgdc.gov/what-we-do/develop-geospatial-shared-services/interoperability/gira), an architecture that provides practical guidance that is aligned with U.S. federal policy, principles and practices for enterprise architecture to make geographic information discoverable, accessible, and usable to all stakeholders, including the general public. The GIRA provides a framework for the management, design, and development of geospatial systems and solutions and recommends performance measures for validating and reporting results. It provides guidance for governance, business, data, applications, services, infrastructure, standards, and security of geospatial information.

The GIRA includes reference models to guide development of business, data, application/service, infrastructure, security, geospatial interoperability, and performance components for the NSDI. A reference model describes common practices to describe, use, and share information or methods across organizations. After presenting a geospatial baseline assessment matrix for data inputs, the data reference model describes a structure that standardizes data description, data context, and data sharing. Detailed descriptions of all the reference models are furnished in the GIRA document (see also GIRA 2015). Use of International Standards Organization (ISO) metadata (ISO 19115-1:2014) is required as a consistent method for describing data for all federally stewarded geospatial data. Data content is augmented with any information that provides additional meaning to data or the purpose for which the data were created, which can also refer to categorization methods, such as taxonomies or ontologies (OMB 2013). Data content should be documented through a web accessible language or service to allow search and discovery of the data and metadata. Lastly, data sharing requires agreed-upon content models that enable participating organizations to create interfaces to access or to distribute data.

23.2.2 Geospatial Platform

Through coordination by the FGDC, an internet portal for a Geospatial Platform was developed for the NSDI allowing access to tools, applications, products, services and communities, to promote collaboration and use of NSDI resources (FGDC 2013). The current platform (https://www.geoplatform.gov/) includes featured content about new and forthcoming datasets or services, such as the Ocean Reports web tool and the Amazon public domain dataset of USGS 3-D Elevation Program (3DEP) lidar point cloud data. The site includes access to the National Geospatial Data Asset (NGDA) themes, originally including 34 themes (OMB 2010) that have been aggregated into 17 primary themes (OMB 2017) including the core framework data themes (geodetic control, orthoimagery, elevation and bathymetry,

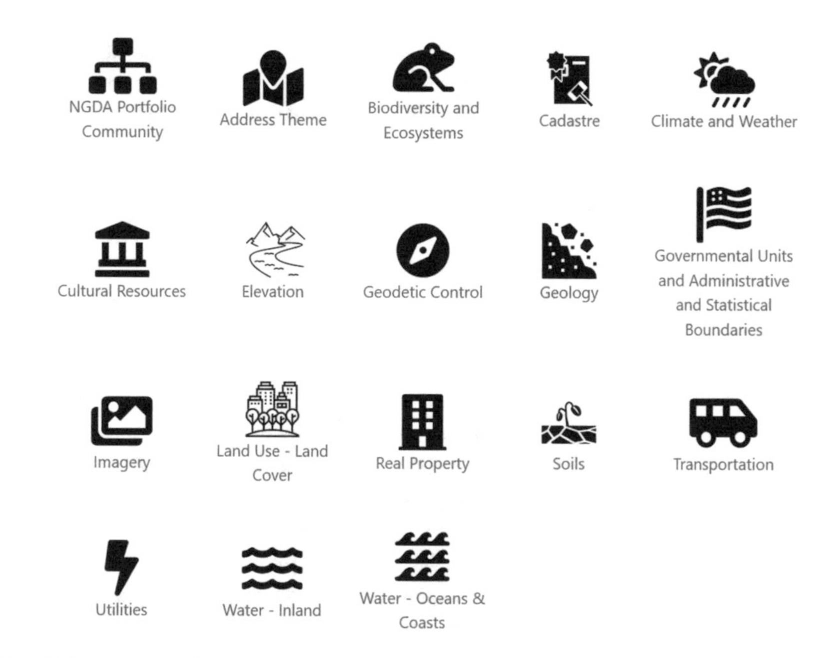

Fig. 23.2 National Geospatial Data Asset (NGDA) Themes on the Geospatial Platform (https://www.geoplatform.gov/)

transportation, hydrography, cadastral, and governmental units) (OMB 2010). An NGDA Portfolio Theme is also available, which coordinates additional NGDA themes (Fig. 23.2). The portal enables users to search or explore resources for data, services, maps, galleries, and communities through portfolio pages.

23.2.3 Cloud Computing

Cloud computing utilizes a network of remote servers hosted on the Internet to store, manage and process data. The technology provides an alternative for building data services on the internet rather than building them on personal or local computers. A big advantage is that it provides immediate storage and processing services with a full range of processing speeds and storage capacities to handle a range of user needs that may not be available locally. Cloud computing can provide organizations with cost effective alternatives for high data volume or processing services. Common examples of cloud services include generic social media platforms such as Facebook, Instagram, and Twitter, and online media services such as Pandora, Netflix, and Amazon. Commercial Geographic Information System (GIS) platforms, such as

Esri, Google Maps, and Bing Maps rely heavily on cloud computing architectures to promote products and services.

Cloud computing capabilities can also be supported by freely distributed open source software, such as GRASS (Geographic Resources Analysis Support System), GDAL (Geospatial Data Abstraction Library), PDAL (Point Data Abstraction Library), QGIS, R, and Python, along with a myriad of specialized modules such as pandas, NumPy, Fiona, SciPy, Spectral, etc. As envisioned by the Open Source Geospatial Foundation (OSGeo, https://www.osgeo.org/about/), groups or individuals can contribute to cloud-based infrastructures without proprietary software licensing restrictions that can limit free sharing of computing resources. Thus, many customizable open source computing capabilities are being distributed through data and software repositories, such as GitHub and Bitbucket. One example is the geospatial data analysis platform called Whitebox Tools (https://github.com/jblindsay/whitebox-tools). Another example is the proposed EarthMAP initiative at USGS, intended as an integrated predictive science capability to provide actionable intelligence that integrates data and interpreted results in support of earth system characterization science spanning disciplinary boundaries. Such cloud computing and open-source capabilities align with best practices for reproducible science (Stodden et al. 2018; Petras et al. 2015). The goal is to publish linked and executable data, code, and results in open-source formats, through publicly accessible online portals, to support both replicability as well as capacity-building that will nurture more efficient advances in knowledge as well as public trust in science (Peng 2011).

A related effort at USGS has seen further development, by making 3DEP data accessible through the cloud. The USGS is making all National Map products and services available through a cloud-based Amazon Web Services (AWS) infrastructure. Data available through the cloud includes 3DEP data, derived products and services. Since 2016, USGS 3DEP has collected airborne-based lidar point cloud data with the goal of providing very fine resolution elevation data for the entire country by 2023 (https://www.usgs.gov/core-science-systems/ngp/3dep/what-is-3dep). Airborne-based Interferometric Synthetic Aperture Radar (IfSAR) data are used for Alaska where weather conditions hinder lidar collection. Early in 2017, nearly 12 trillion lidar data points (about 98 terabytes) in LAS zip (LAZ) file format were made available through a requestor-pays Amazon cloud service. These LAZ files are still available, however, beginning in 2019, through a collaborative effort between Hobu, Inc., the U.S. Army Corps of Engineers, and AWS. LAZ files were converted to Entwine Point Tile files (Fig. 23.3), which is a lossless, streaming octree-based LAZ file encoding structure. These cloud-optimized data are now part of the Open Data registry provided by AWS (https://registry.opendata.aws/usgs-lidar/) and freely downloadable and instantly accessible for processing within the cloud. LidarExplorer tools are under development by the USGS to implement typical processing requests in the cloud for the entwined lidar data, such as clip, project, filter and transform.

USGS / Entwine

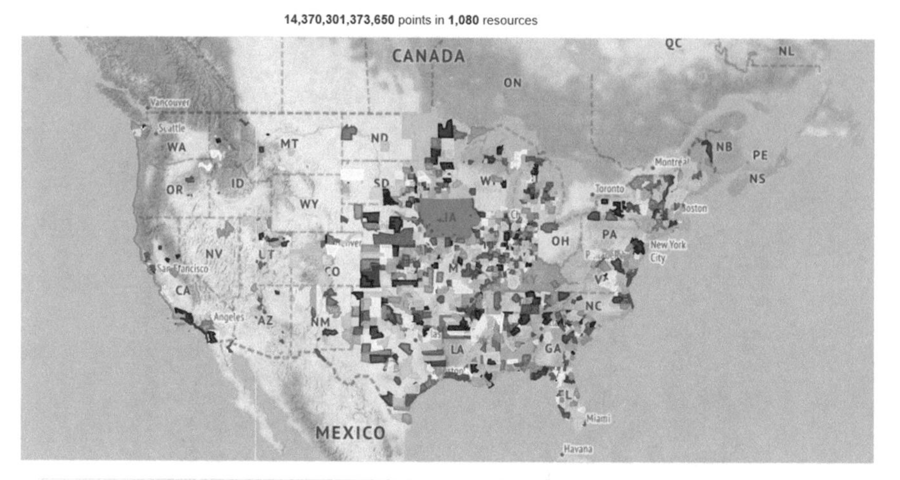

Filter

Name ↓	Points	Potree	Plasio
AK_BrooksCamp_2012	529,285,317	●	●
AK_Coastal_2009	55,711,772	●	●
AK_Fairbanks-NSBorough_2010	1,266,097,458	●	●
AK_Juneau_2012	2,211,557,952	●	●
AK_Kenai_2008	14,054,081,685	●	●

Fig. 23.3 USGS lidar projects with data that are currently available in Entwine file encoding. https://usgs.entwine.io/

23.2.4 Multiagency Geospatial Acquisition

The fourth FGDC foundation supporting national shared services for the NSDI promotes collaboration among agencies and organizations. Collaboration generates cost savings by leveraging government purchasing power for acquisition of geospatial data and services, and by distributing costs among contributing organizations. In the United States, collaboration on data production, maintenance and exchange is coordinated by the FGDC steering committee and executive committee, each composed of representatives from many government agencies, private and state organizations. The two committees oversee activities and distribute responsibilities among thematic subcommittees and coordination groups, each focused upon various geospatial data themes and services (https://www.fgdc.gov/organization). An overview of the organizational structure and components of the FGDC is shown in Fig. 23.4. A good example of collaborative efforts supported by the FGDC is the collection of lidar data by the USGS 3DEP, as described

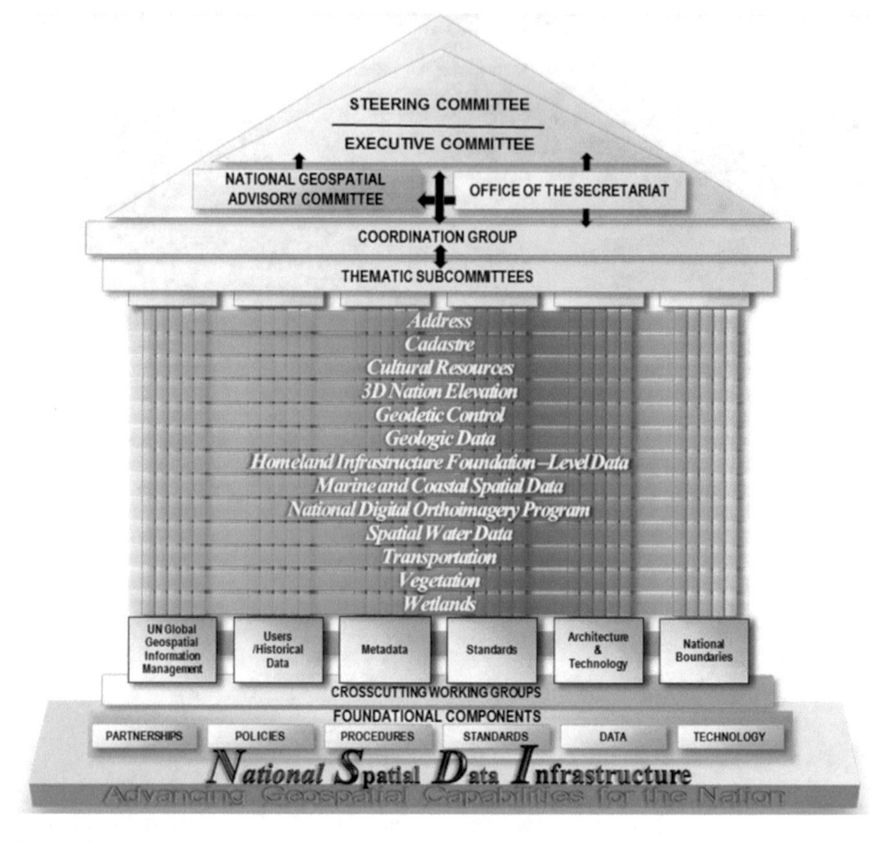

Fig. 23.4 Organizational structure and components of the Federal Geospatial Data Committee. (From https://www.fgdc.gov/organization)

above, whereby costs and work have been distributed among multiple government and private agencies. This lidar program uses an online geospatial analysis tool called "SeaSketch" to gather and prioritize data requirements among multiple agencies. Details on how this part of the NSDI strategy is being implemented are discussed on the project homepage (https://www.seasketch.org/#projecthomepage/5272840f6ec5f42d210016e4/about). A cost-benefit analysis coordinated by FGDC triggered the incentive for multiple federal, state, and private organizations to collaborate on lidar data collection for 3DEP.

It should be clear that a fully operational NSDI for a country as vast and diverse as the United States could not be undertaken without the collaborative efforts of multiple agencies and partnerships, an enterprise reference architecture that encourages interoperability and tools to support data discovery, as well as technological advances such as a publicly accessible geospatial platform, and cloud computing that make open access and freely distributed data products feasible.

To summarize, the four foundations of NSDI provide a firm and reliable base upon which to establish comprehensive horizontal and vertical integration, utilizing emerging approaches such as machine learning and advanced data structures to sustain a workable trajectory for using NSDI to address the pressing societal, environmental and economic issues facing the United States today. The next section of this chapter presents a conceptual framework that could support the vision for utilizing NSDI to greatest national and international benefit.

23.3 The Role of the Fourth Paradigm

In 2007, Microsoft researcher Jim Gray presented an argument (Gray 2007) to the National Research Council's Computer Science and Telecommunications Board arguing for a major transformation of scientific research to focus more directly on data-intensive systems and open access to data, processing and scientific communication. Gray termed the transformation a Fourth Paradigm arguing that it evolved from three historical paradigms of experimentation (a focus on what can be directly observed and measured), theory building (a focus on what can be formally hypothesized or inferred in order to lead to generalizable explanations), and simulation (relying on maturing computational power to reconstruct past conditions or to forecast future states that may be probable but which are "... too complicated to solve analytically" (Gray as quoted in Hey et al. 2009: xvii). The Fourth Paradigm focuses upon data-intensive science consisting of data capture and validation, curation including organization, cataloging and metadata creation, analysis, and open communication and publication of data and findings.

Roots of Gray's argument were founded on the widespread acceptance by several scientific disciplines that the increasing volume of data forming the basis for important scientific advances were stored on physical media that was not publicly accessible, that were vulnerable to magnetic decay and/or could be misplaced or even discarded due to impermanent information provenance. The National Science Foundation's (NSF's) 1993 Digital Libraries Initiative marked an early federal effort to catalog, cross-reference and link data collections in networked environments, and eventually led to development of search engines such as Google (https://www.nsf.gov/discoveries/disc_summ.jsp?cntn_id=100660). In 2005, NSF's National Science Board solicited research (NSB 2005) about preserving data over longer timeframes to support advancements in science and education, defining a new transdisciplinary community of what they termed "data scientists". This community included information scientists, computer scientists, database engineers, data curators and archivists, and domain experts.

Data science, also called data-intensive science, e-science, and data-driven science by various researchers, emerged at the nexus of three domains: computer and information science, mathematics and statistics, and topical domains including business and management science. These support advanced methods such as machine learning and artificial intelligence (AI), software and algorithm development and deployment, technologic progress in distributed computing and cyberprocessing, and societal and sociotechnical implications of applying such methods to problem domains including but not limited to medicine and genetics, earth and atmospheric sciences, economics, social sciences and digital humanities. The objective is to take full advantage of existing data, instead of formulating theory and then trying to establish analytic results to confirm or deny it (Steadman 2013). Data science strategies set up research frameworks that search for and elicit possibly latent patterns in very large data archives, essentially letting the data speak for themselves. As Gould (1981: 176) stated prophetically more than a decade before these efforts and initiatives were formalized:

> If we are to pay reverent heed, and write *theoria* for a Science that is the theory of the real, then we must let relations between things, the connections between elements of sets, be stated in such a language that the complexity of data that is trying to speak to us is not crushed out before we even start.

Culmination of the threads summarized here underscores a national need for a vertically and horizontally linked, fully transparent, and fully operational spatial data infrastructure, and raises the central question posed in this chapter, specifically, what would it take to achieve an NSDI (or even a federated global spatial data infrastructure "GSDI") that can support data driven science using linked and integrated big data and following best practices of reproducible science? The remainder of this chapter presents a selection of activities and research domains that could move the nation closer to that objective.

23.4 Activities for Short- and Longer-Term NSDI Implementation

Based upon the principles espoused in Jim Gray's Fourth Paradigm concept, we propose activities to further integrate big data and national spatial infrastructures from three temporal perspectives, taking into account obstacles and challenges that might need to be resolved. Some solutions are already developed or in testing. Other obstacles are either more challenging or cannot be achieved without other advances in technology or knowledge, and will require longer term attention to become a part of NSDI.

One obstacle relates to the demand for geospatial data at a continuous or near-continuous range of scales and resolutions, where data producers are generally able to generate data products only at a small number of scales. Another obstacle relates to the need to explore and compare among the large number of available databases for any theme (hydrography, soils, road networks, vegetation, terrain, etc.) in order to make informed decisions about which database from which producer will serve the immediate display or analytical purpose, the required level of data quality, attribution, file size and form and dimensionality. A third obstacle relates to changes in technology that have allowed more sophisticated data types, such as streaming data, immersive video, and augmented reality. The challenge for NSDI is to provide geospatial data to support user creation of such products. There is every reason to envision that in the longer term, NSDI can move towards direct provision of these advanced data types as a foundation for users to create even more advanced analyses and data products. Solutions will be proposed that might be implemented in the short term (1–4 and 5–7 years) and in the longer term (more than 7 years to implementation).

23.4.1 Short-Term Goals: Integrate NSDI Across Spatial and Temporal Scales

In the short term (1–5 years), tools to increase NSDI multiscale functionality will provide data with continuous or near-continuous feature scales. At present, national databases are compiled within a single spatial scale or localized scale range, with each version isolated from others. The National Hydrography Dataset High-Resolution (NHD HR) version is somewhat of an exception. Through collaborative efforts, the USGS produces the NHD HR, which includes the best available hydrographic data for the country (https://www.usgs.gov/core-science-systems/ngp/national-hydrography/national-hydrography-dataset?qt-science_support_page_related_con=0#qt-science_support_page_related_con). NHD HR is compiled from multiscale data sources to meet the needs of all collaborators. A visibility attribute allows users to filter NHD HR feature content to eight scales (1:24,000; 1:50,000; 1:100,000; 1:250,000; 1:500,000; 1:1,000,000; 1:2,000,000, and 1:5,000,000) for cartographic or analysis purposes (https://www.usgs.gov/core-science-systems/ngp/national-hydrography/visibilityfilter). Separate versions of the NHD are also available at 1:100,000-scale (Gary et al. 2010) and 1:10,000,000 (Instituto Nacional de Estadística Geografía e Informática et al. 2006).

In recent years the NHD Plus Version 2, based on 1:100,000-scale NHD, has been used by scientists to model phenomena ranging from lake eutrophication to water supply stress and fish species habitat at state, regional and national scales (Hill et al. 2018; Merriam et al. 2019). In some cases a variety of data versions are applied within the same research project due to variations in jurisdictional area and processing capabilities of contributing organizations (e.g., Martin 2018). In such cases, generalized data may be preferred over more detailed datasets if the aim is to standardize reporting and facilitate program consistency. The lack of data integration impedes users attempting to understand what products are available and to obtain data that is appropriate to the needs of a specific research or mapping project. A single-source, customizable multiscale product would facilitate data exploration, data modeling and analysis, development of derivative data, map production in paper and virtual forms and data redistribution by individuals, governments and private organizations.

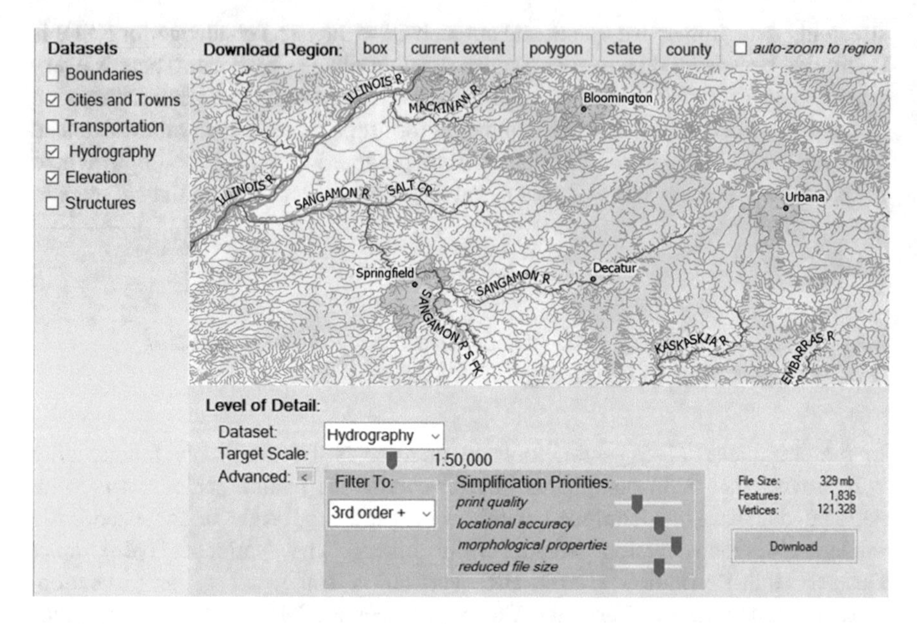

Fig. 23.5 A vision of a user interface for a multiscale data portal targeting a lay audience or K-12 educational communities

One might argue that data production could be limited to the finest possible resolution (i.e., the largest possible scale), but a single product scale cannot possibly meet the full range of display and analytic tasks without provision of software tools to modify levels of detail. For example, searching for patterns of isostatic rebound following retreat of continental glaciers would be sheer folly in a centimeter resolution database, just as a search for erosion and deposition patterns in a continental or global scale database would come up with few results.

A second goal that is immediately achievable relates to opening access to NSDI data to users of all levels of expertise. Figure 23.5 shows a hypothetical interface for an NSDI multiscale data portal as it might look to an end user. The purpose of this interface would be to browse among a suite of data themes and identify appropriate amounts of detail presented in each theme prior to download. The interface shown here might be suitable for lay users and/ or K-12 students who are either unaccustomed to more technical data jargon or just learning about online data portals. The interface design is straightforward and includes common elements such as tools to select datasets (left) and to identify a region of interest (top). Options to select a level of detail (LoD) appear at the bottom but data filtering options are presented in terms of output quality, accuracy, stream order, and file size. As noted above, visibility filtering algorithms are already in place for NHD (https://www.usgs.gov/core-science-systems/ngp/national-hydrography/visibilityfilter). Filters for other data themes could be developed by any data producer and tailored to the specifics of the data theme. Options also might be provided for advanced users. This recognizes that the aims of data users are not monolithic, and that data are needed to support clear linework on paper maps, visualization in digital interactive environments, numerical analysis and modeling, or construction of derivative data products. Lastly, feedback is provided to the user prior to download about the file characteristics.

Achieving an NSDI that supports continuous- or near-continuous data resolutions will require the integration of multiple theoretical advances in data generalization and processing as well as considerable experimentation to control data quality and estimate data processing times. At least four areas of research are needed to make this vision a reality. First, feature visibility attributes must be tied to a continuous range of target scales. This is feasible using methods of automated scale inference such as multi-criteria decision systems and/or an empirical formula for the number of features at a given scale (Jiang et al. 2013; Touya and Reimer 2015; Stauffer et al. 2016). Second, progressive or hierarchical geometric reduction techniques are needed to simplify features to any given target scale. Methods for hierarchical line Simplification are well-known for line simplification algorithms that process vertices sequentially (Cromley 1991) or progressively (Saalfeld 1999) and implementations are publicly available (e.g., Bloch and Harrower 2006).

Third, a comprehensive multiscale product must avoid topological errors. Maintaining topological consistency is a computationally difficult problem, but numerous advances occurred in this area over the past few decades. Early work developed simple modifications of common algorithms to preserve topology (Saalfeld 1999; Visvalingam and Whyatt 1993). The process can be made computationally efficient by using a point indexing structure such as a 2-dimensional tree, and can be extended to segment collapse algorithms (e.g., Tutić and Lapaine 2009; Kronenfeld et al. 2019) by checking for line segments that intersect collapsed triangles using an R-tree indexing structure. Triangulated irregular networks (TINs) provide another approach to maintaining topological consistency (Gold 1994; Ai et al. 2016). Recently there has been much interest in preserving spatial relations between data themes as for example between streams and elevation surfaces (Gaffuri et al. 2008; Sinha and Silvasesrith 2012), addressing a longstanding challenge to support vertical integration. Further work is needed in this area.

Fourth, research is needed to preserve specific data characteristics that are important to users with needs ranging from cartographic display to domain-specific modeling (Sinha and Silavisesrith 2012). General metrics of accuracy/error such as linear and areal displacement have long been recognized as important for quality control (McMaster 1986), but semantic and computational challenges are only beginning to be fully understood (Hangouet 1995; Kronenfeld and Deng 2019). More sophisticated models are being developed that preserve higher-order characteristics such as line density (Stanislawski et al. 2012, 2015) and sinuosity (Kronenfeld et al. 2019). Preservation of user-specified characteristics will require domain-specific approaches (e.g. Christophe and Ruas 2002; Tutić et al. 2016) as well as adaptive techniques that utilize local measures of shape and complexity to determine the best algorithm and parameterization (Buttenfield et al. 2011).

While several techniques to implement a single source, multiscale NSDI data portal are available today, the effort required to create a continuous scale data infrastructure should not be underestimated. The solution likely will require intermediate phases with pre-built options that involve some degree of manual intervention and/or quality control. However, in the long run a fully automated system offers greater control for the end user as well as eliminating the need for mapping agencies to maintain multiple versions of each data theme.

The challenge of maintaining single as opposed to multiple versions is tightly coupled with the need to establish standardized item-level identifiers within each version. Continuing with the example application domain of hydrography, many stream features in each version of the NHD are linkable through reach codes, but because of logistical limitations, finer resolution codes are not always transferred to coarser resolution versions, and thus database updates cannot be propagated automatically across some scale versions, introducing temporal disparities. The same is true of other national databases maintained by other agencies, but for brevity the discussion here will continue to focus on hydrography.

The lack of item-level linkages results from the sheer volume of data contained in each hydrography database at each scale. Early solutions reconstituted smaller scale versions from a finer resolution compilation (e.g., Buttenfield et al. 2013; Stauffer et al. 2016), but these were limited to selected features or to smaller scale target databases. Current strategies partition data into tiles, cross-referencing linkages within tiles to reduce search times. Chaudry and Mackaness (2007) adopt a hierarchical approach to establish partonomic relations. Even with faster parallel processing and full automation available today, the absence of strategies to standardize item-level identifiers in data themes of a size required for the NSDI means that comprehensive integration of database versions compiled at isolated scales may push the goal of a fully multi-scale NSDI into the longer-term.

One promising avenue for implementing item-level linkages will apply machine learning and neural networks for feature extraction of vector data from imagery or from raster terrain. This strategy directly follows the data-intensive practices of the Fourth Paradigm, namely to set up research frameworks that search for and elicit latent patterns in very large data archives. Machine learning and deep learning encompass the development, testing and examination of algorithms and statistical models that automatically perform a specific task without explicit instructions, relying on pattern recognition, training algorithms, and inference. Machine learning is considered an important component of AI. To date, it has not been used widely in generalization (Kang et al. 2019; Touya et al. 2019) but is playing an important role in current work on automated feature extraction from historical maps (Uhl et al. 2018; Uhl et al. 2017), conflation of vector information with historical archives (Duan et al. 2017), removal of road artifacts to extract drainage networks from digital terrain (Stanislawski et al. 2018), and use of training strategies to identify first-order tributaries from lidar point clouds (Shavers and Stanislawski 2018).

Wider adoption of learning methods and neural network strategies could provide many benefits and advantages to spatial data compilation and data integration for the NSDI, specifically landscape classification, spatial reasoning through regression trees, data fusion, aggregation and dimension reduction, reinforcement learning and possibly performance evaluation in conflation studies and algorithm validation. There are apparent similarities between machine learning and earlier conventional approaches to classification, clustering, regression and data reduction (factor analysis). It is important to note that the newer machine learning approaches obviate the need for meeting conventional method assumptions simply by means of the very large amounts of input data. The advantage of working on big data with machine learning is the iterative benefits, because models exposed to new data can adapt independently of manual intervention. This aspect alone makes the management of big data feasible for pattern recognition, similarity assessments, feature extraction and database alignment, all of which will be required to achieve a fully integrated NSDI.

Data alignment proceeds using several types of analytics, especially ontologic analysis and semantic knowledge discovery. Arpinar et al. (2004) argue these alignment methods enable capabilities to automatically extract metadata from unstructured or structured data, and to support reasoning about spatial and temporal proximity. Examples of promising work on data alignment focus upon matching feature types in gazetteer ontologies (Zhu et al. 2016), gathering place semantics from diverse sources (Hu 2017). Varanka and Cheatham (2016) couple semantic or RDF (Resource Description Framework) triples with relative spatial relations to align hydrographic ontologies. Arundel and Usery (2020) extract terrain and vegetation features from raw DEMs and from imagery. Duckam and Worboys (2007) demonstrate how reduction of uncertainty can improve semantic alignment and thus improve data fusion. Delgado del Hoyo et al. (2013) highlight problems arising from ontologies with only partial alignment in a case study matching the CityGML standard for urban city objects with other data domains such as digital building information.

23.4.2 Longer-Term Goals: Aligning NSDI with User Needs and Demands

Looking forward to the longer term of 5–10 years, geospatial science will need to adapt to other trends in technology and research related to engaging with big data. In addition to machine learning strategies, two approaches that stand out as important to address for multiscale representation involve the use of high-dimensional geovisualization and making data readily accessible and valuable to diverse user groups. Geovisualization tools are already available for lower-dimensional data with animated maps, data brushing, and augmented reality. High-dimensional data offers more sophisticated capabilities to interact with data in multiple spatial and statistical dimensions, and the vision for NSDI is to couple these types of functionality to improve efficiency and effectiveness of data intensive discovery.

There are many perspectives on what data dimensionality means. Shneiderman (1996) gives seven dimensions for geospatial data: 1D (lists or text); 2D (traditional maps and plots); 3D (volumetric data with x, y, and z coordinates); multi-dimensional (features or areas linked across spatial scales or more complex statistical dimensions such as those resulting from factor analyses, hierarchical clustering, or advanced regression methods); temporal (animations and also space-time trajectories including individual time lines up through space-time aquaria as proposed early on by Huisman and Forer (1998) and Kwan (2004)); trees (node data connected to parent or child nodes in a hierarchy); and network data (nodes and edges with a varied arrangement). Potential dimension multipliers also include mixed or augmented reality such as linked data maps aiding urban navigation using a headset or lenses (McKendrick et al. 2016). Real-time and near-real-time streaming data will be seen in NSDIs across the globe in the very near future to capture meteorological information, camera and drone surveillance of routine and exceptional movement behaviors of animals and humans. Hybrid data models integrating spatial and temporal data with uncertainty metrics are also being proposed and implemented (Qiang et al. 2018).

Defining dimensions as well as detailing the when, where and how to use high-dimensional, or n-dimensional (nD), visualization in the near future is a challenge. Yet, complex dimensionality will increasingly be unavoidable for timely data acquisition and use in scenarios such as defense and disaster response and is also essential to taking full advantage of big geospatial data and analyzing temporal and interdisciplinary trends. The increasing complexity in data visualization and analysis is compounded when addressing data modeling across spatial and temporal scales. Specific challenges in multiscale representation for nD data relate to technology, data and metadata standards, and making data accessible and valuable to diverse communities of interdisciplinary users.

Technical challenges such as adapting spatial data to untethered augmented and mixed reality, where users rapidly change view and detail level, are clear and made more significant when considering national databases and the need for related standards. The development of national structures and standards for Shneiderman's (1996) high-level tasks of visual information processing (overview, filter, then detailed analysis) on nD datasets of yet to be fathomed size remains a moving target. Nonetheless progress is needed for the sake of scientific analysis and development of the NSDI that will be expected to support such activities.

Advances such as mobile devices and web hosting require new strategies and push the bounds of traditional geospatial principles. Recent advances in nD scene generalization include GPU feature simplification (DeCoro and Tatarchuck 2007) which has been pioneered by developers working in the gaming industry. Adapting real-time high-speed simplification developed for gaming into research quality geospatial data with feature context preservation and semantic links is in early stages (Vollmer et al. 2018). Geovisualization in nD strategies and standards will require continued attention to intentional level of detail scaling for: viewing region, format, and user (Forberg 2007) as well as the automated generalization principles of adjacency, displacement, and topology, ideas not necessarily of precedence for gaming.

A second longer term goal is to make big data more accessible to as many user communities as possible. One requirement for ready access is to standardize data and its associated metadata, so that users can inform themselves about data content, format and provenance prior to download. The NSDI advocates the use of ISO (2014) standards on metadata. Organizations similar to the ISO addressing the above-mentioned challenges include the Open Geospatial Consortium (OGC), an international organization hosting stakeholder committees focused on open-source geospatial standards and processes, and the American Society for Photogrammetry and Remote Sensing (ASPRS) standards committee, responsible for development of United States lidar quality standards among others. Organizations such as OGC and ASPRS are advancing standards for new visualization platforms and data structures that offer direction for a national data infrastructure. For example, by addressing data advances such as the OGC Indexed 3D Scene Layers (I3S) that defines strategies for 3D scene structure (OGC 2017), the organizations offer guidelines and potential forms of national standards.

In Europe, the Centrum Wiskunde & Informatica (CWI) is a national research organization in the Netherlands with teams working on computer science and mathematics development (https://www.cwi.nl/). An example of the CWI development projects is MonetDB, an open-source column store system for big data storage and processing, making advances in structures for moving science towards big data versatility. Methods and technology for gleaning value from big data are being advanced by governments, academia, and industry, whether in collaboration in the case of the OGC or as independent groups such as CWI.

On-demand processing ability in nD for diverse user communities using tools such as data brushing and interpolation will require the data structures and standards discussed above as well as an understanding of user needs. The ongoing trend towards interdisciplinary research will complicate definition of user groups and increasingly require data interoperability for groups in disparate fields, likely extending the number of years needed to understand and respond to user community needs and purposes for spatial data. To shorten this longer-term goal, the unique processing demands and data types that are being explored require tools and structures that are adaptable by users, supporting the move towards open-source processing methods and adaptable interfaces such as a multiscale data portal.

Diverse user needs shape technology and analytical strategies as can be seen with current machine learning models being built and adapted for disparate disciplines such as cell biology, autonomous vehicle navigation, and social media. Geospatial user needs can be broadly grouped into users most often engaged in navigation, learning, or linking data and users who are interested in analysis for business, scientific research, or governance. Designing a national data infrastructure that is scalable and appropriately generalized is a complex and dynamic goal that would be more efficiently addressed by taking advantage of crowd sourcing and open-source programming, as discussed earlier. The open-source tool and web-hosted data model employed by Google Earth Engine (Gorelick et al. 2017) and Open Data Cube (http://www.opendatacube.org) is a potential strategy. The web format, with common themes such as satellite images and elevation data, navigated through an intuitive and simple map view is accessible and adaptable for the general public. Standard analysis, such as overlay or spatial statistics, is tailored by the user with some basic tools and operations for use with web-hosted data. More complex and discipline specific algorithms geared towards researchers can be coded in using languages such as Python.

And obviously, one would be remiss by neglecting to discuss the future of nD visualization in the context of AI. While AI strategies inform many aspects of data processing and nD visualization, spatial data modeling and generalization for the most part is still governed by defined strategies. This is certainly the case with national datasets where standards on representation and attribute preservation are well defined and required for scientific research at the national scale. While not yet feasible, development of efficient data storage, rapid cloud based processing, and clear standards for nD visualization in a linked environment set the stage for AI models to access, processes, and visualize data in ways currently unimaginable.

The goal of optimized multiscale representation is on-demand processing using built-in algorithms, linked data, and data brushing in nD for diverse user communities and ushering in a research environment where stakeholders can employ these tools to answer current and future questions. Looking forward, as machine learning models advance, one can expect much of the work of symbol and scene simplification to be automated. There will be a point when the models determine a choice among alternative simplification strategies and make decisions as to what feature characteristics to preserve and represent at differing scales for specific user audiences by tracking user research records or traits of the users' interaction with a GUI. At that point one might anticipate system designers and administrators learning from the machines and asking questions now about how to interpret the decisions and methods of models such as neural networks to prepare for even larger leaps in the not-too-distant future. This point is argued also by the Fourth Paradigm community – that eventually (in the longer term of 10 or more years) we can anticipate that machine automation will provide us interpretations of big data patterns, and may guide the sequences of questions we ask.

23.5 Implications and Prospects of the Fourth Paradigm for the NSDI

The question raised at the beginning of this chapter is how data-driven science might overcome obstacles challenging the present United States spatial data infrastructure, to modernize an NSDI characterized by a unified and interoperable data framework with sufficient horizontal and vertical integration to support natural and social science applications, delivering data across a wide range of user-specified spatial resolutions, with item-level metadata, linked versions of features, and landscape descriptors that can inform decisions about the most appropriate algorithms and operators for specific modeling and mapping tasks.

There is little doubt that NSDI serves a significant national imperative, and that emerging data analytics and data-intensive science offer promising opportunities for reframing science research, education, and literacy. But conditions in the United States differ from those in smaller countries; and some of these continue to impede accelerated progress on a fully integrated NSDI. One reason mentioned previously in this chapter impacts many expansive nations, namely, the sheer size of the United States and its territories results in a larger data volume, and the highly diverse landscapes, terrain, and settlement patterns complicate any "one-size-fits-all" data management, archival and cataloging. No matter the advances in current processing technologies, updates cannot be implemented exhaustively for any national level layer at a single point in time. This forces agencies to prioritize updates for fast-changing regions, thus fragmenting currentness in unpredictable ways (as for example following environmental crises or major storm events).

Variation across federal agency missions as well as across federal, state, tribal and municipal policies governing data access and use further confound creation of a comprehensively unified NSDI. One example of this can be seen in differing policies about county-level access to cadastral data, for which some counties charge access fees, others charge only specified target user groups, and still others charge only for finer resolution data. Another example relates to varying data definitions whose semantics become difficult to align among various jurisdictions. Take for example the definition of an "address", which for the US Postal Service refers to the coordinate location of a mailbox, but for 911 services and emergency dispatch refers to the location of a front door. In urban areas, the two are usually proximal, but in rural areas they may differ by kilometers. Conflating the two databases is problematic since it is impossible to distinguish database errors from discrepancies in the data dictionaries. All of these geographic conditions, data conditions, technical and policy issues lead to infrastructure fragmentation, problems with horizontal and vertical feature and attribute integration. The efforts by national and international standards organizations has come a long way toward resolving such variations, but there is much more work to be done, and scientific communities are urged to adhere to their guidance, even if adherence increases the time required to move research to practice or to the marketplace.

As discussed in the chapter, the community of data producers has been able to implement some solutions fully or partially, following technical advances as well as conceptual and theoretical developments. The point argued here is that continued progress can be streamlined in large part by following precepts espoused in the Fourth Paradigm. Focusing science and technology explicitly on data-driven science and open access to data, processing and scientific communication is intended to invoke several benefits. First, by harnessing the increased data volume with supercomputing and distributed processing, it becomes possible to train algorithms (through machine learning and AI) to elicit otherwise latent patterns in the data. Second, by distributing methods, data and results openly and transparently, it becomes possible for others to replicate and confirm results. Open source tools offer processing capabilities that obviate the need for centralized software solutions to meet the diversity of data management and analytics. Off-loading software development to the private sector or open-sourcing makes it possible to meet demands unforeseen by data producers. This in turn raises the need to have users create metadata according to OGC and other standards organizations, further democratizing big data curation and archival.

Of course, adoption of the Fourth Paradigm is not a panacea. For example, Kitchin (2014) and Gahegan (2020) caution that data are not by definition free of bias, but instead are collected and curated within technologies and mandates that "... actively shape its constitution" (Kitchin 2014:5). Kitchin also argues, validly, that not all patterns or products of data discovery are meaningful. Some are random or carry high correlations due simply to data volume, and to conclude causality from mere association can generate ecological fallacies (Freedman 2001). The responsible strategy for big-data-driven science is to use it for data exploration and to provide a context within which to ask rather than to definitively answer empirical and theoretical questions. Taking this approach will allow scientists to add value and "... to make sense of massive, interconnected data sets, fostering interdisciplinary research that conjoins domain expertise ... " (Kitchin 2014:6).

An important component of the vision for NSDI is a framework supporting linked multiscale representation. While this meets the aim of simplifying ease of use, data users will prioritize different objectives, such as enhancing visual clarity, reducing data volume, or preserving particular analytic characteristics of various component datasets. Many nations produce only a small number of single resolution data products, in some cases integrating features across resolutions, but overall directing data production for specific agency mandates (topographic mapping, civil engineering, strategic purposes, or cataloging natural and human resources). NSDI development in the United States has followed a different strategy, obtaining and integrating highest quality multi-resolution datasets to support scientific analysis, while enriching data attributes (e.g., visibility filters), and building tools to support generalization by collaborators and third parties. This follows open and democratized data access and processing as advised by Fourth Paradigm proponents. At roughly the same time as the emergence of the Fourth Paradigm, Buddhathoki et al. (2008) pointed to the need for shifting away from totally centralized data production and supply to a passive user base, towards greater reliance on volunteered geographic information, to reconceptualize NSDI within which users take a more proactive role by adding value to data, adding tools and advancing knowledge more effectively for scientists and for the nation as a whole.

As the United States confronts the complexity of an integrated NSDI for a single (albeit very expansive) region, the European Union (EU) has initiated and made great progress on creating a suite of federated national SDIs, integrating spatial data and tools for multiple nations to exchange data (see the INSPIRE chapter in this volume). Coordination of SDIs for member states of the EU are overseen by the Infrastructure for Spatial Information in Europe (INSPIRE) Directive (European Parliament and Council 2007). Contributions of EU member states to INSPIRE are guided by standards, with the incentive to implement services and data through INSPIRE for possible business or economic growth and resource management through geospatial applications. INSPIRE contains a smaller volume of data than

NSDI, with roughly the same number (17) data themes. But the effort to integrate multiple national missions for geospatial data demonstrates a starting point for an internationally federated GSDI.

The integration of international spatial data infrastructures into a GSDI will obviously mandate a level of coordination beyond any of the visions described here. Quite possibly, the experience and lessons learned by the United States in modernizing its own very large NSDI can inform such a globalized effort, opening the way to the Fourth Paradigm vision of open and transparent data availability to address the environmental and societal issues confronting every nation and in so doing, help to sustain the planet for future generations. Possibly the vision is quixotic, and yet current technology and changing environmental conditions create important opportunities for the global cooperation that would make such a vision real.

Disclaimer *Any use of trade, firm, or product names is for descriptive purposes only and does not imply endorsement by the U.S. Government.*

References

Ai T, Shu K, Yang M, Li J (2016) Envelope generation and simplification of polylines using Delaunay triangulation. Int J Geogr Inf Sci 31(2):297–319. https://doi.org/10.1080/13658816.2016.1197399

Arpinar IB, Sheth A, Rmakrishnan C, Usery EL, Azami M, Kwan MP (2004) Geospatial ontology development and semantic analytics. In: Wilson JP, Fotheringham AS (eds) Handbook of geographic information science. Wiley-Blackwell, Hoboken

Arundel ST, Usery EL (2020) Spatial data reduction through element-of-interest (EOI) extraction (this Springer volume)

Bloch M, Harrower M (2006) MapShaper.org: a map generalization web service. Proceedings, AutoCarto 2006. The 16th International Research Symposium on Computer-based Cartography. June 26-28, 2006, Vancouver, Washington, USA

Bialecka E, Dukaczewski D, Janczar E (2018) Spatial data infrastructure in Poland - lessons learnt from so far achievements. Geodesy and Cartography 67(1):3–20. https://doi.org/10.24425/118702

Buddhathoki NR, Bruce B, Nedovic-Budi Z (2008) Reconceptualizing the role of the user of spatial data infrastructure. Geojournal 72:149–160

Buttenfield BP, Stanislawski LV, Brewer CA (2011) Adapting generalization tools to physiographic diversity for the United States National Hydrography Dataset. Cartogr Geogr Inf Sci 38(3):289–301. http://www.tandfonline.com/doi/abs/10.1559/15230406382289

Buttenfield BP, Stanislawski LV, Anderson-Tarver C, Gleason MJ (2013) Automatic enrichment of stream networks with primary paths for use in the united states national atlas. Proceedings, International Cartographic Conference (ICC2014), Dresden Germany

Chaudry O, Mackaness WA (2007) Utilizing partonomic information in the creation of hierarchical geographies. Proceedings 11th ICA Workshop on Generalisation and Multiple Representation, Moscow, Russia. https://kartographie.geo.tu-dresden.de/downloads/ica-gen/workshop2007/Chaudhry-ICAWorkshop.pdf

Christophe S, Ruas A (2002) Detecting building alignments for generalisation purposes. Symposium on Geospatial Theory, Processing and Applications, Ottawa, Canada

Craglia M, Roglia E, Tomas R (2014) INSPIRE public consultation 2014: report of findings. Ispara Italy: European Commission Joint Research Centre Institute for Environment and Sustainability. 64pp. http://www.jrc.ec.europa.eu/
Coetzee S, Du Preez J, Behr F, Cooper AK, Odijk M, Vanlishout S, Buyle R, Jobst M, Cauke M, Fourie N, Schmitz P, Erwee F (2019) Collaborative custodianship through collaborative cloud mapping: challenges and opportunities. Proceedings International Cartographic Association 29th Int. Cartographic Conference. July 15–20, 2019, Tokyo, Japan
Cromley RG (1991) Hierarchical methods of line simplification. Cartogr Geograph Inf Syst 18(2):125–131
DeCoro C, Tatarchuk N (2007) Real-time mesh simplification using the GPU. Proceedings 2007 symposium on Interactive 3D graphics and games. ACM
Delgado del Hoyo F, Martínez-González MM, Finat J (2013) An evaluation of ontology matching techniques on geospatial ontologies. Int J Geogr Inf Sci 27:2279–2301. https://doi.org/10.1080/13658816.2013.812215
Duan WW, Chiang YY, Knoblock CA, Jain V, Feldman D, Uhl JH, Leyk S (2017) Automatic alignment of geographic features in contemporary vector data and historical maps. Proceedings 1st Workshop on Artificial Intelligence and Deep Learning for Geographic Knowledge Discovery (GeoAI '17), pp 45–54. https://doi.org/10.1145/3149808.3149816
Dübel S, Röhlig M, Tominski C, Schumann H (2017) Visualizing 3D terrain, geo-spatial data, and uncertainty. Informatics 4(1):6. Multidisciplinary Digital Publishing Institute
Duckham M, Worboys M (2007) Automated geographic information fusion and ontologyalignment. Chapter 6. In: Belussi A, Catania B, Clementini E, Ferrari E (eds) Spatial data on the web: modelling and management. Springer, Berlin, pp 109–132. http://hdl.handle.net/11343/33603
European Parliament and Council (2007) Directive 2007/2/EC of the European Parliament and of the Council of 14 March 2007 establishing an Infrastructure for Spatial Information in the European Community (INSPIRE). Off J Eur Union L108(50):1–14
FGDC (Federal Geographic Data Committee) (2013) National spatial data infrastructure strategic plan 2014–2016. Federal Geographic Data Committee, Reston, Virginia, USA, p 19
Forberg A (2007) Generalization of 3D building data based on a scale-space approach. ISPRS J Photogramm Remote Sens 62(2):104–111
Freedman DA (2001) Ecological inference and the ecological fallacy. In: International encyclopedia for the social and behavioral sciences, vol 6. Pergamon Press, Oxford, pp 4027–4030
Gaffuri J, Duchêne C, Ruas A (2008) Object-field relationships modelling in an agent-based generalisation model. 11th ICA Workshop on Generalisation and Multiple Representation, June 20–21, Montpellier, France
Gahegan M (2020) Fourth paradigm GIScience? Prospects for automated discovery and explanation from data. Int J Geograph Inf Sci 34(1):1–21. https://doi.org/10.1080/13658816.2019.1652304
Gary RH, Wilson ZD, Archuleta CM, Thomson FE, Vrabel J (2010) Production of a national 1:1,000,000-scale hydrography dataset for the United States—feature selection, simplification, and refinement. scientific investigations report 2009-5202. Reston, Virginia, USA: US Geological Survey. https://s3.amazonaws.com/nhdplus/NHDPlusV21/Documentation/History/Making_the_Digital_Water_Flow.pdf. Accessed 23 Sept 2019
GIRA (2015) Geospatial Interoperability Reference Architecture (GIRA): increased information sharing through geospatial interoperability, Washington, DC: April 2015, p 212. https://www.dni.gov/files/ISE/documents/DocumentLibrary/GIRA.pdf. Accessed 24 Sept 2019
Gold CM (1994) Three approaches to automated topology, and how computational geometry helps. In: Proceedings, Sixth International Symposium on Spatial Data Handling: Advances in GIS Research, Edinburgh, Scotland, pp 145–158
Gorelick N, Hancher M, Dixon M, Ilyushchenko S, Thau D, Moore R (2017) Google Earth Engine: planetary-scale geospatial analysis for everyone. Remote Sens Environ 202:18–27
Gould P (1981) Letting the data speak for themselves. Ann Assoc Am Geogr 71(2):166–176. https://doi.org/10.1111/j.1467-8306.1981.tb01346.x

Gray J (2007) e-Science: a transformed scientific method. Transcript of a presentation to NRC Computer Science and Telcommunications Board, 11 Jan 2007. http://research.microsoft.com/en-us/um/people/gray/JimGrayTalks.htm

Hangouet JF (1995) Computation of the Hausdorff distance between plane vector polylines. In: Proceedings of the 12th International Symposium on Computer-Assisted Cartography, Charlotte, NC, USA, 13–14 March 1995, pp 1–10

Hey T, Tansley S, Tolle K (2009) The fourth paradigm: data-intensive scientific discovery. Microsoft Research, Redmond

Hill RA, Weber MH, Debbout RM, Leibowitz SG, Olsen AR (2018) The Lake-Catchment (LakeCat) Dataset: characterizing landscape features for lake basins within the conterminous USA. Microsoft Corporation, Redmond. http://research.microsoft.com/en-us/um/people/gray/talks/NRC-CSTB_eScience.ppt

Hu Y (2017) Geospatial semantics. In: Huang B, Cova TJ, Tsou MH (eds) Comprehensive geographic information systems. Elsevier, Oxford, UK. https://doi.org/10.1016/B978-0-12-409548-9.09597-X

Huisman O, Forer P (1998) ComLife Spaces: a preliminary realisation of the time geography of student lifestyles. Proceedings Geocomputation 1998, Paper # 68. http://www.geocomputation.org/1998/68/gc_68a.htm

Instituto Nacional de Estadística Geografía e Informática, Government of Canada, Natural Resources Canada, Canada Centre, and US Geological Survey (2006) North american atlas - hydrography: Government of Canada, Ottawa, Ontario, Canada

ISO (International Standards Organization) (2014) ISO 19115-1:2014: geographic information—metadata, part 1: fundamentals. https://www.iso.org/standard/53798.html. Accessed Sept 2019

Jiang B, Liu X, Jia T (2013) Scaling of geographic space as a universal rule for map generalization. Ann Assoc Am Geogr 103(4):844–855

Kang Y, Gao S, Roth RE (2019) Transferring multiscale map styles using generative adversarial networks. Int J Cartogr 5(203):115–141. https://doi.org/10.1080/23729333.2019.1615729

Kitchin R (2014) Big data, new epistemologies and paradigm shifts. Big Data and Society April – June 2014, pp 1–12. https://doi.org/10.1177/2053951714528481

Kronenfeld BJ, Deng J (2019) Between the lines: measuring areal displacement in line simplification. Adv Cartogr GIScі Int Cartogr Assoc 1(9):8. https://doi.org/10.5194/ica-adv-1-9-2019

Kronenfeld BJ, Stanislawski LV, Buttenfield BP, Brockmeyer T (2019) Simplification of polylines by segment collapse: minimizing areal displacement while preserving area. Int J Cartogr 6(1):22–46. https://doi.org/10.1080/23729333.2019.1631535

Kwan MP (2004) GIS methods in time-geographic research: geocomputation and geovisualization of human activity patterns. Geogr Ann 86B(4):267–280

Lindsay JB (2016) Whitebox GAT: a case study in geomorphometric analysis. Comput Geosci 95:75–84. https://doi.org/10.1016/j.cageo.2016.07.003

Mackaness WA, Periklesous S, Chaudry O (2007) Representing forested regions at small scales: automatic derivation from the very large scale. Proceedings International Cartographic Congress (ICC 2007), Moscow, Russia

Martin EH (2018) Assessing and prioritizing barriers to aquatic connectivity in the eastern United States. J Am Water Resour Assoc. https://doi.org/10.1111/1752-1688.12694

McKendrick R, Parasuraman R, Murtza R, Formwalt A, Baccus W, Paczynski M, Ayaz H (2016) Into the wild: neuroergonomic differentiation of hand-held and augmented reality wearable displays during outdoor navigation with functional near infrared spectroscopy. Front Hum Neurosci 10:216

McMaster RB (1986) A statistical analysis of mathematical measures for linear simplification. Am Cartogr 13(2):103–116

Merriam ER, Todd Petty J, Clingerman J (2019) Conservation planning at the intersection of landscape and climate change: brook trout in the Chesapeake Bay watershed. Ecosphere 10(2):e02585. https://doi.org/10.1002/ecs2.2585

Nebert D (Ed.) (2004) Developing spatial data infrastructures: the SDI Cookbook. Global Spatial Data Infrastructure. V. 2.0 Technical Working Group monograph,

Global Spatial Data Infrastructure Secretariat, p 171. Retrieved 15 Dec 2019 from https://edisciplinas.usp.br/pluginfile.php/371105/mod_resource/content/4/2-%20Livro%20sobre%20Developing%20Spatial%20Data%20Infrastructures.pdf

Newman DR, Lindsay JB, Cockburn JMH (2018) Evaluating metrics of local topographic position for multiscale geomorphometric analysis. Geomorphology 312:40–50. https://doi.org/10.1016/j.geomorph.2018.04.003

NRC (National Research Council Mapping Science Committee) (1993) Toward a coordinated spatial data infrastructure for the nation. National Academy Press, Washington, D.C. 171 pp. https://www.nap.edu/catalog/2105/toward-a-coordinated-spatial-data-infrastructure-for-the-nation

NRC (National Research Council) (2012) Advancing strategic science: a spatial data infrastucture roadmap for the U.S. Geological Survey. The National Acadamies Press, Washington DC, p 115

NSB (National Science Board) (2005) Long-lived digital data collections: enabling research and education in the 21st century. Technical report NSB-05-40, Washington DC: National Science Foundation, September 2005. www.nsf.gov/pubs/2005/nsb0540/nsb0540.pdf

OGC (2017) OGC indexed 3d scene layer (I3S) and scene layer package format specification. http://docs.opengeospatial.org/cs/17-014r5/17-014r5.html. Accessed 20 Oct 2019

Oguchi T (2019) Geomorphological mapping based on DEMs and GIS: a review. Proceedings ICC 2019 Tokyo Japan

OMB (2010) OMB circular A-16 and supplemental guidance Washington DC: OMB memorandum M-11-03 to Heads of Executive Departments and Agencies. https://www.whitehouse.gov/wp-content/uploads/2017/11/Circular-016.pdf. Accessed 10 Sept 2019

OMB (2013) Federal enterprise architecture framework appendix c: data reference model. Washington DC Version 2.0, January 29, 2013. https://obamawhitehouse.archives.gov/sites/default/files/omb/assets/egov_docs/fea_v2.pdf. Accessed 10 Sept 2019

OMB (2017) OMB supplemental guidance—appendix E—NGDA data themes, definitions, and lead agencies: OMB circular A-16 and supplemental guidance. https://www.fgdc.gov/policyandplanning/a-16/appendixe/20170324-ngda-themes-fgdc-sc-revised-appendixe.pdf. Accessed 9 Sept 2019

OMB (Office of Management and Budget) (2002) Circular No. A-16 revised. Washington DC: The White House Office of Management and Budget August 19, 2002

Peng RD (2011) Reproducible research in computational science. Science 6060:1226–1227. https://doi.org/10.1126/science.1213847

Petras V, Petrasova A, Harmon B, Meentemeyer RK, Mitasova H (2015) Integrating free and open source solutions into geospatial science education. ISPRS Int J Geo Inf 4(2):942–956. https://doi.org/10.3390/ijgi4020942

Qiang Y, Buttenfield BP, Lam NS-N, Van de Weghe N (2018) Novel models for multiscale spatial and temporal analyses. Proceedings GIScience 2018 Melbourne Australia, LIPIcs 114. https://doi.org/10.4230/LIPIcs.GISCIENCE.2018.55

Saalfeld A (1999) Topologically consistent line simplification with the Douglas-Peucker algorithm. Cartogr Geogr Inf Sci 26(1):7–18. https://doi.org/10.1559/152304099782424901

Sangireddy H, Stark CP, Passalacqua P (2017) Multiresolution analysis of characteristics length scales with high resolution topographic data. J Geophys Res Earth 122:1296–1324. https://doi.org/10.1002/2015JF003788

Shavers E, Stanislawski LV (2018) Streams do work: measuring the work of low-order streams on the landscape using point clouds. Int Arch Photogram Remote Sens Spat Inf Sci XLII-4:573–578. https://doi.org/10.5194/isprs-archives-XLII-4-573-20

Shneiderman B (1996) The eyes have it: a task by data type taxonomy for information visualizations. Proceedings 1996 IEEE symposium on visual languages, pp 336–343

Sinha G, Silavisesrith W (2012) Multicriteria generalization (MCG): a decision-making framework for formalizing multiscale environmental data reduction. Int J Geogr Inf Sci 26(5):899–922

Stanislawski LV, Buttenfield BP (2011) Hydrographic generalization tailored to dry mountainous regions. Cartogr Geogr Inf Sci 38(2):117–125. https://doi.org/10.1559/15230406382117

Stanislawski LV, Doumbouya AT, Miller-Corbett CD, Buttenfield BP, Arundel ST (2012) Scaling stream densities for hydrologic generalization: Proceedings 7th International Conference on Geographic Information Science, September 18–21, 2012, Columbus, Ohio, p 6

Stanislawski LV, Falgout J, Buttenfield BP (2015) Automated extraction of natural drainage density patterns for the conterminous United States through high performance computing. Cartogr J 52(2):185–192

Stanislawski L, Brockmeyer T, Shavers E (2018) Automated road breaching to enhance extraction of natural drainage networks from elevation models through deep learning. Int Arch Photogram Remote Sens Spat Inf Sci XLII-4:597–601. https://doi.org/10.5194/isprs-archives-XLII-4-597-2018

Stauffer AJ, Finelli E, Stanislawski LV (2016) Moving from generalization to the 'Visibility Filter Attribute': Leveraging database attribution to support efficient generalization decisions. American Water Resources Association 2016 Summer Specialty Conference, GIS & Water Resources IX, July 11–13, 2016, Sacramento, California

Steadman I (2013) Big data and the death of the theorist. Wired, 25 January 2013. Available at: http://www.wired.co.uk/news/archive/2013-01/25/big-data-end-of-theory

Stodden V, Seiler J, Ma Z (2018) An empirical analysis of journal policy effectiveness for computational reproducibility. Proc Natl Acad Sci 115(11):2584–2589. https://doi.org/10.1073/pnas.1708290115

The White House (1994) Coordinating geographic data acquisition and access: the national spatial data infrastructure. Federal Register 59(71). https://www.archives.gov/files/federal-register/executive-orders/pdf/12906.pdf. Accessed 19 Sept 2019

The White House (2019) Executive order on maintaining American leadership in artificial intelligence. U.S. Mission to the Organization for Economic Cooperation and Development, Washington, D.C 11 Feb 2019. https://usoecd.usmission.gov/executive-order-on-maintainingamerican- leadership-in-artificial-intelligence/

Touya G (2008) First thoughts for the orchestration of generalisation methods on heterogeneous landscapes. Proceedings, 12th ICA Workshop on Generalizations. Montpellier, France. https://kartographie.geo.tu-dresden.de/downloads/ica-gen/workshop2008/01_Touya.pdf

Touya G, Reimer A (2015) Inferring the scale of OpenStreetMap features. In: Arsanjani JJ, Zipf A, Mooney P, Helbich M (eds) OpenStreetMap in GIScience: experiences, research, and applications. Springer Lecture Notes in Geoinformation and Cartography, Berlin

Touya G, Zhang X, Lokhat I (2019) Is deep learning the new agent for map generalization? Int J Cartogr 5(2–3):142–157. https://doi.org/10.1080/23729333.2019.1613071

Tutić D, Lapaine M (2009) Area preserving cartographic line generalization. J Croat Cartogr Soc 8(11):84–100

Tutić D, Štanfel M, Jogun T (2016) Automation of cartographic generalisation of contour lines. 10th ICA Mountain Cartography Workshop, April 28, 2016, Berchtesgaden, Germany

Uhl JH, Leyk S, Chiang Y, Duan WW, Knoblock CA (2017) Extracting human settlement footprint from historical topographic map series using context-based machine learning. Proceedings 8th Int'l Conference on Pattern Recognition Systems (ICPRS-2017). https://doi.org/10.1049/cp.2017.0144

Uhl JH, Leyk S, Chiang Y, Duan WW, Knoblock CA (2018) Exploring the potential of deep learning for settlement symbol extraction from historical map documents. Proceedings 22nd International Research Symposium on Computer-based Cartography and GIScience (Auto-Carto 2018), Madison, WI. https://www.ucgis.org/assets/docs/AutoCarto-2018Proceedings.pdf

Usery EL, Varanka D, Finn MP (2009) A 125 year history of topographic mapping and GIS in the US Geological Survey 1884–2009, Part 2: 1980—2009. USGS, Reston, Virginia, USA. Retrieved September 20, 2019 https://pubs.er.usgs.gov/publication/70004689

Varanka D, Cheatham M (2016) Spatial concepts for hydrography ontology alignment. Proceedings AUTOCARTO 2016, Albuquerque, New Mexico. https://cartogis.org/docs/proceedings/2016/Varanka_and_Cheatham.pdf

Visvalingam M, Whyatt JD (1993) Line generalisation by repeated elimination of points. Cartogr J 30(1):46–51

Vollmer JO, Trapp M, Schumann H, Döllner J (2018) Hierarchical spatial aggregation for level-of-detail visualization of 3D thematic data. ACM Trans Spat Algorith Syst 4(3):9

Zhu R, Hu Y, Janowicz K, McKenzie G (2016) Spatial signatures for geographic feature types: examining gazetteer ontologies using spatial statistics. Technical Paper, STKO Lab, Dept. Geography, UCSB. https://geog.ucsb.edu/~jano/spatialsignatures.pdf

Chapter 24
INSPIRE: The Entry Point to Europe's Big Geospatial Data Infrastructure

Marco Minghini, Vlado Cetl, Alexander Kotsev, Robert Tomas, and Michael Lutz

The views expressed are purely those of the authors and may not in any circumstances be regarded as stating an official position of the European Commission.

24.1 Introduction

Technological advancements in the last decade have enabled governments, businesses and citizens to produce and collect increasingly larger amounts of data. The availability of personal digital devices with built-in sensors increased while their price significantly decreased, bringing the chance to collect multitudes of data in a simple and fast way to everyone's reach. This unprecedented large and heterogeneous amount of data collected at exceptional scales and speeds has subsequently led to the establishment of the term Big Data, which has currently become ubiquitous in many areas. Multiple definitions of Big Data are available, which bring together different concepts such as volume, variety, cloud, technology, storage, analytics, processing, information, and transformation. According to the common formal definition proposed by De Mauro et al. (2015) Big Data represents the "Information assets characterised by such a High Volume, Velocity and Variety to require specific Technology and Analytical Methods for its transformation into Value". The exploitation of Big Data is often connected to cloud computing platforms, which are composed of data infrastructures put in place in order to store and manage data, high-bandwidth networks to transport data, and high-performance computers to process data (European Commission 2016).

It is clear that the data revolution that is underway is already reshaping how knowledge is produced, business conducted and governance enacted (Kitchin 2014). The main components of this data revolution are digitalisation, big data, open data and data infrastructures. These components, whose impact is already visible in

M. Minghini (✉) · V. Cetl · A. Kotsev · R. Tomas · M. Lutz
European Commission, Joint Research Centre (JRC), Ispra, Italy
e-mail: marco.minghini@ec.europa.eu

M. Werner, Y.-Y. Chiang (eds.), *Handbook of Big Geospatial Data*,
https://doi.org/10.1007/978-3-030-55462-0_24

science, business, government and civil society are addressed by new and emerging fields, such as data science, social computing, and artificial intelligence.

In the geospatial arena, Information and Communication Technology (ICT) developments have been continuously adopted as well. Geospatial data management started with the development of Geographic Information Systems (GIS) which in turn evolved into Spatial Data Infrastructures (SDIs), and consequently in Spatial Knowledge Infrastructures (SKIs). SDIs developments started on the one hand with legally binding governmental initiatives and on the other hand with more business-oriented initiatives driven by the private sector. Typical examples of the former are National Spatial Data Infrastructures (NSDIs) such as the one in the US (Clinton 1994) and European Spatial Data Infrastructures (ESDI), which in the European Union (EU) are driven by the INSPIRE[1] (Infrastructure for Spatial Information in Europe) Directive (European Parliament and Council 2007). In addition to the EU Member States (MS) and European Free Trade Association (EFTA) countries, candidate and potential candidate countries (e.g. Western Balkans) and some European neighbourhood countries (e.g. Ukraine and Moldova) are also building their NSDIs in accordance with INSPIRE (Cetl et al. 2014). The latter group of SDIs is represented by the mapping frameworks from commercial surveying companies including Google Maps, Microsoft Bing Maps, and HERE Maps. There is also a third group of SDIs driven by crowdsourced initiatives such as OpenStreetMap,[2] which has built the largest, most diverse and most detailed open geospatial database to date (Mooney and Minghini 2017) and whose quality can equal that of authoritative data (see e.g. Haklay 2010; Girres and Touya 2010; Fan et al. 2014; Brovelli et al. 2016). There is no doubt that all these initiatives have triggered the creation of Big Geospatial Data. There is also no doubt that all of them are interrelated and their user bases are becoming more and more similar (see e.g. Köbben and Graham 2009; Minghini et al. 2019). Those heterogeneous initiatives combined, have huge potential for creating synergies in Big Geospatial Data management.

In this chapter our emphasis is on governmental, i.e. authoritative geospatial data in the EU. These are officially recognised, quality-certified data provided by authoritative sources such as Environmental Protection Authorities (EPAs) and National Mapping and Cadastral Agencies (NMCAs). Authoritative geospatial data in the EU are managed and shared through NSDIs, interlinked in a European Big Geospatial Data infrastructure that is to a large extent shaped by the legal provisions of the INSPIRE Directive. Structurally, the chapter is organised as follows. Section 24.2 offers an overview of the current Big Data initiatives started by the European Commission (EC), with special focus on those characterised by a geospatial component and their relation with INSPIRE. A more detailed introduction to the INSPIRE legal, technical and organisational framework as well as the state of play of INSPIRE is provided in Sect. 24.3. Particular attention is placed on the INSPIRE

[1] https://inspire.ec.europa.eu

[2] https://www.openstreetmap.org

Geoportal, which is the entry point of the whole infrastructure. This is followed by Sect. 24.4, which maps the main characteristics of Big Data, expressed by the popular six Vs, to the main features of the INSPIRE infrastructure, thus proving its nature of a Big (Geospatial) Data infrastructure. A number of open issues in making full use of the INSPIRE infrastructure from a user perspective are then listed. Finally, Sect. 24.5 concludes the chapter by reflecting on those issues and the lessons learnt from the INSPIRE implementation from a Big Data perspective, and outlining some potential evolutions from a traditional SDI to a modern data ecosystem.

24.2 Big Data in the EU

According to the Digital Single Market strategy of the European Commission,[3] data represents a key asset for the economy and society similar to the traditional categories of human and financial resources. The need to make sense of Big Data – regardless of their nature (geospatial, statistics, weather, research, transport, energy, or health) and source (public services, connected objects, private sector, citizens, research) – is leading to innovations in technology, development of new tools and new skills.

At the EU level, the response to the potential of using the cloud as a platform for Big Data exploitation resides in the European Cloud Initiative (European Commission 2016), which aims to interconnect the existing EU data infrastructures and coordinate their support, ensuring that the sharing of data and the capacity to exploit them are maximised. The European Cloud Initiative is based on the Digital Single Market strategy and several other EU initiatives addressing Big Data, including the 2012 European Cloud Strategy (European Commission 2012a), the High Performance Computing (HPC) Strategy (European Commission 2012b), and the policy developed in the Communication on Big Data (European Commission 2014). A number of reasons are listed why Europe has not yet exploited the full potential of data: non-openness of publicly-funded research data, lack of data interoperability, data fragmentation (i.e. infrastructures scattered across countries and domains), and lack of a European world-class HPC infrastructure. Regarding the lack of data interoperability, the geospatial domain is explicitly mentioned as an exception thanks to the INSPIRE Directive.

The tool envisioned to give Europe a global lead in scientific data infrastructure is the European Open Science Cloud (EOSC; Koski et al. 2015), an open, interoperable, distributed, service-oriented, publicly funded and publicly governed platform connecting networks, data, computing systems, software, tools and services of EU MS to enable gathering, management, analysis, sharing and discovery of scientific data according to the principles of Open Science (European Commission 2015) to

[3]https://ec.europa.eu/digital-single-market/en/big-data

ultimately lead to economic and societal innovation. The EOSC will be research-centric, though not specific to any discipline or field, and will aim to make all scientific data produced by the Horizon 2020 Programme open by default according to the principles of Findability, Accessibility, Interoperability and Reusability (FAIR; Wilkinson et al. 2016). However, turning the FAIR principles into reality would require an effort at the level of both the technological infrastructure and the research culture (European Commission Expert Group on FAIR Data 2018). Finally, the European Cloud Initiative already foresees quantum computing as the next breakthrough in supercomputing and secure networking and anticipates the need for Europe to make significant investments to be at the forefront.[4]

Looking more into the geospatial dimension of Big Data in the EU, in addition to the INSPIRE framework which is separately described in Sect. 24.3 there are several other initiatives worth to be mentioned. Copernicus is the EU's Earth Observation (EO) Programme, looking at our planet and its environment for the ultimate benefit of all European citizens.[5] It offers service providers, public authorities and other international organisations a number of Services – focused on atmosphere, marine, land, climate change, security and emergency – based on satellite EO and in situ (non-space) data. The processed data and the information disseminated, both freely and openly accessible, put Copernicus at the forefront of the Geospatial Big Data paradigm. More than five million products have been published in the Sentinel repository managed by the European Space Agency (ESA) and more than 100.000 users have downloaded more than 50 PB of data since the system became operational, with 1 PB of data corresponding to about 750.000 datasets (Koubarakis et al. 2019). This volume, as well as the velocity at which data are collected and processed, will increase in the future with the launch of new Sentinel satellites. Copernicus is closely connected with INSPIRE since Copernicus Services need access to openly available, up-to-date and harmonised geospatial information across Europe for production and validation purposes. In turn, many geospatial datasets and services produced by Copernicus are exposed according to the INSPIRE guidelines to maximize their interoperability. In the field of positioning and navigation, a partnership between the European Space Agency (ESA) and the EC has resulted into Galileo, the European Global Navigation Satellite System (GNSS) offering high-quality positioning, navigation and timing services to users across the world.[6] Galileo is fully compatible with the American GPS and Russian GLONASS, thus offering enhanced combined performance; in contrast to them, it is specifically designed to remain under civilian control. Galileo's full operational constellation – still under construction – will consist of 24 operational satellites plus six spares circling Earth in three circular medium-Earth

[4]https://ec.europa.eu/digital-single-market/en/news/quantum-technologies-opportunities-european-industry-report-round-table-discussion-and

[5]https://www.copernicus.eu

[6]https://www.gsa.europa.eu/european-gnss/galileo/galileo-european-global-satellite-based-navigation-system

orbits, at an altitude of about 23,000 km. Galileo builds upon the success of the European Geostationary Navigation Overlay Service (EGNOS),[7] operational since 2009 to provide safety of life navigation services to aviation, maritime and land-based users over most of Europe. Finally, the EC is participating in the Group on Earth Observations (GEO),[8] a global network of governmental institutions, research organisations, data providers, scientists and experts working together to build a Global Earth Observation System of Systems (GEOSS)[9] aimed at strengthening the monitoring of the Earth and improving decision making. Thanks to the knowledge acquired through Copernicus, Galileo, EGNOS and other programmes, Europe has positioned itself as a global force in the field of EO and in 2019 the European portion of GEO was renamed EuroGEO, and EuroGEOSS was established as the European component of GEOSS, yet again with a clear link to INSPIRE (European Commission 2017).

In the context of a general strategy to build a European data economy,[10] and within a framework for digital trust granted by the General Data Protection Regulation (European Parliament and Council 2016), a recent initiative dedicated to the establishment of a common European data space is addressing challenges pertaining to data value chains in the era of Big Data. The related EC Communication defines a data space as "a seamless digital area with the scale that will enable the development of new products and services based on data" and positions public data at the centre of data-driven innovation (European Commission 2018). Additional emphasis is put on ensuring access to publicly-funded data held by private companies, and different data-flows (business-to-business, business-to-government, etc.) that are beneficial for all actors involved. This is further addressed by the recent Open Data Directive (European Parliament and Council 2019), which encourages the FAIR management of EU public and publicly funded data and recognises INSPIRE as a good practice. The Directive also introduces the concept of *high-value datasets*, i.e. datasets with "the potential to (i) generate significant socio-economic or environmental benefits and innovative services, (ii) benefit a high number of users, in particular SMEs, (iii) assist in generating revenues, and (iv) be combined with other datasets"; hence, it requires that such datasets are made available free of charge, in machine-readable formats and provided via Application Programming Interfaces (APIs) and as a bulk download where relevant. The Directive does not provide a full list of such datasets – which is left for future work – but only defines categories of datasets, one of which is the geospatial one.

---

[7] https://www.gsa.europa.eu/egnos/what-egnos

[8] http://www.earthobservations.org/index.php

[9] https://www.earthobservations.org/geoss.php

[10] https://ec.europa.eu/digital-single-market/en/policies/building-european-data-economy

24.3 INSPIRE State of Play

24.3.1 Legal, Technical and Organisational Framework

The legal framework for INSPIRE has been set by the Directive 2007/2/EC (European Parliament and Council 2007) and related interdependent legal acts, which are called Implementing Rules, in the form of Commission Regulations and Decisions. By design, the INSPIRE infrastructure is built upon the NSDIs established and operated by the EU MS and EFTA countries that are then made compliant with the Implementing Rules, covering its core components: metadata, network services, interoperability of spatial datasets and services, data sharing and monitoring and reporting (Tomas et al. 2015; Cetl et al. 2019; Minghini et al. 2020). The Implementing Rules for metadata, the interoperability of data themes, the network services (that help to share the infrastructure's content online) and the data sharing are complemented by non-legally binding Technical Guidance documents. These guidelines explain a possible technical approach to fulfill the legal requirements and embed additional recommendations that may help data providers in their implementation for a range of use cases.

The thematic scope of INSPIRE includes 34 cross-sectoral categories, named data themes (see Fig. 24.1), listed in the three annexes of the Directive and reflecting two main types of data: baseline geospatial data (presented in Annex I and partly in Annex II), which define a location reference that the remaining data themes (in Annex III and partly in Annex II) can then refer to.

Data and metadata are shared through web-based services, referred to as network services (European Commission 2009), based on a Service Oriented Architecture

Fig. 24.1 INSPIRE themes, organised in three Annexes. (Source: European Commission, Joint Research Centre)

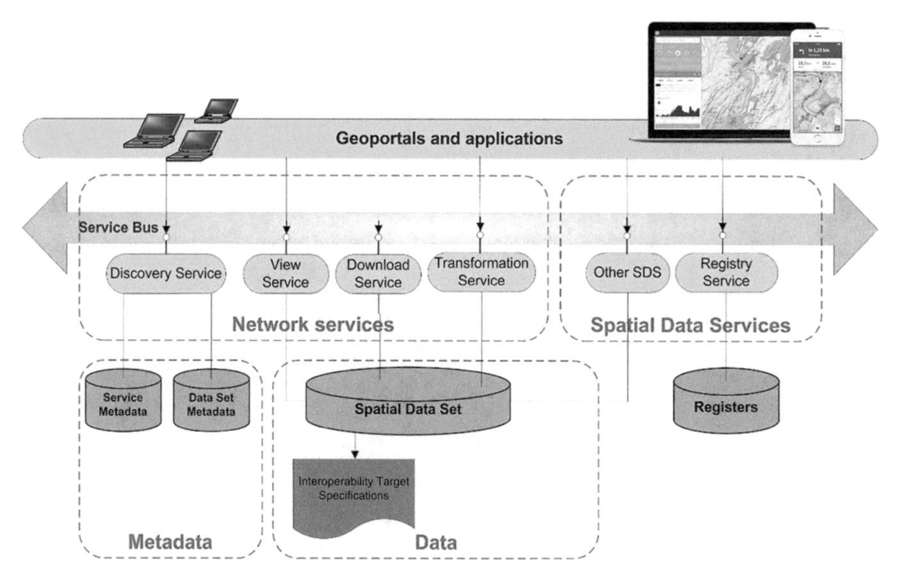

Fig. 24.2 Distributed Service Oriented Architecture of INSPIRE. (Source: European Commission, Joint Research Centre)

(SOA)[11] approach (see Fig. 24.2). Network services are implemented through well-established international standards for geospatial interoperability, mainly developed by the Open Geospatial Consortium (OGC).[12] Technical Guidance documents illustrate how data providers can establish access to metadata for Discovery Services through the Catalogue Service for the Web (CSW).[13] Similarly for View Services, the interactive visualisation of georeferenced content involves guidelines using the Web Map Service (WMS)[14] and Web Map Tile Service (WMTS)[15] standards. Download Services also have guidelines that recommend the use of Atom feeds,[16] Web Feature Service (WFS),[17] Web Coverage Service (WCS)[18] and Sensor Observation Service (SOS),[19] for appropriate types of data. There are also various Transformation Services defined, which can support coordinate and data transformations. In addition to all the above, there are generic services (registry and

[11]https://www.opengroup.org/soa/source-book/soa/p1.htm

[12]http://www.opengeospatial.org/

[13]https://www.opengeospatial.org/standards/cat

[14]https://www.opengeospatial.org/standards/wms

[15]https://www.opengeospatial.org/standards/wmts

[16]https://validator.w3.org/feed/docs/atom.html

[17]https://www.opengeospatial.org/standards/wfs

[18]https://www.opengeospatial.org/standards/wcs

[19]https://www.opengeospatial.org/standards/sos

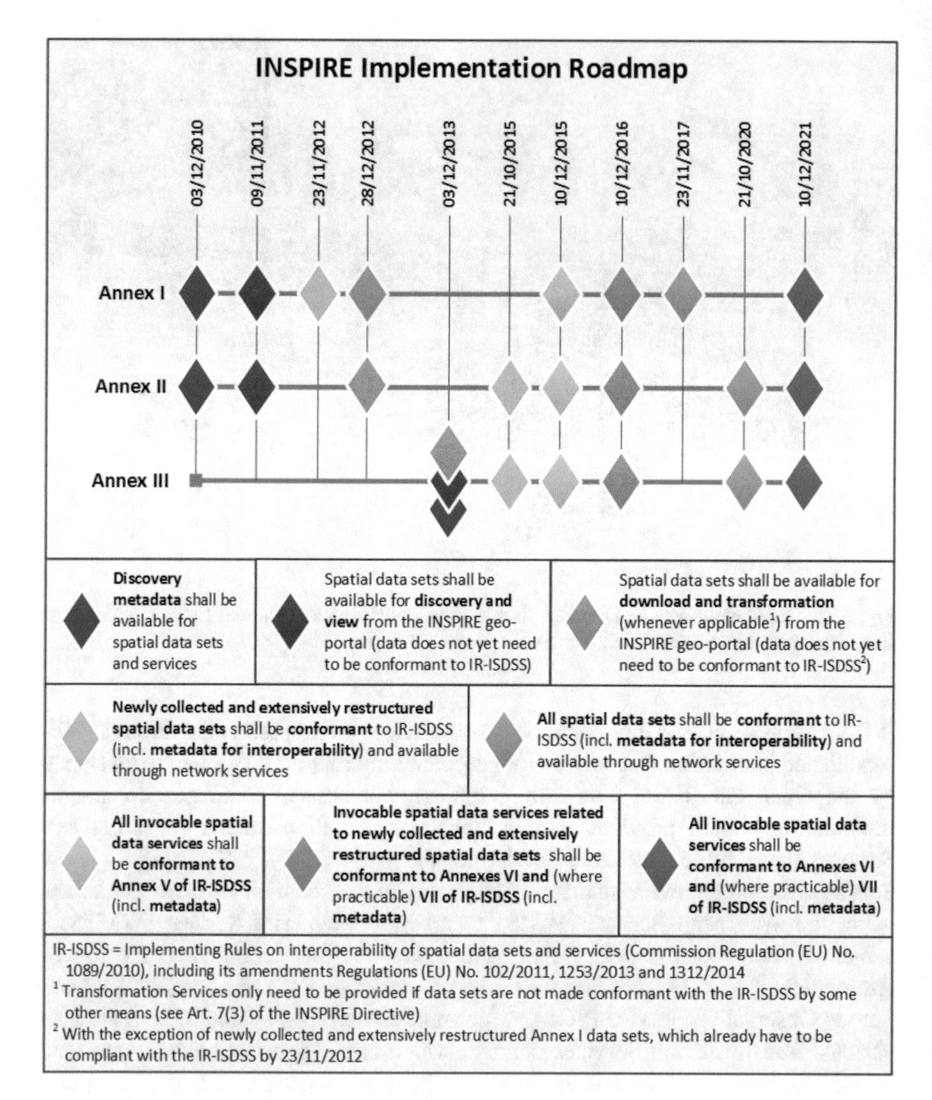

Fig. 24.3 INSPIRE Implementation Roadmap. (Source: European Commission, Joint Research Centre)

other spatial data services), that are implemented on a national as well as European level.

Deadlines for the implementation of the different components of the infrastructure are defined by the INSPIRE roadmap (see Fig. 24.3), which foresees different milestones till 2021 according to the Annexes and the type of resources or services.

A number of important milestones have been already reached, however there are still activities to be completed, especially regarding data harmonisation and conformity which is crucial for the overall interoperability of the infrastructure. It goes in particular for Annexes II and III where the deadline for data harmonisation is set for the end of 2020. This means that, at the time of writing (beginning of 2020), data sets falling under related data themes are still made available mostly in a non-harmonised manner.

24.3.2 INSPIRE Geoportal

The entry point to the INSPIRE infrastructure is the INSPIRE Geoportal.[20] It serves as a central access point to the data and services from public organisations in the EU MS and EFTA countries which fall under the scope of INSPIRE. The INSPIRE Geoportal enables cross-border data discovery, access, visualisation and download. It does not store any geospatial data, but it simply acts as the main client application of the whole INSPIRE infrastructure by exposing data through the harvesting of the CSW endpoints made available by MS. Alongside the INSPIRE Geoportal, which is operated by the EC, there are also national geoportals operated by single countries. Links to national geoportals are available in the INSPIRE Knowledge Base (IKB) section entitled *INSPIRE in your country.*[21]

The first operational Geoportal Pilot was developed by the Joint Research Centre (JRC)[22] of the EC and released in 2011. In September 2018, a redesigned version was published (see Figs. 24.4, 24.5, and 24.6) offering easier access to geospatial data in the EU. The new Geoportal was developed by the JRC in collaboration with and support from the EC Directorate-General for Environment,[23] Eurostat[24] and the European Environment Agency.[25] It builds on the experience of running the Geoportal Pilot and supports several actions of the INSPIRE Maintenance and Implementation Work Programme,[26] especially regarding improving the accessibility of data sets through Network Services and improving the availability of priority data sets for environmental reporting.[27] The redesigned Geoportal is a one-stop shop for public authorities, businesses and citizens to find, access and use geospatial data sets related to the environment in Europe. It also provides overviews of the availability of data sets by country and thematic area, and provides ready-to-use data either through interoperable web services or by direct download, to maximize their exploitation in third-party GIS clients and applications.

The Geoportal landing page provides access to three main applications:

---

[20]http://inspire-geoportal.ec.europa.eu

[21]https://inspire.ec.europa.eu/INSPIRE-in-your-Country

[22]https://ec.europa.eu/jrc/en

[23]https://ec.europa.eu/dgs/environment

[24]https://ec.europa.eu/eurostat

[25]https://www.eea.europa.eu

[26]https://webgate.ec.europa.eu/fpfis/wikis/pages/viewpage.action?pageId=268249090

[27]http://inspire.ec.europa.eu/metadata-codelist/PriorityDataset/

Fig. 24.4 Landing page of the INSPIRE Geoportal. (Source: European Commission, Joint Research Centre)

1. *Priority Data Sets Viewer*, that displays the availability and provides access to the priority datasets used for environmental reporting[28];

[28]https://ies-svn.jrc.ec.europa.eu/projects/2016-5/wiki

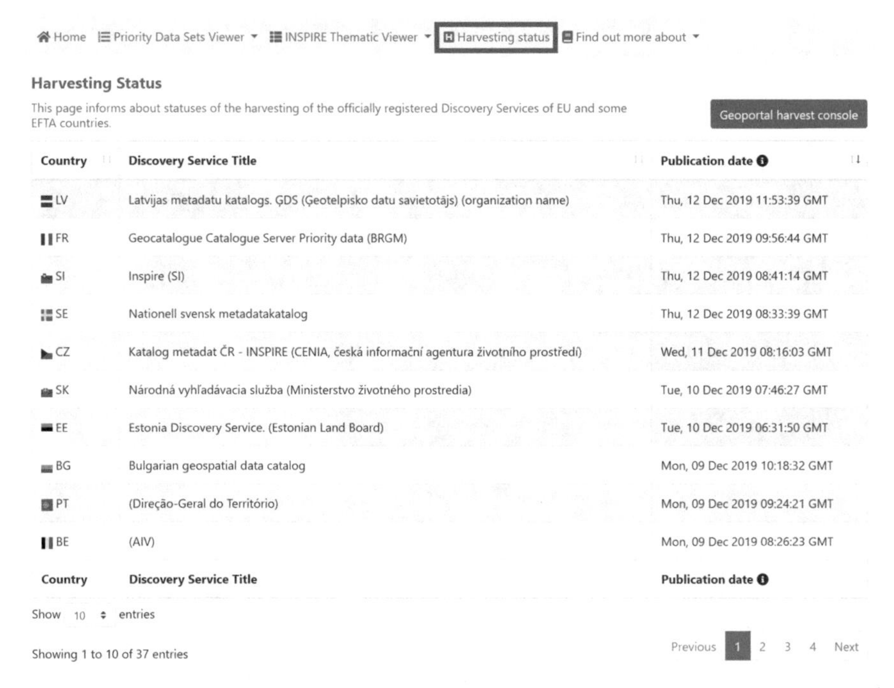

Fig. 24.5 Harvesting status on 15/12/2019. (Source: European Commission, Joint Research Centre)

2. *INSPIRE Thematic Viewer*, that displays the availability and provides access to all EU MS and EFTA countries datasets falling under the scope of the INSPIRE Directive, filtered by data themes and/or countries;
3. *INSPIRE Reference Validator*, a separate application that helps data providers check whether their data sets, services and metadata meet the INSPIRE requirements.

As mentioned above, the input source to the INSPIRE Geoportal is the harvesting of metadata from the officially registered Discovery Services of EU MS and EFTA countries. At the time of writing (January 2020) 37 Discovery Services are harvested on a regular basis (most of them weekly or monthly, however this is fully decided by the administrators of each Discovery Service), as shown in the *Harvesting status* section[29] of the Geoportal (see Fig. 24.5).

Insights into the current implementation status of the infrastructure are provided by the INSPIRE Thematic Viewer, which offers two possibilities for browsing datasets: by individual EU MS & EFTA country and by INSPIRE data theme. Figure 24.6 shows the availability of datasets in EU MS and EFTA countries as of October 2019.

[29]https://inspire-geoportal.ec.europa.eu/harvesting_status.html

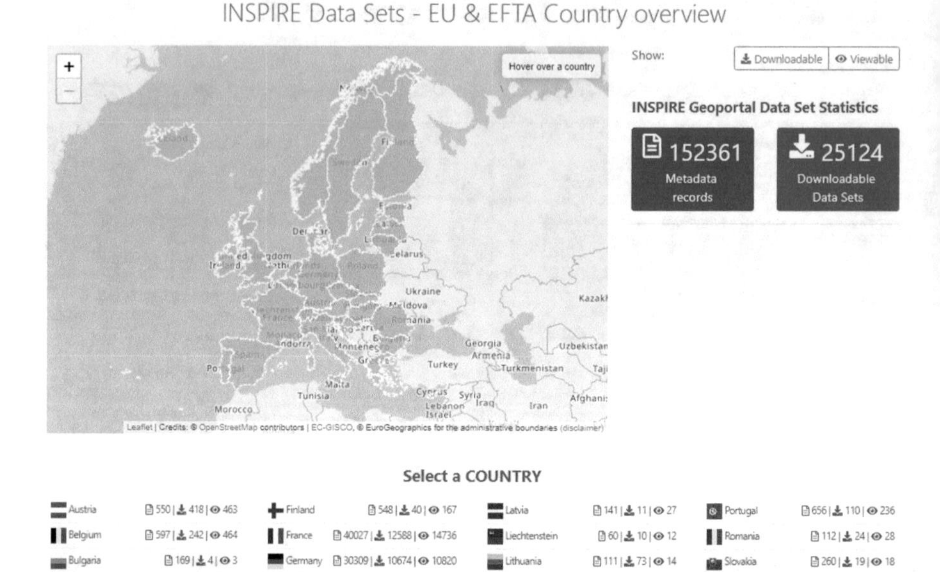

Fig. 24.6 Availability of INSPIRE datasets in EU MS and EFTA countries in October 2019. (Source: European Commission, Joint Research Centre)

The three numbers related to each country correspond to the number of available metadata records, downloadable datasets (i.e. data sets for which a Download Service is available) and viewable datasets (i.e. datasets for which a View Service is available). In the INSPIRE Geoportal about 150k datasets are available with metadata, of which about 24k are also viewable and about 13k downloadable (see Fig. 24.6). The differences between the number of metadata records and the number of viewable and downloadable datasets demonstrate that the full implementation of INSPIRE is yet to be achieved. Similarly, for each EU MS and EFTA country the Priority Data Sets Viewer displays the availability of metadata, viewable and downloadable datasets for the priority datasets used for environmental reporting, which can be filtered by country, environmental domain or environmental legislation.

24.4 Inspire as a Big Data Infrastructure

24.4.1 Characteristics of INSPIRE in Terms of Big Data

As already outlined, Big Data represents the information assets characterized by such a volume, velocity and variety to require specific technology and analytical methods for its transformation into value (De Mauro et al. 2015). In addition to volume, velocity, variety and value, characteristics of Big Data are veracity and visualisation, thus leading to the six Vs of Big Data. The Big Data landscape, including both the technologies (i.e. data lakes[30]) and infrastructures (i.e. databases and analytical tools) is evolving quickly. Geospatial data has always been considered to have more complex and larger datasets relevant for many other applications (McDougall and Koswatte 2018). The term Big Geospatial Data typically refers to spatial datasets exceeding the capacity of widespread computing systems. Many evidences have witnessed that a significant portion of Big Data is, in fact, Big Geospatial Data (Lee and Kang 2015). Spatial data comes from many sources and is used within many domains. According to the business model of the INSPIRE Directive, an efficient use of government resources requires that "spatial data are stored, made available and maintained at the most appropriate level" and that "it is possible to combine spatial data from different sources and share them between several users and applications" (European Parliament and Council 2007) – thus envisioning INSPIRE as a digital platform bringing together spatial data holders, analytics providers and users in sharing, combining and exploiting data. In Table 24.1, we map the six Vs of Big Data to the characteristics of data within the INSPIRE infrastructure.

Table 24.1 shows that the INSPIRE infrastructure shares the characteristics of Big Data and thus it could be considered in all respects a Big Geospatial Data infrastructure. The INSPIRE geospatial interoperability principles offer a flagship model for integrating huge amounts of heterogeneous sources of public sector information originating from a variety of providers, domains, administrative levels and cultural borders.

24.4.2 Challenges from the User Perspective

While several opportunities emerge from the establishment of the INSPIRE Big Geospatial Data infrastructure, at the same time there are also many challenges which especially pertain to the usability of such infrastructure. The most pressing ones are elaborated in more detail in the following subsections.

[30]https://en.wikipedia.org/wiki/Data_lake

Table 24.1 Characteristics of INSPIRE from the perspective of Big Data

Big data characteristics	INSPIRE characteristics
Volume – refers to the size of data	The infrastructure includes geospatial data falling under 34 data themes from all EU MS, EFTA countries and some candidate and potential candidate countries. Size of data differs but many are large data sets covering up to the whole countries, commonly acquired and used in the form of raster imagery, point cloud data, sensor observations, etc.
Variety – refers to heterogeneous sources and the nature of data, both structured and unstructured	Geospatial data in the infrastructure are produced at different levels (municipalities, cities, regions and the whole countries). Some of them are harmonised and structured in an INSPIRE conformant way, but many of them are "as is" data sets. There are several different types of data, e.g. imagery data, geotagged text data, spatio-temporal observation data, structured and unstructured data, raster and vector data – many with complex structures. The INSPIRE *Find your scope* application[a] lists 338 spatial object types, 121 data types and 294 code lists/enumerations.
Velocity – refers to the speed of generation of data	The amount of geospatial data in the infrastructure is increasing on a daily basis as long as new resources are collected/produced and made available in MS Discovery Services. Data velocity can be monitored through the INSPIRE Geoportal.
Veracity – refers to quality of data and data sources	The quality of geospatial data in the infrastructure varies from source to source and is expressed in relevant metadata elements e.g. lineage.
Visualisation – refers to presentation of data of almost any type in a graphical format that makes it easy to understand and interpret them	While the INSPIRE geoportal embeds a simple map-based visualisation, the datasets available in the infrastructure can be analysed and presented through any other visualisation tool suitable for geospatial datasets, including not only traditional ones such as charts, maps and imageries but also 3D models, animations, hotspot and change detection maps. The use of OGC interoperability standards to serve INSPIRE data potentially allows any client application to access and visualise them.
Value – refers to the worth of the data being extracted or used	The extraction of information and the actual use of INSPIRE data has direct economic benefits for Europe. MS implementation reports clearly show such benefits (Cetl et al. 2017).

[a]https://inspire-regadmin.jrc.ec.europa.eu/dataspecification/FindYourScope.action

24.4.2.1 Discoverability of Datasets

Fostered by INSPIRE, the development of NSDIs in Europe has made more and more geospatial resources (datasets and services) available on the web. The first visible component for users, which crucially enables them to search and retrieve resources is metadata (Cetl et al. 2016). Metadata records in INSPIRE are split into two, with individual records being created for (i) geospatial data and (ii) geospatial services, both served in the standardised way by using Discovery Services (usually

OGC CSW). Users then search for the resources by using discovery clients, for example the INSPIRE Geoportal as well as popular GIS clients such as QGIS.[31] This works fine in an SDI environment where both data producers and users are aware of SDI principles, in particular the corresponding services, the specialised GIS tools and how to use them. However, many users (both mainstream ICT developers and end users) are not at all aware of SDIs. Those non-expert users typically search for geospatial resources through standard web search engines such as Google and Bing. In addition to that, there are still a lot of geospatial data producers who – instead of creating and publishing metadata in a predefined structure to make their resources discoverable through an SDI – simply make them available on the web without documentation and standardised publication. The solution to that could be the use of metasearch enhanced crawlers to collect online accessible geospatial resources published by OGC services. There are also experiences and good practices with landing pages of datasets and services that could be generated from catalogue metadata (rather than maintaining the information in different systems). Linked data also constitutes a fast emerging trend, with clear potential to benefit SDIs (Bucher et al. 2020), leveraging a way to interconnect related data resident on the web, and deliver it in a more effective manner to increase its value for users. The resulting "Web of data" has recently started being populated with geospatial data.[32] There are other ongoing discussions dedicated to web developers, spatial data publishers and search engine optimisation (SEO) experts related to search engines indexing optimisations when publishing geospatial data. A first, notable effort to create a search engine for geospatial resources is Google's Dataset Search[33] launched in 2018.

24.4.2.2 Combining National Datasets to Create Pan-European Products

There is a growing demand for more and better quality data both within the EC and the EU MS to support a number of key policies related, but not limited to the environment. Often, several data sources are available at MS level representing the same spatial objects which differ in various characteristics and quality criteria such as cartographic scale, level of detail, positional accuracy, timeliness, update frequency and licensing conditions. The ultimate goal of INSPIRE is to have harmonised national datasets from MS that can be seamlessly used at cross-border and transnational levels and facilitate the creation of consistent pan-European datasets without or with limited additional processing. However, the latter is not among the original INSPIRE objectives. The latter should be provided based on performant and stable services in line with the INSPIRE requirements. The challenges in creation

[31]https://qgis.org

[32]http://ggim.un.org/meetings/GGIM-committee/8th-Session/documents/Standards_Guide_2018.pdf

[33]https://datasetsearch.research.google.com

of such pan-European datasets are twofold, i.e. related to a technical and a non-technical harmonisation. In terms of technical harmonisation, reference data are not yet available for all MS from the INSPIRE Geoportal with harmonised physical data models and functioning services. Further challenges include harmonisation in terms of level of detail, scale and edge-matching; in addition, the scope of the datasets provided by MS in the INSPIRE infrastructure might differ a lot, ranging from single national datasets to multiple regional or even local datasets. From the non-technical perspective, despite the efforts made to harmonise the access conditions to spatial datasets and services (European Commission 2010, 2013), there is still a high diversity of licensing conditions between countries and in some cases even among national public authorities of the same country that overall might create serious legal obstacles to data access and reuse. Data licenses range from open data licenses such as those from Creative Commons[34] to more restrictive and in some cases custom or national-specific licenses. This diversity clearly hampers the joint use of datasets as well as the development of consistent and harmonised EU-wide products derived from national datasets. Even in the circumstance when multiple datasets from different MS or data providers are published under open data licenses, the diversity between these licenses might pose legal obstacles to their combination and joint use from third-party actors. In addition, sometimes MS organisations adopt different technical means to restrict access to data services (e.g. through authentication mechanisms or by imposing hefty access fees), thus further impacting on data accessibility and usability.

24.4.2.3 Data Access and Consumption by Clients

When implementing INSPIRE, MS have adopted a number of different strategies for the implementation of network services. Often the heterogeneity of the European data landscape had led to a lack of agreement between data providers on how to organise Big Data for effective utilisation. Examples include the number of data sets grouped together within one (or few) View and Download Services as well as the criteria used to group such data sets (e.g. by data theme, geographic area, scale, use case or national provider/organisation). A first drawback of this approach is the difficulty for client applications to easily find the desired resources. In some cases Big Geospatial Data such as huge databases or coverages covering whole countries (e.g. national registries of addresses or national orthophotos), which typically correspond to files of gigantic size, are also served through one single service. The consequence on the client side is the overall difficulty in accessing such datasets served through Download Services such as WFS or Atom feeds, in particular the extremely long waiting times required for download. Even in the case that the download is successful, the ultimate consequence is still the issue of consuming (i.e. visualising, analysing and processing) such data for end users.

[34]https://creativecommons.org

24.4.2.4 Cloud Infrastructures

The components of an SDI can be integrated into the cloud as value-added services (Schäffer et al. 2010). Cloud infrastructures are now available as a cost-effective and efficient alternative to on-premises provision of INSPIRE services from providers such as Amazon and Microsoft (Bragg 2017). Only few MS have started to publish their INSPIRE resources in a cloud infrastructure and it can be expected that more data providers and organisations will adapt the same approach in the future, although cloud computing typically raises several concerns due to lack of trust and transparency. INSPIRE can ultimately benefit from the ability of cloud infrastructures to handle large amount of requests and deliver data in a robust and performing manner. By migrating services to the cloud, the geospatial resources provided by these services would be immediately available in a scalable fashion for on-demand use. The central components of the INSPIRE infrastructure, which are technically managed by the JRC, have also not yet been migrated to the cloud. The only exception is the above mentioned INSPIRE Reference Validator,[35] which is deployed on the cloud since spring 2019 to address the increased user base while at the same time providing satisfactory performances.

24.5 Conclusions and Outlook

Since its adoption in 2007, the INSPIRE Directive has been the driver behind the development of an EU-wide Spatial Data Infrastructure based on the interoperability principles ultimately aiming at the creation of a single European (geo)data space. The existence of this SDI has been initially considered of primary importance in support of EU environmental policies and activities impacting the environment. However, location has become pervasive across multiple policy and societal domains and so is the relevance and potential of geospatial data. Accordingly, INSPIRE holds the potential to enable a full use of spatial information across the public sector, allow multiple stakeholders (including not only governmental agencies but also private companies, researchers and citizens) to access spatial data across Europe, assist cross-border policy-making and support better integrated public eGovernment services (Cetl et al. 2017). Thanks to the numbers, increasing on an almost daily basis, of metadata records, datasets and services shared by European countries, INSPIRE is gradually evolving into a reference European SDI whose data possess all the characteristics of Big Data (see Sect. 24.4). Given that INSPIRE as an SDI has been a pioneer of the European digital society and economy, and that many efforts to build SDIs even beyond Europe have looked at INSPIRE as a model, after more than 10 years since INSPIRE inception a number of lessons from a Big Data perspective have been learnt. These include opportunities, threats

[35]http://inspire.ec.europa.eu/validator/about

and dependencies coming from the emerging technological, societal and economic trends, which should guide the future steps so that INSPIRE – subject to the necessary political support and mandate – can continue to play a key role within the European geospatial digital data revolution.

From our perspective the notion of SDI as it was originally defined (Clinton 1994) is evolving. The processes which are typically described by SDIs are mainly linear, i.e. data is first collected (often only once) by specific actors (usually trained professionals), harmonised and published by governmental data providers and finally consumed by data users. Based on this, there has been the belief that once such an infrastructure was in place, people would simply use it. But, in the era of Big Data, reality has become much more complex. First, new actors such as private companies, citizens and researchers have become key players in terms of data collection, thus leading to the term 'produsers' to denote their blurred role of being both producers and users of data (Coleman et al. 2009). Nonetheless, the public sector remains a major actor, whose main advantage is the fact that large portions of its datasets are quality controlled and often rooted in the formalised data value chain processes that are legislatively defined. Second, data are currently collected at unprecedented speeds, and from sources such as drones, smartphones, in situ sensor networks and Internet of Things (IoT) devices which did not even exist when the term SDI was first coined. This increased amount of data brings obvious complexities when it comes to defining ownership, privacy, and licensing. In this new context, it is crucial that traditional SDIs evolve into modern data ecosystems. These can be defined as complex systems of people, organisations, technology, policies, and data in a specific area that interact with each other and their surrounding environment for a specific purpose. Such ecosystems evolve and adapt through a cycle of data creation and sharing, data analytics, and value creation in the form of new products, services, or knowledge, which, when used, produce new data feeding back into the ecosystem (Pollock 2011; UN Environment Assembly 2019; Oliveira et al. 2019). Thus, the key difference of data ecosystems compared to SDIs is the cyclical flow that links the processes of data creation and sharing, data analytics, value creation and use, in turn generating new data in a continuous feedback loop between the stakeholders involved. In other words, the processes within data ecosystem are mainly driven by specific use cases, address specific users and are described by dynamic and non-linear processes. From a Big Data perspective, data analytics combined with artificial intelligence are particularly important for improving policymaking and service delivery (Lisbon Council 2019).

We therefore anticipate the evolution of INSPIRE into a new data ecosystem for environment (Kotsev et al. 2020). In line with the recently-published priorities of the EC, which set out a European Green Deal for the EU and its citizens (European Commission 2019) as well as the strategy to establish a common European data space (European Commission 2018), INSPIRE should act as an integrated data ecosystem which, sitting on top of this horizontal EU data space and interconnected with the EOSC (i.e. with data and services from many different sources and actors, not necessarily geospatial), can deliver efficient solutions to ensure good policymaking in the environmental domain and beyond. Using another term to express the same idea, INSPIRE should evolve from a highly distributed and fragmented infrastructure into a centralized platform (Reitz 2019).

The transition towards a successful data ecosystem implies transitions in a number of dimensions. From the technological perspective, it is still crucial that such a data ecosystem is built on open standards. However, while a SOA based on traditional OGC web standards was at the forefront when INSPIRE was conceived, such standards no longer reflect the modern practices of data exchange through the web (Open Geospatial Consortium 2019). A new family of API-based standards, collectively called OGC APIs,[36] is under development through a user-centric, data-driven approach to maximise the benefit for the future users. The first and currently only standard published, the OGC API - Features,[37] has been already identified as a candidate standard on which a proposal for a specification for setting up INSPIRE Download Services is under development.[38] The OGC API - Features is a REST API that quickly and easily accesses geospatial features on the web, potentially allowing to overcome the issues described in Sect. 24.4.2.3. Being designed as a modern web standard, it is GeoJSON-oriented (although other encodings are also supported) and thus it goes in the same direction of defining alternative encodings to simplify and flatten the INSPIRE complex data models, an activity that the JRC has already started in 2018.[39] An increased flexibility, which – at least for selected spatial object types – relaxes some semantic requirements and only secures a basic level of interoperability, might definitely help improve the overall usability of the infrastructure. Similarly, the OGC SensorThings API standard[40] is proposed as an INSPIRE Download Service (Kotsev et al. 2018). SensorThings API is also based on REST principles and provides a simple yet powerful means for retrieval of observation data.

In the same direction of simplifying data access and usability, which would address the issue described in Sect. 24.4.2.1, APIs are also identified by the Open Data Directive (European Parliament and Council 2019) as the required tool to publish high-value datasets. APIs have real potential to make the new generation of data ecosystems user-friendly and usable by developers and end-users, often not familiar with spatial web services, as building blocks to create third-party applications to generate additional value. In recent years, several European countries such as France, Germany, Sweden, Ireland and Croatia have developed an API-based approach as an integral part of their SDI or INSPIRE developments.

Non-technical transitions are also key for the success of INSPIRE as a data ecosystem. An open platform model characterised by cyclical data flows between the actors and stakeholders involved only works if it generates value for all of them. Thus, it will be increasingly important to move away from a traditional

[36]http://www.ogcapi.org

[37]https://www.opengeospatial.org/standards/ogcapi-features

[38]https://github.com/INSPIRE-MIF/gp-ogc-api-features

[39]https://github.com/INSPIRE-MIF/2017.2

[40]https://www.opengeospatial.org/standards/sensorthings

vision which only looks at the interface between data providers and users, and include other key stakeholders such as partners, software providers, companies selling value-added services or data analytics, and intermediary organisations that facilitate interactions within the platform. All of this without forgetting that the same stakeholders can play multiple roles. Finally, horizontal aspects such as data management, governance, protection and sharing issues will need to be prioritised. The latter includes addressing the INSPIRE licensing scheme which is currently a serious obstacle for a full exploitation of the infrastructure. Taking the Open Data Directive – which requires that high-value datasets are "made available for reuse with minimal legal restrictions and free of charge" – as a reference, a possible path towards an increased usability of the INSPIRE infrastructure could be to require the publication of specific datasets without any access obstacle (e.g. authentication or payment of a license fee) and under an open license that allows for re-use for any purpose, including commercial. In turn, this would facilitate the cross-border combination and the creation of pan-European products, at least for these specific datasets, thus solving the issue described in Sect. 24.4.2.2. Last but not least, the currently slow and only partial implementation of INSPIRE (already discussed in Sect. 24.3.2) should be addressed through a well-thought combination of regulatory interventions and other non legal measures, the latter including incentives, benefits and constructive competitions (Lisbon Council 2019).

References

Bragg K (2017) Leveraging cloud infrastructure for INSPIRE services. INSPIRE Conference, Strasbourg, 6–8 September 2017. https://inspire.ec.europa.eu/sites/default/files/presentations/Leveraging_Cloud_Infrastructure_for_INSPIRE.pdf. Accessed: 2020-02-04

Brovelli MA, Minghini M, Molinari ME, Zamboni G (2016) Positional accuracy assessment of the OpenStreetMap buildings layer through automatic homologous pairs detection: The method and a case study. The International Archives of the Photogrammetry, Remote Sensing and Spatial Information Sciences, Volume XLI-B2, 2016 XXIII ISPRS congress, 12–19 July 2016, Prague. https://doi.org/10.5194/isprs-archives-XLI-B2-615-2016

Bucher B, Tiainen E, Brasch TEV et al (2020) Conciliating perspectives from mapping agencies and web of data on successful European SDIs: toward a European geographic knowledge graph. ISPRS Int J Geo Inf 9(2):62. https://doi.org/10.3390/ijgi9020062

Cetl V, Tóth K, Smits P (2014) Development of NSDIs in Western Balkan countries in accordance with INSPIRE. Surv Rev 46(338):316–321

Cetl V, Kliment T, Kliment M (2016) Borderless Geospatial Web (BOLEGWEB). The International Archives of the Photogrammetry, Remote Sensing and Spatial Information Sciences, Volume XLI-B4, XXIII ISPRS Congress, 12–19 July 2016, Prague. https://doi.org/10.5194/isprsarchives-XLI-B4-677-201

Cetl V, de Lima VN, Tomas R, Lutz M, D'Eugenio J, Nagy A, Robbrecht J (2017) Summary report on status of implementation of the INSPIRE directive in EU. JRC Technical Report. https://doi.org/10.2760/143502. http://publications.jrc.ec.europa.eu/repository/bitstream/JRC109035/jrc109035_jrc109035_jrc_inspire_eu_summaryreport_online.pdf. Accessed: 2020-02-06

Cetl V, Tomas R, Kotsev A, de Lima VN, Smith RS, Jobst M (2019) Establishing common ground through INSPIRE: the legally-driven European spatial data infrastructure. In: Service-oriented mapping. Cham, Springer, pp 63–84

Clinton W (1994) Coordinating geographic data acquisition and access: the National Spatial Data Infrastructure. *Executive Order 12906*. Fed Regist 59(71):17671–17674

Coleman D, Georgiadou Y, Labonte J (2009) Volunteered geographic information: the nature and motivation of producers. Int J Spatial Data Infrastruc Res 4(4):332–358. https://doi.org/10.2902/1725-0463.2009.04.art16

De Mauro A, Greco M, Grimaldi M (2015) What is big data? A consensual definition and a review of key research topics. AIP Conf Proc 1644:97. https://doi.org/10.1063/1.4907823

European Commission (2009) Commission regulation no 976/2009 of 19 October 2009 implementing directive 2007/2/EC of the European Parliament and of the council as regards the network services. Official Journal of the European Union L 264, 9–18, 20.10.2019. https://eur-lex.europa.eu/legal-content/EN/TXT/PDF/?uri=CELEX:32009R0976&from=EN. Accessed: 2020–01–16

European Commission (2010) Commission regulation (EU) no 268/2010 of 29 march 2010 implementing directive 2007/2/EC of the European Parliament and of the council as regards the access to spatial data sets and services of the member states by community institutions and bodies under harmonised conditions. Official Journal of the European Union L 83, 8–9, 30.3.2010. https://eur-lex.europa.eu/legal-content/EN/TXT/PDF/?uri=CELEX:32010R0268&from=EN. Accessed: 2020-02-07

European Commission (2012a) Communication from the Commission to the European Parliament, the Council, the European Economic and Social Committee and the Committee of the Regions. Unleashing the Potential of Cloud Computing in Europe. COM(2012) 529 final. 27.9.2012. https://eur-lex.europa.eu/LexUriServ/LexUriServ.do?uri=COM:2012:0529:FIN:EN:PDF. Accessed: 2020-01-20

European Commission (2012b) Communication from the Commission to the European Parliament, the Council, the European Economic and Social Committee and the Committee of the Regions. High-Performance Computing: Europe's place in a Global Race. COM(2012) 45 final. 15.2.2012. https://eur-lex.europa.eu/LexUriServ/LexUriServ.do?uri=COM:2012:0045:FIN:EN:PDF. Accessed: 2020-01-20

European Commission (2013) Guidance on the 'Regulation on access to spatial data sets and services of the Member States by Community institutions and bodies under harmonised conditions'. https://inspire.ec.europa.eu/documents/Data_and_Service_Sharing/DSSGuidanceDocument_v5.0.pdf. Accessed: 2020-02-07

European Commission (2014) Communication from the Commission to the European Parliament, the Council, the European Economic and Social Committee and the Committee of the Regions. Towards a thriving data-driven economy. COM(2014) 442 final. 2.7.2014. https://ec.europa.eu/transparency/regdoc/rep/1/2014/EN/1-2014-442-EN-F1-1.PDF. Accessed: 2020-01-21

European Commission (2015) Final report of public consultation on Science 2.0 / open science. https://ec.europa.eu/digital-single-market/news/final-report-science-20-public-consultation. Accessed: 2019-03-20

European Commission (2016) Communication from the Commission to the European Parliament, the Council, the European Economic and Social Committee and the Committee of the Regions. European Cloud Initiative - Building a competitive data and knowledge economy in Europe. COM(2016) 178 final, 19.4.2016. https://ec.europa.eu/newsroom/dae/document.cfm?doc_id=15266. Accessed: 2019-03-20

European Commission (2017) EuroGEOSS Concept Paper. 22 November 2017. https://ec.europa.eu/info/sites/info/files/eurogeoss/eurogeoss_concept_paper-2017.pdf. Accessed: 2020-01-31

European Commission (2018) Communication from the Commission to the European Parliament, the Council, the European Economic and Social Committee and the Committee of the Regions. Towards a Common European Data Space. COM(2018) 232 final, SWD(2018) 125 final. https://eur-lex.europa.eu/legal-content/EN/TXT/PDF/?uri=CELEX:52018DC0232&from=EN. Accessed: 2020-01-21

European Commission (2019) Communication from the Commission to the European Parliament, the Council, the European Economic and Social Committee and the Committee of the Regions. The European Green Deal. COM(2019) 640 final. https://eur-lex.europa.eu/resource.html?uri=cellar:b828d165-1c22-11ea-8c1f-01aa75ed71a1.0002.02/DOC_1&format=PDF. Accessed: 2020-02-06

European Commission Expert Group on FAIR Data (2018) Turning FAIR into reality: Final Report and Action Plan from the European Commission Expert Group on FAIR Data. https://doi.org/10.2777/1524. Accessed: 2020-01-21

European Parliament and Council (2007) Directive 2007/2/EC of the European Parliament and of the Council of 14 March 2007 establishing an Infrastructure for Spatial Information in the European Community (INSPIRE). Official Journal of the European Union L 108/1, Volume 50, 25.4.2007. https://eur-lex.europa.eu/legal-content/EN/TXT/PDF/?uri=OJ:L:2007:108:FULL&from=EN. Accessed: 2020–01–16

European Parliament and Council (2016) Regulation (EU) 2016/679 of the European Parliament and of the council of 27 April 2016 on the protection of natural persons with regard to the processing of personal data and on the free movement of such data, and repealing directive 95/46/EC (general data protection regulation). Official Journal of the European Union L 119, 4.5.2016. https://eur-lex.europa.eu/legal-content/EN/TXT/PDF/?uri=CELEX:32016R0679&from=EN. Accessed: 2020-01-23

European Parliament and Council (2019) Directive (EU) 2019/1024 of the European Parliament and of the council of 20 June 2019 on open data and the re-use of public sector information (recast). Official Journal of the European Union L 172, 56–83, 26.6.2019. https://eur-lex.europa.eu/legal-content/EN/TXT/PDF/?uri=CELEX:32019L1024&from=EN. Accessed: 2020-01-23

Fan H, Zipf A, Fu Q, Neis P (2014) Quality assessment for building footprints data on OpenStreetMap. Int J Geogr Inf Sci 28(4):700–719

Girres JF, Touya G (2010) Quality assessment of the French OpenStreetMap dataset. Trans GIS 14(4):435–459

Haklay M (2010) How good is volunteered geographical information? A comparative study of OpenStreetMap and ordnance survey datasets. Environ Plan B Plan Design 37(4):682–703

Kitchin R (2014) The data revolution: big data, open data, data infrastructure and their consequences. Sage Publications, London

Köbben B, Graham M (2009) Maps and mash-ups: the national atlas and Google Earth in a geodata infrastructure. In Proceedings of the 12th AGILE International Conference on Geographic Information Science: Grid technologies for geospatial applications, Hannover

Koski K, Hormia-Poutanen K, Chatzopoulos M, Legré Y, Day B (2015) Position paper: European open science cloud for research. https://eudat.eu/sites/default/files/PositionPaperEOSCcard.pdf. Accessed: 2020-01-20

Kotsev A, Schleidt K, Liang S et al (2018) Extending INSPIRE to the internet of things through SensorThings API. Geosciences 8(6):221. https://doi.org/10.3390/geosciences8060221

Kotsev A, Minghini M, Tomas R, Cetl V, Lutz M (2020) From spatial data infrastructures to data spatial technological perspective on the evolution of European SDIs. ISPRS International Journal of Geo-Information, 9(3):176

Koubarakis M, Bereta K, Bilidas D et al (2019) From copernicus big data to extreme earth analytics. Proceedings of the 22nd international conference on Extending Database Technology (EDBT), March 26–29, 2019

Lee J-G, Kang M (2015) Geospatial big data: challenges and opportunities. Big Data Res 2:74–81, Elsevier Inc.

Lisbon Council (2019) The public-data opportunity: why governments should share more. https://lisboncouncil.net//index.php?option=com_downloads&id=1474. Accessed: 2020-02-06

McDougall K, Koswatte S (2018) The Future of Authoritative Geospatial Data in the Big Data World – Trends, Opportunities and Challenges. FIG Commission 3 Workshop 2018. http://fig.net/resources/monthly_articles/2019/Mcdougall_etal_January_2019.asp. Accessed: 2020-02-06

Minghini M, Kotsev A, Lutz M (2019) Comparing INSPIRE and OpenStreetMap: how to make the most out of the two worlds. The International Archives of the Photogrammetry, Remote Sensing & Spatial Information Sciences, Volume XLII-4/W14, 2019 FOSS4G 2019 – academic track, 26—30 August 2019, Bucharest

Minghini M, Cetl V, Ziemba L, Tomas R, Francioli D, Artasensi D, Epure A, Vinci F (2020) Establishing a new baseline for monitoring the status of EU Spatial Data Infrastructure

Mooney P, Minghini M (2017) A review of OpenStreetMap data. In: Foody G, Fritz S, Mooney P, Olteanu-Raimond A-M, Fonte CC, Antoniou V (eds) Mapping and the citizen sensor. London, Ubiquity Press, pp 37–59

Oliveira MIS, Lima GDFB, Lóscio BF (2019) Investigations into data ecosystems: a systematic mapping study. Knowl Inf Syst:1–42. https://doi.org/10.1007/s10115-018-1323-6

Open Geospatial Consortium (2019) OGC APIs and the evolution of OGC standards. https://www.opengeospatial.org/blog/2996. Accessed: 2020-02-06

Pollock R (2011) Building the (Open) Data Ecosystem. https://blog.okfn.org/2011/03/31/building-the-open-data-ecosystem. Accessed: 2020-02-06

Reitz T (2019) INSPIRE 2030: a successful open platform. Inspire Helsinki 2019, Helsinki, 22–24 October 2019. https://www.youtube.com/watch?v=2svoCE1CE9w&list=PL3ZPoPTgFIRFr68ehmLDVmQRms68txhMn&index=6. Accessed: 2020-02-06

Schäffer B, Baranski B, Foerster T (2010) Towards spatial data infrastructures in the clouds. In: Painho M, Santos M, Pundt H (eds) Geospatial thinking. Lecture notes in geoinformation and cartography. Berlin, Springer, pp 399–418

Tomas R, Harrison M, Barredo JI, Thomas F, Isidro ML, Pfeiffer M, Čerba O (2015) Towards a cross-domain interoperable framework for natural hazards and disaster risk reduction information. Nat Hazards 78(3):1545–1563. https://doi.org/10.1007/s11069-015-1786-7

UN Environment Assembly (2019) The case for a digital ecosystem for the environment: bringing together data, algorithms and insights for sustainable development. Discussion paper: 5 March 2019. https://un-spbf.org/wp-content/uploads/2019/03/Digital-Ecosystem-final-2.pdf. Accessed: 2020-02-06

Wilkinson MD, Dumontier M, Aalbersberg I et al (2016) The FAIR guiding principles for scientific data management and stewardship. Sci Data 3:160018. https://doi.org/10.1038/sdata.2016.18

Printed in the United States
by Baker & Taylor Publisher Services